Outlook and Challenges of Nano Devices, Sensors, and MEMS

Ting Li • Ziv Liu
Editors

Outlook and Challenges of Nano Devices, Sensors, and MEMS

Springer

Editors
Ting Li
State Key Lab of Electronic Thin Film & Integrated Device
University of Electronic Science & Technology of China
Chengdu, China

Ziv Liu
Microelectronics & Solid-state Electronics College
University of Electronic Science & Technology of China
Chengdu, China

ISBN 978-3-319-50822-1 ISBN 978-3-319-50824-5 (eBook)
DOI 10.1007/978-3-319-50824-5

Library of Congress Control Number: 2017931961

Printed on acid-free paper

This Springer imprint is published by Springer Nature
The registered company is Springer International Publishing AG
The registered company address is: Gewerbestrasse 11, 6330 Cham, Switzerland

Foreword

This book mainly covers the current status, challenges, and future directions on design, fabrication, simulation, reliability, and applications of nanoscale semiconductor devices, MEMS, and sensors. And these novel devices are the backbone for realizing the next-generation Internet of things. This book provides readers important information related to the recent development on these nanoelectron apparatuses, which represents that this book will become an indispensable handbook for anyone in a relevant field.

This book contains 17 chapters, and all the chapters can be divided into 4 parts: MOS, sensors (including photoelectric detection and gas detection), nanostructured oxide materials, and other nano devices. In spite of the large coverage of related fields, the focus of this book is explicitly placed on the interaction of nanoscale structure and electrical and/or optical behavior of novel semiconductor devices, MEMS, and sensors. Besides, it enables detailed understanding on the difficulties and challenges of realizing the nano devices moving forward.

In this single book, the reader can access up-to-date and critical knowledge on the design, fabrication, simulation, and reliability of the increasingly important nanoscale electron apparatuses, including semiconductor devices, MEMS, and sensors.

University of Electronic Science and Technology
Chengdu, China — He Yu

University of Electronic Science and Technology
Chengdu, China — Lanhui Wu

Preface

This book contains 17 chapters, which is written by selected experts who have present high-quality papers on topics closely relevant to the theme of nano devices, sensors, and MEMS. We can understand current status, challenges, and future directions on design, fabrication, simulation, reliability, and applications of nanoscale semiconductor devices, MEMS, and sensors from this book.

The field of nanoelectronic is evolving rapidly, and novel nanoscale devices are being investigated and developed at many leading research labs and companies worldwide. Nanoscale semiconductor devices, MEMS, and sensors are the backbone for realizing the next-generation Internet of things. The research of design, fabrication, reliability, modeling, simulation, and applications of these nanoscale electron components is becoming more and more important to achieve further technology breakthrough. Learning how the nanoscale structure interacts with the electrical and/or optical performances, how to find optimal solutions to achieve the best outcome, how these apparatuses can be designed via models and simulations, how to improve the reliability, and what are possible challenges and roadblocks moving forward is necessary to any relative researchers.

This is a book that describes the dielectric and noise analysis of MOS devices, devices modeling and CMOS technology, and medical application in MEMS and sensors, as well as relative fabrication and optimizing technology. It is really a book which provides readers important information related to the recent development on these nanoelectron apparatuses and challenges of realizing the nano devices moving forward that shows processes and materials needed for fabrication and the interaction of nanoscale structure and electrical and/or optical behavior, as well as the performance optimization and reliability analysis by modeling and simulating. Springer is highly desirable and can offer engineers, research scientists, and students a much needed reference in this field.

Last but not the least, we want to express our deep appreciation to the endless support from writers of this book. Also, we really hope that the readers can access up-to-date and critical knowledge on the design, fabrication, simulation, and reliability of the increasingly important nanoscale electron apparatuses from a single book. Know more about nowadays and think more about the future. Let Springer be a window through which we can discover and explore deeper in this field.

Chengdu, China

Ting Li
Ziv Liu

Acknowledgements

Sincere thanks must be expressed to all authors who have devoted considerable time and effort to prepare excellent comprehensive works of high scientific quality and value. Without their help, it would be impossible to prepare this book in line with the high standards that we have set from the very beginning of this project.

We would also like to thank our Springer editorial contacts. It would not have been possible to bring out this book within such a short span of time without her concerted encouragement. We very much appreciate the continued support extended by Springer.

Lastly, I would like to express my thanks to Professor He Yu, student Haicheng Yao, and other students of the University of Electronic Science and Technology of China, who assisted in this book.

Contents

Contributors

Yan Bo School of Communication and Information Engineering, University of Electronic Science and Technology of China, Chengdu, China

Mansun Chan Peking University Shenzhen Graduate School, Shenzhen, China

Rishu Chaujar Department of Engineering Physics, Delhi Technological University, (Formerly Delhi College of Engineering), Delhi, India

Yuanfu Chen State Key Lab Elect Thin Film and Integrated Device, University of Electronic Science and Technology of China, Chengdu, China

Zhi Chen School of Optoelectronic Information, University of Electronic Science and Technology of China, Chengdu, China

Department of Electrical and Computer Engineering and Center for Nanoscale Science and Engineering, University of Kentucky, Lexington, KY, USA

Hui Cheng State Key Laboratory for Electronic Thin Films and Integrated Devices, University of Electronic Science and Technology of China, Chengdu, China

Wanling Deng College of Information Science and Technology, Jinan University, Guangzhou, China

Fu Dong-bing Sichuan Institute of Solid-State Circuits, China Electronics Technology Group Corp, Chongqing, People's Republic of China

Jielin Fang College of Information Science and Technology, Jinan University, Guangzhou, China

Xuan P.A. Gao Department of Physics, Case Western Reserve University, Cleveland, OH, USA

Waqas Amin Gill Department of Electrical and Computer Engineering, Masdar Institute of Science and Technology, Abu Dhabi, UAE

Li Guangjun School of Communication and Information Engineering, University of Electronic Science and Technology of China, Chengdu, China

Shaofeng Guo Institute of Microelectronics, Peking University, Beijing, China

Zhaojia Guo State Key Lab Elect Thin Film and Integrated Device, University of Electronic Science and Technology of China, Chengdu, China

Mengmeng Hao School of Optoelectronic Information, University of Electronic Science and Technology of China, Chengdu, China

Fei Hou State Key Laboratory for Electronic Thin Films and Integrated Devices, University of Electronic Science and Technology of China, Chengdu, China

Ru Huang Institute of Microelectronics, Peking University, Beijing, China

Chengjun Huang Institute of Microelectronics of Chinese Academy of Sciences, Beijing, China

Zhang Jun-an School of Communication and Information Engineering, University of Electronic Science and Technology of China, Chengdu, China

Key Laboratory of Analog Integrated Circuit, Chongqing, China

Ajay Giri Prakash Kottapalli School of Mechanical and Aerospace Engineering, Nanyang Technological University, Singapore, Singapore

Ting Li State Key Lab Elect Thin Film and Integrated Device, University of Electronic Science and Technology of China, Chengdu, China

Shenzhen Geno-Immune Medicine Institute, Shenzhen, Guangdong Province, China

Zebin Li State Key Lab Elect Thin Film and Integrated Device, University of Electronic Science and Technology of China, Chengdu, China

Shenzhen Geno-Immune Medicine Institute, Shenzhen, Guangdong Province, China

Li Li Institute of Microelectronics of Chinese Academy of Sciences, Beijing, China

Xinnan Lin Peking University Shenzhen Graduate School, Shenzhen, China

Honggang Liu Key Laboratory of Microelectronics Devices and Integrated Technology, Institute of Microelectronics, Chinese Academy of Sciences, Beijing, China

Jizhi Liu State Key Laboratory for Electronic Thin Films and Integrated Devices, University of Electronic Science and Technology of China, Chengdu, China

Zhiwei Liu State Key Laboratory for Electronic Thin Films and Integrated Devices, University of Electronic Science and Technology of China, Chengdu, China

Haijun Lou Peking University Shenzhen Graduate School, Shenzhen, China

Sun Woh Lye School of Mechanical and Aerospace Engineering, Nanyang Technological University, Singapore, Singapore

Jaya Madan Department of Engineering Physics, Delhi Technological University, (Formerly Delhi College of Engineering), Delhi, India

Jianmin Miao School of Mechanical and Aerospace Engineering, Nanyang Technological University, Singapore, Singapore

Zakriya Mohammed Department of Electrical and Computer Engineering, Masdar Institute of Science and Technology, Abu Dhabi, UAE

Xingkun Ning Hebei Key Lab of Optic-electronic Information and Materials, The College of Physics Science and Technology, Hebei University, Baoding, People's Republic of China

Leslie K. Norford Center for Environmental Sensing and Modeling (CENSAM), Singapore-MIT Alliance for Research and Technology (SMART) Centre, Singapore, Singapore

Department of Architecture, Massachusetts Institute of Technology, Cambridge, MA, USA

Boan Pan State Key Lab Elect Thin Film and Integrated Device, University of Electronic Science and Technology of China, Chengdu, China

Luo Pu Sichuan Institute of Solid-State Circuits, China Electronics Technology Group Corp, Chongqing, People's Republic of China

Ling-Xuan Qian School of Microelectronics and Solid-State Electronics, University of Electronic Science and Technology of China, Chengdu, China

Mahmoud Rasras Department of Electrical and Computer Engineering, Masdar Institute of Science and Technology, Abu Dhabi, UAE

Zhang Rui-tao Key Laboratory of Analog Integrated Circuit, Chongqing, China

Kai Tao School of Mechanical and Aerospace Engineering, Nanyang Technological University, Singapore, Singapore

Rui Tian State Key Laboratory for Electronic Thin Films and Integrated Devices, University of Electronic Science and Technology of China, Chengdu, China

Wenbo Wan Peking University Shenzhen Graduate School, Shenzhen, China

Runsheng Wang Institute of Microelectronics, Peking University, Beijing, China

Z.H. Wang Shenyang National Laboratory for Materials Science, Institute of Metal Research, Chinese Academy of Sciences, Shenyang, People's Republic of China

Chapter 1
High-k Dielectric for Nanoscale MOS Devices

Ling-Xuan Qian

1.1 Introduction

The Integrated Circuit (IC) industry has successfully followed the guidance of the Moore's law in the past decades: the number of transistors in a dense IC doubles approximately two years. Increasing integration level, reducing cost and power consumption while improving product performance has always been the primary goal of IC industry. In order to keep pace with the Moore's Law, it has been decades that device scaling is one of the essential focuses of the development of MOS technology.

After becoming the most important device for forefront high-density integrated circuits such as microprocessors and semiconductor memories, Metal-Oxide-Semiconductor Field-Effect Transistor (MOSFET) is becoming an important power device as well. Since 1970, the gate-length dimension of MOSFET in production ICs has been decreasing at a steady pace and will continue to scale down in the future, as shown in Fig. 1.1 [1].

However, as the scaling down of the MOS devices, the departures from long-channel behaviors, as known as the short-channel effects, are inevitable, leading to varieties of undesirable electrical phenomena. Up till now, the most widely used scaling rule to avoid short-channel effects is the "constant-field scaling", i.e. scale down all dimensions and voltage of a long-channel MOSFET so that the internal electric fields are kept the same [2]. In Fig. 1.2, the dimensions of MOSFET before and after scaling are compared, all dimensions including channel length (L), oxide thickness (T_{ox}), and junction depth (r_j) are ideally shrunk by a same scaling factor α. Noted that the doping level of the substrate increases by α and the applied voltages

L.-X. Qian (✉)
School of Microelectronics and Solid-State Electronics, University of Electronic Science and Technology of China, Chengdu, China
e-mail: lxqian@uestc.edu.cn

T. Li, Z. Liu (eds.), *Outlook and Challenges of Nano Devices, Sensors, and MEMS*,
DOI 10.1007/978-3-319-50824-5_1

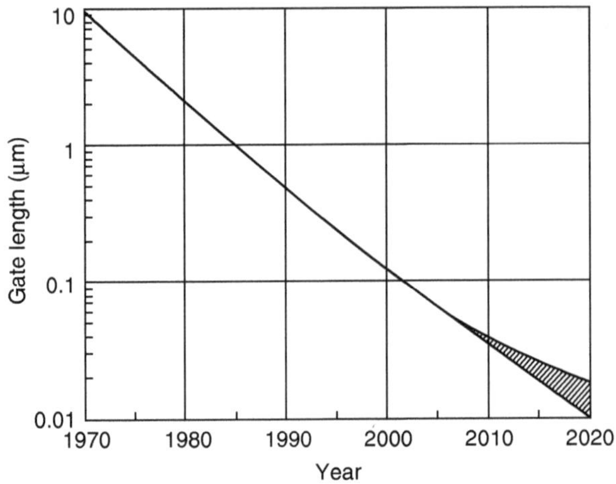

Fig. 1.1 Minimum gate length of MOSFET in commercial IC vs. the year of production [1]

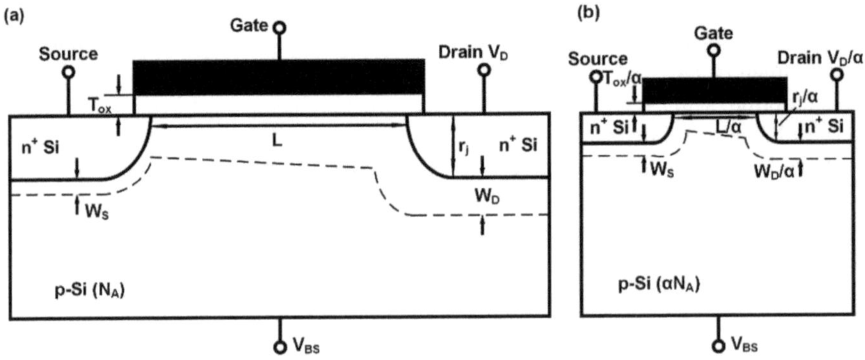

Fig. 1.2 Schematic diagram of MOSFET before and after scaling [2]. (**a**) Original device, (**b**) Scaled device

decreases by α, thus a reduction of the junction depletion width (W_D) by α occurs beneath the drain area. Unfortunately, the scaling factor in actual cases cannot achieve the ideal value because of the limitation of some other factors. Table 1.1 summarizes the ideal and actual scaling factors together with the limitation in scaling [2, 3].

Since the oxide thickness should also reduce with the device scales down according to Fig. 1.2 and Table 1.1, an oxide thickness of less than 3 nm is needed for MOS devices with channel length of 100 nm or less. Theoretically, the minimum physical thickness of SiO_2 is 0.7 nm with at least two layers of neighboring oxygen atoms to prevent the gate/SiO_2 and the SiO_2/Si interfaces from overlapping with each other, and thus make the gate insulator conductive [4]. But in fact, because of the quantum mechanical effects that carriers are capable of tunneling through the ultra-thin gate dielectric directly and leading to very large leakage

Table 1.1 Scaling factors and limitation factors of MOSFET

Parameters	Ideal scaling factor	Actual scaling factor	Limitation
Electric field	1	>1	–
Threshold voltage V_{th}	$1/\alpha$	$>>1/\alpha$	Leakage current
Gate length L	$1/\alpha$	$1/\alpha$	–
Oxide thickness T_{ox}	$1/\alpha$	$>1/\alpha$	Defects, tunneling
Junction depth r_j	$1/\alpha$	$>1/\alpha$	Resistance
Drain bias V_D	$1/\alpha$	$>>1/\alpha$	System and V_{th}
Substrate doping	α	$<\alpha$	Junction breakdown

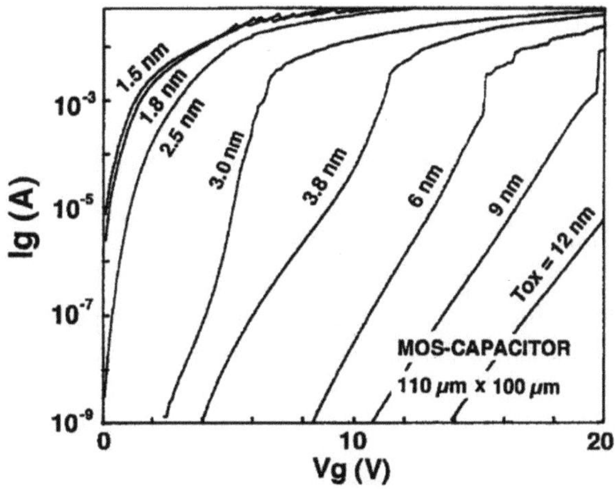

Fig. 1.3 Dependence of gate leakage current of MOS capacitors on the thickness of gate oxide [7]

current, the performance of the MOS devices are not suitable for actual application when the SiO_2 dielectric is thinner than 3 nm [5, 6]. Figure 1.3 shows the leakage current of MOS capacitors with different gate oxide thicknesses range from 12 nm down to 1.5 nm, it is obvious that capacitors with gate oxide thinner than 2.5 nm have a significant leakage component in low gate bias region caused by direct tunneling. Especially when $T_{ox} = 1.5$ nm, the leakage current reaches the order of 1×10^{-3} A/cm^2, which is not tolerable [7].

Apart from leakage current, device reliability is another essential factor that should not be ignored, the 10-year reliability criterion of CMOS technology cannot be guaranteed in devices with gate oxide thinner than 2 nm [8]. For example, a series of physical and chemical phenomena caused by dielectric degradation due to the emission of hot electrons at the Si/SiO_2 interface will eventually lead to the breakdown of the gate dielectric.

Back to 2007, in the reports of International Technology Roadmap of Semiconductors (ITRS), it was mentioned that equivalent gate oxide thickness (EOT) is the most difficult challenge associated with the future device scaling. For high-performance applications, EOT of less than 1 nm with adequate reliability is needed, while on the other hand, for low-power applications, leakage current caused by ultrathin gate oxide must be decreased. Moreover, in the executive summary of 2013 ITRS, the reduction of EOT was mentioned still to be a difficult challenge in the near term.

With the advantages of thermodynamically and electrically stability, superior electrical isolation properties, low mid-gap interface state densities with Si (10^{10}/cm^2), high dielectric strength (15 MV/cm), etc. [9], thermally grown amorphous SiO_2 as gate dielectric has kept its predominance in the past decades. However, as the existence of the limiting factors mentioned above, measures must be taken in order to keep the scaling trend of integrated devices without sacrificing the performance of the devices. Therefore, the idea of replacing conventional SiO_2 with material with high dielectric constant is proposed.

1.2 High-k Dielectrics and Electrical Performance

1.2.1 Dielectric Constant

Dielectric constant (k), as known as relative permittivity, is a factor that reflects decreasing of the electric field between charges relative to vacuum, which can be defined by Eq. (1.1):

$$k = \frac{k(\omega)}{k_0} \tag{1.1}$$

where k(ω) is the complex frequency-dependent absolute permittivity of the material, and k_0 is the permittivity of vacuum. Since the accumulation capacitance per unit area (C_{ox}) of an MOS capacitor equals to:

$$C_{ox} = \frac{kk_0}{T_{ox}} \tag{1.2}$$

the k value of a material is usually calculated by Eq. (1.3):

$$k = \frac{C_{ox}T_{ox}}{k_0} \tag{1.3}$$

It is well known that the saturation current (I_D) of a MOSFET can be written as:

$$I_D = \frac{W}{2L}\mu C_{ox}\left(V_g - V_{th}\right)^2 \tag{1.4}$$

in which W and L are the channel width and length, μ is the carrier mobility, Vg is the gate bias and V_{th} is the threshold voltage of the MOSFET. Therefore, during

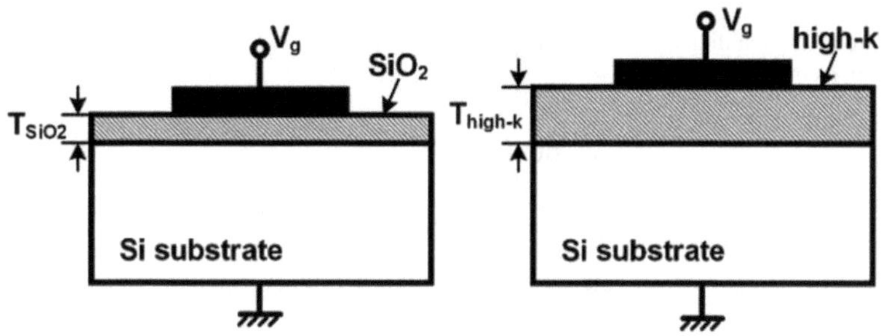

Fig. 1.4 Illustration of high-k material replacing SiO_2

scaling down, in order to maintain a certain level of drive current that can make the device functional, C_{ox} must be increased or at least be kept constant. As discussed in the last section, it is almost impossible to further decrease the thickness of SiO_2, replacing SiO_2 with other material with higher k value is the only way to increase C_{ox}. After replacement, the equivalent oxide thickness (EOT) can be calculated as:

$$EOT = \frac{k_{SiO2}}{k_{high-k}} T_{high-k} \tag{1.5}$$

meaning that the device with high-k material as gate dielectric is capable of achieving the same capacitance as the conventional SiO_2 based device but with larger physical thickness, which can also solve the problem of the unacceptable leakage current caused by direct tunneling, as shown in Fig. 1.4.

The first high-k material studied was silicon oxynitride (SiO_xN_y), K.S. Krisch et. al found that nitrogen can be effectively incorporated into SiO_2 by annealing the film in NH_3 or other gas contains nitrogen, and thus the reliability of the dielectric is improved [10]. Besides, doping N into SiO_2 is also reported to be able to reduce oxide charges and border traps [11]. However, if the N doping concentration is too high, the positive charges induced will cause large threshold voltage shift and interface degradation. In 2007 ITRS reports, it was mentioned that silicon oxynitride no longer meet the strict leakage current requirement anymore, therefore, material with high k value should be studied.

1.2.2 *Selection of High-k Dielectric*

Figure 1.5 demonstrates the k values together with band gaps of the promising candidates in high-k materials, however, it is never a simple job to determine which one is the best choice because there is always a trade-off between k value and the band offset that band gap of metal oxides tends to vary inversely with k value [12]. For example, as shown in Fig. 1.5, TiO_2 with k value of 80 has a band gap of only 3.3 eV, and the k value of one of the perovskite oxide $SrTiO_3$ can reach around

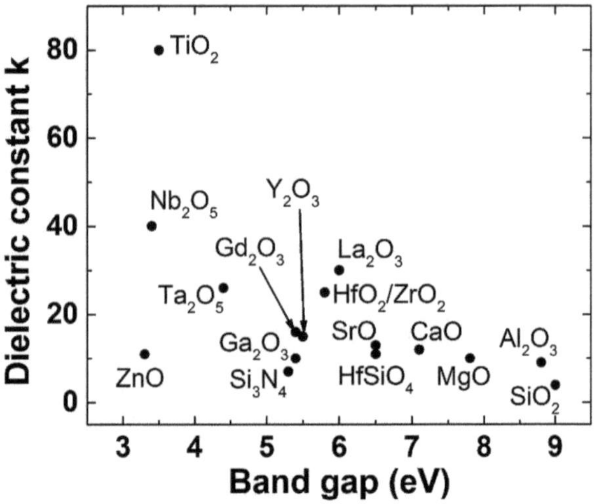

Fig. 1.5 k value vs. band gap of promising candidates as gate dielectric

2000, but its band gap is even smaller (3.2 eV). Therefore, apart from k value, band gap is also an important criterion of choosing dielectric material. Band gap should be large enough (usually over 5 eV) to make sure the dielectric material is an insulator, besides, the band offset with Si substrate should also be large enough (usually over 1 eV) so that the Schottky emission of carriers into the oxide band can be prevented [13].

Moreover, since the interface between high-k dielectric and Si substrate is the dominant factor that that influence the electrical performance of MOS devices, kinetic and thermodynamic stability should be considered as another important criterion. In ideal devices, there should be no reactions between the dielectric film and the substrate, and no interfacial layer should be formed either. Unfortunately, most high-k materials are compatible with Si at some certain processing conditions [14]. Assuming the high-k dielectric is M_xO_y, then there are basically four types of reactions between M_xO_y and Si [15–17]:

1. Silicon oxidation

$$Si + M_xO_y \rightarrow SiO_2 + xM \tag{1.6}$$

2. Silicide formation

$$Si + M_xO_y \rightarrow SiO_2 + M_xSi \tag{1.7}$$

$$Si + M_xO_y \rightarrow M_xSiO_y + M \tag{1.8}$$

$$Si + M_xO_y \rightarrow M_xSiO_y + M_xSi_y \tag{1.9}$$

3. Silicon oxide layer might be formed on the Si surface during the growth of the dielectric film or annealing, which enables the formation of silicates

$$SiO_2 + M_xO_y \rightarrow M_xSiO_y \tag{1.10}$$

4. SiO gas formation at low oxygen pressures and diffuse through the oxide film

$$Si + M_xO_y \rightarrow SiO \uparrow + M_xSi_y \tag{1.11}$$

These reactions will induce large amount of dangling bonds, oxygen vacancies and traps that negatively impact the device performance, and the SiO_2 layer formed will decrease the k value of the gate stack and thus nullify the benefit of high-k dielectric. Therefore, it is essential to choose a dielectric with large Gibbs free energy and low oxygen diffusion coefficient to inhibit the reactions with Si and the formation of the low-k SiO_2 interfacial layer during annealing at high temperature [18, 19].

1.2.3 *Defects and Interfaces of High-k Dielectrics*

When compared with SiO_2, high-k oxides contain more defects because firstly, off-stoichiometry defects such as oxygen vacancies in SiO_2 are rare because of the high heat during its formation [20]. Secondly, unlike the covalent bonding SiO_2, high-k oxides have ionic bonding and have larger coordination number, making high-k oxides poor glass formers. Therefore, it is much easier for them to crystallize during high temperature annealing, as a result, it is difficult for the crystal network to relax and rebond to remove the defects [21].

In ionic oxides, the intrinsic defects are oxygen vacancies and oxygen interstitials. Forster et al. calculated the energy levels of the defects in HfO_2 and ZrO_2 as shown in Fig. 1.6, in which V^+ and V^{2+} represent positively charged oxygen vacancies and O^0, O^{20}, O^-, O^{2-} are oxygen interstitials which are negatively charged [22, 23]. It is found that the defects are capable of trapping electrons from Si substrates, but further work is still needed to study the mechanism of the defects.

Besides, hydrogen is another type of defect which comes from forming gas ($N_2 + H_2$) annealing or the deposition of high-k film using atomic layer deposition (ALD). Reports show that hydrogen acts as shallow donor in most of the oxide materials (HfO_2, ZrO_2, La_2O_3, Y_2O_3, TiO_2, $LaAlO_3$ and $SrTiO_3$, etc.), while in silicates, SiO_2 and Al_2O_3 hydrogen is deep donor and therefore a source of fixed charge [24, 25]. The sign of the fixed charges in different oxides depends on the position of H_0 energy level with respect to their band gaps. For example, in HfO_2, ZrO_2, La_2O_3, Y_2O_3, level of H_0 lies above the band gap of Si, so the hydrogen in these oxides tends to ionize as H^+ and donate electrons to Si, and thus positive fixed

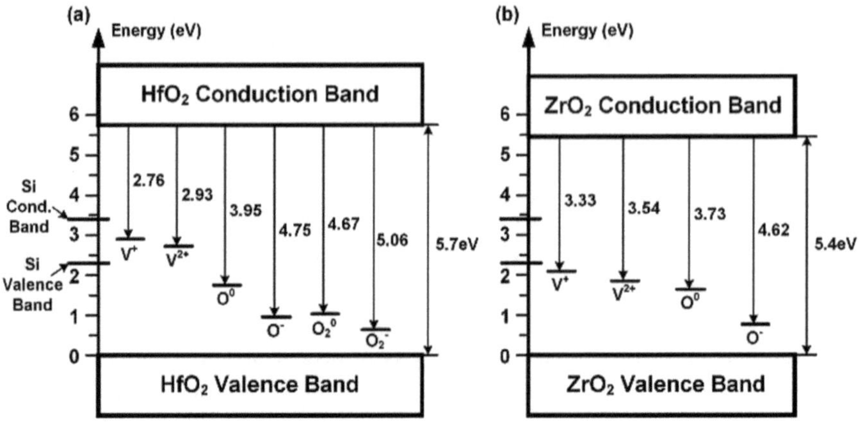

Fig. 1.6 Energy levels of the defects in (**a**) HfO_2 and (**b**) ZrO_2 [19, 20]

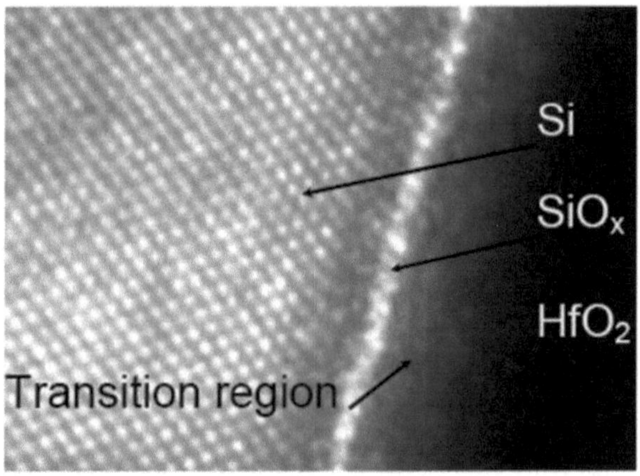

Fig. 1.7 TEM graph of HfO_2/Si interface [24]

charges are formed. Whereas, the H_0 level in Al_2O_3 locates below the band gap of Si and tends to capture electrons to form positive fixed charge. Therefore, in consistent with experimental results, the fixed charge in HfO_2 and ZrO_2 is usually positive but in Al_2O_3 it is negative [14, 26].

As mentioned in the previous section, most of the high-k oxides have poor kinetic and thermodynamic stability, leading to the formation of unstable and undesirable interfacial layer, and thus worsen the interface quality. For example, the low Gibbs free energy of Ta_2O_5 and TiO_2 makes them easy to react with Si at 1000 K [16], also, and interfacial layer SiO_x can be formed after depositing HfO_2 directly on Si (Fig. 1.7) [27]. These will cause a high interface charge density Dit. Reports show that most high-k dielectrics have high D_{it} (~10^{11}–10^{12} cm^{-2} eV^{-1}) and result in mobility degradation and flatband voltage shift [14]. Therefore, in order to improve the interface quality, the idea of adding a buffer layer between Si and high-k material is proposed [28]. According to Gutowski et al. a stable interface silicate layer can be formed between HfO_2/ZrO_2 and Si if a thermal treatment is implemented after the deposition of the dielectric, and this silicate layer is frequently used as buffer layer [29].

Apart from the interface between high-k and Si substrate, the high-k/gate interface is also an important factor that influence the device properties. Since the capacitance degradation in the depleted polysilicon electrode cannot be ignored in ultrathin gate dielectric, plus most high-k materials are thermally unstable and easy to crystallize, polysilicon is no longer a suitable gate material and metal gates are needed [30]. The selecting criteria of metal gate includes work function, ease of processing, thermal stability, etc. Up till now, single metal (Al, Ti, W, Hf, Ta, Mo, Ru, Au, Ni and Pt), metal nitrides (TaN, TiN, MoN and WN), metal silicide (CoSi and NiSi) and metal oxide (RuO_2) have already been investigated.

Figure 1.8 shows the work functions of some common metals. It is known that mid-gap metals are not capable of providing suitable work function on high-k dielectrics and thus negatively impact the performance of the transistor. Therefore, mid-gap metals like W, Fe, Cr, Mo, etc. cannot be used as the gate of MOS devices. In order to achieve relatively low and symmetry threshold voltage for nMOS and pMOS devices, alloy materials with tunable work function are preferable [31]. Moreover, the work function can be further tuned by adding different content of nitrogen such as TiAlN [32, 33].

It is well known that a high-temperature annealing is usually needed for metal gates so that ohmic contact can be formed. For metal nitride gate, during the annealing, nitride tends to diffuse into the dielectric bulk, which is capable of reduce the density of oxide charges by passivating oxygen vacancies in the high-k layer and also decrease D_{it} at high-k/Si interface [34]. However, the intermixing of the metal atom and the high-k metal atom may happen at the same time, which will negatively influence the characteristic of the gate stack [34].

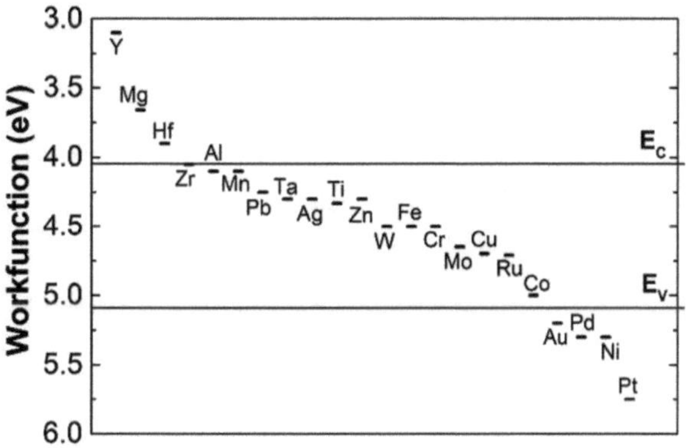

Fig. 1.8 Work functions of different metals

1.2.4 Mobility Degradation and Threshold Voltage Shift

Because of the existence of the defects in dielectric bulk and the interface states mentioned in the previous section, carrier mobility will be degraded due to various scattering mechanisms at both Si bulk and high-k/Si interface. The effective channel carrier mobility (μ_{eff}) is varies as a function of gate effective field and determined by Coulombic scattering (μ_{coul}), phonon scattering (μ_{ph}), and surface roughness (μ_{rough}) at low, moderate, and high fields respectively, which can be calculated by Eq. (1.12) according to Matthiessen's rule.

$$\frac{1}{\mu_{eff}} = \frac{1}{\mu_{coul}} + \frac{1}{\mu_{ph}} + \frac{1}{\mu_{rough}} \tag{1.12}$$

MOSFETs with high-k gate oxide exhibit lower electron mobility than FETs with SiO_2 gate oxide, as shown in Fig. 1.9 [26]. The exact cause of the degradation in mobility still needs more works to make sure, but several reasons can be deduced according to the studies up till now. Firstly, the higher oxide trap and interface trap of high-k oxide than that of SiO_2 enhances the Coulomb scattering. Secondly, a large amount of optical phonons in high-k oxides have low-lying soft polar mode, leading to a remote scattering of carriers, and thus limit the carrier mobility [35]. It has been reported that the soft phonon mechanism can be inhibited by using $HfSiO_4$ ($ZrSiO_4$) or adding a SiO_2 buffer layer to keep the high-k dielectric away from the channel [36, 37], however, this will inevitably increase EOT. In addition, high-k oxides have larger surface roughness than SiO_2 due to the longer bond length of metal-O and metal-Si than that of Si-Si.

Charge-trapping centers can be induced by the large amount of fixed charges locate in high-k materials, leading to a threshold voltage (V_{th}) shift which is another

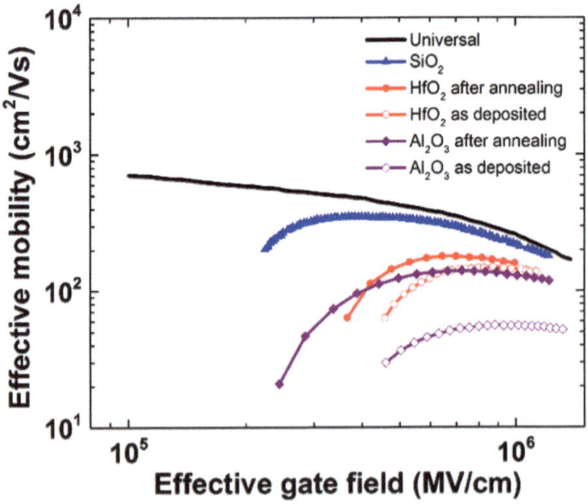

Fig. 1.9 Comparison of electron mobility in devices with high-k and SiO_2 dielectric [26]

concern of applying high-k oxides in MOSFETs. Besides, the reasons for V_{th} shift also include Fermi-level pinning effect at high-k/Si interface which can be modeled by the metal-induced gap states (MIGS) theory [38, 39]. In MIGS theory, the Schottky barrier height is given by:

$$\Phi_n = S_p(\Phi_m - \Phi_{CNL}) + (\Phi_{CNL} - \chi_S) \tag{1.13}$$

where Φ_{CNL} is the charge neutrality level, Φ_M and χ_S are the work function and the electron affinity of the semiconductor substrate, then the derivational Fermi-level pinning parameter S can be obtained as:

$$S = \frac{1}{1 + q^2 N_{it}\delta / \varepsilon_0 \varepsilon_{Si}} \tag{1.14}$$

in which Nit is the interface states density per unit area and δ is their extent into the semiconductor. In ideal case, there is no interface states, so the parameter S = 1 and no Fermi-level pinning happens. On the contrary, if S approaches to 0, high density of states exist at/near the high-k/Si interface and the device will suffer from sever Fermi-level pinning, leading to the variation of Schottky barrier height and thus the shift of V_{th}.

Experimentally, take HfO_2 as an example, Fig. 1.10 is the comparison between C-V characteristics of MOS capacitors measured at different temperatures with HfO_2 as gate dielectric with and without nitrogen incorporation [40]. The large negative flat band voltage (V_{fb}) found in the sample without nitrogen proves the existence of positive fixed charges as discussed in Sect. 1.2.3. With nitrogen incorporation, the C-V curve shift to the positive direction and the stretch-out is decreased, besides, the temperature-dependent flat band shift is also decreased to a large extent, implying a significant improvement of the interface quality due to the

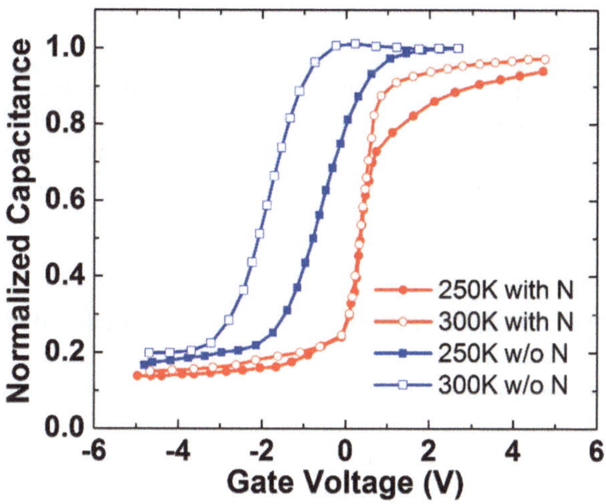

Fig. 1.10 High-frequency (1 MHz) C-V characteristics of HfO_2 films at different temperatures with and without nitrogen incorporation [40]

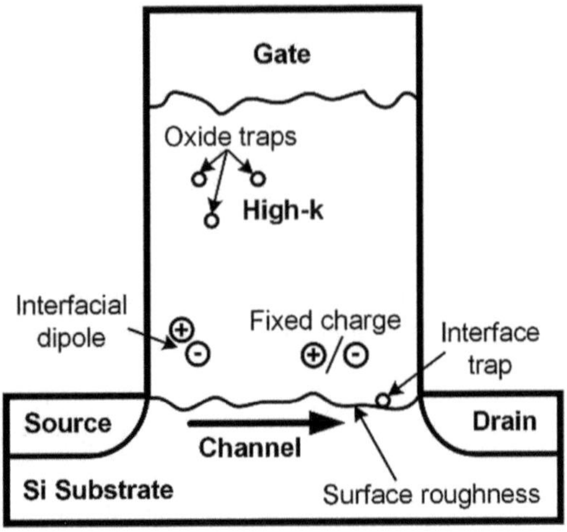

Fig. 1.11 Factors that cause mobility degradation and Vth shift [41]

converting of the Hf–Si bond to Hf–N. Apart from nitrogen implantation, thermal treatment and Al incorporation are two other methods to weaken the shift of Vfb by reducing the defects in the oxide bulk and/or at the high-k Si interface.

To sum up, the large amount of defects in the high-k oxide and the poor interface between high-k dielectric and Si are the main reasons that cause mobility degradation and V_{th} shift, as shown in Fig. 1.11. Therefore, the main challenge is to improve the bulk and interface quality of high-k dielectrics in order to optimize the performance of MOS devices.

1.3 Deposition

Being an essential factor that influence the densities of oxide defects and interface states, the quality of the high-k dielectric film must be paid sufficient attention. In order to guarantee acceptable performance of the resulting devices, grown dielectric films with good thickness uniformity, thermal stability and interfacial properties are required, therefore, correlating with the film quality and properties, deposition methods should be chosen carefully. Several deposition techniques have already been developed, the comparison of which are listed in Table 1.2 [41]. Chemical vapor deposition (CVD), Atomic layer deposition (ALD), and Physical vapor deposition (PVD) will be introduced in this section.

Basically, the chemical vapor deposition (CVD) is the transportation of the required materials from metallic-organic and some other precursors to the surface of the substrate. Through thermal, plasma or photo treatment, the precursors are decomposed with the materials needed deposit on the substrate. For example, trisdipyvaloylmethanato lanthanum ($La(C_{11}H_{19}O_2)^3$) can be used as the precursor

Table 1.2 Comparison of different deposition technologies

Method	ALD	MBE	CVD	Sputter	Evaporation	PLD
Uniformity	Good	Fair	Good	Good	Fair	Fair
Film density	Good	Good	Good	Good	Poor	Good
Step coverage	Good	Poor	Varies	Poor	Poor	Poor
Interface quality	Good	Good	Varies	Poor	Good	Varies
Multi-materials	Fair	Good	Poor	Good	Fair	Poor
Low-temp. deposition	Good	Good	Varies	Good	Good	Good
Deposition rate	Fair	Poor	Good	Good	Good	Good
Industrial applicability	Good	Fair	Good	Good	Good	Poor

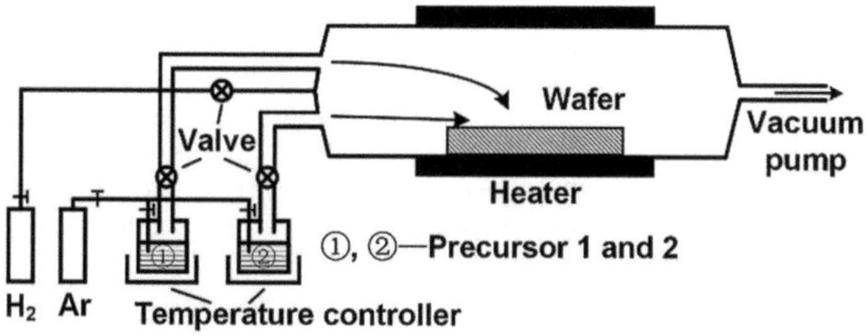

Fig. 1.12 Schematic diagram of a simplified ALD system

of growing La_2O_3, with a deposition temperature of 400–650°C in the ambient of O_2. However, uniformity problem is found in the La_2O_3 film and a silicate interfacial layer is formed at the La_2O_3 interface [42]. The major advantage of CVD is it is able to simply control the ratio of elements in tenary oxides, as reported by Y. Ohshita, et al. the percentages of Hf and Si in $Hf_{1-x}Si_xO_2$ can be controlled by varying the pressure, gas flows and temperature [43]. On the other hand, step coverage is one of the disadvantages of CVD, perfect coverage of steps cannot be obtained with aspect ratios larger than 10:1. Besides, relatively high temperature (400–850°C) is necessary during deposition, which is not good for materials with poor thermal stability such as rare earth oxides.

First known as atomic layer epitaxy (ALE) proposed in 1975 [44], it is developed to atomic layer deposit (ALD) with a wider range of materials and became the most widely used deposition method in actual application nowadays. The idea of ALD modifies the reaction of CVD by separating the two precursors into different containers. Figure 1.12 demonstrates a simple schematic diagram of an ALD system, high speed valves are used to strictly control the gas flow so that one single atomic layer can be deposited at a time. Therefore the thickness of the film can be determined by the number of transition cycles precisely, resulting in high quality film with excellent uniformity. The major disadvantages of ALD are (1) an interfacial low-k layer is still inevitable, (2) the poor stoichiometric control capability

Table 1.3 Different types of ALD

Type of ALD	Precursors	Target material	Temperature	Application
Catalytic ALD	(Metal)Cl_4, H_2O	Metal oxides	>32°C	High k dielectric layers, protective layers, anti-reflective layers
Al_2O_3 ALD	MCl_4, H_2O, Ti $(OiPr)_4$, $M(Et)_2$	Al_2O_3, Metal oxides	30–300°C	Dielectric layers, insulating layers, passivation layers
Metal ALD (Thermal chemistry)	$M(C_5H_5)_2$, $(CH_3C_5H_4)M$ $(CH_3)_3$, $Cu(thd)_2$, $Pd(hfac)_2$, Ni $(acac)_2$, H_2	Metal fluorides, organometallics, catalytic metals	175–400°C	Conductive pathways, catalytic surface, MOS devices
ALD on particles	Various gases	BN, ZrO_2, CNT, polymer particles	25–100°C (polymer), 100–400°C (metal/ alloy)	Insulative coating, optical and mechanical property modification, composite structures, conductive mediums
Plasma/ radical-enhanced ALD	Organometallics, MH_2Cl_2, TBTDET, NH_3	Pure metals, metal nitrides	450–800°C	DRAM, MOSFET, capacitors

makes it difficult to deposit complex oxides with required doping concentration. Up till now, many types of ALD are available due to different mechanisms, some representative examples are listed in Table 1.3 [45–47].

Most of the precursors used in CVD and ALD tend to induce contamination to the dielectric films, besides, the reaction between precursors and Si substrate is inevitable, resulting in the formation of unstable low-k materials. Another technique called physical vapor deposition (PVD) including sputtering, evaporation, and molecular beam epitaxy (MBE) is proposed to deposit transition metal/rare earth metal oxides [48]. Argon or electron beam are used in PVD to sputter the required material from the metal or oxide target to the wafer. Since no other chemical is applied in PVD, much lower contamination from carbon, hydrogen and chlorine can be achieved when compared with CVD and ALD. When a metal target is used, oxygen ambient is necessary in order to deposit oxide dielectrics, resulting in the formation of interfacial SiO_x layer [49]. In addition, as long as the target is metal oxide, after being sputter away from the target, the elements of required material deposited on the wafer cannot be guaranteed to be the expected ratio, i.e. the film obtained might be Y_xO_y instead of Y_2O_3, leading to a high density of oxygen vacancies [50] that negatively impact the performance of the device as discussed in the previous sections. Summing up, PVD has the advantages of low contamination and compositional profile of the film, but owns the drawbacks of poor conformality for large aspect ratio steps and high traps and/or interface states density.

1.4 Current Alternative High-k Dielectrics

1.4.1 Hf-based Oxides

As discussed, SiO_2 have already reached its physical limit as gate dielectric, therefore, with relatively high k value, suitable band gap, and fewer reliability problems than other high-k materials, HfO_2 gate dielectrics with metal gate is now being applied in most of complementary metal-oxide-semiconductor field effect transistors (CMOSFETs). However, as the defects and interface states still exist in HfO_2 dielectric, plus the formation of crystalline structure of HfO_2 at low temperature can provide pathways for leakage current and lead to dielectric breakdown eventually, taking measures to make improvement in order to meet the requirement of further scaled MOSFETs is necessary.

The first strategy proposed is nitrogen incorporation. As discussed in Sect. 1.2.4, nitrogen is capable of passivating oxygen vacancies in high-k bulk and reducing interface states, this is consistent with experimental results that better characteristics including higher crystallizing temperature have been found in HfO_2 film, and thus the performance of MOS devices has been improved significantly [51, 52]. However, since the N 2p states lies above O 2p states, doping nitrogen also means decreasing band gap [53], besides, nitrogen-related traps might be induced [54], so the concentration of nitrogen should be considered and well controlled.

A second method is doping other oxides to optimize the film and interface quality, band gap offset, k value, as well as crystallization of HfO_2. SiO_2, Al_2O_3, Ta_2O_5, and La_2O_3 as dopants are discussed in this section according to the existing reports.

Since an unstable low-k SiO_x interfacial layer always appears in the HfO_2/Si gate stack and cause negative effects on interface quality and k value, doping SiO_2 into HfO_2 during the dielectric deposition is a feasible solution that inhibit the formation of the interfacial layer and give rise to the crystallization temperature of HfO_2 as well [55]. Also, doping nitrogen (HfSiON) can further improve the thermal stability as mentioned above. Unfortunately, reduction of k value is an unescapable consequence of doping SiO_2, therefore, this trade-off should be considered when selecting SiO_2 concentration.

Frequently, Al_2O_3 plays the role of improving the thermal stability of high-k oxides. In the case of HfO_2, the crystallization of HfO_2 and the out-diffusion of Si from the substrate to the dielectric can be effectively inhibited by Al_2O_3 incorporation [56]. As confirmed by Ref. [56], with Al_2O_3 concentration of 33%, the $HfAlO_x$ remains amorphous after post-deposition annealing at 900°C. Besides, according to M. S. Joo, et al. HfAlO also has good compatibility with metal nitride gate after PDA treatment at high temperature, as shown in Fig. 1.13 [57]. Another effect of Al_2O_3 is to passivate oxygen vacancies, T. J. Park reported that oxygen vacancies can be decreased by around 4% after Al incorporation [58]. Moreover, the bandgap structures of HfO_2 and HfAlO are compared in Fig. 1.14 according to Ref. [58], with Al incorporation, shallow defect states can be removed and thus the

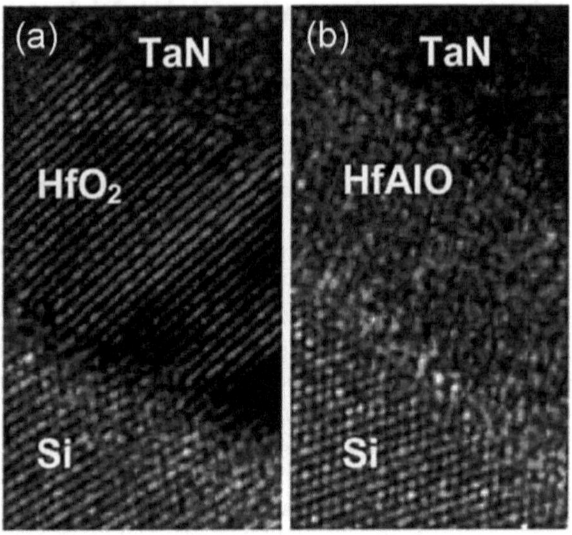

Fig. 1.13 TEM images of (**a**) HfO_2 and (**b**) HfAlO on Si after PDA treatment at 700°C [57]

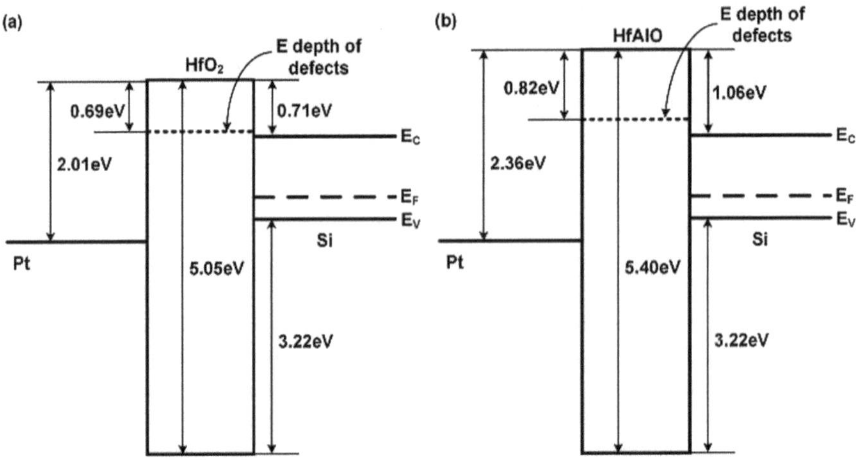

Fig. 1.14 Bandgap structures of gate stacks with (**a**) HfO_2 and (**b**) HfAlO as dielectric [58]

bandgap energy is increased, and the leakage current can be decreased accordingly. However, similar as SiO_2, the k value of Al_2O_3 (~9) is relatively low. In addition, Bae et al. pointed out negative fixed charges that degrades carrier mobility are introduced due to the accumulation of Al at the HfAlO/Si interface [59].

In order to improve the thermal stability of HfO_2 but meanwhile keep its relatively high k value, Ta_2O_5 (k ~26) incorporation is proposed. It has been proved by X. F. Yu, et al. that doping Ta_2O_5 effectively enhances the crystallization temperature of HfO_2. Figure 1.15 compares the cross sectional images of HfO_2

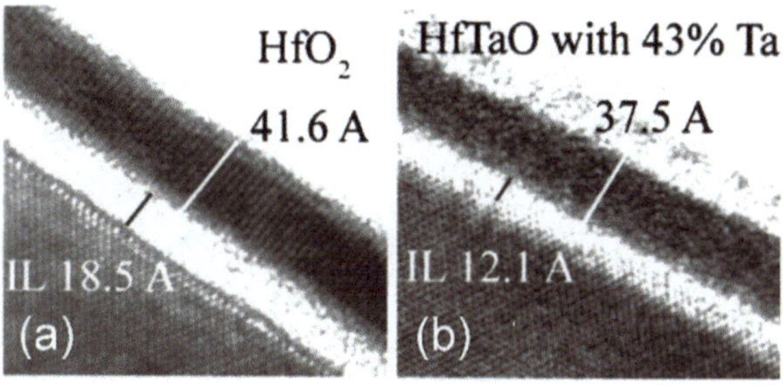

Fig. 1.15 TEM images of (**a**) HfO_2 and (**b**) HfTaO films after high temperature treatments [60]

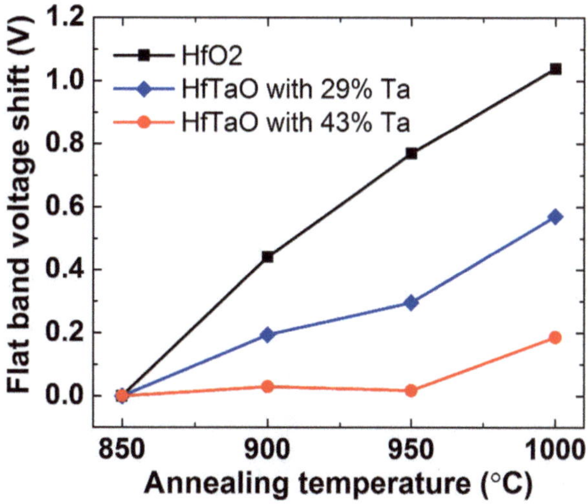

Fig. 1.16 Comparison of flat band voltage shift of pMOS capacitors with HfO_2 and HfTaO as gate dielectric after annealing at different temperatures [60]

and HfTaO after a PDA treatment at 700°C and an activation annealing at 950°C, obviously the crystal lines of HfO_2 are formed whereas HfTaO still remains amorphous [60]. Besides, in Fig. 1.15, the low-k interfacial layer formed on the surface of the HfO_2 sample is thicker than that of the HfTaO sample, implying that Ta_2O_5 incorporation is able to inhibit the intermixing of the dielectric and the Si substrate. Furthermore, in Ref. [60], the flat band voltage shift after high temperature annealing is also studied as shown in Fig. 1.16. The boron penetration-induced flat band voltage shift in HfO_2 film is dramatically suppressed by Ta_2O_5 incorporation, indicating less defects and interface states.

Rare earth elements like La is also a promising candidate as dopant to improve the properties of Hf-based high-k dielectric. Same as Ta_2O_5, doping La_2O_3 into HfO_2 can also increase crystallization temperature meanwhile keep the relatively high k value due to the high permittivity of La_2O_3 (above 20). A decrease of fixed charge density in the oxide film can be found after introducing La_2O_3 into HfO_2 by Y. Yamanoto et al. because the dependence of flat band voltage (V_{fb}) on capacitance equivalent thickness (CET) is very week according to Fig. 1.17 [61]. As

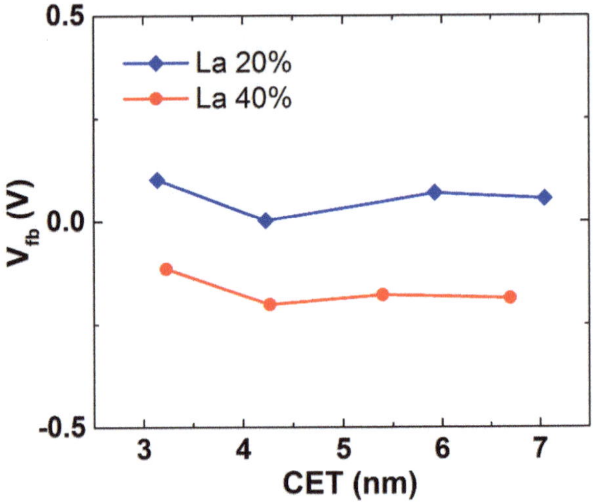

Fig. 1.17 Vfb vs. CET of MOS capacitors with $HfLaO_x$ as gate dielectric [61]

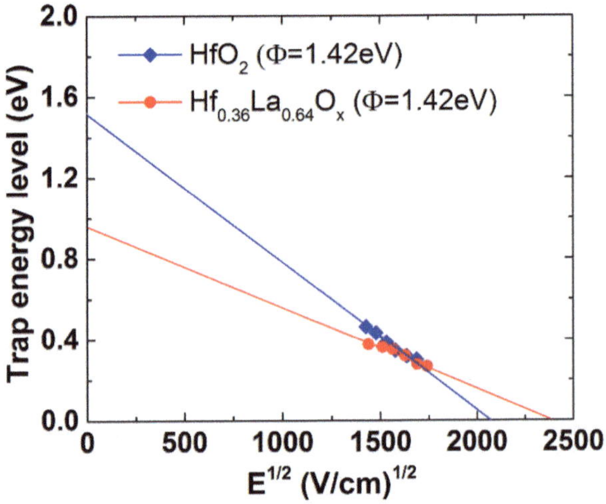

Fig. 1.18 Trap energy level vs. applied electric field of HfO_2 and HfLaO films [62]

confirmed by C. H. An, et al. in Fig. 1.18, the intrinsic trap energies in HfLaOx films are 1.42 eV, 1.34 eV, 1.03 eV, and 0.98 eV with La/(Hf + La) ratios of 0%, 42%, 57%, 64%, respectively, indicating a removing of shallow traps after doping La_2O_3 [62]. In addition, the metal work function of MOSFETs with TaN/$HfLaO_x$ or HfN/$HfLaO_x$ gate stacks can be tuned by varying the La concentration [63], which is preferred in achieving relatively low Vth as mentioned in Sect. 1.2.3. However, the major drawback of La_2O_3 is its hygroscopic nature, if the La_2O_3 possesses extremely high ratio in HfO_2, the film will absorb moisture easily and negatively influence the roughness of the dielectric and thus impact the carrier mobility.

1.4.2 Rare-Earth Oxides

Rare-earth oxide films are considered as promising candidates in varieties of applications such as luminescent materials, catalysts, buffer and protecting layer, constituents in oxide superconductors, and solid oxide fuel cell [64]. As dielectric material, rare-earth oxides own advantages of relatively high k value (La_2O_3 ~ 24–27, Y_2O_3 ~ 12–18, Lu_2O_3 ~ 12–19, etc.) excellent chemical stability on Si, and large band offset with Si. Besides, one other advantage that cannot be found in the commonly used HfO_2 is the higher electron effective mass and barriers in some rare-earth oxides can suppress tunneling current [65]. Unfortunately, the main concern of rare-earth oxides is the hygroscopicity, leading to poor stability of the film. In addition, formation of silicon-metal oxide mixtures or crystallized silicate phases easily happens due to the diffusion of Si from the substrate, which is a sever deterioration of the interface [66, 67].

Up till now, the most mature study on rare-earth oxides as gate dielectric is about La_2O_3 which will be discussed in this section as a representative of rare-earth oxides. As mentioned, the hygroscopic nature of the rare-earth oxides is the main concern in real application because during CMOS fabrication, wet processes and exposing to air are inevitable, therefore, the effects of moisture absorption is an important factor that influents the electrical properties of MOS devices. A moisture absorption experiment of La_2O_3 was carried out and it was found that after being exposed in the air for 12 h, La_2O_3 was totally replaced by $La(OH)_3$ [68]. Roughness of La_2O_3 film before and after being exposed in air is compared in Fig. 1.19, the surface roughness increases from 0.5 nm to 2.4 nm, one possible reason for which is that the density of hexagonal $La(OH)_3$ is smaller than that of hexagonal La_2O_3, resulting in volume expansion, therefore, nonuniform volume expansion happens due to the nonuniform moisture absorbing, leading to an increased roughness [68].

As for the influence of moisture absorption on the electrical properties of MOS devices, it is controversial on the direction of V_{fb} shift. S. Guha et al. hold the opinion that the replacement of O^{2-} by OH^- means an inducing of positive charges, resulting in a negative shift of V_{fb} [69]. However, in Ref. [68], V_{fb} of MOS capacitor with La_2O_3 as gate dielectric exhibits shift to the positive direction as shown in Fig. 1.20, which is explained by the formation of OH^- that contains negative charges during absorbing moisture. Moreover, Fig. 1.20 also illustrate the

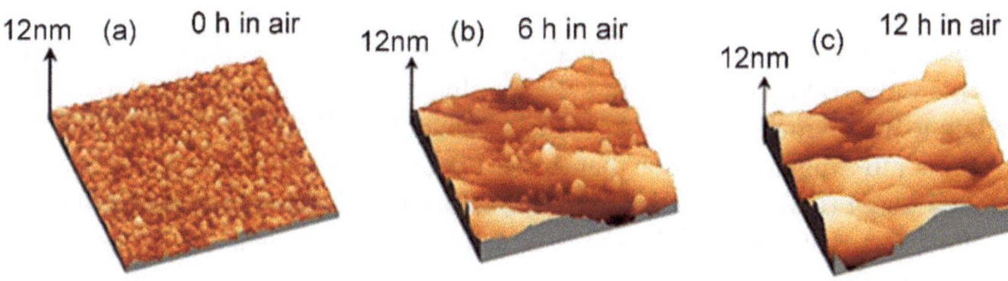

Fig. 1.19 Roughness of La_2O_3 film before and after being exposed in air [68]

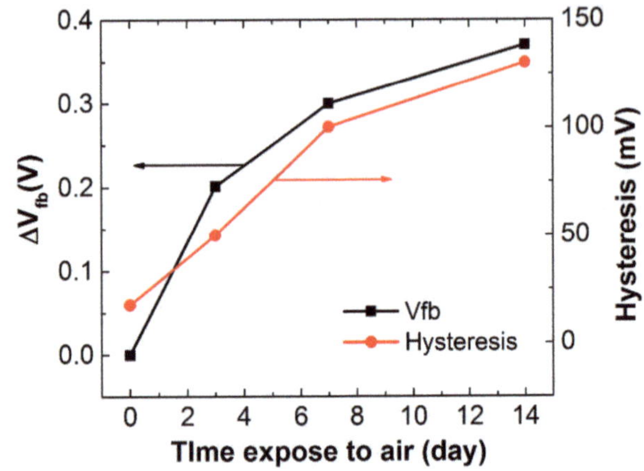

Fig. 1.20 V_{fb} shift and hysteresis of La_2O_3 film before and after exposed to air [65]

effect of moisture absorption on the hysteresis of C-V characteristic. Obvious increment of hysteresis can be found after being exposed in air, indicating the introduction of traps. Besides, leakage current of La_2O_3 is also found to be increased by several orders after moisture absorbing [68].

Apart from the hygroscoticity of La_2O_3, the absorption of carbon dioxide during post-deposition ambient exposure has also been reported [70], with reaction as followed:

$$La_2O_3 + 3CO_2 \rightarrow La_2(CO_3)_3 \tag{1.15}$$

However, the effect of CO_2 absorption is found much weaker than that of moisture absorption film, and thus negligible.

Another concern of La_2O_3 as dielectric that should be paid attention to is its k value degradation due to the hygroscopicity of La_2O_3. Theoretically, the k value of La_2O_3 should be 20–30, but many literatures has reported much lower k value than the theoretical value. As mentioned, $La(OH)_3$ is formed after moisture absorption, the k value of $La(OH)_3$ can be calculated by the Clausius-Mossotti relationship:

$$k = \left(3V_m + 8\pi\alpha^T\right) / \left(3V_m - 4\pi\alpha^T\right) \tag{1.16}$$

in which Vm is the molar volume and αT is the total polarizability. According to the Shannon's additivity rule, αT equals to 12.81Å3 [71], and 71 Å3 can be plugged in as the value of V_m [72]. Therefore, the k value of $La(OH)_3$ is estimated to be ~10, leading to an obvious degradation in the k value of La_2O_3.

In order to explain the reason that causes La_2O_3 absorbs moisture, the concept of Lattice Energy (LE) should be introduced here. LE is the energy required to break the ionic bonds in an ionic lattice of a solid ionic compound and separate one molecule into gaseous ions completely. It has been found that the LE of ionic oxides is inversely proportional to the metal ion radius [73]. Meanwhile, smaller LE tends to promote the reaction of moisture absorption [74]. Therefore, since La owns the largest ion radius among of the rare-earth metals, the LE of La_2O_3 is the smallest, and thus make La_2O_3 has very weak hygroscopic tolerance, indicating that enhancing the LE of La_2O_3 is the key point of suppressing its moisture absorption.

Doping other elements or oxides is the first idea, the dopant should own larger LE. Being in the same group with La, Y has many properties similar to La except that Y_2O_3 exhibits much lower crystallization temperature than La_2O_3 and thus able to achieve much larger LE. It is expected that Y_2O_3 might be a promising dopant and studies have been made on the properties of LaYO film. It has been reported that $La_{2-x}Y_xO_3$ films with 40–70% Y concentration has much larger lattice energy than La_2O_3 and high k value can be obtained with nearly no degradation after exposure, besides, MOS capacitor with $La_{2-x}Y_xO_3$ as gate dielectric shows better C-V behavior and smaller leakage current [68], indicating Y_2O_3 is indeed a good choice for dopant to enhance the hygroscopic tolerance of La_2O_3. However, the low crystallization temperature is usually not preferred. Apart from Y, Hf [75], Ta [76], and Nb [77] as dopant to La_2O_3 have also been tried with improved properties obtained.

Another idea is ultraviolet (UV) ozone treatment proposed by Ref. [68]. An experiment has been carried out comparing the influence of different treatment after La_2O_3 depositing on surface roughness with result shown in Fig. 1.21. The as-deposited sample and the 0.1% O_2 + N_2 sample show better moisture resistance than the N_2 annealing sample is possibly due to the decrease of oxygen vacancies in O_2 related ambient. Implying that the moisture absorbing tendency of La_2O_3 might has some dependency on the density of oxygen vacancies in the film. It has been reported that oxygen vacancies in oxide films can be passivated by UV ozone treatment at room temperature [78]. It is well known that charge transfer between La and O atoms tends to happen with the existence of oxygen vacancies and make

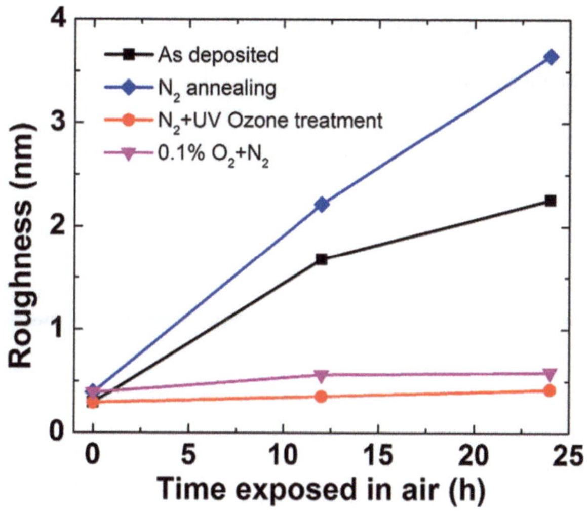

Fig. 1.21 RMS surface roughness of La_2O_3 vs. exposure time with different post deposition treatment [78]

the La–O bond more ionic, resulting in a decrease LE, so passivating oxygen vacancies can suppress the moisture absorbing at the same time. Therefore the N_2 + UV ozone treatment sample in Fig. 1.21 exhibit the best moisture resistance. Besides the formation of low-k interfacial layer can also be prevented because there is no need to provide high temperature.

1.4.3 Perovskite Structured Oxides

In recent years, perovskite structured oxide has received much attention for further device scaling due to their extremely high k values, especially $SrTiO_3$ (STO) [79–81]. In Ref. [79], relatively good results are obtained with the experimental STO k value of 45.2 and EOT of 5.4 Å. However, the main concern of STO is about leakage current due to its small band gap (3.2 eV) and conduction band offset with Si (~0 eV) [82]. The first idea to solve this problem is adding an interfacial layer with wider band gap, but usually a SiO_x layer is formed at the STO/Si interface after the deposition. This oxide layer with much larger band gap can be considered as a buffer layer to decrease the leakage current, which has been confirmed by R. Droopad, et al [83]. Based on the drain current curves in Fig. 1.22, the mobility of nMOS- and pMOSFET are 220 $cm^2\,V^{-1}\,s^{-1}$ and 62 $cm^2\,V^{-1}\,s^{-1}$ respectively, which are impressive. And the leakage current of these devices are also acceptable. But since SiO_x is not as stable as SiO_2 and might contains dangling bonds, the interface quality cannot be guaranteed and thus impact the carrier mobility.

In order to further decrease the leakage current of MOS device with STO as gate dielectric, the idea of doping Al into STO has been proposed to enlarge the band gap and meanwhile passivate oxygen vacancies. It has been proved that after doping Al

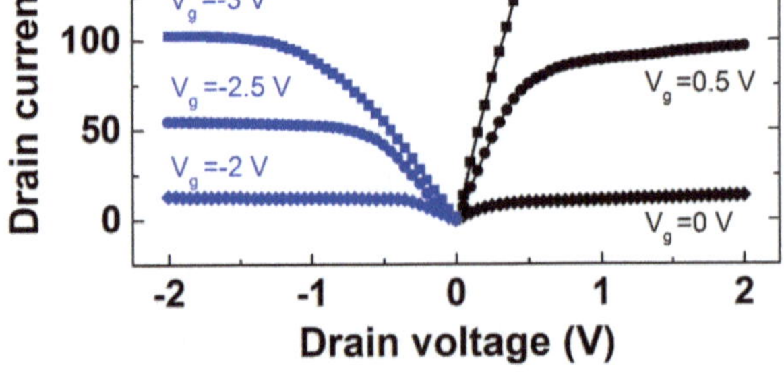

Fig. 1.22 Drain current curves of nMOS- and pMOSFET with STO as gate dielectric [83]

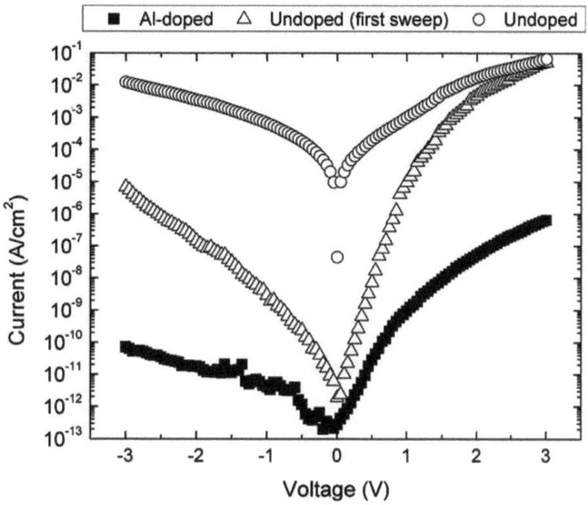

Fig. 1.23 Leakage current of 90Å-thick undoped and 20% Al-doped STO dielectric on Si [84]

with a ratio of 20%, the band gap of STO increases ~0.3 eV, and the leakage current decreases significantly as shown in Fig. 1.23 [84]. In addition, after the first voltage sweep, the leakage current of the undoped sample increases from 4×10^{-6} A/cm^2 to 8×10^{-4} A/cm^2 at 1V, whereas no increment is found in the Al doped sample (7×10^{-10} A/cm^2), indicating the oxygen vacancies are effectively passivated by Al. However, the poor thermal stability of STO is a fatal drawback that will impact the reliability of MOS device, so more studies should be made to achieve further improvement.

$LaAlO_3$ is another promising candidate with perovskite structure due to its large band gap (5.6 eV), relatively high k value (~24), large band offset and thermal stability with Si. High crystallization temperature of $LaAlO_3$ has already been confirmed (above 850°C) [85], besides, the formation of interfacial $LaAlSiO_x$ layer can be effectively suppressed if the $LaAlO_3$ film is deposited in N_2 ambient (LaAlON) [86]. Being consistent with the discussion of nitrogen incorporation in Sect. 1.4.1, LaAlON performs better than $LaAlO_3$ as gate dielectric of MOS device as proved by W. F. Xiang et al. in Fig. 1.24 [87]. Obvious improvement including higher k value, lower Vfb and leakage current is achieved by the LaAlON sample, which should be attributed to the reducing of interface states density by nitrogen and the more stable SiON interfacial layer than SiO_x formed at the high-k/Si interface.

The electrical performance and reliability of MOSFET with $LaAlO_3$ as gate dielectric has been studied by I. Y. Chang, et al [88]. Mobility of nearly 700 cm^2 V^{-1} s^{-1} and V_{th} of 0.35 V can be considered as impressive results. The temperature reliability of the device is illustrated by Fig. 1.25. It is well known that V_{th} is affected by the charges in the dielectric oxide and the states at/near the oxide/Si substrate as expressed in Eq. (1.16).

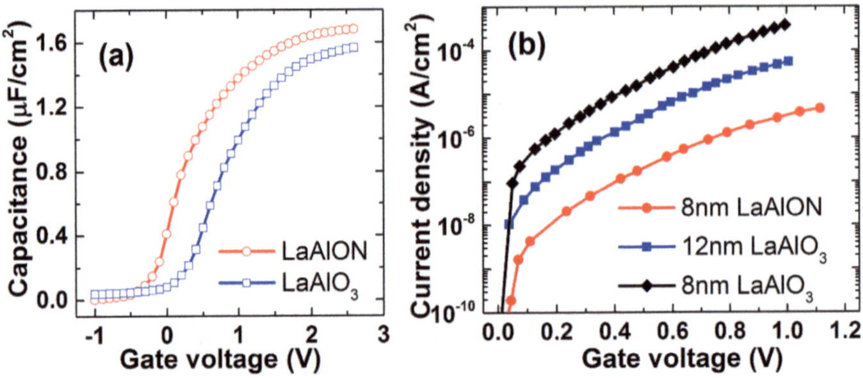

Fig. 1.24 Comparison of (**a**) C-V and (**b**) leakage current characteristics of MOS capacitors with $LaAlO_3$ and LaAlON as gate dielectric [87]

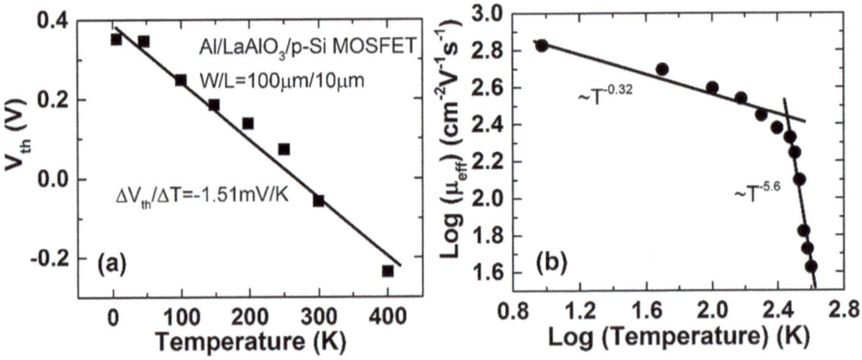

Fig. 1.25 Variation of (**a**) V_{th} and (**b**) mobility with the increase of temperature [89]

$$V_{th} = \Phi_{ms} + 2\psi_B + \frac{\sqrt{4\varepsilon_s q \psi_B N_A}}{C_{ox}} - \frac{Q_f + Q_{ot} + Q_m + Q_{it}}{C_{ox}} \tag{1.16}$$

where Φ_{ms} is the work function difference between the gate and the Si substrate, Q_f, Q_{ot}, and Q_m are the densities of fixed charges, oxide traps, and mobile ionic charges in the oxide layer, respectively, while Qit is the density of interface trapped charges. Since Φ_{ms}, Q_f, Q_m, and Q_{it} are independent with temperature, the variation of Vth¬ with temperature can be obtained by differentiating Eq. (1.17):

$$\frac{dV_{th}}{dT} = -\frac{1}{C_{ox}}\frac{dQ_{ot}}{dT} + \frac{d\psi_B}{dT}\left(2 + \frac{1}{C_{ox}}\sqrt{\frac{\varepsilon_s q N_A}{\psi_B}}\right) \tag{1.17}$$

Therefore, in the case of this section, the oxide density Q_{ot} is the determining factor of ΔV_{th}. In Fig. 1.25a, the temperature-dependent change rate of V_{th} ($|\Delta V_{th}/\Delta T|$) is 1.51 mV/K, larger than that of MOSFET with SiO_2 as gate oxide (1 mV/K), indicating larger amount of oxide charges in $LaAlO_3$. On the other hand, the electron mobility is proportional to $T_{-5.6}$ at high temperature, since the comparative value for device with SiO_2 is $T_{-1.75}$, conclusion of $LaAlO_3$ induces severer phonon scattering than SiO_2 can be drawn [89].

To sum up, perovskite structured oxides have the drawbacks of small bandgap and relatively poor thermal stability, nevertheless, they are attracting broad attention recently due to their extraordinary electronic and magnetic properties such as extremely high k value, high-T_c superconductivity, colossal magnetoresistance, and multiferroic behavior. Apart from the STO and $LaAlO_3$ introduced in this section, materials like $BaTiO_3$, $BaTi_{0.5}(Fe_{0.33}Mo_{0.17})O_3$, Sr_2MgMoO_6, etc. are being studied in varieties of areas such as memory, solar cell, and solid oxide fuel cell. Therefore, more studies needs to be done to make further improvement in the applications with perovskite oxides.

1.5 Applications

1.5.1 DRAM Cell Capacitors

It is well known that it is capacitor that DRAM uses to determine a bit value of 0 or 1 by whether the capacitor is charged or not. Since the capacitance of a parallel capacitor with an oxide layer as insulator is calculated by:

$$C = \frac{k\varepsilon_0 A}{T_{ox}} \tag{1.18}$$

with the scaling down of the device, the capacitance (C) will inevitably decreases with the decreasing of electrode area (A). However, a DRAM capacitor must guarantee a minimum capacitance of ~ 25 f. per cell to provide enough sensing margin and data retention time, so measures should be taken to increase C. In order to achieve an EOT of less than 1 nm, high-k materials have to be introduced as the insulating layer. Metal-insulator-metal (MIS) structured DRAM with Al_2O_3 insulator deposited by ALD is the first case with enhanced capacitance and relatively low leakage current [90, 91]. After that, Ta_2O_5 has become the most widely used high-k material and has been put in production due to its high k value, high breakdown fields, good compatibility with Si and strong ability of charge storing [92].

Besides, perovskites structured materials such as STO and $BaSrTiO_3$ have also been studied for the purpose of significantly increasing k value, however, as discussed in Sect. 1.4.3, the inevitable interfacial SiO_x layer is the fatal factor

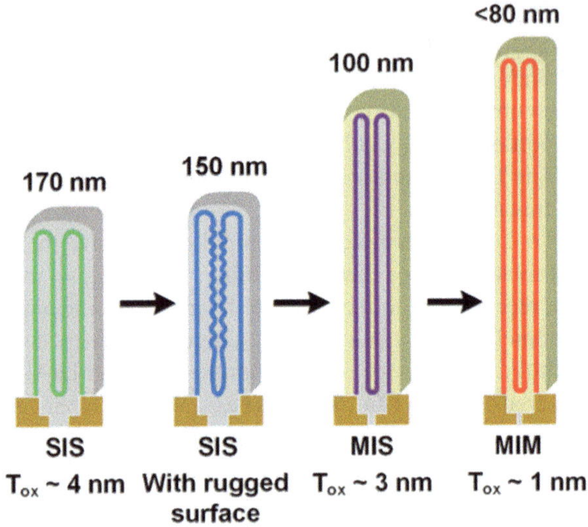

Fig. 1.26 DRAM capacitor evolution [95]

that impact the capacitance. Therefore, the replacement of MIS structure with metal-insulator-metal (MIM) structure is proposed to prevent the formation of the low-k interfacial layer. A record low leakage current density of 10^{-6} A/cm^2 with EOT of 0.4 nm have been obtained from the MIM DRAM with ALD deposited STO as insulator and Ru as bottom electrode, which enables the scalability of DRAM to the 3X nm mode [93]. Consequently, Al_2O_3 doped ZrO_2 and sandwich structured ($ZrO_2/Al_2O_3/ZrO_2$) stack with TiN as electrode MIM DRAM are reported to be able to achieve excellent reliability [94]. Figure 1.26 is a summary of the development of DRAM, more new materials and structures are still being studied [95].

1.5.2 Nonvolatile Memories

Nonvolatile memory typically plays the role of secondary storage or long-term persistence storage in modern electronic systems and evolves from read only memory (ROM), electrically programmable ROM (EPROM), electrically erasable programmable ROM (EEPROM) to flash memory. Flash memory has kept its dominating position in recent years due to its small cell size and the ability of operating program/erase more than 10^5 times with high reliability.

The first successful type is floating gate flash memory with control gate/inter-poly dielectric (IPD)/floating gate (FG)/tunneling layer (TL)/Si substrate stack. With the advantages of high program/erase speeds, low operating voltage and low power consumption, floating gate flash memory has continuously scaling down to increase the data-storage density, however, a relatively thick tunneling oxide (6–7 nm) and IPD layer (9–10 nm) are needed to guarantee the reliability due to the leakage caused by the possibility that charges stored in the floating gate tunneling through the IPD and TL with the assistant of defects. Oxide/nitride/oxide (ONO) stack or high-k materials as TL and/or IPD have been studied for a period to solve the leakage issue [96]. Unfortunately, temperature instability is found in ONO stack and the ionic defects (e.g. oxygen vacancy) in high-k materials tend to give rise to varieties of reliability issues. Therefore, novel structures and/or materials have to be investigated for further scaling.

Ferro-electric random-access memory (FeRAM) defines '0' and '1' by the remnant polarization of ferroelectric dielectrics through the control of an external field [97]. $Pb(Zr_xTi_{1-x})O_3$ (PZT) and $SbBi_2Ta_2O_9$ (SBT) are the most widely used ferroelectrics for FeRAM. High random access operating speed and low power consumption are the major advantages for FeRAM, but the rewriting operation after each read process is a severe problem that cause extra consumption [97].

Another type of flash memory that makes use of high-k material is discrete charge-trapping memory (CTM), in which '0' or '1' is defined by removing or adding charges from the charge-storage layer. The typical structure of CTM is based on the metal/oxide/nitride/oxide/Si (MONOS) gate stack, corresponding to gate/blocking layer (BL)/charge trapping layer (CTL)/tunneling layer (TL)/substrate stack, as shown in Fig. 1.27. The conventional SiN-based CTM has a trade-off between high program/erase speeds and good data retention, so band engineering is needed. The blocking layer contact directly with Si substrate plays the role of preventing the diffusion of carriers in the charge trapping layer back to the Si substrate, so the choice of high-k material has criteria of large band gap, few defects and trap in the bulk and at the interface, and good stability and compatibility with CMOS technology. With large band gap (8.8 eV) and excellent thermal stability, Al_2O_3 becomes the best choice for the blocking layer [14]. In order to further

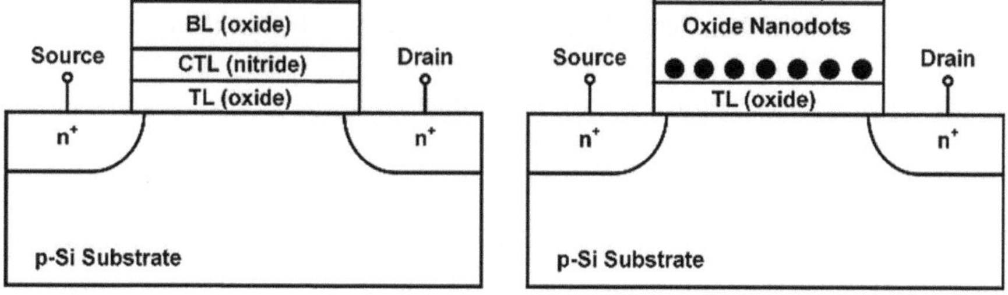

Fig. 1.27 Schematic diagrams of (**a**) nitride-based and (**b**) nanocrystal-based charge trapping memory

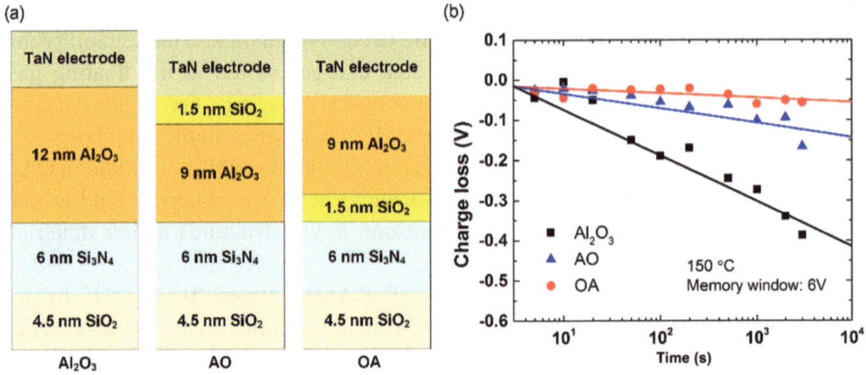

Fig. 1.28 (**a**) Gate stacks of CTMs with Al_2O_3 and Al_2O_3/SiO_2 as blocking layer. (**b**) Charge loss vs. time duration for the devices in (**a**) [101]

increase the k value, compound dielectrics such as Gd^- [98], La^- [99], and Y^- [100] doped Al_2O_3 have been studied with improvements in program speed, saturation window and breakdown characteristic have been made. An Al_2O_3/SiO_2 double layer is another idea for the blocking layer to improve the device properties due to the lower density of defects and larger barrier height of SiO_2, which is proved by H. Park et al in Fig. 1.28b, the SiO_2 layer effectively weaken the loss of stored data [101].

In terms of the charge trapping layer, shallow-level trap density should be as low as possible to maintain enough conduction band offset with the tunneling layer. Various high-k dielectrics (e.g. La_2O_3, Y_2O_3, ZrO_2, Ta_2O_5, etc.) have been investigated to replace conventional Si_3N_4 due to their deep-level traps and stronger scaling ability [102–105]. Recently, compound high-k materials as charge trapping layer are widely being studied such as LaTiON [106], Nb doped La_2O_3 [107], fluorinated STO [108], and Zr doped BTO [109].

Nanocrystals (NC) including Si, Ge, Pt, Au, etc. embedded in a dielectric matrix is another promising medium for trapping charges in CTMs as shown in Fig. 1.28b. Metal NCs own higher deep-level trap density and wider range of work function than their semiconductor NCs counterparts, but their poor thermal stability and contamination are fatal disadvantages [110, 111]. Therefore, high-k materials start to be applied in NC based memories due to their high deep-trap density and compatibility with MOS process, devices with Ga_2O_3 [112], Al_2O_3 [113], $TiAl_2O_5$ [114], etc. as NC charge trapping layer exhibit excellent memory properties.

Furthermore a bandgap-engineered (BE) barrier with an oxide/nitride/oxide stack as TL consists of low-k and high-k has been investigated in order to maintain good retention by enhancing the tunneling of carriers during the program/erase operating and thus inhibit the charge loss in retention mode [115]. BE barriers using Si_3N_4/SiO_2 [116] and HfO_2/SiO_2 [117] etc. have already been tried.

1.5.3 Novel Channel Materials

1.5.3.1 Ge

According to Eq. (1.4) in Sect. 1.2.1, the channel carrier mobility μ is another important factor that affects the drive current of MOS devices. Besides, as mentioned in Sect. 1.2.4, mobility degradation is inevitable when high-k dielectric is applied on Si. The measure of replacing conventional Si substrate has been taken so that the mobility issue can be solved, and thus further enhance the drive capability of MOS devices. This replacement begins with strained Si and SiGe, but limitations in mobility and V_{th} still exists for the high performance and low power consumption requirements in the future, and that is when Ge becomes attractive because of its superior intrinsic electron and hole mobility. The comparison between Si and Ge listed in Table 1.4 illustrates the advantage of Ge in mobility over Si.

High-k materials including HfO_2 [118], HfTiON [119], and LaTiON [120], etc. have been tried with the highest k value of over 30 obtained, however, the main challenges of Ge substrate with high-k dielectrics are poor high-k/Ge interface, Fermi-level pinning near valance band, and the easily formed unstable and water-soluble native oxide, so the interface quality still need to be improved. Post deposition treatment is always a good choice to decrease the defects and/or traps in the dielectric bulk and at the interface, which has been confirmed by Ref. [119] and [121] using wet N_2 annealing and fluorine incorporation respectively. Besides, it has been found that pulsed laser annealing is also capable of positively influence the quality of the high-k film and the interface (Fig. 1.29) [122]. With no need of high temperature, the risk of crystallization is also eliminated.

An ultrathin passivation interfacial layer is another method that is being widely investigated. It is important to choose materials with large bandgap and good thermal stability as passivation layer. Up till now, varieties of high-k oxides/oxynitrides have been applied with improvement obtained, including Y_2O_3 [123], TaON [124], LaTaON [125], and $GeSnO_x$ [118] etc.

1.5.3.2 III-V Compound Semiconductors

According to Table 1.4, Ge has much higher carrier mobility than Si, but the lower breakdown field is not suitable for high-voltage applications. Meanwhile, although the smaller band gap is preferable for low-voltage operation devices, large leakage

Table 1.4 Comparison of properties between Si and Ge

Property	Mobility ($cm^2V^{-1}s^{-1}$) Electron	Hole	Breakdown field (V/cm)	Band gap (eV)	k value	Saturation velocity (10^7 cm/s)	Melting point (°C)
Si	1350	500	3×10^5	1.12	11.9	1.0	1415
Ge	3900	1900	10^5	0.66	16.0	0.5	937

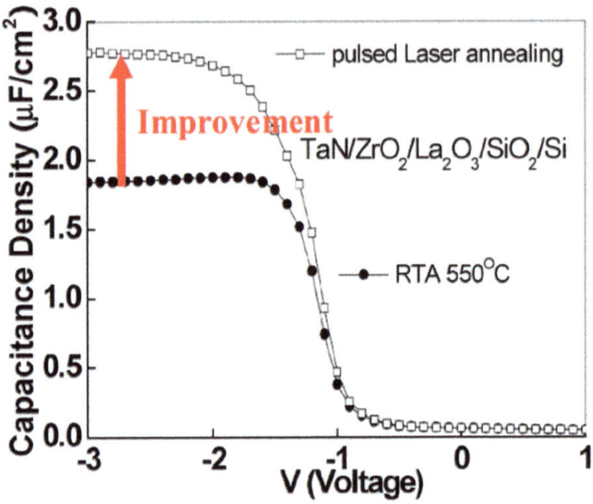

Fig. 1.29 C-V characteristics of Ge MOS capacitors with pulsed laser annealing and conventional thermal annealing [122]

Table 1.5 Basic parameters of major III-V compound semiconductors

Materials	Electron mobility ($cm^2V^{-1}s^{-1}$)	Hole mobility ($cm^2V^{-1}s^{-1}$)	Band gap (eV)	Lattice constant (Å)
Si	1400	450	1.12	5.43
GaAs	8500	400	1.42	5.65
InAs	4×10^4	500	0.37	6.03
$Ga_{0.53}In_{0.47}As$	1.2×10^4	850	0.85	5.87
InSb	7.7×10^4	300–400	0.17	6.48
InP	5400	200	0.35	5.87

current is also caused which will increase power consumption. Therefore, III-V compound semiconductors with even higher mobility have been proposed and paid much attention recently (Table 1.5 [41]). Among of the III-V compounds, GaAs with electron mobility around five times than Si and larger band gap and breakdown field has received most attention and is the most representative member up till now.

When high-k oxides deposited on GaAs (e.g. HfO_2), the Ga– and As– dangling bonds cannot fully saturated, mid-gap states will be induced by the partially saturated bonds and cause Fermi-level pinning, meanwhile, the dangling bonds at the interface might form unstable Ga–O, As–O, As–As, Ga–Hf, and As–Hf interfacial bonds during fabrication process. Therefore, similar with Ge, GaAs also suffers from the easily formed unstable native oxide and high density of interface states (D_{it}), mobility degradation and instability of devices, which makes surface passivation of the GaAs surface the most essential factor of optimizing the performance of MOS devices.

It has been found by F. S. Aguirre-Tostado et al. that sulfur passivation by treating GaAs surface with $(NH_4)_2S$ is capable of reducing interfacial space charges

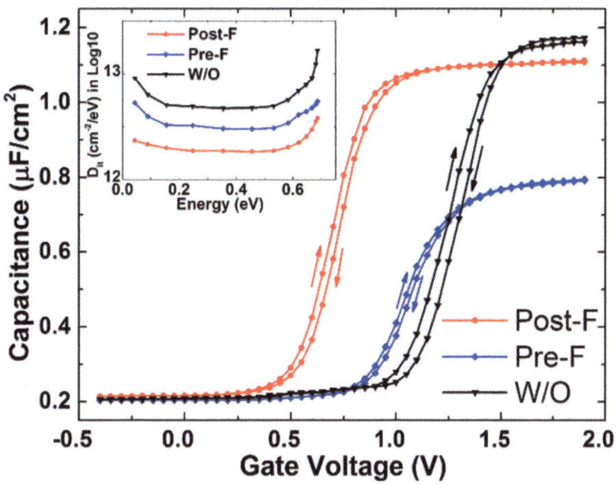

Fig. 1.30 C-V curves of GaAs capacitors with and without fluorine treatment [128]

[126]. With reactions [1.19] and [1.20] happens, Ga_2S_3 and As_2S_3 are formed at the GaAs surface, thus decreases the Ga- and As-related vacancies [127].

$$(NH_4)_2S + H_2O \rightarrow 2NH_4{}^+ + S^{2-} + H^+ + OH^- \tag{1.19}$$

$$2GaAs + 12H^+ + 6S^{2-} \rightarrow Ga_2S_3 + As_2S_3 + 6H_2 \tag{1.20}$$

In addition, post-deposition fluorine treatment can also effectively passivate oxygen vacancies in high-k oxides and suppress the formation of GaAs native oxide [128]. Figure 1.30 compares the C-V characteristics of GaAs MOS capacitors with post- and pre-deposition fluorine treatment, the Vfb and Dit of the post treatment sample is decreased significantly, overweighs a slightly loss in k value probably due to the high bonding energy between metal atom and incorporated F atoms [128].

Similarly, passivation layer is another effective solution of the poor interface quality. With the same criteria of passivation layer on Ge, Si, Ge, AlON, TaON, nitride $Ga_2O_3(Gd_2O_3)$, and LaTaON have been studied with improved performance obtained [129, 130].

1.5.3.3 Metal Oxide Thin Film Transistor (TFT)

Tremendous efforts have been made in the past few decades on thin film transistors (TFTs) because of its application in flat-panel display. Since the invention of TFT in 1962 using polycrystalline CdS as semiconductor material, the active layer of TFT has developed from a-Si: H, poly Si, SnO_2, and ZnO to meet the requirement of the increasing display panel size, consequently, in order to further increase the mobility, TFT with amorphous InGaZnO (a-IGZO) deposited at room temperature as channel material has been reported by Nomura et al. in 2004 with a mobility of

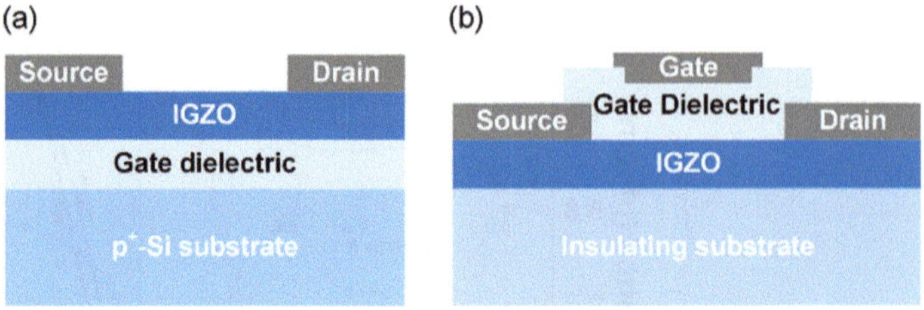

Fig. 1.31 Schematic diagrams of (**a**) top contact bottom gate and (**b**) top contact top gate structured IGZO TFT

~8.3 $cm^2\ V^{-1}\ s^{-1}$ [131]. Up till now, due to its advantages including high carrier mobility compatible with large panel size display, good stability, low-temperature fabrication process and high transparency, a-IGZO is still placed in the dominant position as the channel material of TFT.

Figure 1.31 shows the two typical structures of IGZO TFT. The top contact bottom gate structure (Fig. 1.31a) has simpler fabricating process with only one mask needed for the source/drain electrodes. As for the top gate in Fig. 1.31b, since there is no exposure of the IGZO surface, the device degradation can be significantly inhibited. In order to meet the future requirement of low power consumption, the subthreshold slope (SS) (defined and calculated by Eq. (1.21) should be decreased so that the operating voltage range can be narrowed down [132]. That is why the replacement of SiO_2 with high-k materials with larger Cox as gate dielectric layer in TFT happens.

$$SS = \frac{dV_{GS}}{d(\log I_D)} = \frac{kT\ln 10}{q}\left(1 + \frac{qN_t}{C_{ox}}\right) \tag{1.21}$$

Dielectrics such as Al_2O_3 [133], HfO_2 [134], Y_2O_3 [135] and BaSrTiO [136] have been studied but the degradation in carrier mobility mentioned in Sect. 1.2.4 is still inevitable due to the large polarization of high-k oxides. In Eq. (1.21), Nt is the trap density at/near the dielectric/a-IGZO interface which directly influences SS. Hence, methods of suppressing current leakage, formation of defect-related traps and interfacial reaction are still being explored to optimize the performance of IGZO TFT. It is well known that the quality of high-k bulk and interface varies with different conditions of post deposition treatment, and so does the performance of the TFT. Experimental results show that both post deposition annealing (PDA) and fluorine treatment are capable of optimizing the properties of the devices. L. X. Qian et al. have done investigations on different annealing time and ambient of a-IGZO TFT with HfLaO as gate dielectric, with results listed in Tables 1.6 and and 1.7 [137, 138].

According to Ref. [137] (Table 1.6), negative oxide charges can be effectively reduced by PDA in O_2, with acceptor-like border and interface traps removed as

Table 1.6 Electrical properties of a-IGZO TFTs with different annealing time in O_2

Annealing time (min)	0	10	30	60	120
μ_{sat} ($cm^2V^{-1}s^{-1}$)	4.3	15.7	35.2	25.7	24.9
SS (V/dec)	0.310	0.231	0.292	0.378	0.350
V_{th} (V)	4.3	3.5	2.8	1.7	1.7
ΔV_{th} (V)	1.8	0.4	−1.1	−2.1	−1.7
I_{on}/I_{off}	7.5×10^5	2.5×10^6	5.2×10^6	6.9×10^6	6.2×10^6
C_{ox} ($\mu F/cm^2$)	0.266	0.264	0.267	0.294	0.286

Table 1.7 Electrical properties of a-IGZO TFTs with different annealing ambient

Annealing ambient	w/o	O_2	N_2	NH_3
μ_{sat} ($cm^2V^{-1}s^{-1}$)	4.3	15.7	35.1	19.3
SS (V/dec)	0.310	0.231	0.206	0.315
V_{th} (V)	4.3	3.5	3.3	1.9
ΔV_{th} (V)	1.8	0.4	−0.03	−0.82
I_{on}/I_{off}	7.5×10^5	2.5×10^6	5.1×10^6	6.1×10^6
C_{ox} ($\mu F/cm^2$)	0.266	0.264	0.241	0.275

well, so the electrical properties of the device is improved significantly after PDA with saturation mobility increases from 4.3 $cm^2\ V^{-1}\ s^{-1}$ to 35.2 $cm^2\ V^{-1}\ s^{-1}$. However, the influence of annealing becomes negligible beyond 60 min and a decrease of mobility is found. As for Table 1.7, N_2 shows the most effective effect of oxygen vacancies are filling together with electron concentration decreasing in a-IGZO, and thus obtains the best electrical performance.

In addition, multicomponent dielectric is always a good choice according to the discussion in several previous sections. Doping SiO_2 into Ta_2O_5 for leakage reducing and oxygen vacancy passivating has been investigated in Ref. [139], as shown in Fig. 1.32, improvement in drive current, V_{th}, leakage current and SS can be found in the TaSiO sample when compared with the Ta_2O_5 one. Furthermore, multicomponent high-k dielectric such as HfLaO [137], LaTaO [140], and NbLaO [77] have all been investigated and proved to be promising candidates. For example, the saturation mobility as high as 39.8 $cm^2\ V^{-1}\ s^{-1}$ has been realized in the fluorinated a-IGZO/HfLaO TFT [141], as exhibited in Fig. 1.33.

1.5.3.4 Other Novel MOS Devices

In planar MOSFET, even though the replacement of SiO_2 with high-k materials can solve the leakage issue, problems caused by short channel such as drain-induced barrier lowering (DIBL) is no longer negligible with the relentless scaling. Hence the idea of multi-gate field effect transistor (MGFET) is proposed to increase the effective channel area. FinFET, one type of MGFET has now been considered as

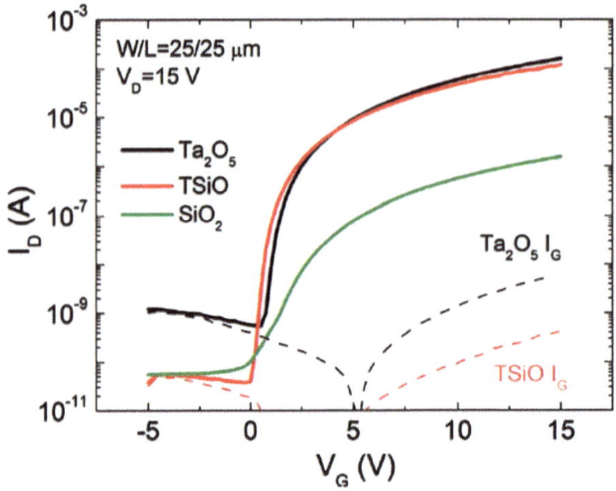

Fig. 1.32 Comparison of I-V characteristics of IGZO TFTs with Ta_2O_5, TaSiO, and SiO_2 as gate dielectric [139]

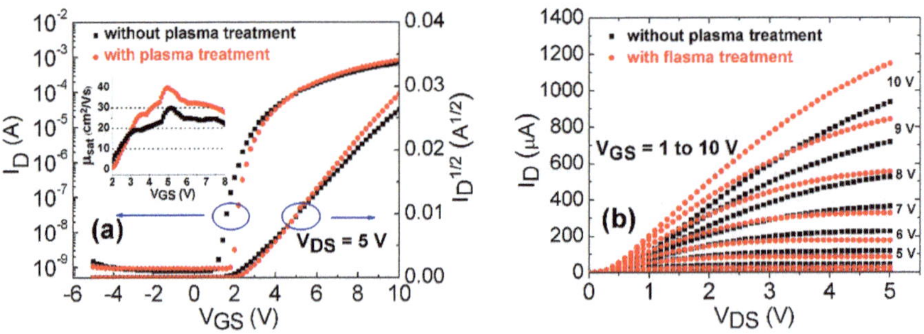

Fig. 1.33 Transfer (**a**) and output (**b**) characteristics of the a-IGZO/HfLaO TFTs with or without CHF_3/O_2 plasma treatment [141]

the most desirable alternative to planar MOSFET due to its simple structure and fabrication process with the simplified structure shown in Fig. 1.34 [142] and has already been adopted in 22 nm technology node by companies like Intel and TSMC.

Same as planar MOSFET, high-k gate dielectrics are also needed to suppress the increasing of leakage current as the devices scale down. FinFET with HfO_2 as dielectric and Mo as metal gate with and without nitrogen incorporation into Mo has been investigated by D. Ha, et al. with SEM image and transfer characteristic demonstrated in Fig. 1.35 [143]. Low leakage current density is achieved for EOT of 1.72 nm, and impressive Vth and SS are obtained. When comparing the I-V curve of the samples with and without nitrogen implant into Mo, a V_{th} shift of 0.45 V is found, indicating a work function modification of Mo, however, the reduction in work function by nitrogen of Mo on HfO_2 is lower than that of Mo on SiO_2 which is probably due to the Fermi-level pinning and nitrogen diffusion into HfO_2 [143], which needs future works to improve.

Fig. 1.34 Simplified structure of FinFET [142]

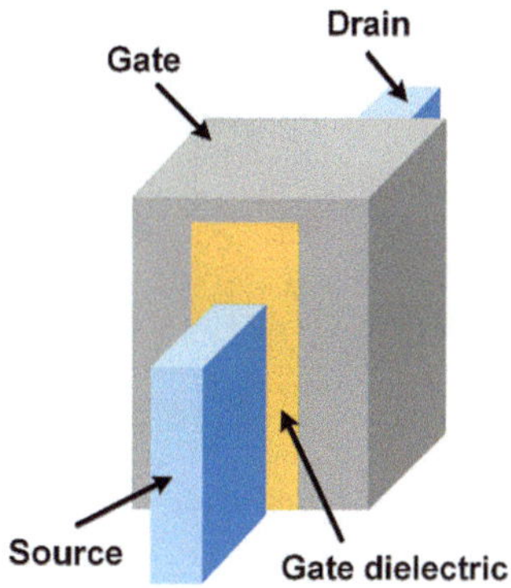

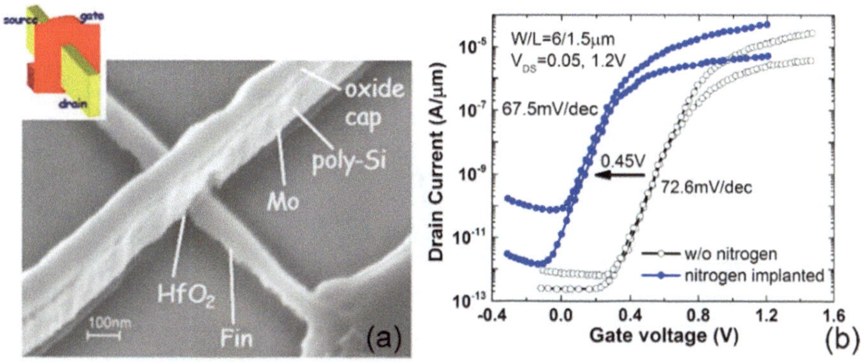

Fig. 1.35 (a) SEM image and (b) I–V characteristic of HfO_2 FinFET [143]

III-V compound semiconductors have also been applied as the substrate of FinFET. For example, GaAs FinFET with Al_2O_3 as gate dielectric is fabricated and evaluated in Ref. [144]. As shown in Fig. 1.36b, relatively large $I_{on/off}$ ratio (2.54×10^5), small V_{th} (−0.25V), low SS (80 mV/dec at $V_{DS} = 1V$), and weak DIBL are exhibited by the sample, implying great potential of GaAs or other III-V compounds based FinFET in device scaling in the future.

Recently, nano-materials such as graphene, carbon nanotubes, nanowires based MOSFETs have attracted much attention due to their even higher carrier mobility [145, 146]. As a representative, nanowire based FET will be introduced in the section. ZnO nanowire FET with Al_2O_3 as dielectric with the device structure shown in Fig. 1.37a is investigated [147]. Carrier mobility can reach 127 $cm^2V^{-1}\ s^{-1}$ when the thickness of Al_2O_3 is 21.2 nm and the value decreases with the increase of Al_2O_3 thickness which is probably caused by the decrease of transconductance. The relatively low V_{th} and acceptable leakage current plus the high carrier mobility indicates the potential of ZnO nanowire FET with high-k as gate dielectric, which has been proved by D. Yeom, et al. through the application in NOT and NAND logic circuits [148].

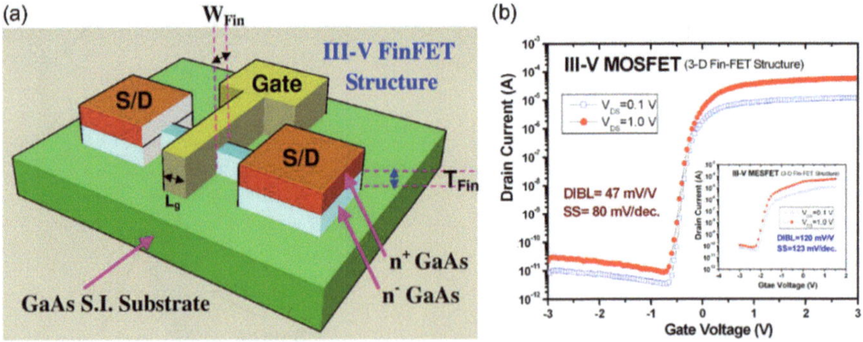

Fig. 1.36 (**a**) Schematic diagram and (**b**) I–V characteristic of GaAs FinFET with Al_2O_3 as gate dielectric [144]

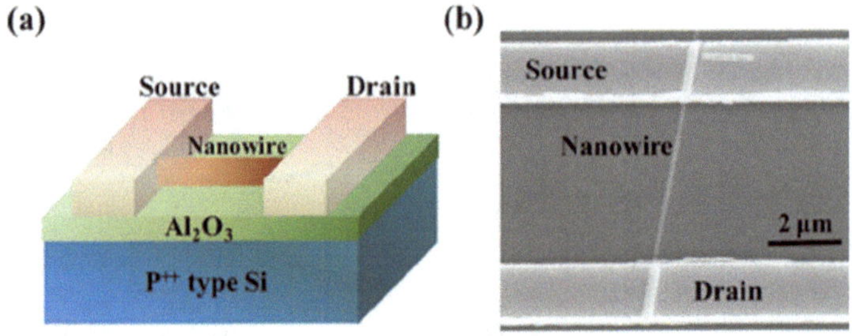

Fig. 1.37 (**a**) Schematic diagram and (**b**) SEM image of ZnO nanowire FET [147]

Fig. 1.38 Cross sectional SEM images of the twin nanowire FET [149]

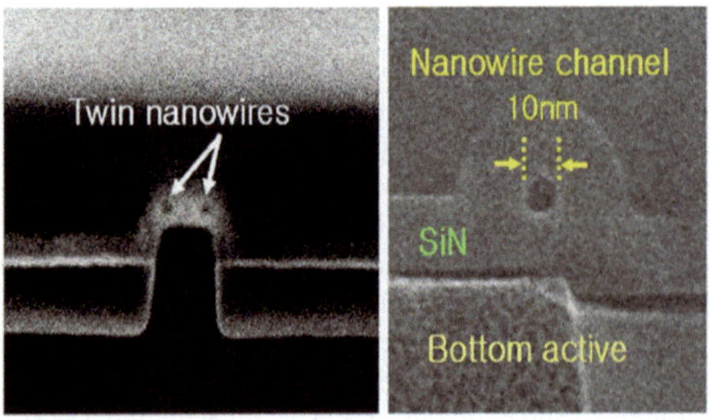

In addition, impressive electrical properties have been achieved from the idea of gate-all-around twin silicon nanowire MOSFET with the gate length of 30 nm [149]. According to the experimental results, an on/off ratio of 10^6 has been achieved, SS and DIBL are 71 mV/dec and 13 mV/V, implying the good immunity of nanowire to short channel effect. Besides, the testing on hot carrier lifetime and

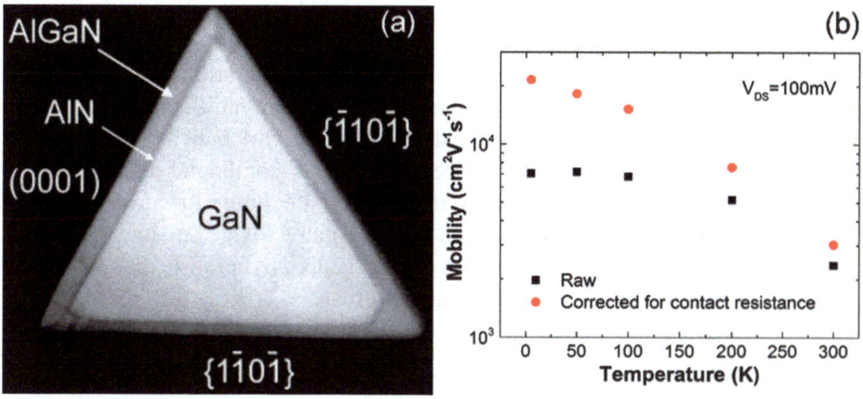

Fig. 1.39 (**a**) TEM image and (**b**) carrier mobility at different temperature of GaN/AlN/AlGaN nanowire [151]

the gate induced drain leakage current indicates its better reliability than that of planar MOSFET. Therefore the twin nanowire is a promising structure and might be improved by applying materials with higher k value. Similarly, III-V compounds can also be considered as the alternative of Si substrate in nanowire based MOSFET, for example, the vertical wrap-gated InAs/high-k nanowire gate stack exhibits low D_{it} near the semiconductor conduction band [150], which is necessary for achieving high carrier mobility.

Furthermore, III-V compound nanowires are suitable for high electron mobility transistors (HEMT). Y. Li, et al. has achieved extremely high mobility using GaN/AlN/AlGaN radial nanowire [151]. MOCVD is used as the method of the formation of the radial structure and ZrO_2 plays the role of gate dielectric. In this work, the carrier mobility, on/off ratio and SS are 21000 cm^2 V^{-1} s^{-1}, 10^7, and 68 mV/dec, respectively. These excellent results opens up new opportunities for nanoelectronics and provide new thoughts of the development of MOS devices.

1.6 Summary

In order to keep the trend of device scaling down, high-k is proposed as the alternative of SiO_2 due to the approaching of the physical limit of SiO_2. Since a reversely proportional relationship exists between k value and band gap of materials, apart from k value, many factors including band gap, band offset with the substrate, thermal and kinetic stability, etc. have to be considered when choosing the suitable one. Hf- and La-based high k oxides have attracted most attention recently, especially HfO_2 has already been put in to commercial applications. However, similar to other high k materials, high density of defects and traps in the bulk and at/near the high-k/substrate interface and low crystallizing temperature are still the main challenges of HfO_2 and La_2O_3. Therefore, post deposition treatments such as thermal annealing and fluorine incorporation are necessary to reduce the defects and traps. Besides, the idea of applying multicomponent high-k materials is an effective way to improve the quality of high-k bulk and the interface, and thus optimize the device performance. Ge and III-V compound semiconductors like GaAs, InGaAs, etc. are expected to replace Si in the future for further mobility increasing. However, their easily formed unstable native oxide and high density of interface states when contact with high-k oxides leading to Fermi-level pinning and mobility degradation still needs more future works.

References

1. S.M. Sze, K.K. Ng, *Physics of Semiconductor Devices* (Wiley, Hoboken, NJ, 1981)
2. R.H. Dennard, F.H. Gaesslen, H. Yu, V.L. Rideout, E. Bassons, A.R. LeBlanc, IEEE J. Solid State Circuits **SC-9**, 256 (1974)
3. D.J. Frank, R.H. Dennard, E. Nowak, P.M. Solomon, Y. Taur, H.P. Wong, Proc. IEEE **89**, 259 (2001)
4. H. Wong, H. Iwai, Microelectron. Eng. **83**, 1867 (2006)
5. M. Depas, B. Vermeire, P.W. Mertens, R.L. Van Meirhaeghe, M.M. Heyns, Solid State Electron. **38**, 1465 (1995)
6. G. Chindalore, S.A. Hareland, S. Jallepalli, A.F. Tasch, C.M. Maziar, V.K.F. Chia, S. Smith, IEEE Electron Device Lett. **18**, 206 (1997)
7. H.S. Monose, M. Ono, T. Yoshitomi, T. Ohguro, S. Nakamura, M. Saito, H. Iwai, IEEE Trans. Electron Devices **43**, 1233 (1996)
8. M. Cao, P.V. Voorde, M. Cox, W. Greene, IEEE Electron. Device Lett. **19**, 291 (1998)
9. M.L. Green, E.P. Gusev, R. Degraeve, E.L. Garfunkel, J. Appl. Phys. **90**, 2057 (2001)
10. K.S. Kirsch, C.G. Sodini, J. Appl. Phys. **76**, 2284 (1994)
11. D.A. Buchanan, A.D. Marwick, D.J. DiMaria, L. Dori, J. Appl. Phys. **76**, 3595 (1994)
12. J. Robertson, High dielectric constant oxides. Eur. Phys. J. Appl. Phys. **28**, 265 (2004)
13. R.L. Puurumen, J. Appl. Phys. **97**, 121301 (2005)
14. G.D. Wilk, R.M. Wallace, J.M. Anthony, J. Appl. Phys. **89**, 5243 (2001)
15. K.J. Hubbard, D.G. Schlom, J. Mater. Res. **11**, 2757 (1996)
16. D.G. Schlom, J.H. haeni, Mater. Res. Soc **27**, 198 (2002)
17. S. Stemmer, Z.Q. Chen, P.S. Lysagt, J.A. Gisby, J.R. Taylor, Proc. Electrochem. Soc. 2003-2 (Silicon Nitride and silicon Dioxide Thin Insulating films) 119, (2003)
18. M. Houssa, L. Pantisano, L. Ragnarsson, R. Degraeve, T. Schram, G. Pourtois, S.D. Gendt, G. Groeseneken, Mater. Sci. Eng. Rep. **51**, 37 (2006)
19. R.M.C. Almeida, I.J.R. Baumvol, Surf. Sci. Rep **49**, 1 (2003)
20. J. Robertson, Solid-States Electron. **49**, 283 (2005)
21. G.M. Rignanese, X. Gonze, A. Pasuarello, Phys. Rev. B **63**, 104305 (2001)
22. A.S. Foster, F.L. Gejo, A.L. Shluger, R.M. Nieminen, Phys. Rev. B **65**, 174117 (2002)
23. A.S. Foster, V.B. Sulimov, F.L. Gejo, A.L. Shluger, R.M. Nieminen, Phys. Rev. B **64**, 224108 (2001)
24. P.W. Peacock, J. Robertson, Appl. Phys. Lett. **83**, 2025 (2003)
25. A. Yokozawa, J. Robertson, Phys. Rev. B **55**, 13783 (1997)
26. E.P. Gusev, D.A. Buchanan, E. Cartier, A. Kumar, D. DiMaria, S. Guha, A. Callegari, S. Zafar, P.C. Jamiso, D.A. Neumayer, M. Copel, M.A. Gribelyuk, H. Okorn-Schmidt, C. D'Emic, P. Kozlowski, K. Chan, N. Bojarczuk, L.A. Ragnarsson, P. Ronsheim, K. Rim, R.

J. Fleming, A. Mocuta, and A. Ajmera, IEEE international electron devices meeting 20.1.1, (2001)
27. O. Engstrom, B. Raeissi, J. Piscator, I.Z. Mitrovic, S. Hall, H.D.B. Gottlob, M. Schmidt, P.K. Hurley, K. Cherkaoui, J. Telecommun. Inf. Technol. **1**, 10 (2010)
28. T. Yasuda, Y. Ma, S. Habermehl, G. Lucovsky, Appl. Phys. Lett. **60**, 434 (1992)
29. M. Gutowski, J.E. Jaffe, C.L. Liu, M. Stoker, R.I. Hegde, R.S. Rai, P.J. Tobin, Appl. Phys. Lett. **80**, 1897 (2002)
30. T. Nabatame, M. Kadoshima, K. Iwamoto, N. Mise, S. Migita, M. Ohno, H. Ota, N. Yasuda, A. Ogawa, K. Tominaga, H. Satake, A. Toriumi, IEEE Trans. Electron Devices Meet. **83** (2004)
31. T.L. Li, C.H. Hum, W.L. Ho, H.C.H. Wag, C.Y. Chang, IEEE Trans. Electron Devices **52**, 1172 (2005)
32. H. Wakabayashi, Y. Saito, K. Takeuchi, T. Mogami, T. Kunio, IEEE Trans. Electron Devices Meet. **253** (1999)
33. T.H. Cha, D.G. Park, T.K. Kim, S.A. Jang, I.S. Yeo, J.S. Roh, J.W. Park, Appl. Phys. Lett. **82**, 4192 (2002)
34. J. Lu, Y. Kuo, S. Chatterjee, J.Y. Yewg, J. Vac. Sci. Technol. B **24**, 349 (2006)
35. M.V. Fischetti, D.A. Neumayer, E.A. Cartier, J. Appl. Phys. **90**, 4587 (2001)
36. M.H. Chowdhury, M.A. Mannan, S.A. Mahmood, Int. J. Emerg Technol. Sci. Eng. **2**, 1 (2010)
37. A. Oshiyama, Jpn. J. Appl. Phys. **2**, 37 (1998)
38. W. Monch, Phys. Rev. Lett. **58**, 1260 (1987)
39. J. Robertson, J. Vac. Sci. Technol. B. **18**, 1785 (2000)
40. H. Wong, B. Sen, B.L. Yang, A.P. Huang, P.K. Chu, J. Vac. Sci. Technol. B **25**, 1853 (1858)
41. G. He, Z. Sun, *High-k gate dielectrics for CMOS technology* (Wiley-VCH, Germany, 2012)
42. H. Yamada, T. Shinizu, T. Suzuki, Jpn. J. Appl. Phys. **41**, L368 (2002)
43. Y. Ohshita, A. Ogura, M. Isikawa, T. Kada, H. Mahida, Jpn. J. Phys. **42**, L578 (2003)
44. T. Suntola, J. Antson, U. S. Patent 4058430, (1977)
45. S.M. George, Chem. Rev. **110**, 111 (2010)
46. F. Werner, W. Stals, R. Gortzen, B. Veith, R. Brendel, J. Schnidt, Energy Procedia **8**, 1301 (2011)
47. H. Kim, J. Vac. Sci. Technol. B **21**, 2231 (2003)
48. P. Darmawan, P.S. Lee, Y. Setiawan, J.C. Lai, P. Yang, J. Vac. Sci. Technol. B **25**, 1203 (2007)
49. N. Zhan, M.C. Poon, C.W. kok, K. La, Ng, H. Wong, J. Electrochem. Soc **150**, F200 (2003)
50. H. Wong, K.L. Ng, N. Zhan, M.C. Poon, C.W. Kok, J. Vac. Sci. Technol. B **22**, 1094 (2004)
51. C.S. Kang, H.J. Cho, K. Onishi, R. Nieh, R. Choi, S.K. Gopalan, J.H. Han, J.C. Lee, Appl. Phys. Lett. **81**, 2539 (2002)
52. K.J. Choi, J.H. Kim, S.G. Yoon, W.C. Shin, J. Vac. Sci. Technol. B **22**, 1755 (2004)
53. P. Kroll, Phys. Rev. Lett. **90**, 125501 (2003)
54. Y. Yonamoto, Y. Inaba, N. Akamatsu, Appl. Phys. Lett. **98**, 232906 (2011)
55. C. Yang, L. Hang, J. Xu, Mater. Sci. Semicond. Process **29**, 321 (2015)
56. H.Y. Yu, N. Wu, M.F. Li, C. Zhu, B.J. Cho, D.L. Kwong, W.D. Wang, D.Z. Chi, C.H. Ang, J. Z. Zheng, S. Ramanathan, Appl. Phys. Lett. **81**, 3618 (2002)
57. M.S. Joo, B.J. Co, C.C. Yeo, S.H. Cha, S.J. Whoang, S. Mathew, B.L. Kanta, N. Balasubramanian, D.L. Kwong, IEEE Trans. Electron Devices **50**, 2088 (2003)
58. T.J. Park, J.H. Kim, J.H. Jang, C.K. Lee, K.D. Na, S.Y. Lee, H.S. Jung, M. Kim, S. Han, C.S. Hwang, Chem. Mater. **22**, 4175 (2010)
59. S.H. Bae, C.H. Lee, R. Clark, D.L. Kwong, IEEE Electron Device Lett. **24**, 556 (2003)
60. X. Yu, C. Zhu, M.F. Li, A. Chin, A.Y. Du, W.D. Wang, D.L. Kwong, Appl. Phys. Lett. **85**, 2893 (2004)
61. Y. Yamamoto, K. Kita, K. Kyuno, A. Toriumi, Appl. Phys. Lett. **89**, 032903 (2006)
62. C.H. An, M.S. Lee, J.Y. Choi, H. Kim, Appl. Phys. Lett. **94**, 262901 (2009)

63. M.F. Li, C.X. Zhu, C.Y. Shen, F.F. Xiong, P. Yuan, Y.C. Ywo, A. Chin, D.L. Kwong, S.H. Wang, A.Y. Du, G. Samudra, ECS Trans. **1**, 717 (2006)
64. M. Leskela, K. Kukli, M. Ritala, J. Alloys Compd. **418**, 27 (2006)
65. Y.C. Yeo, T.J. King, C. Hu, Appl. Phys. Lett. **81**, 2091 (2002)
66. G. Lippert, J. Dabrovski, V. Melnik, R. Sorge, C. Wenger, P. Zaumseil, H.J. Mussig, Appl. Phys. Lett. **86**, 042902 (2005)
67. K. Kukli, M. Ritala, T. Pilvi, T. Sajavaara, M. Leskela, A.C. Jones, H.C. Aspinall, D.C. Gilmer, P.J. Tobin, Chem. Mater. **16**, 5162 (2004)
68. Y. Zhao, Material **5**, 1413 (2012)
69. S. Guha, E. Cartier, M.A. Gribelyuk, N.A. Bojarczuk, M.C. Copel, Appl. Phys. Lett. **77**, 2710 (2000)
70. T. Gougousi, D. Niu, R.W. Ashcraft, G.N. Parsons, Appl. Phys. Lett. **83**, 3543 (2003)
71. R.D. Shannon, J. Appl. Phys. **73**, 348 (1993)
72. W.C. Koehler, E.O. Wollan, Acta. Crystallog. **6**, 741 (1953)
73. A.F. Kapustinskii, Q. Rev, Chem. Soc. **10**, 283 (1956)
74. Y. Yokogawa, M. Yosimura, S. Somiya, J. Mater. Sci. Lett. **10**, 509 (1991)
75. N.C. Su, S.J. Wang, A. Chin, IEEE Electron Device Lett. **30**, 1317 (2009)
76. C.Y. Han, W.M. Tang, C.H. Leung, C.M. Che, P.T. Lai, Org. Electron. **15**, 2499 (2014)
77. J.Q. Song, C.Y. Han, P.T. Lai, Comparative study of Nb2O5, NbLaO, and La_2O_3 as ate dielectric of InGaZnO thin film transistor. IEEE Trans. Electron Devices **63**, 1928 (2016)
78. W.J. Song, S.K. So, D.Y. Wang, Y. Qiu, L.L. Cao, Appl. Surf. Sci. **177**, 158 (2001)
79. S. Jeon, F.J. Walker, C.A. Billman, R.A. McKee, H. Hwang, IEEE Electron Device Lett. **24**, 218 (2003)
80. G.J. Norga, C. Marchiori, C. Rossel, A. Guiller, J.P. Locquet, H. Siegwart, D. Caimi, J. Fompeyrine, J.W. Seo, C. Dieker, J. Appl. Phys. **99**, 084102 (2006)
81. J.H. Hao, J. Gao, Z. Wang, D.P. Yu, Appl. Phys. Lett. **87**, 131908 (2005)
82. S.A. Chambers, Y. Liang, Z. Yu, R. Droopad, J. Ramdani, K. Eisenbeiser, Appl. Phys. Lett. **77**, 1662 (2000)
83. R. Droopad, Z. Yu, L. Hilt, J. Curless, C. Overgaard, J.L. Edwards, J. Finder, K. Eisenbeise, W. Ooms, Mater. Sci. Eng. B **87**, 292 (2001)
84. A.B. Posadas, C.W. Lin, A.A. Demkov, S. Zollner, Appl. Phys. Lett. **103**, 142906 (2013)
85. X.B. Lu, Z.G. Liu, X. Zhang, R. Huang, H.W. Zhou, X.P. Wang, B.Y. Nguyen, J. Phys. D. Appl. Phys. **36**, 3047 (2003)
86. X.B. Lu, Z.G. Liu, G.H. Shi, H.Q. Ling, H.W. Zhou, X.P. Wang, B.Y. Nguyen, Appl. Phys. A Mater. Sci. Process **78**, 921 (2003)
87. W.F. Xiang, H.B. Lu, B.H. Chen, M. He, X.B. Lu, L.F. Liu, H.Z. Guo, Z.Y. Liang, Chin. Phys. Lett. **22**, 182 (2005)
88. I.Y. Chang, S. You, M. Chen, P. Juan, C. Chen, J.Y. Lee, J. Appl. Phys. **105**, 104512 (2009)
89. S. Takagi, A. Toriumi, M. Iwase, H. Tango, IEEE Trans. Electron Devices **41**, 2357 (1994)
90. Y.K. Kim, S.H. Lee, S.J. Choi, H.B. Park, Y.D. Seo, K.H. Chin, D. Kim, J.S. Lim, W.D. Kim, K.J. Nam, M.H. Cho, K.H. Hwang, Y.S. Kim, S.S. Kim, Y.W. Park, J.T. Moon, S.L. Lee, M. Y. Lee, IEEE International Electron Devices Meeting, Technical Digest, (2000)
91. J. Lutzen, A. Bimer, M. Goldbach, M. Gutsche, T. Hecht, S. Jakschik, A. Orth, A. Sanger, U. Schroder, H. Seidl, B. Sell, D. Schumann, Symposium on VLSI Technology, Digest of Technical Papers, (2002)
92. E. Atanassova, A. Paskaleva, Microelectron. Reliab. **47**, 913 (2007)
93. N. Pawlak, M. Popovici, J. Swerts, K. tomida, M.S. Kim, B. Kaczer, K. Posomer, M. Schaekers, P. Favia, H. Bender, IEEE International Electron Devices Meeting, 929, (2010)
94. D. Zhou, U. Schroeder, J. Xu, J. Heitmann, G. Jegert, W. Weinreich, M. Kerber, S. Knebel, E. Erben, T. Mikolajick, J. Appl. Phys. **108**, 124104 (2010)
95. S.K. Kim, G.J. Choi, S.Y. Lee, H. Seo, S.W. Lee, J.H. Han, H.S. Ahn, S. Han, C.S. Hwang, Adv. Mater. **20**, 1429 (2008)

96. S. Mori, Y. Araki, M. Sato, H. Meguro, H. Tsunoda, E. Kamiya, K. Yoshikawa, N. Arai, E. Sakagami, IEEE Trans. Electron Devices **43**, 47 (1996)
97. A. Sheikholeslami, P.G. Gulak, Proc. IEEE **88**, 677 (2000)
98. J. Pu, S.J. Kim, B.J. Cho, IEEE Trans. Electron Devices **56**, 2739 (2009)
99. W. He, J. Pu, S.H. Chan, B.J. Cho, IEEE Trans. Electron Devices **56**, 2746 (2009)
100. X.D. Huang, L. Liu, J.P. Xu, P.T. Lai, IEEE Trans. Device Mater. Reliab. **13**, 490 (2011)
101. H. Park, G. Bersuker, D. Gilmer, K.Y. Lim, M. Jo, H. Hwang, A. Padovani, L. Larcher, P. Pavan, W. Taylor, P.D. Kirsch, Proc. IMW **175** (2010)
102. J.K. Hyo, S.Y. Cha, D.J. Choi, Mater. Sci. Semcond. Process **13**, 9 (2010)
103. Y.M. Pan, W.W. Yeh, IEEE Trans. Electron Devices **55**, 2354 (2008)
104. Y.H. Wu, L.L. Chen, M.L. Wu, C.C. Lin, C.H. Chang, IEEE Electron Device Lett. **31**, 1008 (2010)
105. X.G. Wang, J. Liu, W.P. Bai, D.L. Kwong, IEEE Trans. Electron Devices **51**, 597 (2004)
106. X.D. Huang, P.T. Lai, J.K.O. Sin, Appl. Phys. A Mater. Sci. Process **108**, 229 (2012)
107. R.P. Shi, X.D. Huang, C.H. Leung, J.K.O. Sin, P.T. Lai, IEEE Trans. Device Mater. Rev. **15**, 123 (2015)
108. X.D. Huang, J.K.O. Sin, P.T. Lai, IEEE Tran. Electron Devices **58**, 4235 (2011)
109. X.D. Huang, P.T. Lai, J.K.O. Sin, Solid State Electron. **79**, 285 (2013)
110. Z. Liu, C. Lee, V. Narayanan, G. Pei, E.C. Kan, IEEE Trans. Electron Devices **49**, 1606 (2002)
111. J.C. Wang, C.T. Lin, C.S. Lai, J.L. Hsu, Appl. Phys. Lett. **97**, 023513 (2010)
112. X.D. Huang, P.T. Lai, J.K.O. Sin, ECS Solid State Lett. **1**, Q45 (2012)
113. J.H. Chen, W.J. Yoo, D.S.H. Chan, L.J. Tang, Appl. Phys. Lett. **86**, 073114 (2005)
114. Y. Zhou, J. Yin, H. XU, Y. Xia, Z. Liu, A. Li, Y. Gong, L. Pu, F. Yan, Y. Shi, Appl. Phys. Lett. **97**, 143504 (2010)
115. H.T. Lue, S.Y. Wang, E.K. Lai, Y.H. Shih, S.C. Lai, L.W. Yang, K.C. Chen, J. Ku, K.Y. Hsieh, R. Liu, C.Y. Lu, in IEEE International Electron Devices Meeting, 547, (2005)
116. Y.Q. Wang, W.S. Hwang, G. Zhang, G. Samudra, Y.C. Yeo, W.J. Yoo, IEEE Trans. Electron Devices **54**, 2699 (2007)
117. B. Chakrabarti, H. Kang, B. Brennan, T.J. Park, K.D. Cantley, A. Pirkle, S. Mcdonell, J. Kim, R.M. Wallace, E.M. Vogel, IEEE Trans. Electron Devices **58**, 4189 (2011)
118. M. Zhao, R. Liang, J. Wang, J. Xu, Appl. Phys. Lett. **102**, 142906 (2013)
119. C.X. Li, X. Zou, P.T. Lai, J.P. Xu, C.L. Chan, Microelectron. Reliab. **48**, 526 (2008)
120. H.X. Xu, J.P. Xu, C.X. Li, C.L. Chan, P.T. Lai, Appl. Phys. A Mater. Sci. Process **99**, 903 (2010)
121. C.X. Li, C.D. Wang, C.H. Leung, P.T. Lai, J.P. Xu, Microelectron. Eng. **86**, 1596 (2009)
122. W.B. Chen, B.S. Shie, C.H. Cheng, K.C. Hsu, A. Chin, IEEE International Electron Devices meeting, 420, (2010)
123. C.X. Li, H.X. Xu, P.T. Lai, J.P. Xu, Appl. Phys. Lett. **93**, 022910 (2009)
124. F. Ji, J.P. Xu, J.G. Liu, C.X. Li, P.T. Lai, Appl. Phys. Lett. **98**, 182901 (2011)
125. F. Ji, J.P. Xu, Y. Huang, L. Liu, P.T. Lai, IEEE Trans. Electron Devices **61**, 3608 (2014)
126. F.S. Aguirre-Tostado, M. Milojevic, K.J. Choi, H.C. Kim, C.L. Vogel, J. Kim, T. Yang, Y. Xuan, P.D. Ye, R.M. Wallace, Appl. Phys. Lett. **93**, 061907 (2008)
127. M.K. Lee, C.F. Yen, Appl. Phys. A Mater. Sci. Process **116**, 2051 (2014)
128. L.N. Liu, H.W. Choi, P.T. Lai, J.P. Xu, J. Vac. Sci. Technol. B **33**, 050601 (2015)
129. L.S. Wang, J.P. Xu, L. Liu, W.M. Tang, P.T. Lai, Appl. Phys. Express **7**, 061201 (2014)
130. L.N. Liu, H.W. Choi, J.P. Xu, P.T. Lai, Appl. Phys. Lett. **107**, 213501 (2015)
131. K. Nomura, H. Ohta, A. Takagi, T. Kamiya, M. Hirao, H. Hosono, Nature **432**, 488 (2004)
132. J.S. Park, W.J. Maeng, H.S. Kim, J.S. Park, Thin Solid Films **520**, 1679 (2012)
133. L.F. Lan, J.B. Peng, IEEE Trans. Electron Devices **58**, 1452 (2011)
134. Y.H. Lin, J.C. Chou, J. Nanomater. **2014**, 347858 (2014)
135. H. Yabuta, M. Sano, K. Abe, T. Aiba, T. Den, H. Kumomi, K. Nomura, T. Kamiya, H. Hosono, Appl. Phys. Lett. **89**, 112123 (2006)

136. J. Kim, C. Fuentes-Hernandez, B. Kippelen, Appl. Phys. Lett. **93**, 242111 (2008)
137. L.X. Qian, P.T. Lai, IEEE Trans. Device Mater. Reliab. **14**, 177 (2014)
138. L.X. Qian, P.T. Lai, Microelectron. Reliab. **54**, 2396 (2014)
139. P. Barquinha, L. Pereira, G. Goncalves, R. Martins, D. Kuscer, M. Kosec, E. Fortunato, J. Electrochem. Soc. **156**, H824 (2009)
140. L.X. Qian, P.T. Lai, W.M. Tang, Appl. Phys. Lett. **104**, 123505 (2014)
141. L.X. Qian, P.T. Lai, IEEE Electron Device Lett. **35**, 363 (2014)
142. D Ghattacharya, N.K. Jha, 2014, 365689, (2014).
143. D. Ha, H. Takeuchi, Y.K. Choi, T.J. King, W.P. Bai, D. L. Kwong, A. Agarwal, M. Ameen, IEEE International Electron Devices meeting, 643, (2004)
144. S.H. Chen, W.S. Liao, H.C. Yang, S.J. Wang, Y.G. Liaw, H. Wang, S. Gu, M.C. Wang, Nanoscale Res. Lett. **7**, 431 (2012)
145. F. Schwierz, Nat. Nanotechnol. **5**, 487 (2010)
146. P.A. Alvi, K.M. Lal, M.J. Siddiqui, S.A.H. Naqvi, Indian J. Pure Appl. Phys. **43**, 899 (2005)
147. M. Choe, G. Jo, J. Maeng, W.K. Hong, M. Jo, G. Wang, W. Park, B.H. Lee, H. Hwang, T. Lee, J. Appl. Phys. **107**, 034504 (2010)
148. D. Yeom, K. Keem, J. Kang, D.Y. Jeong, C. Yoon, D. Kim, S. Kim, Nanotechnology **19**, 265202 (2008)
149. S.D. Suk, S.Y. Lee, S.M. Kim, E.J. Yoon, M.S. Kim, M. Li, C.W. Oh, K.H. Yeo, S.H. Kim, D. S. Shin, K.H. Lee, H.S. Park, J.N. Han, C.J. Park, J.B. Park, D.W. Kim, D. Park, B.I. Ryu, IEEE International Electron Devices meeting, 717, (2005)
150. J. Wu, K. Jansson, A.S. Badadi, M. Berg, E. Lind, L.E. Wernersson, IEEE Trans. Electron Devices **63**, 584 (2016)
151. Y. Li, J. Xiang, F. Qian, S. Gradecak, Y. Wu, H. Yan, D.A. Blom, C.M. Lieber, Nano Lett. **6**, 1468 (2006)

Chapter 2
Challenge of High Performance Bandgap Reference Design in Nanoscale CMOS Technology

Zhang Jun-an, Li Guangjun, Zhang Rui-tao, Yang Yu-jun, Li Xi, Yan Bo, Fu Dong-bing, and Luo Pu

2.1 Introduction

Reference circuit is a critical block in analog and mixed-signal circuits such as A/D converters, D/A converters or SOCs, and it can be used as an accurate reference signal or a steady bias source. Based on the development of semiconductor technology, IC's power supply voltage is becoming lower, so as the analog signal swing. But the resolution demands (by data converters for an example) may be constant, and that means the demand for reference signal is actually more critical. Most of the reference circuit concentrates on its DC performance than its AC performance, so the fast speed benefit from technology improvement had done nothing to a reference circuit design. On the contract, there are many negative effects in bandgap reference design caused by technology shrinking, such as low source voltage, large temperature coefficient of on-chip resistors, low EARLY

Z. Jun-an (✉)
School of Communication and Information Engineering, University of Electronic Science and Technology of China, Chengdu 611731, China

Key Laboratory of Analog Integrated Circuit, Chongqing 400060, China
e-mail: zhangjunan2000@sina.com

L. Guangjun • Y. Bo
School of Communication and Information Engineering, University of Electronic Science and Technology of China, Chengdu 611731, China

Z. Rui-tao
Key Laboratory of Analog Integrated Circuit, Chongqing 400060, China

Y. Yu-jun • L. Xi • F. Dong-bing • L. Pu
Sichuan Institute of Solid-State Circuits, China Electronics Technology Group Corp, Chongqing 400060, P.R. China

T. Li, Z. Liu (eds.), *Outlook and Challenges of Nano Devices, Sensors, and MEMS*, DOI 10.1007/978-3-319-50824-5_2

voltage, large drain leakage current, etc., that will cause a great challenge to achieve a high performance bandgap reference circuit among a large temperature variation range.

This chapter is arranged as follows. Sect. 2.2 discusses five negative effects of nanoscale CMOS technology in bandgap reference design. Sect. 2.3 presents a bandgap reference circuit implementation with measurement results and comparison. Sect. 2.5, conclusion.

2.2 Negative Effects of Nanoscale CMOS Technology in Bandgap Reference Design

2.2.1 Voltage Headroom Decrease

As CMOS technology is shrinking, MOSfet's power supply voltage is becoming lower, as shown in Fig. 2.1, there are not enough voltage headroom to achieve the conventional bandgap reference circuit.

To deal with this problem, many low voltage reference circuit schemes are presented in recent published literatures [1–18]. Figure 2.2 shows the first kind of low voltage bandgap reference circuit named current-mode bandgap reference [1] which utilizes a reference current (temperature independent) and a resistor to generate a reference voltage. The value of reference current and resistor can be choosed freely, so the reference voltage output will not be limited by the source

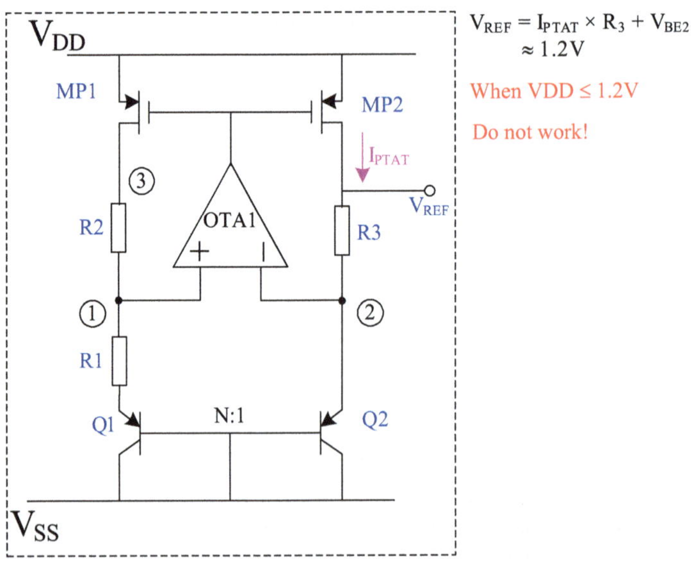

Fig. 2.1 Conventional bandgap circuit

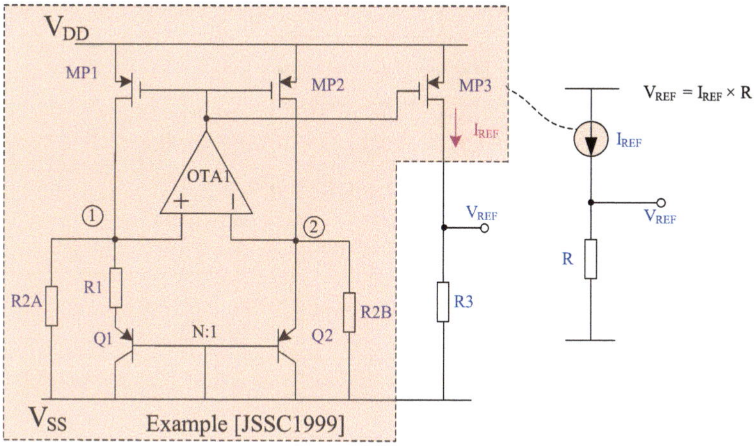

Fig. 2.2 Curent-mode low voltage bandgap reference circuit

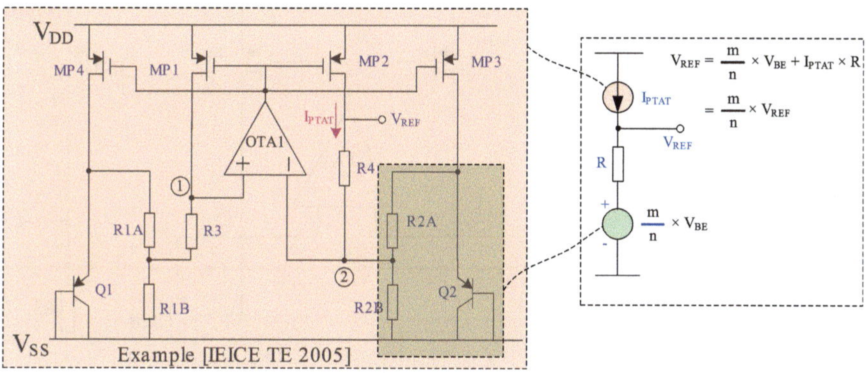

Fig. 2.3 Fractional low voltage bandgap circuit

voltage headroom. The lowest source voltage of the example circuit is about $V_{BE} + V_{DSAT}$ which can be lower than 1 V (regardless OTA's limitation).

Figure 2.3 shows the second kind of low voltage bandgap reference circuit named fractional bandgap reference [2] which utilizes a fraction of the P-N junction voltage (V_{BE}) added a PTAT voltage (which is achieved by a PTAT current with a resistor) to generate the reference voltage. The value of reference voltage is a fraction of the conventional bandgap voltage (≈1.2 V), and it also can be lower than 1 V. The lowest source voltage of the example circuit [13] is the minimum of $V_{REF} + V_{DSAT}$ with $V_{BE} + V_{DSAT}$ (regardless OTA's limitation).

Figure 2.4 shows the third kind of low voltage bandgap reference circuit which principle likes the output common voltage extraction in full differential operational amplifier [3]. The common voltage of a PTAT voltage and a CTAT voltage will achieve a temperature independent reference voltage if the opposite temperature coefficient can be counteracted perfectly. The value of reference voltage is 1/2 of

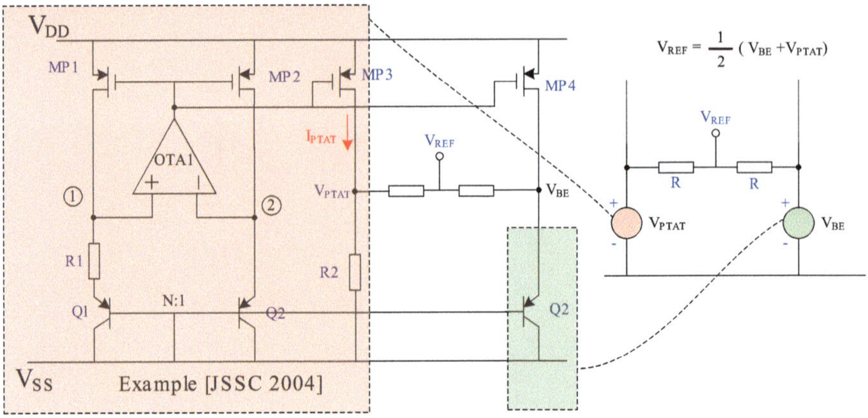

Fig. 2.4 Common voltage extraction method bandgap circuit

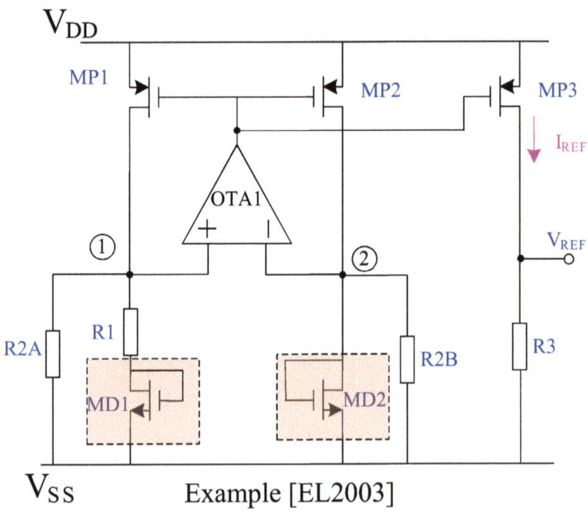

Fig. 2.5 Current-mode bandgap reference circuit with MOSfet

conventional bandgap voltage (≈0.6 V). The lowest source voltage of the example circuit is the minimum of $V_{PTAT} + V_{DSAT}$ with $V_{BE} + V_{DSAT}$ (regardless OTA's limitation).

MOSfet can be used to replace the bipolar transistor, because the threshold voltage (V_{TH}) has the similar temperature characters as P-N junction voltage (V_{BE}). As shown in Fig. 2.5, MD1 and MD2 have been used to replace the bipolar transistor [4]. Due to the threshold voltage of MOSfet is lower than P-N junction voltage, the source voltage of reference circuit can be even lower. Figure 2.5 shows that the current-mode bandgap reference circuit's lowest source voltage is $V_{TH} + V_{DSAT}$ (regardless OTA's limitation). Comparing with P-N junction voltage, the threshold voltage of MOSfet is considered as unsteady, the variation range of process corner is large, so some extra process step must be utilized to adjust it.

There are many other architectures and methods of low voltage reference circuit. Some method utilizes special devices (DTMOS [5]) and some others utilize special technology (CMOS/SOI [6], hybrid with Germanium diode [7]). In now days, the level of source voltage decreasing is slowed down, but the trend is inevitable. In the future, some new architectures and methods would be innovated continuously.

2.2.2 *Large Temperature Coefficient of On-Chip Resistors*

As shown in Fig. 2.1, the conventional bandgap reference circuit output a reference voltage directly, and it can be expressed as Eq. (2.1):

$$V_{REF} = \frac{R_3}{R_1} V_T \ln N + V_{BE2} \tag{2.1}$$

When R_1 and R_3 use the same type of resistor, temperature coefficient of the resistors will not affect the reference voltage (V_{REF}).

But the current-mode bandgap reference circuit can only get a reference voltage through a reference current, and it can be expressed as Eq. (2.2) and Eq. (2.3):

$$I_{REF} = \frac{V_T \ln N}{R_1} + \frac{V_{BE2}}{R_{2A}} \tag{2.2}$$

$$V_{REF} = I_{REF} \times R_3 \tag{2.3}$$

Temperature coefficient of the R_3 will directly affect the reference voltage (V_{REF}). Current-mode bandgap reference circuit cannot obtain a high performance reference voltage if the used resistor's temperature coefficient are a large. On the other hand, when utilize the conventional bandgap circuit's reference voltage and a resistor to generate a reference current, this reference current will be affected by the resistor's temperature coefficient too.

Unfortunately, all the on-chip resistors provided by nanoscale CMOS technology have a large temperature coefficient. As shown in Fig. 2.6, the temperature sweep simulation results of all the resistors which is provided by a 65 nm CMOS technology.

As shown in Fig. 2.7, the smallest temperature coefficient on-chip resistor the variation value is $\Delta R = 2.4\%$ among −55 to 125°C which temperature coefficient is about 133 ppm/°C. This temperature coefficient is still large compared with a one-order bandgap reference.

As shown in Fig. 2.2, if R_3 set as only the smallest temperature coefficient on-chip resistor cannot achieve a high performance reference voltage.

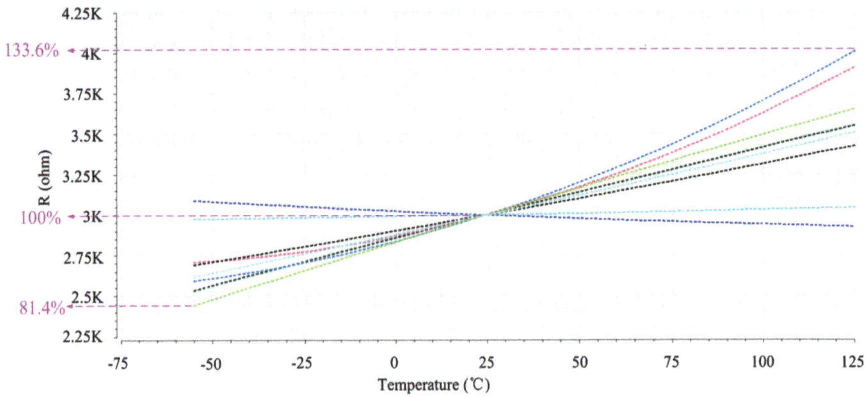

Fig. 2.6 Temperature performance simulation result of on-chip resistors

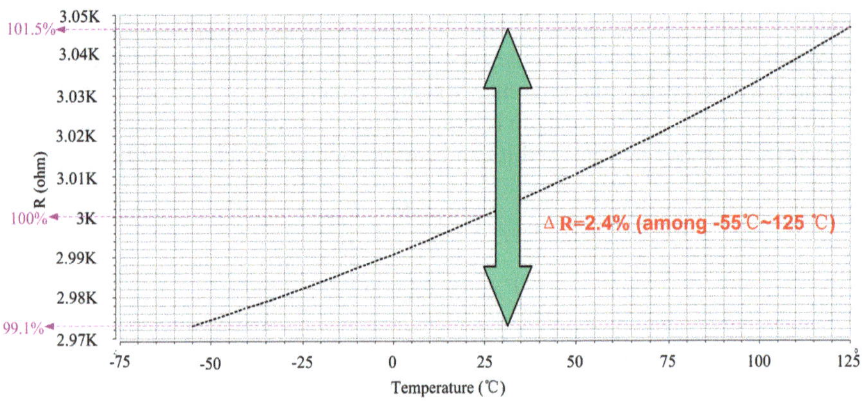

Fig. 2.7 The smallest temperature coefficient on-chip resistor

2.2.3 EARLY Voltage Decreasing

As CMOS technology is shrinking, MOSfet's EARLY voltage is becoming lower. As shown in Figs. 2.8 and 2.9, a PMOSfet V-I curve comparison between a 65 nm CMOS technology and a 0.18μm CMOS technology. It can be seen from Fig. 2.8 that in 65 nm CMOS technology EARLY voltage of PMOSfet with the smallest gate length decreased sharply. It can be seen from Fig. 2.9 that in different technology, even choosing the same transistor size the EARLY voltage is still different.

The negative effect of EARLY voltage decreasing to a reference design is when using one stage current mirror [8, 9, 14] (even cascode current mirror [13] which needs more voltage headroom) with small gate length can not achieve an accurate current replication. Even large gate length MOSfet (which means large V_{TH}

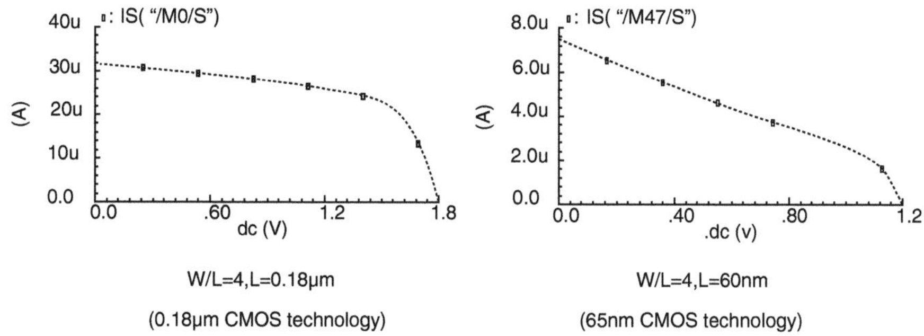

Fig. 2.8 PMOSfet V-I curve of different CMOS technology (smallest gate length)

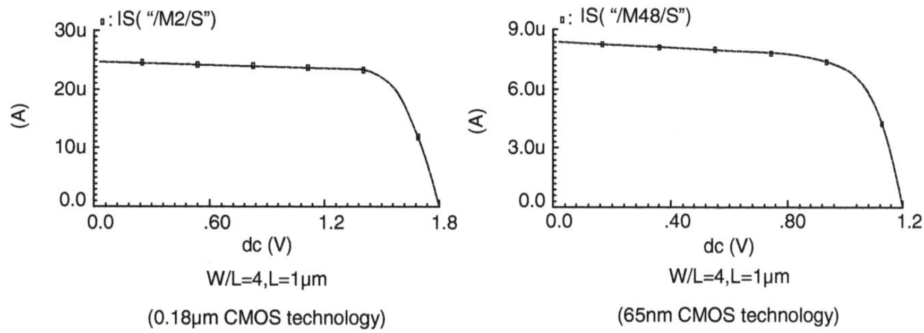

Fig. 2.9 PMOSfet V-I curve of different CMOS technology (same gate length)

consumed more voltage headroom) also cannot achieve an accurate current replication here. The reason is that the drain voltage of MP2 is a P-N junction voltage (V_{BE}), and the drain voltage of MP3 is expected to be a temperature independent voltage (V_{REF}). Due to V_{BE} has a large variation range among −55°C to 125°C, and with the EARLY voltage decreasing, one stage current mirror can not copy a current accurately.

Figure 2.10 shows the method which utilizes one stage current mirror to generate V_{REF}. Figure 2.11 shows the MP2 and MP3's drain current simulation result of Fig. 2.10, the length of MP2 and MP3 is set as 2 μm. The replication error is very large.

As shown in Fig. 2.12, a cascode current mirror is used for I_{REF} current replication. MP4's gate length is set as 2 μm too. The bias voltage of MP4 is generated from a diode-connected PMOSfet with a drain current which has a certain ratio with I_{D2}.

Figure 2.13 shows the MP2 and MP3's drain current simulation result of Fig. 2.12. The replication error is smaller than one stage current mirror, but for a high performance reference circuit this accuracy level is not enough.

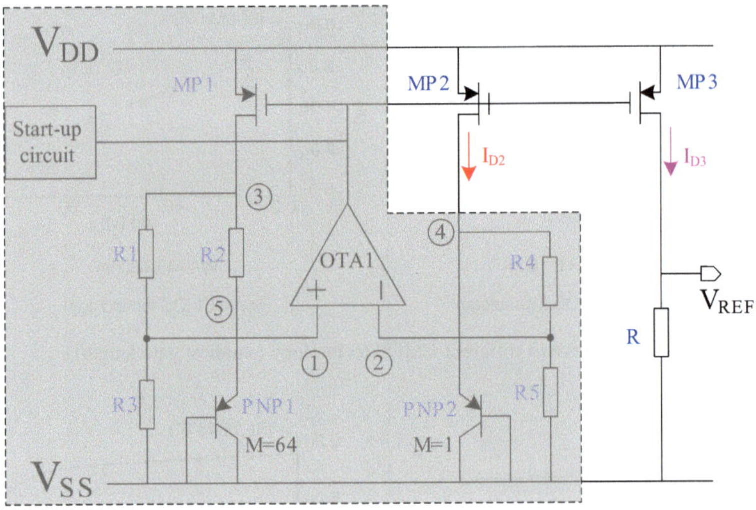

Fig. 2.10 One stage current mirror for V_{REF}

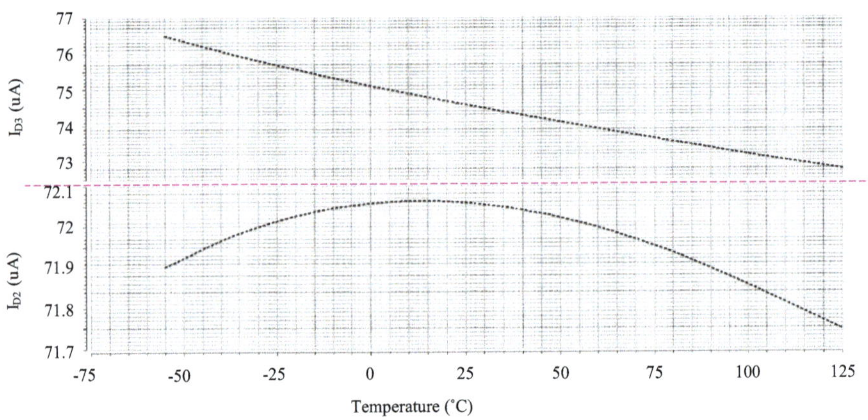

Fig. 2.11 Current replication performance by one stage current mirror

2.2.4 *Effect to DC Operation Points at Extreme Low Temperature*

Due to P-N junction voltage (V_{BE}) and MOSfet's threshold voltage (V_{TH}) all have a negative temperature coefficient, and a low voltage design is difficult to maintain suitable DC operation points in an extremely low temperature condition. Most of the published literatures' temperature range were not lower than −40°C [1–18].

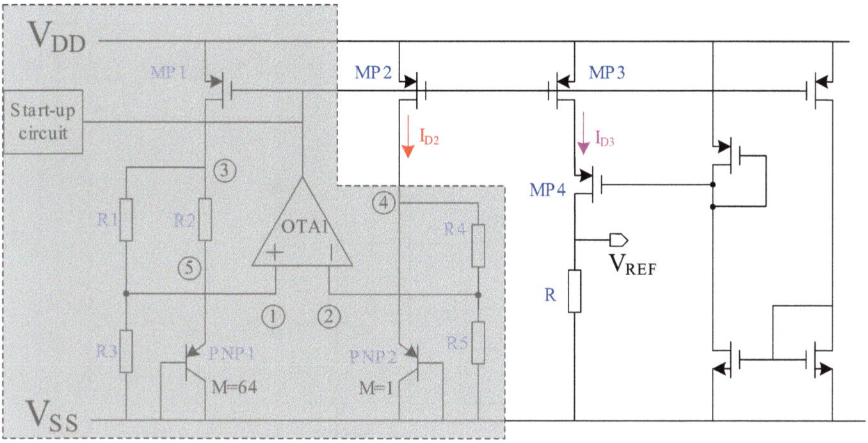

Fig. 2.12 Cascode current mirror for V_{REF}

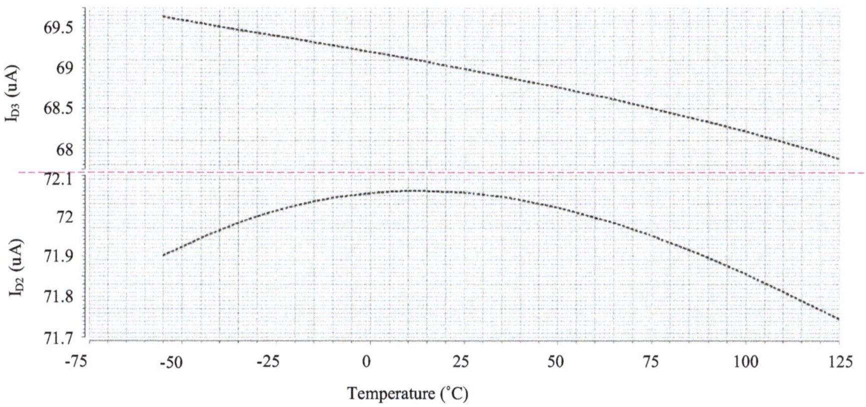

Fig. 2.13 Current replication performance by cascode current mirror

As mentioned in Sect. 2.2.1, all the lowest source voltages of the example reference circuits are regardless OTA's limitation. As Fig. 2.14 shows the lowest source voltage must fulfil Eq. (2.4):

$$V_{BE} + V_{TH\_P4} + V_{DSAT\_P4} + V_{DSAT\_P6} \leq V_{DD} \tag{2.4}$$

This is much larger than V_{BE} + V_{DSAT} (the lowest source voltage limitation regardless of OTA), and the MOSfets of OTA may exceed the saturation zone in the extreme low temperature, so the OTA must be carefully designed.

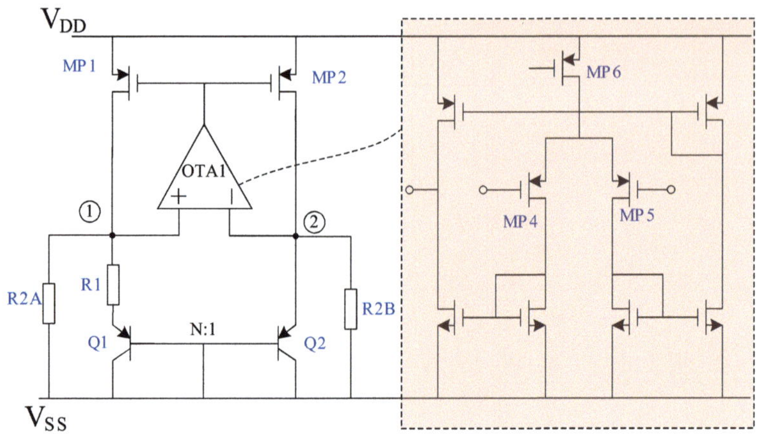

Fig. 2.14 Curent-mode low voltage bandgap reference circuit

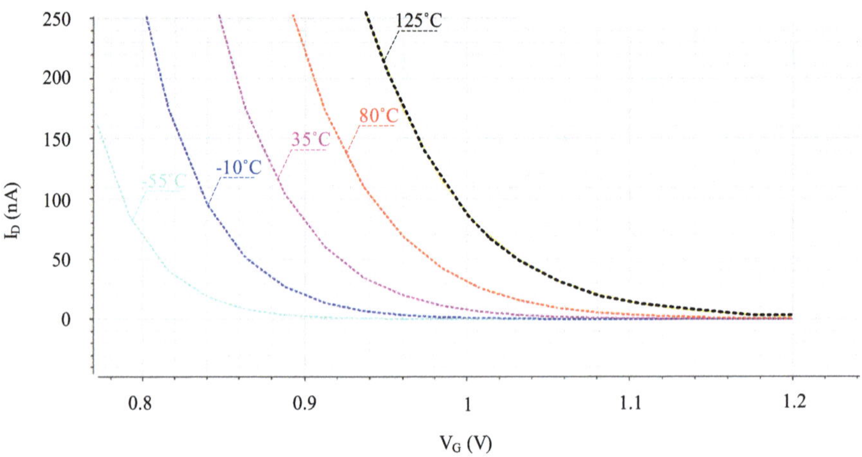

Fig. 2.15 Curent-mode low voltage bandgap reference circuit

2.2.5 *Large Drain Leakage Current*

Another negative effect in nanometer CMOS technology is the MOSfet's large drain leakage current. As shown in Fig. 2.15, a PMOSfet with a size as 300 nm/ 60 nm, and the source voltage (V_S) is 1.2 V, and the drain voltage (V_D) is 0 V. The gate voltage is swept from 0 V to 1.2 V in different temperature. The simulation result shows that in the temperature of 125°C V_{GS} is 0.2 V, the drain leakage current is beyond 100 nA.

The negative effect of large drain leakage current to a reference design is mainly the startup circuit. That because the start-up circuit will be designed as lower power as possible, the operation current of the startup circuit may be less than 1 uA. If the leakage current and start-up circuit's operation current are at the same quantity level, the start-up circuit may lose its effectiveness, or the output state may be flipped and then affect the reference circuit's normal operation.

2.2.6 Summary

The discussed five effects caused by nanoscale CMOS technology are negative for a high performance reference circuit design. The only positive fact caused by CMOS technology shrinking is the higher manufacture accuarcy which means a smaller element mismatch. But the quantitative level of mismatch improvement cannot compensate all the negative effects. Furthermore, when a large temperature variation range is demanded, many reference circuits may not maintain a suitable DC operating point at every process corner and source voltage corner, and that will affect the circuit's yield. So, to design a high performance bandgap reference circuit among a large temperature variation range in nanoscale CMOS technology is very difficult.

2.3 A Bandgap Reference Circuit in 65 nm CMOS

2.3.1 Circuit Implementation

Figure 2.16 shows the schematic of the presented bandgap reference circuit. The bandgap core uses a classic low voltage bandgap architecture as literature [8], but due to low threshold voltage NMOSfets are utilized, the OTAs can use a simple architecture.

The principle of bandgap core can be explained as follows: MP1 and MP2 have the same W/L, and the emitter area ratio of PNP1 to PNP2 is set as 64. The node voltage $V_1 \approx V_2$ because of the feedback of OTA1, and lets $R_1 = R_4$, $R_3 = R_5$, so it can get $V_3 \approx V_4 \approx V_{BE2}$. The drain current of MP1(and MP2) can be expressed as Eq. (2.5):

$$I_D = \frac{V_3 - V_5}{R_2} + \frac{V_3}{R_1 + R_3} = \frac{V_{BE2} - V_{BE1}}{R_2} + \frac{V_3}{R_1 + R_3} = \frac{V_T \ln 64}{R_2} + \frac{V_{BE2}}{R_1 + R_3} \quad (2.5)$$

When a suitable value of R_1–R_5 is set, I_D can be temperature independent. It needs to be considered that the temperature coefficient of R_1–R_5 can not affect I_D's temperature characteristic if all the resistors are the same type (this will be verified by the measured results in Sect. 2.3.2).

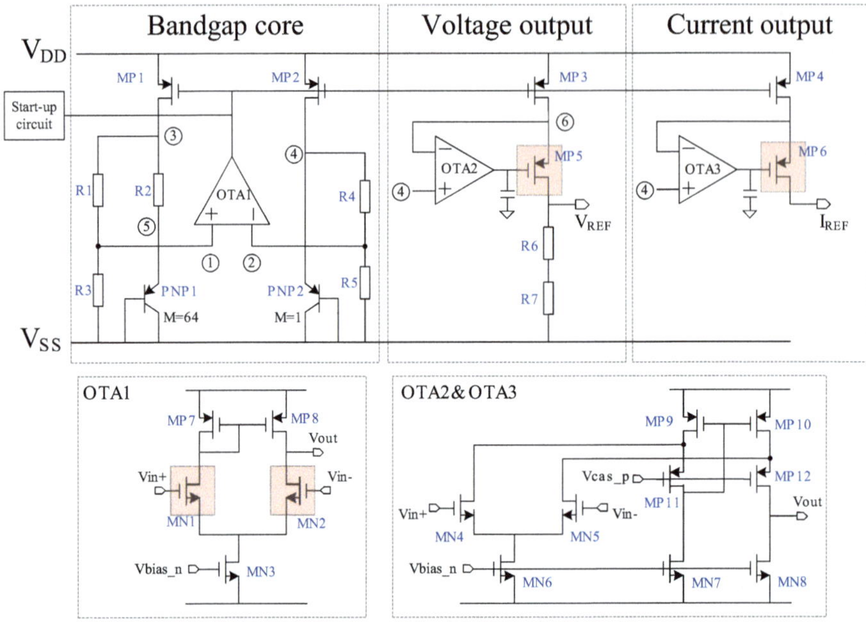

Fig. 2.16 The presented bandgap reference architecture

The input common-mode voltage of OTA1 (V_1/V_2) is a fraction of V_{BE2}, and this voltage must fulfil the Eq. (2.6):

$$\begin{gathered} V_1 \geq V_{DSAT\_N3} + V_{TH\_N2} + V_{DSAT\_N2} \\ \frac{R_3}{R_1 + R_3} V_{BE2} \geq V_{DSAT\_N3} + V_{TH\_N2} + V_{DSAT\_N2} \end{gathered} \tag{2.6}$$

V_{DSAT} is the drain-source saturation voltage of MOSfet. V_1/V_2 may not fulfil Eq. (2.6) under extreme low temperature, to avoid this case, the input MOSfets of OTA1 (MN1/MN2) are choosed as low threshold voltage NMOSfet. PMOSfet input OTA also can be used here as literature [8], but in order to fulfil the output voltage range of OTA, two stage architecture must be utilized.

As Fig. 2.16 shown, the presented reference voltage output and current output circuit utilize the familiar architecture. OTA2 and OTA3 are used to obtain an accurate current replication and high DC PSRR performance. The principle is that let the drain voltage of MP3 and MP4 tracking MP1 and MP2 (and the gate voltage and source voltage of MP1–MP4 are equal) can obtain an accurate current replication under every PVT corner even the gate length of MP1–MP4 are small. Figure 2.17 shows the simulation result of current replication performance, compared with Fig. 2.11 and Fig. 2.13, the accuracy of the current replication are significantly improved.

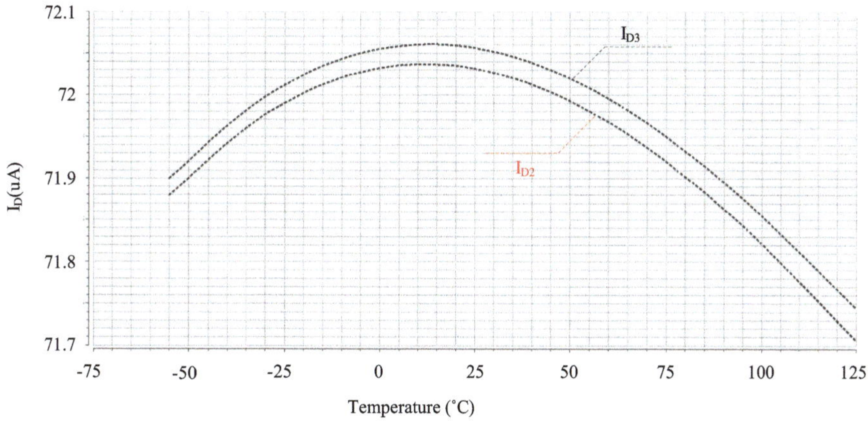

Fig. 2.17 Current replication performance by OTAs are utilized

The input common-mode voltage of OTA2 and OTA3 (V_4/V_6) is V_{BE2}, so MN4 and MN5 can utilize normal threshold voltage NMOSfet. The output voltage of OTA2&OTA3 must fulfil the Eq. (2.7) and Eq. (2.8):

$$V_6 \geq V_{TH\_P5} + V_{DSAT\_P5} + V_{DSAT\_N8}$$
$$V_{BE2} \geq V_{TH\_P5} + V_{DSAT\_P5} + V_{DSAT\_N8} \tag{2.7}$$
$$V_6 - V_{REF} \geq V_{DSAT\_P5}$$
$$V_{REF} \leq V_{BE2} - V_{DSAT\_P5} \tag{2.8}$$

V_{REF} is the reference output voltage. In order to fulfil the output voltage range of OTA2&OTA3, the MP5 and MP6 are choosed as low threshold voltage PMOSfet.

As shown in Fig. 2.18, R_6 and R_7 have the opposite temperature coefficient which is connected in series with suitable ratio to obtain a one-order temperature independent characteristic. This series connection's temperature curve also achieve a curvature which is inverse to the curvature of reference current's, as shown in Fig. 2.19, so a weak curvature compensation effect for voltage reference can be achieved.

Comparing with normal MOSfets, low threshold voltage MOSfets have many disadvantages such as large parameter variation range under PVT corners, large leakage current under very high temperature, inaccurate simulation model, etc. In this design, low threshold voltage MOSfets are only used in the feedback loop, and all those negative effects will be alleviated by OTA's close-loop feedback.

2.3.2 *Measurement Results and Comparison*

Figure 2.20 shows the photograph of this bandgap reference, and the chip is implemented in a 65nm CMOS technology, occupies 0.75 mm × 0. 67 mm

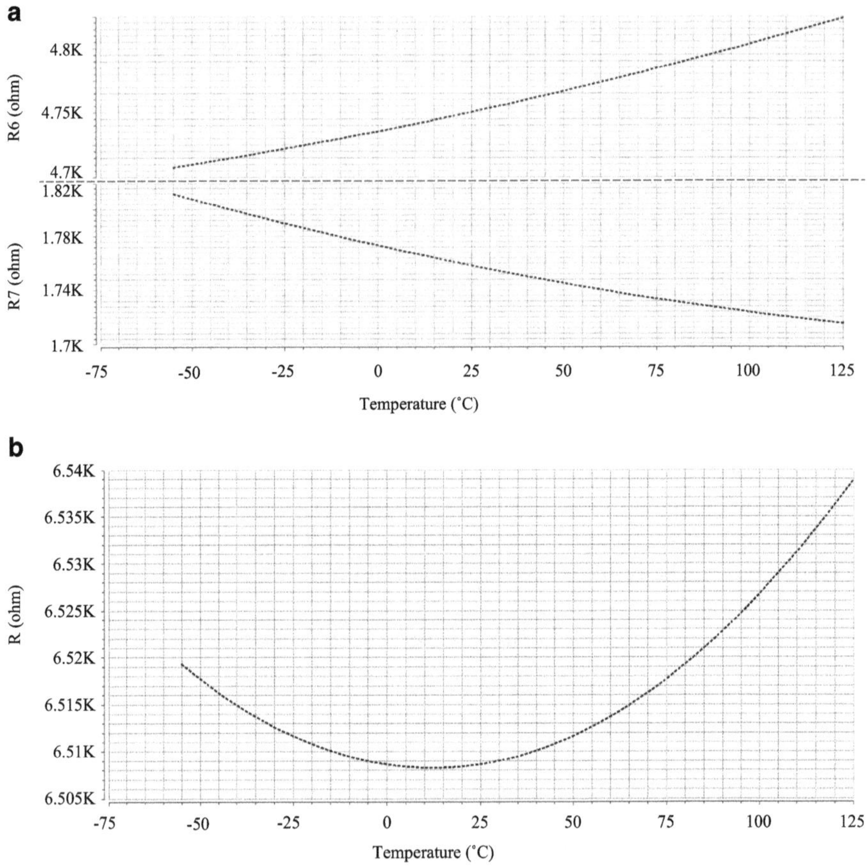

Fig. 2.18 One-order temperature independent resistor. (**a**) R6 and R7 with opposite TC. (**b**) TC of R6 and R7 in series connection

including bond pads, bonded in a D16S package. Test equipments include: Agilent 34401 multimeter, Agilent N6705B DC power analyzer, and THERMOJET-ES environmental testing apparatus.

Figure 2.21a shows the measured result of sampler1's current output, and Fig. 2.21b shows sampler1's voltage output. When measuring the current output, an external resistor is utilized to set the drain voltage of MP6 as similar as MP5's, and the external resistor is put outside of the temperature control box, so the temperature coefficient of external resistor will not affect the temperature performance of current output, and the measurement of current output's PSRR (DC) also utilizes this external resistor.

Figure 2.22a shows the measured DC power supply rejection performance of samplers1' current output under three temperature, and Fig. 2.22b shows the measured DC power supply rejection performance of samplers1' voltage output under three temperature.

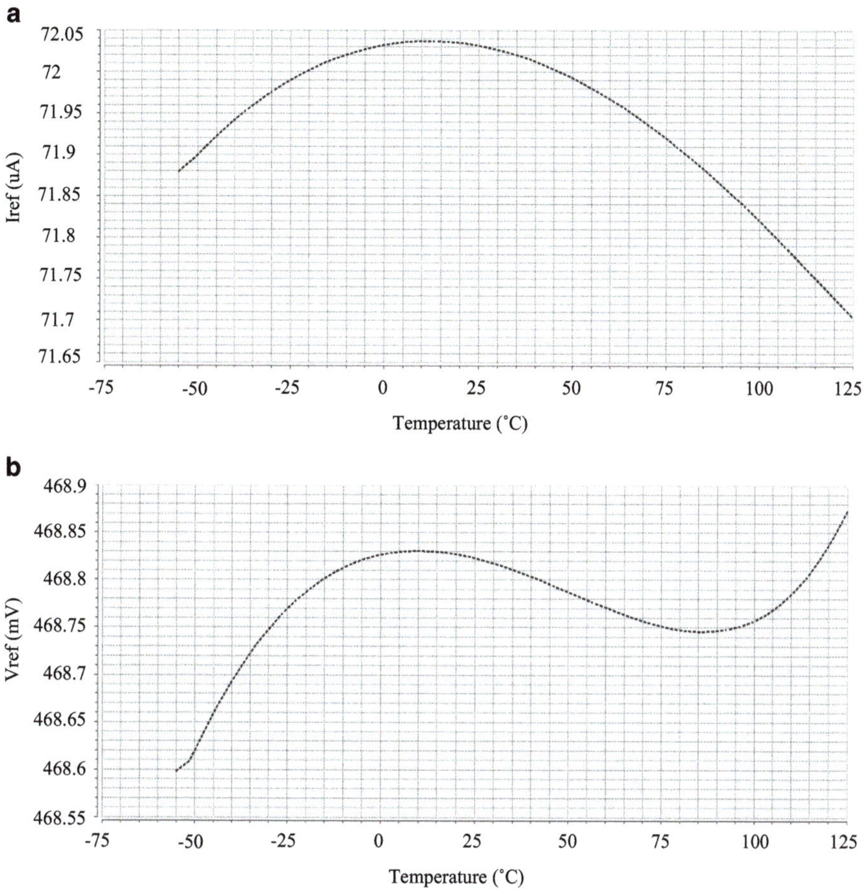

Fig. 2.19 Temperature performance of iref and vref 's simulation result. (**a**) Temperature Performance of IREF. (**b**) Temperature Performance of VREF

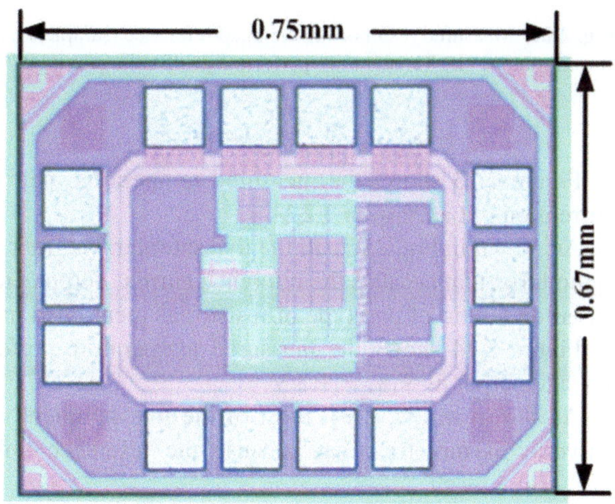

Fig. 2.20 Photograph of the bandgap reference chip

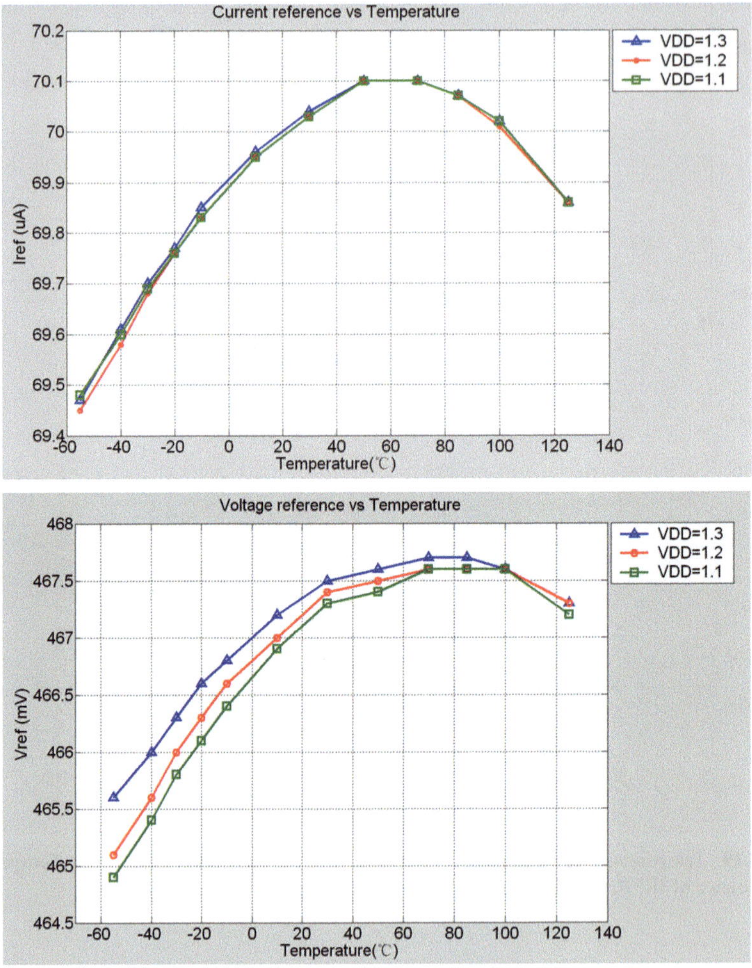

Fig. 2.21 Measured result of the bandgap reference (sampler1). (**a**) Temperature performance of current output. (**b**) Temperature performance of voltage output

Figure 2.23 shows the measured temperature performance of all 5 samplers' current output as $V_{DD} = 1.2$ V.

As shown in Fig. 2.23, all the five samplers' current output temperature curve is symmetric, that because the reference current output's temperature characteristic is based on a ratio of one type of resistor as mentioned in Sect. 2.3.1.

Figure 2.24 shows the measured temperature performance of all 5 samplers' voltage output as $V_{DD} = 1.2$ V.

As shown in Fig. 2.24, most of the five samplers' voltage output temperature curve is asymmetric, that because the reference voltage output's temperature

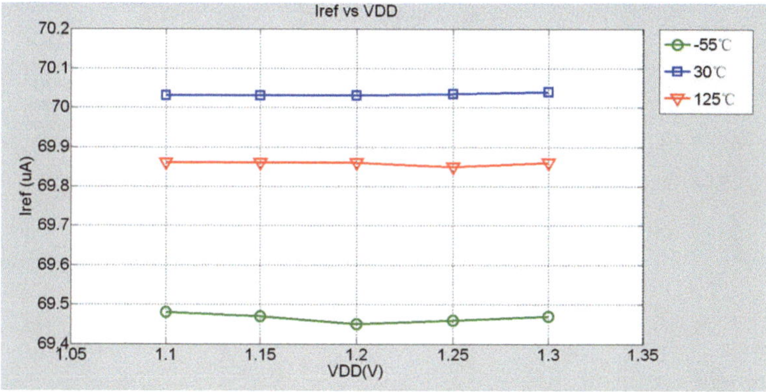

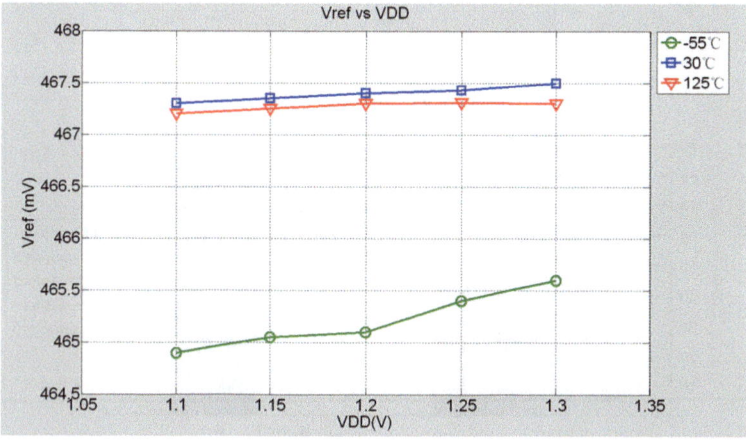

Fig. 2.22 Measured result of the bandgap reference (sampler1). (**a**) PSSR performance of current output. (**b**) PSSR performance of voltage output

characteristic is based on two types of resistors in series connection. Good matching for two types of resistors at only one time is nearly impossible, so a layout optimization can be done in the next manufacture to trimming this temperature curve, and an improved temperature performance can be expected.

Figure 2.25 shows the measured DC power supply rejection performance of all 5 samplers' current output as T = 30°C.

Figure 2.26 shows the measured DC power supply rejection performance of all 5 samplers' voltage output as T = 30°C.

Figure 2.27 shows the measured temperature performance of all 5 samplers' current and voltage output as $V_{DD} = 1$ V.

From Fig. 2.27, when $V_{DD} = 1$ V, all the 5 samplers start becoming invalid as the temperature below 0°C, and this is due to the unsuitable DC operation points.

Table 2.1 summarizes the measured performance.

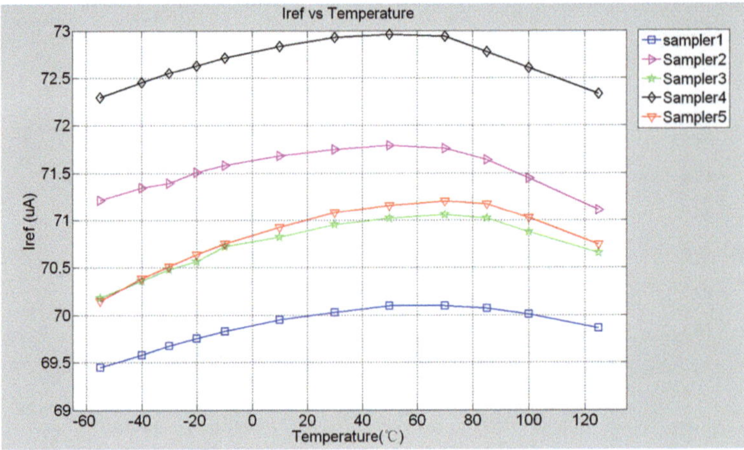

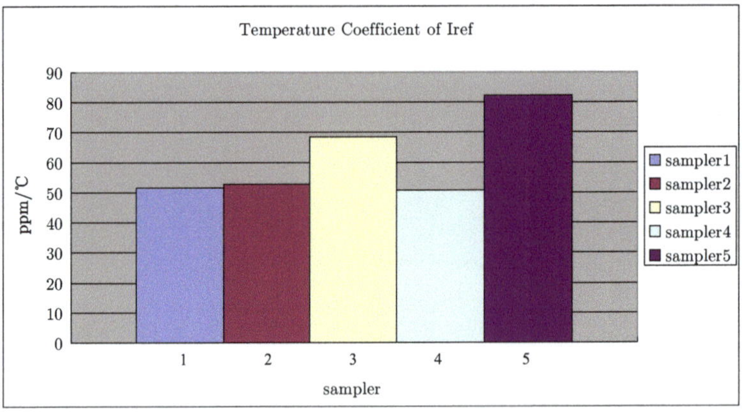

Fig. 2.23 Measured temperature performance of current output (5 samplers, VDD = 1.2 V). (**a**) Temperature performance of current output. (**b**) Temperature coefficient of current output

Table 2.1 compares the measured performance of the bandgap reference with several other published literatures'.

As shown in Table 2.2, because of low threshold voltage MOSfets are utilized, temperature range of the presented bandgap reference is larger than all of the listed literatures, and it can operate at the extreme low temperature (−55°C). The temperature coefficient of voltage output is better than current output, that due to the weak curvature compensation effect. DC PSRR also achieves a high performance as OTAs are utilized.

2.3.3 Summary

The presented bandgap reference circuit is implemented in 65 nm CMOS. To achieve a larger temperature range and a higher DC power supply rejection rate

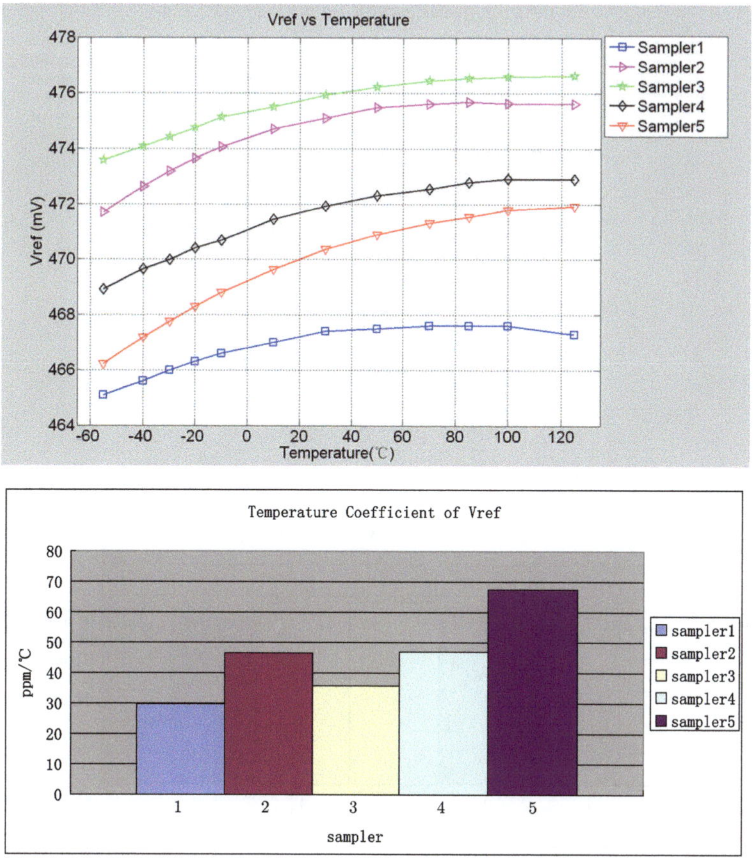

Fig. 2.24 Measured Temperature performance of voltage output (5 samplers, VDD = 1.2 V). (**a**) Temperature performance of voltage output. (**b**) Temperature coefficient of voltage output

(PSRR), low threshold voltage MOSfet and operational transconductance amplifiers have been utilized. A weak curvature compensation effect for reference voltage's generation have been achieved by two opposite temperature coefficient resistors in series connection. The measured results shows that the bandgap reference achieves all the design targets.

2.4 Conclusion

In this chapter, the authors discuss five negative effects caused by CMOS technology shrinking in high performance bandgap reference design, and then present a bandgap reference circuit implemented in a 65 nm CMOS with both voltage output and current output. The design includes three tips: first, low threshold voltage

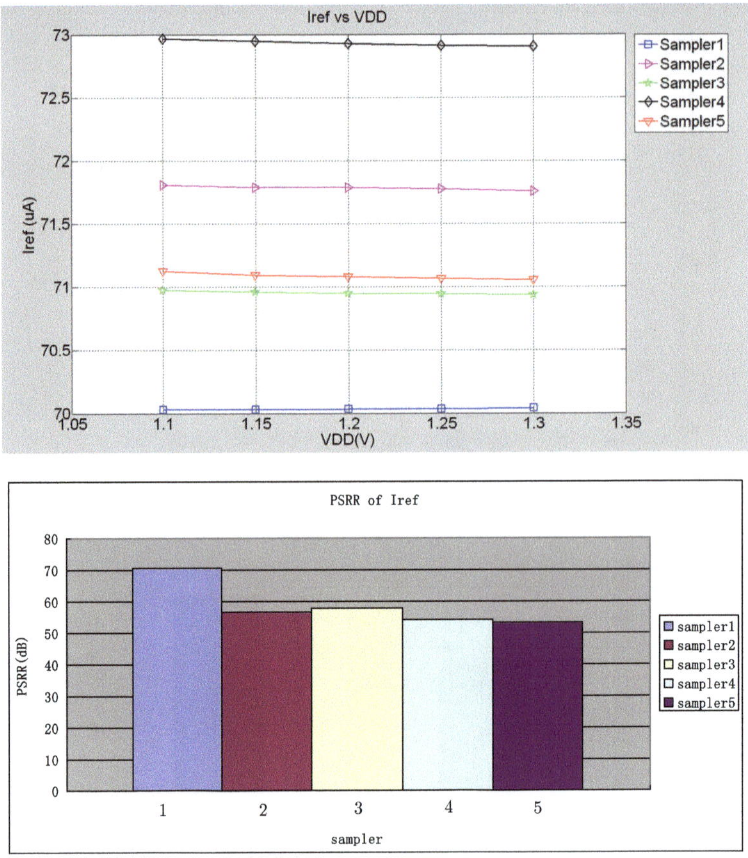

Fig. 2.25 Measured DC power supply rejection performance of current output (5 samplers, T = 30°C). (**a**) Reference current output versus VDD. (**b**) DC PSRR performance of current output

MOSfet have been utilized to obtain a suitable DC operating point at low temperature. Second, OTA has been used to obtain an accurate current copy and a high DC PSRR. Third, two opposite temperature coefficient resistors have been connected in series to obtain a one-order temperature independent resistor which also achieves a weak curvature compensation effect for reference voltage's generation. The measured results shows that the reference circuit achieves a higher DC PSRR and a smaller temperature coefficient, and the temperature range is −55°C to 125°C which is larger than most of the published literatures.

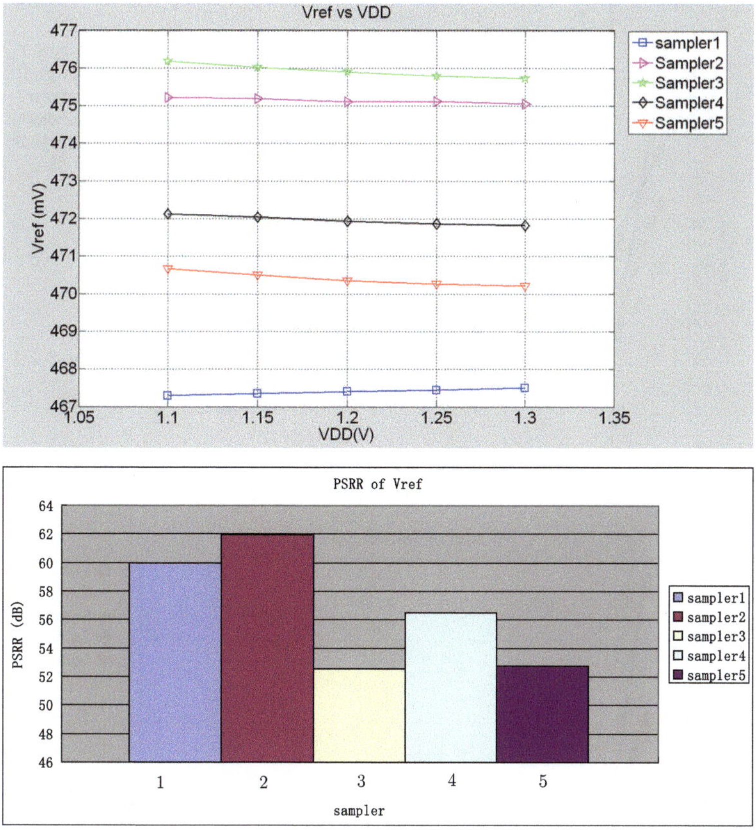

Fig. 2.26 Measured DC power supply rejection performance voltage output (5 samplers, T = 30°C). (**a**) Reference voltage output versus VDD. (**b**) DC PSRR performance of voltage output

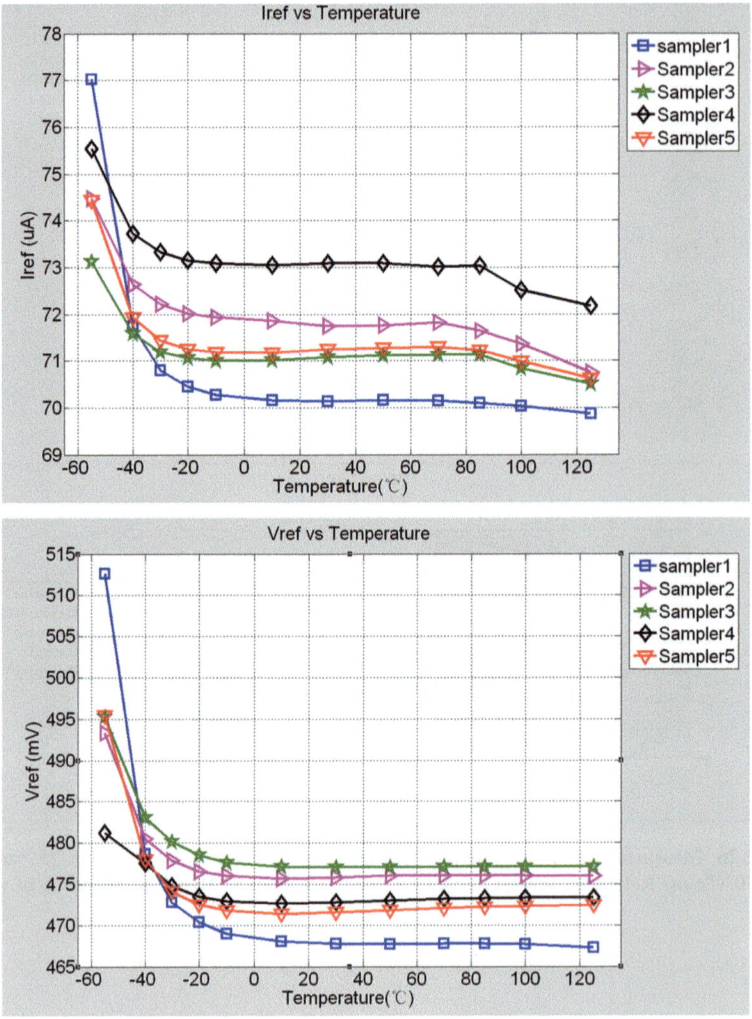

Fig. 2.27 Measured temperature performance of current and voltage output (5 samplers, VDD = 1 V). (**a**) Temperature performance of current output. (**b**) Temperature performance of voltage output

Table 2.1 Measured performance of the bandgap reference

Specification	Measured results
Technology	65 nm CMOS (with low V_{th} MOSfet)
Power supply	1.1–1.3 V (1.2 V typical)
Temperature range	−55–125°C
V_{REF}	468 mV
I_{REF}	70 uA
temperature coefficient	As shown in Table 2.2
PSRR (DC)	As shown in Table 2.2
Power consumption (1.2 V)	0.488 mW
Die area (including bond pads)	0.75 mm × 0.67 mm

Table 2.2 Comparison

Literature	V_{DD} V	TR °C	TC ppm/°C	Process nm	PSRR (DC) (room temp) dB
[8]	0.98–1.5	0–100	15	600	≈41[a]
[9]	1.2	−25–125	119	1200	40@5KHz
[10]	1.4	−20–100	12.4	350	68@100Hz
[11]	1	−40 to 125	11	500	≈86.4[b]
[12]	0.75	−20 to 85	40	130	≈86[b]
[13]	1.2	−40 to 120	4.5[c]	180	≈61[a]
[14]	0.9	0–100	19.5	250	≈30[a]
[15]	1.2/0.7	−40 to 120	147/114	180	≈33/43[a]
[16]	0.9	0–120	≈1030[d]	32	N/A
[17]	1.1	−40 to 125	6	28	≈81.5[b]
This work (V_{REF})	1.1–1.3	−55 to 125	30 (better)	65	60 (better)
			45 (average)		56 (average)
This work (I_{REF})	1.1–1.3	−55 to 125	52 (better)	65	70 (better)
			61 (average)		58 (average)

[a]Calculated from the literatures' temperature curve vs supply voltage
[b]Calculated from the literatures' line regulation, measurement result
[c]After trimming
[d]Calculated from the literatures' temperature curve

Acknowledgments This work is supported by National Natural Science Foundation of China (No.61006027, No.61176030), Research Foundation of Key Laboratory of Analog Integrated Circuit (No.0c09YJTJ1505, No.9140c090105140C09041, No.0c09YJTJ1602), and Fundamental Research Funds for the Central Universities of China (No.ZYGX2012J003).

References

1. H. Banba, H. Shiga, A. Umezawa, et al., A CMOS bandgap reference circuit with sub 1-V operation. IEEE J. Solid State Circuits **34**(5), 670–674 (1999)
2. M.D. Ker, J.S. Chen, C.Y. Chu, A CMOS bandgap reference circuit for sub 1-V operation without using extra low-threshold voltage device. IEICE Trans. Electron. **E88**(11), 2150–2155 (2005)
3. J. Doyle, Y.J. Lee, Y.B. Kim, et al., A CMOS subbandgap reference circuit with 1-V power supply voltage. IEEE J. Solid-State Circuits **39**(1), 252–254 (2004)
4. T. Ytterdal, CMOS bandgap voltage reference circuit for supply voltages down to 0.6 V, IEEE. Electron. Lett. **39**(20), 1427–1428 (2003)
5. A.-J. Annema, Low-power bandgap references featuring DTMOSTs. IEEE J. Solid State Circuits **34**(7), 949–955 (1999)
6. M. Ugajin, K. Suzuki, T. Tsukahara, A 0.6 V supply, voltage-reference circuit based on threshold-voltage summation architecture in fully depleted CMOS/SOI. IEICE Trans. Electron. **E85**(8), 1588–1595 (2002)
7. J.W. Kim, B. Murmann, R. Dutton, in *International Symposium on Quality Electronic Design (ISQED)*. Hybrid Integration of bandgap reference circuits using silicon ICs and germaninum devices (Stanford University, San Jose, CA, 2008), pp. 429–432
8. K.N. Leung, P.K.T. Mok, A Sub-1-V 15-ppm/°C CMOS bandgap voltage reference without requiring low threshold voltage device. IEEE J. Solid State Circuits **37**(4), 526–530 (2002)
9. G. Giustolisi, G. Palumbo, M. Criscione, F. Cutrì, A low-voltage low-power voltage reference based on subthreshold MOSFETS. IEEE J. Solid State Circuits **38**(1), 151–154 (2003)
10. T. Raymond, Perry, S.H. Lewis, A. Paul Brokaw, T.R. Viswanathan, A 1.4 V supply CMOS fractional bandgap reference. IEEE J. Solid State Circuits **42**(10), 2180–2186 (2007)
11. K. Sanborn, D. Ma, A. Paul Brokaw, V. Ivanov, A sub-1-V low-noise bandgap voltage reference. IEEE J. Solid State Circuits **42**(11), 2466–2481 (2007)
12. V. Ivanov, R. Brederlow, J. Gerber, An ultra low power bandgap operational at supply from 0.75 V. IEEE J. Solid State Circuits **47**(7), 1515–1523 (2012)
13. B. Ma, F. Yu, A novel 1.2–V 4.5-ppm/°C curvature-compensated CMOS bandgap reference. IEEE Trans. Circuits Syst. I Regul. Pap. **61**(4), 1026–1035 (2014)
14. M.-D. Ker, J.-S. Chen, New curvature-compensation technique for CMOS bandgap reference with sub-1-V operation. IEEE Trans. Circuits Syst. II Express Briefs **53**(8), 667–671 (2006)
15. Y. Osaki, T. Hirose, N. Kuroki, M. Numa, 1.2-V Supply, 100-nW, 1.09-V bandgap and 0.7-V supply, 52.5-nW, 0.55-V subbandgap reference circuits for nanowatt CMOS LSIs. IEEE J. Solid State Circuits **48**(6), 1530–1538 (2013)
16. A.J. Annema, P. Veldhorst, G. Doornbos, B. Nauta, A sub-1V bandgap voltage reference in 32 nm FinFET technology. ISSCC. 331–333 (2009)
17. D.F. Bowers, E.J. Modica, Curvature-corrected low-noise sub-bandgap reference in 28 nm CMOS technology. Electron. Lett. **50**(5), 396–398 (2014)
18. V.V. Ivanov, K.E. Sanborn, I.M. Filanovsky, Bandgap voltage references with 1 V supply. Proceedings of the 32nd European Solid-State Circuits Conference. ESSCIRC 2006, pp. 311–314 (2006)

Chapter 3
Metal Oxide Semiconductor Thin-Film Transistors: Device Physics and Compact Modeling

Wanling Deng, Jielin Fang, Xixiong Wei, and Fei Yu

3.1 Introduction

Metal oxide semiconductor thin-film transistor (TFT) is poised to become a prominent next-generation technology due to its high carrier mobility, good uniformity, low process temperature, compatibility with rollable transparent electronic applications, etc. Similar to the conventional MOSFET devices, metal oxide semiconductor TFTs can be also used in IC designs, for example, AMOLED line driver [1], logical circuits [2], digital to analog converters (DAC) [3], RFID or near field communication (NFC) applications [4], etc. Today, some new applications with metal oxide semiconductor TFTs are emerging in electronic device markets, such as intelligent wearable and textile integrated systems [5, 6], epidermal devices [7, 8], artificial skins [9, 10], medical implants [11, 12], active matrix liquid-crystal displays [13], and so on. These new applications are revolutionizing our daily life.

Table 3.1 [14] gives a comparison of some important device properties for the established TFTs, including amorphous silicon (a-Si), polycrystalline silicon (poly-Si), organic semiconductors, and metal oxide semiconductors. As shown in Table 3.1, metal oxide semiconductors TFTs present some outstanding advantages on device properties like smaller manufacturing cost, larger scalability, lower process complexity and temperature, and so on [14].

The first metal oxide semiconductor TFT was manufactured by Klasens and Koelmans [15]. However, metal oxide semiconductor TFTs did not catch much attention at that time. Until 1996, working as active layers in ferroelectric memory TFTs [16], metal oxide semiconductors could reclaim new lives. The antimony-doped SnO_2 (SnO_2:Sb) TFTs were presented by Prins et al. [16], and at the same

W. Deng (✉) • J. Fang • X. Wei • F. Yu
College of Information Science and Technology, Jinan University, Guangzhou 510630, China
e-mail: dwanl@126.com

T. Li, Z. Liu (eds.), *Outlook and Challenges of Nano Devices, Sensors, and MEMS*,
DOI 10.1007/978-3-319-50824-5_3

Table 3.1 Comparison between metal-oxide-based TFTs and other counterparts [14]

Property	a-Si TFTs	Poly-Si TFTs	Organic TFTs	Metal-oxide-based TFTs
Mobility (cm^2/Vs)	0.5–1	50–100	0.1–10	10–100
Integration of circuits	Poor	Good	Poor	Good
Uniformity	Good	Poor	Good	Good
Process temperature (°C)	150–300	350–500	RT-250	RT-350
Manufacturing cost	Low	High	Low	Low

time, the first indium oxide (In_2O_3) non-volatile memory TFT with ferroelectric gate dielectric was shown by Seager et al. [17]. Up to 2003, the increasing attention was paid to ZnO TFTs [18, 19], owing to the carrier mobility around 1–40 $cm^2/(Vs)$. The high carrier mobility guarantees metal oxide TFTs can be employed into display backplanes. Furthermore, Nomura et al. [20] proposed an indium gallium zinc oxide (InGaZnO) TFT in 2003, which demonstrated the electron mobility of 80 $cm^2/(Vs)$. Recently, some other multicomponent metal oxide semiconductor TFTs are prepared and excellent electrical performance is obtained, such as zinc tin oxide (ZTO) TFTs [21], indium zinc oxide (IZO) TFTs [22], and so on.

Comparing to the traditional MOSFET technology, metal oxide semiconductor TFTs are mainly fabricated on insulating substrates (e.g. glass and plastic). Of course, the use of the different techniques would lead to different device properties. For example, Fig. 3.1 presents the schematic cross-section image of the bottom-gate-structure a-IZO TFT on the polyimide (PI) substrate [23]. A molybdenum (Mo) gate electrode was fabricated on the buffer layer with 200 nm thickness. Then, a stacked buffer layer of SiN_x/SiO_2 as gate insulator film was deposited by PECVD at 310°C with a total thickness of 300 nm. The indium zinc oxide (In:Zn = 1:1, IZO) active layer was grown onto gate dielectric by RF magnetron sputtering, having a thickness of 30 nm. After that, a SiO_2 layer (200 nm) as etch stop layer (ESL) was deposited and patterned by dry-etch. The Mo/Al/Mo stacked layers as the source/drain (S/D) electrodes were then sputtered at room temperature and patterned by photolithography and wet etching. Finally, the device was post-annealed at 350 °C for 30 min in O_2 atmosphere [23]. The PI substrate was attached to glass carrier during the entire TFT fabrication first and de-bonded mechanically from the carrier only when the device formation was completed.

As we mentioned above, the flexible metal oxide semiconductor TFTs have been considered to be one of the most important devices because of their widely applications in flexible AMOLED displays, e-Papers, soft electronic skins, wearable and textile integrated devices, etc. The field of flexible electronics has attracted much attention and presents us with a new mode of daily life. For the future flexible electronic applications, more and more complex digital and analog circuits with high TFT count will be included. Given the fact that the flexible TFT technology is

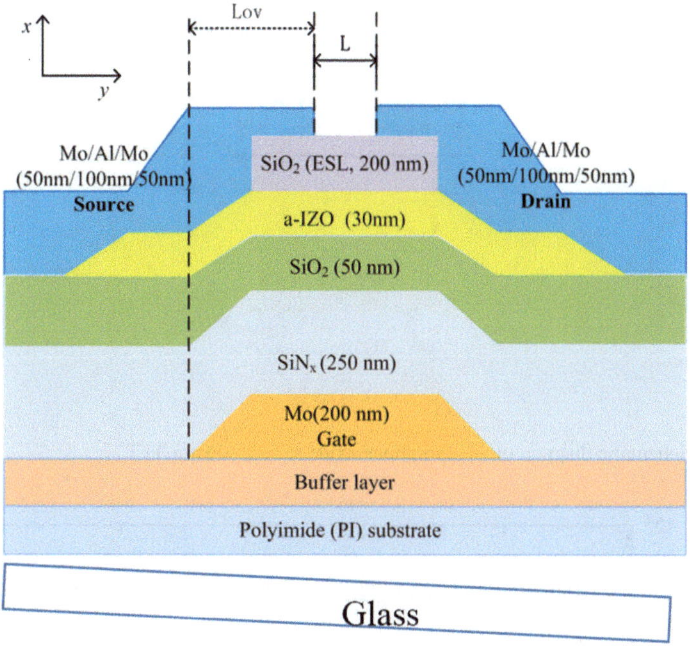

Fig. 3.1 Cross-sectional view of the bottom gate a-IZO TFT on PI substrate [23]

going main-stream, compact model adequate for circuit simulators is urgently needed to reduce the design cycle for developing industry standard devices and future use of these devices in integrated circuits. Compact model is the interface between circuit designers and device technology. A good compact model is required to accurately capture all real-device effects and maintain high computational efficiency. In this chapter, a compact model of metal oxide semiconductor TFTs is introduced, which is a proper balance between accuracy and complexity. Surface-potential-based approach is employed here, which is single-piece and give an accurate and continuous description of surface potential and current in all operation regions.

3.2 Fundamentals of Metal Oxide-Based TFTs

Figure 3.2 presents the schematic structure of an n-type and un-doped metal oxide semiconductor TFT, in which x is the perpendicular direction across the channel thickness and y is the parallel direction along the channel. The device in Fig. 3.2 has an inverted staggered bottom gate. In the amorphous oxide semiconductor thin film, the trap states are classified as acceptor-like and donor-like states. Since the device is assumed as n-type, donor-like trap densities can be neglected. As depicted in Fig. 3.3, the energy-dependent acceptor-like density of states (DOS) over the bandgap, i.e., $g(E)$, is modeled by the superposition of deep and tail states in exponential forms as [24]

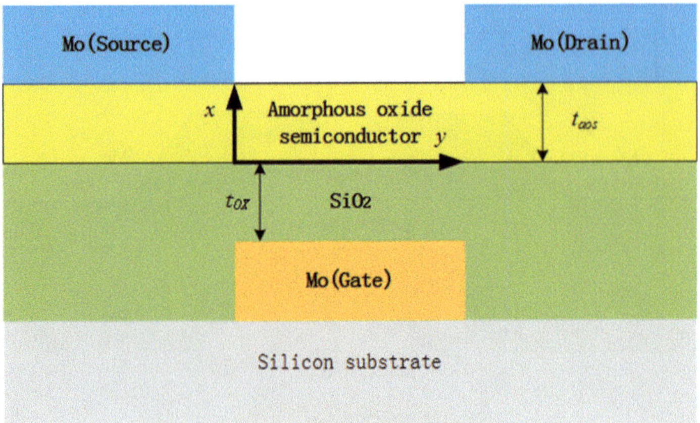

Fig. 3.2 Schematic diagram of an amorphous oxide semiconductor TFT

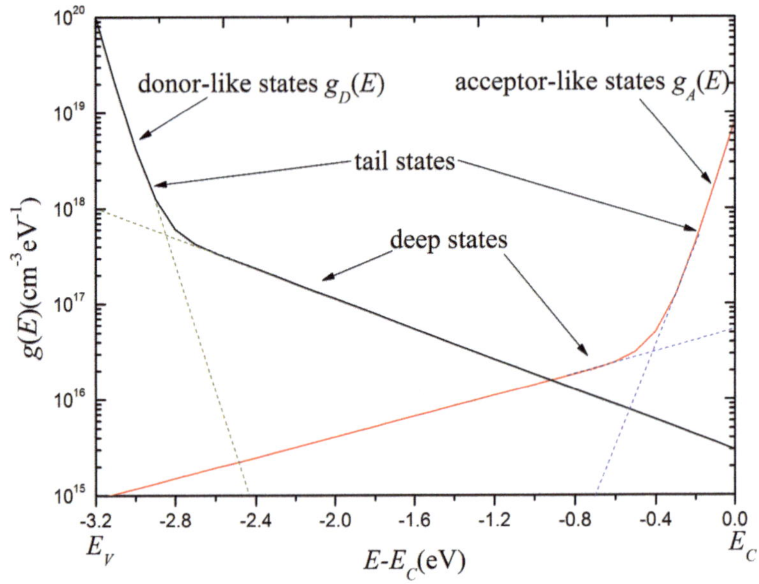

Fig. 3.3 Distribution of DOS in logarithmic scale. The donor-like and acceptor-like states are each approximated by the sum of two exponential

$$g(E) = N_{D0} \exp\left(\frac{E - E_C}{E_d}\right) + N_{T0} \exp\left(\frac{E - E_C}{E_t}\right) \tag{3.1}$$

where E is the energy, N_{D0} and N_{T0} are the deep- and tail-state densities at the conduction band edge E_C respectively, k is the Boltzmann constant, E_d ($E_d = kT_d$) is the characteristic energy for the deep trap states, and E_t ($E_t = kT_t$) is the characteristic energy for the tail trap states. Note that, although the energy distribution of deep state density may be either an exponential or a Gaussian form depending on the fabrication processes, an exponential distribution is adopted here [25].

The density of ionized acceptor traps (n_{trap}) is given as a function potential by

$$n_{trap} = \int_{E_V}^{E_C} g(E)f(E)dE = n_{deep} + n_{tail} \tag{3.2}$$

where E_V is the valence-band maximum, n_{deep} and n_{tail} are the ionized trap components for deep- and tail-states respectively. The function $f(E)$ is the occupation probability:

$$f(E) = \frac{1}{1 + \exp\left(\frac{E-E_f}{kT}\right)} \tag{3.3}$$

herein, E_f is the Fermi level and T is the temperature.

Following the gradual channel approximation, the one-dimensional Poisson's equation for potential $\phi(x)$ along the vertical direction is described by [25]

$$\frac{\partial^2 \phi}{\partial x^2} = \frac{q}{\varepsilon_{aos}} \left(n_{free} + n_{deep} + n_{tail}\right) \tag{3.4}$$

where q is the electron charge, ε_{aos} is amorphous oxide semiconductor permittivity, and n_{free} is the density of free charges. In Eq. (3.4), the impacts of free carriers, localized deep states and tail states are taken into account.

We also assume that the potential $\phi(0)$ and electric field at the back interface ($x = 0$) are zero. Using Gauss' law at the front interface ($x = t_{aos}$), we get

$$\varepsilon_{aos} \frac{\partial \phi}{\partial x}\bigg|_{x=t_{aos}} = -\varepsilon_{aos} F_s = C_{OX}(\phi_G - \phi_s) \tag{3.5}$$

where $\phi_G = V_{gs} - V_{fb}$, C_{OX} is the gate capacitance per unit area, V_{gs} is the gate-to-source voltage, V_{fb} is the flat-band voltage, F_s is the vertical electric field at the surface, and ϕ_s is the surface potential with $\phi_s = \phi(t_{aos})$. In Eq. (3.5), the implicit relation between ϕ_s and V_{gs} can be obtained. However, due to the inclusion of complicated functions in Eq. (3.4), a closed form of Eq. (3.5) with the analytical approach is impossible. To accurately calculate the surface potential, a numerical computation should be used.

Furthermore, to evaluate the drain current of metal oxide semiconductor TFTs, it is possible to extend the Pao-Sah model [26] developed for crystalline silicon MOSFETs. It uses a single expression accounting for both the drift and diffusion components, and is valid in the all operating regions when the gate voltages are larger than flat-band voltage. Hence, following such an approach, the drain current (I_{ds}) is expressed as:

$$I_{ds} = \mu \frac{W}{L} \int_{\phi_{ss}}^{\phi_{sL}} \frac{Q_i(\phi, V_{ch})}{d\phi/dV_{ch}} d\phi \tag{3.6}$$

where μ is the effective mobility, V_{ch} is the channel potential (viz., quasi-Fermi potential), W is the channel width, L is the channel length, Q_i is the mobile charge density per unit area depending on the surface potential, and V_{ds} is the drain-to-source voltage. In addition, ϕ_{ss} and ϕ_{sL} are the surface potentials at source and drain ends, respectively, which are the solutions of Eq. (3.5) with $V_{ch} = 0$ and $V_{ch} = V_{ds}$, respectively.

3.3 A Review of Core Compact Models for Metal Oxide-Based TFTs

As the demands for novel applications of metal oxide-based TFTs aforementioned increase rapidly, physics-based compact models are highly desirable to well predict the performance in TFT-based circuits. The models must be physically based and offer better computational efficiency than numerical alternatives. In this section, we mainly provide a unified review of the recent compact models developed for metal oxide semiconductor TFTs. Several research groups have important contributions in this field, and their models will be introduced with a brief description of the physical mechanisms.

In recent years, considerable attention has been drawn to the semiconductor device models based on surface potential formulations. However, the surface potential (ϕ_s) is given by an implicit relation on the terminal voltages, which can only be solved iteratively and leads to long computation times. Consequently, regional approach with the approximate analytical solution of a local domain is an efficient way and is widely adopted in modeling algorithm to derive the asymptotic solution of ϕ_s or drain current. For example, A. Tsormpatzoglou et al. [27] regarded DOS in the active layer as a Gaussian distribution of tail states near the energy band, and neglected the free charge density (n_{free}). Note that the subgap DOS used in [27] differs from Eq. (3.1). By reformulating Gaussian function as two exponential formulas, the derivation of the surface potential can be distinguished in two separated regions.

Unlike the work described above [27], M. Ghittorelli et al. [28] and L. Colalongo [29] took n_{free} into account and assumed DOS as a tail exponential function, but n_{deep} was negligible. Accounting for both multiple-trapping-and-release and percolation mechanisms, the drain current expression in terms of surface potential in different regions was developed. Moreover, the work of [29] relies on a new symmetric quadrature method called SQM, similar to the modulation of symmetric linearization method (SLM) used in PSP model [30]. The basic of SQM is the calculation of the induced mobile charge Q_i, which is approximated by a second order polynomial in terms of surface potential at the center, i.e., $\phi_{\mathrm{m}} = (\phi_{ss} + \phi_{sL})/2$. Finally, based on the charge sheet model, the drain current can be easily obtained.

To explicitly solve the surface potential, an efficient method or known as the effective carrier density approach was proposed in [24, 31]. Nevertheless, the effective carrier density in the bottom of conduction band (see N_{EFF} in [31]) is fixed to a constant in all the operation regions and independent to the gate voltage, which is less physical for metal oxide-based TFTs. The work in [24] indicates that N_{EFF} is equal to tail state density limited to the subthreshold region and free carrier density limited to the above-threshold region. Using this regional approach and the effective density method, the drain current can be deduced and unified through Matthiessen's rule.

From the above summary, it clearly indicates that a better model should consider both localized trapped charge and free carrier in all the operation regions without iterative evaluation. Our group [32] has utilized the effective charge density approach to develop an analytical drain current model based on the explicit calculation of surface potential with both deep and tail states. The proposed model is inherently single-piece, and gives an accurate and continuous description of surface potential and current in all operation regions including the transition. Detailed derivation will be given in the next section.

Up to now, our discussion assumes that the metal oxide film is not degenerated, and thus, the Fermi-Dirac statistic can be simplified by Boltzmann. However, considering a reduced DOS of metal oxide materials, when the gate bias is above turn-on voltage, it can push the Fermi level above the band edge. In this case, Boltzmann approximation is not valid, and the charge concentration should be described in a new form. In literature, only a few TFT models focus on the degenerate conduction. M. Ghittorelli et al. [33–35] at the first time proposed an analytical model that described the charge concentration in the degenerate regime using Lambert *W* function.

Due to drain current is most sensitive to the small errors of surface potential, an accurate and physic-based surface potential model is pressing needed. In the following sections, an efficient solution to the surface potential is derived. Meanwhile, the drain current model is built based on the regional approach, which combines the short calculation time of regional models and the high accuracy of implicit single-piece models.

3.4 Drain Current Model for Metal Oxide Based TFTs

3.4.1 Models Accounting for the Non-degenerate Conduction

In this section, an explicit and closed-form scheme for the surface potential calculation is developed by including both exponential deep and tail states. As we mentioned above, Eq. (3.5) cannot be solved in a closed form. High accuracy of surface potential ϕ_s can only be acquired by using an iterative numerical approach such as the Newton-Raphson algorithm. For compact TFT models, this poses a severe computational burden. As a consequence, the effective charge density approach is adopted here. The resulting DC and surface potential models give accurate descriptions with single-piece formulas, and provide a better platform to develop the advanced surface-potential-based model for the circuit simulation.

3.4.1.1 Derivation of Surface Potential

In amorphous oxide semiconductor (AOS) thin film, spherical overlapping orbitals form the conduction band bottom, leading to a smaller DOS which is 2 or 3 orders of magnitude lower than that of amorphous covalent semiconductors. Hence, Fermi level pinning effect is eliminating [36] and the Fermi level is entering into the conduction band. Carriers percolate among potential barriers in delocalized states. Therefore, the transport mechanism is governed by the percolation conduction. This also explains the high field-effect mobility values measured in metal oxide semiconductor TFTs [28]. Nonetheless, whenever we need a simple and efficient model, the conduction band structure can be described by non-degenerate free-electron-like band and the film is assumed to be non-degenerated [37]. In this case, the calculation of the free carrier density and ionized acceptor traps in Eq. (3.4), can be simplified by Boltzmann statistics. As a consequence, the integral in Eq. (3.2) reads [32]

$$n_{tail} = N_{TO}\frac{\pi kT}{\sin(\pi kT/E_t)}\exp\left(\frac{E_f + q\phi - qV_{ch} - E_C}{E_t}\right) = N_T\exp\left(\frac{q\phi}{E_t}\right) \tag{3.7}$$

$$n_{deep} = N_{DO}\frac{\pi kT}{\sin(\pi kT/E_d)}\exp\left(\frac{E_f + q\phi - qV_{ch} - E_C}{E_d}\right) = N_D\exp\left(\frac{q\phi}{E_d}\right) \tag{3.8}$$

where $N_D = N_{DO}\frac{\pi kT}{\sin(\pi kT/E_d)}\exp\left(\frac{E_f - E_C - qV_{ch}}{E_d}\right)$ when $kT < E_d$, and $N_T = N_{TO}\frac{\pi kT}{\sin(\pi kT/E_t)}\exp\left(\frac{E_f - E_C - qV_{ch}}{E_t}\right)$ when $kT < E_t$.

The free-electron density is given by

$$n_{free} = n_0\exp\left(\frac{\phi}{V_T}\right) \tag{3.9}$$

where V_T is the thermal voltage kT/q, $n_0 = N_C\exp\left(\frac{E_f - E_C - qV_{ch}}{kT}\right)$, and N_C is the effective density of states in the conduction band edge E_C. Substituting Eqs. (3.7)–(3.9) into (3.4), we acquire the 1-D Poisson's equation as [32].

$$\frac{\partial^2\phi}{\partial x^2}=\frac{q}{\varepsilon_{aos}}\left[n_0\exp\left(\frac{\phi}{V_T}\right)+N_D\exp\left(\frac{\phi}{E_d/q}\right)+N_T\exp\left(\frac{\phi}{E_t/q}\right)\right]. \tag{3.10}$$

Integrating Poisson's equation once with the boundary condition $(\partial\phi/\partial x)_{x=0}=0$, yields

$$F(\phi)=\frac{\partial\phi}{\partial x}=\sqrt{\frac{2q}{\varepsilon_{aos}}}\left\{n_0V_T\left[\exp\left(\frac{\phi}{V_T}\right)-1\right]+N_D\phi_{deep}\left[\exp\left(\frac{\phi}{\phi_{deep}}\right)-1\right]\right.$$
$$\left.+N_T\phi_{tail}\left[\exp\left(\frac{\phi}{\phi_{tail}}\right)-1\right]\right\}^{1/2} \tag{3.11}$$

where F is the vertical electric field, $\phi_{deep}=E_d/q$ and $\phi_{tail}=E_t/q$.

In Eq. (3.11), the calculation of electric field at the surface (F_s) can be obtained. The electric field in the accumulation layer can be expressed as a function of the electrostatic potential ϕ and of the channel voltage V_{ch}. Consequently, combining Eqs. (3.5) and (3.11), we get an implicit expression of the surface potential ϕ_s, but it cannot be solved analytically due to the inclusion of three exponential functions in Eq. (3.11). To solve this problem, the effective charge density (n_{eff}) approach [31] can be used. Therefore, Eq. (3.11) is rewritten as

$$F_{eff}(\phi)=\sqrt{\frac{2q}{\varepsilon_{aos}}n_{eff}\phi_{eff}\left[\exp\left(\frac{\phi}{\phi_{eff}}\right)-1\right]}\approx\sqrt{\frac{2q}{\varepsilon_{aos}}n_{eff}\phi_{eff}}\exp\left(\frac{\phi}{2\phi_{eff}}\right)=A\exp\left(\frac{\phi}{2\phi_{eff}}\right) \tag{3.12}$$

where $\phi_{eff}=(V_T+\phi_{deep}+\phi_{tail})/3$, $A=\sqrt{2qn_{eff}\phi_{eff}/\varepsilon_{aos}}$, and

$$n_{eff}(\phi)=\frac{F^2(\phi)}{\frac{2q}{\varepsilon_{aos}}\phi_{eff}\exp\left(\frac{\phi}{\phi_{eff}}\right)}. \tag{3.13}$$

Note that, $n_{eff}(\phi_s)$ [cm^{-3}] is the basis of this model, which is a function of ϕ_s and includes both the localized trapped charges and the free electrons. Applying Eq. (3.12), the implicit surface potential function (3.5) can be re-expressed as [32]

$$\phi_G-\phi_s=\frac{\varepsilon_{aos}}{C_{OX}}A\exp\left(\frac{\phi_s}{2\phi_{eff}}\right). \tag{3.14}$$

Normalized form of Eq. (3.14) in terms of variables $x_s=\phi_s/\phi_{eff}$ and $x_G=\phi_G/\phi_{eff}$ is

$$x_G - x_s = \Delta_{TFT}\exp\left(\frac{x_s}{2}\right) \tag{3.15}$$

where $\Delta_{TFT} = \frac{\sqrt{2q\varepsilon_{aos}n_{eff}/\phi_{eff}}}{C_{OX}}$.

The normalized form of surface potential can be explicitly solved by using the Lambert W function to yield the following expression:

$$x_s = x_G - 2W_0\left[\frac{\Delta_{TFT}}{2}\exp\left(\frac{x_G}{2}\right)\right] \tag{3.16}$$

where W_0 is the principal branch of the Lambert W function [38], a special function which is defined as the solution to the equation $W_0(z) \times \exp[W_0(z)] = z$. The Lambert W function has already proved its usefulness in numerous physics applications. It can be found already incorporated into some circuit simulation tools [39].

In fact, the calculation of n_{eff} within Δ_{TFT} involves an initial estimate of surface potential denoted by ϕ_{s0}. The following algorithm is used to obtain ϕ_{s0}. As done in [25], a distinction can be made between two different regions, namely the strong accumulation region and the subthreshold region. In the subthreshold region, the deep trapped charge density dominates Eq. (3.4). Using the regional approach, Eq. (3.5) can be approximated as

$$C_{OX}^2(\phi_G - \phi_s)^2 = 2q\varepsilon_{aos}N_D\phi_{deep}\exp\left(\frac{\phi_s}{\phi_{deep}}\right). \tag{3.17}$$

When the device enters into the strong accumulation region, all traps are occupied and the density of free charges dominates Eq. (3.4). We can get an asymptotical expression of the surface potential as

$$C_{OX}^2(\phi_G - \phi_s)^2 = 2q\varepsilon_{aos}n_0V_T\exp\left(\frac{\phi_s}{V_T}\right). \tag{3.18}$$

Equations (3.17) and (3.18) can be also solved by using Lambert W function to get the explicit expression of surface potential restricted to different operational regions

$$\phi_{sd} = \phi_G - 2\phi_{deep}W_0\left[\frac{\Delta_{0d}}{2}\exp\left(\frac{\phi_G}{2\phi_{deep}}\right)\right] \tag{3.19}$$

$$\phi_{sn} = \phi_G - 2V_TW_0\left[\frac{\Delta_{0n}}{2}\exp\left(\frac{\phi_G}{2V_T}\right)\right] \tag{3.20}$$

herein, ϕ_{sd} and ϕ_{sn} are the asymptotical results of surface potential in the subthreshold and strong accumulation regions, respectively. Moreover, $\Delta_{0n} = \sqrt{2q\varepsilon_{aos}n_0/V_T}/C_{OX}$

and $\Delta_{0d} = \sqrt{2q\varepsilon_{aos}N_D/\phi_{deep}}/C_{OX}$. Although Eq. (3.19) is indeed an exact solution of (3.17), having neglected the "−1" term as shown in Eq. (3.11) introduces a small error around the flat-band voltage. This simplification was made before attempting the solution of (3.19). If an accurate solution is really needed, it is easy to include the "−1" term straightforwardly by modifying the solution to

$$\phi_{sd} = \phi_G + \Delta_{0d}\phi_{deep} - 2\phi_{deep}W_0\left[\frac{\Delta_{0d}}{2}\exp\left(\frac{\phi_G/\phi_{deep} + \Delta_{0d}}{2}\right)\right]. \tag{3.21}$$

Moreover, an additional term of Δ_{0d} is added to ϕ_G in Eq. (3.21) to guarantee the surface potential becomes zero when ϕ_G approaches zero, which describes the ϕ_s behavior more accurately in the vicinity of V_{fb}.

In the transition region between the above two operation regions, both terms of free charge and trap states contribute to the band bending, and neither approximate results work as they should be. Here, we use the following smoothing function to link the different operation regions. The unified initial valuation of surface potential used in n_{eff} is obtained as

$$\phi_{s0} = \frac{1}{m}\ln\left[\frac{1}{1/\exp(m\phi_{sd}) + 1/\exp(m\phi_{sn})}\right]. \tag{3.22}$$

Here, m (for typical devices, $m > 10$) is a weight parameter to connect the different asymptotical results.

It should be noted that when taking both deep and tail states into account, the calculation of surface potential in Eq. (3.16) is explicit and accurate as a function of gate voltage, but some millivolt errors are given due to the introduction of ϕ_{s0}. However, for a surface-potential-based compact model, high accuracy is desirable. To further improve the accuracy in circuit simulation, the solution given by Eq. (3.16) is refined by adding a correction term ω, which is deduced from Eqs. (3.5) and (3.11), i.e.,

$$\omega = \frac{K_0}{1 + \frac{y_W''}{2y_W'}K_0} \tag{3.23}$$

where $K_0 = -y_W/y_W'$, $y_W = \left(\phi_G - x_s\phi_{eff}\right)^2 - \left[\frac{\varepsilon_{aos}}{C_{OX}}F\left(x_s\phi_{eff}\right)\right]^2$, y_W' and y_W'' are the first- and second- order derivatives of y_W, respectively.

Finally, the complete expression for the surface potential becomes

$$\phi_s = x_s\phi_{eff} + \omega\left(y_W, y_W', y_W''\right). \tag{3.24}$$

Note that, the solution of surface potential from Eq. (3.24) is explicit and accurate over a wide range of the gate voltage.

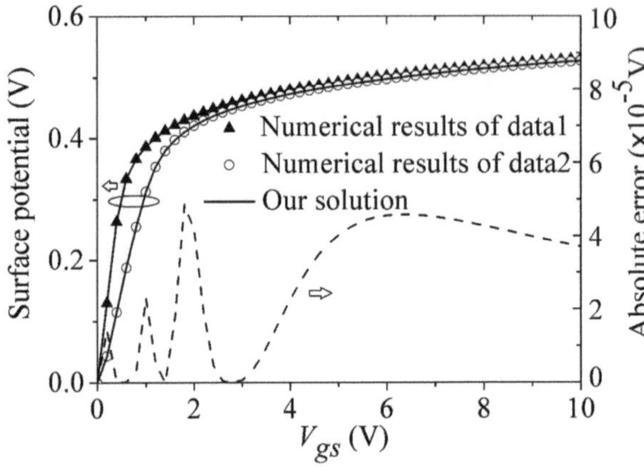

Fig. 3.4 Comparison of the surface potential solution (*lines*) with the numerical method (*markers*) for various DOS distributions. Absolute error of the surface potential solution compared with the numerical results of data2 is also shown

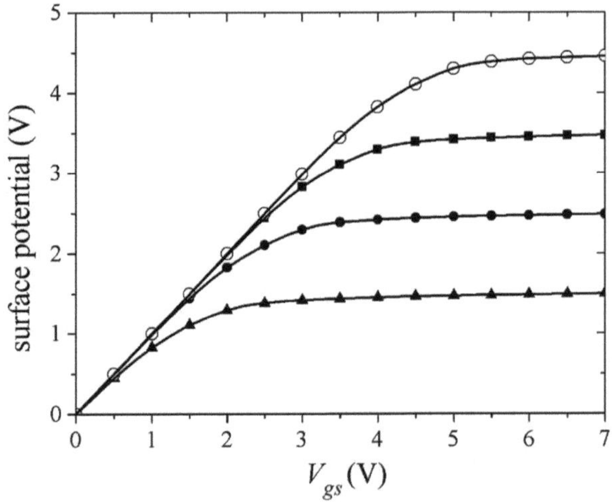

Fig. 3.5 Comparison of the surface potential solution (*lines*) with the numerical method (*markers*) for different V_{ch}, which are 1, 2, 3, 4 V. Trap state parameters are used as data2 in Table 3.2

To verify the proposed model, we have compared the surface potential solution of (3.24) with numerical simulation of Eqs. (3.5) and (3.11) in Figs. 3.4 and 3.5, where the parameters used are listed in Table 3.2. As shown in Fig. 3.4, for the smaller density of states (data1), the rise of curve becomes steeper, and the device enters the accumulation mode in a lower gate voltage, i.e., threshold voltage becomes smaller, which is consistent with the surface potential characteristics of a-Si:H TFTs and poly-Si TFTs. In Fig. 3.4, the absolute error of the proposed solution is also demonstrated, with the maximum absolute error about 5×10^{-5} V. The n_{eff} in Δ_{TFT} computed by Eqs. (3.11) and (3.13) takes all the charge densities into account. Therefore, an excellent accuracy is achieved for the subthreshold, above threshold and transition regions over a wide range of the gate voltage.

In addition, Fig. 3.5 shows the surface potential characteristics with different V_{ch} that corresponds to different points in the channel, and good agreements with numerical results can be also found.

Table 3.2 Parameters used in simulation of Figs. 3.4 and 3.5

Symbols (Units)	Values	Symbols (Units)	Values
V_{fb} (V)	0	N_C (cm^{-3})	4×10^{18}
t_{OX} (nm)	100	$E_f - E_C$ (eV)	−0.5
ϕ_{deep} (V)	0.17	ϕ_{tail} (V)	0.03
N_{D0} (cm^{-3} eV^{-1})	1×10^{16} (data1)	N_{T0} (cm^{-3} eV^{-1})	1×10^{17} (data1)
N_{D0} (cm^{-3} eV^{-1})	2×10^{17} (data2)	N_{T0} (cm^{-3} eV^{-1})	3×10^{18} (data2)
V_{ch} (V)	0 (Fig. 3.4)		

To show the relative dominance of each contribution in the electric field, according to Eq. (3.11), F_s is simplified as

$$F_s^2 = F^2(\phi_s) = \frac{2q}{\varepsilon_{aos}}\left(N_{free} + N_{tail} + N_{deep}\right) \tag{3.25}$$

where $N_{free} = n_0 V_T\left[\exp\left(\frac{\phi_s}{V_T}\right) - 1\right]$, $N_{deep} = N_D \phi_{deep}\left[\exp\left(\frac{\phi_s}{\phi_{deep}}\right) - 1\right]$, and $N_{tail} = N_T \phi_{tail}\left[\exp\left(\frac{\phi_s}{\phi_{tail}}\right) - 1\right]$. Figure 3.6 shows N_{free}, N_{deep}, and N_{tail} versus ϕ_s, where parameters are the same as in Fig. 3.4 except for $N_{D0} = 2 \times 10^{17}$cm^{-3}eV^{-1}, $N_{T0} = 5 \times 10^{18}$cm^{-3}eV^{-1} and ϕ_{tail}. When ϕ_s is large enough (i.e., the above-threshold region), N_{free} related to the free charge density is the dominant term in the electric field of Eq. (3.25). When the surface potential is small (i.e., the subthreshold region), in Fig. 3.6 (a), N_{deep} associated with deep states has a significant effect on (3.25), but in Fig. 3.6b where ϕ_{tail} increases to 0.08 V, $N_{deep} + N_{tail}$ is dominant which relates to both deep and tail states. Furthermore, in the transition region, all the components of charges, including free charge and trapped charge in deep and tail states, contribute to the electric field. As a result, the dominant terms in the transcendental Eq. (3.5) are different depending on the trap state parameters and the applied voltage. The application of the effective charge density approach (3.14) is much more accurate to reflect the device physics, because it takes the free charge, the ionized deep and tail states into account in all the operation regions.

3.4.1.2 Derivation of Drain Current Model

As we mentioned above, Eq. (3.6) can be used to derive the I-V model. Therefore, combining the effective charge density approach, the induced free charge in the channel, i.e., Q_i in Eq. (3.6), can be expressed as

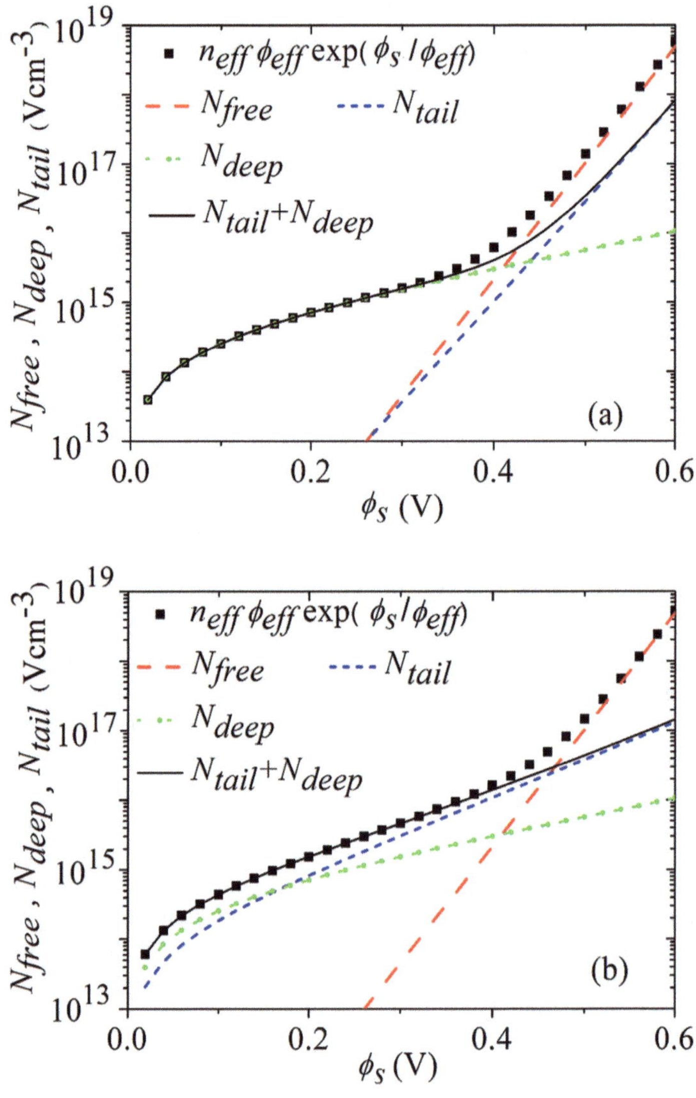

Fig. 3.6 The plot of N_{free}, N_{deep}, and N_{tail} as a function of ϕ_s with (**a**) $\phi_{\text{tail}} = 0.03$ V and (**b**) $\phi_{\text{tail}} = 0.08$ V

$$Q_i(\phi_s) = q\int_0^{\phi_s} \frac{n_0 \exp\left(\frac{\phi}{V_T}\right)}{\frac{d\phi}{dx}} d\phi = q\int_0^{\phi_s} \frac{n_0 \exp\left(\frac{\phi}{V_T}\right)}{A\exp\left(\frac{\phi}{2\phi_{eff}}\right)} d\phi$$
$$\approx \frac{2n_0 q\phi_{eff} V_T}{A(2\phi_{eff} - V_T)} \exp\left[\left(\frac{1}{V_T} - \frac{1}{2\phi_{eff}}\right)\phi_s\right] \cdot \quad (3.26)$$

Substituting Eq. (3.14) into (3.26), yields

$$Q_i(\phi_s) = \frac{2n_0 q\phi_{eff} V_T}{A(2\phi_{eff} - V_T)} \left(\frac{\phi_G - \phi_s}{A\frac{\varepsilon_{igzo}}{C_{OX}}} \right)^{\frac{2\phi_{eff}}{V_T} - 1} = \alpha \left(\frac{\phi_G - \phi_s}{\beta} \right)^{\frac{2\phi_{eff}}{V_T} - 1} \tag{3.27}$$

where $\alpha = \frac{2n_0 q\phi_{eff} V_T}{A(2\phi_{eff} - V_T)}$, and $\beta = A\frac{\varepsilon_{igzo}}{C_{OX}}$.

Meanwhile, differentiating Eq. (3.14) with respect to ϕ_s and using Eq. (3.14), we get

$$\frac{dV_{ch}}{d\phi_s} = 1 + \frac{2\phi_{eff}}{(\phi_G - \phi_s)} \tag{3.28}$$

Substituting Eqs. (3.27) and (3.28) into Eq. (3.6), the following expression for drain current can be obtained [32]

$$I_{ds0} = -\frac{W}{L} \frac{\mu\alpha}{\beta^{\frac{2\phi_{eff}}{V_T} - 1}} [P(\phi_{sL}) - P(\phi_{ss})] \tag{3.29}$$

$$P(\phi_s) = \frac{V_T}{2\phi_{eff}} (\phi_G - \phi_s)^{\frac{2\phi_{eff}}{V_T}} + \frac{2\phi_{eff} V_T}{2\phi_{eff} - V_T} (\phi_G - \phi_s)^{\frac{2\phi_{eff}}{V_T} - 1}. \tag{3.30}$$

The drain current formula of Eq. (3.29) is a single-piece surface-potential-based model available for all the operation regions above the flat-band voltage. Moreover, if the influence of non-saturation characteristics [40] in saturation region is considered, from Eq. (3.29), the total drain current becomes

$$I_{ds} = (1 + \lambda V_{deff}) I_{ds0} \tag{3.31}$$

where λ is a fitting parameter accounting for non-saturation, and V_{deff} is the effective drain-to-source voltage, viz., $V_{deff} = V_{ds} - (\phi_{sL} - \phi_{ss})$.

The drain current model of Eq. (3.31) was directly verified by comparing with the measured output and transfer characteristics of an a-IGZO TFT [41] with W/L = 200 μm/2 μm in Fig. 3.7. Parameters used in simulation are listed in Table 3.3. It should be noted that, similar to a-Si:H TFTs, the behavior of the effective mobility of a-IGZO TFTs can be reproduced by a power function as shown in Table 3.3. In Fig. 3.7, it can be clearly observed that this model gives an accurate description for all values of gate voltage above V_{fb}, even in the transition between subthreshold and above-threshold regions, where both the drift and the diffusion components contribute equally to the total drain current. This accurate description is attributed to fully taking the single-piece ϕ_s *vs* . V_{gs} relation into account.

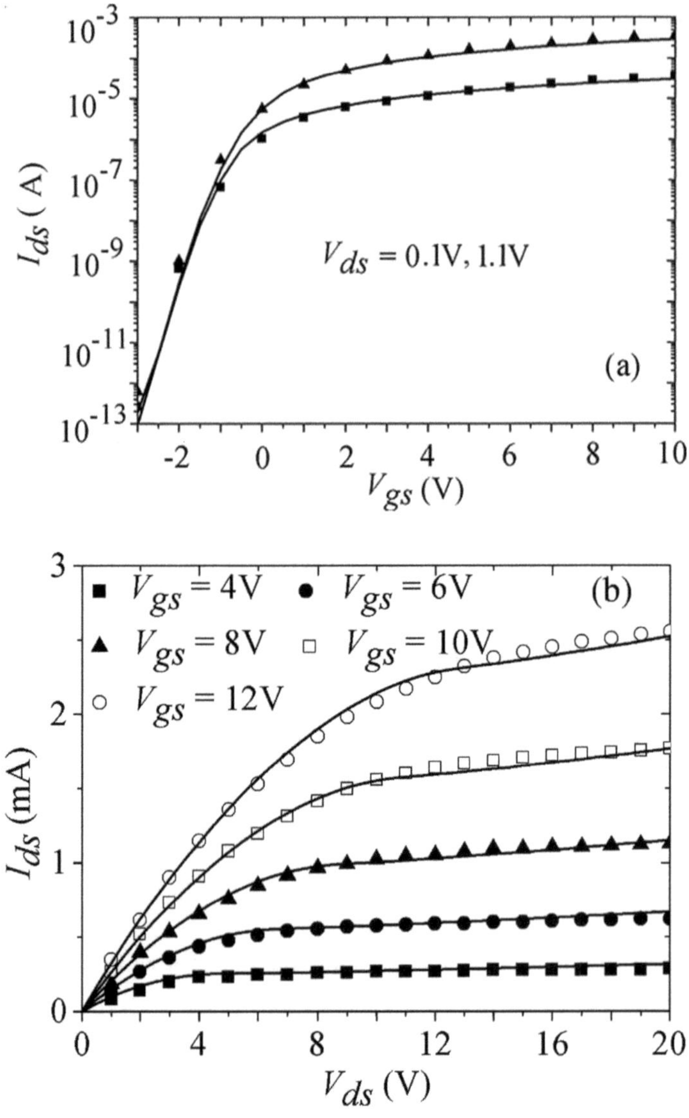

Fig. 3.7 Comparison of (**a**) transfer characteristics and (**b**) output characteristics, between model results (*curves*) and experimental data [41] (*markers*)

Table 3.3 Parameters used in simulation of Fig. 3.7

Symbols (units)	Values	Symbols (units)	Values
V_{fb} (V)	-3	N_C (cm^{-3})	4×10^{18}
t_{OX} (nm)	100	$E_f - E_C$ (eV)	-0.5
ϕ_{deep} (V)	0.2	ϕ_{tail} (V)	0.1
μ (cm^2 V^{-1} s^{-1})	$10.8\phi_G^{0.3}$	λ (V^{-1})	0.02
N_{D0} (cm^{-3} eV^{-1})	4×10^{17}	N_{T0} (cm^{-3} eV^{-1})	2×10^{18}

3.4.2 Models Accounting for the Degenerate Conduction

As the previous section described, with different chemical bonding, Fermi level pinning effect is eliminated in metal oxide TFTs. The *s* orbitals spherical symmetry leads to a reduced density of band tail states. The Fermi level enters into the conduction band with the rise of gate bias, enabling electrons to transport in the nonlocalized band states. Considering such intrinsic material nature, in this section, Fermi statistic is applied instead of Boltzmann approximation, because the latter is only valid for non-degenerate assumption. Meanwhile, for simplification, in contrast to the two exponential in Eq. (3.1), DOS distribution in the upper half of the gap is assumed as a single exponential function, i.e., $g(E) = N_{T0} \exp\left(\frac{E-E_C}{kT_t}\right)$.

For n-type semiconductors, the relationship between the free-electron density and the Fermi level is given by

$$\frac{E_f - E_C}{kT} = \ln\left(\frac{n_{free}}{N_C}\right) + 2^{-3/2}\frac{n_{free}}{N_C}. \tag{3.32}$$

Taking degeneration into account, Eq. (3.32) can be straightforwardly rewritten by means of W_0 function and reads

$$n_{free}(\phi, V_{ch}) = N_C 2\sqrt{2} W_0\left[R_0 \exp\left(\frac{\phi}{V_T}\right)\right] \tag{3.33}$$

where $R_0 = \exp\left(\frac{E_f-E_C-V_{ch}}{V_T}\right)/2^{1.5}$. Analogously, using Fermi-Dirac statistic, the integral of (3.2), that is the trapped states density, can be expressed in a similar form as [35]

$$n_{trap}(\phi, V_{ch}) = N_{T0}\theta_t\left\{2\sqrt{2}W_0\left[R_0 \exp\left(\frac{\phi}{V_T}\right)\right]\right\}^{T/T_t} \tag{3.34}$$

where $\theta_t = \frac{\pi kT}{\sin(\pi T/T_t)}$. Equations (3.33) and (3.34) give a physics-based description of charge concentration valid for both non-degeneration and degeneration. Model verification is shown in Fig. 3.8. Parameters used in simulation are summarized in Table 3.4. As we can see, the total value n_{all}, viz., the sum of n_{free} and n_{trap}, is in good agreement with the overall concentration which is numerically solved by Fermi-Dirac integral equation. Moreover, the concentration based on Boltzmann statistic is also plotted. Compared with the numerical results, we can clearly find that although Boltzmann approximation is valid in non-degeneration, it overestimates n_{all} when E_f exceeds E_C.

With the relationship $W_0(x)\exp(W_0(x)) = x$, one can easily obtain $dW_0(x)/dx = W_0(x)/(x(1+W_0(x)))$, so the surface electric field can be derived from integrating Eqs. (3.33) and (3.34) with respect to ϕ, i.e.,

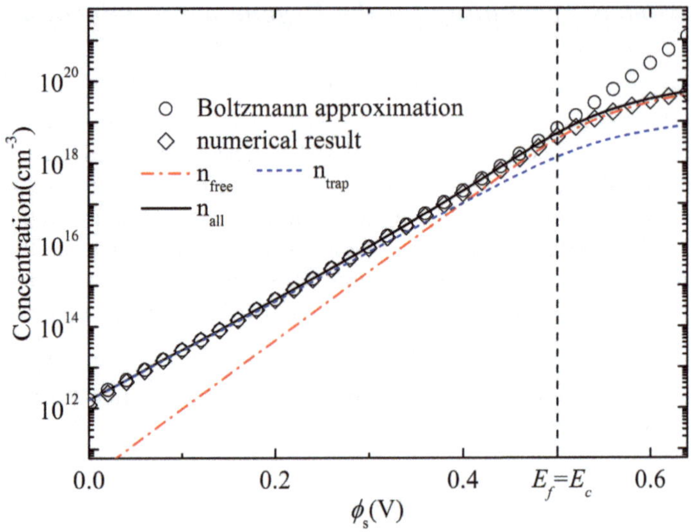

Fig. 3.8 The plot of charge concentrations as a function of ϕ_s

Table 3.4 Parameters used in simulation of Figs. 3.8, 3.9, 3.10, and 3.11

Symbols (Units)	Values	Symbols (Units)	Values
V_{fb} (V)	−0.5	N_{T0} (cm^{-3} eV^{-1})	1.6×10^{19}
$E_f - E_C$ (eV)	−0.5	T (K)	300
N_C (cm^{-3})	5×10^{18}	T_t (K)	418

$$F(\phi_s, V_{ch}) = \sqrt{\int_0^{\phi_s} \frac{2q}{\varepsilon_{aos}}(n_{free} + n_{trap})d\phi}$$
$$= \sqrt{G_f W_\phi(1 + 0.5W_\phi) + G_t W_\phi{}^{T/T_t}[1 + TW_\phi/(T + T_t)]} \quad (3.35)$$

herein $G_f = 2^{5/2} N_C q V_T / \varepsilon_{aos}$, $G_t = 2^{1+3T/2T_t} N_{T0} \theta_t E_t / \varepsilon_{aos}$, and $W_\phi = W_0\left[R_0 \exp\left(\frac{\phi_s}{V_T}\right)\right]$.

3.4.2.1 Derivation of Surface Potential

Combining Eq. (3.35) with Eq. (3.5), the implicit equation between ϕ_s with applied voltage can be obtained. Similar to the case in non-degeneration, it cannot be analytically solved because there is a square root of a polynomial about W_ϕ. Thus, further investigation is required. Fig. 3.8 points out the relative dominant terms in the Eq. (3.35). As shown in Fig. 3.8, n_{trap} is much larger n_{free} when ϕ_s is small, viz. a low gate voltage is applied and the transistor operates in the subthreshold region. On the contrary, when ϕ_s is large enough, corresponding to the strong accumulation region, n_{free} has a higher concentration than n_{trap}. As a result, the regional approach can be adopted to derive the asymptotic solution of ϕ_s.

In the subthreshold region, most of charge resides in the localized states, n_{trap} becomes dominant and n_{free} can be negligible. Furthermore, with moderate band bending, $E_f < E_C - 2kT$ is satisfied in this region, and thus $W_0(x)$ is approximated as x. Equation (3.34) can be simplified as Eq. (3.7), viz.,

$$n_{trap}(\phi, V_{ch}) = N_{T0}\theta_t \exp\left(\frac{\phi - V_{ch} + \phi_{f0}}{kT_t/q}\right) \tag{3.36}$$

where $\phi_{f0} = (E_f - E_C)/q$.

After the same calculation procedure mentioned above, one obtains

$$\phi_{sub} = \phi_G + \Delta_T\phi_{trap} - 2\phi_{trap}W_0\left[\frac{\Delta_T}{2}\exp\left(\frac{\phi_G/\phi_{trap} + \Delta_T}{2}\right)\right] \tag{3.37}$$

where $\Delta_T = \sqrt{2q\varepsilon_{aos}N_{T0}\theta_t \exp\left[\left(-V_{ch} + \phi_{f0}\right)/\phi_{trap}\right]/\phi_{trap}}/C_{OX}$ and $\phi_{trap} = kT_t/q$.

In the accumulation region, the Fermi level moves toward the conduction band bottom E_C with the increasing gate biases. Therefore, the degenerate conduction occurs and Boltzmann approximation becomes invalid. Neglecting n_{trap} in Eq. (3.35) and combing with Eq. (3.5), yields

$$\phi_G - \phi_s = \varepsilon_{aos}\sqrt{G_f W_\phi\left(1 + 0.5W_\phi\right)}/C_{OX}. \tag{3.38}$$

Equation (3.38) can be treated as a quadratic equation with variable W_ϕ. With a simple mathematical procedure, one obtains

$$W_\phi = -1 + \sqrt{1 + 2C_{OX}^2(\phi_G - \phi_s)^2/\left(\varepsilon_{aos}^2 G_f\right)}. \tag{3.39}$$

Unfortunately, Eq. (3.39) is still too complicated to derive an analytical evaluation of ϕ_s. In order to make some reasonable approximations, we introduce $\Delta\phi = \phi_s - \phi_{sn}$, where ϕ_{sn} is the surface potential only accounting for free carriers in non-degeneration with the expression as Eq. (3.20). In this way, Eq. (3.39) can be rearranged as

$$\begin{aligned} W_\phi &= -1 + \sqrt{1 + 2C_{OX}^2(\phi_G - \phi_{sn} - \Delta\phi)^2/\left(\varepsilon_{aos}^2 G_f\right)} \\ &\approx -1 + \sqrt{1 + \frac{2C_{OX}^2(\phi_G - \phi_{sn})^2}{\varepsilon_{aos}^2 G_f} - \frac{4C_{OX}^2(\phi_G - \phi_{sn})\Delta\phi}{\varepsilon_{aos}^2 G_f}} \\ &= -1 + \sqrt{1 + \frac{2C_{OX}^2(\phi_G - \phi_{sn})^2}{\varepsilon_{aos}^2 G_f}}\sqrt{1 - \frac{4C_{OX}^2(\phi_G - \phi_{sn})\Delta\phi}{\varepsilon_{aos}^2 G_f + 2C_{OX}^2(\phi_G - \phi_{sn})^2}}. \end{aligned} \tag{3.40}$$

Here we neglect $\Delta\phi^2$ term since it is a relative small value. Further, only considering the first order term, the second square root in Eq. (3.40) can be simplified by using the Taylor expansion. Hence, one obtains

$$W_\phi \approx -1 + A_0\left[1 - \frac{2C_{OX}^2(\phi_G - \phi_{sn})\Delta\phi}{\varepsilon_{aos}^2 G_f A_0^2}\right] \tag{3.41}$$

where $A_0 = \sqrt{1 + 2C_{OX}^2(\phi_G - \phi_{sn})^2/(\varepsilon_{aos}^2 G_f)}$. Following the discussions above, the difference between ϕ_s and ϕ_{sn} can be expressed as

$$\Delta\phi = \frac{\varepsilon_{aos}^2 G_f A_0(A_0 - 1 - W_\phi)}{2C_{OX}^2(\phi_G - \phi_{sn} + b)}. \tag{3.42}$$

Noteworthy, ϕ_{sn} is roughly equal to ϕ_G when $V_{ch} > \phi_G$, and thus, a tiny constant b is introduced here to avoid the denominator term being zero. This approximation is reasonable because in the accumulation region, $\phi_G - \phi_{sn} \gg b$ is satisfied and b will not actually modify the final results. In other words, the existence of b is only used to refine the algorithm. Finally, one can obtain the normalized surface potential as

$$x_s = \frac{\varepsilon_{aos}^2 G_f A_0(A_0 - 1)}{2V_T C_{OX}^2(\phi_G - \phi_{sn} + b)} - \frac{\varepsilon_{aos}^2 G_f A_0 W_\phi}{2V_T C_{OX}^2(\phi_G - \phi_{sn} + b)} + x_{sn} \tag{3.43}$$

where $x_s = \phi_s/V_T$, and $x_{sn} = \phi_{sn}/V_T$. We apply the exponential function and multiply R_0 to both sides of Eq. (3.43), i.e.,

$$R_0 \exp(x_s) = R_0 \exp(B)/\exp\left[\frac{\varepsilon_{aos}^2 G_f A_0 W_\phi}{2V_T C_{OX}^2(\phi_G - \phi_{sn} + b)}\right] \tag{3.44}$$

here $B = \frac{\varepsilon_{aos}^2 G_f A_0(A_0 - 1)}{2V_T C_{OX}^2(\phi_G - \phi_{sn} + b)} + x_{sn}$. Substituting $R_0 \exp(x_s) = W_\phi \exp(W_\phi)$ into Eq. (3.44), results in

$$W_\phi \exp\left[W_\phi\left(1 + \frac{\varepsilon_{aos}^2 G_f A_0}{2V_T C_{OX}^2(\phi_G - \phi_{sn} + b)}\right)\right] = R_0 \exp(B). \tag{3.45}$$

Thanks to this, Eq. (3.45) can be explicitly solved by using the Lambert W function to yield the following expression

$$W_{\phi} = \frac{W_0\left[R_0 \exp(B)\left(1+\frac{\varepsilon_{aos}^2 G_f A_0}{2V_T C_{OX}^2(\phi_G - \phi_{sn}+b)}\right)\right]}{1+\varepsilon_{aos}^2 G_f A_0/\left[2V_T C_{OX}^2(\phi_G - \phi_{sn}+b)\right]}. \tag{3.46}$$

Hence we get the explicit expression of W_{ϕ}, and according to (3.46), the surface potential restricted to accumulation operation region is obtained as

$$\phi_{ab} = V_T\left(\ln W_{\phi} + W_{\phi} - \ln R_0\right). \tag{3.47}$$

To get a single piece formula, we use the following smoothing function to link the different operation regions

$$\phi_s = \frac{\phi_{sub}}{1+m_1\exp\left[\left(V_{gs}-V_k\right)/m_2\right]} + \frac{\phi_{ab}}{1+m_1\exp\left[\left(V_k-V_{gs}\right)/m_2\right]} \tag{3.48}$$

where m_1 and m_2 are weight parameters to connect the different asymptotical results. V_k is gate voltage corresponding to the condition of equal values of n_{free} and n_{trap}, viz.

$$\begin{aligned} V_k = {} & V_{fb} + V_T(\ln W_{Vk} + W_{Vk} - \ln R_0) \\ & + \frac{\varepsilon_{aos}}{C_{OX}}\sqrt{G_f W_{Vk}(1+0.5W_{Vk}) + G_t {W_{Vk}}^{T/T_t}[1+TW_{Vk}/(T+T_t)]} \end{aligned} \tag{3.49}$$

where

$$W_{Vk} = (N_C/N_{T0}\theta_t)^{\frac{T_t}{T-T_t}}/2\sqrt{2}. \tag{3.50}$$

To further improve the accuracy of obtained surface potential, corrections based on the same exact formulation by using Schroder series [42] can be applied as mentioned in Sect. 3.4.1.

To verify the proposed model, comparison between surface potential given by Eq. (3.48) with numerical simulation is shown in Fig. 3.9, where model parameters are the same as in Fig. 3.8. An excellent agreement is achieved with the maximum absolute error about 2×10^{-5} V in the transition region. Moreover, the model with non-degenerate assumption shows a discrepancy due to overestimation of charge concentrations by using Boltzmann statistics when $E_f > E_C - 2kT$. This discrepancy becomes larger with the rise of gate voltage. Thus, it indicates a more accurate and physics-based surface potential model aforementioned is important, because the current model is most sensitive to the small errors of surface potential.

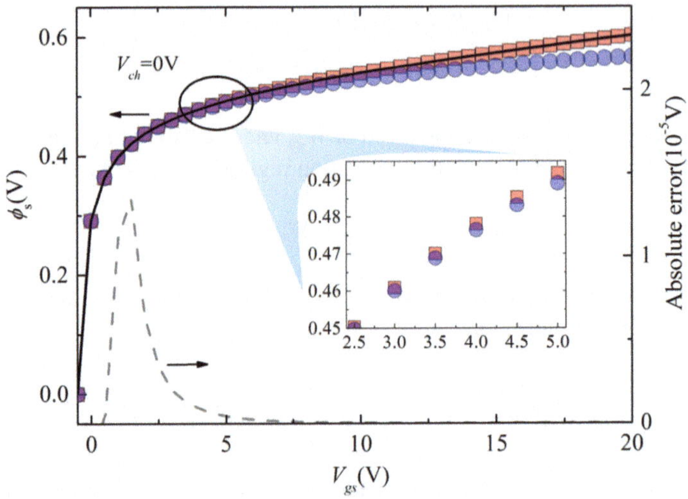

Fig. 3.9 Comparison of surface potential solution (*solid lines*) with numerical results (*square*) and the results based on non-degenerate assumption (*circle*). Absolute error (*dashed lines*) is also shown

3.4.2.2 Derivation of Drain Current Model

By applying Gauss' law to the insulator-semiconductor interface and only accounting for n_{trap}, Eq. (3.5) becomes

$$\phi_G - \phi_s = \varepsilon_{aos}\sqrt{G_t W_\phi{}^{T/T_t}\left[1 + TW_\phi/(T + T_t)\right]}/C_{OX}. \tag{3.51}$$

Hence, the induced free charge inside the active layer of metal oxide TFTs in the subthreshold region, i.e., Q_i, is given by

$$Q_i = q\int_0^{\phi_s} \frac{N_C 2\sqrt{2}W_\phi}{\sqrt{G_t W_\phi{}^{T/T_t}\left[1 + TW_\phi/(T + T_t)\right]}} d\phi. \tag{3.52}$$

The integral in Eq. (3.52) cannot be solved explicitly. When multiplying the denominator by $\left[1 + TW_\phi/(T + T_t)\right]^{1-T_t/T}$, only a very small error may be induced because W_ϕ is always a relatively small value compared to 1 in the subthreshold region. As a result, Eq. (3.52) can be rewritten as

$$\begin{aligned} Q_i &\approx \frac{2\sqrt{2}qN_C}{\sqrt{G_t}} \int_0^{\phi_s} \frac{W_\phi}{W_\phi{}^{T/2T_t}\left[1 + TW_\phi/(T + T_t)\right]^{3/2-T_t/T}} d\phi \\ &= \frac{4\sqrt{2}N_C E_t T}{\sqrt{G_t}(2T_t - T)} \left\{W_\phi{}^{T/T_t}\left[1 + TW_\phi/(T + T_t)\right]\right\}^{\frac{2T_t - T}{2T}}. \end{aligned} \tag{3.53}$$

Differentiating Eq. (3.51) with respect to ϕ_s, one obtains

$$\frac{dV_{ch}}{d\phi_s} = 1 + \frac{2E_t C_{OX}\sqrt{W_\phi{}^{T/T_t}[1 + TW_\phi/(T + T_t)]}}{q\sqrt{G_t}\varepsilon_{aos}W_\phi{}^{T/T_t}}. \tag{3.54}$$

Substituting Eqs. (3.53) and (3.54) into Eq. (3.6), the drain current can be expressed as

$$\begin{aligned} I_{sub} = \; & 4\sqrt{2}\mu\frac{WN_C E_t T}{L\sqrt{G_t}(2T_t - T)} \times \int_{\phi_{ss}}^{\phi_{sL}} \left\{W_\phi{}^{T/T_t}[1 + TW_\phi/(T + T_t)]\right\}^{\frac{2T_t - T}{2T}} \\ & + \frac{2E_t C_{OX}\left\{W_\phi{}^{T/T_t}[1 + TW_\phi/(T + T_t)]\right\}^{T_t/T}}{q\sqrt{G_t}\varepsilon_{aos}W_\phi{}^{T/T_t}} d\phi_s. \end{aligned} \tag{3.55}$$

The second term in the integral is too complicated to evaluate the currents. Therefore, multiplying the second term by $[1+TW_\phi/(T+T_t)]^{-2}$ and with some simple mathematical procedures, the drain current expression in the subthreshold region eventually reads

$$I_{sub} \approx 4\sqrt{2}\mu\frac{WN_C E_t C_{OX}^{\gamma-1}}{LG_t^{\gamma/2}\varepsilon_{aos}^{\gamma-1}(\gamma - 1)}[f_{sub}(\phi_{sL}) - f_{sub}(\phi_{ss})] \tag{3.56}$$

where $\gamma = 2T_t/T$ and

$$f_{sub}(\phi_s) = -\frac{1}{\gamma}(\phi_G - \phi_s)^\gamma + \frac{4E_t{}^2}{q^2(\gamma - 2)}(\phi_G - \phi_s)^{\gamma-2}. \tag{3.57}$$

In the accumulation region, the free charge contributes to the band bending near the surface. Neglecting n_{trap} term and using Eq. (3.38), the induced free charge Q_i is written as

$$Q_i = C_{OX}(\phi_G - \phi_s) = \varepsilon_{aos}\sqrt{G_f W_\phi(1 + 0.5W_\phi)}. \tag{3.58}$$

Differentiating Eq. (3.58) with respect to ϕ_s, yields

$$\frac{dV_{ch}}{d\phi_s} = 1 + \frac{2V_T C_{OX}\sqrt{W_\phi[1 + 0.5W_\phi]}}{\sqrt{G_f}\varepsilon_{aos}W_\phi}. \tag{3.59}$$

Analogously, substituting Eqs. (3.58) and (3.59) into Eq. (3.6), the drain current in the accumulation region can be expressed as

$$I_{acc} = 4\sqrt{2}\mu \frac{qWN_C V_T C_{OX}}{LG_f \varepsilon_{aos}} [f_{ab}(\phi_{sL}) - f_{ab}(\phi_{ss})] \tag{3.60}$$

where

$$f_{ab}(\phi_s) = \left(\frac{V_T^2 C_{OX}^2}{\varepsilon_{aos}^2 G_f} - \frac{1}{2} \right) (\phi_G - \phi_s)^2 + 2V_T \phi_s. \tag{3.61}$$

As discussed above, the drain current in the subthreshold and accumulation regions is derived by different expressions. The limiting regions as well as the transition can be connected by using a smoothing function, or known as Matthiessen's model [43]. Thus, I_{ds} reads

$$I_{ds} = [(1/I_{acc})^{m_3} + (1/I_{sub})^{m_3}]^{\frac{-1}{m_3}}. \tag{3.62}$$

According to [34], m_3 is a weight parameter which involves the influence of the temperature on the transition between non-degeneration and degeneration. Also it [34] indicates that m_3 is a function of both T and T_t, which offers the relation as

$$m_3 = 0.36\gamma(\gamma - 2). \tag{3.63}$$

The current model is applied to an a-IZO TFT with inverted staggered bottom gate structure, of which $W/L = 20$ μm/2 μm, $C_{OX} = 41$ nF/cm^2 and other extracted model parameters are summarized in the Table 3.4. As demonstrated in Figs. 3.10 and 3.11, the new model results are well consistent with the measured transfer and

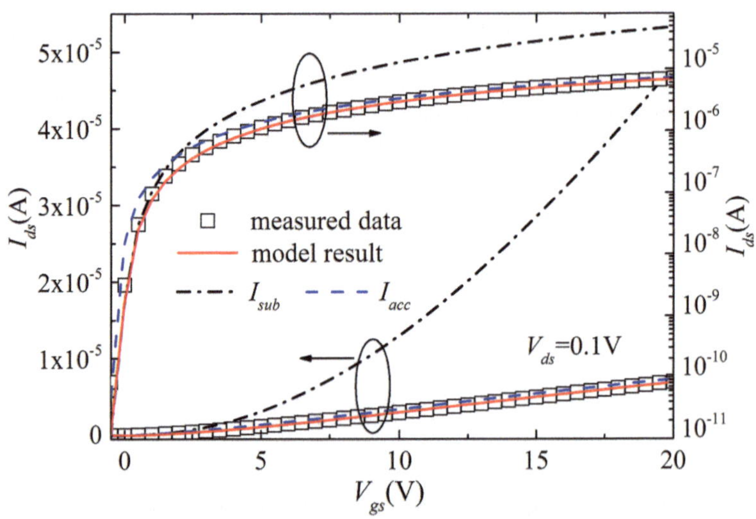

Fig. 3.10 Comparison of transfer characteristics between model results (*curves*) and measured data (*markers*) with $V_{ds} = 0.1$ V both in logarithmic and linear scales

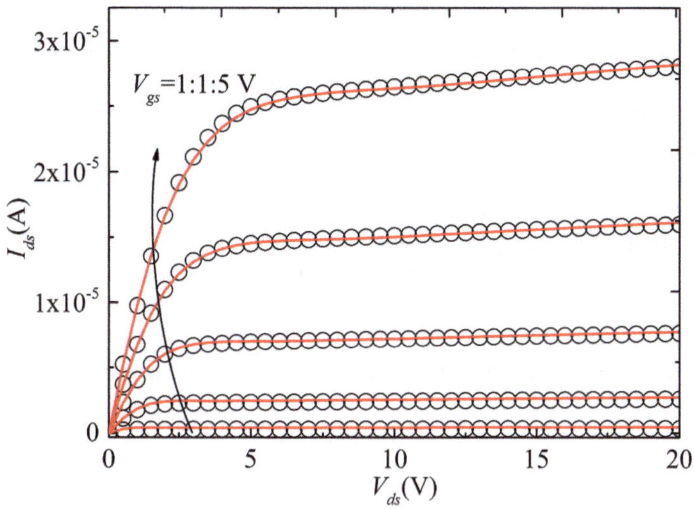

Fig. 3.11 Comparison of output characteristics between model results (*curves*) and measured data (*markers*) with different V_{gs}

output characteristics over a wide range of voltage bias. Meanwhile, in Fig. 3.10, it shows the contributions of I_{sub} and I_{acc} on the total drain current. One can see that, I_{acc} overestimates in the subthreshold region, because in the limit of low voltage bias, E_f resides within the tail states and carrier transport is controlled by multiple trap and release events (MTR), viz. trap-limited conduction. Hence, in the subthreshold region, neglecting n_{trap} will bring overestimation of free electrons concentration, and thus a larger I_{acc} is observed. Analogous analysis is applied to the strong accumulation region. At higher gate voltages, E_f moves toward the E_C, and percolation becomes dominant. As a result, I_{sub} is far above measured data since it only accounts for n_{trap} and ignores the contribution of n_{free}.

3.5 The Effective Mobility Models

The determination of the mobility is one of the primary issues for the development of metal oxide TFTs models. Similar to a-Si:H TFTs, in many publications [44], the characteristics of the effective mobility of metal oxide semiconductor TFTs can be reproduced by a power function. As we mentioned before, in the current-voltage compact models of Eqs. (3.29) and (3.62), the effective mobility (μ) obeys a power law. However, in contrast to a-Si:H TFTs, it is believed that the charge transport mechanisms are governed by the multiple-trapping-and-release (MTR) in the subthreshold regime and the percolation in the strong accumulation regime [28]. In both cases, the field effect mobility follows a power law [45].

In the strong accumulation region, i.e., when E_f exceeds the conduction band edge, the carrier density is degenerate and has a week temperature dependence. In this case, μ is controlled by percolation. The basic concept of percolation conduction [46] is that there are different electron conduction paths that have different carrier transport properties. Adler and co-workers assume that there is a potential fluctuation above the mobility edge in amorphous semiconductor [46]. Therefore, the potential barrier height leads to different conductance for electron transferring over the potential barrier. As a result, the carriers in metal oxide semiconductor move by finding the path of least resistance among the potential barriers [45]. Percolation conduction can be modeled by the potential barrier height and the coherence ratio of average barrier width (W_B)-to-distance (D_B) [46–48].

When percolation conduction dominates, the gate voltage dependent μ can be approximated by [45]

$$\mu = \mu_0 \exp\left[-\frac{q\phi_{B0}}{kT} + \frac{(q\sigma_B)^2}{2(kT)^2}\right] B^*\left(V_{gs} - V_P\right)^{4\left(\frac{D_B - W_B}{D_B}\right)} \tag{3.64}$$

where μ_0 is the band mobility and $B^* = \left[C_{OX}^2/\left(2^{2.75}\varepsilon_{aos}N_C kT\right)\right]^{2[(D_B - W_B)/D_B]}$. As shown in Fig. 3.12, ϕ_{B0} is the average potential barrier height and σ_B is the variance. In addition, V_P is a percolation threshold voltage when $E_f = E_C$, which can be calculated by Poisson's equation and Gauss' law [45], i.e.,

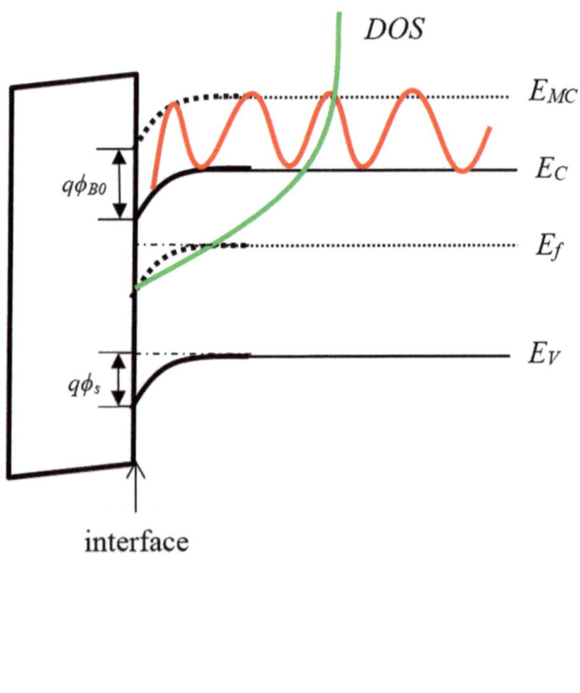

Fig. 3.12 Energy-band diagram of a metal oxide semiconductor TFT [45]

$$V_P \approx \frac{\left(2q\varepsilon_{aos}\int_{E_f}^{E_C}\left[n_{free}(E_f)+n_{trap}(E_f)\right]dE_f\right)^{1/2}}{C_{OX}}. \tag{3.65}$$

On the other hand, when the Fermi level resides within the localized tail states, i.e., in the subthreshold region, carrier transport is dominated by MTR, which is also called trap-limited conduction (TLC). In this case, the effective mobility can be analytically expressed as [45]

$$\mu = \mu_0 \exp\left[-\frac{q\phi_{B0}}{kT}+\frac{(q\sigma_B)^2}{2(kT)^2}\right]A^*\left(V_{gs}-V_{TH}\right)^{2\left(\frac{T_t}{T}-1\right)} \tag{3.66}$$

where A^* is a constant related to the density of tail states and V_{TH} is the threshold voltage.

Comparing with Eqs. (3.64) and (3.66), we can find that these two equations are very similar, which can be unified by a power law [45], i.e.,

$$\mu = K\left(V_{gs}-V_{T,P}\right)^{\eta} \tag{3.67}$$

where K and η are the fitting parameters. It is interesting that, from Eq. (3.67), in both trap-limited conduction dominance and percolation dominance, the gate voltage dependent mobility follows a power law but with different coefficients.

In Fig. 3.13, Eq. (3.67) is used to fit the measured mobility data from the IZO TFTs with $W/L = 50$ μm/30 μm, the detailed fabrication of which is described in Sect. 3.1. When TLC is dominant, the values of K_1 and η_1 are extracted as 26.5 and 0.2, respectively. Meanwhile, at a higher V_{gs} where percolation is dominant, the parameters of K_2 and η_2 are extracted as 46.4 and 0.02, respectively. Furthermore, the transition point can be computed by using Eq. (3.65), viz., $V_P = 7.5$ V. In other words, when V_{gs} is decreased below V_P, TLC prevails. In contrast, when V_{gs} increases to V_P, the percolation becomes important.

3.6 Conclusions

This chapter introduces the physics-based compact models for metal oxide semiconductor TFTs. The fundamental issues related to the surface potential and current characteristics are discussed and modeled. These compact models can be generally categorized into either non-degeneration or degeneration based, but some similarities exist among these models developed from the two different assumptions. The accuracy of the models has been verified from the surface potential and drain current comparisons with numerical simulations and experimental results. This

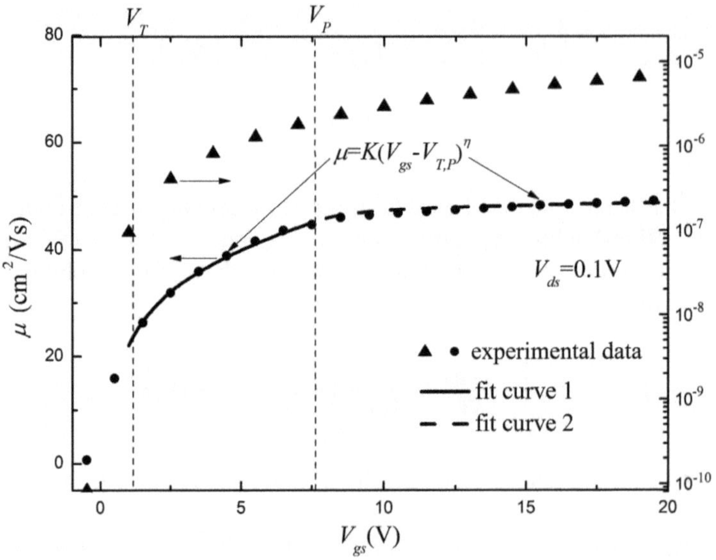

Fig. 3.13 Drain-source current (I_{ds}) and the effective mobility (μ) as a function of V_{gs}

chapter may provide a useful knowledge to readers who are engaged or interested in the device physics, modeling, and circuit design of the increasingly important metal oxide semiconductor TFTs.

References

1. W.J. Wu, L. Zhou, R.H. Yao, J.B. Peng, A new voltage-programmed pixel circuit for enhancing the uniformity of AMOLED displays. IEEE Electron Device Lett. **32**(7), 931–933 (2011)
2. R.E. Presley, D. Hong, H.Q. Chiang, C.M. Hung, R.L. Hoffman, J.F. Wager, Transparent ring oscillator based on indium gallium oxide thin-film transistor. Solid State Electron. **50**(3), 500–503 (2006)
3. D. Raiteri, F. Torricelli, K. Myny, et al., A 6b 10MS/s current-steering DAC manufactured with amorphous Gallium-Indium-Zinc-Oxide TFTs achieving SFDR>30 dB up to 300kHz. Int. Solid State Circuits Conf., 314–316 (2012)
4. H. Ozaki, T. Kawamura, H. Wakana, T. Yamazoe, H. Uchiyama, Wireless operations for 13.56-MHz band RFID tag using amorphous oxide TFTs. IEICE Electron. Express **8**(4), 225–231 (2011)
5. A. Nathan, A. Ahnood, M.T. Cole, et al., Flexible electronics: the next ubiquitous platform. Proc. IEEE **100**, 1486–1517 (2012)
6. D. Makarov, M. Melzer, D. Karnaushenko, O.G. Schmidt, Shapeable magnetoelectronics. Appl. Phys. Rev. **3**(1), 011101 (2016)
7. S. Wagner, S.P. Lacour, J. Jones, et al., Electronic skin: architecture and components. Physica E Low Dimens. Syst. Nanostruct. **25**(2–3), 326–334 (2004)

8. D.-H. Kim, N. Lu, R. Ma, Y.-S. Kim, R.-H. Kim, et al., Epidermal electronics. Science **333**(6044), 838–843 (2011)
9. V.J. Lumelsky, M.S. Shur, S. Wagner, Sensitive skin. IEEE Sens. J. **1**(1), 41–51 (2001)
10. M.L. Hammock, A. Chortos, B.C.K. Tee, J.B.H. Tok, Z. Bao, 25th anniversary article: The evolution of electronic skin (e-skin): a brief history, design considerations, and recent progress. Adv. Mater. **25**(42), 5997–6037 (2013)
11. D.-H. Kim, J. Viventi, J.J. Amsden, J. Xiao, L. Vigeland, et al., Dissolvable films of silk fibroin for ultrathin conformal bio-integrated electronics. Nat. Mater. **9**(6), 511–517 (2010)
12. S.-W. Hwang, H. Tao, D.-H. Kim, H. Cheng, J.-K. Song, et al., A physically transient form of silicon electronics. Science **337**(6102), 1640–1644 (2012)
13. J. Genoe, K. Obata, M. Ameys, K. Myny, T.H. Ke, et al., Integrated line driver for digital pulse-width modulation driven AMOLED displays on flex. IEEE J. Solid State Circuits **50**(1), 282–290 (2015)
14. L. Petti, N. Münzenrieder, C. Vogt, H. Faber, L. Büthe, et al., Metal oxide semiconductor thin-film transistors for flexible electronics. Appl. Phys. Rev. **3**(2), 021303 (2016)
15. H.A. Klasens, H. Koelmans, A tin oxide field-effect transistor. Solid State Electron. **7**(9), 701–702 (1964)
16. M.W.J. Prins, K.O. Grosse-Holz, G. Müller, J.F.M. Cillessen, et al., A ferroelectric transparent thin-film transistor. Appl. Phys. Lett. **68**(25), 3650–3652 (1996)
17. C.H. Seager, D.C. McIntyre, W.L. Warren, B.A. Tuttle, Charge trapping and device behavior in ferroelectric memories. Appl. Phys. Lett. **68**(19), 2660–2662 (1996)
18. R.L. Hoffman, B.J. Norris, J.F. Wager, ZnO-based transparent thin-film transistors. Appl. Phys. Lett. **82**(5), 733–735 (2003)
19. P.F. Carcia, R.S. McLean, M.H. Reilly, Oxide engineering of ZnO thin-film transistors for flexible electronics. J. Soc. Inf. Disp. **13**(7), 547–554 (2005)
20. K. Nomura, H. Ohta, K. Ueda, T. Kamiya, M. Hirano, H. Hosono, Thin-film transistor fabricated in single-crystalline transparent oxide semiconductor. Science **300**(5623), 1269–1272 (May 2003)
21. H.Q. Chiang, J.F. Wager, R.L. Hoffman, J. Jeong, D.A. Keszler, High mobility transparent thin-film transistors with amorphous zinc tin oxide channel layer. Appl. Phys. Lett. **86**(1), 013503 (2005)
22. N.L. Dehuff, E.S. Kettenring, D. Hong, H.Q. Chiang, J.F. Wager, et al., Transparent thin-film transistors with zinc indium oxide channel layer. J. Appl. Phys. **97**(6), 064505 (2005)
23. Z. Xu, M. Li, M. Xu, et al., Flexible amorphous oxide thin-film transistors on polyimide substrate for AMOLED. Proc. SPIE **9270**, 92700A (2014)
24. M. Bae, Y. Kim, D. Kong, et al., Analytical models for drain current and gate capacitance in amorphous InGaZnO thin-film transistors with effective carrier density. IEEE Electron Device Lett. **32**(11), 1546–1548 (2011)
25. M. Bae, K.M. Lee, E.S. Cho, et al., Analytical current and capacitance models for amorphous Indium-Gallium-Zinc-Oxide thin-film transistors. IEEE Trans. Electron Device **60**(10), 3465–3473 (2013)
26. H.C. Pao, C.T. Sah, Effects of diffusion current on characteristics of metal-oxide (insulator) -semiconductor transistors. Solid State Electron. **9**(10), 927–937 (1966)
27. A. Tsormpatzoglou, N.A. Hastas, N. Choi, F. Mahmoudabadi, M.K. Hatalis, C.A. Dimitriadis, Analytical surface-potential-based drain current model for amorphous InGaZnO thin film transistors. J. Appl. Phys. **114**(18), 184502 (2013)
28. M. Ghittorelli, F. Torricelli, L. Colalongo, et al., Accurate analytical physical modeling of amorphous InGaZnO thin-film transistors accounting for trapped and free charges. IEEE Trans. Electron Device **61**(12), 4105–4112 (2014)
29. L. Colalongo, Compact model of amorphous InGaZnO thin film transistors based on symmetric quadrature of accumulation charge. IEEE Electron Device Lett. **37**(4), 416–418 (2016)

30. G. Gildenblat, X. Li, W. Wu, H. Wang, A. Jha, R. van Langevelde, et al., PSP: an advanced surface-potential-based MOSFET model for circuit simulation. IEEE Trans. Electron Devices **53**(9), 1979–1992 (2006)
31. J.H. Park, Y. Kim, S. Kim, et al., Surface-potential-based analytic DC I-V model with effective electron density for a-IGZO TFTs considering the parasitic resistance. IEEE Electron Device Letters **32**(11), 1540–1542 (2011)
32. W. Deng, J. Huang, X. Ma, T. Ning, An explicit surface-potential-based model for amorphous IGZO thin-film transistors including both tail and deep states. IEEE Electron Device Lett. **35** (1), 78–80 (2014)
33. M. Ghittorelli, F. Torricell, J.-L. Van Der Steen, C. Garripoli, A. Tripathi, et al., Physical-based analytical model of flexible a-IGZO TFTs accounting for both charge injection and transport. IEDM Tech. Dig., 28.2.1–28.2.4 (2015)
34. M. Ghittorelli, F. Torricelli, Z.M. Kovács-Vajna, Analytical physical-based drain-current model of amorphous InGaZnO TFTs accounting for both non-degenerate and degenerate conduction. IEEE Electron Device Lett. **36**(12), 1340–1343 (2015)
35. M. Ghittorelli, F. Torricelli, Z.M. Kovács-Vajna, Physical modeling of amorphous InGaZnO thin-film transistors: the role of degenerate conduction. IEEE Trans. Electron Devices **63**(6), 2417–2423 (2016)
36. J. Robertson, Disorder and instability processes in amorphous conducting oxides. Phys. Status Solidi B **245**(6), 1026–1032 (2008)
37. T. Kamiya, K. Nomura, H. Hosono, Origins of high mobility and low operation voltage of amorphous oxide TFTs: electronic structure, electron transport, defects and doping. J. Disp. Technol. **5**(7), 273–288 (2009)
38. R.M. Corless, G.H. Gonnet, D.E.G. Hare, et al., On the Lambert W function. Adv. Comput. Math. **5**, 329–359 (1996)
39. A. Ortiz-Conde, F.J.G. Sanchez, M. Guzman, Exact analytical solution of channel surface potential as an explicit function of gate voltage in undoped-body MOSFETs using the Lambert W function and a threshold voltage definition therefrom. Solid State Electron. **47**(11), 2067–2074 (2003)
40. K. Abe, N. Kaji, H. Kumomi, et al., Simple analytical model of on operation of amorphous InGaZnO thin-film transistors. IEEE Trans. Electron Device **58**(10), 3463–3471 (2011)
41. S. Lee, J.-H. Park, K. Jeon, et al., Modeling and characterization of metal-semiconductor-metal-based source-drain contacts in amorphous InGaZnO thin film transistors. Appl. Phys. Lett. **96**(11), 113506 (2010)
42. J. He, M. Chan, X. Zhang, et al., A physics-based analytic solution to the MOSFET surface potential from accumulation to strong-inversion region. IEEE Trans. Electron Devices **53**(9), 2008–2016 (2006)
43. O. Marinov, M.J. Deen, U. Zschieschang, H. Klauk, Organic thin-film transistors: Part I-Compact DC modeling. IEEE Trans. Electron Devices **56**(12), 2952–2961 (2009)
44. P. Servati, D. Striakhilev, A. Nathan, Above-threshold parameter extraction and modeling for amorphous silicon thin-film transistors. IEEE Trans. Electron Devices **50**(11), 2227–2235 (2003)
45. S. Lee, K. Ghaffarzadeh, A. Nathan, et al., Trap-limited and percolation conduction mechanisms in amorphous oxide semiconductor thin film transistors. Appl. Phys. Lett. **98**(20), 203508 (2011)
46. T. Kamiya, K. Nomura, H. Hosono, Electronic structures above mobility edges in crystalline and amorphous In-Ga-Zn-O: percolation conduction examined by analytical model. J. Disp. Technol. **5**(12), 462–467 (2009)
47. C.G. Lee, B. Cobb, A. Dodabalapur, Band transport and mobility edge in amorphous solution-processed zinc tin oxide thin-film transistors. Appl. Phys. Lett. **97**(20), 203505 (2010)
48. T. Kamiya, K. Nomura, H. Hosono, Origin of definite Hall voltage and positive slope in mobility-donor density relation in disordered oxide semiconductors. Appl. Phys. Lett. **96**(12), 122103 (2010)

Chapter 4
AC Random Telegraph Noise (AC RTN) in Nanoscale MOS Devices

Jibin Zou, Shaofeng Guo, Ru Huang, and Runsheng Wang

4.1 Introduction

As the aggressive device scaling down, the random telegraph noise (RTN) has become as one of the most critical issues that hamper the development of future VLSI technology [1–5]. This kind of low-frequency noise, which is attributed to capture and emission of channel carriers by individual oxide traps in gate dielectrics, has been thoroughly studied with respects to: its origination [6–8], "atomistic" device simulation [9], physical model [10, 11], process dependence [12], device structural dependence [13, 14], etc.

Meanwhile, due to the long-tail distribution of RTN amplitude, RTN has already been considered as an important source of dynamic variations in VLSI design [15]. Many recent works indicate RTN can seriously impact on the performance of VLSI circuits, such as SRAM, DRAM, PPL, etc. [16–18]. Still, except for several experimental demonstrations on very particular circuits such as ring oscillators [19, 20], most studies that try to combine the RTN interference to digital circuits only rely on understandings from measuring RTN under conventional constant voltage condition (DC RTN), which is also within a very limited device gate voltage (VG) range. However, it should be pointed out that devices in digital circuits actually operate under AC signals, which generally cover full-swing range from GND to VDD. Thus, how RTN practically impacts on digital circuits should be further studied based on full-swing AC experimental investigations. Yet, this still remains a challenging question, due to the absence of a powerful technique which enables RTN characterization under such AC operations.

In this chapter, a novel statistical technique for characterizing RTN under GND–VDD full-swing digital operations, named as AC RTN [21–23], is presented and

J. Zou • S. Guo • R. Huang • R. Wang (✉)
Institute of Microelectronics, Peking University, Beijing 100871, China
e-mail: r.wang@pku.edu.cn

T. Li, Z. Liu (eds.), *Outlook and Challenges of Nano Devices, Sensors, and MEMS*, DOI 10.1007/978-3-319-50824-5_4

discussed with AC experimental details. This newly proposed AC RTN technique demonstrates its unique capability of detecting trap behaviors and characterizing RTN statistics exceeding the narrow RTN "detectable" window by the conventional DC RTN method, which covers full-swing operation range from GND to VDD. That is, even device is under VG voltage at GND or VDD, where RTN or trap behavior is very challenging to be directly detected by DC RTN method, this AC RTN technique can precisely monitor these possibly concealed trap behaviors. As a result, this technique reveals the nature of RTN or trap behaviors under full-swing AC operations, and further enables the study of practical RTN impact on digital circuits working with actual AC signals.

This chapter is organized as follows. Firstly, how RTN can practically influence digital circuit and why RTN should be studied with respect to AC signals are discussed in Sect. 4.2. In order to characterize RTN under AC signals, the novel AC RTN technique is presented in Sect. 4.3. Then, AC RTN statistics is discussed based on experimental observations in Sect. 4.4. In Sect. 4.5, the implication of practical RTN impact is discussed in terms of trap activity and trap occupancy rate, as two important FoMs in the AC RTN study. Next, AC RTN technique is further applied to different types of MOS devices including scaled multi-gate Fin-FETs and planar MOSFETs in Sect. 4.6. The universal AC RTN characteristics are found regardless of the different device structures and gate dielectric materials, revealing the universal RTN or trap behaviors under AC full-swing operations covering GND–VDD. Further in Sect. 4.7, the universality of AC RTN characteristics is explained by a simple model based on interpretations of transient trap occupancy rate under practical AC signals. Moreover, the details on the device-structure and gate-dielectric-material dependence of AC RTN characteristics are compared in Sect. 4.8. The discussions implement two important FoMs for circuit predictions: trap occupancy rate (TOR) and trap activity (ACT). Simulations with experimental results reveal the sensitivity of circuit performance to TOR and ACT. With the new understanding of the physical trap properties under AC signals, practical RTN impact on circuit functionality and signal-integrity are revisited by simulations of SRAM and RO in Sect. 4.9. Finally is the summary and other open issues.

4.2 Potential Impacts of RTN on Digital Circuits

Before discussing the RTN characterization method under AC full-swing operations, how RTN can practically impact on digital circuits and how the impact links to AC signals are discussed in the following two aspects.

4.2.1 On the Circuit Functionality

The most intuitive way that RTN influence on the digital circuit is disturbing the logic criteria of '0' and '1' by its large noise amplitude. In some cases towards the distribution tail, the RTN amplitude can beyond 50%, which leads to the difficulty in identifying logic levels of '0' and '1', and consequently results in the interference on circuit functionality. It should be noticed that under AC conditions of devices frequently switched between statuses of 'ON' and 'OFF' in digital circuits, RTN interference during device 'ON' is inevitably modulated by the status of device 'OFF'. It is because: firstly, the electrical field changes in the gate dielectrics can have non-ignorable influence on the RTN or trap behaviors; and secondly, even the device is biased at "OFF" status, it does not mean that the RTN or trap behaviors are "OFF". Thus, both "ON" and "OFF" statuses of the device that enabled by AC signals affect the practical RTN impact on the functionality aspect of circuit performance.

4.2.2 *On the Circuit Signal Integrity*

In majority cases, the RTN amplitude may not be large enough to cause direct logic failure, and it would have little direct interference on the circuit functionality. When device works at high gate voltage, i.e., in strong inversion region, RTN amplitude can become quite small, compared to weak inversion or sub-threshold operating regions [12]. However, it does not mean this RTN has no impact on circuit performance at all. For instance, as a typical digital circuit of inverter buffer shown in Fig. 4.1, although no direct or distinguishable RTN interference is observed when device (NMOS) is biased at high voltage of VDD, RTN does affect

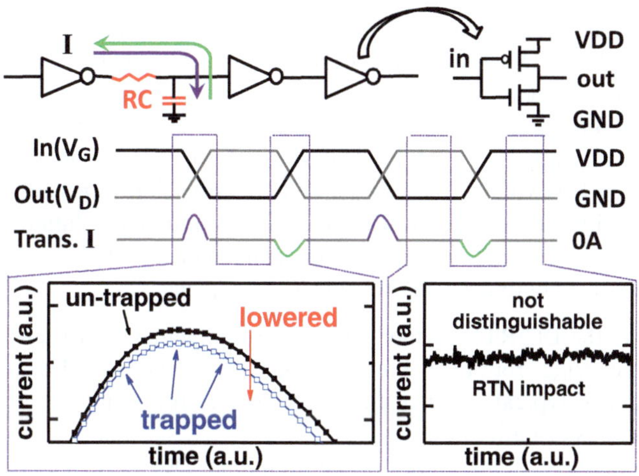

Fig. 4.1 Although no distinguishable RTN behavior can be detected at high voltage of VDD and low voltage of GND, RTN does affect digital circuit operation by reducing the transient current during the switches of device operating conditions

the transient or fan-out current during logic status changing. The lowered transient current between circuit stages due to the presence of RTN results in enlarged RC delay in circuitry. Also, as a result of the discontinuous current lowering by traps capture and emissions, large jitter noise on signal rises and falls is expected. These effects will further induce failures such as transient bit error in digital circuits.

In short, except for influence on circuit functional failure, RTN leads to influence on the signal integrity aspect of circuit performance as well. Generally speaking, this signal integrity interference is mainly modulated by both voltages of GND and VDD in the AC signals, at which condition most of the trapping and detrapping events are controlled.

Thus, for either two possible aspects of RTN impact on digital circuits, the full-swing AC signal covering GND and VDD voltages play an important role in the study of practical RTN impact on circuit performance.

4.3 AC RTN as a Characterization Technique

The traditional RTN measurement under constant voltage condition, named as DC RTN here, is inadequate to characterize RTN or trap behaviors under AC signals. Firstly, the DC bias cannot imitate the switching electrical field under AC operations, which definitely affects actual trap behaviors as previously discussed. Secondly, DC RTN method shows inefficiency in following conditions:

1. RTN or trap behaviors are not detectable at zero (GND) gate voltage by DC RTN method, because drain current already approaches 0 A. This statement holds true for almost all the possible RTNs and traps under discussion.
2. For most cases, RTN or trap behaviors cannot be clearly extracted at higher gate voltage such as VDD, because the RTN amplitude drops to be either smaller than the minimum resolution of current monitoring techniques, or to be small enough buried in other noise at device "ON" status, making it hard to distinguish captures and emissions of RTN behaviors. Also, there is another possibility that at high bias voltage the capture and emission time constants of the RTN differs significantly in time scale, challenging both time resolution and memory depth of the monitoring oscilloscope or semiconductor analyzer, and making it hard to extract continuous and reliable RTN sequences.

Therefore, the DC RTN method can be only implemented with a very limited VG range, covering from over threshold voltage to somewhere less than VDD in most cases. The typical value of this RTN "detectable" window is found to be around hundreds of millivolt or less.

To overcome the shortcomings of the traditional DC RTN, the AC RTN technique that characterizes RTN behaviors under AC operation signals is proposed as illustrated in Fig. 4.2. VG of device under test (DUT) is changed from DC constant voltage to alternating AC signals as shown in Fig. 4.2a. In order to separately investigate the influence of GND and VDD voltages on the trap behaviors, a

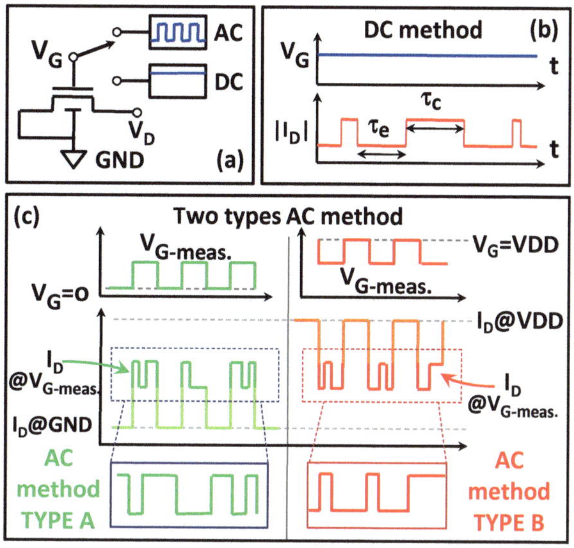

Fig. 4.2 So as to compensate the inefficiency of traditional DC RTN method, the two-type of AC RTN characterization technique is proposed. (**a**) Schematics of DC and AC measurement set-ups. (**b**) Typical DC RTN measurement. (**c**) Two-type AC RTN method: Type A: gate voltage from GND to VG-meas.; Type B: gate voltage from VG-meas. to VDD

measure or probe voltage of VG-meas. is simply inserted between the full-swing range of GND–VDD to separate voltages. The VG-meas. is chosen to be within the DC RTN "detectable" window, where the RTN behaviors can be easily observed with distinguishable RTN amplitude and reasonable time constants. This particular inserted voltage VG-meas. indirectly monitors RTN and reflects trap behaviors at voltages of GND and VDD, which exceeds the "detectable" window in conventional DC RTN measurements.

Therefore, as schematically shown in Fig. 4.2c upper figures, two types of AC signals are generated for characterizing AC RTN. Type A: VG switches between GND and VG-meas.; Type B: VG switches between VG-meas. and VDD. Since the rise and fall times are chosen to be short enough for both types, RTN are mostly expected to be observed at condition of VG = VG-meas. The schematic of RTN embedded in the AC response of drain current (I_D) is shown in the middle part of Fig. 4.2c. Based on this approach, effective RTN information can be further extracted in lower part of Fig. 4.2c.

Figure 4.3 is a typical experimental demonstration of this approach. The DUT is a SOI FinFET. With both types of AC RTN tested at different frequencies, the raw drain current outputs of DUT under AC signals are shown in the middle parts, where the embedded RTN behaviors can be clearly observed at VG-meas. for both types. The upper and lower parts of Fig. 4.3 correspond to the de-embedded I_D at higher voltage and lower voltage of the AC signals separately. For Type A (B) AC RTN, upper (lower) parts of Fig. 4.3 represents the extracted I_D at VG-meas., where reliable RTN information including RTN amplitude, effective capture time constant τ_c and emission time constant τ_e can be extracted for further study. Also, as expected, within the lower (upper) parts of Type A (B), which represent the I_D extracted at GND (VDD), no distinguishable RTN is detected. It should also be

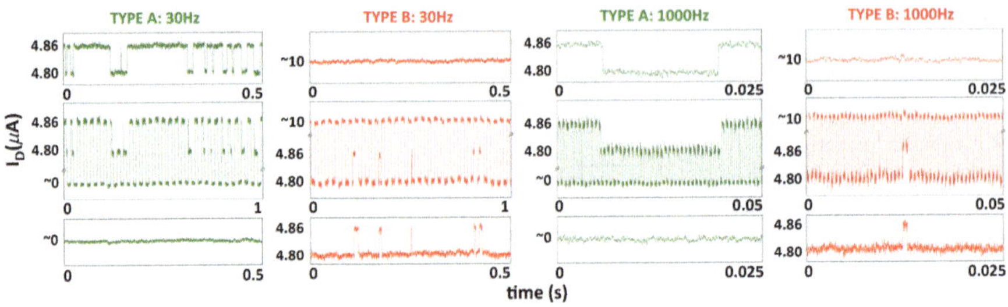

Fig. 4.3 Typical experimental results of two types AC RTN methods at different frequencies. Upper figures denotes the de-embedded current at higher voltage of AC signals (for Type A AC RTN, it is VG-meas.; while for Type B, it is VDD). Lower figures are the de-embedded current measured at lower voltages (type A AC RTN as GND; while type B AC RTN as VG-meas.). It is found that only at middle voltage of VG-meas., RTN behaviors can be observed clearly, for both types

noticed that although RTN not detected during periods of GND and VDD, trapping and detrapping behavior would not stop at such conditions. The similar results of AC RTN method can also be obtained in planar MOSFETs, which will be also discussed and compared in this chapter.

4.4 Experimental Results and Discussions of AC RTN

After extracting RTN information by the proposed two-type method at different frequencies, AC RTN results can be obtained. As shown in Fig. 4.4, for the same RTN under discussion, RTN information at VG-meas. is extracted for both types of AC RTN method. Apparently, the AC RTN statistics largely derive from DC RTN statistics as the frequency increasing. For both types, time constants of RTN change dramatically with frequency. The trends of time constant changes in AC RTN are symmetric for Type A and Type B. That is, at higher frequency, emission time constant τ_e in Type A becomes smaller significantly; while in Type B, the one that decreases is τ_c. Yet, as shown in Fig. 4.5, RTN amplitude does not show clear frequency dependence.

The results imply that, with unchanged RTN amplitude, the RTN or trap behaviors can be modified symmetrically by the voltages of GND and VDD in AC signals, in terms of capture and emission time constants. This effect should be considered in the studies that try to link RTN and circuit impact together.

In order to investigate the detailed RTN behavior changes under the AC RTN method, RTN time constants of two types are extracted as a function of frequency, as illustrated in Fig. 4.6. For Type A AC RTN results, τ_c seems to be independent of frequency (Fig. 4.6a). And as expected from Fig. 4.4, τ_e drops dramatically with increasing frequency (Fig. 4.6b). The results of Type B exhibit opposite or symmetric trends with τ_c-dependent (Fig. 4.6c) and τ_e-independent (Fig. 4.6d) of the

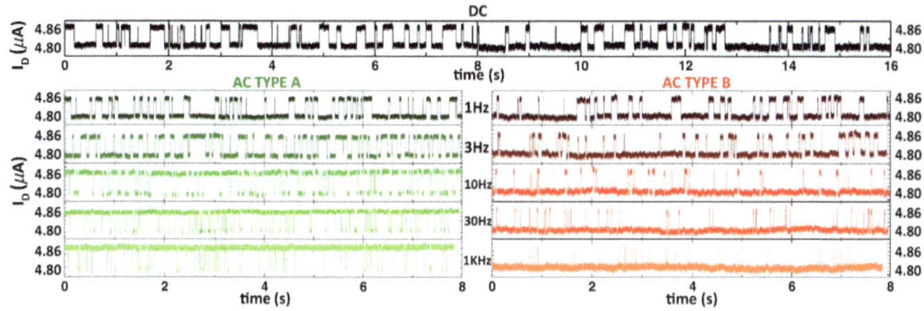

Fig. 4.4 *Upper figure*: DC RTN measurement results. *Lower figures*: de-embedded RTN information from two types of AC RTN methods. It is clearly detected that as the increasing frequency, type A and type B exhibit different trends with regards to the capture and emission times

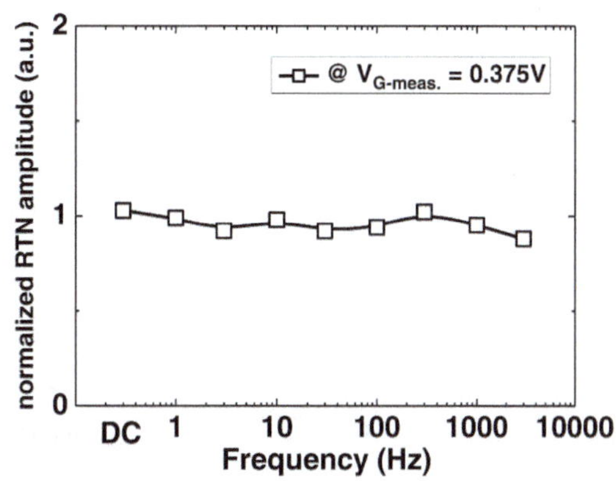

Fig. 4.5 RTN amplitude shows weak frequency dependence. RTN under

AC frequency. Also, it is clearly observed that τ_e of Type A and τ_c of Type B shows much weaker voltage dependence at higher frequency than at DC and low-frequency cases.

These results can be understood by detailed observations on the trapping and detrapping events embedded in the AC RTN measurements. Figure 4.7 shows the typical trapping and detrapping events embedded in drain current at VG-meas. of two-type AC signals, which can be directly observed. However, when the device is biased at GND or VDD, there may also be trap behaviors which cannot be directly detected, but can be indirectly inferred by comparing two adjacent cycles of AC signals. If there is status discontinuity at the end of the first AC cycle and the beginning of the second AC cycle, there must be concealed trapping and detrapping events between these two adjacent cycles, either on GND or VDD period in-between. The actually existent trap behaviors here under discussion do affect the impact of RTN on circuits. And the principles of trapping and detrapping on GND and VDD should be studied carefully.

With various capture and emission scenarios defined by up/down and blue/red arrows in the caption of Fig. 4.7, it can be statistically concluded from Fig. 4.7 as:

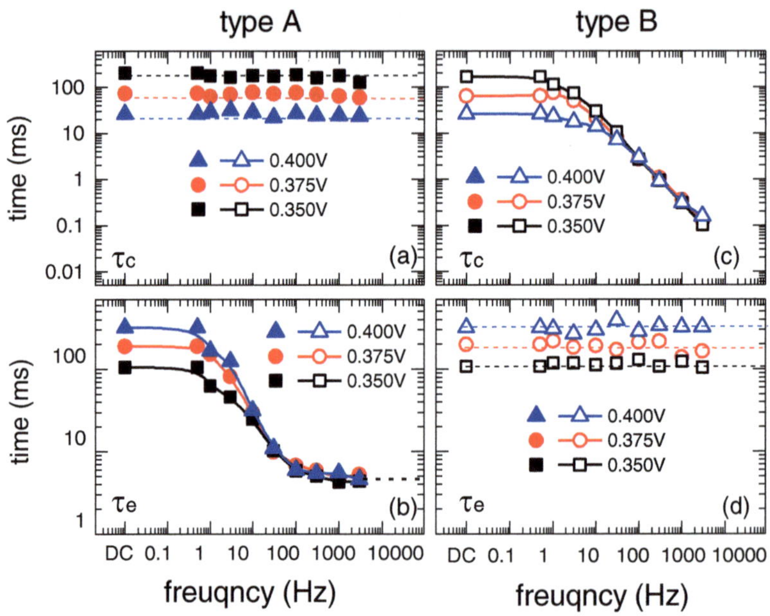

Fig. 4.6 Extracted capture and emission time constants as a function of operating frequency. Asymmetric trends are found for type A and B

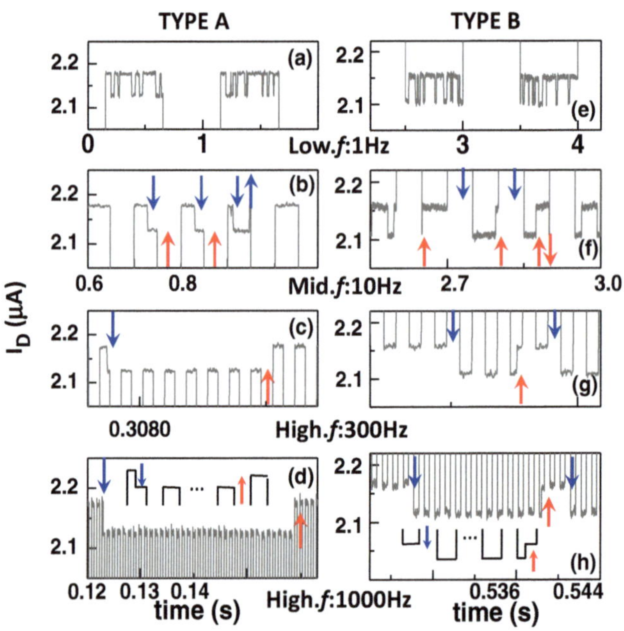

Fig. 4.7 Detailed observations of AC RTN experimental results. Drain current at GND or VDD are not shown in this figure, which is same as the middle part of Fig. 4.3. *Down arrows*: captures (trapping). *Up arrows*: emissions (detrapping). For simplicity, the bottom voltage in AC signals (GND in Type A and VG-meas. in Type B) are named as the low voltage (LV), while the top voltage (VG-meas. in Type A and VDD in Type B) are named as the high voltage (HV) of the AC signals. Trapping and detrapping events that occur at HV are marked by *blue arrows*, while events at LV are *red arrows*. Statistical results can be concluded at different frequencies

1. At low frequency, nearly all capture and emission processes effective for gathering RTN information, are controlled by voltage of VG-meas. for both types.
2. At middle frequency, capture and emission processes can be both controlled by either high voltage (HV) or low voltage (LV) of the AC signals. (All kinds of arrows can be found for both Type A and B at middle frequency).
3. At high frequency, HV of AC signals dominates the capture process (all down arrows are blue), while LV controls the emission (all up arrows are red). Therefore, the statistical changes of time constants observed in Figs. 4.4 and 4.6 can be explained as follows:

   (a) Covering all frequency range, HV (VG-meas.) in Type A dominates all the cap-ture events. Thus, capture time τ_c extracted from HV (VG-meas.) keeps unchanged with frequency as previously discovered in Fig. 4.6a.
   (b) HV (VG-meas.) in Type A only dominates emission time τ_e at low frequency. It is the LV (GND) in Type A that really controls τ_e at high frequency (which is a much smaller τ_e), resulting in the τ_e drop as a function of frequency in Fig. 4.6b.
   (c) For Type B AC RTN results, the symmetric trapping and detrapping principles are found compared to Type A, explaining the symmetric time constant trends as a function of frequency in Fig. 4.6c, d. The only difference in Type B is: HV is VDD; while LV is VG-meas.
   (d) Based on (1)–(3), because the constant voltage of GND (VDD) controls τ_e (τ_c) of Type A (B) at higher frequency, these τ_e (τ_c) of Type A (B) will have no de-pendence on middle voltage of VG-meas. That's why τ_e (τ_c) of Type A (B) shows little gate voltage dependence at higher frequency, as shown in Fig. 4.6a, d.

To further verify these explanations, Fig. 4.8 put the extracted τ_e (τ_c) from type A (B) AC RTN at high frequency, into the VG-dependence plot of τ_e and τ_c extracted from DC RTN method. The DC RTN "detectable window" is tested to be as wide as possible. However, it is clearly seen that the window around several hundred of millivolt cannot cover the full signal swing of GND–VDD.

As expected, the τ_e (τ_c) of AC RTN matches the extrapolation of DC RTN win-dow at the GND (VDD) point. This confirms the conclusion that the τ_e (τ_c) of AC RTN at high frequency obtained by the proposed AC RTN characterization method is actually the τ_e (τ_c) at GND (VDD), which is blind from conventional method. Thus it strongly supports that, the AC RTN statistics is the combination of VDD-controlled-τ_c and GND-controlled-τ_e in practical digital circuits with full-swing high-frequency AC signals.

Therefore, the proposed AC RTN technique demonstrates its unique capability of characterizing RTN statistics out of the RTN "detectable" window. Actually, other than GND and VDD, this trap behavior detecting technique is compatible with any other voltages out of DC "detectable" window.

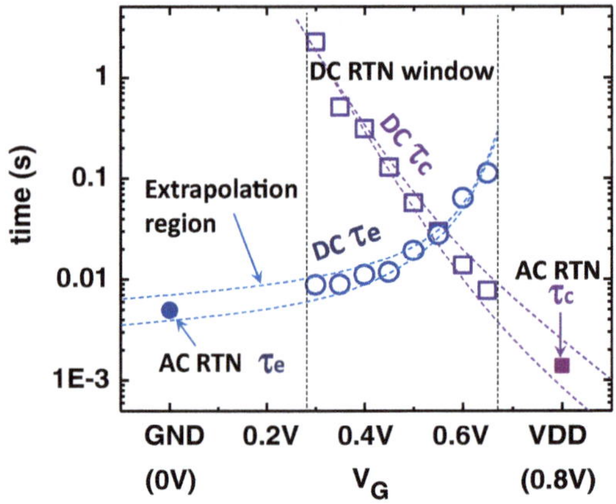

Fig. 4.8 Emission time extracted from high-f Type A AC RTN and capture time extracted from high-f Type B AC RTN, plotted in traditional DC RTN measurement results as a function of operating voltage. AC RTN τ_e and τ_c perfectly locates the reasonable extrapolation region of DC results, confirming the capability of AC RTN for extracting RTN behaviors out of traditional DC RTN observable window

4.5 RTN FoMs for Digital Circuit Applications

From the above discussions, it is well understood that RTN behavior derivate largely from AC to DC case: under actual full-swing GND-VDD operations, both time constants of τ_c and τ_e will be changed from the value extracted by DC RTN. These changes also reveal that under AC operations, the trap physical properties of trap occupancy rate and trap activity will be much different from DC case. Thus, trap occupancy rate and activity, as two figures of merit (FoMs) of RTN characteristics, can be extracted and implemented for studying the practical RTN impact on circuits, as discussed in the following part.

4.5.1 Effective Trap Occupancy Rate

The effective trap occupancy rate (TOR, or named trap occupancy probability) is defined as the following equation, literally means the average rate of trap being occupied:

$$TOR = \frac{\tau_e}{\tau_e + \tau_c} \tag{4.1}$$

Intuitively, if a trap has larger TOR, the impact of this trap on digital circuits will be consequently larger. The circuit will suffer more functional failures under the condition that the trap averagely spends more time on the “trapped” status.

From the understanding of the results above, it can be inferred that the TOR of RTN under AC signal depends on the actual capture time constant at HV (VDD) and emission time constant LV (GND) of this AC signal, which can be very different from the DC case:

$$TOR_{AC} = \frac{\tau_{e\_GND}}{\tau_{e\_GND} + \tau_{c\_VDD}} \neq TOR_{DC} = \frac{\tau_{e_{VDD}}}{\tau_{e_{VDD}} + \tau_{c_{VDD}}} \tag{4.2}$$

Thus, the TOR obtained by the AC RTN technique has practical significance for evaluating RTN impact on circuit functionality performance.

4.5.2 Apparent Trap Activity

The characteristic time constant of trap (or, RTN) can be expressed as:

$$\tau_{trap} = \frac{\tau_c \tau_e}{\tau_e + \tau_c} \tag{4.3}$$

Then, the trap activity (ACT, or named RTN activity) is defined as how many trapping and detrapping events per unit time (second):

$$ACT = \frac{1}{\tau_{trap}} = \frac{\tau_e + \tau_c}{\tau_c \tau_e} \tag{4.4}$$

From Fig. 4.4 we can understand there is a huge difference for trap activity at DC and AC cases, and the ACT would be much higher for AC cases. Mostly, from the AC RTN measurement, the difference is larger than ten times. And for some cases, the difference is approaching 100 times. (As will be shown in the Sect. 4.8). This means that in the real case of AC operations for digital circuits, the trap is much more active than the previous thoughts. And it will definitely make differences for circuit evaluations. Generally, this trap activity is an important factor when analyzing circuit signal integrity performance as a function of time. For example, the jitter noise characteristics of digital circuits is actually much worse than predictions based on DC RTN, due to the faster trapping and detrapping behaviors under AC operations.

The circuit-level applications of these two FoMs of AC RTN will be revisited with further discussed in Sect. 4.9.

4.6 Universal AC RTN Characteristics

With help of the inserted measuring voltage of VG-meas. inside the full-swing GND–VDD AC signals, the AC RTN technique can be separated into two types: Type A with AC signals from GND to VG-meas.; and Type B with AC signals from VG-meas. to VDD.

It has been shown above that this two-type AC RTN characterization method has the capability to characterize RTN or trap behaviors out of the conventional DC RTN "detectable" window, especially under practical AC operating conditions of switching between GND (zero) voltage and VDD voltage, where the RTN or trap behaviors are dominated but cannot be directly detected.

In this section, the proposed technique is applied to devices with different structures and different gate-dielectric materials for further comparison and investigations. As shown in Fig. 4.9, τ_c and τ_e from AC RTN measurement are extracted, with three kinds of devices as DUTs: multi-gate FinFETs with SiO_2 gate dielectrics; planar MOSFETs with high-κ metal-gate (HKMG) stack; planar MOSFETs with SiON gate dielectrics. It is found that all these time constants exhibit the same trends:

1. τ_c (τ_e) extracted from Type A (B) shows weak AC frequency dependence, while they exhibits strong voltage dependence regardless of the frequency.
2. τ_e (τ_c) extracted from Type A (B) are strongly frequency dependent, while they show clear voltage dependence only under low frequency.

These typical observations of AC RTN characteristics can be phenomenologically explained based on detailed observations of each trapping and detrapping processes as described in Sect. 4.4. In short, it can be summarized as: at low frequency, both capture and emission are controlled by the middle voltage of VG-meas.; while at high frequency, emissions are dominated by low voltage (LV) of GND, captures are mostly controlled by high voltage (HV) of VDD. Actually, in AC RTN technique, the capture (emission) time constant extracted at

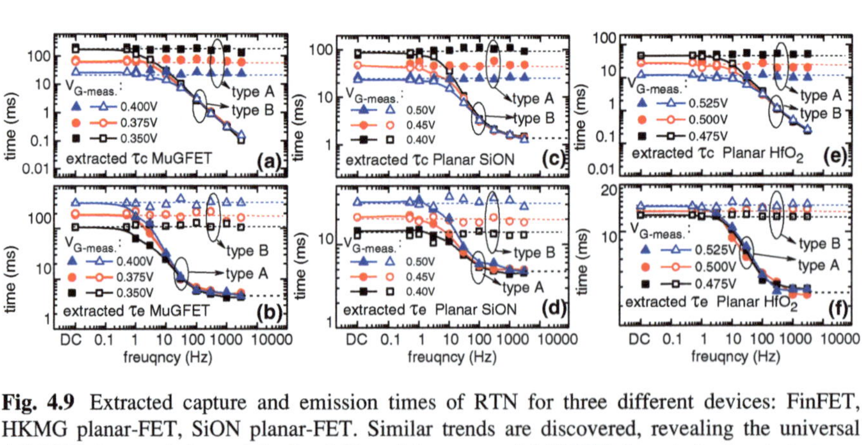

Fig. 4.9 Extracted capture and emission times of RTN for three different devices: FinFET, HKMG planar-FET, SiON planar-FET. Similar trends are discovered, revealing the universal AC RTN characteristics and the feasibility of the proposed AC RTN characterization technique

high frequency can be regarded as the capture (emission) time constant at HV (LV). As for middle frequency case, the results will be somewhere in-between low and high frequencies. These similar trends obtained from different device structures with different gate dielectric materials prove the wide applicability of the proposed AC RTN technique. Meanwhile, the universality of the AC RTN characteristics shown in Fig. 4.9 also indicate the unified physical mechanisms of the trap response to AC signals, which are independent of the device types.

4.7 A Simple Model for AC RTN Statistics

Although the universal AC RTN characteristics in these devices can be experimentally explained by the detailed observations, deep understandings with a unified physical model on trap response to AC signals is required to describe the AC RTN or trap behaviors regardless of the device types.

Firstly, DC trap occupancy rate of a single trap under a constant voltage of VX is defined as the average rate of a trap being trapped, and it can be expressed as:

$$p_{Vx} = \frac{\tau_{eVx}}{\tau_{eVx} + \tau_{cVx}} \tag{4.5}$$

where τ_{eVx} and τ_{cVx} represents the emission and capture time of RTN under constant voltage of V_X.

Also, under VX, the averaged time constant of RTN can be denoted as:

$$\tau_{Vx} = \frac{\tau_{eVx}\tau_{cVx}}{\tau_{eVx} + \tau_{cVx}} \tag{4.6}$$

Given that the RTN behaviors, i.e. the trapping and detrapping behaviors follow the Markov process [24, 25], the transient occupancy rate (defined as the trap occupancy rate at a certain time, p(t)) under switching AC signals (as shown in Fig. 4.10) can be described with an exponential decay towards the DC trap occupancy rate under high voltage (HV) case or low voltage (LV) case:

$$p(t) = \begin{cases} p_{HV} + (p_1 - p_{HV})\exp\left(-\frac{t}{\tau_{HV}}\right), 0 \le t < \frac{T}{2} \\ p_{LV} + (p_2 - p_{LV})\exp\left(-\frac{t}{\tau_{LV}}\right), -\frac{T}{2} \le t < 0 \end{cases} \tag{4.7}$$

where T (= 1/f) denotes the AC period. p_1 represents the transient occupancy rate at –T, 0, T, 2T, . . ., 2nT; while p_2 is the transient occupancy rate at –T/2, T/2, 3T/2, . . ., (2n+1)T/2.

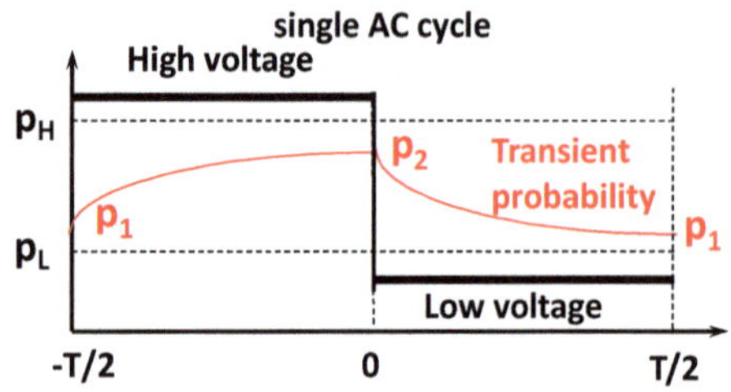

Fig. 4.10 The trap occupancy rate at a certain time is defined as transient trap occupancy rate p(t). It is a function of the DC trap occupancy rate at high voltage (PHV), DC trap occupancy rate at low voltage (PLV) and the AC frequency f or period T

Together with the boundary conditions of AC cycles in Fig. 4.10:

$$p\left(-\frac{T}{2}\right) = p\left(\frac{T}{2}\right), p(0_-) = p(0_+) \tag{4.8}$$

The averaged transient trap occupancy rate TOR at HV and LV can be expressed as a function of frequency or period:

$$TOR_{HV} = 2f\int_0^{T/2} p(t)dt, \ \ TOR_{LV} = 2f\int_{-T/2}^{T0} p(t)dt \tag{4.9}$$

which is a monotonous function that saturates at high frequency.

At high frequency, if T is much smaller than τ_{HV}, τ_{LV}, then from Eq. (4.7)), the probability at boundaries can be derived as:

$$\begin{cases} p\left(\frac{T}{2}\right) = p_{HV} + (p_1 - p_{HV})\exp\left(-\frac{T}{2\tau_{HV}}\right) = p_2 \\ p\left(-\frac{T}{2}\right) = p_{LV} + (p_2 - p_{LV})\exp\left(\frac{T}{2\tau_{LV}}\right) = p_1 \end{cases} \tag{4.10}$$

It can be obtained that p_1 equals to p_2 at high frequency case.

Thus, at high frequency, Eq. (4.9)) can be expressed as:

$$TOR_{HV} = TOR_{LV} = p_1 = p_2 \tag{4.11}$$

Equations (4.10) and (4.11) implies that at high frequency, the transient trap occupancy rate degenerates to a constant value, which is independent of frequency, and will be modulated only by HV and LV voltages under discussion. Here the degenerated transient trap occupancy rate is in consistence with the trap occupancy rate defined in Eq. (4.1). From the derived TOR in Eq. (4.11), the trap can be described as entering a stable status, which is just similar as the DC RTN case with a stable or constant trap occupancy rate. AC stable status and DC stable status of the trap can have different TOR values, according to Eq. (4.2). And we define this difference as ratio of TOR:

$$TOR_{ratio} = \frac{TOR_{AC(HV\&LV)}@high\,f}{TOR_{DC}} \tag{4.12}$$

Further based on Eqs. (4.8) and (4.10), TOR at high frequency is solved as:

$$TOR_{AC(HV\&LV)}@high\,f = p_1 = p_2 = \frac{\frac{p_{LV}}{\tau_{LV}} + \frac{p_{HV}}{\tau_{HV}}}{\frac{1}{\tau_{LV}} + \frac{1}{\tau_{HV}} = \frac{1}{\tau_{cHV}} + \frac{\frac{1}{\tau_{cLV}}}{\frac{1}{\tau_{cHV}} + \frac{1}{\tau_{cLV}} + \frac{1}{\tau_{eHV}} + \frac{1}{\tau_{eLV}}}} \tag{4.13}$$

For the majority of traps with reducing capture time and increasing emission time as gate voltage increases, τ_{cHV} and τ_{eLV} is much smaller than τ_{cLV} and τ_{eHV}, and Eq. (4.13) can be simplified as:

$$TOR_{AC(HV\&LV)}@high\,f = \frac{\tau_{eLV}}{\tau_{eLV} + \tau_{cHV}} \tag{4.14}$$

From this equation, the time constants of this RTN under discussion can be understood as: emission time constant equals to τ_{eLV} capture time constant equals to τ_{cHV}. Therefore, this physical model based on averaged transient occupancy rate calculation is in consistence with the phenomenological conclusion obtained by experimental observations in Sect. 4.4: under high frequency AC operations, capture and emission processes are controlled by HV and LV separately, and the corresponding time constant equals to τ_{cHV} and τ_{eLV}.

It should be noticed that, for Type A AC RTN, VG-meas. is HV; while for Type B AC RTN, VG-meas. is LV. Different parameters in TOR should be chosen to understand the trend of extracted time constants in Fig. 4.9, especially for the high frequency cases.

In order to further verify the developed physical model, a RTN with trap behavior distinguished at both HV and LV is investigated as the extracted time constants shown in Fig. 4.11. In this particular case, the same RTN can be extracted and analyzed on both HV and LV at the same time. From HV (LV) we can get a constant τ_{cHV} (τ_{eLV}), while the τ_{cLV} (τ_{eHV}) apparently depends on frequency. What is interesting and also as expected is that the capture time τ_{cLV} (emission time τ_{eHV}) extracted from LV (HV) finally equals to τ_{cHV} (τ_{eLV}) at high frequency. This observation confirms the results predicted by both the phenomenological explanations (Sect. 4.4) and the physical model (Eq. (4.10)) that the τ_c (τ_e) extracted from AC RTN technique @high-f just equals to τ_{cHV} (τ_{eLV}) extracted from HV (LV).

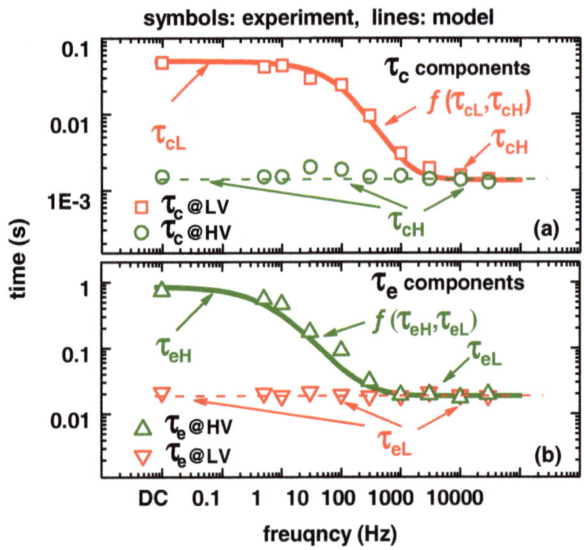

Fig. 4.11 Measurement for verifying the physical explanation based on transient trap occupancy rate. RTN behaviors are observable at both HV and LV. The theoretical physical model agrees well with the time constants extracted from both HV and LV voltages, confirming the understandings of the universal AC RTN characteristics

Furthermore, only with DC and the high-frequency extracted results as initial parameters to calculate the transient trap occupancy rate, this physical model agrees well with experiments for the whole range of frequency covering from the DC to high-f cases. The overlaps of the model and data lines in Fig. 4.11 ensure the validity of this physical model.

4.8 Technology Dependence of AC RTN Characteristics

The universal behavior of AC RTN characteristics can be well understood from both phenomenological explanation and the above physical model. Yet, the AC RTN characteristics for different device structures and gate-dielectric materials should be discussed more carefully in terms of two important FoMs for circuit applications based on the physical RTN or trap property: trap occupancy rate TOR and trap activity ACT, as defined above.

The trap occupancy rate represents the average rate of a trap being trapped during DC or AC operations. TOR here is calculated by the actual emission and capture time measured under DC or AC signals. And the trap activity is defined as the number of trapping and detrapping events per unit time (second), shown in Eq. (4.2). Similar as TOR_{ratio} defined in Eq. (4.12), ACT_{ratio} is defined as the ratio of ACT between AC and DC cases.

ACT and TOR for three device types are extracted and shown in Fig. 4.12 for further comparison. Firstly, it is observed that although ACT at AC condition all increases dramatically compared to ACT at DC condition, the extent of the increase, i.e., ACT_{ratio} shows significant differences (Fig. 4.12a). ACT_{ratio} of

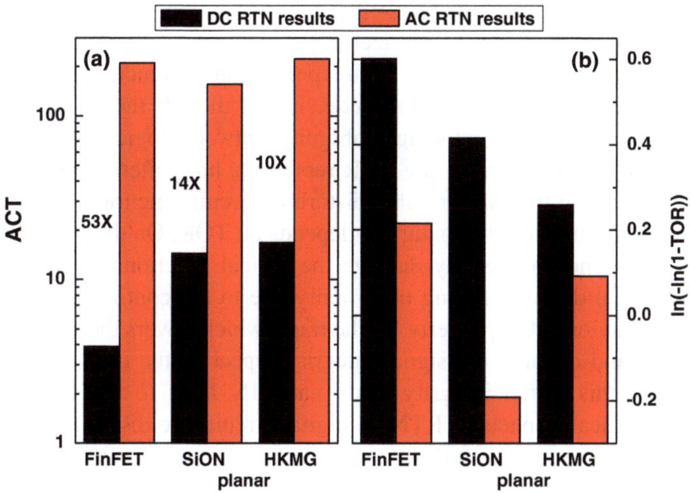

Fig. 4.12 Although the AC RTN behaviors are universal in spite of device types, the physical trap properties in terms of trap activity and occupancy rate can be much different. (**a**) Comparison of trap activity. (**b**) Comparison of trap occupancy rate

FinFET is much larger than planar MOSFETs. From the results in Fig. 4.9, we can understand the reason for this observation: τ_c and τ_e of RTN in FinFET drops faster as a function of frequency than the planar MOSFET cases, resulting in larger differences between AC-high-f and DC ACT. And the reason for the relatively smaller τ_c (controlled by HV) and τ_e (controlled by LV) of RTN in FinFETs at higher frequency can be understood by the stronger HV and LV voltage dependence of time constants in FinFETs, which is due to the thin-body structure that enhances the quantum confinement effects [14, 26].

On the other hand, as shown Fig. 4.12b, although TOR of three cases all drops, TOR_{ratio} does not show distinguished difference for FinFETs. Actually, TOR_{ratio} may vary within a large range, due to the diversity of the traps under discussion [27], which will also be addressed in the future work. Unless the time constants of the trap are not sensitive to HV and LV of the AC signals at all, the TOR_{ratio} will be smaller than 1 for most cases.

Thus, although the AC RTN characteristics generally shows same trends, the important FoMs for evaluating practical RTN impact on digital circuits could differ a lot in terms of TOR and ACT for different device types.

4.9 Understanding RTN Impacts on Digital Circuits

In this section, the practical RTN impacts on digital circuits are revisited, with the two FoMs of TOR and ACT.

From the above results, we know that the motivation and application of the proposed AC RTN characterization technique is to practically evaluate the RTN impacts on two aspects of digital circuit performance: functionality and signal integrity. If a RTN as a large amplitude, it can disturb the original designed functionality seriously by inducing ambiguity between logic state '0' and '1'. Practically, the trap occupancy rate TOR represents how often the trap is trapped or to what extent the trap has possible disturb to circuit functions. Thus, the RTN impact on circuit functionality naturally depends on TOR. On the other hand, even if the RTN does not necessarily damage the logical functions of circuits, it will induce additional delay or timing uncertainty due to the continuous trapping and detrapping behavior (ACT property) of the traps, which lowers the current between circuit stages and disturbs the signal integrity aspect of the circuit performance. Consequently, this signal integrity study naturally links to the FoMs of ACT. Therefore, practical impacts of RTN on digital circuits are discussed regarding to TOR and ACT separately as follows.

4.9.1 Understanding RTN Impacts on Circuit Functionality with TOR

Based on the new observations and the derived physical trap property of TOR, firstly we examine the practical RTN impact on functionality aspect of digital circuits. A typical 6-T SRAM cell (Fig. 4.13a) is implemented in this approach with 16 nm PTM model [28] for evaluations at advanced technology node.

Because the random nature of RTN, even though one RTN at a single device in the SRAM structure is not large enough to be responsible for a function failure, lots of possible RTNs at different devices will. In this section, every device in the SRAM structure is considered to possess one RTN. And the model of RTN is a Voltage Controlled Voltage Source (VCVS) with RTN time constant specifications according to the practical AC RTN results as shown in Fig. 4.13b. This schematic in Fig. 4.13c describes that the circuit failure of SRAM could be mainly caused by the trapping processes of multiple RTNs at the same time. Also, Fig. 4.13c schematically shows that the TOR of traps under discussion will manipulate the circuit functional failures significantly.

Functional impact of RTN is discussed in term of read failure probability of the standard 6-T SRAM cell. According to the previous work [24], there will be a constant probability of read-failure induced by RTN within a wide range of power supply VDD, which is defined as the read failure probability of SRAM. After testing more than 105 read-after-write cycles in total, the read failure probability is plotted as a function of TOR selected for modeling the RTNs. Apparently, the read failures of SRAM has a strong dependence on the TOR of RTN or traps. This result further indicates that there would be large error in circuit functional performance prediction if only applying the TOR results obtained from conventional DC

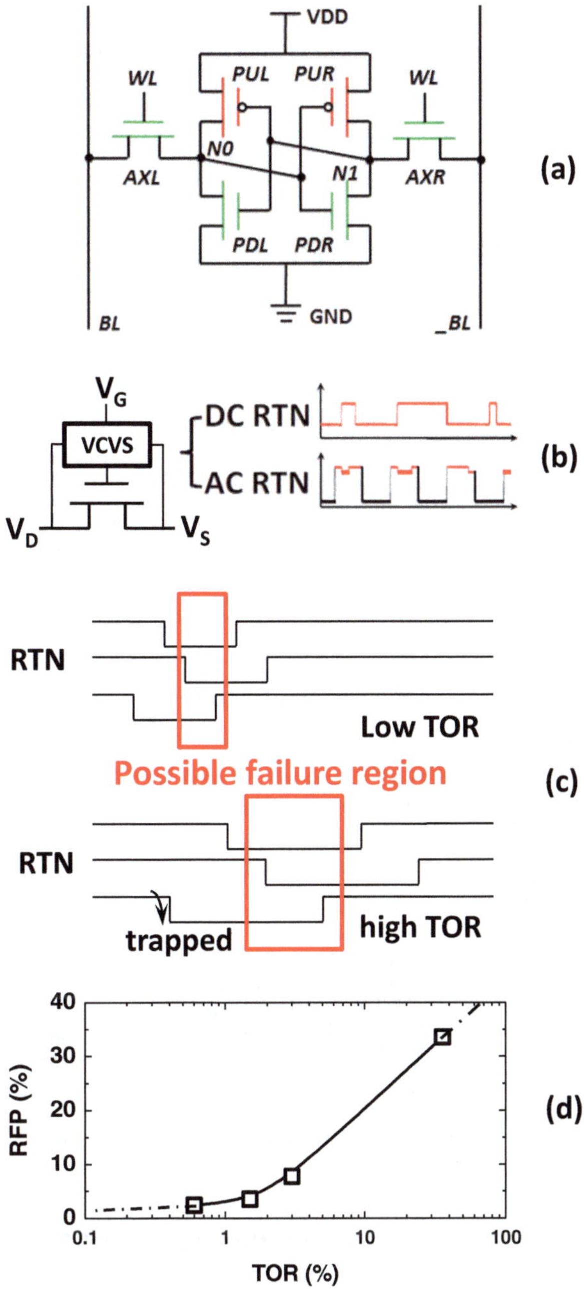

Fig. 4.13 (**a**) 6-T SRAM structure used in this study. (**b**) Equivalent VCVS RTN model, considering both DC and AC cases. (**c**) Multiple RTN behaviors will lead to circuit failure. (**c**) schematically illustrates that with high TOR, the circuit failure probability may be higher. (**d**) simulation results of SRAM read failure probability as a function of TOR

RTN methods. For example, if the tested TOR_{ratio} is 0.1, the actual read failure probability under AC operations could be much differed from the DC RTN based predictions. The circuit performance predictions are quite sensitive to the TOR under discussion.

4.9.2 Understanding RTN Impacts on Circuit Integrity with ACT

As discussed in Sect. 4.2, under conditions that RTNs do not exhibit large amplitude that causes functional failure, they will still impact the signal integrity aspect of circuit performance. And this aspect of impact is evaluated by investigating the frequency performance of a 5-stage ring oscillator (Fig. 4.14a). For simplification, each inverter is designed with a RTN on the NMOS. Because the timing or frequency performance of the ring oscillator will be strongly related to the RTN amplitude during the signal switching between GND and VDD, the VCVS model

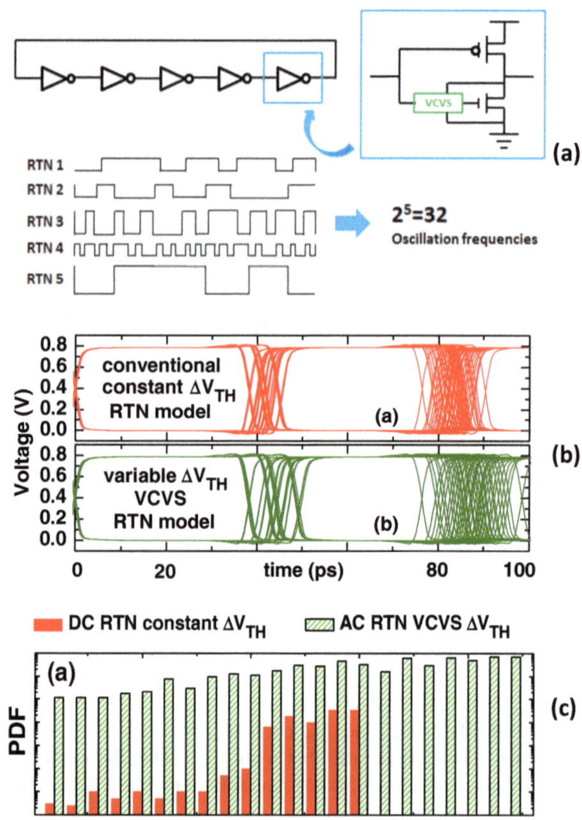

Fig. 4.14 (**a**) 5-stage Ring Oscillator (RO) structure used in this study. Every NMOS is considered to be with a RTN, randomly selected from experimental data pool. There would be 32 oscillating frequencies. (**b**) Comparison of previous DC RTN model and the newly proposed VCVS RTN model with AC considerations. (**c**) Jitter noise comparisons for DC and AC RTN cases. It is observed that jitter noise will be seriously underestimated if only taking DC RTN results into account

applied in this paper considers both the dynamic voltage dependence of RTN amplitudes and the larger trap activity under AC signals in digital circuits.

Since five different RTNs are chosen from the experimental result pool for practical simulations, there could be total $2^5 = 32$ oscillating frequencies as shown in Fig. 4.13a. And the Fig. 4.13b are the eye diagram of for all possible cases the can be generated from this ring oscillator, with the comparison of conventional constant RTN model and the proposed VCVS RTN model. From Fig. 4.13b, it can be observed that jitter variation range is larger for the AC RTN case due to the dynamic VCVS RTN model considerations. Actually, the larger jitter noise range quantitatively depends on RTN amplitude profile as a function of device operation condition, which will be comprehensively studied in the future work.

Furthermore, the normalized probability density functions (PDF) of the jitter noise as a function of time are extracted in Fig. 4.13c. Other than the larger jitter variation range which is in accordance with Fig. 4.13b, the probability density function of jitter noise with the AC RTN results (higher trap activity) is much larger than that based on DC RTN (lower trap activity). (ACT_{ratio} is chosen to be around 50X as shown in Fig. 4.12). This confirms that during GND–VDD full-swing operations, there will be much more frequent trapping and detrapping processes, and result in larger jitter noise in terms of PDF, due to the highly activated trap response to AC signals. Therefore, physical trap property of ACT reflects the practical RTN impact on signal integrity aspect of circuit performance, which may be predicted with large error if only considering DC RTN characteristics.

This section only provide a simple introduction on the RTN impacts on digital circuits. For more comprehensive studies, please refer to [29, 30].

4.10 Summary and Other Open Issues

In this chapter, a special AC RTN characterization technique is proposed to detect RTN or trap behaviors which are usually concealed out of the conventional DC RTN "detectable" window. Results obtained by AC RTN technique reveal that RTN statistics under practical AC operations largely derivate from constant voltage measurement (DC RTN) case. Although RTN amplitude is not frequency dependent, the RTN time constants drop dramatically as frequency increasing, which should be considered in the evaluation of RTN impact on digital circuits. Based on detailed observation of experimental results, principles of trapping and detrapping behaviors are revealed: the capture and emission time constants of the RTN under discussion are controlled by high voltage (VDD) and low voltage (GND) of the AC signals separately. The results also indicate that, evaluating practical RTN impacts on logic circuits is not reliable solely based on DC RTN characteristics. With the important FoMs of RTN trap occupancy rate and trap activity obtained from AC RTN technique, it is possible to predict RTN impacts on functionality and signal integrity aspects of digital circuits.

With the newly proposed AC RTN characterization technique, AC RTN characteristics are investigated with different devices. With the physical model based on transient trap occupancy rate, the universality of AC RTN characteristics, which is regardless of device structures and gate-dielectric materials, is well explained. The detailed differences for RTN response to AC signals are discussed in terms of trap occupancy rate TOR and trap activity ACT. The results based on AC RTN understandings and simulations indicate: circuit performances are quite sensitive to TOR and ACT; and evaluating RTN impacts on digital circuits is not reliable if only considering DC RTN characteristics. With the AC RTN characterization technique and the corresponding results, it is possible to predict the practical RTN impact on functionality and signal integrity of the digital circuits.

There are still some open issues remained for RTN, such as, understanding RTN amplitude from microscopic modeling (e.g., [8, 9, 11, 31–33]), origin of the switching oxide traps (e.g., [34, 35]), metastable defect states (e.g., [34–37]), connections between RTN and BTI (e.g., [8, 37, 38]), careful data analysis of RTN measurements (e.g., [24, 25]), design solutions against RTN (e.g., [39]), complex RTN (e.g., [40–42]), etc., which require further studies.

Acknowledgments This work was partly supported by NSFC (Nos. 61522402 and 61421005). The authors would like to thank Mulong Luo for the input, and thank Tibor Grasser and K. P. Cheung for the helpful discussions.

References

1. K.K. Hung, P.K. Ko, C. Hu, Y.C. Cheng, Random telegraph noise of deep-submicrometer MOSFETs. IEEE Electron Device Lett. **11**(2), 90–92 (1990)
2. M. Tsai, T.-P. Ma, The impact of device scaling on the current fluctuations in MOSFETs. IEEE Trans. Electron Devices **41**(11), 2061–2068 (1994)
3. N. Tega, H. Miki, F. Pagette, D.J. Frank, A. Ray, M.J. Rooks, W. Haensch, and K. Torii, Increasing threshold voltage variation due to random telegraph noise in FETs as gate lengths scale to 20 nm. 2009 Symposium on VLSI Technology, 2009, pp. 50–51
4. T. Nagumo, K. Takeuchi, S. Yokogawa, K. Imai, and Y. Hayashi, New analysis methods for comprehensive understanding of random telegraph noise. 2009 I.E. International Electron Devices Meeting (IEDM), 2009, pp. 759–762
5. J.P. Campbell, L.C. Yu, K.P. Cheung, J. Qin, J.S. Suehle, A. Oates, and K. Sheng, Random telegraph noise in highly scaled MOSFETs. 2009 I.E. International Reliability Physics Symposium. 2009, pp. 382–388
6. M.J. Uren, D.J. Day, M.J. Kirton, 1/f and random telegraph noise in silicon metal-oxide-semiconductor field-effect transistors. Appl. Phys. Lett. **47**(11), 1195–1197 (1985)
7. G. Ghibaudo, T. Boutchacha, Electrical nosie and RTS fluctuations in advanced CMOS devices. Microelectron. Reliab. **42**(4–5), 573–582 (2002)
8. T. Grasser, H. Reisinger, W. Goes, T. Aichinger, P. Hehenberger, P.-J. Wagner, M. Nelhiebel, J. Franco, and B. Kaczer, Switching oxide traps as the missing link between negative bias temperature instability and random telegraph noise. 2009 I.E. International Electron Device Meeting (IEDM), 2009, pp. 729–732
9. A. Asenov, R. Balasubramaniam, A.R. Brown, J.H. Davies, RTS amplitudes in de-ca-nanometer MOSFETs: 3-D simulation study. IEEE Trans. Electron Devices **50**(3), 839–845 (2003)

10. E. Simoen, B. Dierichx, C.L. Claeys, G.J. Declerck, Explaining the amplitude of RTS noise in submicrometer MOSFETs. IEEE Trans. Electron Devices **39**(2), 422–439 (1992)
11. R.G. Southwich, K.P. Cheung, J.P. Campbell, S.A. Drozdov, J.T. Ryan, J.S. Suehle, and A.S. Oates, Physical model for random telegraph noise amplitudes and implications. 2012 I. E. Silicon Nanoelectronics Workshop (SNW), 2012, pp. 1–2
12. J. Chen, Y. Higashi, I. Hirano, and Y. Mitani, Experimental study of channel doping concentration impacts on random telegraph noise and successful noise suppression by strain induced mobility enhancement. 2013 Symposium on VLSI Technology, 2013, pp. T184–T185
13. Y.F. Lim, Y.Z. Xiong, N. Singh, R. Yang, Y. Jiang, D.S.H. Chan, W.-Y. Loh, L.K. Be-ra, G.Q. Lo, N. Balasubramanian, D.L. Kwong, Random telegraph noise in gate-all-around Si-FinFET with ultranarrow body. IEEE Electron Device Lett. **27**(9), 765–768 (2006)
14. C. Liu, Y. Wang, J. Zou, R. Huang, C. Fan, L. Zhang, J. Fan, Y. Ai, and Y. Wang, New understanding of the statistics of random telegraph noise in Si nanowire transistors—the role of quantum confinement and non-stationary effects. 2011 I.E. International Electron Devices Meeting (IEDM), 2011, pp. 521–524
15. R. Wang, and Y. Cao, Impact of temporal transistor variations on circuit reliability. IEEE ISCAS, 2015, pp. 2453–2456
16. Y. Tsukamoto, S.O. Toh, C.Shin, A. Mairena, T.J.K. Liu, and B. Nikolic, Analysis of the relationship between random telegraph signal and negative bias temperature instability. Proceedings of IEEE International Reliability Physics Symposium, May 2010, pp. 1117–1121
17. S.O. Toh, Y. Tsukamoto, Z. Guo, L. Jones, T.J.K. Liu, and B. Nikolic, Impact of random telegraph signals on Vmin in 45 nm SRAM. Proceedings of IEEE IEDM, Dec 2009, pp. 767–770
18. J.S. Kolhatkar, Steady-state and cyclo-stationary RTS noise in MOSFETS. Ph.D. dissertation, Department of Electrical Engineering, Mathematics and Computer Science, University of Twente, Twente, The Netherlands, 2005
19. Q. Tang, X. Wang, J. Keane, and C.H. Kim, RTN induced frequency shift measurements using a ring oscillator based circuit. 2013 Symposium on VLSI Technology, 2013, pp. T188–T189
20. S. Fujimoto, I.A.K.M. Mahfzul, T. Matsumoto, H. Onodera , Inhomogeneous ring oscillator for WID variability and RTN characterization. 2012 I.E. International Conference on Microelectronic Test Structures (ICMTS), 2012, pp. 25–30
21. R. Wang, J. Zou, X. Xu, C. Liu, J. Liu, H. Wu, Y. Wang, and R. Huang, New observations on the AC random telegraph noise (AC-RTN) in nano-MOSFETs. 2012 International Symposium on VLSI Technology, System and Applications (VLSI-TSA), 2012, pp. 1–2
22. J. Zou, R. Wang, N. Gong, R. Huang, X. Xu, J. Ou, C. Liu, J. Wang, J. Liu, J. Wu, S. Yu, P. Ren, H. Wu, S.-W. Lee, and Y. Wang, New insights into AC RTN in scaled high-k/metal-gate MOSFETs under digital circuit operations. 2012 Symposium on VLSI Technology, 2012, pp. 139–140
23. J. Zou, R. Wang, M. Luo, R. Huang, N. Xu, P. Ren, C. Liu, W. Xiong, J. Wang, J. Liu, J. Wu, W. Wong, S. Yu, H. Wu, S.-W. Lee, and Y. Wang, Deep understanding of AC RTN in MuGFETs through new characterization method and impacts on logic circuits. 2013 Symposium on VLSI Technology, 2013, pp. T186-T187
24. H. Miki, N. Tega, M. Yamaoka, D.J. Frank, A. Bansal, M. Kobayashi, K. Cheng, C.P. D'Emic, Z. Ren, S. Wu, J. Yau, Y. Zhu, M.A. Guillorn, D. Park, W. Haensch, E. Leo-bandung, and K. Torii, Statistical measurement of random telegraph noise and its impact in scaled-down high-k/metal-gate MOSFETs. 2012 I.E. International Electron Devices Meeting (IEDM), 2012, pp. 19.1.1–19.1.4
25. M. Li, R. Wang, J. Zou, R. Huang, Characterization of random telegraph noise in scaled high-k/metal-gate MOSFETs with SiO_2/HfO_2 gate dielectrics. ECS Trans. **52**(1), 974–946 (2013)
26. R. Wang et al., in *Towards Quantum FinFET*, ed by W. Han, Z. M. Wang. Random telegraph noise in multi-gate FinFET/nanowire devices and the impact of quantum confinement (Springer, Cham, 2013), pp. 205–225

27. H. Miki, M. Yamaoka, N. Tega, Z. Ren, M. Kobayashi, C.P. D'Emic, Y. Zhu, D.J. Frank, M.A. Guillorn, D.-G. Park, W. Haensch, and K. Torii, Understanding short-term BTI behavior through comprehensive observation of gate-voltage dependence of RTN in highly scaled high-k metal-gate pFETs. 2011 Symposium on VLSI Technology, 2011, pp. 148–149
28. Predictive Technology Model (PTM), available online at http://ptm.asu.edu/
29. M. Luo et al., Impacts of random telegraph noise (RTN) on digital circuits. IEEE Trans. Electron Devices **62**(6), 1725–1732 (2015)
30. D. Mao et al., Deep understanding of random telegraph noise (RTN) effects on SRAM stability, 2016 International Symposium on VLSI Technology, System and Applications (VLSI-TSA), 2016, pp. 1–2
31. J. Franco et al., Impact of single charged gate oxide defects on the performance and scaling of nanoscaled FETs. IEEE IRPS, 2012, pp. 5A.4.1–5A.4.6
32. L. Gerrer et al., Modelling RTN and BTI in nanoscale MOSFETs from device to circuit: a review. Microelectron. Reliab. **54**, 682–697 (2014)
33. Z. Zhang et al., Investigation on the amplitude distribution of random telegraph noise (RTN) in nanoscale MOS devices. IEEE INEC, 2016
34. T. Grasser et al., On the microscopic structure of hole traps in pMOSFETs. IEDM Technical Digest, 2014, pp. 530–533
35. J. Ji et al., New framework for the random charging/discharging of oxide traps in HfO_2 gate dielectric. IEDM Technical Digest, 2014, pp. 542–545
36. D. Mao et al., Impacts of metastable defect states on gate oxide trapping in nanoscale MOS devices. IEEE INEC, 2016
37. T. Grasser et al., A unified perspective of RTN and BTI. IEEE IRPS, 2014, pp. 4A.5.1-4A.5.7
38. R. Wang et al., Understanding NBTI-induced dynamic variability in the nano-reliability era: from devices to circuits. IEEE IPFA, 2015, pp. 119–121
39. S. Guo et al., DTMOS mode as an effective solution of RTN suppression for robust device/circuit co-design. IEDM Technical Digest, 2014, pp. 319–322
40. P. Ren et al., New observations on complex RTN in scaled high-k/metal-gate MOSFETs—the role of defect coupling under DC/AC condition. IEDM Technical Digest, 2013, pp. 778–781
41. P. Ren et al., Experimental study on the oxide trap coupling effect in metal oxide semiconductor field effect transistors with HfO_2 gate dielectrics. Appl. Phys. Lett. **104**, 263512 (2014)
42. J. Zou et al., New understanding of state-loss in complex RTN: statistical experimental study, trap interaction models, and impact on circuits. IEDM Technical Digest, 2014, pp. 832–835

Chapter 5
Passivation and Characterization in High-k/III–V Interfaces

Shengkai Wang and Honggang Liu

5.1 Merit of III–V Channels

The tremendous growth in semiconductor technology has been based on the well-behaved Si-SiO_2 system. In 1965, Gordon Moore, co-founder of Intel, saw the future. His prediction, now popularly known as Moore's Law, states that the number of transistors on a chip doubles approximately every 2 years. Up until now this "law" has been followed by evolutionary progress of the basic technology. For the last decade, pure geometrical scaling has failed to deliver the expected benefits in terms of performance and power consumption, and industry has gradually moved to innovation-driven scaling, bringing to market chips based on the likes of strained silicon-on-insulator, high-κ/metal gate technology and tri-gate devices. Now researchers everywhere are foreseeing that in the coming years, silicon—regardless of its form—will probably fail to meet the ultra-low power consumption targets imposed by the exploding demand of 'Mobile-Everywhere' applications. This progress has reached its limit. The main lever for power scaling is the operating voltage of the chips. The target is to trim this from 0.8 to 0.9 V, which is where it stands today, down to 0.5 V. This cut in operating voltage must go hand-in-hand with a maintaining of the drive-current for the transistors, in order to ensure no reduction in performance. But realizing this will not be easy. It will require the charge carriers in the transistor's channel—either electrons or holes, depending on the particular transistor—to travel far faster from the source to the drain. Today, increases in the charge carrier velocity in silicon often result from the application of very high levels of strain in the material, but the opportunities for further gains are now minimal.

S. Wang (✉) • H. Liu
Key Laboratory of Microelectronics Devices and Integrated Technology, Institute of Microelectronics, Chinese Academy of Sciences, Beijing 100029, China
e-mail: wangshengkai@ime.ac.cn; liuhonggang@ime.ac.cn

T. Li, Z. Liu (eds.), *Outlook and Challenges of Nano Devices, Sensors, and MEMS*,
DOI 10.1007/978-3-319-50824-5_5

Table 5.1 Bandgap and mobility for Si and typical III–V materials

Materials	Bandgap	μe (cm^2/Vs)	μ_h (cm^2/Vs)
Si	1.12	1500	450
GaAs	1.42	8500	400
InGaAs	0.74	>10,000	400
InP	1.34	5400	200
GaSb	0.73	3000	1000
InSb	0.17	>70,000	800

What is possible is that the next revolution in the semiconductor industry will come from the introduction of compound semiconductors as channel materials. As listed in Table 5.1, electron and hole mobility in silicon are just 1500 $cm^2\ V^{-1}\ s^{-1}$ and 450 $cm^2\ V^{-1}\ s^{-1}$, respectively, and far higher values are promised by switching to carefully selected III–V compounds. For example, the electron mobility in InGaAs is in excess of 10,000 $cm^2\ V^{-1}\ s^{-1}$, while hole mobility in GaSb can hit 1000 $cm^2\ V^{-1}\ s^{-1}$ [1]. Using these materials to form the n-type and p-type transistors for CMOS chips would be a promising solution for future high performance logic application.

5.2 Brief History of III–V MOSFET

III–V MOSFETs concept has been considered for over 50 years. In 1965, Becke and White from Radio Corporation of America reported the first GaAs MOSFETs [2]. When III–V MOSFETs first developed, the major issue was lack of high quality thermodynamic stable gate dielectrics. Unlike Si-SiO_2 interface, the native oxides of III–V produce much more interface traps, which makes it difficult to bend Fermi level into inversion mode. In the early 1960, some researchers attempted to use the native oxides of GaAs as gate dielectric. However, due to the compound nature of GaAs, its native oxide is a mixture of Ga_2O_3, Ga_2O, As_2O_3, As_2O_5, strongly increasing the complexity in using its native oxide as dielectric. In order to find some alternative oxides as gate dielectric, researchers tried to use SiO_2, SiN_x, SiON, and Al_2O_3 [2–5]. However, limited by the deposition method, most dielectric materials at that time required high temperature, causing severe interface degradation by forming arsenic oxides or vacancies.

In late 1970s, besides physical vapor deposition (PVD) and chemical vapor deposition (CVD), molecule beam epitaxy (MBE) technique has been used for dielectric films deposition on GaAs, such as AlGaAs [6]. In 1978, Dingle et al. made the first enhancement mode high mobility III–V transistor by using AlGaAs/GaAs superlattice [7] and lead to the invention of high electron mobility transistors (HEMT) in 1980 [8]. Then, GaAs HEMT has been commercialized and widely used in telecommunication, aerospace technology and military purpose. Compared with mature HEMT technology, the unsolved dielectric/III–V interface problem pushes the research of III–V MOSFETs technology away from the main stream until 1987,

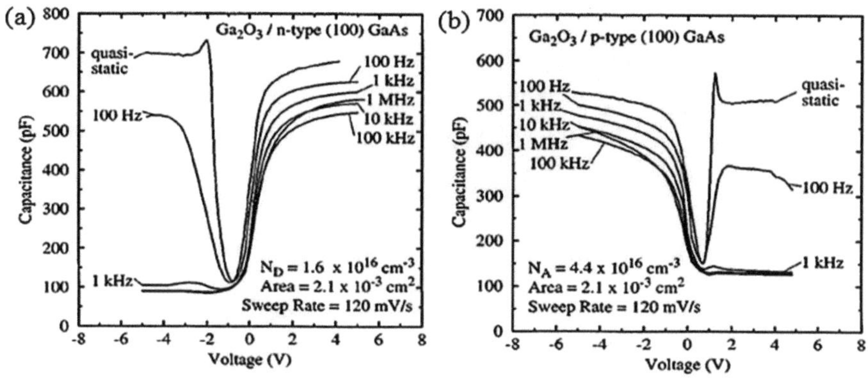

Fig. 5.1 Typical C–V characteristics of (**a**) n-type and (**b**) p-type samples measured in quasi-static mode and at various frequencies between 100 Hz and 1 MHz

when the sulfur passivation technology was verified to be effect on GaAs by Bell Lab [9]. This breakthrough brought III–V MOSFETs back, and many researchers started to find solutions for dielectric/III–V systems.

In 1995, a milestone was achieved by Passlack and Hong et al. from Bell Lab. They have made the field-break work by depositing crystalline $GaGdO_3$ (GGO) onto GaAs (100) using MBE [10]. For the first time, the Fermi level of GaAs has been bent to inversion mode, as depicted in Fig. 5.1. Since low interface trap density (D_{it}) has been obtained by GGO epitaxy, encouraging people to continue the research in crystalline oxides on III–Vs (not limited to GGO). Now, there are still several teams are working on crystalline oxides on III–Vs, such as Hong's team in National Tsinghua University, the team from Freescale, and Ye and Gorden teams in Purdue University and Harvard University [11–13].

In 2001, another milestone has been reached in the history of III–V MOSFETs. Wilk and Ye et al. from Bell Lab first delivered atomic layer deposition (ALD) technology into the formation of high-k dielectrics on GaAs, opening a new era for III–V MOSFETs. Later after that, Ye et al., for the first time, demonstrated a III–V compound semiconductor MOSFET with the Al_2O_3 gate dielectric grown by ALD [14]. By using alternating pulses of $Al(CH_3)_3$ and H_2O precursor in a carrier N_2 gas flow, the thickness and uniformity of deposited oxide layer was well controlled at angstrom level. Growth of ALD-Al_2O_3 process results in an abrupt interface with the GaAs substrate, as illustrated by the high-resolution transmission electron microscopy (Fig. 5.2). The oxide layer appears as a desirable amorphous form, while the GaAs exhibits clear lattices. The ALD process removes the native oxide and excess As on GaAs surface, resulting in a very thin Ga-Oxide interfacial layer. This process is now commonly acknowledged as "self-cleaning" effect. Compared with traditional sputtering, MBE, and CVD methods, ALD has many advantages, such as good uniformity, accurate in thickness control, low cost, and compatible for III–V processes. The above-mentioned advantages of ALD method attracts more and more researchers into the field of III–V MOSFETs, such as researchers from

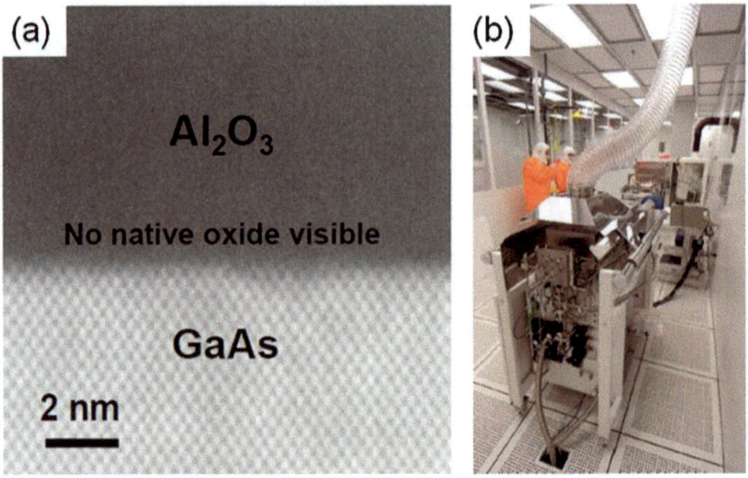

Fig. 5.2 (**a**) ALD deposited Al_2O_3 on GaAs, no native oxide visible at the Al_2O_3/GaAs interface due to the "self-cleaning" reaction during ALD deposition. (**b**) A typical ALD deposition system, which is small in size and low cost in thin film deposition

MIT, Intel, IMEC, Purdue, U. Tokyo, UT-Dallas, UT-Austin, Harvard, NUS, IMECAS, and so on, making ALD to be the most popular method for high-k deposition on III–V substrates and strongly pushing the performance of III–V MOSFETs to higher level. For example, in 2008, Ye's group demonstrated a high-performance inversion-type enhancement-mode InGaAs MOSFET with maximum drain current exceeding 1 A/mm [15]. In 2013, Gu et al. demonstrated high performance gate-all-around (GAA) nanowire MOSFETs with equivalent oxide thickness (EOT) of 1.2 nm using ALD [16]. In 2013–2014, Kim et al. have made progress in high performance extremely thin-body (ETB) InAs-OI MOSFETs using ALD high-k dielectrics, obtaining high I_{on} and $G_{m,max}$ of >2 mA/μm and ~2 mS/μm [17, 18].

Besides dielectrics/III–V interface problems, III–V MOSFETs also confronts with some other challenges from the aspects of (i) integration with Si, (ii) metal-contact resistance reduction, (iii) limited in density of state for channel carrier, (iv) building of non-planar structure, (v) device reliability, etc. In the past 10 years, albeit significant progresses have been achieved in the above aspects, this chapter cannot cover all the details about them, and only focuses on the passivation and characterization of high-k/III–V interfaces.

5.3 Passivations in High-k/III–Vs

Building a high quality high-k/III–V gate stack is regarded as the most important issue in III–V MOSFETs study. In this section, we want to share our experiences in high-k/III–V gate stack engineering by taking some specific examples in passivations and characterizations of high-k/III–V interfaces.

5.3.1 Al_2O_3/InP MOS Capacitors with Pre-gate Sulfur/Nitrogen Passivations

To evaluate the gate stack quality of a MOSFET, in most cases, it is not necessary to fabricate a real MOSFET device. As an alternative, fabricating a simple Metal-oxide-semiconductor (MOS) capacitor structure is enough for use and much convenient in process flow.

In this subsection, MOS capacitors were fabricated on n-InP substrates (3–5 × 10^{15} cm^{-3}, supplied by AXT). After organic cleaning using acetone, and alcohol (5 min each), wafers were cleaned by sequential immersion in 10% HCl solution for 1 min, 25% NH_4OH solution for 5 min. After that, some samples were rinsed in 20% $(NH_4)_2S_x$ solution for 15 min at room temperature to remove the native oxide and passivate the surface, denoted as S-passivation. While for some other samples, after the cleaning process and skipped the S-passivation process, these samples, denoted as N-passivation, were directly transferred into the ALD deposition chamber to receive in-situ 50 W plasma nitridation at 200°C using activated NH_3 gas [19]. Immediately after S-passivation or N-passivation, a low temperature process was then applied for gate dielectric deposition: 3 nm of Al_2O_3 was deposited on the substrate at a temperature of 200°C using a Beneq TFS-200 ALD system with successive cycles of tri-methyl-aluminum and H_2O precursors. After that, in situ post deposition annealing (PDA) at 200°C for about 10 min in vacuum was performed for further dielectric densification and purification. Al metal gate contacts (~200 nm) were e-beam evaporated to form MOS capacitors having an area of 7.85 × 10^{-5} cm^2. After that, post metallization annealing (PMA) treatments were performed at several temperatures ranging from 250 to 350°C in N_2 for 30 s, respectively. Then, backside ohmic contact was fabricated by Al evaporation. 3 nm ALD Al_2O_3 were deposited at 200°C followed by the deposition of Al as gate electrode. The fabrication process flow of the MOS capacitors is schematically shown in Fig. 5.3.

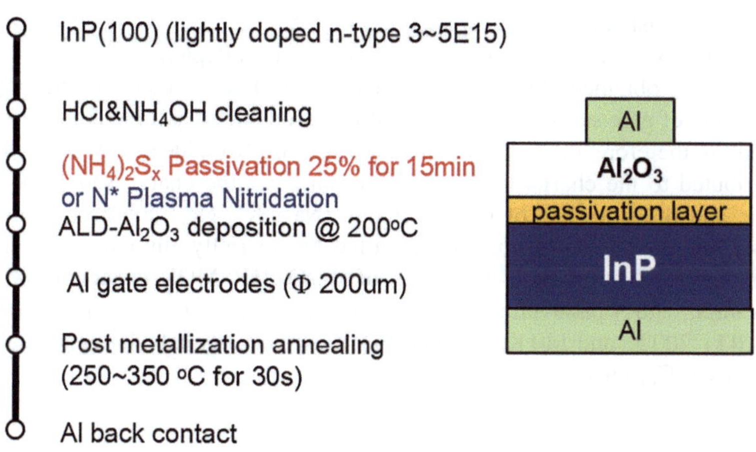

Fig. 5.3 The fabrication process flows of the MOS capacitors with S-passivation and N-passivation

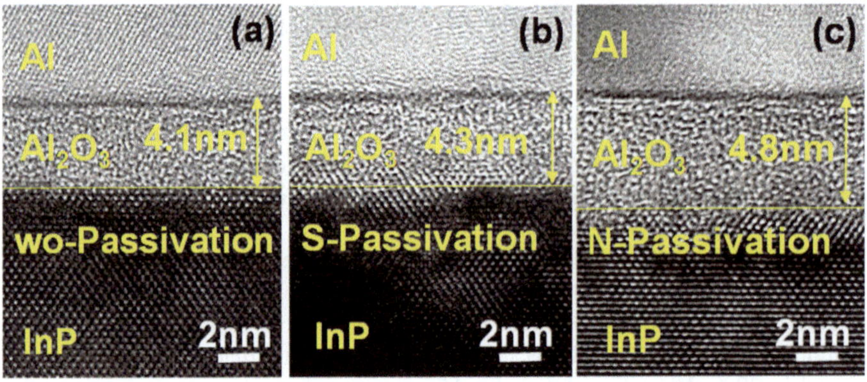

Fig. 5.4 Cross sectional TEM images of $Al/Al_2O_3/InP$ MOS capacitors with different interface passivation processes (**a**) w/o Passivation, (**b**) S-Passivation, (**c**) with N-Passivation

To further confirm the structure of the above MOS capacitors, the cross-sectional TEM images of the above samples with the detailed dielectric thickness are shown in Fig. 5.4.

Although 3 nm Al_2O_3 was initially deposited, the total dielectric thicknesses are all over 4 nm. The overgrowth in dielectric thickness is attributed to the existence of 1.1 nm-thick AlO_x layer between Al top electrode and ALD-Al_2O_3 due to the redox reaction by assuming that the interfacial layer thickness between Al_2O_3 and InP for the without passivation one is negligible [20]. Therefore, by subtracting 1.1 nm from the total thickness, the interfacial layer thickness for S-passivation and N-passivation ones are estimated to be around 0.2 nm and 0.7 nm.

Among various electrical characterizations of a MOS capacitor, capacitance-voltage (C–V) measurement is the most popular and effective one in probing the oxide charge types, interface trap density and distribution, amount and types of bulk traps, dielectric reliability, and so on. For a typical C–V measurement, it contains three steps. The measurement sequence of C–V curves of Al_2O_3/InP stacks is schematically listed below, as depicted in Fig. 5.5.

In the first step, by sweeping the gate voltage from depletion to accumulation, fresh C–V is obtained. In C–V measurement, fresh C–V contains the original information of gate stack. In fresh C–V, it is reasonable to assume that most traps are neutral; therefore, the flatband voltage (V_{FB}) shift from ideal C–V curve should be attributed to the charge in the stack. In our case, the charge should be fixed charge, and the influence of mobile charge (Na^+, K^+, etc.) could be neglected, because the mobile charge contamination has been strictly eliminated. Figure 5.6 shows the 1 MHz fresh C–V curves of Al_2O_3/InP MOS capacitors without passivation, with S-passivation and with N-passivation measured at room temperature (RT), 200 K, and 140 K, in which V_{FB} is extracted by calculating the flatband capacitance (C_{FB}) using the following equations:

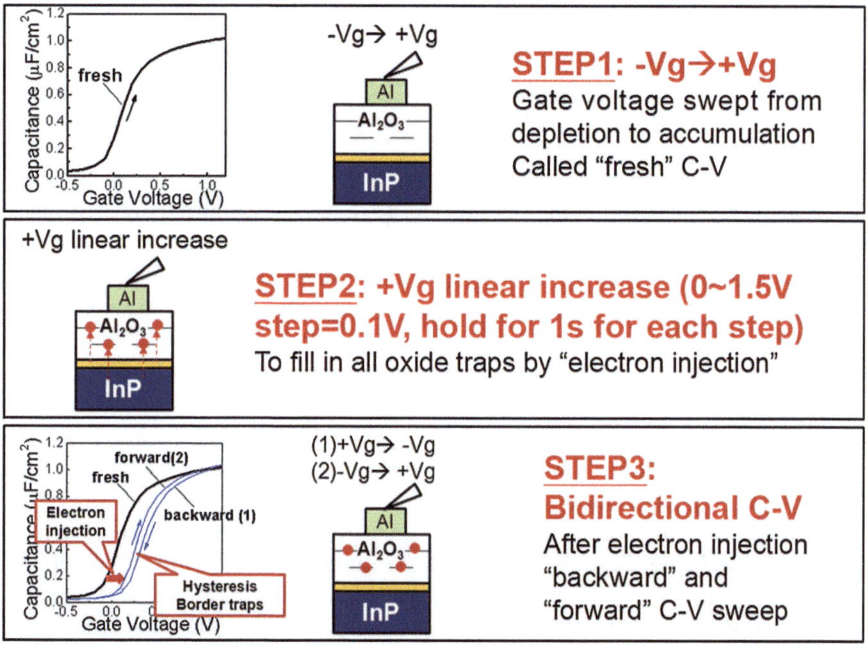

Fig. 5.5 The measurement sequence of C–V curves

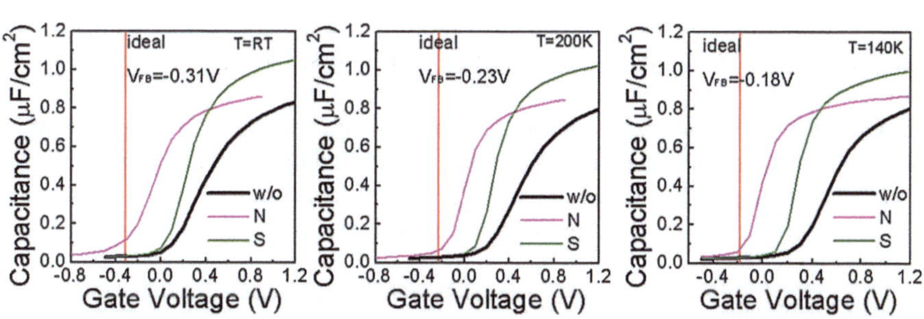

Fig. 5.6 Fresh C–V curves of Al_2O_3/InP MOS capacitors without passivation, with S-passivation and with N-passivation measured at RT, 200 K, and 140 K

$$C_{FB} = \frac{1}{\frac{t_{ox}}{\varepsilon_0 \varepsilon_{ox}} + \frac{L_D}{\varepsilon_0 \varepsilon_s}} = \frac{C_{ox}}{1 + \frac{\varepsilon_{ox} L_D}{\varepsilon_s t_{ox}}} \tag{5.1}$$

$$L_D = \sqrt{\frac{\varepsilon_0 \varepsilon_s k_B T}{q^2 N_A}} \tag{5.2}$$

Combine Eqs. (5.1) and (5.2) and assuming $C_{ox} \approx C_{max}$, we have

$$C_{FB} = \frac{C_{max}}{1 + C_{max}/\sqrt{N_A q^2 \varepsilon_s \varepsilon_0 / k_B T}} \tag{5.3}$$

where C_{ox} is the dielectric capacitance, N_A is the substrate doping concentration, L_D is the Debye length, e is the electron charge, ε_0 is the permittivity of free space, t_{ox} is the dielectric thickness, ε_{ox} is the oxide dielectric constant, ε_s is the semiconductor dielectric constant, k_B is Boltzmann's constant, and C_{max} is the maximum capacitance at 1 MHz in accumulation. Note that the Cox here is not simply determined by assuming $t_{ox} = 3$ nm and $\varepsilon_{ox} = 8$ for Al_2O_3, because of the AlO_x layer between Al_2O_3 and the Al electrode, which reduces the total C_{ox}. By applying Eq. (5.3), it is possible for us to calculate C_{FB} directly without considering ε_{ox} and t_{ox}. Concerning the assumption of $C_{ox} \approx C_{max}$ in Eq. (5.3), C_{max} can never reach C_{ox} values because InP has a low conduction band density of states, which leads to a deviation in V_{FB} calculation [21].

In RT case, compared with the V_{FB} value from the ideal C–V characteristic (−0.31 V by theoretical calculation), all the samples show positive shift of V_{FB} value. In specific, the capacitor with N-passivation shows least V_{FB} shift, then the S-passivation one, and the one without interface passivation show the largest V_{FB} shift. If we do not consider the impact of interfacial dipole, the positive shift could be mainly ascribed to the existence of negative fixed charge. By changing the temperature to 200 K and 140 K, although V_{FB} of ideal C–V is positively shifted, the trend of V_{FB} shift remains identity. To further quantitatively investigate V_{FB} shift against temperature among the three capacitors, the data are re-plotted in Fig. 5.7.

On the basis of the results in Fig. 5.7, the V_{FB} shift seems to be independent of temperature, verifying that such V_{FB} shift should mainly result from the negative fixed charge originally existed in the dielectric but not from the trap, because trap/de-trap behavior is a temperature dependent process.

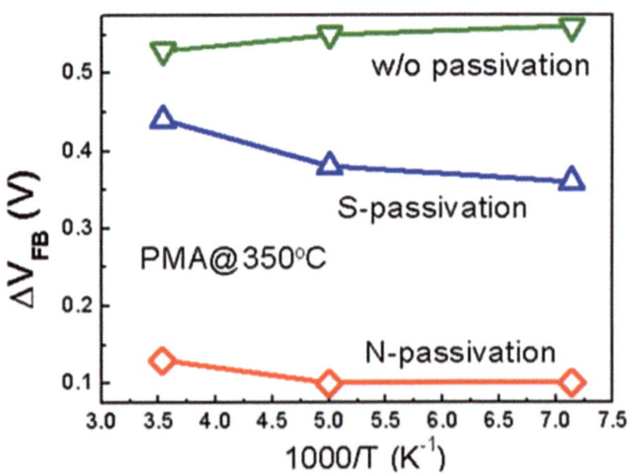

Fig. 5.7 Temperature dependence of the V_{FB} shift among the capacitors without passivation, with S-passivation and with N-passivation

In the second step, by linearly sweeping the voltage from 0 to 1.5 V with step of 0.1 V, and holding at each point for 1 s, all the traps are filled with electrons. In our case, the electrical field across the Al_2O_3 is estimated to be about 1.5 V/4 nm = 1.5 × 10^{-6} MV/4 × 10^{-7} cm = 3.75 MV/cm. Compared with the breakdown filed of our Al_2O_3 (>7 MV/cm), this field is just nearly half of that, thus could be reasonably assume to be enough for electron injection and sufficiently low without causing additional traps. Therefore, in the third step, on the basis of the above assumption, the total oxide traps (N_{tot}) and border traps (N_{bt}) could be roughly extracted from the difference between fresh C–V and the backward C–V, and the difference between forward C–V and the backward C–V (hysteresis) using the following equations, respectively:

$$N_{tot} = \frac{1}{q}\int C_{back}(V) - C_{fresh}(V)dV \tag{5.4}$$

$$N_{bt} = \frac{1}{q}\int C_{back}(V) - C_{for}(V)dV \tag{5.5}$$

where C_{back} (V), C_{fresh} (V) and C_{for} (V) are capacitance from backward C–V, fresh C–V and forward C–V curves, respectively.

Figure 5.8a, b show the temperature dependence of total traps (N_{tot}) and border traps (N_{bt}) for the above samples, N-passivation is proved to show lowest trap amount than S-passivation and w/o passivation ones, in which similar trend are observed in both total traps and border traps. In detail, the amount of both total traps and border traps are reduced after interface passivation, and the capacitor receiving N-passivation show better effect in reducing oxide traps. Concerning the interface trap density (D_{it}) in the MOS capacitors, there are several method to extract that, such as conductance method, charge pumping, Terman method, and Castagné-Vapaille method. For high-k/InP system, since the bandgap of InP is large enough, response of minority carrier can be easily distinguished from the

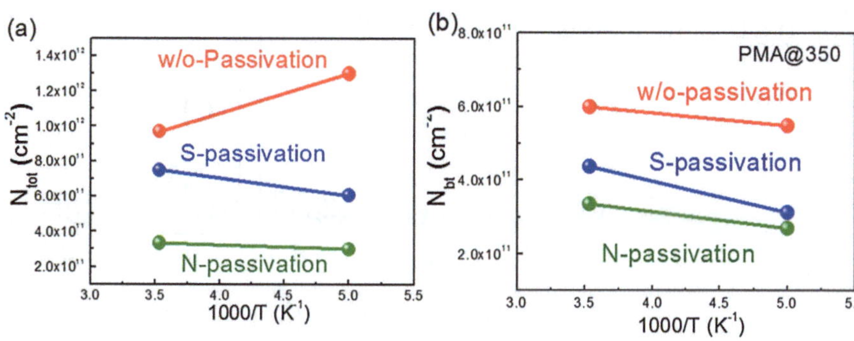

Fig. 5.8 Temperature dependence of (**a**) total traps (N_{tot}) and (**b**) border traps (N_{bt}) for Al/Al_2O_3/n-InP MOS capacitors with S- and N-passivation and without passivation

D_{it} response due to their different time constant. Therefore, in this case, Castagné-Vapaille is one of the most convenient and effective method [22]. The D_{it} distributions as a function of energy within semiconductor bandgap were derived from the formula

$$D_{it} = \frac{C_{ox}}{Aq^2}\left(\frac{C_{lf}/C_{ox}}{1 - C_{lf}/C_{ox}} - \frac{C_{hf}/C_{ox}}{1 - C_{hf}/C_{ox}}\right) \tag{5.6}$$

where Dit is in $cm^{-2}\,eV^{-1}$, A is the electrode area, C_{ox} is the oxide capacitance, C_{lf} is low frequency capacitance, C_{hf} is high frequency capacitance, and q is electron charge. By measuring the C–V curves at high-low frequencies (1 MHz–1 kHz), the corresponding D_{it} could thus be extracted.

Figure 5.9a–c illustrate the multi-frequency bi-directional C–V curves at room temperature. Compared with the without passivation sample, both S-passivation and N-passivation ones show improved C–V characteristics with less frequency dispersion and hysteresis, indicating both S- and N-passivation are effective in reducing the Dit and oxide traps. Figure 5.10 shows the D_{it} distribution profiles extracted by Castagné-Vapaille method for S-passivation and N-passivation samples. For both samples, the extracted D_{it} shows a peak at E_i+0.3–0.4 eV and a tail, which extends into the InP conduction band. This trend is similar to the observations by Galatage et al. N-Passivation shows higher D_{it} in the bandgap but lower D_{it} inside the conduction band (CB). At present, the reason of such distribution is mainly attributed to the difference in anion coordination number of N and O, and further investigation is still needed to understand the detailed mechanism.

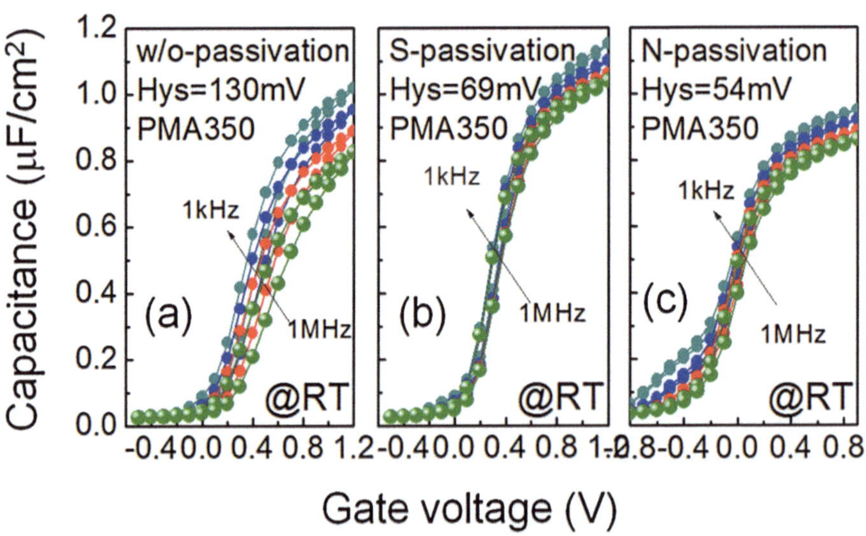

Fig. 5.9 The bi-directional C–V curves for samples with different interface passivation processes at room temperature (**a**) w/o passivation. (**b**) S-passivation (**c**) N-passivation

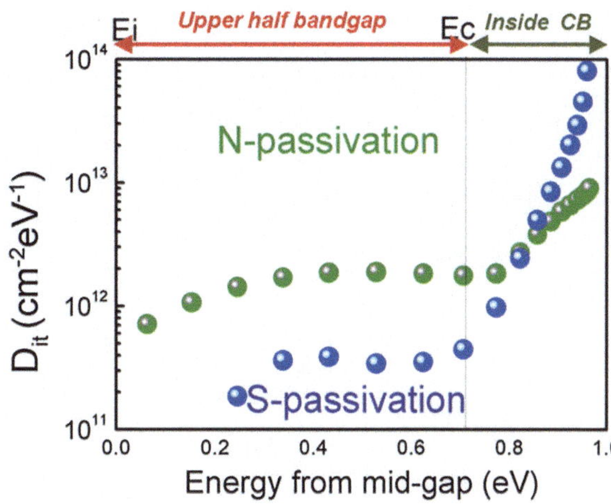

Fig. 5.10 Dit distribution for Al/Al_2O_3/n-InP MOSCAPs with S- and N-passivation

If we evaluate the results in Fig. 5.10 based on the standard of SiO_2/Si interface, the minimal D_{it} in the order of 2×10^{11} cm^{-2} eV^{-1} for S-passivation sample is acceptable when compared with the D_{it} value of N-passivation one. However, D_{it} under high electrical field, which is mainly extracted from the capacitance frequency dispersion in accumulation region, still needs further improvement.

Concerning the origin of the frequency dispersion in accumulation in high-k/III–V system, explanations are many, such as series resistance, border traps, disorder induced gap states or Maxwell–Wagner effect [12, 23, 24]. In our case, it is able to draw conclusion that series resistance effect is not the main reason because frequency dispersion with exact the same magnitude has been observed from a 350°C PMA treated Al/Al_2O_3/n-InP capacitor with high doping level (~5×10^{17} cm^{-3}), and also no ω^2 angular frequency dependence can be observed here, as depicted in Fig. 5.11.

To probe the mechanism of capacitance frequency dispersion in accumulation region, besides passivating the interface with various termination species, thermodynamic control of the process is also an effective way. Therefore, Al/Al_2O_3/InP MOS capacitors with PMA temperatures ranging from 250 to 350°C in N_2 for 30 s were prepared. Figure 5.12 illustrates room temperature frequency dependent bi-directional capacitance-voltage curves for samples with 3 nm-Al_2O_3 dielectrics without PMA and with PMA temperature of 250, 300, and 350°C, respectively. The corresponding V_{FB} are extracted to be 0.51, 0.29, 0.27 and 0.44 V from the C–V characteristics measured at 1 MHz. The C–V characteristics for PMA samples measured at 140 K are shown in the insets. Note 100 Hz C–V was not reliable for the capacitor without PMA due to a large dissipation factor (>1) during the measurement, while for all the other measurements, dissipation factor are less than 0.1.

Compared with the initial stack without PMA treatment, the ones with PMA treatment show negative shift of V_{FB}, indicating that PMA treatment is effective in reducing the negative charge. Moreover, compared with the capacitor without

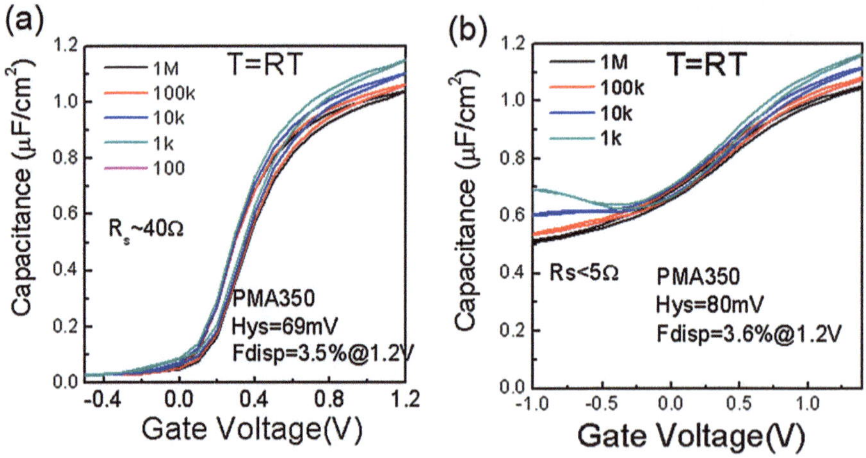

Fig. 5.11 A comparison between two 350°C PMA treated Al/Al_2O_3/n-InP capacitor with low (~5 × 10^{15} cm^{-3}) and high doping level (~5 × 10^{17} cm^{-3}), the capacitance frequency dispersion in accumulation regime are quite closer

Fig. 5.12 Room-temperature multi-frequency C–V characteristics of Al/Al_2O_3 (3 nm)/n-InP MOS capacitors without and with PMA treatment at 250, 300, and 350°C. The corresponding V_{FB} is extracted to be 0.51, 0.29, 0.27, and 0.44 V from the C–V characteristics measured at 1 MHz. The C–V characteristics of the PMA samples measured at 140 K are shown in the insets

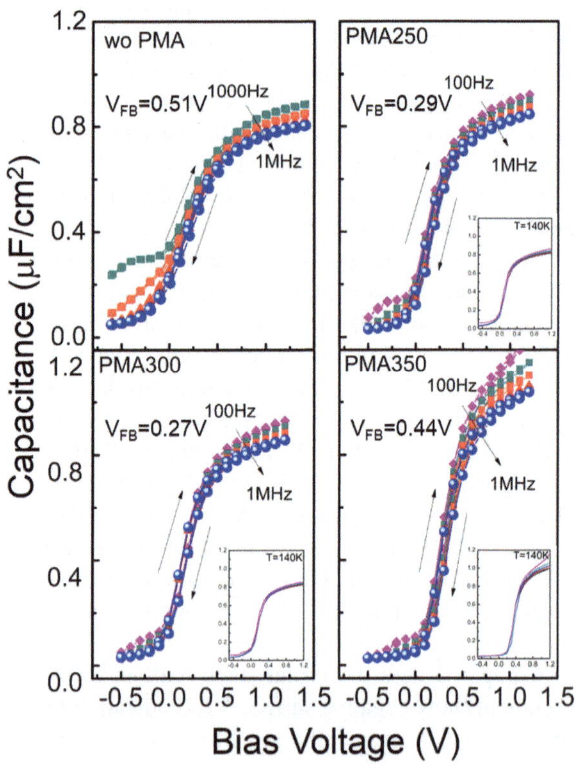

PMA, the ones with PMA treatment show small capacitance frequency dispersion near V_{FB}, indicating PMA treatment is effective in reducing the D_{it}. However, frequency dispersion in accumulation capacitance is different for each sample. Apparently from the insets, similar C–V behavior with relatively small frequency dispersion are observed at accumulation region for 250 and 300°C samples, and almost no frequency dispersion is observed for all the three samples in the depletion region. Concerning the accumulation capacitance difference between the 350°C PMA sample and the other ones at 1 MHz, we are inclined to believe that the increasing of accumulation capacitance could be attributed to the coupling of some border traps distributed inside the oxide with very small time constant and thus even cannot be fully frozen out at 140 K, which agrees well with the distributed circuit model proposed by Chen et al. [23]. For 250 and 300°C PMA cases, low temperature measurements indicate that these devices exhibit true accumulation, and the room temperature C–V measurements are not dominated by trap response.

To quantitatively discuss the effect of PMA treatment against D_{it}, Castagné-Vapaille method is employed. The energy distributions of D_{it} extracted for no PMA and 250, 300, 350°C PMA for the Al_2O_3/InP interface are depicted in Fig. 5.13. Moreover, compared with the no PMA capacitor, which shows a flatten distribution from mid-gap to conduction band edge with magnitude of 1.5–2.5 × 10^{12} cm^{-2} eV^{-1}, the ones received PMA treatment show obvious improvement in interface quality with Dit magnitude of 1.2–6 × 10^{11} cm^{-2} eV^{-1} across the upper bandgap, where the 300°C PMA sample demonstrates the lowest average D_{it} with a minimum value of 1.2 × 10^{11} cm^{-2} eV^{-1} near the mid-gap. For the trap density inside the conduction band extracted from the accumulation region, the 350°C PMA sample show much higher D_{it} than the 250 and 300°C ones.

On the basis of the above results, low PMA temperature has been proved to be effective to reduce the capacitance frequency dispersion as well as keep the minimal D_{it} in the range of 1–4 × 10^{11} cm^{-2} eV^{-1}. In order to further understand the detailed mechanism about the above trend, thermal desorption spectroscopy

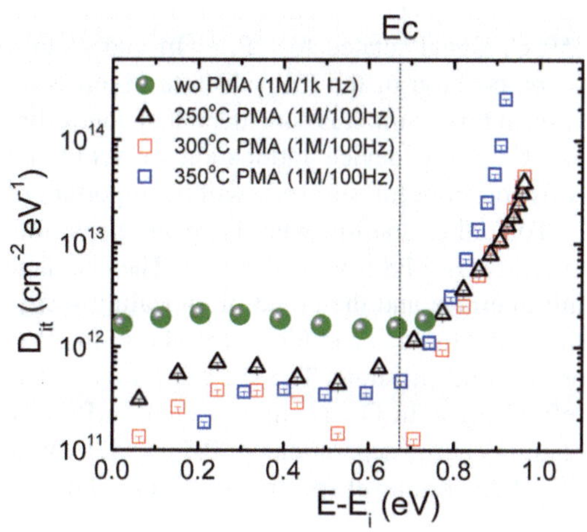

Fig. 5.13 D_{it} distribution for Al/Al_2O_3 (3 nm)/n-InP without and with PMA treatment at 250, 300, and 350°C, respectively. The extracted D_{it} shows a peak at E_i+0.3 eV and a tail, which extends into the InP conduction band. 300°C PMA sample demonstrates the lowest average D_{it} with a minimum value of 1.2 × 10^{11} cm^{-2} eV^{-1} near the mid-gap

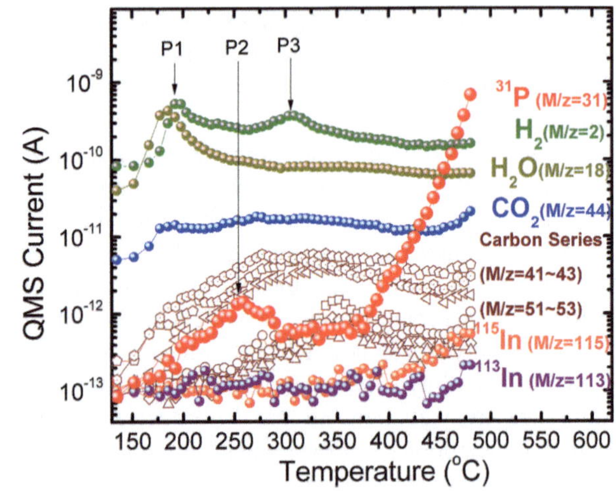

Fig. 5.14 TDS spectra of various desorption species from the Al/Al_2O_3 (3 nm)/n-InP stack without PMA treatment. Sweeping rate is kept to be 60°C/min during the whole measurement. Note mass number corresponding to 41–43 and 51–53 here are ascribed to the signal from carbon series contamination (denoted by open symbols), M/z=2, 18, 31, 44, 113 and 115 are regarded as H_2, H_2O, ^{31}P, CO_2, ^{113}In, and ^{115}In, respectively

(TDS) and x-ray photoelectron spectroscopy (XPS) measurements were performed. Figure 5.14 illustrates the TDS spectra of various desorption species (mass numbers of 2, 18, 31, 41–43, 51–53, 113, 115) from the Al/Al_2O_3 (3 nm)/InP stack without PMA treatment. Note mass numbers corresponding to 41–43 and 51–53 here are ascribed to the signal from carbon series contaminations (denoted by open symbols), and M/z=2, 18, 31, 44, 113 and 115 are regarded as H_2, H_2O, ^{31}P, CO_2, ^{113}In, and ^{115}In, respectively. Although 200°C in situ PDA treatment has been performed immediately after Al_2O_3 deposition, water desorption with a peak temperature (denoted as P1 in Fig. 5.14) still exists. This is attributed to the water absorption in terms of Van der Waals way during the exposure before metal evaporation. By elevating the annealing temperature to 250°C, a peak related to ^{31}P desorption exists (denoted as P2 in Fig. 5.14), together with desorption of carbon series contaminations. When annealing temperature goes up to 300°C, a peak which comes from H_2 (denoted as P3 in Fig. 5.14) appears, together with desorption of carbon series contaminations. While by further increasing the temperature to over 350°C, signal related to ^{31}P, ^{113}In and ^{115}In drastically increases, implying the decomposition of the InP substrate in terms of atom emission. Therefore, on the basis of this result, it is concluded that controlling the processing temperature below 350°C during device fabrication to avoid the negative effect caused by atom emission from InP substrate will be important to realize high-performance devices.

To further confirm what happens inside the stack, looking into the Al_2O_3/InP interface by XPS is very effective. The chemical composition of the Al_2O_3 (3 nm)/InP interface and the effect of annealing were investigated by XPS. Figure 5.15a shows the In $3d_{5/2}$ and Fig. 5.15b shows the P 2p core level spectra for 3 nm Al_2O_3 on an n-InP substrate. The In 3d spectra are deconvoluted into four parts including InP substrate, In_2O_3, a combination of $In(PO_3)_3$ and $InPO_4$, and an InP-O_x state with binding energy separations of 0.5 eV, 1.1 eV, and 1.7 eV from the bulk InP peak. The P 2p core level spectra are deconvoluted into $InPO_4$, $In(PO_3)_3$ and P_2O_5 with

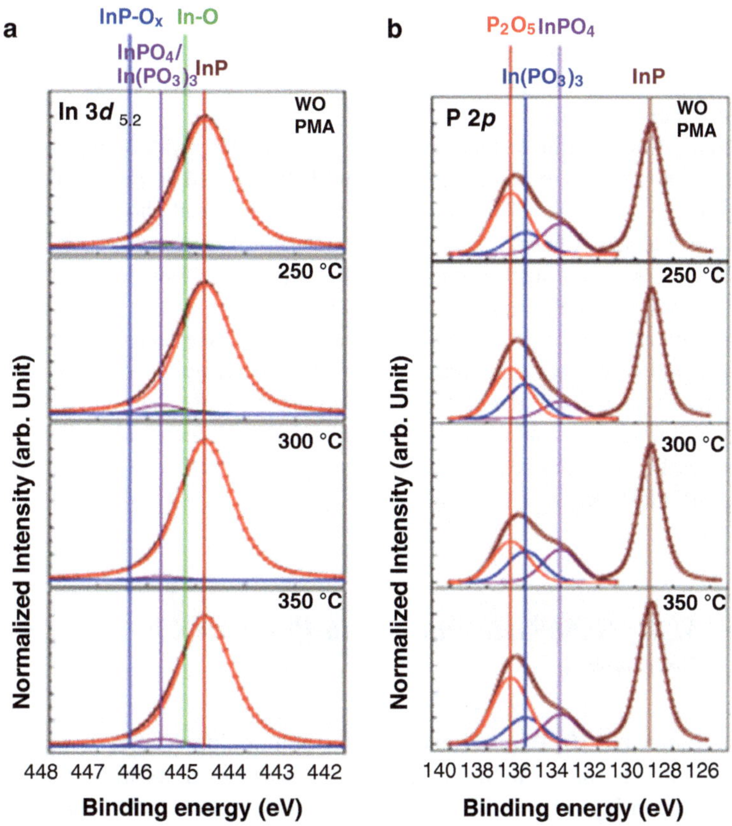

Fig. 5.15 (**a**) In 3d and (**b**) P 2p core level spectra for samples with varying ultra-high vacuum annealing temperature for 30 s, respectively. Smoothed raw data are plotted in dotted lines. The In 3d spectra are deconvoluted into four parts including InP substrate, In_2O_3, a combination of In $(PO_3)_3$ and $InPO_4$, and an InP-O_x state with binding energy separations of 0.5 eV, 1.1 eV, and 1.7 eV from the bulk InP peak. The P 2p core level spectra are deconvoluted into $InPO_4$, $In(PO_3)_3$ and P_2O_5 with binding energy separations of 4.7 eV, 6.6 eV and 7.4 eV from the InP bulk peak, respectively

binding energy separations of 4.7 eV, 6.6 eV and 7.4 eV from the InP bulk peak, respectively [25]. In Fig. 5.15a, by elevating the annealing temperature, the peak related to In-O decreases and fully eliminated after 300°C, meanwhile the signal from $InPO_4$ and $In(PO_3)_3$ is also found to be the smallest when compared with the other annealing temperatures. Moreover in Fig. 5.15b, although the trends of $InPO_4$ and $In(PO_3)_3$ are hard to be described individually, the total intensity of the oxidized states shows first a decrease and reaches its minimum at 300°C, followed by an increase with further increasing the annealing temperature. In specific, this trend could be obviously reflected from the shift of intensity of P_2O_5 against annealing temperature. By taking D_{it} characteristics as well as the TDS spectra above into consideration, these observations are found to be quite consistent with each other and could be well explained by considering the trade-off between the phosphorus-related desorption below 300°C and oxide generation above 350°C from interfacial reaction.

At first, after the high-k deposition and 200°C in situ PDA treatment, there are still some carbon contamination and residual P_2O_5 existed at the surface, interface and inside the dielectrics. These contamination and residual phosphorus oxides are harmful to electrical behavior of the MOS capacitor in the direction of causing capacitance frequency dispersion, large D_{it}, and higher leakage current density. When PMA treatment is applied at relatively low temperature (250–300°C), the carbon contamination and residual phosphorus oxides are effectively removed from the stack while the oxide generation from the interfacial reaction is negatively small. However, when PMA temperature becomes higher than 350°C, compared with the reduction of contamination and residual phosphorus oxides, the generation of interfacial oxide becomes much dominant, resulting in the net degradation of the high-k/InP interface.

5.3.2 *Al_2O_3/GaSb Gate Stacks with Post-gate CF_4 Passivation*

Among all of the III–V materials, gallium antimonide (GaSb) is an attractive channel material for p-MOSFET application because of its high hole mobility over Si. Recently, various pre-gate interface passivation techniques have been developed in GaSb gate stack, such as in-situ hydrogen plasma exposure, sulfur passivation, and amorphous Si passivation. These surface passivation techniques can effectively reduce the interface state densities [26–28]. To further eliminate the bulk defects, fluorine (F) incorporation into the high-κ gate dielectric on Si, Ge, and III–V semiconductors, has been widely investigated. By forming strong metal-F bonds, high-k dielectric on Si, Ge, InGaAs and InP substrates are effectively passivated. At the same time, the formation of high binding energy bonds, such as Si-F and Ge-F can improve the interface quality [29, 30]. Therefore, following the scheme in Sect. 5.3.1, in this subsection, the impact of post-gate CF_4 passivation on Al_2O_3/GaSb is investigated.

N-type GaSb (100) wafers with a doping concentration of ~2×10^{17} cm^{-3} were used for MOS capacitor fabrication. All wafers were sequentially immersed in acetone, ethanol and de-ionized (DI) water to remove organic contamination, finally dried by compressed N_2 blowing. The native oxide was removed by a cyclic rinsing between diluted HCl and DI water. After that, an Al_2O_3 gate dielectric layer of about 10 nm was then deposited by ALD at a substrate temperature of 300°C. Some samples were transferred to reactive ion etching (RIE) chamber, a mixed flow of CF_4 (50sccm) and O_2 (5sccm) gas was introduced to fluorinate the sample for 3 minutes. Control samples without CF_4 plasma treatment were also fabricated as

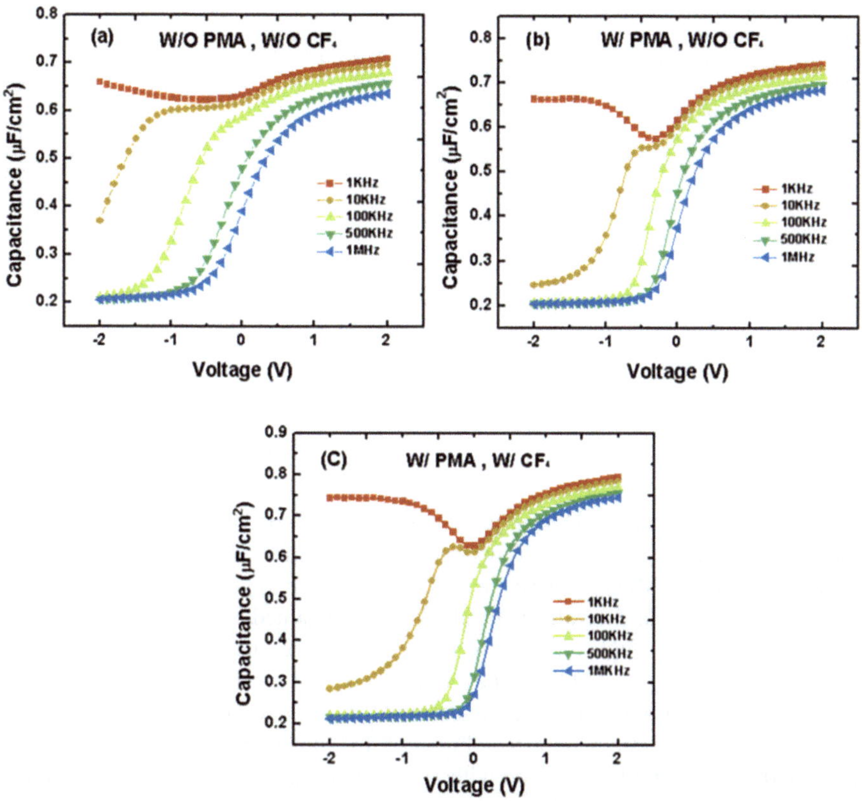

Fig. 5.16 C–V characteristics of MOS capacitors for samples (**a**) with neither CF_4 plasma nor PMA treatment (W/O CF_4, W/O PMA), (**b**) with PMA treatment only (W/O CF_4, W/ PMA) and (**c**) with both CF_4 plasma and PMA treatment (W/CF_4, W/PMA)

references. Post deposition annealing (PDA) was then performed at 350°C in N_2 ambient for 30 s. After electron beam evaporation of Ti/Au gate electrodes, PMA was selectively performed at 300°C in N_2 ambient for 30 s. Finally, about 200 nm Al film was deposited on the back side of GaSb substrates for ohmic contact.

Figure 5.16a–c show the room temperature frequency dependent C–V curves of the samples without any treatment, with PMA only, and with both CF_4 plasma and PMA, respectively. As shown in Fig. 5.16a, bumps are observed in the inversion region even at high frequency and large frequency dispersion exists in accumulation region. After PMA, as depicted in Fig. 5.16b, the response is observed in the moderate frequency and the frequency dispersion decreases, indicating improved interface quality. As shown in Fig. 5.16c, the C–V characteristics are further improved for the samples with both CF_4 plasma and PMA treatment. Compared with the samples without any post-gate treatment, the frequency dispersion in accumulation region has reduced from 10.3% to 4.9%, which indicates a better interface.

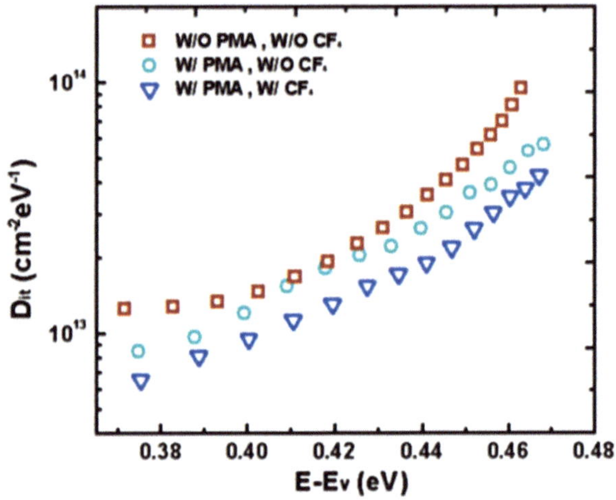

Fig. 5.17 Dit distribution for samples without any post-gate treatment, with PMA only, and with both F passivation and PMA

In order to evaluate the D_{it} distribution, Castagné-Vapaille method was used. As shown in Fig. 5.17, reduction of D_{it} was achieved by PMA and post-gate CF_4 plasma treatment. For the MOS interface with both post-gate treatments, a D_{it} value of 6.5×10^{12} cm^{-2} near the valence band edge was obtained, while 8.5×10^{12} cm^{-2} for with PMA only sample and 1.24×10^{13} cm^{-2} for the one without any treatment. The reduction of D_{it} by PMA can be explained by the mechanism, that an enough thermal energy rearranges the bonds at the Al_2O_3/GaSb interface and dangling bonds throughout the film are passivated. Figure 5.17 shows the gate leakage current characteristics of samples with and without CF_4 plasma treatment. In the accumulation region, samples with CF_4 plasma treatment exhibit lower leakage currents, indicating the reduction of trap assisted tunneling. With the F atoms incorporation, defects states at Al_2O_3/n-GaSb interface were further passivated.

The high frequency (1 MHz) C–V hysteresis characteristics of 10 nm Al_2O_3/n-GaSb gate stack with and without post-gate treatments are illustrated in Fig. 5.18. A large hysteresis (~310 mV) is observed for sample with neither PMA nor CF_4 plasma treatment. After PMA, the hysteresis is reduced to ~160 mV, and sample with both PMA and CF_4 plasma treatment shows a smaller hysteresis (~110 mV). The reduction of hysteresis for sample with post-gate treatment is related to the "border traps", which can reveal the slow traps in high-κ film near the interface region with the substrate and/or interface region between the high-k film and the interfacial oxide. The ΔN_{bt} of the MOS capacitors with different postgate treatments are shown in Fig. 5.19. It is clearly observed that PMA and postgate CF_4 plasma treatment can effectively decrease the ΔNbt. The sample with neither CF_4 plasma nor PMA treatment shows a largest ΔN_{bt} of 6.25×10^{11} cm^{-2}, while 5.26×10^{11} cm^{-2} for samples with both postgate treatments. Therefore, the slow traps in Al_2O_3 are decreased by using the PMA and postgate CF_4 plasma treatments.

In order to understand the effect of post-gate CF_4 plasma treatment on the properties of Al_2O_3 gate dielectric, XPS analysis was performed. Figure 5.20a, b

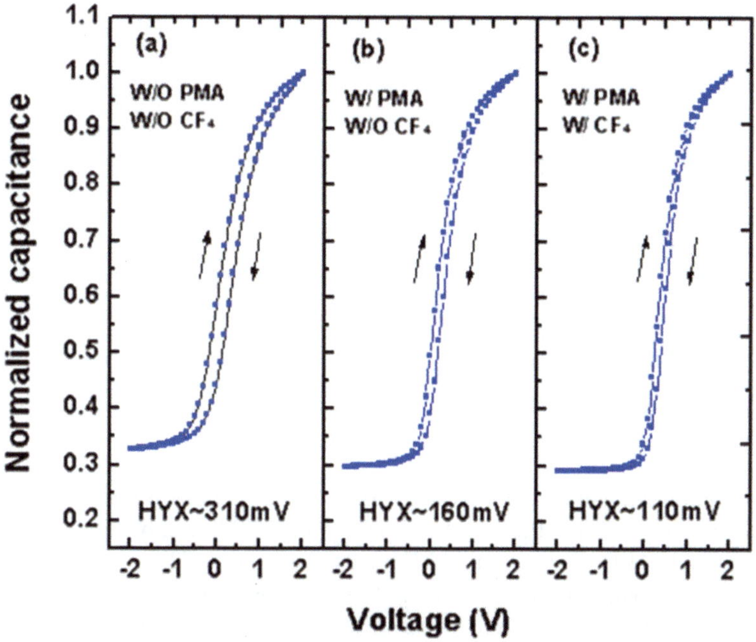

Fig. 5.18 C–V hysteresis characteristics of 10 nm Al_2O_3/n-GaSb gate stacks (**a**) W/O CF_4, W/O PMA, (**b**) W/O CF_4, W/PMA and (**c**) W/CF_4, W/PMA

Fig. 5.19 Total effective border trap density (ΔN_{bt}) for samples without any post-gate treatment, with PMA only, and with both F incorporation and PMA

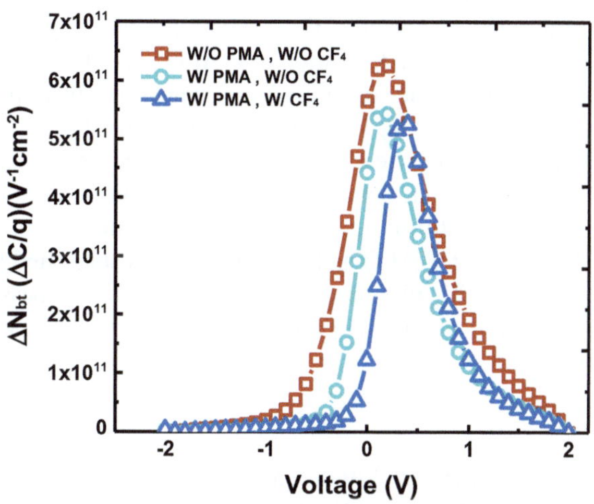

plot the XPS spectra obtained from 3 nm-thick Al_2O_3/n-GaSb structures. Figure 5.20a shows the F 1 s spectra of samples without and with 3 min CF_4 plasma treatment, respectively. The F 1 s peak located at ~686.5 eV corresponds to the F-Al bonds in the bulk Al_2O_3, indicating that F is incorporated into the gate stack after

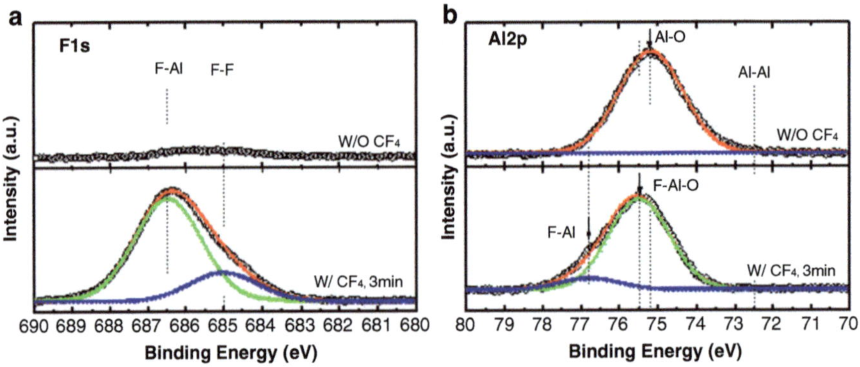

Fig. 5.20 F 1 s and Al 2p XPS spectra for Al_2O_3/n-GaSb stacks without and with 3 min CF_4 plasma treatment

CF_4 plasma treatment and none of F 1 s peak is observed for sample without CF_4 plasma treatment. The Al 2p spectra shown in Fig. 5.20b can be divided into four peaks: Al-Al, Al-O, F-Al-O and Al-F bonds. The one at 72.5 eV is attributed to typical Al-Al bonds and the other at 75.2 eV is assigned to Al-O bonds. With CF_4 plasma treatment, F-terminated bonding at Al atoms forms F-Al-O bonds. For sample with 3 min plasma treatment the binding energy moves to 75.5 eV and Al-F peak is also observed at binding energy of 76.8 eV, which agree well with the values reported by K. M. Fan, et al. [31]. Therefore, with the CF_4 plasma treatment, Al-O bonds were replaced by F-Al-O and Al-F bonds. Judging from these results, the further improvement of GaSb MOS interfaces by post-gate CF_4 plasma treatment can be owe to the formation of F-Al-O and Al-F bonds. Oxygen vacancy in Al_2O_3 bulk, and interface dangling bonds were effectively passivated by F incorporation.

In summary, we have systematically investigated the effects of PMA and post-gate CF_4 plasma treatment on Al_2O_3/n-GaSb MOS capacitors. Experimental results have shown that both methods could improve the interface properties. The XPS analysis conforms the F incorporation into the gate dielectric and proves the formation of F-Al-O and F-Al bonds. It is suggested that PMA in N_2 ambient and post-gate CF_4 plasma treatment can effectively passivate the interface and bulk oxide traps and they are very promising to improve the quality of the gate dielectric/ GaSb MOS stacks

5.3.3 Raman Spectroscopy in Al_2O_3/GaAs Interface Characterization

Compared with the studies in InP and GaSb, there are many groups working on high-k/GaAs and high-k/InGaAs gate stacks using electrical characterizations. For example, Prof. P. D. Ye's group from Purdue University has published results on ALD-grown dielectrics on GaAs with various substrate orientations; while Prof.

S. Takagi's group from the University of Tokyo has reported many research progresses in the field of surface channel InGaAs MOSFETs with ALD gate dielectrics [15–18, 20]. Therefore, as compensation to the investigation in high-k/GaAs system, in this subsection, we take Al_2O_3/GaAs stack as an example to introduce the application of Raman spectroscopy for interface characterization.

For Al_2O_3/GaAs stacks, due to the existence of oxide and interface charges, the energy band of GaAs surface will be bent. The band bending is defined as "surface barrier", which will shift the V_{FB} from its ideal value. As we have demonstrated in Sect. 5.3.1, the amount of charge could be extracted from C–V measurements. Different from electrical characterization, the surface barrier can be also measured by Raman spectroscopy.

Intrinsic GaAs(100) and GaAs(111) wafers were used in this study. After organic cleaning using acetone, and alcohol (5 min each), wafers were cleaned by sequential immersion in 10% HCl solution for 1 min, 25% NH_4OH solution for 5 min. After that, some samples were rinsed in 20% $(NH_4)_2S_x$ solution for 15 min at room temperature followed by 50 nm ALD-Al_2O_3 deposition. Then, PDA treatments were applied at 450–650°C for 10 min in N_2 ambient. Finally, Raman spectroscopy measurements were performed using 473 nm laser source, and the penetration depth is about 70 nm.

Figure 5.21a–d show the Raman spectra of GaAs(100) and GaAs(111) with various PDA treatments. It is obvious from Fig. 5.20a, b that the intensities of transverse optical (TO) mode (291.3 cm^{-1}) and longitude optical (LO) mode (268.6 cm^{-1}) are different for the substrates with (100) and (111) orientations [32, 33]. And the intensity ratio between LO mode and TO mode, I(LO)/I(TO), becomes larger when PDA temperature increases, as depicted in Fig. 5.20c, d. The surface depletion thickness of GaAs(100) and GaAs(111) δ could be calculated by the following equation [34, 35]:

$$\frac{I(LO)}{I(TO)} = \frac{I_0(LO)}{I_0(TO)} \times \frac{\left(1 - e^{-2\delta/D}\right)}{e^{-2\delta/D}} \tag{5.7}$$

where $I_0(LO)$ and $I_0(TO)$ are the intensities of LO mode and TO mode without any surface passivation, D is the penetration depth of laser. In our case, we use the data of $I_0(LO)/I_0(TO)$ from Ref. [34] for calculation, and the calculation results of depletion thickness are shown in Fig. 5.22a, b.

Therefore, the surface barrier height of GaAs(100) and GaAs(111) at the Al_2O_3/GaAs interface could thus be calculated by Eq. (5.8),

$$V_I = \frac{eN_D\delta^2}{2\varepsilon_0\varepsilon_s} \tag{5.8}$$

where e is the unit charge, N_D is the doping concentration of GaAs wafer, ε_0 is the permittivity of free space, ε_s is the semiconductor dielectric constant. For intrinsic wafer, the intrinsic carrier density of GaAs is used for calculation.

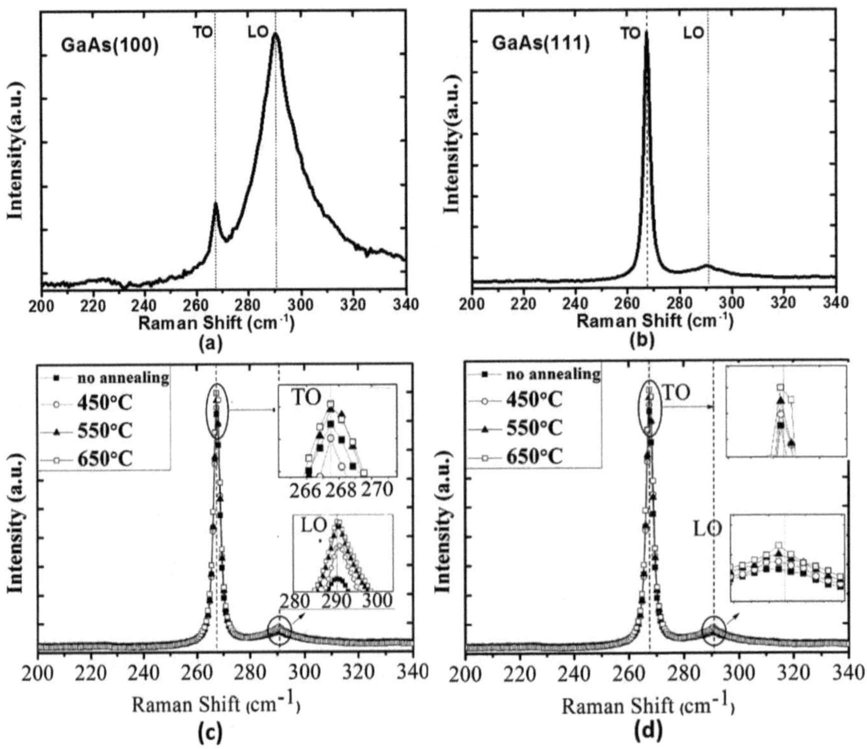

Fig. 5.21 Raman spectra of GaAs:(**a**) for unannealed GaAs(100), (**b**) for unannealed GaAs(111), (**c**) for annealed GaAs(100) with different temperature and (**d**) for annealed GaAs(111) with different temperature where the inset is the enlarged view of the peak

Figure 5.23a, b show the surface barrier heights for GaAs after PDA with different temperatures on the reciprocal of thermodynamic temperature for GaAs (100) and GaAs(111). On the basis of the results in Figs. 5.22 and 5.23, the depletion thickness and surface barrier height at Al_2O_3/GaAs interface become larger when PDA temperature increases. Since the oxide charge is proportional to the surface barrier height, therefore, the results in Fig. 5.23 indicate that additional charge is generated in Al_2O_3/GaAs system when PDA temperature is increased. We infer that although S-passivation is effective in terminating the surface dangling bonds and repairing the vacancies, high temperature PDA process destroys the S-passivation layer and forms new defects at the Al_2O_3/GaAs interface. Moreover, compared with GaAs(100), GaAs(111) case shows lower surface barrier height in each PDA temperature, indicating a better interface quality. This trend is consistent with the reports by Ye's group [36]. Figure 5.24a, b show the schematics of the surface atoms configuration of GaAs samples after S-passivation for GaAs(100)

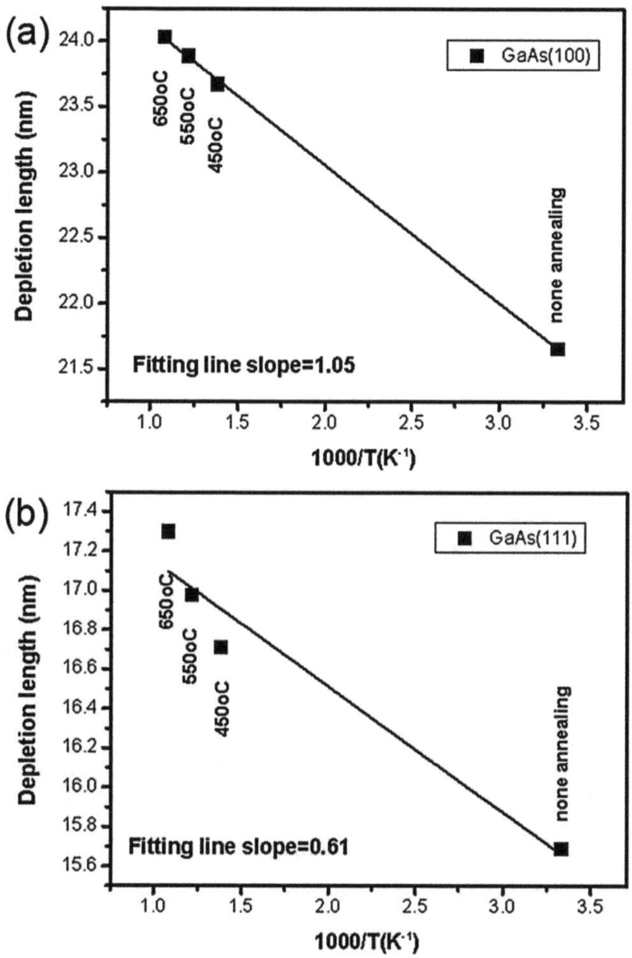

Fig. 5.22 Dependence of the depletion layer thickness for GaAs after annealing with different temperature on the reciprocal of thermodynamic temperature; (**a**) for GaAs(100) and (**b**) for GaAs (111)

and GaAs(111), respectively. We infer the better interface quality for GaAs(111) is closely related to the surface area density of dangling bonds. As illustrated in Fig. 5.24, the GaAs(100) surface contains more dangling bonds than GaAs(111), which cannot be fully terminated by S-passivation due to the existence of As on the surface. The interfacial As may form As-S bonds, which is believed to be less stable than Ga-S bond upon thermal treatment [37], resulting in a larger interface charge density and surface barrier height.

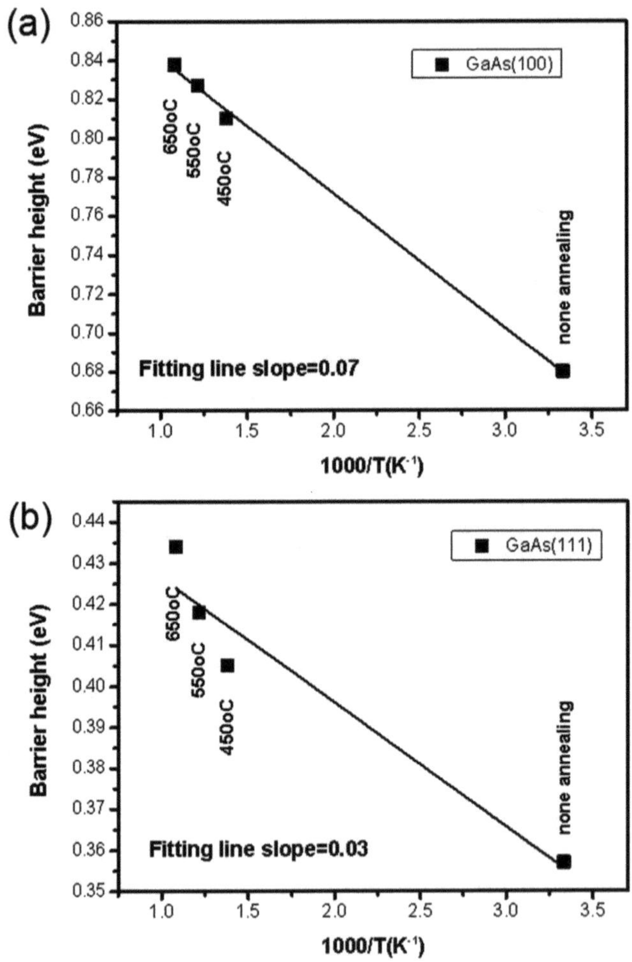

Fig. 5.23 Dependence of the surface barrier heights for GaAs after annealing with different temperature on the reciprocal of thermodynamic temperature; (**a**) for GaAs(100) and (**b**) for GaAs(111)

5.4 Summary

In this chapter, impacts and mechanisms of sulfur-passivation, nitrogen-passivation, CF_4 post-gate passivation on high-k/III–V stacks are introduced and discussed. Sulfur-passivation with low PMA temperature has been proved to be effective in reducing D_{it} as well as the traps related to capacitance frequency dispersion for Al_2O_3/InP system, while nitrogen-passivation shows better impacts on reducing fixed charge and border traps in the oxide. For high-k/GaSb system, CF_4 post-gate treatment is demonstrated to be positive in both interface and border trap reduction. For high-k/GaAs system, compared with GaAs(100), GaAs(111) shows less

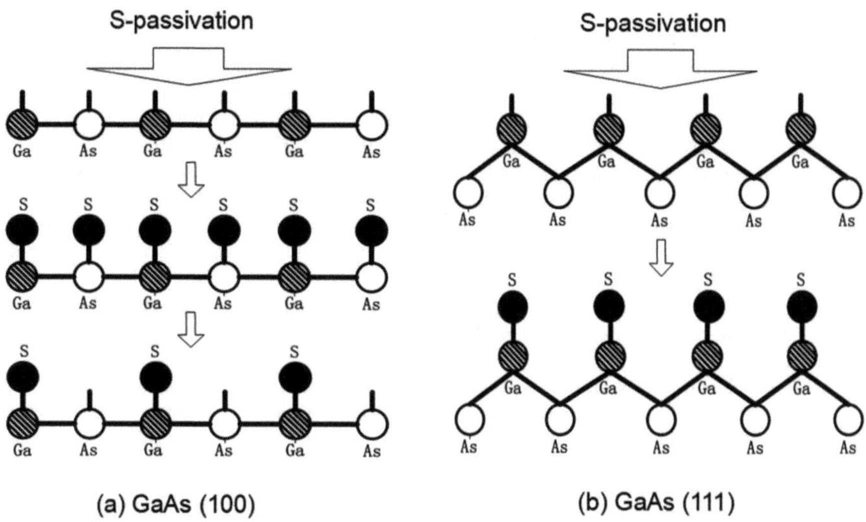

Fig. 5.24 Schematic of the surface atoms configuration of GaAs samples after S-passivating; (**a**) for GaAs(100) and (**b**) for GaAs(111)

interface charge density upon post deposition annealing, which could be explained by considering the different surface geometry between GaAs(100) and (111). Moreover, C–V measurements for extraction of fixed charge, border trap, interface trap in high-k/III–V stacks are systemically introduced and investigated. At last, application of Raman spectroscopy in evaluation of high-k/III–V interface in terms of surface barrier height is also given.

Acknowledgments This work was supported by the National Science and Technology Major Project of China under Grant No. 2011ZX02708-003, and the Opening Project of Key Laboratory of Microelectronics Devices and Integrated Technology, Institute of Microelectronics, Chinese Academy of Sciences and the authors are grateful to Miss Yuyu Su for her revision.

References

1. G.E. Moore, Cramming more components onto integrated circuits (Reprinted from Electronics, pg 114–117, April 19, 1965). Proc. IEEE **86**(1), 82–85 (1998)
2. H. Becke, R. Hall, J. White, Gallium arsenide MOS transistors. Solid State Electron. **8**(10), 813in7819–818in8823 (1965)
3. H. Becke, J. White, Gallium arsenide FET's outperform conventional silicon MOS devices (GaAs epitaxial layer in transistors with silicon nitride insulator surpasses silicon MOS device in power gain, frequency response and temperature range). Electronics **40**, 82–85 (1967)
4. L. Messick, Power gain and noise of InP and GaAs insulated gate microwave FETs. Solid State Electron. **22**(1), 71–76 (1979)
5. K. Kamimura, Y. Sakai, The properties of GaAs-Al_2O_3 and InP-Al_2O_3 interfaces and the fabrication of MIS field-effect transistors. Thin Solid Films **56**(1–2), 215–223 (1979)

6. H. Casey Jr., A. Cho, E. Nicollian, Use of oxygen-doped AlxGa1- xAs for the insulating layer in MIS structures. Appl. Phys. Lett. **32**(10), 678–679 (1978)
7. R. Dingle, H. Störmer, A. Gossard, et al., Electron mobilities in modulation-doped semiconductor heterojunction superlattices. Appl. Phys. Lett. **33**(7), 665–667 (1978)
8. T. Mimura, S. Hiyamizu, T. Fujii, et al., A new field-effect transistor with selectively doped GaAs/n-AlxGa1-xAs heterojunctions. Jpn. J. Appl. Phys. **19**(5), L225 (1980)
9. E. Yablonovitch, C. Sandroff, R. Bhat, et al., Nearly ideal electronic properties of sulfide coated GaAs surfaces. Appl. Phys. Lett. **51**(6), 439–441 (1987)
10. M. Passlack, M. Hong, J. Mannaerts, Quasistatic and high frequency capacitance–voltage characterization of Ga_2O_3–GaAs structures fabricated by insitu molecular beam epitaxy. Appl. Phys. Lett. **68**(8), 1099–1101 (1996)
11. T. Lin, H. Chiu, P. Chang, et al., High-performance self-aligned inversion-channel In0.53Ga 0.47As metal-oxide-semiconductor field-effect-transistor with Al_2O_3/Ga_2O_3 (Gd_2O_3) as gate dielectrics. Appl. Phys. Lett. **93**, 033516 (2008)
12. M. Passlack, N. Medendorp, R. Gregory, et al., Role of Ga_2O_3 template thickness and gadolinium mole fraction in GdxGa0.4-xO0.6/Ga_2O_3 gate dielectric stacks on GaAs. Appl. Phys. Lett. **83**, 5262 (2003)
13. L. Dong, X. Wang, J. Zhang, et al., GaAs enhancement-mode NMOSFETs enabled by atomic layer epitaxial as dielectric. IEEE Electron Device Lett. **34**(4), 487–489 (2013)
14. P. Ye, G. Wilk, J. Kwo, et al., GaAs MOSFET with oxide gate dielectric grown by atomic layer deposition. IEEE Electron Device Lett. **24**(4), 209–211 (2003)
15. Y. Xuan, Y. Wu, P. Ye, High-performance inversion-type enhancement-mode InGaAs MOSFET with maximum drain current exceeding 1 A/mm. IEEE Electron Device Lett. **29** (4), 294–296 (2008)
16. J.J. Gu, X. Wang, H. Wu, et al., Variability improvement by interface passivation and EOT scaling of InGaAs nanowire MOSFETs. IEEE Electron Device Lett. **34**(5), 608–610 (2013)
17. S.H. Kim, M. Yokoyama, R. Nakane, et al., High Performance Tri-Gate Extremely Thin-Body InAs-On-Insulator MOSFETs With High Short Channel Effect Immunity and Vth Tunability, IEEE Trans. Electron Devices **61**(5), 1354–1360 (2014)
18. S. Kim, M. Yokoyama, R. Nakane, et al., High-performance InAs-on-insulator n-MOSFETs with Ni-InGaAs S/D realized by contact resistance reduction technology. IEEE Trans. Electron Devices **60**(10), 3342–3350 (2013)
19. M. Cao, Z. Lin, S.K. Wang, et al., Impact of InP surface nitridation on interface properties and leakage current properties of Al_2O_3/InP metal-oxide-semiconductor capacitors. Chin. J. Vac. Sci. Technol. **36**(1), 110–116 (2016)
20. T. Haimoto, T. Hoshii, S. Nakagawa, et al., Fabrication and characterization of metal-insulator-semiconductor structures by direct nitridation of InP surfaces. Appl. Phys. Lett. **96** (1), 012107 (2010)
21. S.-K. Wang, M. Cao, B. Sun, et al., Reducing the interface trap density in Al_2O_3/InP stacks by low-temperature thermal process. Appl. Phys. Express **8**(9), 091201 (2015)
22. R. Castagne, A. Vapaille, Apparent interface state density introduced by the spatial fluctuations of surface potential in an MOS structure. Electron. Lett. **22**(6), 691–694 (1970)
23. H.-P. Chen, J. Ahn, P.C. McIntyre, et al., Comparison of bulk-oxide trap models: lumped versus distributed circuit. IEEE Trans. Electron Devices **60**(11), 3920–3924 (2013)
24. T. Sawada, H. Hasegawa, Anomalous frequency dispersion of MOS capacitors formed on n-type GaAs by anodic oxidation. Electron. Lett. **18**(12), 471–473 (1976)
25. R. Galatage, H. Dong, D. Zhernokletov, et al., Effect of post deposition anneal on the characteristics of HfO_2/InP metal-oxide-semiconductor capacitors. Appl. Phys. Lett. **99**(17), 172901 (2011)
26. L.B. Ruppalt, E.R. Cleveland, J.G. Champlain, et al., Atomic layer deposition of Al_2O_3 on GaSb using in situ hydrogen plasma exposure. Appl. Phys. Lett. **101**(23), 231601 (2012)

27. Z. Lian-Feng, T. Zhen, W. Jing, et al., Improved interfacial and electrical properties of GaSb metal oxide semiconductor devices passivated with acidic $(NH4)_2S$ solution. Chin. Phys. B **23** (7), 078102 (2014)
28. P. Nagaiah, V. Tokranov, M. Yakimov, et al., In situ deposited HfO_2 with amorphous-Si passivation as a potential gate stack for high mobility (In) GaSb-based p-MOSFETS. ECS Trans. **41**(3), 223–230 (2011)
29. W.C. Wu, C.-S. Lai, T.-M. Wang, et al., Carrier transportation mechanism of the structure with silicon surface fluorine implantation. IEEE Trans. Electron Devices **55**(7), 1639–1646 (2008)
30. R. Xie, W. He, M. Yu, et al., Effects of fluorine incorporation and forming gas annealing on high-k gated germanium metal-oxide-semiconductor with GeO2 surface passivation. Appl. Phys. Lett. **93**(7), 3504 (2008)
31. K.M. Fan, C.S. Lai, H.K. Peng, et al., Improvements on interface reliability and capacitance dispersion of fluorinated ALD-Al_2O_3 gate dielectrics by CF_4 plasma treatment. J. Electrochem. Soc. **155**(3), G51–G55 (2007)
32. D. Spirkoska, G. Abstreiter, A.F.I. Morral, Size and environment dependence of surface phonon modes of gallium arsenide nanowires as measured by Raman spectroscopy. Nanotechnology **19**(43), 435704 (2008)
33. P. Puech, G. Landa, R. Carles, et al., Strain effects on optical phonons in $< 111 >$ GaAs layers analyzed by Raman scattering. J. Appl. Phys. **82**(9), 4493–4499 (1997)
34. L. Farrow, C. Sandroff, M. Tamargo, Raman scattering measurements of decreased barrier heights in GaAs following surface chemical passivation. Appl. Phys. Lett. **51**(23), 1987 (1931–1933)
35. J. Geurts, Analysis of band bending at III–V semiconductor interfaces by Raman spectroscopy. Surf. Sci. Rep. **18**(1), 1–89 (1993)
36. Y. Wu, O. Koybasi, T. Shen, et al., Metal-oxide-semiconductor field-effect transistors on GaAs (111) A surface with atomic-layer-deposited Al_2O_3 as gate dielectrics. Appl. Phys. Lett. **94** (21), 212104 (2009)
37. Z. Lu, M. Graham, X. Feng, et al., Structure of S on passivated GaAs (100). Appl. Phys. Lett. **62**(23), 2932–2934 (1993)

Chapter 6
Low Trigger Voltage and High Turn-On Speed SCR for ESD Protection in Nanometer Technology

Jizhi Liu, Zhiwei Liu, Fei Hou, Hui Cheng, Liu Zhao, and Rui Tian

6.1 Introduction

As CMOS technology scales down to the nanometer technology, the ICs are more vulnerable to be damaged from ESD because of the thinner gate oxide, shorter channel length, shallower drain/source junction. ESD has became a main reliability concern for integrated circuit (IC) in the nanometer technology [1, 2].

There are four type ESD protection devices, such as diodes, MOSFETs, bipolar transistors and silicon controlled rectifiers (SCRs). The SCR has long been used as an on-chip ESD protection devices [3, 4], because it has much higher ESD robustness level within a smaller layout area comparing to the other ESD protection devices.

The SCR has "snapback" characteristics. The typical I–V curve and ESD design window of the SCR are shown in Fig. 6.1. When the ESD pulse voltage is lower than the trigger voltage (Vt1) of the SCR, the current of the SCR keeps small and increases slowly. Once the ESD pulse voltage is higher than Vt1, the anode voltage of the SCR starts to reduce, but the current of the SCR increases. The SCR is in the snapback state. Then, the anode voltage reduces to the holding voltage (Vh), the snapback state ends, and the anode voltage increases with the current again. When the current arrives the second breakdown current (It2), the SCR runs away. So the trigger voltage, holding voltage and second breakdown current of the SCR are important parameters for designing the ESD protection circuits.

The ESD protection circuits should be restricted in the ESD design window. In the nanometer technology, the lower limit and the upper limit of the ESD design

J. Liu (✉) • Z. Liu • F. Hou • H. Cheng • L. Zhao • R. Tian
State Key Laboratory for Electronic Thin Films and Integrated Devices,
University of Electronic Science and Technology of China, Chengdu, China
e-mail: jzhliu@uestc.edu.cn

T. Li, Z. Liu (eds.), *Outlook and Challenges of Nano Devices, Sensors, and MEMS*,
DOI 10.1007/978-3-319-50824-5_6

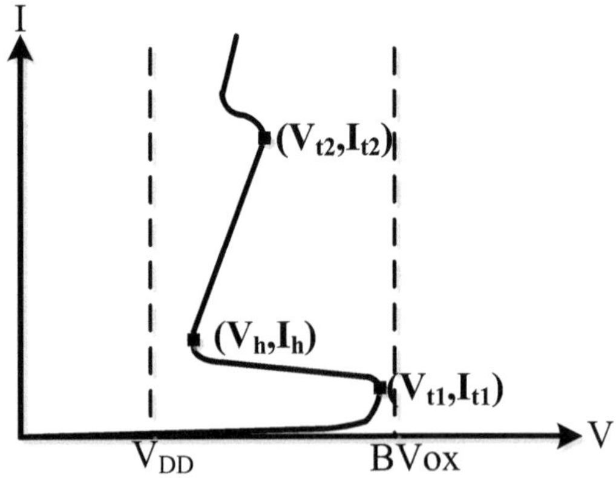

Fig. 6.1 The type I–V curve and the ESD design window of SCRs

window are the VDD operation voltage and the gate oxide breakdown voltage of the internal MOSFET devices, respectively. When the trigger voltage of the ESD protection circuit is larger than the VDD operation voltage, the ESD protection circuit is off under normal circuit operation condition and cannot effect the normal circuit operation. When the trigger voltage of the ESD protection circuit is lower than the gate oxide breakdown voltage of the internal MOSFET devices, the ESD protection circuit can protect the internal circuits without gate oxide damage. In addition, the holding voltage of the ESD protection circuit which is higher than the VDD operation voltage can prevent the latchup under normal circuit operation condition [5].

The SCR has some drawbacks, such as higher trigger voltage and low holding voltage. In the nanometer technology, the VDD operation voltage of the IC system is low enough and the low holding voltage of the SCR can be ignored. However, as the CMOS technology develops to the nanometer nodes, the gate-oxide thickness has been scaled down. The breakdown voltage of the gate oxide reduces to under 5 V shown in Fig. 6.2. The efficient ESD protection design is becoming increasingly difficult. The trigger voltage should be small enough. So how to reduce the trigger voltage of the SCR is a challenge to design the effective SCR device for ESD protection. Section 6.2 will present a novel GCSCR structure to reduce the trigger voltage.

Furthermore, the turn-on speed of the SCR is slow which can cause the overshoot voltage at the beginning of the ESD event. In the nanometer technology, the gate-oxide thickness has been scaled down, and its time-to-breakdown (t_{BD}) or charge-to-breakdown (Q_{BD}) has also be decreased in the same process. Hence, it is imperative to enhance the turn-on speed of SCR for efficiently protecting the ultrathin gate oxide from latent damage or rupture [7], especially against the fast charged device-model (CDM) ESD events [8]. In Sect. 6.3, the novel effective VSCR structure is proposed to improve the turn-on speed under the ESD event.

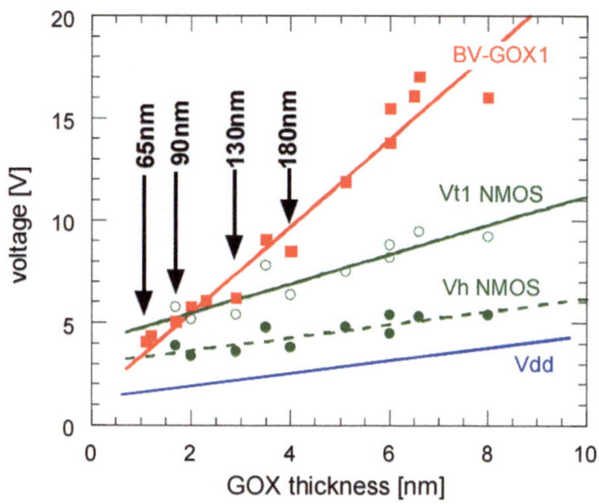

Fig. 6.2 Evolution of the breakdown voltage of the gate oxide in Bulk CMOS technologies [6]

6.2 Low Trigger Voltage SCR for ESD Protection

In order to reduce the trigger voltage of the SCR, there are two conventional methods. One is to lower the breakdown voltage of the N/P junction in the SCR structure, such as MLSCR [9] and LVTSCR [10]. But this approach has a main problem that the trigger voltage is decided by the process parameters, the trigger voltage can be only adjusted a little. The other is to inject a trigger current through external trigger circuitry, such as the diode triggered [11], substrate triggered [12], inductor triggered [13], and MOS triggered [14]. This method can effectively reduce the trigger voltage. However, the method need additional layout area to implement the external trigger circuitry. It is non area-efficient particularly in the nanometer ICs.

This section presents a robust gate-coupled silicon-controlled rectifier (GCSCR) device that provides a low trigger-voltage solution without using any external trigger circuitry and mask layer.

6.2.1 Device Structure of the GCSCR

The device cross-sectional view and equivalent circuit of the conventional LVTSCR structure are shown in Fig. 6.3. The trigger voltage of the conventional LVTSCR is decided by the breakdown voltage of the n+/pwell junction which is larger than that of the gate oxide in the nanometer technology. So this structure cannot be used to protect the interior IC circuit in the nanometer technology.

Figure 6.4 shows the device cross-sectional view and equivalent circuit of the novel GCSCR. Compared to the conventional LVTSCR, there is an additional poly-silicon gate structure on the silicon between the N+ and P+ in the Nwell. The gate is

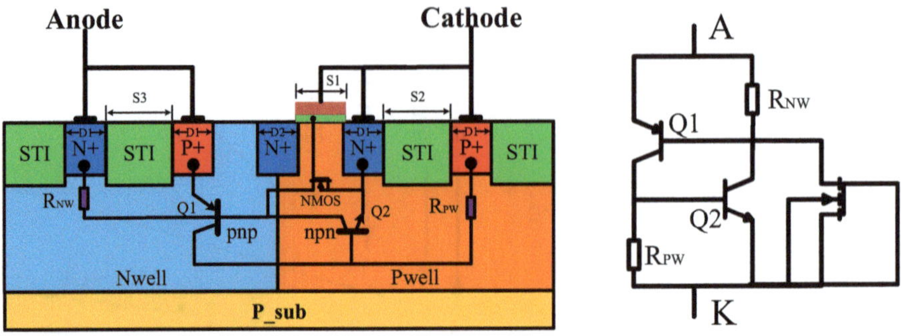

Fig. 6.3 Device cross-section of the LVTSCR structure and equivalent circuit

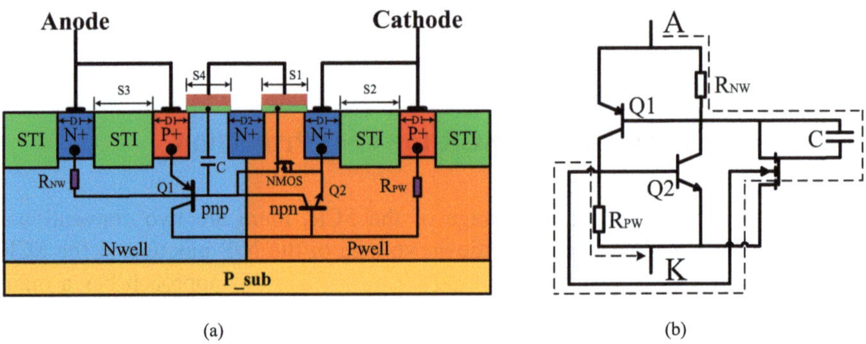

Fig. 6.4 Device cross-section of the novel GCSCR structure and equivalent circuit

connected to the gate of the parasitic NMOS. So there is a parasitic RC sub-network between the anode and the cathode in the structure of the GCSCR. The RC sub-network consists of the RNW, C, the gate capacity of the NMOS and RPW.

Once the ESD pulse arrives the anode of the GCSCR, the gate potential of the NMOS increases due to the parasitic RC sub-network. When the gate voltage of the NMOS is larger than the threshold voltage, the NMOS turns on. The channel current flows through RNW, which builds enough potential to turn on the emitter junction of the Q1, then the Q2, and finally the SCR to discharge ESD pulse. Hence, the trigger voltage of the GCSCR is decided by the parasitic RC network without external trigger circuitry.

6.2.2 *Simulation Results and Discussion*

6.2.2.1 Turn-On Process Simulation

To gain insights of GCSCR's operations, a 2-D, TLP-like numerical simulations is conducted. The rise time and pulse width of the TLP pulse is 10 and 100 ns,

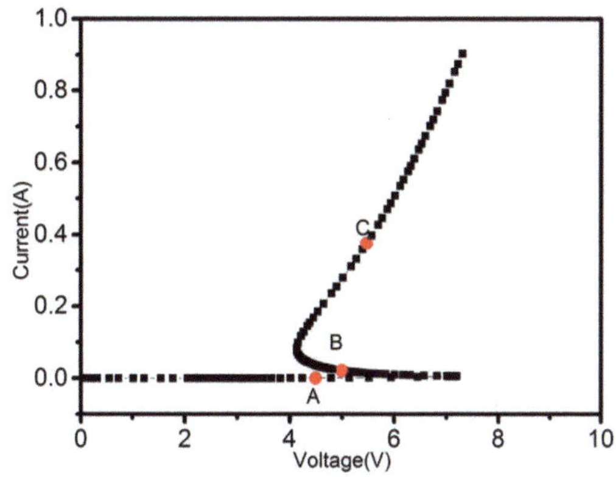

Fig. 6.5 Simulated transient ESD discharging I–V characteristic for the novel GCSCR

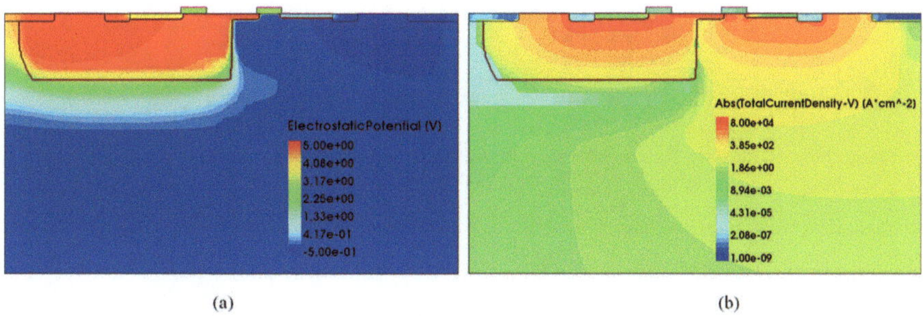

Fig. 6.6 Electric potential distribution and Current density simulation of the case A

respectively. The simulated transient ESD discharging I–V characteristic for this novel GCSCR is shown in Fig. 6.5. The turn-on physical mechanisms of the proposed GCSCR can be illustrated with the simulated electrostatic potential, and current distributions of the structure under the before turn-on, snapback and post-snapback conditions, which are correspond to cases A, B and C marked on the I–V curve as shown in Fig. 6.5, respectively.

At case A, the anode voltage is 4.5 V, the electrostatic potential distribution is shown in Fig. 6.6a. The gate voltage of NMOS is about 2.9 V obtained by the parasitic RC network. This gate voltage of the parasitic NMOS is higher than the threshold voltage, and the parasitic NMOS turns on. Figure 6.6b shows the current density distribution at case A. The anode current flows to the drain of parasitic NMOS through N-well, and then flows to the cathode of the GCSCR mainly through channel.

Figure 6.7a, b shows the electrostatic potential distribution and current density simulation of the case B, respectively. With the anode current increasing, there are enough potential difference at the emitter junction of Q1 and Q2 because of the current through the R_{NW} and R_{PW}. So the parasitic Q1 and Q2 transistors turn on, the GCSCR goes into the snapback state called as case B. Meanwhile, the potential

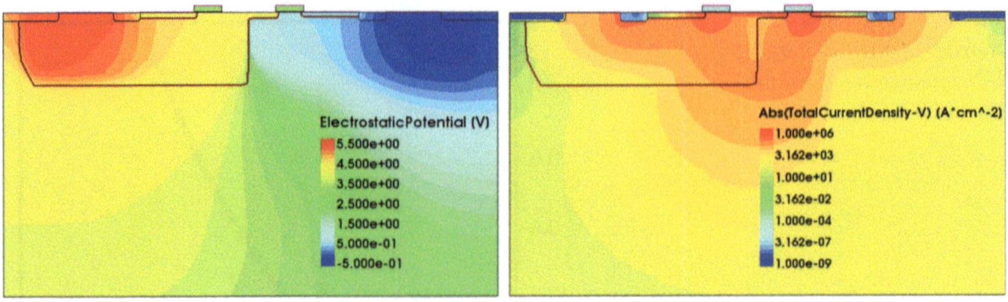

(a) (b)

Fig. 6.7 Electric potential distribution and Current density simulation of the case B

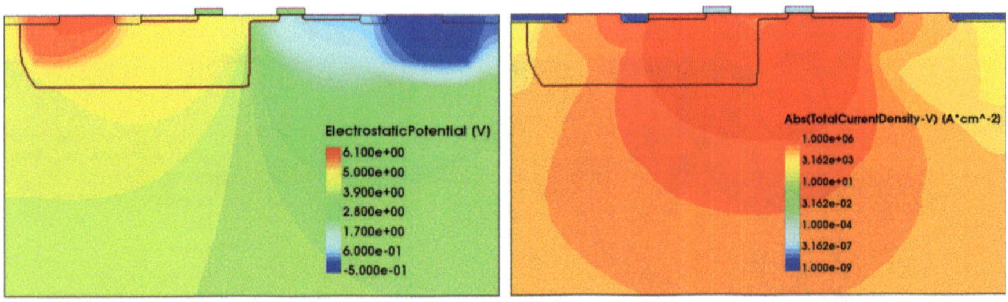

(a) (b)

Fig. 6.8 Electric potential distribution and Current density simulation of the case C

difference between the gate and substrate of the NMOS reduces. In case B, the SCR turn on, the current can discharge through both SCR and parasitic NMOS. But a large proportion of the anode current flows through the NMOS, and a small proportion of the anode current flows through the SCR.

Figure 6.8a shows the electrostatic potential distribution simulation results at case C. The potential difference between the gate and substrate of the NMOS continue to reduce. The current increases rapidly. Almost all current discharges through the SCR and there is few current in the NMOS channel shown in Fig. 6.8b.

6.2.2.2 DC Leakage Current Simulation

In order to ensure that there is few current in the GCSCR under the normal working state, the simulation for the leakage current is implemented. Figure 6.9 shows DC leakage curve of the novel GCSCR with a voltage range from 0 to 5 V. When a DC voltage applies on the anode, the parasitic RC sub-network does not work. Hence, the leakage current is mainly determined by the reverse saturation current of p-n junction. From the simulation result, when the anode voltage comes to 5 V, the

Fig. 6.9 DC leakage characteristic simulated by TCAD

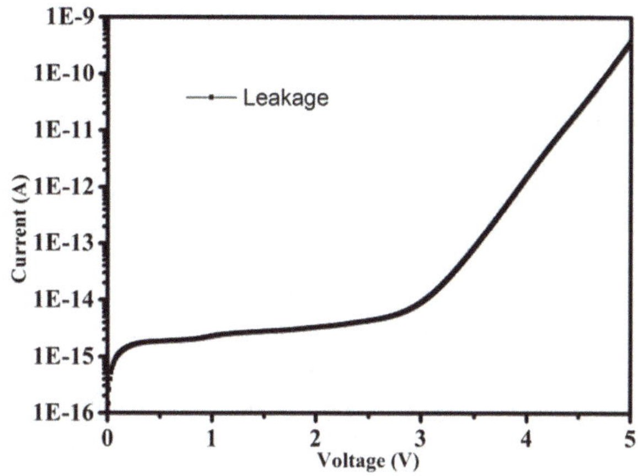

Fig. 6.10 The TLP test result of LVTSCR and GCSCR

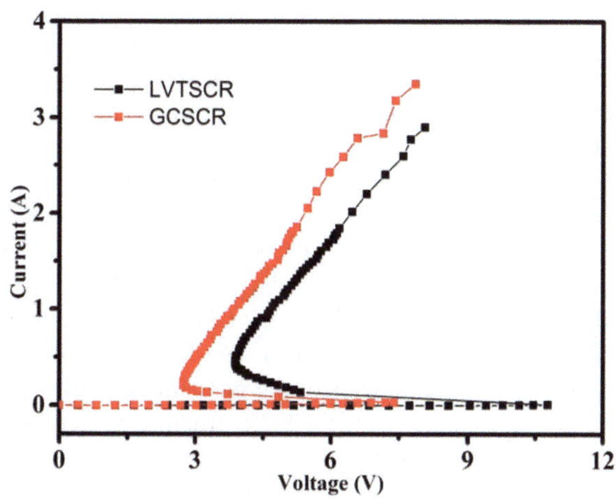

leakage current is as low as 0.4 nA. So the GCSCR cannot affect the normal working of the protected circuit.

6.2.3 Experimental Results and Discussion

The GCSCR structures have been fabricated in a BiCMOS process. For comparison, a conventional LVTSCR is also fabricated in the same process. All of the widths of SCR devices are 50 μm. The TLP current–voltage characteristics of SCRs are shown in Fig. 6.10. From the results, the GCSCR has a lower trigger-voltage (7.39 V) than the conventional LVTSCR (10.8 V). Both of them have a high ESD robustness (60 mA/μm).

The trigger voltage of the GCSCR can be adjusted by the device layout sizes. The dimension S1, S2 and S3 in the GCSCR structure shown in Fig. 6.4 are the key parameters to decide the parasitic RC network, which can adjust the trigger voltage of the GCSCR. The effects on the ESD characteristic due to these parameters are shown as follows.

The TLP test results with the value of the dimension S1 are shown in Fig. 6.11a. When the value of the dimension S1 equals 1, 2 and 4 μm, the trigger voltages of the GCSCRs is 9.08, 9.62 and 10.44 V, respectively. The trigger voltage increases with the increase of the S1 length. Because the gate capacitance of the parasitic NMOS becomes bigger when the value of S1 is increasing, the gate voltage of the parasitic NMOS became smaller, and the NMOS open more difficultly. So the trigger voltage of the GCSCR increases.

Figure 6.11b shows the TLP test results with the value of the dimension S2. By increasing the length of S2, the trigger voltage of the GCSCR varies from 9.08 V to 7.39 V. With the value of S2 increasing, the resistance RPW enlarges, and the current to forward bias the emitter junction of the parasitic Q2 transistors decreases, the SCR turn on easily. So the trigger voltage of the GCSCR decreases.

Figure 6.11c shows the TLP test results with the value of the dimension S3. When the value of the dimension S3 equals 1, 2 and 4 μm, the trigger voltages of the GCSCRs is 9.08 V, 8.71 V and 8.63 V, respectively. Because the resistance RNW enlarges, the current to forward bias the emitter junction of the parasitic Q1 transistors decreases, so the trigger voltage of the GCSCR decreases.

In this section, a novel GCSCR structure for ESD applications has been proposed to reduce the trigger voltage. The GCSCR can be triggered by the parasitic RC sub-network. The simulation and experimental results have been shown that the trigger-voltage can be reduced significantly as compared with the conventional LVTSCR. The parameter of the parasitic RC sub-network can be adjusted by the layout dimensions of the GCSCR structure. Compared with external trigger SCRs, the novel structure don't use any external triggering circuitry and can obtain the low trigger voltage.

6.3 High Turn-On Speed SCR for ESD Protection

The SCR is composed of parasitic n-p-n and p-n-p transistors, and its turn-on time is also determined by the base transit time of the parasitic p-n-p and n-p-n transistors of the SCR [15]. The turn-on mechanism of the SCR is essentially a current trigger event. The turn-on speed of the SCR can be improved by the advanced trigger-assist circuit techniques. The increasing of the triggering current introduced by the external trigger-assist circuit can reduce the trigger voltage of the SCR to improve the turn-on speed of the SCR. Furthermore, the optimization SCR structure with the trigger-assist circuit [16, 17] can also improve the turn-on speed of a SCR device. In these structures, the length of the current path is shortened by the optimization layout and the time that carriers pass from the anode to the cathode

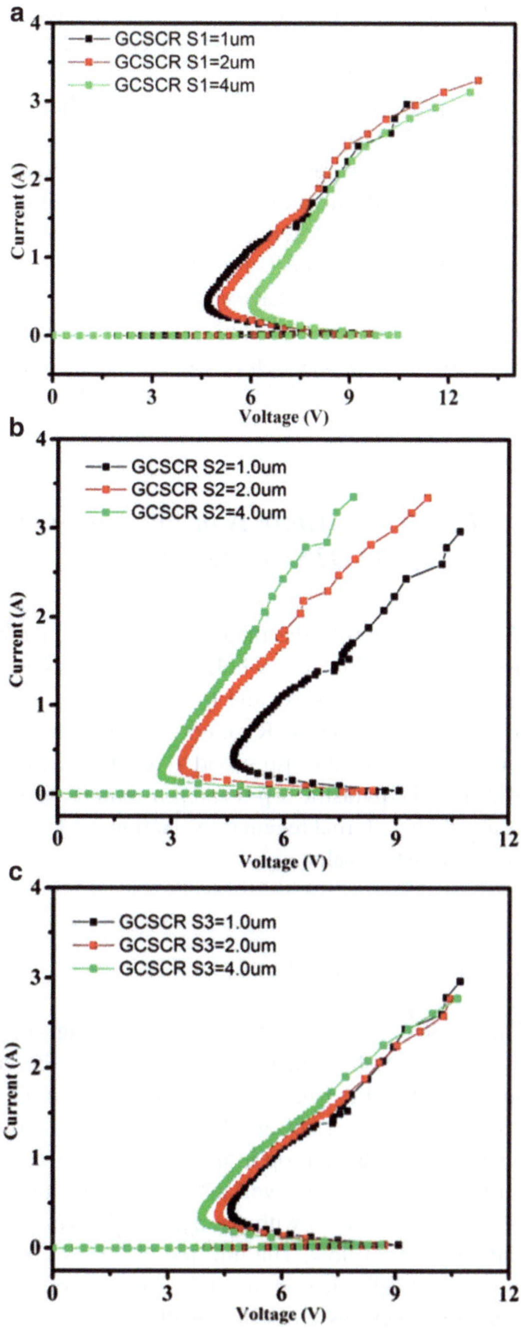

Fig. 6.11 The TLP test results with the various values of the parameters of (**a**) the dimension S1; (**b**) the dimension S2; (**c**) the dimension S3

of the SCR structure is reduced. So the turn-on time of the SCR is reduced. However, the larger triggering current of the external trigger-assist circuit costs the larger chip area.

The previous research on improving the turn-on speed focused on how to introduce the trigger current, little research were on the interior base region structure of the parasitic n-p-n and p-n-p bipolar transistors, and the base doping profiles of SCRs are almost uniform distribution.

In this section, forcing on altering the interior base region structure of the parasitic bipolar transistors, the SCR with the various lateral base doping (VLBD) structure (VSCR) is present to improve the turn-on speed of the SCR without adding extra process masks and increasing the chip area. The VLBD structure in the SCR can introduce an accelerating build-in electric field to reduce the base transit time of the bipolar transistors [16].

6.3.1 *Theory Analysis of the Turn-On Process of the DTSCR Device*

In order to simplify the theoretical analysis, the double trigger SCR (DTSCR) is used in the section. The whole turn-on process of the DTSCR is divided into two parts, the delay process and the rise process. When the trigger signal arrive the trigger nodes, the DTSCR cannot be turned on immediately, but need some time to turn on the parasitic n-p-n and p-n-p transistors. This time is called delay time td. When the parasitic n-p-n and p-n-p transistors turn on, there is the rise time t_r to establish the internal regenerative action, and then the anode current can increase to the steady-state value.

6.3.1.1 Delay Time t_d

The DTSCR device structure is shown in Fig. 6.12, with the current path indicated by the red dashed lines. The DTSCR can be triggered to the on-state by the application of the trigger currents of the PT and NT node. When the NT node is floating, the trigger current flows from the PT node to the GND through the pwell region. The voltage drop in the pwell region due to the pwell trigger current flow forward biases the n+/pwell junction, producing the injection of electrons from the n+ region connected with the cathode K. This does not immediately produce the cathode current flow. The injected electrons transport through the pwell region to the pwell/nwell junction in a finite time interval referred to as the base transit time. Once the electrons cross the pwell/nwell junction, they immediately promote the injection of holes from the p+/nwell. The injection of carriers at the p+/nwell junction initiates current flow through the device. Consequently, the anode current begins to flow after the delay time t_d which is equal to the base transit time for the

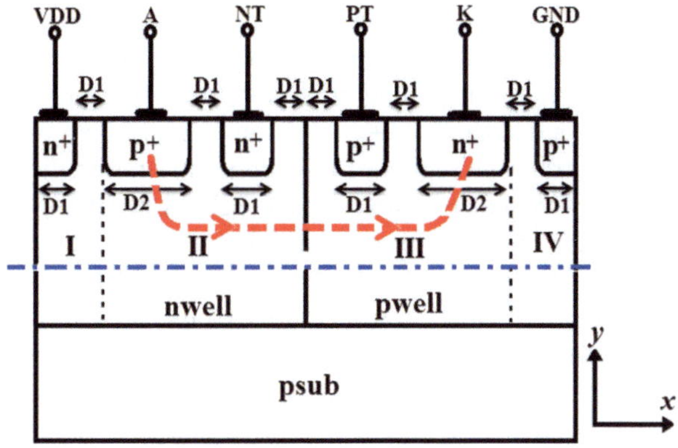

Fig. 6.12 Schematic cross section of DTSCR structure

n-p-n transistor. When the trigger currents are synchronously applied to the NT and PT nodes of the DTSCR, the delay time td equals the maximum value of the base transit times of n-p-n and p-n-p transistor.

According to the injection level of the emitter junction and the base doping profile of the p-n-p and n-p-n transistor, the base transit times can be considered under three conditions.

Low Injection Level and Uniform Base Doping Condition

Under the low injection level and uniform base doping conditions, the base transit times $\tau_{B,npn}$ and $\tau_{B,pnp}$ of the n-p-n and p-n-p transistor are given by [18, 19]

$$\tau_{B,npn} = \frac{W_{BP}^2}{2D_N} \tag{6.1}$$

$$\tau_{B,pnp} = \frac{W_{BN}^2}{2D_P} \tag{6.2}$$

Where D_N and D_P are the diffusion coefficient for electrons and holes, WBP and WBN are the base region width of the n-p-n and p-n-p.

Low Injection Levels and Non-uniform Base Doping Condition

Under the low injection levels and non-uniform base doping condition, supposing the base doping profile of the one-dimensional (1-D) n-p-n is an exponential distribution:

$$N_B(x) = N_B(0)\exp\left(-\frac{x}{C}\right) \tag{6.3}$$

where $N_B(0)$ is the base acceptor concentration at the emitter junction and C is the characteristic length of the exponential distribution of the base doping profile.

When the semiconductor is in thermal equilibrium and the base acceptor concentration decreases in +x direction as (6.3), there will be a diffusion of majority carriers from the region of high concentration to the region of low concentration. The flow of positive holes leave behind negatively charged acceptor ions. The separation of the positive and negative charges induces an electric field that is in –x direction to oppose the diffusion process. When the equilibrium is reached, the mobile carrier concentration is not exactly equal to the fixed impurity concentration and induced electric field prevents any further separation of charge [20].

Under low injection level, $p_B(x) \approx N_B(x)$, the build-in electric field E is given by

$$E = \frac{D_\mathrm{p}}{\mu_\mathrm{p}} \cdot \frac{1}{N_\mathrm{B}(x)} \cdot \frac{\mathrm{d}N_\mathrm{B}(x)}{\mathrm{d}x} \tag{6.4}$$

Combining (6.3) and (6.4), E can be written as

$$E = -\frac{kT}{qC} \tag{6.5}$$

where k is Boltzmann's constant, T is temperature and q is electronic charge. The minus sign represents the director of the electric field is opposite the +x direction

The build-in electric field is inversely proportional to the characteristic length of the exponential distribution of the base doping. The build-in electric field is an accelerating field for the minority in the base region.

Substituting (6.4) into the current density equation, the minority carrier concentration can be written by

$$n_\mathrm{B}(x) = \frac{J_\mathrm{nE} C}{q D_\mathrm{n}} \cdot \left[1 - \exp\left(\frac{x - W_{BP}}{C}\right)\right] \tag{6.6}$$

where J_{nE} is the minority carrier current in the base region.

According to the definition of the base transit time, the transit time of the n-p-n can be written by

$$\tau_\mathrm{B,npn} = \frac{{W_{BP}}^2}{2D_\mathrm{n}} \cdot \frac{2C}{W_{BP}} \left(1 - \frac{C}{W_{BP}} + \frac{C}{W_{BP}} e^{-\frac{W_{BP}}{C}}\right) \tag{6.7}$$

From this equation, τ B,npn can be reduced by the characteristic length of the exponential distribution of the base doping profile. For example, τB,npn of the DTSCR with the non-uniform base doping can be reduced to 56.8% of that with the uniform base doping when the C equals half of WBP.

The base transit time of the p-n-p transistor is the same as the above stated n-p-n transistor under this condition.

High Injection Levels

When the high injection occurs, the minority carrier concentration is not exactly equal to the majority carrier concentration and induced electric field prevents any further separation of charge. Considering an n-p-n transistor as an example, $p_B(x) \approx n_B(x) \gg N_B(x)$, the build-in electric field is given by

$$E = \frac{D_n}{\mu_n} \cdot \frac{1}{n_B(x)} \cdot \frac{dn_B(x)}{dx} \tag{6.8}$$

The build-in electric field is different from the low injection level condition, and is independent of the base doping profile.

The minority carrier current is defined by

$$J_n = qD_n \frac{dn_B(x)}{dx} + q\mu_n n_B(x)E = 2qD_n \frac{dn_B(x)}{dx} \tag{6.9}$$

This is the Webster effect which can reduce the transit time by a factor 2 [21]. So the base transit time under the high injection level are given by

$$\tau_{B,npn} = \frac{W_{BP}{}^2}{4D_N} \tag{6.10}$$

$$\tau_{B,pnp} = \frac{W_{BN}{}^2}{4D_P} \tag{6.11}$$

6.3.1.2 Rise Time t_r

An analytical model for the increase in the anode current during the turn-on process for a 1-D DTSCR structure can be derived based upon the charge control principles. The 1-D device structure of the DTSCR is shown in Fig. 6.2 which is built along the red dash-line in Fig. 6.13. Therein, JA and JK are the anode A and cathode K current of the DTSCR, JPT and JNT are the trigger currents from the NT and PT nodes, and αpnp and αnpn are the common-base current gains of the parasitic p-n-p and n-p-n transistors. This analysis takes into consideration the internal feedback mechanism between the n-p-n and p-n-p transistors within the DTSCR structure to determine the growth of the stored charge within the nwell and pwell region. Due to relatively short time for the turn-on transient compared with the lifetime, the recombination within the nwell and pwell region can be assumed to be negligible during this analysis [22, 23].

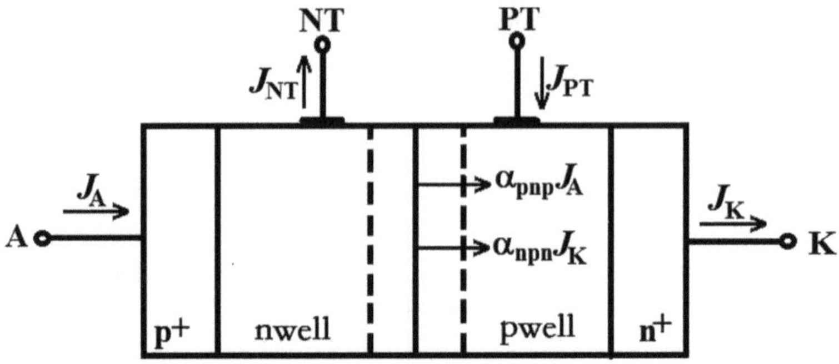

Fig. 6.13 Schematic cross section of 1-D DTSCR

The increase in the stored electrons within the nwell region (Q_{SN}) during the turn-on process occurs due to the electrons collected at the pwell/nwell junction, which are supplied by the injection from the n+ cathode (the emitter of the n-p-n transistor), and the electrons supplied by the trigger current J_{NT},

$$\frac{dQ_{SN}}{dt} = \alpha_{npn}J_K(t) + J_{NT} \tag{6.12}$$

Similarly, the increase in the stored holes within the pwell region due to the holes collected at the pwell/nwell junction and the holes supplied by the trigger current J_{PT},

$$\frac{dQ_{SP}}{dt} = \alpha_{pnp}J_A(t) + J_{PT} \tag{6.13}$$

The collector current for the n-p-n transistor is related to the stored charge in its base region via the transit time

$$J_{C,pnp}(t) = \alpha_{pnp}J_A(t) = \frac{Q_{SN}(t)}{\tau_{Bm,pnp}} \tag{6.14}$$

where $\tau_{Bm,npn}$ is the mean base transit time for the n-p-n transistor. During the turn-on process of the DTSCR, the base transit time changes from the condition under the low injection level to the condition under the high injection level. In order to simplify the derivation, the mean value of the base transit time is used during the rise time.

Similarly, the collector current for the p-n-p transistor is related to the stored charge in its base region via the transit time:

$$J_{C,npn}(t) = \alpha_{npn}J_K(t) = \frac{Q_{SP}(t)}{\tau_{Bm,npn}} \tag{6.15}$$

where $\tau_{Bm,pnp}$ is the mean transit time for the p-n-p transistor.

Combining (6.12), (6.13), (6.14) and (6.15), Q_{SN} can be given

$$Q_{SN}(t) = \sqrt{\tau_{Bm,npn}\tau_{Bm,pnp}} J_{NT} e^{\frac{t}{\sqrt{\tau_{Bm,npn}\tau_{Bm,pnp}}}} - \tau_{Bm,pnp} J_{PT} \tag{6.16}$$

Substituting this expression into (6.14) yields an equation that describes the increase in the anode current during the turn-on process:

$$J_A(t) = \frac{J_{NT}}{\alpha_{pnp}} \cdot \sqrt{\frac{\tau_{Bm,npn}}{\tau_{Bm,pnp}}} e^{\frac{t}{\sqrt{\tau_{Bm,npn}\tau_{Bm,pnp}}}} - \frac{J_{PT}}{\alpha_{pnp}} \tag{6.17}$$

The rise time is defined as the time taken by the anode current density to increase to the steady-state value. Using (6.17), the rise time can be obtained:

$$t_r = \sqrt{\tau_{Bm,pnp}\tau_{Bm,npn}} \ln\left(\frac{\alpha_{pnp} J_{A,SS} + J_{PT}}{J_{NT}} \cdot \sqrt{\frac{\tau_{Bm,pnp}}{\tau_{Bm,npn}}}\right) \tag{6.18}$$

Where $J_{A,SS}$ is the steady-state anode current density.

The rise time of the anode current of the DTSCR is the geometric mean of the mean base transit times for the n-p-n and p-n-p transistors.

6.3.2 Structure and Operation Principle

According to the above analysis, the turn-on speed of the SCR can be improved by the variation lateral base doping (VLBD) structure. The new proposed VSCR is fully process-compatible to the commercial CMOS processes without additional mask or process step in the chip fabrication. The total widths of all test SCR devices in this paper are drawn as 100 μm for performance comparison. All of the test devices occupy the same layout area.

The traditional modified lateral SCR (MLSCR) device [9] is shown in Fig. 6.14a. The MLSCR consists of a pair of parasitic p-n-p and n-p-n bipolar transistors. Figure 6.14b shows the proposed VSCR device which is a MLSCR device with a VLBD structure. Table 6.1 shows the structure parameters of the MLSCR and the VSCRs in the 0.54 μm CMOS process. D1 is the width of the p+ region in the PWELL/NWELL region and D2 is the width of the STI near the PWELL/NWELL junction. The width of the PWELL layer at the PWELL/NWELL junction is D3. The width of the NOSAPW layer is D4. The NOSAPW is self-aligned P-well implant block layer. The VLBD structure of the VSCR is in the base region of the parasitic n-p-n transistor. The VLBD structure can be constructed by using the NOSAPW layer between two PWELL regions to obtain a lightly doped P-type region near the PWELL/NWELL junction.

The simulation results of the doping profiles of the MLSCR and VSCRs along the dash-dot line AA′ by the TCAD software are shown in Fig. 6.15. The lateral

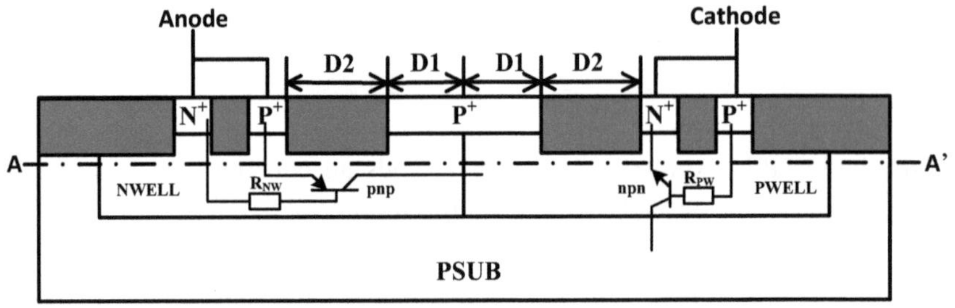

(a)

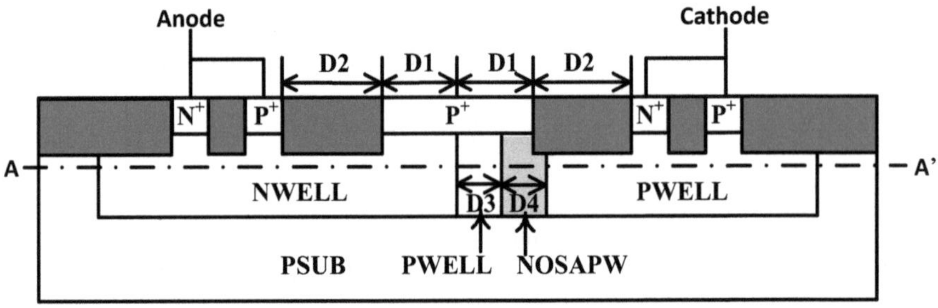

(b)

Fig. 6.14 Devices under study. (**a**) The MLSCR with the traditional uniform base doping (**b**) Proposed VSCR device with the VLBD structure. The VLBD structure can develop by using the PWELL and NOSAPW layer. The NOSAPW layer is drawn between two PWELL regions to block the PWELL implant

Table 6.1 The structure parameters of the MLSCR and the VSCRs

	D1 (μm)	D2 (μm)	D3 (μm)	D4 (μm)
MLSCR	1.5	4		
VSCR1	1.5	4	1	1
VSCR2	1.5	4	1.5	1
VSCR3	1.5	4	2	1
VSCR4	1.5	4	2.5	1

base doping profile of the parasitic n-p-n and p-n-p transistors of the MLSCR is uniform in most of the base region and varies only in a small portion of the base region near the PWELL/NWELL junction by the impurities compensation. Using the VLBD structure in the parasitic n-p-n transistors of the VSCRs, there are the various lateral base doping profiles in most of the base region of the parasitic n-p-n transistors. Once the variation lateral base doping profile of the parasitic n-p-n transistor is formed shown in Fig. 6.15, there is an accelerating build-in electric field in the base region of the parasitic n-p-n transistor to reduce the base transit time of the bipolar transistors. Furthermore, the doping profile in the base region of the parasitic transistors can be changed by the value of D3. With decreasing of the

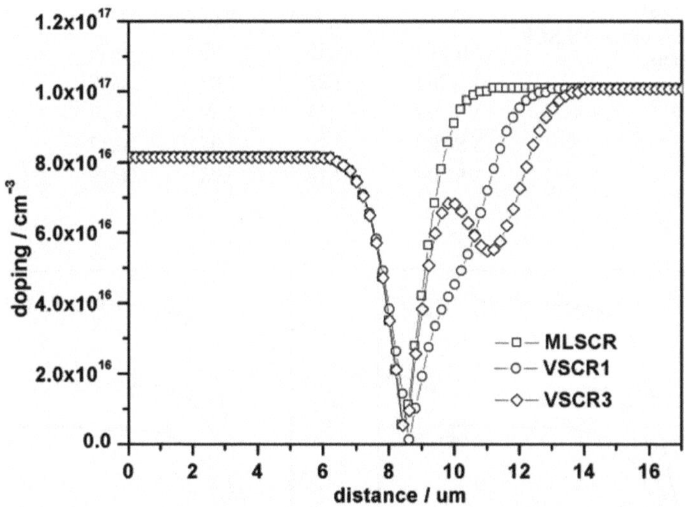

Fig. 6.15 The process simulation results of the doping profile of the MLSCR and VSCRs along the *dash-dot line* AA′ shown in Fig. 6.13 by the TCAD software

value of the D3, the lateral doping profile in the base region becomes steeper near the PWELL/NWELL junction, and then the accelerating build-in electric field can be increased in the parasitic n-p-n transistor.

6.3.3 Experimental Results

6.3.3.1 TLP Test

To investigate the ESD robustness, the samples were stressed using on-wafer Transmission Line Pulsing (TLP) stress system, with 100 ns pulse width and 10 ns rise time. The TLP tester is HED-T5000 by HANWA. The ESD performances of the samples, such as the trigger voltage (Vt1), the holding voltage (Vh), and the second breakdown current (It2), are measured shown in Table 6.2. The failure criterion is defined as the leakage current over 1 μA under the 5 V bias condition. The trigger voltages and holding voltages of the MLSCR and VSCRs are very similar. The ESD current handling capabilities of VSCRs is lower than MLSCR, about a maximum 4% drop in It2, due to the little increase of the value of the on resistance of the SCR devices by reducing the base doping concentration of the parasitic n-p-n transistors.

6.3.3.2 Turn-On Time Test

During the turn-on procedure of the device under test, for a given zap voltage, the current of the device increases from 0 to the value IDUT. The value IDUT equals

Table 6.2 ESD performances of the MLSCR and the VSCRs

	V_{t1} (V)	V_h (V)	I_{t2} (A)	HBM (kV)
MLSCR	17.15	2.99	6.34	>8
VSCR1	17.23	2.87	6.09	>8
VSCR2	17.23	2.89	6.13	>8
VSCR3	17.22	2.97	6.17	>8
VSCR4	17.18	2.98	6.19	>8

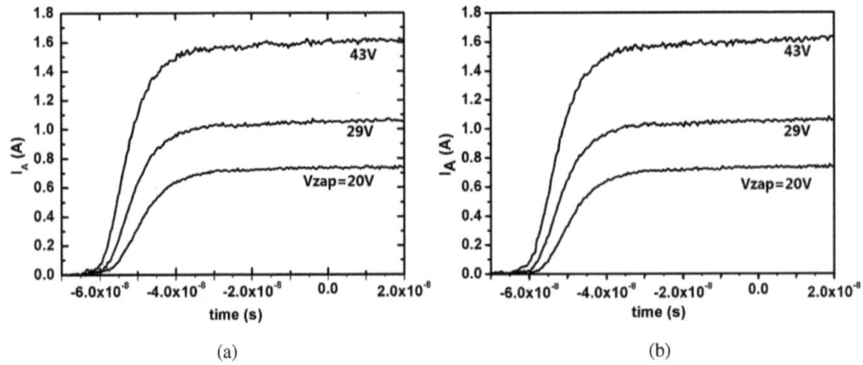

Fig. 6.16 Current waves at the different TLP zap voltage level (**a**) MLSCR; (**b**) VSCR1

the average current from 70 to 90% of the current waveform. The rise time can be defined as the time between 10 and 90% of the value of IDUT [24, 25]. The comparison of turn-on time between a traditional MLSCR and the new VSCRs is made under three zap voltages, 20, 29 and 43 V, respectively. Figure 6.16 shows the extracted turn-on time of the MLSCR and VSCRs under the different rise times and zaps voltages of the TLP pulses.

The measured current waveforms of the MLSCR and the VSCR1 under different TLP zap voltage levels are shown in Fig. 6.16a, b, where the rise time and the width of the applied pulse is 10 and 100 ns, respectively.

Figure 6.17a shows the extracted turn-on time of the SCRs when the raise time and pulse width of the TLP pulse is 10 and 100 ns, respectively. The turn-on time of the VSCRs is smaller than that of the conventional MLSCR by introducing the VLBD structure. With increasing of the TLP zap voltage levels, all of the turn-on times of the SCR devices are reduced because the increasing of the displacement current can reduce the trigger voltage of the SCRs. Furthermore, the turn-on time of the VSCR devices can be reduced with decreasing of the value of the D3. Because the lateral doping profile near the PWELL/NWELL junction becomes steeper with decreasing of the value of the D3, and then the accelerating build-in electric field increases in the base region to reduce the base transit time of the parasitic n-p-n transistor.

The extracted turn-on times of the SCRs are shown in Fig. 6.17b when the raise time and pulse width of the TLP pulse is 2 and 100 ns, respectively. The turn-on time of the VSCRs is smaller than that of the MLSCR due to the VLBD structure,

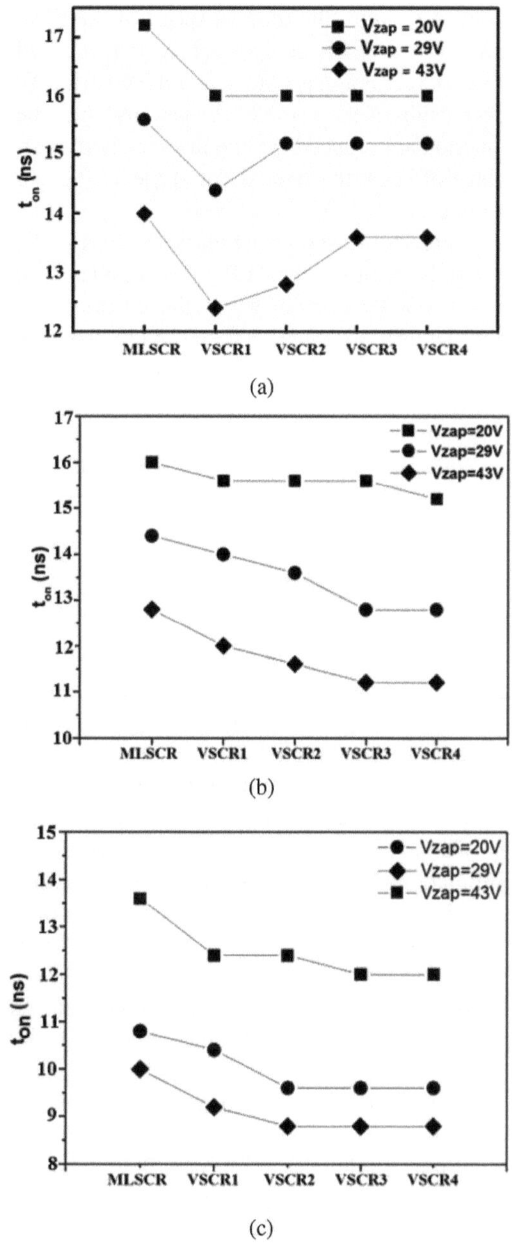

Fig. 6.17 The turn-on times of the MLSCR and VSCRs under different rise time and different zap voltages of the TLP pulses. (**a**) $t_r = 10$ ns and $t_d = 100$ ns, (**b**) $t_r = 2$ ns and $t_d = 100$ ns, and (**c**) $t_r = 0.2$ ns and $t_d = 100$ ns

and the turn-on speed of the SCRs is also improved with increasing of the TLP zap voltage levels. The turn-on times of the SCRs shown in Fig. 6.16b are smaller than that shown in Fig. 6.17a. Because the effect of the displacement current on the turn-on speed of the SCRs cannot be ignored with the increase of the value of the dv/dt of the TLP pulse. The larger displacement current can make the trigger voltage smaller, the turn-on speed of the SCRs can be improved. But the turn-on time of the VSCR devices can be reduced with increasing of the value of the D3. The lateral doping profile near the PWELL/NWELL junction becomes smaller with decreasing of the value of the D3, and then the capacities of the NWELL/PWELL junction are reduced. So the displacement current decreases with decreasing of the value of D3 at the same dv/dt of the TLP pulse, and the turn-on speed of the SCRs reduces.

Figure 6.17c shows the extracted turn-on time of the SCRs when the raise time and pulse width of the TLP pulse is 200 and 100 ns, respectively. The turn-on time of the VSCRs is smaller than that of the MLSCR with using the VLBD structure, and the turn-on speed of the SCRs improves with increasing the TLP zap voltage level. The turn-on times of the VSCR devices can be reduced with increasing of the value of D3.

From the experimental, the turn-on time of the VSCRs with the VLBD structure is 12% less than that of the MLSCR with the traditional uniform base doping.

6.3.4 Discussion

The TCAD simulations are used to investigate the turn-on process of the MLSCR and VSCRs. A well-calibrated device simulation deck is used for the 2-D TCAD simulations. Mobility, high field velocity saturation, and avalanche generation models were calibrated for the turn-on process of the SCRs. Figure 6.18 shows the comparison of the simulation result and the TLP experimental results of the I–V characteristic of the MLSCR when the raise time and pulse width of the TLP pulse is 10 and 100 ns, respectively. The triggering voltage and the holding voltage of the simulation results are same as that of the experimental results.

Figure 6.19 shows the simulated anode voltage and anode current versus time for the MLSCR under the TLP test. The stress pulse has a 25 V amplitude, a 10 ns rise time and a 100 ns pulse width. There are three cases A, B and C. At the case A, the anode voltage is large and the anode current is small, the MLSCR don't be triggered. At the case B, the anode voltage decreases and the anode current increases, the MLSCR is triggered in the snapback state. At the case C, the anode voltage and current both increase, the MLSCR goes into the steady-state. Three cases are three typical work states of the SCR. The turn-on process of the SCR will be discussed by the electronic field distributions and the hole mean velocity distributions at three cases as following.

6.3.4.1 Case A

At the case A, the SCRs are all off. There are only the displacement currents in the devices by the dv/dt of the TLP pulse. When the value of the dv/dt of the TLP pulse is small, the total currents in the SCRs are small. So the electronic fields in the base region of the parasitic n-p-n transistors of SCRs are only decided by the ionized

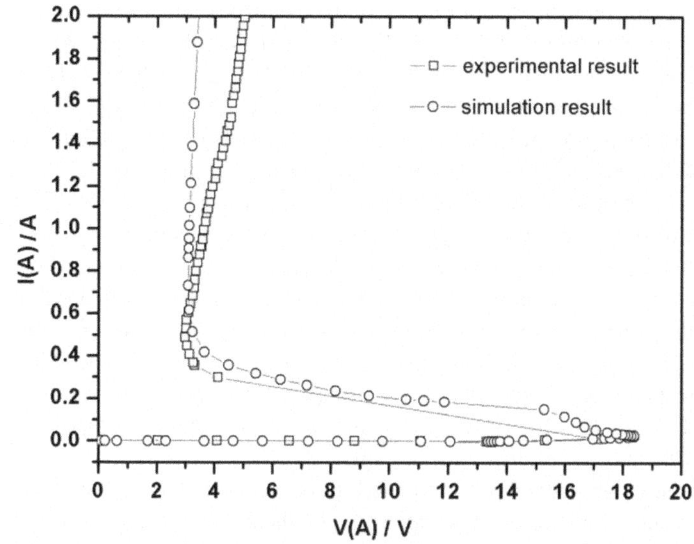

Fig. 6.18 The comparison of the simulation result by MEDICI and the TLP experimental results of the I–V characteristic of the MLSCR. In the TLP test, $t_r = 10$ ns and $t_d = 100$ ns. Calibration of mobility, high field velocity saturation, and avalanche generation models parameters for the MLSCR and VSCRs

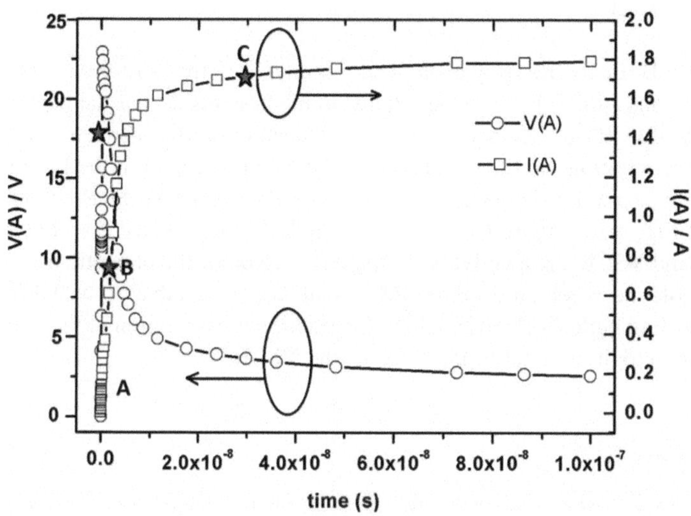

Fig. 6.19 The simulated anode voltage and anode current versus time for the MLSCR under the TLP test. The stress pulse has a 25 V amplitude, a 10 ns rising time and a 100 ns pulse width. There are three cases A, B and C

impurities concentration. From Fig. 6.20, the electronic fields in the base region of the parasitic n-p-n transistors of VSCRs are larger than that of the MLSCR due to the accelerating build-in electric field by the VLDB structure. The VLBD structures in the parasitic n-p-n transistor are marked by the black line frame. The electronic field in the base region of the parasitic n-p-n transistors of the VSCR1 is larger that of the VSCR3. Because the variation trend of the doping profile of the VSCR3 is not the same as that of the VSCR1. The doping profile of the base region of the VSCR3 decreases in the small region and increases in the other region with far away the NWELL/PWELL junction shown in Fig. 6.15. There is the retarding build-in electric field in the fractional base region and the accelerating build-in electric field in the other base region of the parasitic n-p-n transistor of the VSCR3. Figure 6.21 shows the hole mean velocity distributions of the MLSCR and VSCRs, respectively. Since the VLBD structure can introduce the build-in electric field, the hole mean velocity in the base region of the VSCRs are larger than that in the MLSCR, and can reduce with the value of D3.

When the value of the dv/dt of the TLP pulse becomes larger, the total current in the SCRs cannot be ignored. The electronic fields in the base region of the parasitic n-p-n transistors of SCRs are decided by both the ionized impurities concentration and the carrier concentration. So the effect of the accelerating build-in electric field on the turn-on process of the SCRs can be reduced with increase of the value of the dv/dt of the TLP pulse.

6.3.4.2 Case B

At the case B, the SCRs trigger on. There are large currents in the parasitic n-p-n and p-n-p transistors. The electronic fields in the base regions of the parasitic n-p-n transistors of SCRs are decided by both the ionized impurities concentration and the carrier concentration. So the difference of the electronic field in the base region of the parasitic n-p-n transistors in among the VSCRs and the MLSCR becomes small shown in Fig. 6.22. Figure 6.23 shows the hole mean velocity distribution of the MLSCR and VSCRs, respectively. In the base region of the parasitic n-p-n transistors, the hole mean velocities of the VSCRs are larger than that of the MLSCR. Due to the retarding build-in electric field in the fractional base region of the VSCR3, the hole mean velocity is smaller than that of the VSCR3.

6.3.4.3 Case C

At the case C, the currents in the SCRs are very large, the concentrations of the carriers are further larger than that of the ionized impurities. The electronic field in the base region of the parasitic n-p-n transistors of SCRs is decided by only the carrier concentration. So there are few differences of the electric field profile in the base regions among three devices shown in Fig. 6.24. So the VLBD structure cannot introduce the accelerating build-in electric field under this case C. Figure 6.25 shows the hole mean velocity distributions of the MLSCR and

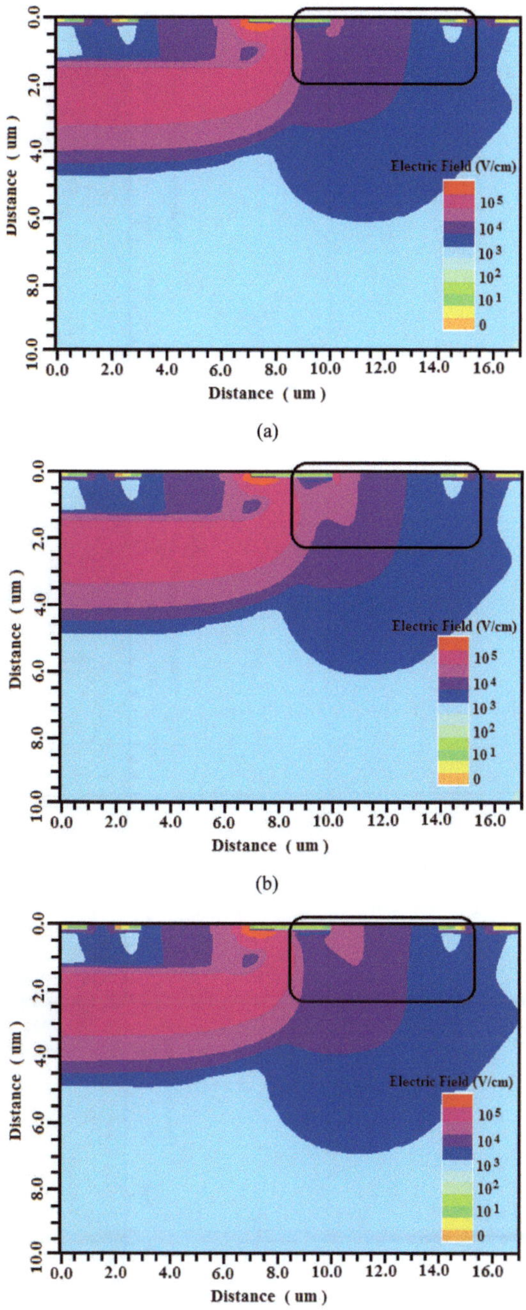

Fig. 6.20 The electronic field distributions of the (**a**) MLDSCR, (**b**) VSCR1 and (**c**) VSCR3 at the case A

Fig. 6.21 The hole mean velocity distributions of the (a) MLDSCR, (b) VSCR1 and (c) VSCR3 at the case A

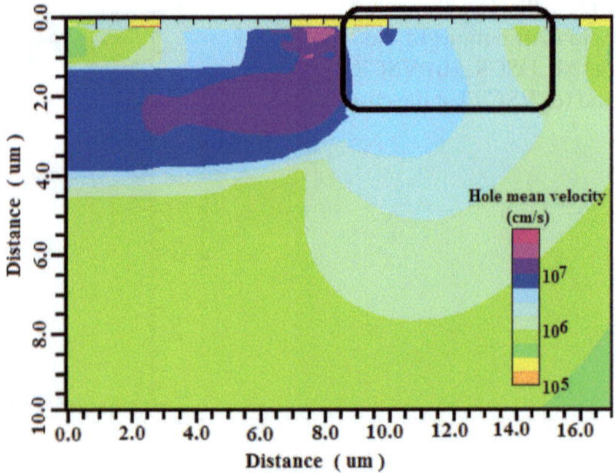

(a)

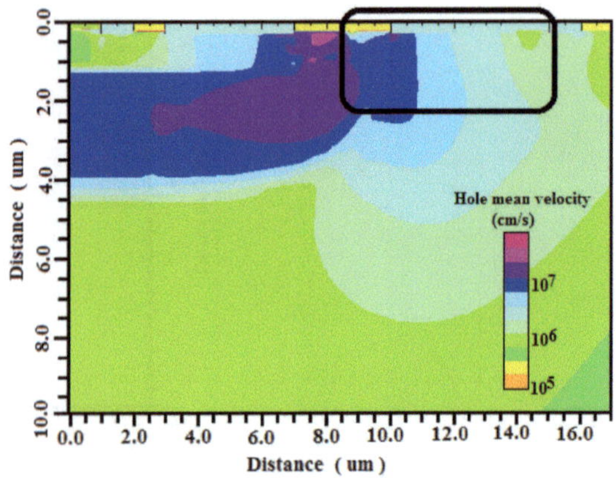

(b)

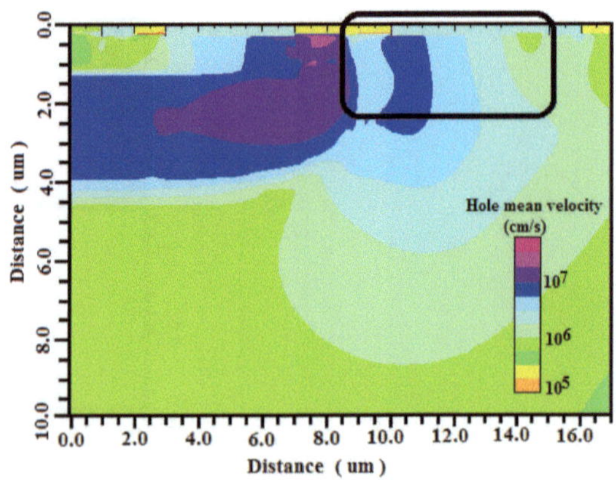

(c)

Fig. 6.22 The electronic field distributions of the (**a**) MLDSCR, (**b**) VSCR1 and (**c**) VSCR3 at the case B

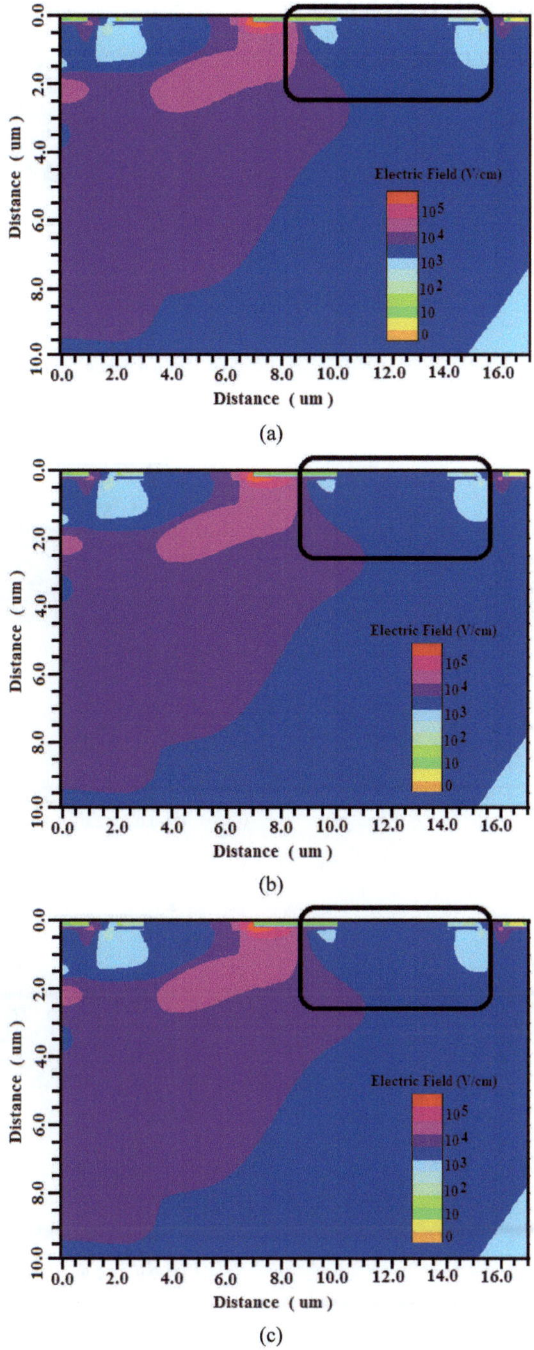

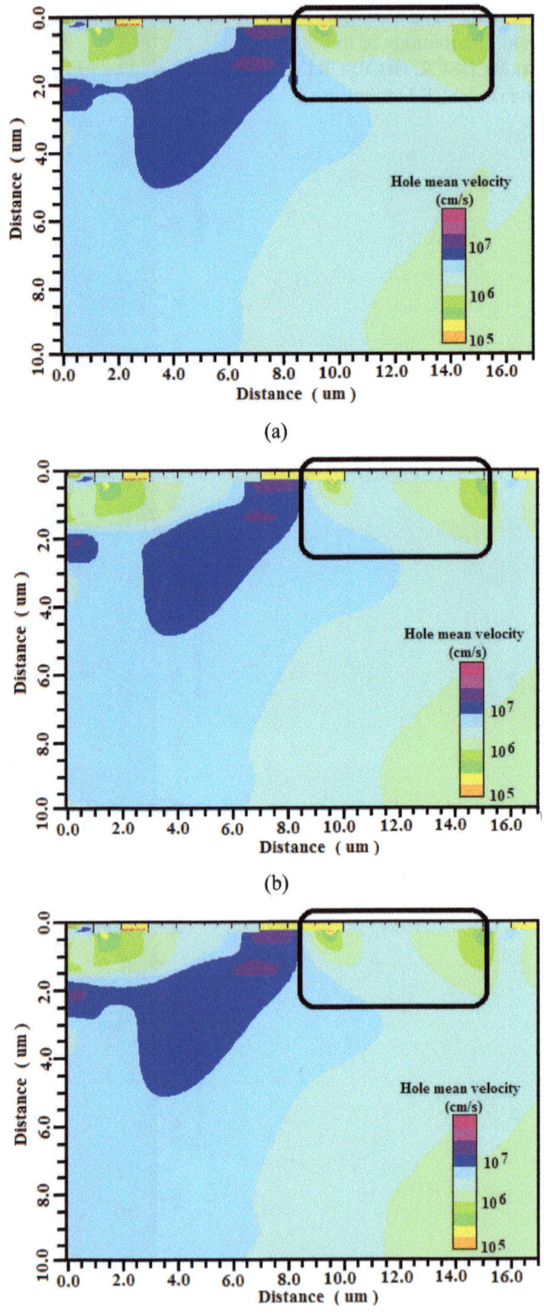

Fig. 6.23 The hole mean velocity distributions of the (**a**) MLDSCR, (**b**) VSCR1 and (**c**) VSCR3 at the case B

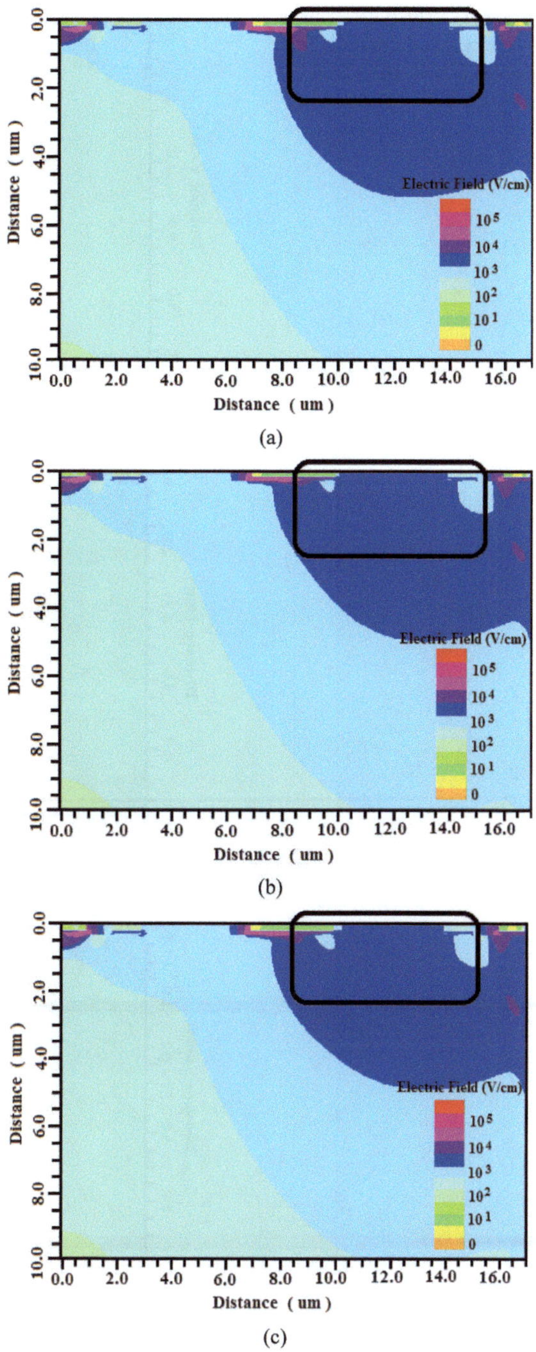

Fig. 6.24 The electronic field distributions of the (**a**) MLDSCR, (**b**) VSCR1 and (**c**) VSCR3 at the case C

Fig. 6.25 The hole mean velocity distributions of the (**a**) MLDSCR, (**b**) VSCR1 and (**c**) VSCR3 at the case C

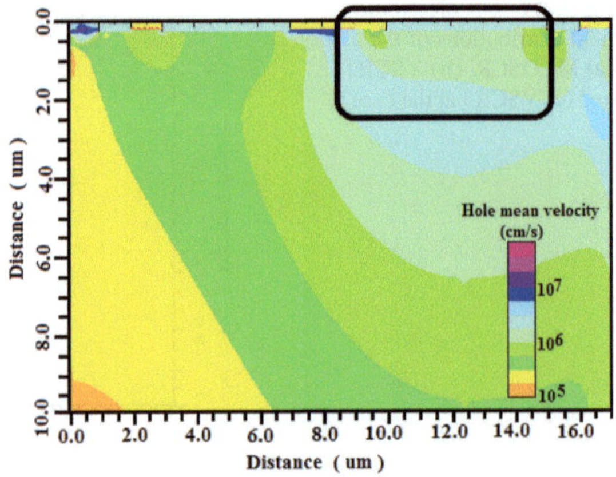

(a)

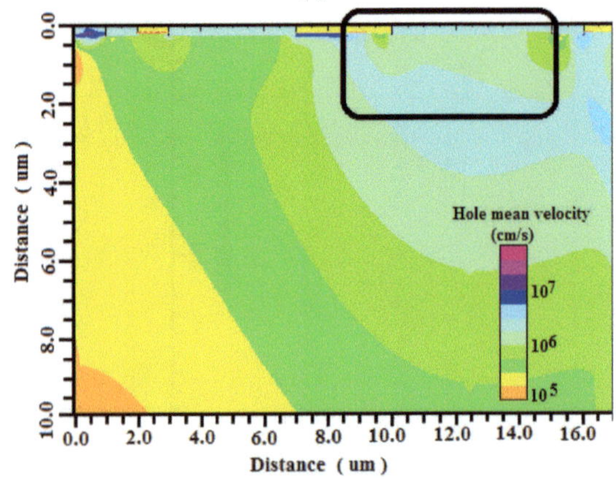

(b)

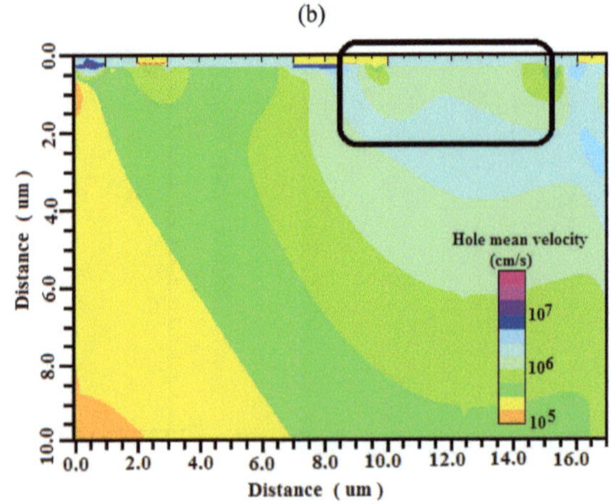

(c)

VSCRs, respectively. All hole mean velocity distributions of three structures are almost same.

The turn-on speed of the VSCR for ESD protection can be improved by the VLBD structure in the base region of the parasitic bipolar transistors of the SCRs. The VLBD structure introduces an accelerating build-in electric field in the base region of the parasitic bipolar transistors during the turn-on process. The accelerating build-in electric field can push the carriers to move faster and reduce the base transit time of the parasitic bipolar transistors. The lateral base doping profile of the parasitic bipolar transistors becomes steeper, and there is larger accelerating build-in electric field in the parasitic n-p-n transistor, so the turn-on time of the SCRs becomes smaller. Furthermore, the lateral base doping profile of the parasitic bipolar transistors can be adjusted by the layout design. In comparison, the turn-on time of the VSCR devices is decreased by maximum 12% than that of the conventional MLSCR with the uniform base doping profile. The VLBD structure can be applied not only to the MLSCR but also all other types of SCR devices for improving the turn-on speed.

6.4 Summary

In this chapter, the novel SCR structures for ESD protection are proposed to reduce the trigger voltage and improve the turn-on speed in the nanometer technology. The novel GCSCR structure for ESD applications has been proposed to reduce the trigger voltage. The GCSCR can be triggered by the parasitic RC sub-network. The simulation and experimental results have been shown that the trigger-voltage can be reduced significantly as compared with the conventional LVTSCR. The parameters of the parasitic RC sub-network can be adjusted by the layout dimensions of the GCSCR structure. Compared with external trigger SCRs, the novel structure don't use any external triggering circuitry and can have the low trigger voltage. The turn-on speed of the VSCR for ESD protection can be improved by the VLBD structure in the base region of the parasitic bipolar transistors of the SCRs. The VLBD structure introduces an accelerating build-in electric field in the base region of the parasitic bipolar transistors during the turn-on process. The accelerating build-in electric field can push the carriers to move faster and reduce the base transit time of the parasitic bipolar transistors. The lateral base doping profile of the parasitic bipolar transistors becomes steeper, and there is larger accelerating build-in electric field in the parasitic n-p-n transistor, the turn-on time of the SCRs becomes smaller. Furthermore, the lateral base doping profile of the parasitic bipolar transistors can be adjusted by the layout design. In comparison, the turn-on times of the VSCR devices are decreased by maximum 12% than that of the conventional MLSCR with the uniform base doping profile. The VLBD structure can be applied not only to the MLSCR but also all other types of SCR devices for improving the turn-on speed.

References

1. L. Lou, J.J. Liou, A novel and robust un-assisted, low-trigger and high-holding voltage SCR (uSCR) for area-efficient on-chip ESD protection. IEEE Conference on Electron Devices and Solid-Static Circuits, 2007, pp. 605–607
2. C. Russ, M.P.J. Mergens et al., GGSCRs:GGNMOS triggered silicon controlled rectifiers for ESD protection in deep sub-micron CMOS processer. Proceedings of EOS/ESD Symposium, 2001, pp. 22–31
3. A. Chatterjee, T. Polgreen, A low-voltage triggering SCR for on-chip ESD protection at output and input pads. IEEE Electron Device Lett. **12**(1), 21–22 (1991)
4. M.D. Ker, K.C. Hsu, Overview of on-chip electrostatic discharge protection design with SCR-based devices in CMOS integrated circuit. IEEE Transactions on Device and Materials Reliability, 2005, pp. 235–249
5. M.-D. Ker, S.-H. Chen, Implementation of initial-on ESD protection concept with PMOS-triggered SCR devices in deep-submicron CMOS technology. IEEE J. Solid-State Circuits **42** (5), 1158 (2007)
6. O. Marichal, G. Wybo, B. Van Camp et al., SCR based ESD protection in nanometer SOI technologies. Electrical Overstress/Electrostatic Discharge Symposium, 2005, pp. 1–8
7. J. Wu, P. Juliano, E. Rosenbaum, Breakdown and latent damage of ultra-thin gate oxides under ESD stress conditions. Electrical Overstress/Electrostatic Discharge Symposium Proceedings, 2000, Anaheim, California, USA, pp. 287–295
8. ESD Association, ESD Association Standard Test Method for Electrostatic Discharge Sensitivity Testing: Charged Device Model-Component Level, ESD STM 5.3.1, 1999
9. C. Duvvury and R. Rountree, A synthesis of ESD input protection scheme. Electrical Overstress/Electrostatic Discharge Symposium Proceedings, 1991, Las Vegas, Nevada, USA, pp. 88–97
10. A. Chatterjee, T. Polgreen, A low-voltage triggering SCR for on-chip ESD protection at output and input pads. IEEE Electron Device Lett. **12**(1), 21–22 (1991)
11. M.P.J. Mergens, C.C. Russ, K.G. Verhaege, et al., Speed optimized diode-triggered SCR (DTSCR) for RF ESD protection of ultra-sensitive IC nodes in advanced technologies. IEEE Trans. Device Mater. Reliab. **5**(3), 532–542 (2005)
12. M.D. Ker, K.C. Hsu, Substrate-triggered SCR device for on-chip ESD protection in fully silicided sub-0.25-μm CMOS process. IEEE Trans. Electron Devices **50**(2), 397–405 (2003)
13. C.-Y. Lin, L.-W. Chu, M.-D. Ker, M.-H. Song et al., ESD protection structure with inductor-triggered SCR for RF applications in 65-nm CMOS process. 2012 I.E. International Reliability Physics Symposium, Anaheim, California, USA, pp. EL.3.1–EL.3.5, 2012
14. S.H. Chen, M.D. Ker, Optimization on MOS-triggered SCR structures for on-chip ESD protection. IEEE Trans. Electron Device **56**(7), 1466–1472 (2009)
15. J. Liu, Z. Liu, Z. Jia, et al., A novel DTSCR with a variation lateral base doping structure to improve turn-on speed for ESD protection. J. Semicond. **35**(6), 064010-1–064010-9 (2014)
16. M.-D. Ker, K.-C. Hsu, SCR device fabricated with dummy-gate structure to improve turn-on speed for effective ESD protection in CMOS technology. IEEE Trans. Semicond. Manuf. **18** (2), 320–327 (2005)
17. B. Song, Y. Han, S. Dong, et al., Compact MOS-triggered SCR with faster turn-on speed for ESD protection. Microelectron. Reliab. **50**(9), 1393–1397 (2010)
18. R.S. Muller, T.I. Kamins, in *Device Electronics for Integrated Circuits*, 3rd edn.. Bipolar transistors (Wiley, New York, 2002), pp. 337–340
19. S.H. Voldman, in *Latchup*. Bipolar transistors (Wiley, Chichester, 2007), pp. 68–135
20. X.B. Chen, Q.Z. Zhang, Y. Chen, in *Microelectric Devices*, 2nd edn.. Bipolar transistors (Electronic Industry Press, Beijing, 2005), pp. 91–95
21. W.M. Webster, On the variation of junction-transistor current-amplification factor with emitter current. Proceedings of IRE, 1954, pp. 914–920

22. B.J. Baliga, in *Fundamentals of Power Semiconductor Devices*. Thyristors (Springer, Chichester, 2008), pp. 662–666
23. S.H. Voldman, in *ESD Physics and Devices*. Semiconductor devices and ESD (John Wiley & Sons, Chichester, 2004), pp. 112–119
24. R.A. Ashton, Extraction of time dependent data from time domain reflection transmission line pulse measurements [ESD protection design]. Proceedings of the 2005 International Conference on Microelectronic Test Structures, Leuven, Belgium, 2005, pp. 239–244
25. M. Huo, Y. Han, Y. Li et al., Study of turn-on characteristics of SCRs for ESD protection with TDR-O and TDR-S TLPs. 17th IEEE International Symposium on the Physical and Failure Analysis of Integrated Circuits, Singapore, 2010, pp. 1–8

Chapter 7
Silicon-Based Junctionless MOSFETs: Device Physics, Performance Boosters and Variations

Xinnan Lin, Haijun Lou, Ying Xiao, Wenbo Wan, Lining Zhang, and Mansun Chan

7.1 Introduction

It is becoming more and more challenging to obtain abrupt junctions in today's CMOS technologies especially in the three dimensional multi-gate MOSFETs. From the physics perspective, electron de-localizations of dopant atoms are in a few nanometers scale. With the transistor volume shrinking to the range of nanometer, even an individual dopant atom in the transistor channel causes a high doping concentration. Lack of control over the number of dopant atoms induces significant doping fluctuations as a source of transistor variation. As a result, the formation of traditional MOSFETs with abrupt junctions impose severe challenges on the doping techniques and thermal budget [1] due to the laws of diffusion. In recent years, a transistor concept called junctionless multi-gate MOSFET (J-MuGFET), with the same doping type and concentration of the source/drain and channel, has been proposed [1]. Due to its non-junction or doping gradients property, the thermal budget issue is waived in its fabrication process. The doping fluctuations are also reduced compared to the junction-based transistors. Junctionless transistors operate just like a simple resistor with its conductivity and carrier density modulated by a gate electrode. Further, it is recognized in the community that junctionless transistors can be prepared with a CMOS-compatible fabrication process, have controllable subthreshold characteristics and superior ability to suppress short channel effects, in comparisons to conventional inversion-mode transistors [2]. Due to these aforementioned advantages, junctionless transistors attract a lot of attentions from the research community [3], and is well accepted as a promising device structure for further scaling of the CMOS technology.

X. Lin (✉) • H. Lou • Y. Xiao • W. Wan • L. Zhang • M. Chan
Peking University Shenzhen Graduate School, Shenzhen, China
e-mail: xnlin@pkusz.edu.cn

T. Li, Z. Liu (eds.), *Outlook and Challenges of Nano Devices, Sensors, and MEMS*,
DOI 10.1007/978-3-319-50824-5_7

Due to the different doping mode in the channel, it is expected that the electrical characteristics of junctionless transistors are quite different from those of the conventional MOSFETs, such as the volume conduction. Theoretical studies are required to reveal their unique device physics. On the other hand, some new problems emerge due to the high channel doping. First, the carrier mobility is reduced significantly with ionized dopant scattering which hurts drive current and transconductance of junctionless MOSFETs [4]. Second, band-to-band tunneling effect plays important roles in junctionless devices causing an increase of the off state current. Geometry engineering is needed to compensate the above negative effects due to high channel doping. In nanoscale devices, variations in these properties cannot be ignored. In the fin-type multi-gate junctionless transistors, vertical non-uniformity and line edge roughness are inevitable, forming two main variation causes. Evaluation of the overall variations in junctionless transistors are necessary for their practical applications.

We provide a review on the junctionless transistors in this chapter, addressing the above mentioned issues. An analytical full range drain current model for long and short channel symmetric junctionless double-gate MOSFETs are developed to understand the device physics of the basic junctionless device. Several performance boosters from the dual-material gate structure to strain engineering are discussed. The dual-material gate improves the current drive, suppresses the tunneling leakage current and enhances the scaling capability [5]. Further optimizations of these parameters including the work function difference and control gate ratio are performed. Finally, variation in the fin-type multi-gate junctionless transistors are reviewed, including the fin's vertical nonuniformity, line edge roughness is investigated based on three-dimensional simulation. A design guideline is proposed to define the optimal parameters of J-MuGFETs for a given technology. Then the high-k spacer is utilized to suppress the subthreshold characteristics variation of J-MuGFETs with non-ideal sidewall angle [6].

7.2 The Development and Fundamental Physics of Junctionless MOSFET

In 1925, J.E.Lilienfied proposed a transistor in his patent, which has a uniform doping concentration and type throughout the channel, source and drain, that is , a real junctionless transistor device prototype. Due to the depletion principal of the device conduction, it needs an extremely thin channel which is impossible to fabricate with the process technology at that time. In recent years, with the rapid development of the integrated circuit manufactory techniques, J. P. Colinge' group has firstly fabricated the junctionless transistor with 1um long, 10 nm high, and 30 nm wide, which could accomplish the basic logic same as traditional inversion MOSFETs [1]. Figure 7.1 shows this fabricated structure of multi-gate junctionless transistor through TEM, and its transfer characteristics is shown in Fig. 7.2.

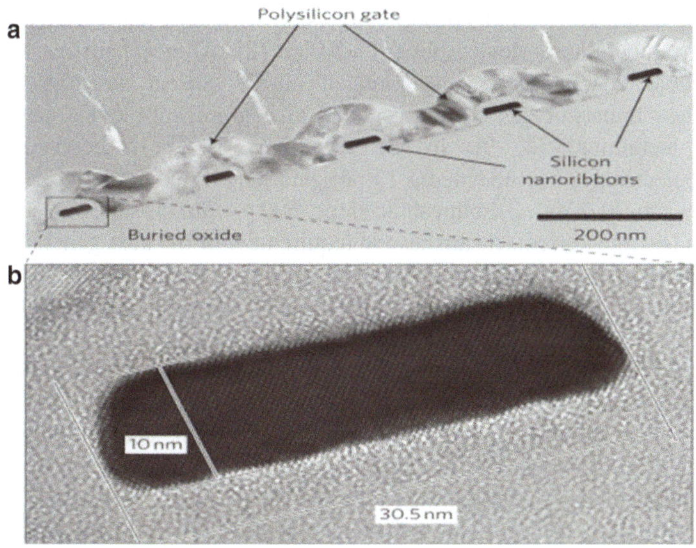

Fig. 7.1 Transmission electron micrograph of silicon gated resistor nanoribbons. (**a**) Five parallel devices with a common polysilicon gateelectrode. (**b**) Magnification of a single nanoribbon device. Individual atomicrows can be seen in the silicon [1]

Fig. 7.2 The current–voltage characteristics of junctionless device [1]

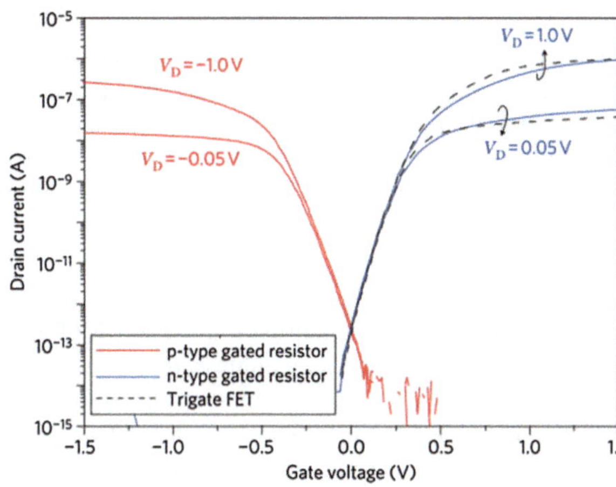

It displays the numerous advantages of junctionless transistor, such as the extremely small leakage current, the superior subthreshold performance, the simple fabrication process and so on. Thus the research of junctionless transistor has become important topic recently [7]. Then, the dimension of junctionless has rapidly scaled down.

In 2011, INTEL has already successfully fabricated the junctionless FINFET of 26 nm gate length [4]. In the same year, the National Nano-materials Technology

Center of South Korea and Pohang University of Science and Technology have scaled the junctionless device down to 20 nm [8]. After exhaustive analysis of subthreshold characteristics, the ratio of on-off current and the noise, the junctionless transistor has been considered as a promising device candidate for 22 nm techniques [8, 9]. In the meantime, LETI in cooperation with STMicroelectronics has invented a 13 nm gate length junctionless device, which possesses sub-70 mV/dec subthreshold slope, 500 μA/μm on-state current and only 3 nA/μm leakage current. It satisfies the requirement of the next generation device [10]. What's more, the AIST from Japan has further scaled down the gate length to 3 nm, and has maintained the 106 on/off ratio, sub 100 mV/dec subthreshold slope [11].

Recently, a lot of governments and organizations are paying great attention to junctionless transistor research. In USA, the DARPA, MARCO and SRC has funded several research programs about its fabrication, characteristic analyses and model development. The European Union has also sponsored the exploration of fabrication of junctionless transistor. Then Japan, South Korea, and Singapore have stepped into the development of junctionless devices, and applied it into memory and thin film transistors based on the national fund support. On the contrary, China started the research of junctionless transistor lately. Several organizations have set to focus on the fabrication and model derivation since 2011. Jin wang et al. from Tsinghua University has investigated and optimized the performance of junctionless transistor [12, 13]. Wang et al. from Fudan University and Hu from Tongji University have proposed a potential-based model for junctionless device [14]. Qing et al. from Shenyang University of Technology has developed a compact current model for surrounding gate junctionless nanowire transistor and then proposed that adding oxide layer under overlap gate inhibits the band to band tunneling current of junctionless [15]. Chen et al. from Xiangtan University has presented a full range potential-based compact current model for junctionless MOSFETs [16].

As junctionless transistor is a novel device with different and special characteristics, J.P.Colinge was the first to analyzed and summarize its physical principles and compared with the traditional inversion MOSFET. There are some obvious distinctions between the junctionless transistor and conventional MOSFET. In traditional MOSFET, the inversion carriers are generated due to the gate voltage effect and confined at the thin surface channel when the device turns on. However, as for the junctionless transistor, the channel is often fully depleted by the high gate metal work function for off-state. Then as the gate voltage changes, the channel is turned to partially depleted and the carriers are flowing through the entire bulk of channel [2]. Figure 7.3 shows the carriers distributions of junctionless transistor versus the gate voltage, which obviously displays the bulk conduction characteristics of junctionless transistor. In other words, the junctionless transistor often works between the fully depletion to partially depletion regime, which is totally different from conventional MOSFET. The inversion-mode MOSFET often operates from depletion to inversion as shown in Fig. 7.4.

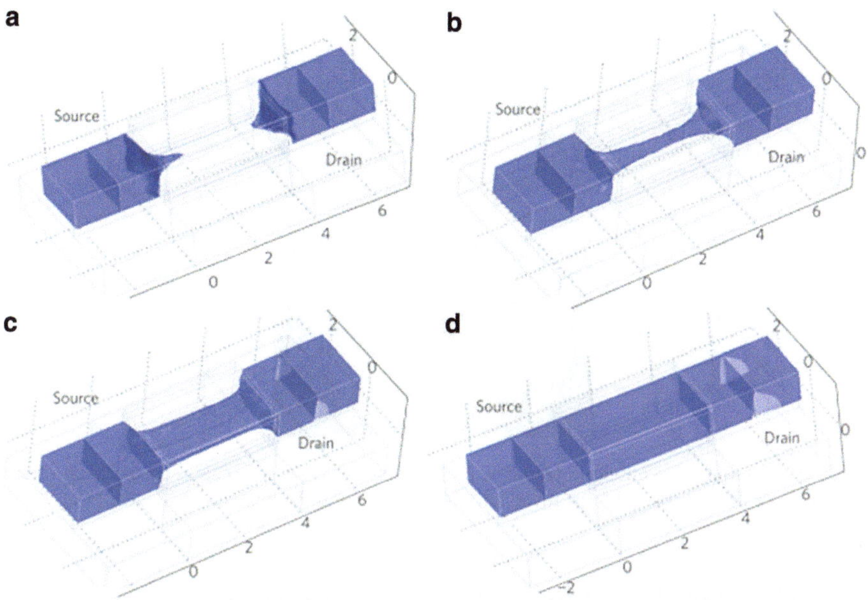

Fig. 7.3 Electron concentration contour plots in an n-type junctionless gated resistor [1]

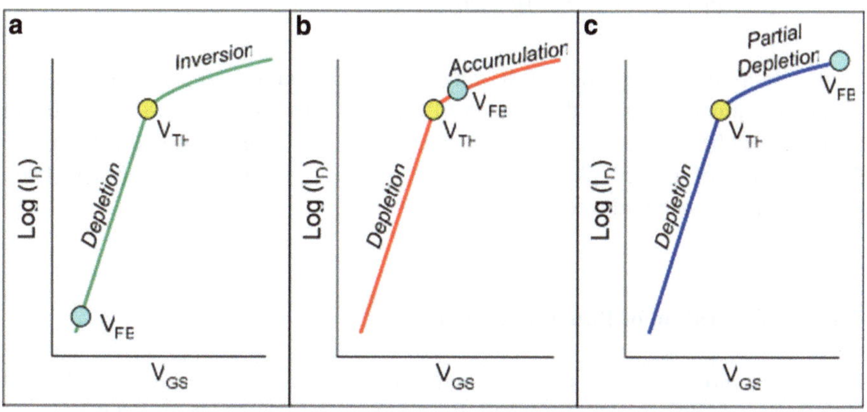

Fig. 7.4 Current in inversion-mode (**a**), accumulation-mode (**b**) and junctionless (**c**) nanowire MuGFETs. Note the very different positions of the flatband voltage, V_{FB} [2]

Furthermore, as for the junctionless transistor, its electric field distribution of channel is also different from traditional MOSFET as shown in Fig. 7.5. The electric field of central point of the channel is lowest and which approaches zero, leading to the highest carrier concentration due to the volume transport [3]. On the contrary, the highest electric field of inversion-mode MOSFET occurs at the surface as the inversion happens, which induces the highest carrier concentration.

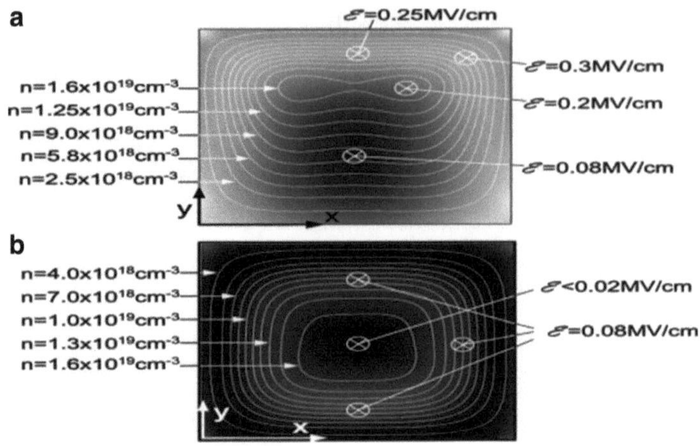

Fig. 7.5 Electron concentration contour lines superimposed to a grayscale representation of the amplitude of the electric field. (**a**) Inversion-mode device and (**b**) junctionless device [3]

7.3 Compact Model for Junctionless Transistor

7.3.1 Full Range Current Model for Long-Channel Junctionless Double-Gate MOSFETs

In this section, an analytical full range drain current model for long channel symmetric junctionless double-gate MOSFETs is proposed [17, 18]. Significant attention has been given to ensure the smoothness of model expressions to ensure the accuracy of device physics, as well as fast convergence which has been validated using standard circuits.

7.3.1.1 Derivation of Potential model

Figure 7.6 shows the cross-section and coordinate system of the long channel junctionless double-gate MOSFET. In the figure, tsi, tox, and L represent the silicon film thickness, gate oxide thickness, and gate length respectively. N_D is the uniform doping concentration in the silicon film. The electron quasi-Fermi potential in the neutral source is defined as zero for the potential reference.

Unlike conventional inversion-mode MOSFETs that works from accumulation region to inversion region, the junctionless transistor operates from the full depletion to accumulation regime. The charge density arising from holes can be ignored in the electrostatic analysis of an n-type junctionless transistor. For a long-channel device, under the gradual-channel-approximation (GCA), the 1-D Poisson-Boltzmann's equation in the channel region can be written as:

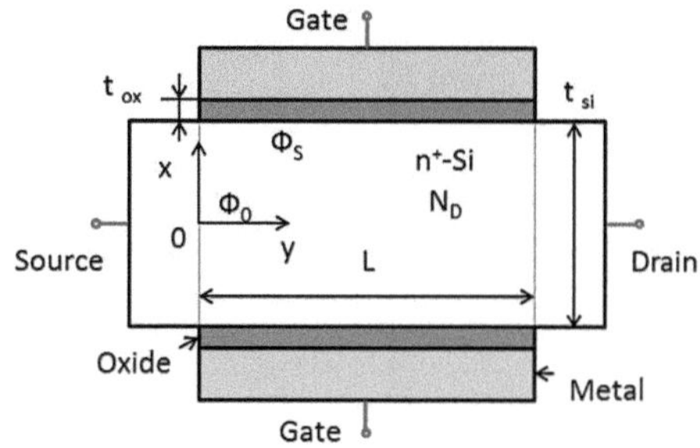

Fig. 7.6 The cross-section and coordinate system of the long channel junctionless double-gate MOSFET

$$\frac{d^2\Phi(x)}{dx^2}=\frac{qN_D}{\varepsilon_{si}}\left[e^{\frac{\Phi(x)-V-V_{bi}}{V_T}}-1\right],\quad -\frac{t_{si}}{2}<x<\frac{t_{si}}{2} \tag{7.1}$$

where q is the elementary charge, ε_{si} is the silicon dielectric permittivity, V is the electron quasi-Fermi potential, V_T is the thermal voltage at 300K, and $V_{bi}=V_T\ln\,(N_D/n_i)$ is the built-in potential. Multiplying both sides of (7.1) by and integrating from to an arbitrary position x in the channel give the electric field E along x-direction

$$E^2-E_0^2=\frac{2qN_D}{\varepsilon_{si}}\left[V_T\left(e^{\frac{\phi-V-V_{bi}}{V_T}}-e^{\frac{\phi_0-V-V_{bi}}{V_T}}\right)-(\phi-\phi_0)\right]. \tag{7.2}$$

Boundary conditions for the symmetric junctionless transistor are:

$$\varepsilon_{si}\frac{\partial\phi}{\partial x}\bigg|_{x=\pm t_{si}/2}=-\varepsilon_{si}E_s=C_{ox}(\phi_G-\phi_s), \tag{7.3}$$

$$\frac{\partial\phi}{\partial x}\bigg|_{x=0}=-E_0=0, \tag{7.4}$$

where is the gate capacitance per unit area and $\phi_G=V_{gs}-V_{fb}$ is the potential on the gate electrode with the flat band voltage $V_{fb}=[W_m-(\chi+E_g/2-K_T\ln(N_D/n_i))]/q$, working function Wm, the silicon affinity χ and the silicon band-gap Eg. Evaluating (7.2) at the silicon-oxide interface and substituting (7.3) and (7.4) into it leads to the channel potential control equation:

$$[C_{ox}(\phi_S-\phi_G)]^2=2qN_D\varepsilon_{si}\left[V_Te^{\frac{\phi-V-V_{bi}}{V_T}}-\phi\right]\Bigg|_{\phi=\phi_0}^{\phi=\phi_S}, \tag{7.5}$$

where ϕ_S and ϕ_0 represent surface and central potential in the channel, respectively.

Since (7.5) contains two unknowns parameters ϕ_S and ϕ_0, another analytical relationship between ϕ_S and ϕ_0 is needed to calculate the channel potential for a given gate bias voltage. From the definition of the electric field with the reference coordinate, the electric field and the electrostatic potential profile in the silicon channel is given by

$$E(v) - E_0 = -\int_0^v \frac{d^2\phi(x)}{dx^2}du, \tag{7.6}$$

$$\phi(x) - \phi_0 = -\int_0^x E(v)dv, \tag{7.7}$$

where u, v are integral intermediate variables.

Electrostatic potential at the arbitrary position in the channel is expressed as a function of central potential and other device parameters by substituting (7.1), (7.4) and (7.6) into (7.7)

$$\phi(x) - \phi_0 = -\frac{qN_D}{2\varepsilon_{si}}\left[x^2 - \eta(x)e^{\frac{\phi_0 - V - V_{bi}}{V_T}}\right], \tag{7.8}$$

where

$$\eta(x) = 2\int_0^x \int_0^v e^{\frac{\phi - \phi_0}{V_T}} dudv. \tag{7.9}$$

By assuming the potential profile in the right hand side of (7.9) is a constant and equal to the central potential, which is defined as a local approximation [19]:

$$\phi(x) = \phi_0. \tag{7.10}$$

After solving Eq. (7.9) analytically, substituting it into (7.8) leads to the closed form expression:

$$\phi(x) - \phi_0 = -\frac{qN_D x^2}{2\varepsilon_{si}}\left[1 - e^{\frac{\phi_0 - V - V_{bi}}{V_T}}\right]. \tag{7.11}$$

The second relationship between surface and central potential can be obtained from (7.11) as

$$\phi_s - \phi_0 = -\frac{qN_D {t_{si}}^2}{8\varepsilon_{si}}\left[1 - e^{\frac{\phi_0 - V - V_{bi}}{V_T}}\right]. \tag{7.12}$$

Now for given gate voltage V_{gs} and quasi Fermi potential V, surface and central potentials in the symmetric junctionless double-gate MOSFET can be derived by solving (7.4) and (7.12) together. A two-order Newton iterative method is used in this work [20]. Analytical solutions of the surface and central potentials with the flat band voltage are used as initial guess, which leads to convergences of the Newton iterations usually within four steps. The local approximation has been used to

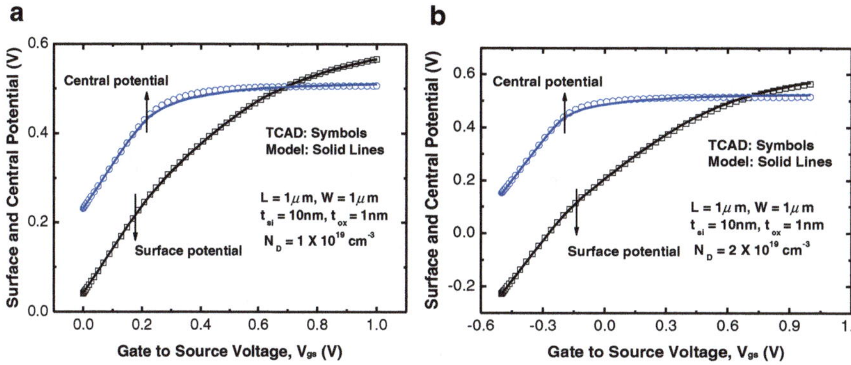

Fig. 7.7 Simulated (symbols) and modeled (solid lines) surface and central potential versus gate voltage at the source terminal. (**a**) $N_D = 1 \times 10^{19}$ cm^{-3}, (**b**) $N_D = 2 \times 10^{19}$ cm^{-3}

model a similar dynamic depletion effect. In the following work, the calculated surface and central potential at the drain are labeled as ϕ_{SL} and ϕ_{0L} with V = Vds, respectively, and those at the source are labeled as ϕ_{S0} and ϕ_{00} with V = 0.

In order to validate the proposed model, the potential model is compared with the 2-D numerical simulations Sentaurus [21]. The channel length and width are set as both 1 um to avoid any secondary order effects, and gate work function is set to be 4.8 eV. A constant electron mobility of 110 cm^2/Vs and Boltzmann statistics are used in the simulation to obtain the potentials and currents.

Figure 7.7 shows model predicted surface and central potentials at the source side versus gate voltage with two channel doping concentrations, in comparisons with numerical results. The electrostatic potential model shows good agreements with device simulations in the whole operation region with smooth transitions. After obtaining the potential distribution, the mobile charge density can be easily calculated by:

$$Q_{mobile} = 2C_{ox}\left[\phi_s - \left(V_{gs} - V_{fb}\right)\right] - qN_Dt_{si}. \tag{7.13}$$

Figure 7.8 shows the normalized mobile charge density ($-Q_{mobile}/qN_Dt_{si}$) versus gate voltage and its comparison with simulation results. There is very good agreement between our model and simulation results in the entire range of operations over a wide range of oxide thicknesses.

7.3.1.2 Derivation of Drain Current Model

As conduction takes place in the middle of the channel in the junctionless MOSFETs under partial depletion and accumulation regions, the charge sheet approximation which assumes conduction through a sheet of charge at the channel

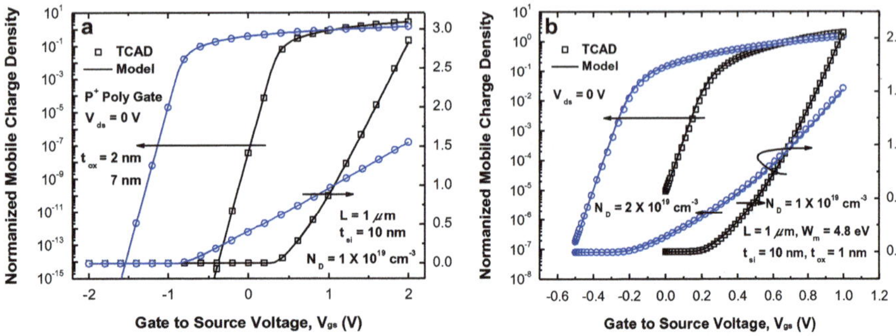

Fig. 7.8 Normalized mobile charge density comparisons between model (solid lines) and simulation results (Symbols) with (**a**) different oxide thickness and (**b**) different doping concentrations

surface close to the insulator interface cannot be used. Based on the potentials obtained in the previous section, the drain current model for JLDG MOSFETs is derived from Pao-Sah's dual integral which contains both the drift and diffusion carrier transport [22, 23]. The drain-current is written as

$$\begin{aligned} I_{ds} &= \mu \frac{W}{L} \int_0^{V_{ds}} \int_{-t_{si}/2}^{t_{si}/2} qndxdV \\ &= -2\mu \frac{W}{L} \int_0^{V_{ds}} \int_{\phi_0}^{\phi_S} \frac{qn}{E} d\phi dV, \end{aligned} \tag{7.14}$$

where u is the constant electronics mobility, W and L are the channel width and channel length, and V_{ds} is the drain to source voltage. Rearranging (7.2) gives an equivalent equation

$$\alpha(V) = E^2(\phi, V) - G^2(\phi, V), \tag{7.15}$$

where

$$G^2(\phi, V) = \frac{2qN_D}{\varepsilon_{si}} \left[V_T e^{\frac{\phi - V - V_{bi}}{V_T}} - \phi \right] \tag{7.16}$$

Following the procedure developed by Pierret and Shields [24], we evaluate the partial derivation of (7.15) and combine it with the definition of n

$$\frac{\varepsilon_{si}}{2E} \frac{d\alpha}{dV} - \varepsilon_{si} \frac{\partial E}{\partial V} = -\frac{\varepsilon_{si}}{2E} \frac{\partial G^2}{\partial V} = \frac{qn_i}{E} e^{\frac{\phi - V}{V_T}} = \frac{qn}{E} \tag{7.17}$$

Then substituting (7.17) into the physical-based drain current expression (7.14) leads to

$$I_{ds} = -2\mu \frac{W}{L} \int_0^{V_{ds}} \int_{\phi_C}^{\phi_S} \left(\frac{\varepsilon_{si}}{2E} \frac{d\alpha}{dV} - \varepsilon_{si} \frac{\partial E}{\partial V} \right) d\phi dV = I_1 + I_2, \tag{7.18}$$

where

$$I_1 = -2\mu\frac{W}{L}\int_0^{V_{ds}}\int_{\phi_C}^{\phi_S}\frac{\varepsilon_{si}}{2E}\frac{d\alpha}{dV}d\phi dV \tag{7.19}$$

$$I_2 = -2\mu\frac{W}{L}\int_0^{V_{ds}}\int_{\phi_C}^{\phi_S}\left(-\varepsilon_{si}\frac{\partial E}{\partial V}\right)d\phi dV. \tag{7.20}$$

Note that is a constant in a given V and does not depend on x, I1 can be calculated as

$$I_1 = \mu\frac{W\varepsilon_{si}}{L}\int_0^{V_{ds}}\frac{d\alpha}{dV}dV\int_0^{t_{si}/2}dx = \mu\frac{W}{L}\frac{t_{si}}{2}\varepsilon_{si}\alpha\Big|_{source}^{drain} \tag{7.21}$$

I_2 can be transformed to a sum of four single integrations as done in the MOSFETs [22, 23].

$$\begin{aligned} I_2 &= 2\mu\frac{W}{L}\varepsilon_{si}\int_0^{V_{ds}}\int_{\phi_C}^{\phi_S}\frac{\partial E}{\partial V}d\phi dV \\ &= 2\mu\frac{W}{L}\varepsilon_{si}\Big[\int_{\phi_{CO}}^{\phi_{CL}}E(\phi,V_C)d\varphi - \int_{\phi_{SO}}^{\phi_{SL}}E(\phi,V_S)d\phi \\ &+\int_{\phi_{CL}}^{\phi_{SL}}E(\phi,V_{ds})d\phi - \int_{\varphi_{CO}}^{\varphi_{SO}}E(\phi,0)d\phi\Big], \end{aligned} \tag{7.22}$$

where V_0 is the value of V specified by ϕ_0, and VS is the value of V specified by ϕ_S. Then, drain current can be calculated as

$$I_{ds} = -\mu C_{ox}\frac{W}{L}\left[\frac{t_{si}}{2C_{ox}}\varepsilon_{si}\alpha - \left(2V_{gf} - \phi_s + 4V_T\right)\phi_s + I'\right]\Bigg|_{source}^{drain} \tag{7.23}$$

where

$$I' = -\frac{4qN_D}{\varepsilon_{si}C_{ox}}\int_0^{t_{si}/2}[\phi(x,L) - \phi(x,0)]dx. \tag{7.24}$$

Substituting (7.11) into (7.24) to calculated the total drain current as

$$\begin{aligned} I_{ds} = -\mu C_{ox}\frac{W}{L}\Big[&\frac{t_{si}}{2C_{ox}}\varepsilon_{si}\alpha - \left(2V_{gf} - \phi_s + 4V_T\right)\phi_s \\ &-\frac{2qN_d t_{si}}{3C_{ox}}(\phi_s + 2\phi_0)\Big]\Bigg|_{source}^{drain} \end{aligned} \tag{7.25}$$

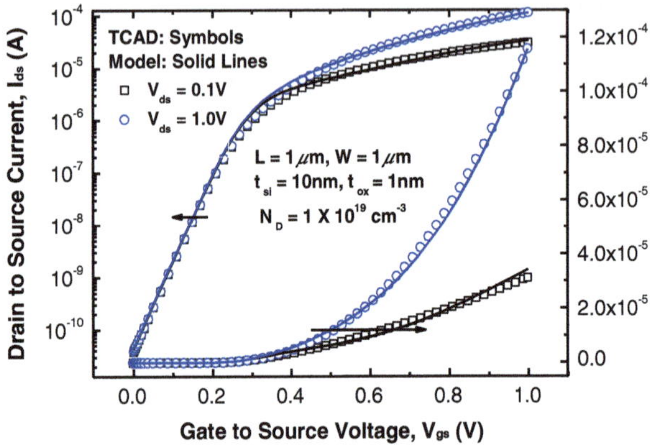

Fig. 7.9 Simulated (symbols) and modeled (solid lines) transfer characteristics at different drain to source voltage

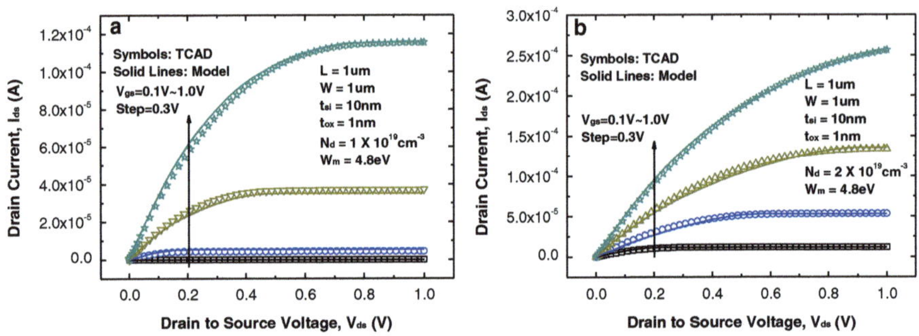

Fig. 7.10 Simulated (symbols) and modeled (solid lines) output characteristics at different gate to source voltage for two doping concentrations JLDG MOSFETs. (**a**) $N_D = 1 \times 10^{19}$ cm^{-3}, (**b**) $N_D = 2 \times 10^{19}$ cm^{-3}

Figure 7.9 shows the model predicted transfer characteristics of a symmetric JLDG MOSFET of two different Vds compared with the numerical simulation results. Good agreement is observed from fully depletion to partially depletion operations. Comparison of the output characteristics from the proposed model and numerical simulations at different gate voltage is shown in Fig. 7.10, with two doping concentrations. It shows good agreement not only in linear region, but also the saturation region.

Figure 7.11 shows the model-predicted drain-current versus gate voltage with different channel thickness, compared with simulation results. The comparison of transfer characteristics between model and simulation results with different gate oxide thickness is shown in Fig. 7.12. Overall, the proposed analytical model for JLDG MOSFETs matches well with numerical results with wide device geometries and different doping levels.

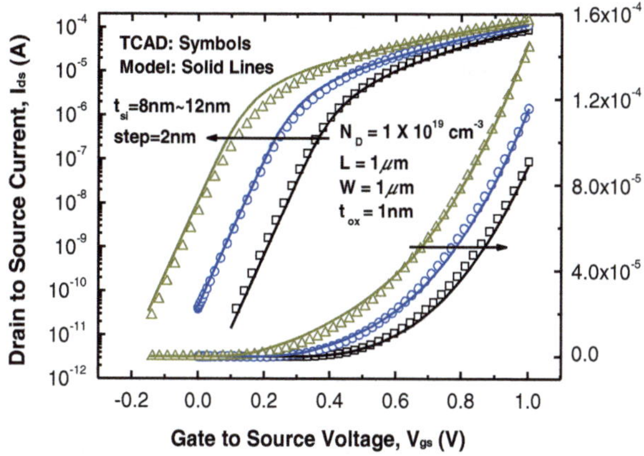

Fig. 7.11 Simulated (symbols) and modeled (solid lines) transfer characteristics for different silicon film thickness JLDG MOSFETs

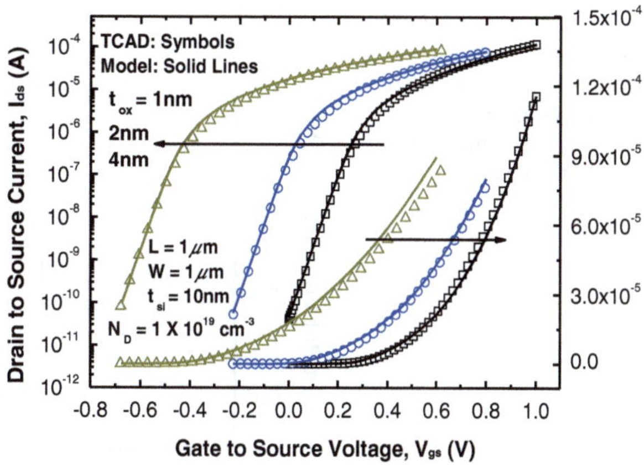

Fig. 7.12 Simulated (symbols) and modeled (solid lines) transfer characteristics for different gate oxide thickness JLDG MOSFETs

7.3.1.3 Application in Circuit Simulation

In order to apply the junctionless MOSFET to practical circuit design, the compatibility of the model to a SPICE platform is essential. To ensure fast convergence, the current derivatives have to be smooth and continuous over the entire operation region of the transistor. Figure 7.13a shows the trans-conductance gm versus the gate voltage characteristics with different oxide thickness. Figure 7.13b shows the output conductance gds versus the drain to source voltage characteristics with different gate to source voltage. Both of them have proved good continuity and smoothness of the full operation region. In addition, the model has been implemented into the circuit simulator HSPICE with Verilog-A code which is available at the i-MOS platform [25]. As a demonstration of the model convergence, an inverter circuit is simulated with different operation voltages and the voltage transfer curves are shown in Fig. 7.14. This benchmark circuit test shows that the developed model can be used for circuit simulation without convergence issues.

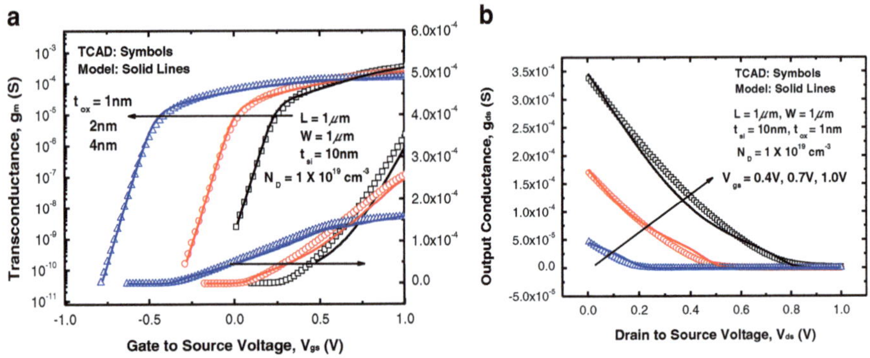

Fig. 7.13 (**a**) Simulated (*symbols*) and modeled (*solid lines*) trans-conductance versus gate to source voltage with different oxide thickness. (**b**) Simulated (*symbols*) and modeled (*solid lines*) Output conductance versus gate to source voltage with different gate to source voltage

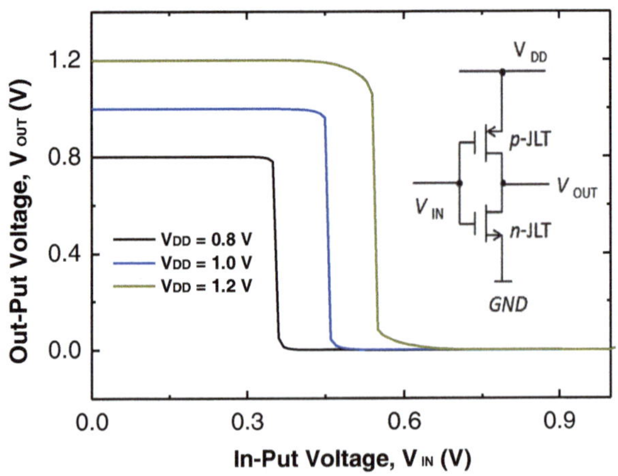

Fig. 7.14 The transfer characteristics at different supply voltage for a simple logic invert circuit, and the inset is the invert circuit configuration

7.3.2 *Short-Channel Subthreshold Current Model for Junctionless MOSFETs with Source/Drain Depletion Effect*

As the dimension of the junctionless transistor has scaled down to sub-32 nm, the short channel effect greatly affects the performance of MOSFETs, however, as for junctionless transistor, the source/drain depletion effect plays increasingly vital role on superior characteristics of junctionless MOSFETs. As for short channel junctionless cylindrical surrounding gate (JLCSG) MOSFET, the fringe field strongly depletes the source/drain(S/D) region of JLCSG MOSFET under the off-state, which reduces the potential of the source, drain and channel, inducing larger difference of current, capacity and threshold voltage. When the transistor becomes smaller, it degrades, so it is really necessary to include the S/D depletion effects in the short channel JLCSG compact model to evaluate the unique performance. Moreover, in order to satisfy the efficiency requirement of large scale circuit design, a fast analytical short channel compact model without numerical iterations is really needed to address this problem.

In this section, a compact potential-based model which include the source/drain depletion effect by using the parabolic approximation in the subthreshold regime for the short channel JLCSG MOSFETs, is developed. For the sake of improving the efficiency in the circuit simulation, a fast analytical drain current model is proposed, which the error is less than 5 %. Furthermore, the threshold voltage (Vth), subthreshold slope (SS), and drain-induced barrier lowering (DIBL) are extracted based on potential expression.

7.3.2.1 Derivation of Potential model

Figure 7.15 shows the cross-section view of the JLCSG MOSFET, where r, tox and L represent the radius of the silicon cylinder,gate oxide thichnese, and gate length respectively. N_D is uniform doping concentration in the silicon film including the source, channel and drain. and represent the source and drain depletion region respectively. The electron quasi-Fermi potential within the neutral source region is defined as zero for the potential reference.

Since we consider the subthreshold regime only, the mobile electronics can be ignored. The 2-D Poisson's equation in the channel is thus written as:

$$\frac{1}{r}\frac{\partial}{\partial r}\left[r\frac{\partial\Phi(r,z)}{\partial r}\right]+\frac{\partial^2\Phi(r,z)}{\partial z^2}=-\frac{qN_D}{\varepsilon_{Si}} \tag{7.26}$$

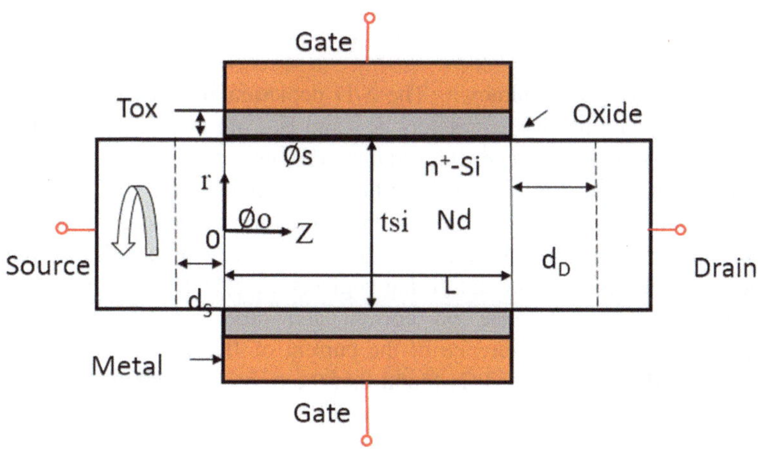

Fig. 7.15 Cross-section view of a JLCSG MOSFET and the coordinate system

where $0 \le r \le R, 0 \le z \le L$, q is the electron charge, and ε_{Si} is the dielectric constant of the silicon. Using the parabolic approximation method and considering the symmetry condition $(\partial\Phi/\partial r)_{r=0} = 0$, hence the 2-D potential along r is assumed as:

$$\Phi(r,z) = \phi_C(z) - [\phi_C(z) - \phi_S(z)]\left(\frac{r}{R}\right)^2 \tag{7.27}$$

where $\phi_C(z)$ and $\phi_S(z)$ are the central and surface potential, respectively.

As we know, the normal component of electrical flux between the silicon film and surrounding oxide should be continuous, thus a boundary condition is added:

$$-\varepsilon_{Si}\frac{\partial\Phi(r,z)}{\partial r}\bigg|_{r=R} = C_{ox}[\phi_S(z) - \phi_G] \tag{7.28}$$

where $C_{ox} = \varepsilon_{ox}/t_{ox\_eff}$ is the gate oxide capacitance per unit area and $t_{ox\_eff} = R \cdot \ln(1 + t_{ox}/R)$ is the effective gate oxide thickness. $\phi_G = V_{GS} - (\Phi_M - \chi - E_g/2)/q$ is the potential on the gate electrode with workfunction Φ_M and gate-to-source bias V_{GS}, χ and Eg are the silicon affinity and band-gap respectively. Substituting (7.27) into (7.26 and using (7.28) to eliminate the surface potential, then the scaling equation for the bulk conduction mode of the JLCSG MOSFETs is expressed as follows:

$$\frac{d^2\phi_C(z)}{dz^2} - \frac{1}{\lambda^2}\cdot\left[\phi_C(z) - \phi_G - \frac{qN_d\lambda^2}{\varepsilon_{Si}}\right] = 0 \tag{7.29}$$

Where $\lambda = \sqrt{(2\varepsilon_{Si}R + R^2C_{ox})/4C_{ox}}$ is the natural length. Then the general solution of the central potential is founded as

$$\phi_C(z) = \frac{\phi_{C2}\sinh\left(\frac{z}{\lambda}\right) + \phi_{C1}\sinh\left(\frac{L-z}{\lambda}\right)}{\sinh(L/\lambda)} + \phi_G + \frac{qN_d}{\varepsilon_{Si}}\cdot\lambda^2 \tag{7.30}$$

where ϕ_{C1} and ϕ_{C2} are unknown. The S/D depletion regimes of the boundary condition for obtaining the parameters ϕ_{C1} and ϕ_{C2} are often ignored. However, this simplified approach reduces the accuracy of compact model.

Figure 7.16a shows the difference in the central potential of JLCSG MOSFET between different boundary condition in the simulations. Due to the fringe electric field and source/drain depletion effect, the potential with source and drain region is obviously smaller than the potential that neglects the S/D depletion region, which will induce a big difference on current and other electric characteristics. Figure 7.16b shows the difference of the current of JLCSG MOSFETs between different boundary condition. It is found that the transistor without S/D depletion region has much larger current than that with S/D depletion region and the difference increases as the scale of device shrinks. The S/D depletion region acts like extent channel which suppresses the SCEs and protects the minimum potential of

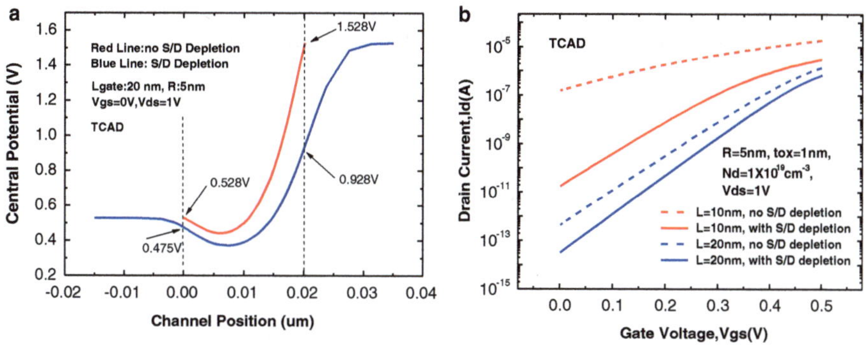

Fig. 7.16 The central potential distribution (**a**) and drain current distribution (**b**) of JLCSG MOSFET with different boundary condition

channel and current from increasing. This unique characteristic of junctionless transistor contributes to its superior short channel performance. The effect of the source/drain depletion region is thus considered in this work. The length of the source and drain depletion region is assumed as and respectively, where and are unknowns to be determined. Therefore, the z-dependent potential at the source and drain, is given by the solution of Poisson's equation. The accurate boundary condition is expressed as:

$$\Phi(r,z) = \phi_{CS}(z) = V_{bi} - \frac{qN_D}{2\varepsilon_{Si}}(z+d_S)^2, \quad -d_S < z < 0 \tag{7.31}$$

$$\Phi(r,z) = \phi_{CD}(z) = V_{DS} + V_{bi} - \frac{qN_D}{2\varepsilon_{Si}}(z-L-d_D)^2, \ L < z < L+d_D \tag{7.32}$$

where $V_{bi} = V_T \ln(N_D/n_i)$ is the built-in potential, VT is the thermal voltage, ni is the intrinsic carrier concentration, and VDS is the drain-to-source bias voltage. According to the continuity of the central potential and its z-derivative at z = 0 and z = L, the equations are obtained as:

$$\phi_{C1} + \phi_G + \frac{qN_D}{\varepsilon_{Si}} \cdot \lambda^2 = V_{bi} - \frac{qN_D}{2\varepsilon_{Si}} \cdot {d_S}^2, \tag{7.33}$$

$$\frac{\phi_{C2} - \phi_{C1}\cosh(L/\lambda)}{\sinh(L/\lambda)} = -\frac{qN_D\lambda}{\varepsilon_{Si}} \cdot d_S, \tag{7.34}$$

$$\phi_{C2} + \phi_G + \frac{qN_D}{\varepsilon_{Si}} \cdot \lambda^2 = V_{DS} + V_{bi} - \frac{qN_D}{2\varepsilon_{Si}} \cdot {d_D}^2, \tag{7.35}$$

$$\frac{\phi_{C2}\cosh(L/\lambda) - \phi_{C1}}{\sinh(L/\lambda)} = \frac{qN_D\lambda}{\varepsilon_{Si}} \cdot d_D. \tag{7.36}$$

These above equations are nonlinear due to the quadratic terms in dS and dD. Considering the nonlinearity of the continuity equation, its exact solution requires iterations to be converged [26]. In order to reduce the iteration time for circuit simulation, an analytical solution of these equations is proposed as follows:

$$d_S = \sqrt{A + (c\lambda)^2 - D} - c\lambda, \tag{7.37}$$

$$d_D = \sqrt{B + (c\lambda)^2 + D} - c\lambda, \tag{7.38}$$

where

$$A = \frac{2\varepsilon_{Si}}{qN_D}\left(V_{bi} - \phi_G - \frac{qN_D}{\varepsilon_{Si}}\lambda^2\right), \tag{7.39}$$

$$B = \frac{2\varepsilon_{Si}}{qN_D}\left(V_{DS} + V_{bi} - \phi_G - \frac{qN_D}{\varepsilon_{Si}}\lambda^2\right), \tag{7.40}$$

$$c = \left(e^{L/\lambda} + 1\right) / \left(e^{L/\lambda} - 1\right) \tag{7.41}$$

$$D = \frac{4b\lambda(B - A)}{\sqrt{B + (c\lambda)^2} + \sqrt{A + (c\lambda)^2} - 8b\lambda}, \tag{7.42}$$

where $b = e^{L/\lambda}/(e^{2L/\lambda} - 1)$. Then from (7.33) and (7.35) we obtain the expression of ϕ_{C1} and ϕ_{C2}. The result of potential model has been compared with the 2-D TCAD simulation. Figure 7.17 shows the central potential ϕ_C distribution at $V_{GS} = 0.05$ V with different channel lengths. The results show that the model results match the simulation well, especially within the channel region. It can be seen that the device with 90 nm channel length exhibits a typical long-channel behavior, which the central potential is flat and almost independent of V_{DS} though the channel region.

As the channel length decreases, the drain potential is further penetrated into the channel that causes the central potential to be no longer flat. Moreover, the minimum central potential also increases with V_{DS}, which reflects the short channel effect. For the junctionless transistors, the threshold voltage is mainly determined by the minimum central potential.

The potential along r for different channel length is shown in Fig. 7.18. The model results are in good agreement with the simulated results, even as the channel length is scaled down to 10 nm. It can be seen that the minimum central potential increases as the channel length decreases. Therefore, the threshold voltage is reduced with the decrease of the channel length.

7.3.2.2 Derivation of Subthreshold Drain Current Model

Based on the potential compact model above, we derived the current expression from the standard drift-diffusion method.

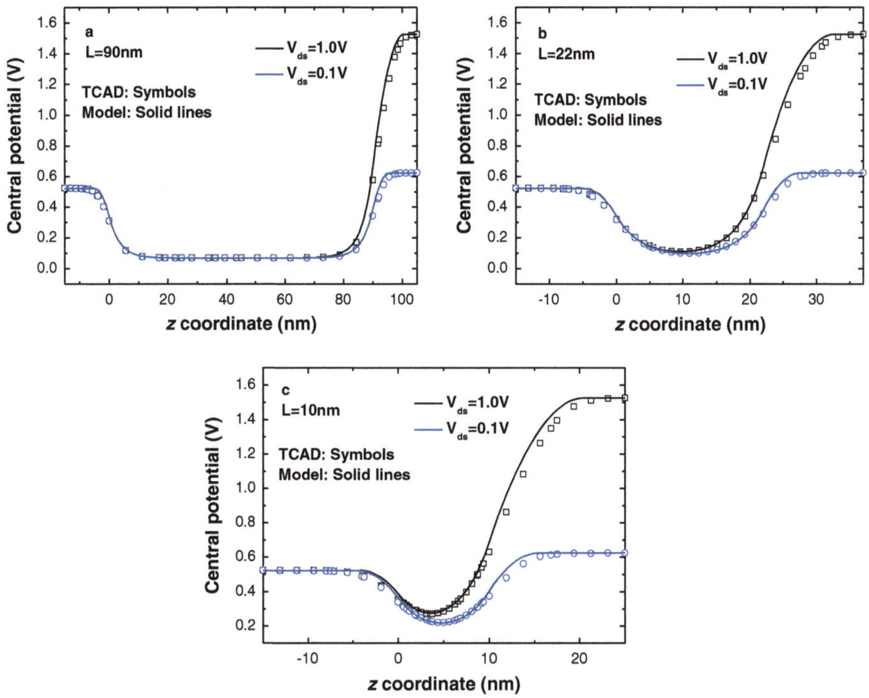

Fig. 7.17 Simulated (*symbols*) and modeled (*solid lines*) central potential for the three devices with (**a**) L = 90 nm, (**b**) L = 22 nm, and (**c**) L = 10 nm at $V_{GS} = 0.05$ V, R = 5 nm, tox = 1 nm, $N_D = 1019\ cm^{-3}$, $\chi = 4.05$ eV, $\Phi_M = 5.2$ eV

Fig. 7.18 The transverse potential along r-direct at z = L/2, for different channel length

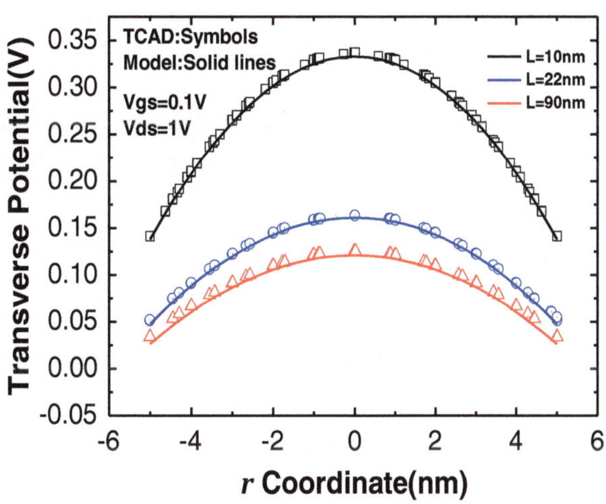

$$J(r,z) = -qun_i \exp\left[\frac{\Phi(r,z) - \varphi_n(r)}{V_T}\right]\frac{d\varphi_n}{dz} \tag{7.43}$$

where u is the electron mobility and assumed as 110 cm^2/V.s [27]. ϕ_n is the electron quasi-Fermi potential, which is assumed to depend only on x, due to the lack of the small current component in the r-direction. Then we integrate current density along the r-direction from 0 to R that leads to the total current equation as follows:

$$I = 2\pi\, qun_i \exp\left[\frac{-\phi_n(r)}{V_T}\right]F(z)\frac{d\phi_n}{dz} \tag{7.44}$$

where W is the device width and

$$F(z) = \int_0^R r \times \exp\left[\frac{\Phi(r,z)}{V_T}\right]dr$$
$$= \begin{cases} \dfrac{R^2}{2}\exp\left[\dfrac{\phi_{cs}(z)}{V_T}\right], & -d_s < Z < 0 \\ \dfrac{2\pi\lambda^2 V_T}{\phi_c(z) - (V_{gs} - V_{fb})}\exp\left[\dfrac{\phi_c(z)}{V_T}\right]\left[1 - \exp\left(-\dfrac{R^2[\phi_c(z) - (V_{gs} - V_{fb})]}{4\lambda^2 V_T}\right)\right], & 0 < Z < L \\ \dfrac{R^2}{2}\exp\left[\dfrac{\phi_{cd}(z)}{V_T}\right], & L < Z < L + d_D \end{cases} \tag{7.45}$$

$\phi_{cs}(z)$and $\phi_{cd}(z)$ can be obtained by (7.31) and (7.32), respectively, and the value of $\phi_c(z)$ is got by (7.30). Then through integrating the Eq. (7.44) by the separation of variable from $z = -d_s$ and $\phi_n = 0$ to $z = L + d_D$ and $\phi_n = V_{DS}$, the final accurate current expression is obtained.

$$I = \frac{2\pi qun_i V_T}{\int_{-d_s}^{L+d_D} F^{-1}(z)dz}\left[1 - \exp\left(\frac{-V_{DS}}{V_T}\right)\right]. \tag{7.46}$$

In order to improve the efficiency of model for simulation, the current model is simplified as an analytical expression. According to the calculation of the integral of bottom of (7.45), the value between 0 and L accounts for 97 %, So the integral of bottom of (7.46) can be approximated as follows [28].

$$\int_{-d_s}^{L+d_D} F^{-1}(z)dz \approx \int_0^L F^{-1}(z)dz \cong F^{-1}(z_{\min})(z_1 - z_2) \tag{7.47}$$

where z_{min} is obtained by (7.48), z1 and z2 are the position for the twice value of the minimum exponential term.

$$\phi_{C\min} = \phi_C(z_{\min}) = \phi_C\left(\frac{L}{2} - \frac{\lambda}{2}\ln\frac{\phi_{C2} - \phi_{C1}e^{-L/\lambda}}{\phi_{C1} - \phi_{C2}e^{-L/\lambda}}\right) \tag{7.48}$$

The whole drain current model can be simplified as follows, with the error less than 5 %:

$$I = \frac{2\pi q u n_i V_T}{F^{-1}(z_{\min})(z_1 - z_2)}\left[1 - \exp\left(\frac{-V_{DS}}{V_T}\right)\right]. \tag{7.49}$$

Figure 7.19a shows the comparison of drain current from the model and the TCAD simulation for the 65-, 22-, and10-nm channel length devices. In the model, the Vgs is varied from 0 to 0.4 V. For all different gate length devices, the current model shows remarkable agreement with TCAD outputs in the subthreshold region, even for the 10-nm devices, which has reveal the strong short channel effects. The results present the high accuracy of our model. For validation of the model for different silicon film thickness, the results are compared with simulation.

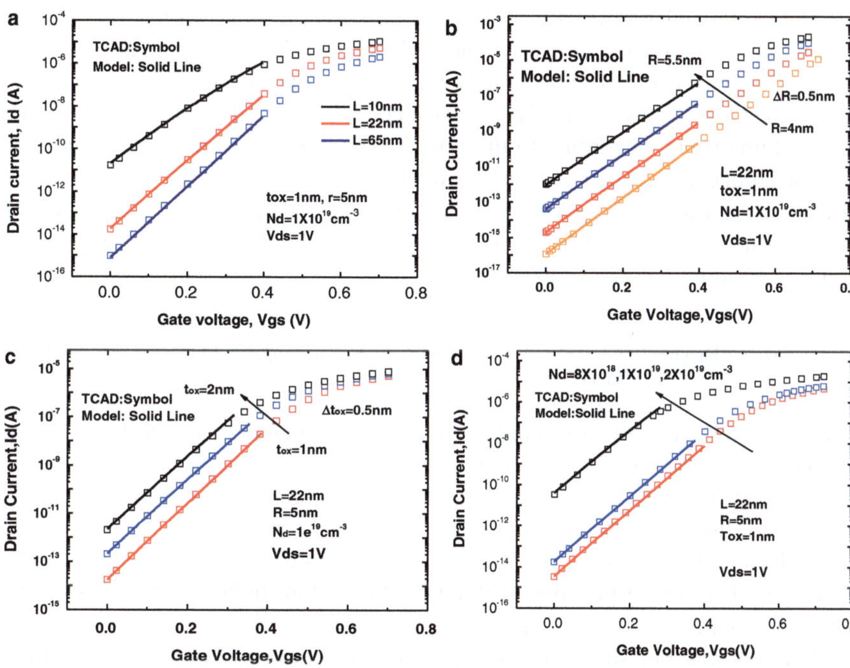

Fig. 7.19 (**a**) The transfer characteristics of junctionless devices for different gate length. (**b**) The transfer characteristics of junctionelss devices for different silicon radius. (**c**) The transfer characteristics of junctionless devices for different oxide thickness. (**d**) The transfer characteristics of junctionless devices for different doping concentration

Figure 7.19b shows the current distribution of four different silicon film radius, which vary from 4 nm to 5.5 nm. The proposed model is seen to work well and exactly for all different silicon film device. From the graph, it can be seen that the transistor is very sensitive to the silicon thickness variation, especially for the threshold voltage. Figure 7.19c shows the comparison between the drain current of the model and simulation results for different gate oxide thickness. The gate oxide varies between 1 nm and 2 nm.When the gate oxide becomes thick, the gate control turns weak, which induces increased short channel and small fringe field. It should be noticed that for the different gate oxide thickness, the result of proposed model can exactly match the simulation results in the subthreshold regime. Figure 7.19d shows the comparison between the drain current of the model and simulation results for different doping concentration. A good agreement is observed between the model and simulation result.

7.3.2.3 Short Channel Effects

The threshold voltage is one of the most important parameters of device for circuit design. As the transistor has been shrunk to nanometers, the shift of threshold voltage becomes greatly severe. Hence it's very necessary to evaluate the shift. The threshold voltage definition of junctionless MOSFETs is based on the minimum of the central potential ϕCmin. Due to the bulk conduction principle that the current flows through the central of bulk silicon majorly, the central potential dominates the whole current expression in subthreshold condition, till the applied outer gate voltage turns it on. According to the Eq. (7.34), the expression of ϕCmin and its location zmin can be calculated analytically by (7.48). The threshold voltage can be defined as the value of V_{GS} which makes $\phi_{\mathrm{Cmin}}^{extr} = V_{bi}$, where $\phi_{\mathrm{Cmin}}^{extr}$ is the value of ϕ_{Cmin} linearly extrapolated from deep subthreshold operation [29]. Thus the expression for threshold voltage is obtained as:

$$V_{TH} = V_{GS1} + \frac{V_{GS2} - V_{GS1}}{\varphi_{\mathrm{Cmin}}(V_{GS2}) - \varphi_{\mathrm{Cmin}}(V_{GS1})}[V_{bi} - \varphi_{\mathrm{Cmin}}(V_{GS1})]. \tag{7.50}$$

Figure 7.20a plots the threshold voltage versus the channel length for different silicon radius. The threshold voltage decreases with the radius and the channel length. While for R = 10 nm, the error becomes bigger with shorter channel length due to the influence of the free carrier in the channel region.

Subthreshold slope is an essential standard to evaluate the short channel characteristics of device, especially working on the subthreshold region. It shows the capacity of gate control and suppressing the short channel effects of the device. Smaller subthreshold slope means transistor could have smaller leakage current with the same on-state current, which is vital for devices under 22-nm. In this paper, we derive a general SS model from the potential model of JLCSG MOSFETs.

$$SS = \frac{\partial V_{gs}}{\partial \log(I_d)} = \frac{kT}{q} \ln(10) \cdot \frac{1}{\partial \phi_{C\min} / \partial V_{gs}}. \tag{7.51}$$

Figure 7.20b shows that SS increases with the channel length and radius. The model results are verified with the simulation, and show good agreement. The DIBL is defined as the threshold voltage variation between the low Vds and high Vds.

$$DIBL = \frac{V_T(V_{dsl}) - V_T(V_{dsh})}{V_{dsh} - V_{dsl}}. \tag{7.52}$$

DIBL is also an essential parameter for small scale device, which reveal the impact of drain voltage on threshold voltage. Figure 7.20c shows that the DIBL with the different channel length and radius. The model results agree well with the simulation.

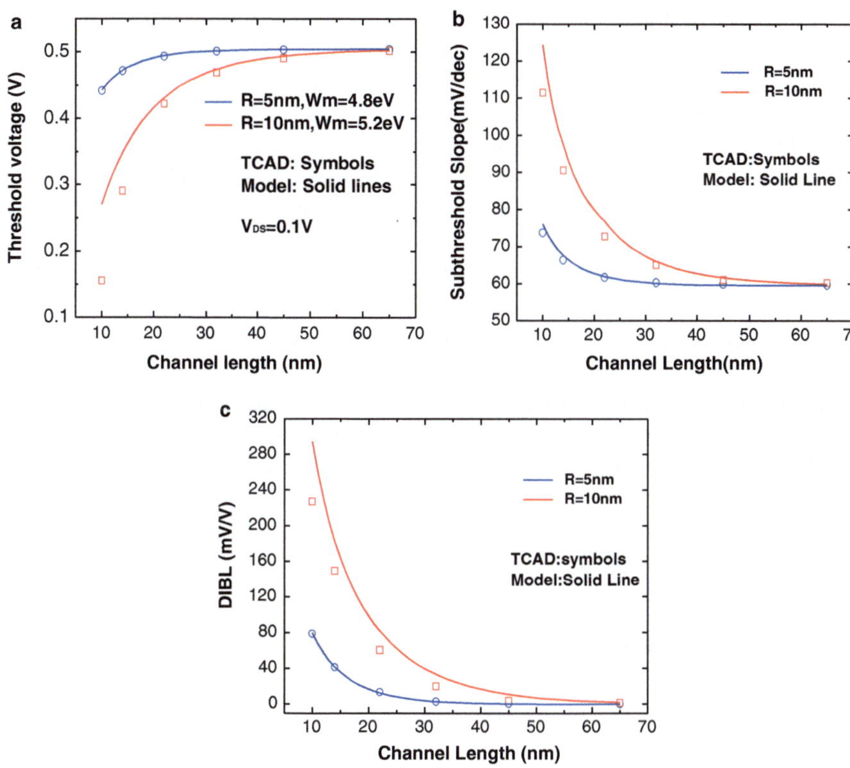

Fig. 7.20 The variation of threshold voltage (**a**), DIBL (**b**) and SS (**c**) with different channel radius

7.4 Performance Optimization

In this section, the dual-material gate structure and high-k spacer are incorporated into junctionless nanowire transistor (JNT) to improve the current drive and suppress the tunneling leakage current, then the key parameters' impact on electrical characteristics and the physical mechanism are discussed in detail.

7.4.1 Junctionless Transistor with a Dual Matetial Gate

As for the junctionless MOSFET, The high doping concentration of the channel would degrade the carrier mobility, which hurts drive current and trans-conductance of it. Hence, the dual-material gate (DMG) devices, which displays good ability to improve the carrier transport efficiency, trans-conductance, and the drain output resistance in conventional MOSFETs by adjusting the channel potential and electric field distributions along the channel have been studied with junctionless transistor to ameliorate its performance.

A dual-material-gate junctionless nanowire transistor (DMG-JNT) is proposed in this section. Its characteristic is demonstrated and compared with a generic single-material-gate JNT using 3-D numerical simulations. The results show that the DMG-JNT has a number of desirable features, such as high ON-state current, a large ON/OFF current ratio, improved trans-conductance Gm, high unity-gain frequency fT, high maximum oscillation frequency fMAX, and reduced drain-induced barrier lowering. The effects of different control gate ratios Ra and varied work-function differences between the two gates are studied. Finally, the optimization of Ra and the work-function difference for the proposed DMG-JNT is presented.

A 3-D schematic view and a cross section along the channel of the DMG-JNT are shown in Fig. 7.21. A single-material-gate JNT (SMG-JNT) with the same channel length L that was proposed in [30] is used as a reference for comparison. The DMG-JNT has two metal gates with different work functions denoted by M1 and M2, which are assigned to be 4.97 and 4.27 eV, respectively, corresponding to

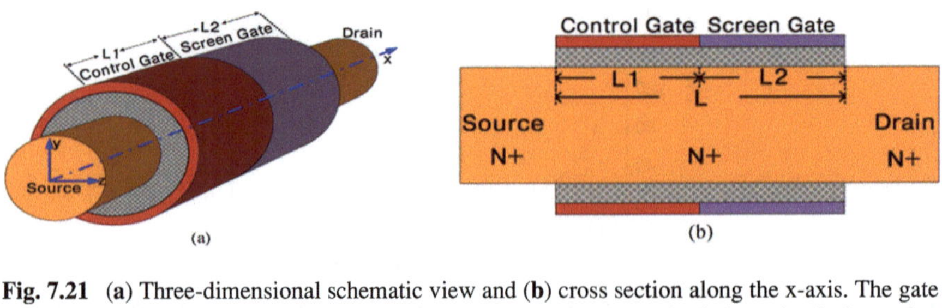

Fig. 7.21 (**a**) Three-dimensional schematic view and (**b**) cross section along the x-axis. The gate near the source region is denoted as "control gate", and the gate near the drain region is denoted as "screen gate"

the values of some common metals (e.g., Co and Mo/SiO2 for M1 and HfSi, Al (111), and NiAl(110) for M2) [31–35]. The values of L1 and L2 are initially set to be 20 nm. The length and work function of the two metal gates can be varied in the simulations. In this work, the simulation is drift–diffusion based without impact ionization, and the mobility model includes the effects of doping concentration and electric field. The density gradient model applied in [36–39] is utilized to account for the quantum effects in this work. The Auger recombination model and the Shockley–Read–Hall model are included for the recombination and generation regimes in the highly doped channel of DMG- and SMG-JNTs.

7.4.1.1 Electrical Characteristics of the DMG-JNT

In Fig. 7.22, the output characteristics of the DMG-JNT and the corresponding SMG-JNT with different gate voltage overdrive V_{GT} defined as $V_{GT} = V_{gs} - V_{th}$ are shown, where Vgs is the gate voltage and Vth is the threshold voltage extracted by the maximum trans-conductance method [40]. In general, the drain-to-source current Ids of the DMG-JNT is higher than that of the SMG-JNT at same V_{GT}. The transfer characteristics of DMG- and SMG-JNTs with different gate work functions at the drain-to-source voltage Vds = 1 V are shown in Fig. 7.23. It is observed that the gate work function is not a physical mechanism responsible for the improved performance of the DMG-JNT. We further investigate the potential and electric field distributions along the channel in both the proposed DMG-JNT and the conventional SMG-JNT with Vds = 2 V and different V_{GT}. As shown in Fig. 7.24a, the potential distribution of a DMG-JNT has an abrupt change near the transition of the two gates, whereas that of the SMG-JNT increases monotonically from the source to the drain. The abrupt change is caused by the difference of gate work function, and this enhances the electric field in the channel of the DMG-JNT. It is also observed that the potential drop across the source/drain extensions in the

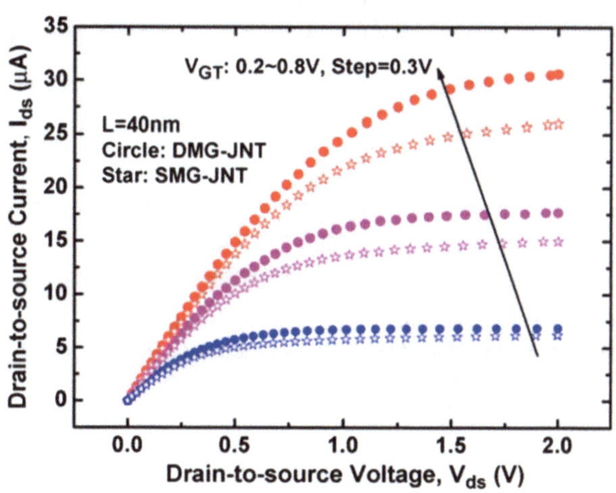

Fig. 7.22 Output characteristics of the DMG-JNT and referencing the SMG-JNT with the channel length L = 40 nm at different V_{GT}

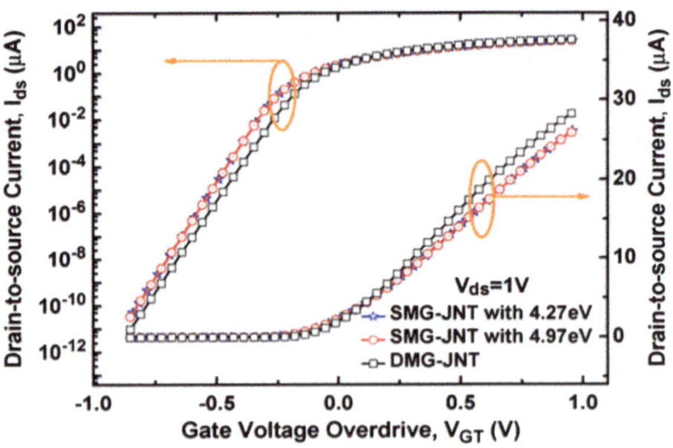

Fig. 7.23 Plot of the drain-to-source current I ds versus V_{GT} for the DMG-JNT and the corresponding SMG-JNT with different gate work functions

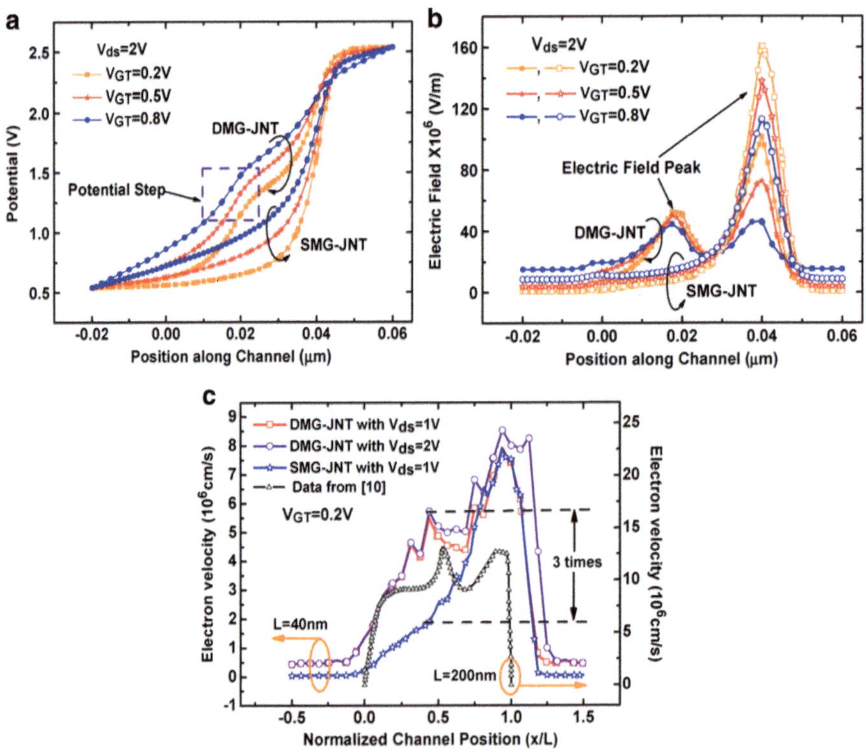

Fig. 7.24 Distributions of (**a**) channel potential, (**b**) channel electric field, and (**c**) channel electron velocity profile along the channel of the DMG-JNT. (**a**) and (**b**) are biased with drain-to-source voltage $V_{ds} = 2$ V and different V_{GT}. (**c**) is biased at $V_{GT} = 0.2$ V and different V_{ds}. The transistor lengths are $L_1 = L_2 = 20$ nm and $L = 40$ nm

DMG-JNT is larger than that in the SMG-JNT, which indicates that the channel ON-state resistance of the DMG-JNT is smaller than that of the SMG-JNT. As shown in Fig. 7.24b, there are two electric field peaks in the DMG-JNT and only one near the drain in the SMG-JNT. Furthermore, the peak value of the electric field near the drain of the DMG-JNT is reduced by 40 % compared with that of the SMG-JNT. As a comparison, the peak values of the electric field near the drain side for the DMG SOI and DMG bulk MOSFETs are only reduced by 20 %, as compared with those of their SMG counterparts. The DMG structure is more effective in reducing the drain channel field, which in turn suppresses the short-channel effect and the hot-carrier effect in JNTs. For ON-state current, the electric field peak near the source region provides more acceleration to the electron in the channel and enhances the presaturation carrier velocity near the source, as shown in Fig. 7.24c. The electron velocity in the DMG-JNT can be increased to three times of that of the SMG-JNT at the same location, and its value is similar with that reported in [36]. In addition, the electron velocity in the source/drain extensions of the DMG-JNT is also increased to further enhance the drive current.

7.4.1.2 Transconductance and High-Frequency Characteristics of the DMG-JNT

In Fig. 7.25a, the transconductance values Gm of the DMG-JNT and the SMG-JNT for comparison are shown. The Gm of the DMG-JNT in the saturation region is higher than that of SMG devices. The maximum values at Vds = 1 V are 0.033 and 0.028 mS, respectively, and those at Vds = 0.4 V are 0.018 and 0.014 mS. The higher Gm in the DMG-JNT is also caused by the abrupt potential step near the source that the same VGT results in high channel potential change, as shown in Fig. 7.24a. Furthermore, the lower Gm for the DMG-JNT below threshold voltage is beneficial for faster turn off.

In Fig. 7.25b, the high-frequency characteristics of the DMG-JNT and the SMG-JNT are shown. The definitions of the unity-gain cutoff frequency fT and

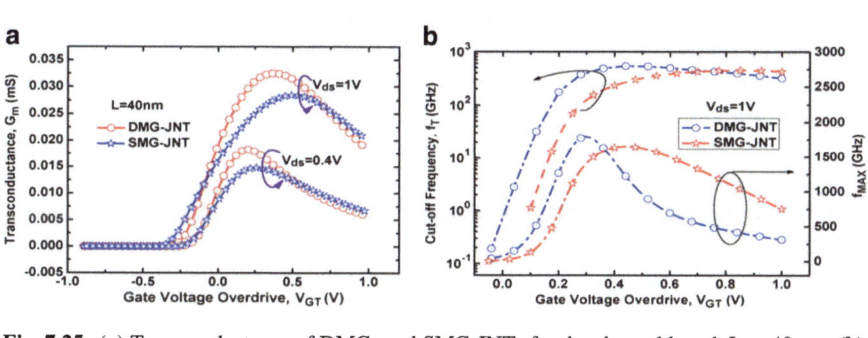

Fig. 7.25 (**a**) Transconductance of DMG- and SMG-JNTs for the channel length L = 40 nm. (**b**) Cut-off frequency f_T and maximum oscillation frequency f_{MAX} of DMG- and SMG-JNTs as a function of V_{GT} at V_{ds} = 1 V

the maximum oscillation frequency fMAX are given in [41]. The DMG device exhibits higher fT and fMAX when VGT is small due to higher Gm compared with its SMG counterpart. At higher VGT, the advantage of higher fT and fMAX in the DMG-JNT gradually disappears due to the increasing influences of the parasitic capacitance. The maximum fT and fMAX of the DMG-JNT are 541 and 1774.97 GHz, which are 20.2 % and 8.1 % improvement, respectively, compared with their SMG counterparts. Both the maximum fT and fMAX appear at VGT $\leq$ 0.4 V. According to the ITRS report, the supply voltage of the device is decreasing, and the device for the high-frequency applications is more likely to work at its maximum fT and fMAX points so that the advantage of DMG-JNT is more obvious.

7.4.1.3 Effects of L1/L Ratio at a Given Channel Length

The output characteristics of DMG-JNTs such as ON-state current Ion, which is extracted at Vds = Vgs = 1 V; OFF-state current Ioff, which is extracted at Vgs = 0 V; saturation current Isat, which is extracted at Vds = VGT = 1 V; subthreshold swing (SS); drain-induced barrier lowering (DIBL); and Gm are studied when the L1/L ratio Ra is varied by changing L1 from 0 to 40 nm with a fixed channel length L = L1 + L2 = 40 nm.

In Fig. 7.26, the current characteristics of DMG-JNTs with different Ra at a given bias are shown. It should be noted that Ra = 0 and Ra = 1 correspond to the SMG-JNT cases with a gate work function of 4.27 and 4.97 eV, respectively. As shown in Fig. 7.26a, Ioff decreases with the increase in Ra. In addition, the effect of Vds on Ioff also decreases with the increasing Ra. It is because the increasing L1 or Ra leads to more depletion charge at the same gate voltages, which screens the effect of Vds. The inset shows that Ion decreases as the Ra increases for different doping concentrations. Ion is larger with higher channel doping due to the shifted threshold voltage. Furthermore, the different work functions of the gates cause the difference of Ion between Ra = 0 and Ra = 1. After subtracting the effect of threshold voltage by using the same VGT, the saturation currents Isat are shown in Fig. 7.26b. In general, Isat of the DMG-JNT is higher than that of the SMG-JNT, whereas IV (extracted at VGT = 0 V) is lower. The higher I sat in the DMG-JNT is a direct result from the high channel electric field at the transition of the two gates, as discussed earlier. With increasing L1 or Ra, the location of the channel peak electric field is moved toward the drain, and the electrons from the source have to travel a longer distance before reaching the same velocity. As a result, the overall trend of Isat decreases with Ra. Isat increases with the doping concentrations due to the extrinsic property such as contact resistance to the metal, whereas the optimal Isat for different channel doping is observed at Ra = 1/4. As for IV, the effect of the screen gate L2 decreases with increasing L1, leading to higher IV due to a stronger influence of drain voltage. The ON/OFF current ratios of the DMG-JNTs are shown in Fig. 7.26c. The ON/OFF current ratio of the DMG-JNT increases with Ra mainly due to the reduced OFF-state current, as shown in Fig. 7.26a. In order to subtract the

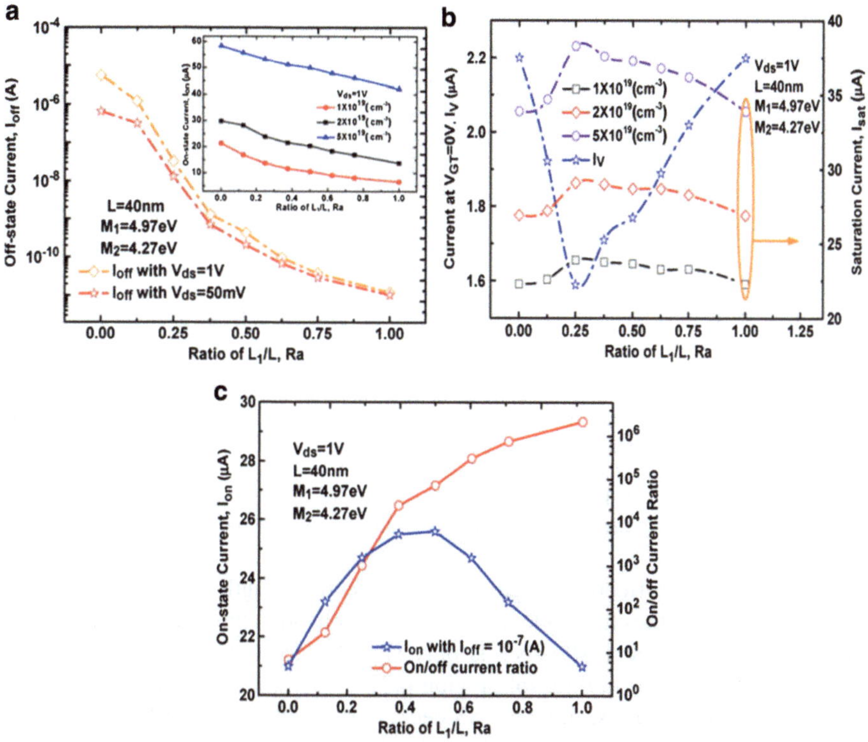

Fig. 7.26 (**a**) Variations of the OFF-state current I_{off}. The inset shows the variations of the ON-state current I_{on} with different channel doping concentrations. (**b**) Variations of the saturation current (I_{sat}) and the current at $V_{GT} = 0$ V (I_V). (c) Variations of the original ON/OFF current ratio and Ion with constant I_{off}. The channel length is fixed at L = 40 nm. Ra = 0 and Ra = 1 correspond to the SMG-JNT with gate work functions of 4.27 and 4.97 eV, respectively.

effect of threshold voltage for a fair comparison, the Ion referencing the same Ioff by changing the gate work function to an appropriate value is also shown in Fig. 7.26c. It indicates that an optimal ON/OFF current ratio can be achieved with Ra = 1/2 and by adjusting the threshold voltage to give Ioff = 10−7A. Such a condition satisfies the requirement for high-performance circuits [42].

In Fig. 7.27a, the characteristics of the SS and DIBL for DMG-JNT devices are shown. The values of DIBL are calculated as the difference between the threshold voltage at Vds = 50 mV and the threshold voltage at Vds = 1 V. It is observed that DIBL has a sharp increase with the introduction of L1 from the SMG-JNT with a gate work function equal to that of L2 due to the introduction of higher channel electric field that leads to a more significant impact of drain voltage. With increasing L1, the peak electric field in the channel is moved toward the drain, thus reducing the impact of the electric field from the drain. As a result, we observe a decreasing DIBL effect with increasing Ra. As for the SS, DMG-JNTs in general have a larger SS compared with corresponding SMG-JNTs. The SS is inversely

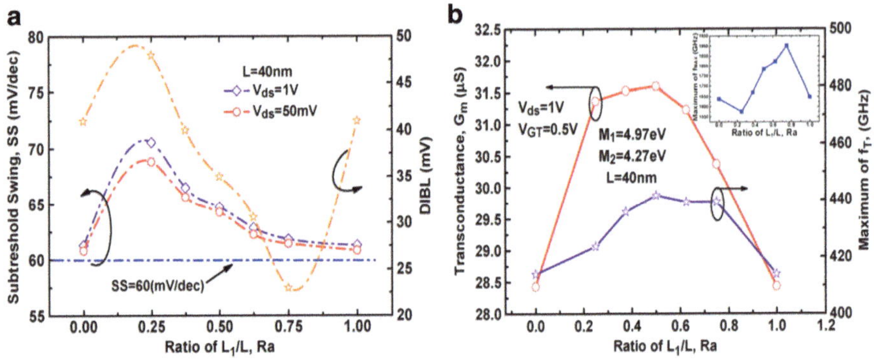

Fig. 7.27 (**a**) Variations of the SS and DIBL effect with varied Ra for the DMG-JNT at a fixed channel length L = 40 nm. Ra = 0 and Ra = 1 correspond tothe SMG-JNT with 4.27 and 4.97 eV, respectively. (**b**) Transconductance G_m and the maximum f_T as a function of Ra with a fixed channel length L = 40 nm. The G_m is extracted at $V_{ds} = 1$ V and $V_{GT} = 0.5$ V, and the maximum f_T is extracted at $V_{ds} = 1$ V. The inset shows the maximum f_{MAX} with the variation of Ra. Ra = 0 and Ra = 1 correspond to the SMG-JNT with 4.27 and 4.97 eV, respectively.

proportional to the effective length, and the effective length of the SMG-JNT is slightly larger than the channel length. On the other hand, it has been shown that the effective length of the DMG-JNT is only slightly larger than the length of the control gate L1 in the subthreshold region [43]. With an increase in L1 or Ra, the effective length increases; thus, the SS decreases. Furthermore, as L1 approaches 0 nm, the screen gate controls the channel, which induces the SS approaching the value of the SMG-JNT.

The Gm at Vds = 1 V and VGT = 0.5 V and the maximum fT at Vds = 1 V with varying Ra are investigated in Fig. 7.27b. DMG-JNTs exhibit higher Gm and, hence, larger maximum fT when compared with SMG-JNTs in the saturation regimes. The optimal Gm and maximum fT are observed at Ra = 1/2. The inset shows that maximum fMAX increases with Ra due to reduction of gate-to-drain capacitance caused by the reduced drain field penetration into the channel.

7.4.1.4 Effects of the Work-Function Difference for DMG-JNTs

For a DMG-JNT with the fixed length of metal gate L1 = L2 = 20 nm, the effect of work-function difference δW = M1 − M2 is investigated by varying M1 and keeping M2 fixed at 4.27 eV, unless otherwise specified.

In Fig. 7.28, the current characteristics with varying work-function difference δW are shown. The δW = 0 eV corresponds to the SMG-JNT. As shown in Fig. 7.28a, both Ion and Ioff decrease with increasing δW due to the enhanced threshold voltage caused by increasing the work function of the control gate. As the doping concentration increases, Ion tends to be larger due to smaller threshold voltage. After the effect of threshold voltage is subtracted, Isat and I_V with varying

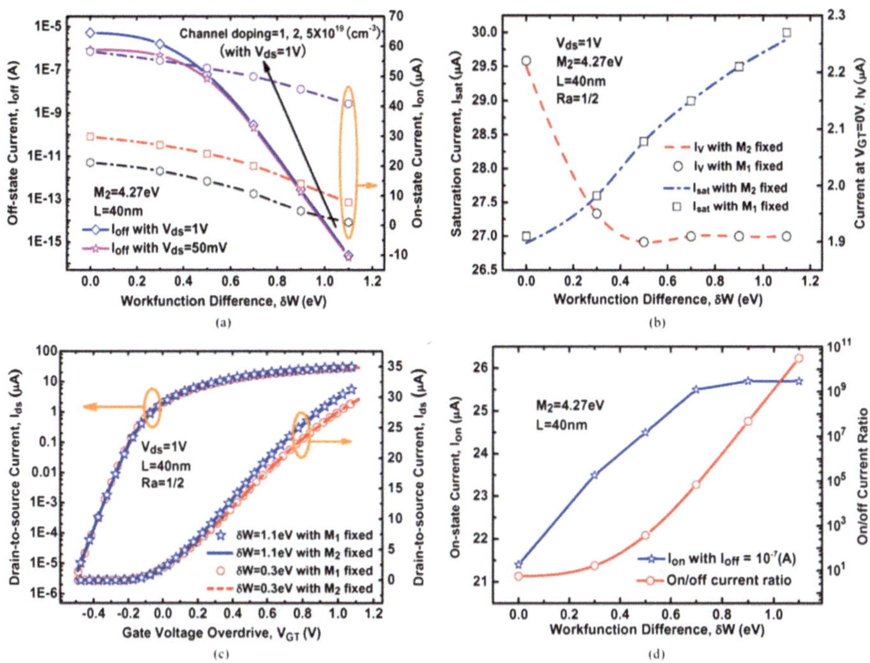

Fig. 7.28 Current characteristics as a function of the work-function difference δW at L = 40 nm. (**a**) Variations of I_{on} and I_{off} . (**b**) Variations of I_{sat} and I_V. (**c**) Transfer characteristics for M_1 or M_2 fixed. (**d**) Variations of the original ON/OFF current ratio and I_{on} with a constant I_{off}. Ra is equal to 1/2 for theDMG-JNT, and δW = 0 eV corresponds to the SMG-JNT

δW are shown in Fig. 7.28b. It is observed that Isat increases with δW due to stronger enhancement of the channel field near the source that improves carrier transport efficiency. On the other hand, I_V decreases with δW and finally flat out at δW = 0.4 eV. Furthermore, the values of Isat and I_V with M1 fixed at 4.97 eV are also shown in Fig. 7.27b with the symbols. For a clearer description of the influences of the work function and δW, the Ids − VGT characteristics of δW = 0.3 eV and δW = 1.1 eV for M1 or M2 fixed are shown in Fig. 7.28c, respectively. The results show that the current properties of the DMG-JNT mainly depend on δW, although the exact values of the work function cause Vth, Ion, and Ioff to vary. In Fig. 7.28d, the ON/OFF current ratio increases with δW. Additionally, an optimal Ion of the DMG-JNT can be obtained at δW = 1.1 eV after adjusting the threshold voltage to obtain $I_{off} = 10^{-7}$A.

In Fig. 7.29, the SS and DIBL of DMG-JNTs, which both increase with increasing δW due to the decreasing effective length of the control gate and the minimum surface potential shift toward the source, are shown. In terms of the SS, an SMG-JNT is superior over that of a DMG-JNT, but the SS of the DMG-JNT can still be maintained at less than 67 mV/dec, which satisfies the circuit applications [44, 45]. For DIBL, the screening effect against the drain voltage due to the screen

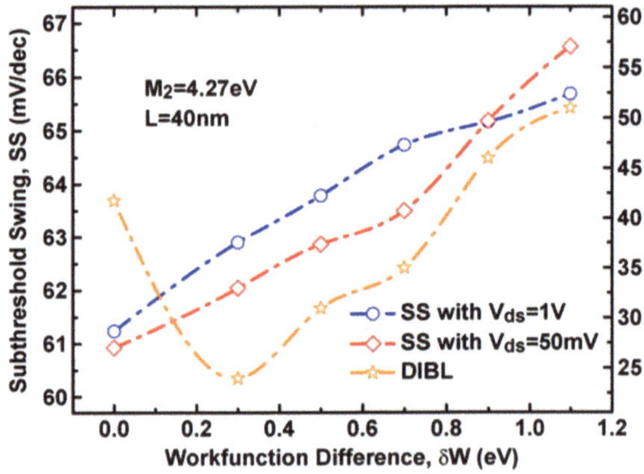

Fig. 7.29 SS and DIBL versus the varying work-function difference δW with $L_1 = L_2 = 20$ nm. δW = 0 eV corresponds to the SMG-JNT

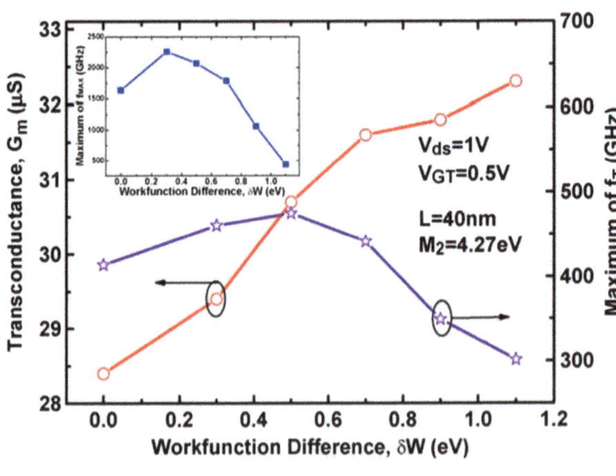

Fig. 7.30 Transconductance G_m and maximum of f_T as a function of work-function difference δW with $L_1 = L_2 = 20$ nm. The G_m is extracted at $V_{ds} = 1$ V and $V_{GT} = 0.5$ V, and the maximum f_T is extracted at $V_{ds} = 1$ V. The inset shows the maximum of f_{MAX} with varied δW. δW = 0 eV corresponds to the SMG-JNT.

gate is reduced as δW approaches 0 eV, which leads to higher DIBL. Compared with that of the SMG-JNT, the DIBL voltage of the DMG-JNT is smaller when δW is smaller than 0.8 eV. The minimum DIBL voltage of the DMG-JNT appears around δW = 0.3 eV, which is only 24 mV and nearly a 45 % improvement over that of the SMG-JNT.

In Fig. 7.30, the Gm and maximum fT with varying δW are shown. In the saturation region, G m increases linearly with δW, which is not observed in other DMG structures such as DMG SOI. It is due to the higher saturation current resulting from increasing δW, as shown in Fig. 7.28b, which has been explained before. The higher saturation current directly translates to higher Gm. For maximum fT, an optimal value is observed at δW = 0.5 eV. Moreover, the

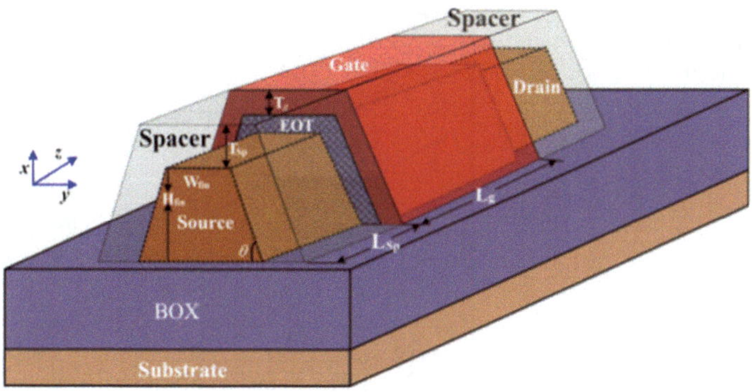

Fig. 7.31 Schematic view of JMT with the trapezoidal cross section

inset in Fig. 7.30 shows that the maximum fMAX decreases with an increasing work-function difference due to higher screening effect, as described before.

7.4.2 Junctionless Transistors Using High-k Spacers

In this section, the high-k spacer is proposed to suppress the subthreshold characteristics variation of junctionless multigate transistor (JMT) with non-ideal sidewall angle for the first time. It is demonstrated that the variation of subthreshold characteristics induced by the changing sidewall angle is efficiently suppressed by high-k spacers due to the enhanced corner effect through the fringe capacitance, and the electrostatic integrity of JMTs is also improved at sub-22 nm gate length. Two key parameters of high-k spacer, the thickness and length, have been optimized in terms of the suppression of subthreshold characteristics variation. Then their optimal values are proposed. The benefit of high-k spacer makes JMTs more scalable.

The proposed device architecture is shown in Fig. 7.31. The fin top width (Wfin) and height (Hfin) are fixed at 10 nm. N-type doping is used and the doping concentration (Nd) is 1×10^{19} cm^{-3}. The gate work function, the supply voltage (Vdd) and the equivalent gate-oxide thickness (EOT) are set to be 5.0 eV, 0.81 V and 1.6 nm [46], respectively. Its vertical non-uniformity is evaluated by the sidewall angle Θ, which is normally kept between 78° and 90° according to the fabrication process [47, 48]. For probing the performance influence induced by the fringe capacitance, the thickness of metal gate is set to be uniform except around the corner.

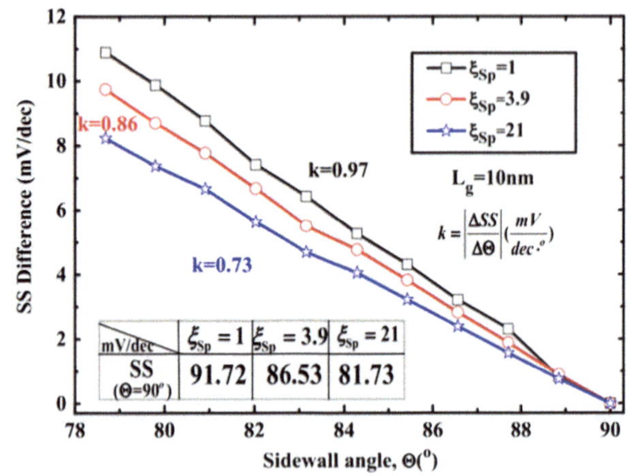

Fig. 7.32 The dependence of SS difference on the sidewall angle Θ

7.4.2.1 The effect of high-k spacer on the subthreshold characteristics variation

The subthreshold swing (SS) difference increases with the decreasing sidewall angle Θ, as shown in Fig. 7.32, where the SS difference is defined as the variation of SS values between the non-ideal sidewall angle case and the ideal case ($\Theta = 90°$). As for the $\Theta = 90°$ case, the SS value in JMTs degrades from 91.72 to 81.73 mV/dec when the dielectric constant of spacer increases from 1 to 21. It is caused by the stronger fringe capacitance applied on the source/drain region through the spacers, especially at the OFF-state owing to the volume depletion mechanism of JMT as the higher dielectric constant of spacer is introduced. Thus, the short channel immunity of JMTs is enhanced. As for the non-ideal sidewall angle case, the SS value in JMTs increases almost linearly with the decrease of the sidewall angle Θ for all the ξSp conditions. Due to its linearity characteristics, the slope k is consequently introduced to evaluate the extent of SS variation as:

$$k = \left|\frac{\Delta SS}{\Delta \Theta}\right| \tag{7.53}$$

The variation slope k is observed to be more suppressed with higher dielectric constant spacers. As shown, when $\xi Sp = 1$, 3.9 and 21, k is equal to 0.97, 0.86 and 0.73, respectively.

Figure 7.33 shows the dependence of the drain-induced barrier lowering (DIBL) voltage on the sidewall angle Θ. For the clarity of the DIBL variation, the DIBL effect is evaluated by the DIBL difference, which is also calculated as the difference in value of the non-ideal sidewall angle case and the ideal case. As shown in the inset table, the DIBL values for the $\Theta = 90°$ case are 145, 115 and 83 mV, respectively, when $\xi Sp = 1$, 3.9 and 21 in JMTs. It is also attributable to the better short channel immunity of JMTs caused by the stronger fringe capacitance with the

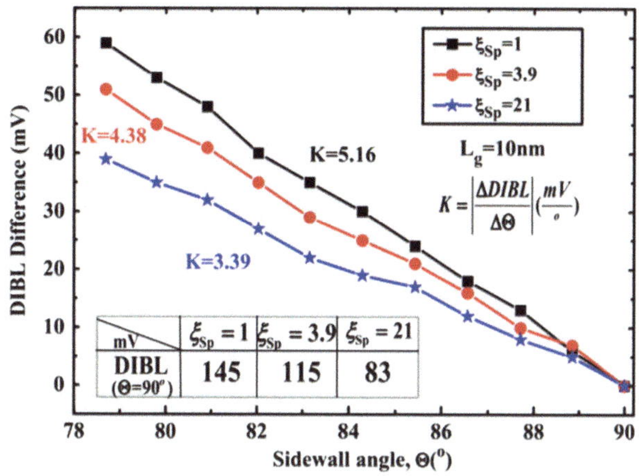

Fig. 7.33 Drain-induced barrier lowering (DIBL) difference versus the sidewall angle Θ. The *inset* table shows the DIBL values for the ideal case.

higher dielectric constant of spacers. Furthermore, the linear relationship between the DIBL value and the Θ is still observed as the SS characteristics do. Hence, the variation of DIBL voltage is measured by using the slope K:

$$K = \left|\frac{\Delta \text{DIBL}}{\Delta\Theta}\right| \tag{7.54}$$

As the dielectric constant ξSp increases from 1 up to 21, the DIBL variation is better suppressed, the slope K is changing from 5.16 to 3.39, respectively. In short, the suppression of subthreshold characteristics variations for JMTs with smaller sidewall angle is better improved by high-k spacers although the SS/DIBL characteristics are degenerated.

As for the triple gate structure of JMT, the weakest control of channel appears approaching the fin bottom owing to the additional control from the top gate. In particular, the electric field in JMT is high when it is in the subthreshold state due to its volume depletion mechanism. The electric field thus strongly corresponds to the gate control, since the subthreshold leakage current often occurs at the channel position with the weakest control [27, 49, 50]. Therefore, the subthreshold characteristics can be characterized by the lowest electrical field in the channel of JMTs, which is dependent with the spacer dielectric constant and the sidewall angle.

Figure 7.34 shows the electrical field distribution along z direction, the value is extracted under the same electron density when it is the minimum in the x-y cross section. Due to the volume depletion in JMTs, the lower electrical field represents the stronger depletion and the better gate control of the channel under the same electron density. As the sidewall angle Θ decreases from 90° to 78.7°, the electrical field increases, thus the gate control is reduced. And as the high-k spacer is introduced, the electrical field is effectively reduced with the broader depletion region, which explores more superior subthreshold characteristics, as shown above. The minimum electrical field in the channel raises from 7.71 × 105 to 9.89 × 105 Vm^{-1}, and from 1.62 × 105 to 2.36 × 105 Vm^{-1} for ξSp = 1 and

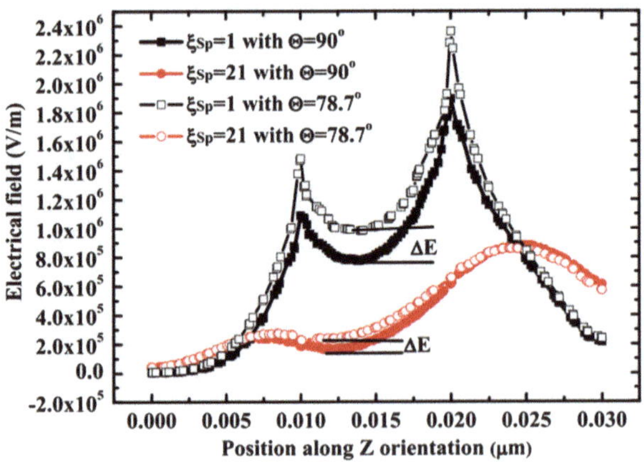

Fig. 7.34 Electric field distribution along z direction

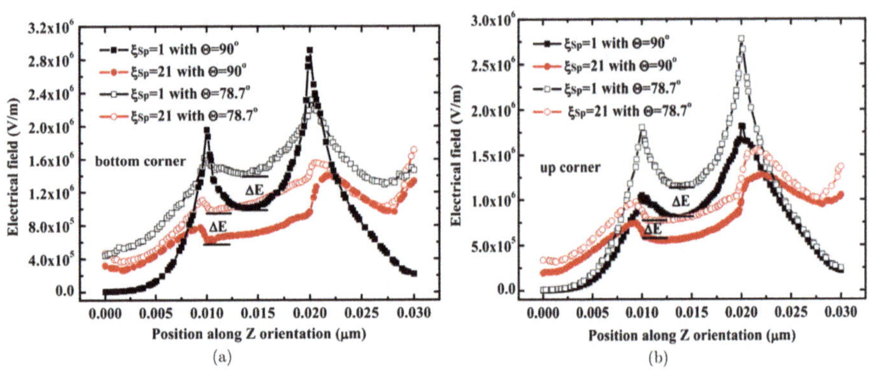

Fig. 7.35 Corner electric field distribution along z direction, (**a**) bottom corner, (**b**) top corner

21, respectively. More importantly, the difference of the minimum electrical field ΔE between Θ = 90° and Θ = 78.7° at ξSp = 21 is smaller than the one at ξSp = 1, which represents the enhanced suppression capability of the subthreshold characteristics variation as the high-k spacer is introduced.

For better understanding of the underlying physical mechanism, the corner electric field distributions under the same electron density are shown in Fig. 7.35. The corner effect enhances the electrical field in the corner of JMTs under the subthreshold state, that causes increased fringe capacitance besides increasing the cross-sectional area of metal gate with the reducing Θ. Consequently, the fringe capacitance depletes the channel more effectively through the corner with the introduction of high-k spacer. As shown in Fig. 7.35a for the bottom corner case, the ΔE between Θ = 90° and Θ = 78.7° is 4.34 × 105 and 3.61 × 105 Vm^{-1} for ξSp = 1 and 21, respectively. Another case is for the up corner as shown in Fig. 7.35b, the minimum electrical field in the channel increases from 7.89 × 105 to 1.14 × 105 Vm^{-1} when ξSp = 1. As the high-k spacer is introduced, the minimum electrical field changes from 5.85 × 105 to 7.67 × 105 Vm^{-1} for

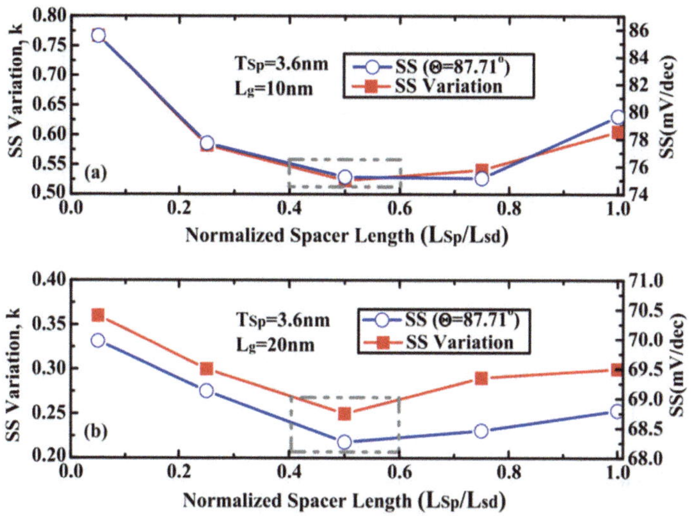

Fig. 7.36 Variation slopes k and SS value versus the spacer length for different channel length L_g, (**a**) $L_g = 10$ nm, (**b**) $L_g = 20$ nm

$\Theta = 90°$ and $\Theta = 78.7°$, whose ΔE is nearly half as much as that at $\xi Sp = 1$. The up corner is more sensitive than the bottom one for changing the difference of the minimum electric field ΔE. It is ascribed to the SOI structure in JMTs, the bottom corner effect of Fin is reduced as compared to the top one. The fringe capacitance through the up corner plays a dominant role in suppressing the subthreshold characteristic variation.

7.4.2.2 The optimization of high-k spacer

Considering two general process methods of spacer, high-k spacer is deposited after the formation of gate dielectric and metal gate, and the chemical mechanical polishing is used. Alternatively, high-k spacer is deposited at first, then gate dielectric and metal gate layers are formed by photo etching and RIE. Hence, either high-k spacer is as thick as the sum of gate dielectric and metal gate or the length of high-k spacer is equal to the dimension of source/drain region. Consequently, the process for the high-k spacer can refer the fabrication of metal gate and source/drain region.

Figure 7.36 shows the impact of the spacer length on the SS variation at Tsp = 3.6 nm. The SS slope k reduces rapidly with the spacer length both for Lg = 10 nm and 20 nm when it is small. As the spacer length increases further, the slope k raises again. Hence, the suppression capability of the SS variation is degenerated. The optimized value is observed at a half length of the source/drain region. The similar tendency of the SS values is also found.

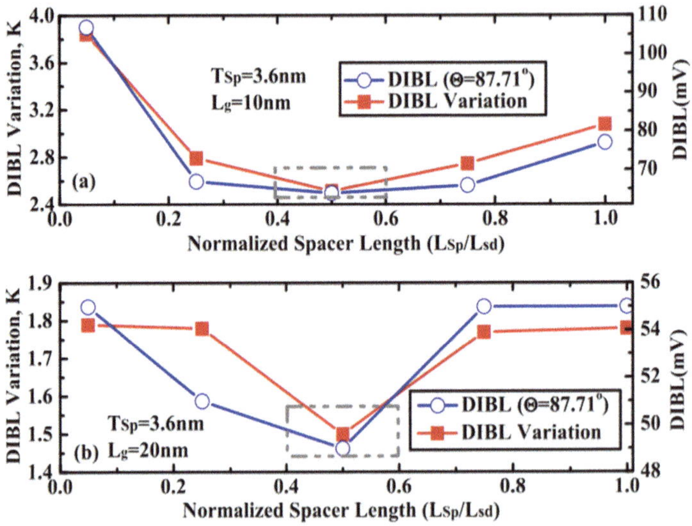

Fig. 7.37 Variation slopes K and DIBL value as a function of the spacer length for different channel length Lg, (**a**) Lg = 10 nm, (**b**) Lg = 20 nm

The dependence of the DIBL value and DIBL variation for different channel length on the spacer length is shown in Fig. 7.37. Similar to the SS value and its variation slope, the DIBL value and its slope K present the consistent tendency. As for the $\Theta = 87.71°$ case at Lg = 10 nm, the DIBL is 107 mV when the normalized spacer length equals 0.05. The DIBL value becomes 64 mV as the normalized spacer length increases to 0.5, while as the spacer covers the whole source/drain region, its DIBL value increases up to 77 mV. As for the DIBL variation case, its tendency with the changing spacer length is similar. The smallest slope K value also occurs at a half length of the source/drain region.

Both the SS/DIBL value and variation slopes change with the spacer length because of the changing impact of the fringe capacitance and the corner effect, which affects the effective gate capacitance (Ceff) in JMTs. The fringe electric field starts from the side of metal gate, then penetrates through the spacers and source/drain region, ending at the source/drain contact. Hence, its fringe capacitance includes the spacer part (Cof) and the source/drain region (Cif). Consequently, the effective gate capacitance Ceff can be expressed as:

$$C_{eff} \approx Series(C_{ox}, C_{Si}) \parallel Series(C_{of}, C_{if}), \tag{7.55}$$

where Cox and C_{Si} are the gate dielectric capacitance and silicon body capacitance, respectively. The changing of fringe capacitance affects the channel depletion induced by the corner when the JMT is in the subthreshold state. At the same time, according to the gate control ability being related to (C_{Si} || Cif) /Ceff, the increase of Cof and the decrease of Cif induces gate control ability enhanced when

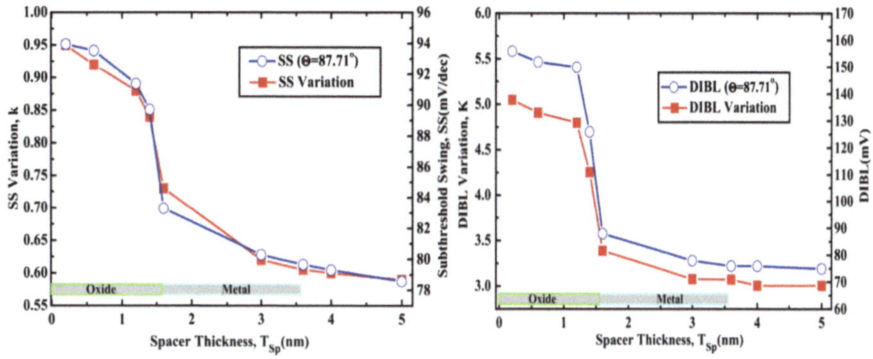

Fig. 7.38 (**a**) Dependences of the SS variation on the spacer thickness. $L_{sp} = L_{sd}$. $L_g = 10$ nm. (**b**) DIBL variation as a function of the spacer thickness. $L_{sp} = L_{sd}$. $L_g = 10$ nm

the spacer length is small. As the spacer length further increases, the gate control capability is turned to be degraded. Correspondingly, the stronger the gate control, the better the suppression of SS/DIBL with the changing sidewall angle, which causes the variation slopes k and K to be optimized around 0.5Lsd, also changing the SS/DIBL values, as shown in Figs. 7.36 and 7.37.

As shown in Fig. 7.38a, the SS and its variation slope k decrease with the spacer thickness Tsp. When the Tsp is much smaller than the thickness of gate dielectric, the fringe electric field cannot efficiently affect the source/drain region through the spacers. Then, the SS and its variation k almost reduce gradually as the Tsp increases. However, as the Tsp approaches the metal gate, the fringe electric field through the spacer rises drastically, thus improving the depletion capability and inducing enhanced suppression of subthreshold characteristics variation. As the Tsp continues increasing, more impact from the edge of the metal gate is applied at the source/drain region through the spacers, which further extends the minimum depletion length and simultaneously reduces the variation of subthreshold characteristics. Until the spacer is raised over the metal gate, the fringe capacitance between the top surface of metal gate and the source/drain extension region is almost negligible as compared with that from sidewall gate because the electric flux strongly depends on the smallest distance between the electrodes [51]. Thus, the SS value and its slope k finally tend to remain constant. In Fig. 7.38b, a similar tendency is shown for the DIBL value and its slope K. When the Tsp increases from 0.2 to 1.2 nm, the DIBL voltage only reduces 6 mV and its slope K changes 0.2. While as the Tsp is between 1.4 and 1.6 nm, the DIBL drastically drops from 126 to 88 mV. At the same time, the DIBL slope K changes from 4.26 to 3.38. when the Tsp increases up to 3.6 nm, the DIBL voltage decreases gradually to 78 mV and its slope is down to 3.08. The DIBL and its slope K finally flatten out as the Tsp becomes larger than 3.6 nm.

7.4.2.3 Junctionless Transistors with Strain Engineering

Mechanical strain is a widely used technique to increase carrier mobility in the channel of silicon conventional MOSFETs. As for junctionless, the high doping concentration of the channel which ranges from 1×10^{18} to 5×10^{19} cm^{-3} would reduce the effective carrier mobility and thus decrease the saturation current. Hence some researches have applied the strain technology to the junctionless to boost the performance.

7.4.2.4 Process of Strain Technology [52]

Figure 7.39 demostrates a fabrication processes flow of the Junctionless MOSFETs on self-aligned n-doped SiNWs grown by electric-field assisted assembly technique. First, source and drain electrodes were made by depositing Ti on the heavily doped p-type Si substrate coated with a 200 nm thermal oxide and patterned by photolithography. Then, the position of Au nanoparticles with 20 nm in diameter was defined by photolithography. Then after spraying the 20 nm Au nanoparticles, the sample was put on the quartz holder and clipped by the stainless steel clamp to perform the VLS growth with applied electric field. A 10 nm thick A12O3 layer that served as the gate dielectric was then deposited on the surface of the SiNWs using atomic layer deposition process, and a top Pt gate was deposited using e-beam evaporation and defined by photolithography.

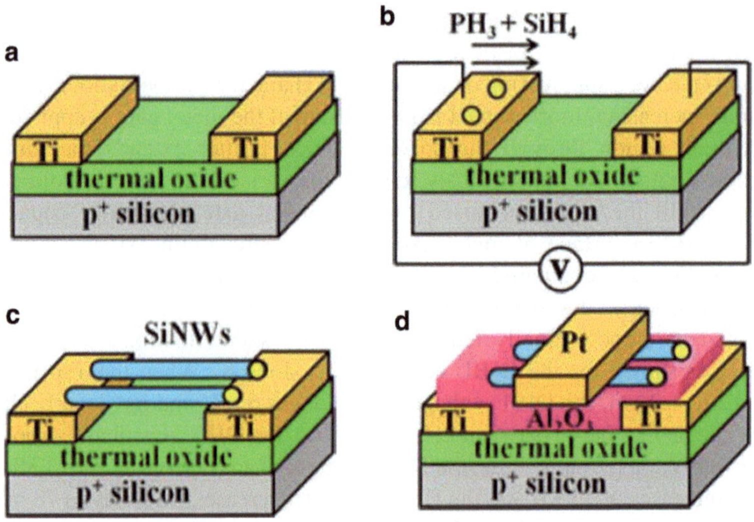

Fig. 7.39 The fabrication process of the position and direction controlled SiNWs JLFET. (**a**) Ti electrodes were patterned on p-type Si wafer with thermaloxide. (**b**) Au catalysts were dripped on pre-defined position. The sample was then clipped by the stainless steel clamp and put on the quartz holder to initiate SiNWs growth. (**c**) Results of the self-aligned SiNWs. (**d**) Dielectric layer of Al2O3 and top gate of Pt were deposited on SiNWs [52]

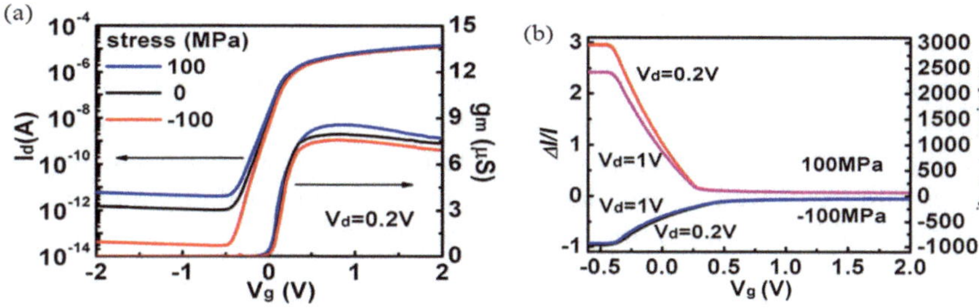

Fig. 7.40 (**a**) Id-Vg characteristics and transconductance, and (**b**) the change indrain current as a function of Vg in the JLFET with and without stress [52]

7.5 Results and Discussion

The mechanical stress applied on the JLFET is assumed to be uniformly distributed in longitude direction. Figure 7.40a shows the Id-Vg curves and transconductances (gm) for the same device at stress values of 0 and 100 MPa and the corresponding current variation ΔI/I0 is shown in Fig. 7.40b. In the linear regime, the stress-induced change in drive current for the JLFET exhibits quite different behavior compared with that for the typical MOSFET case. In stressed MOSFETs, the absolute value of ΔI/I0 decreases with gate bias. It is because carriers flowing along a very thin channel surface layer suffer from scattering by gate field that suppresses the piezoresistance coefficient, and thereby decreasing I0. As Vg increases in the negative direction beyond the threshold voltage, ΔI/I0 varies exponentially as a function of the gate voltage. The underlying mechanism of huge electromechanical response in the JLFET operated in subthreshold regime is relative to the reduction of effective channel width, which is achieved by expanding the depletion region with decreasing gate bias, and thus raises the piezoresistance coefficient. Figure 7.40b shows that when the fully pinch-off condition is reached, the compressive stress of 100 MPa decreases the leakage current, whereas the tensile stress of 100 MPa increases the leakage current. And the absolute value of piezoresistance coefficient is nearly independent of Vd in the linear region and decreases only slightly with increasing Vd in subthreshold region. These experimental results demonstrate the possibility of enhancing mobility in heavily doped silicon junctionless MOSFETs using strain technology.

7.6 Fabrication Variation Impact

The performance of junctionless MOSFET is strongly dependent on the width of channel which is often varied by the fabrication variation. The post-lithography baking processes of the photoresist and the optical absorption in the photoresist would result in the cross section of the photoresist being not ideally rectangular and more trapezoidal [53, 54]. This shape is transferred by fin etching [55]. At the same time, non-ideal anisotropic over-etching, which often occurs during the fabrication, also leads to the cross section of the fin body being trapezoidal [55–57]. Futhermore, the line-edge roughness (LER) is inevitable as a result of the lithography and etching process during fabricating the nanoscale junctionless devices [58]. As is well known, the vertical nonuniformity and the variations of fin width significantly affect the device characteristics [59].

7.6.1 The Impact of Trapezoidal Fin Body on Junctionless MOSFET

A three-dimensional schematic view of a Junctionless MOSFET with trapezoidal fin body is shown in Fig. 7.41. The transfer characteristics of Junctionless with different sidewall angles (Θ) is shown as Fig. 7.42. With a decrease in from 90 to 78.7, the threshold voltage (Vth) shifts from 0.44 to 0.32 V owing to the increase in the fin cross section, which causes the conduction channel to emerge at a small gate voltage [62]. Meanwhile, SS increases from 71 to 76 mV/dec with a reduction in Θ. Figure 7.43 shows plots of the characteristics of DIBL for both the JMT and IM-MOSFET. It can be seen that the DIBL voltage increases almost linearly with the reduction in for both the JMT and IM-MuGFET. As the Hfin of the JMT increases from 5 to 20 nm, DIBL is enhanced and the slope of the DIBL vs Θ increases from 0.5 to 6.83. This is due to the reduced series resistance of the source/drain, which shifts the electric field peak to the channel as Hfin increases and decreases [63], which extends the enhanced effect of the drain voltage in the channel region and enhances the affected area of the drain voltage. Thus, the short-channel effect immunity of the JMT degrades with increasing Hfin and decreasing. The fluctuations in SS and DIBL caused by the variation are significant.

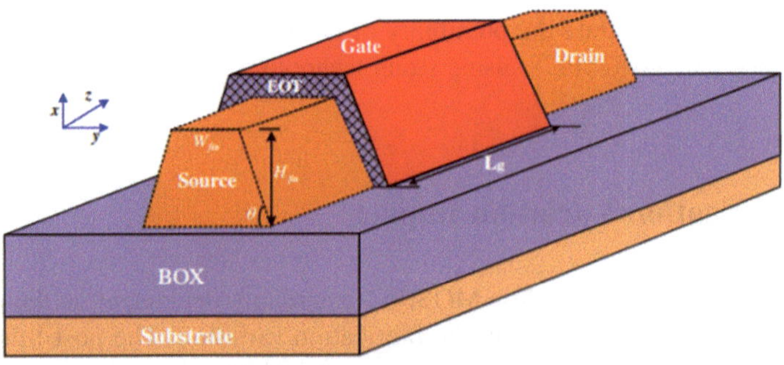

Fig. 7.41 Schematic view of JMT with trapezoidal fin body on BOX substrate [65, 66]

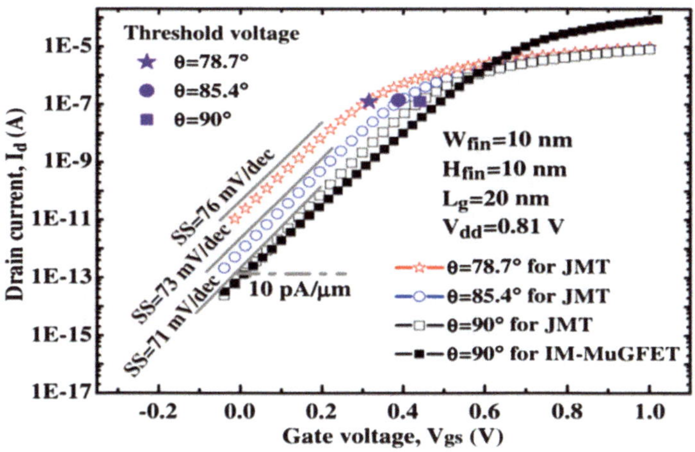

Fig. 7.42 Transfer characteristics with different, $V_{dd} = 0.81$ V, Wfin = Hfin = 10, and Lg = 20 nm [65, 66]

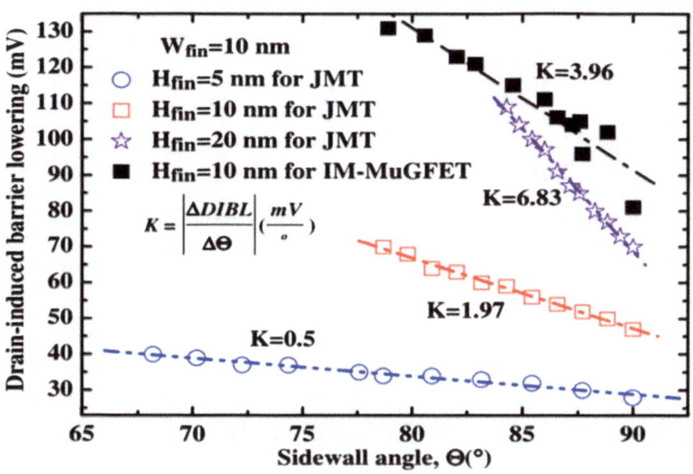

Fig. 7.43 DIBL voltages for both JMT and IM-MuGFET vs sidewall angle (Θ), at Wfin = 10 and Lg = 20 nm. The DIBL voltage is calculated from the difference in the threshold voltage between Vds = 0.05 and 0.81 V [66]

7.6.2 The Impact of LER on Junctionless MOSFET

As mentioned before, the line edge roughness (LER) is inevitable in lithography and etching process, and seriously impairs the performance of Junctionless MOSFET [16]. The LER patterns are generated from Gaussian model (7.55), which is approximate to the experimental line edge roughness as compared with

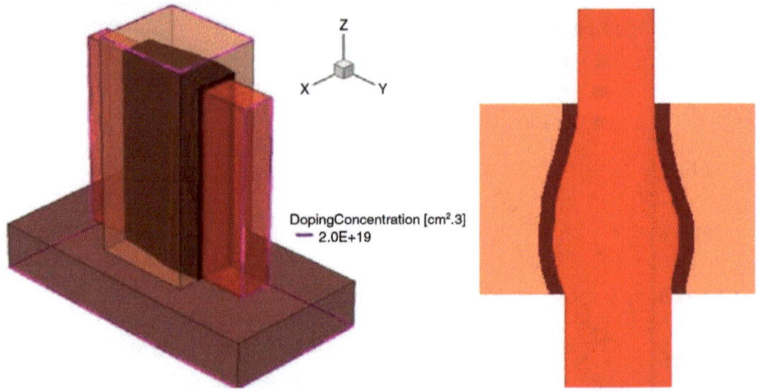

Fig. 7.44 The 3-D schematics and 2-D cross section of simulated junctionless FinFET [61]

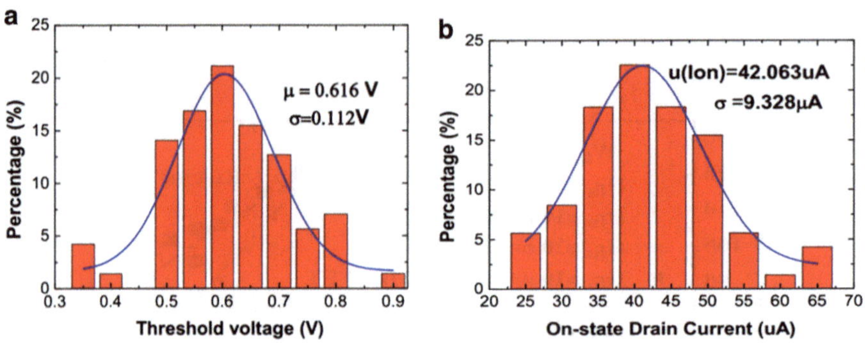

Fig. 7.45 Distributions of Threshold voltage (**a**) and On-state current (**b**) of JL FinFET due to LER effect [61]

the SEM photograph [64]. In order to obtain qualified accuracy of the LER patterns, the two key parameters: root mean square Δ and correlation length Λ of the Gaussian model, are set to 1 and 15 nm respectively.

$$R(x) = \Delta e^{(-(x^2/\Lambda^2))} \quad (7.56)$$

Figure 7.44 presents both three-dimensional schematic and two dimensional cross section of the simulated junctionless FinFET with LER. First, the distributions of the threshold voltage (Vt) and on-state drain current (Ion) in JL-FinFET and IM-FinFET are shown in Figs. 7.45 and 7.46. In JL-FinFET, the Fluctuations exceeds 18 %, with the average threshold voltage of 0.616 V and already surpass over 22 %, with average Ion value of 42.063 uA. Whereas in IM-FinFET the fluctuations are 2.6 % and 16 % respectively. Both of threshold voltage and leakage current suggest the LER effect causes larger variations on JL-FinFET than IM-FinFET. But as for the on-state current, the impact of LER has caused similar fluctuation in both kinds of transistor.

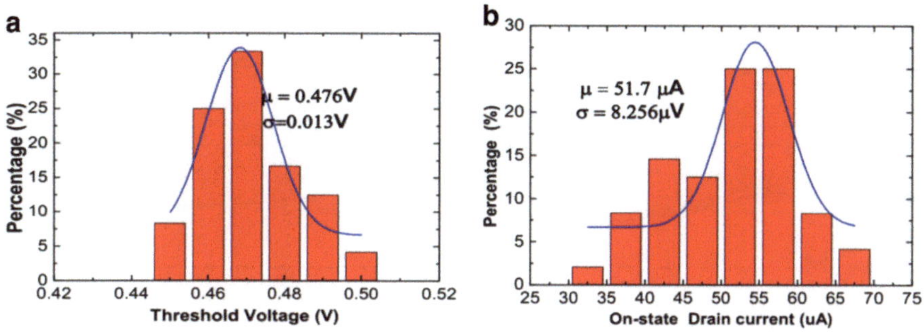

Fig. 7.46 Distributions of Threshold voltage (**a**) and On-state current (**b**) of IM FinFET due to LER effect [61]

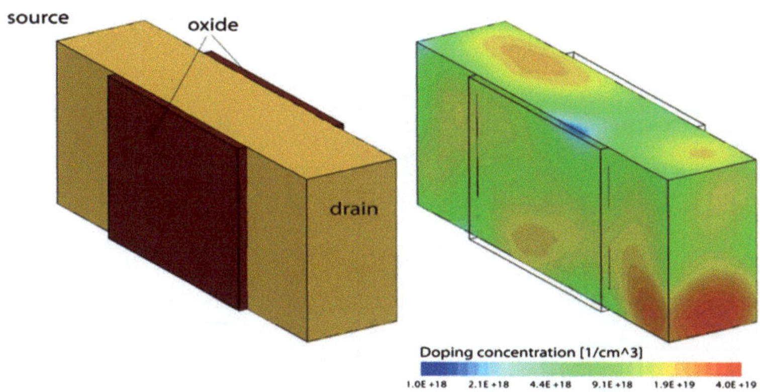

Fig. 7.47 JL-FinFET structures used for device simulations (left) without RDF and (right) with RDF. The nominal doping concentration is a uniform 2×10^{19} cm^{-3} in the left structure, whereas, in the right structure, the local doping concentration is nonuniform and varies between 10^{18}–4×10^{19} cm^{-3} [65]

7.6.3 The Impact of RDF on Junctionless MOSFET [65]

Variability of junctionless field-effect transistor performance due to random dopant fluctuation (RDF) is projected to be a serious problem when scaled to nanoscale dimensions as the total number of dopant atoms becomes increasingly discretized [65, 66]. Three-dimensional n-type double gate JL-FinFETs with RDF shown in Fig. 7.47 were generated using Sentaurus TCAD.

The RDF impact is measured via the standard deviation of $V_{T,lin}$, $V_{T,sat}$, Ion, Ioff, SS, and DIBL (all normalized to their baseline values) with results presented in Fig. 7.48 as a function of fin height Hfin from 10 to 40 nm. Several observations are apparent: (1) the magnitude of device variability is quite large for all cases, particularly in terms of $\sigma V_{T,lin}$, $\sigma V_{T,sat}$, σIon, and σIoff; (2) the curves show an

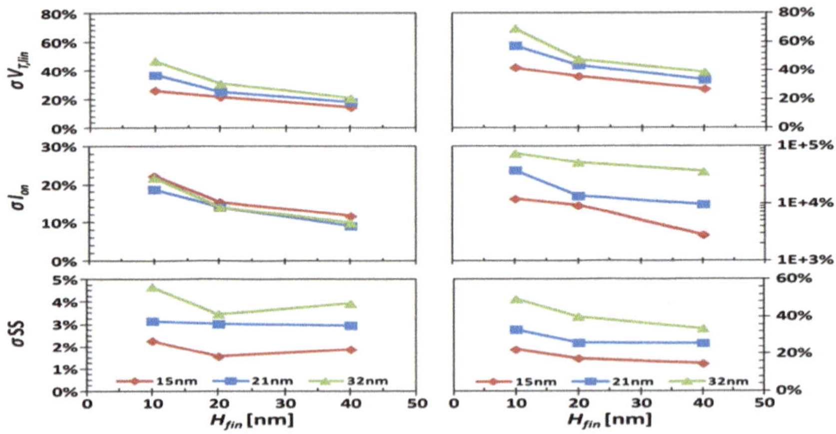

Fig. 7.48 Standard deviations of extracted parameters in JL-FinFETs due to RDF versus fin height and technology node [65]

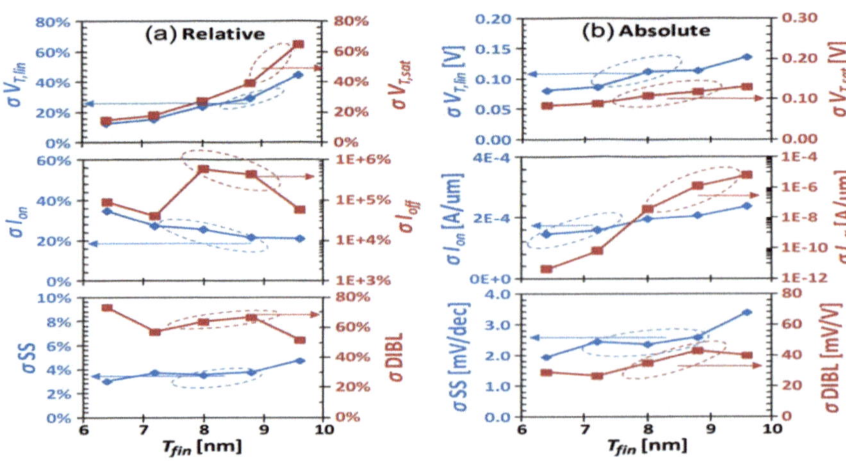

Fig. 7.49 (**a**) Relative variation and (**b**) absolute variation of JL-FinFET performance due to Tfin scaling from 9.6 to 6.4 nm with H_{fin} = 10 nm [67]

inverse dependence with Hfin, which correlates with Pelgrom's law [67]; and (3) more aggressively scaled technologies tend to exhibit less variation with the exception of σIon. The first observation is alarming but not unexpected since V_T and parameters that depend on it are highly sensitive to the actual channel doping profile and, hence, RDF. With σVT,sat ranging between 20 and 60 %, σIon between 10 and 20 %, and σIoff between 103 and 106 %, the effect of RDF is comparable to fin LER up to an amplitude of 1 nm [38]. The second observation is consistent with traditional scaling of the "channel width," (i.e., Hfin in a FinFET).

On the other hand, when Tfin is scaled in Fig. 7.49, we see that smaller fin thickness results in less relative and absolute threshold voltage variation from RDF.

In addition, the magnitudes of $\sigma V_{T,lin}$ and $\sigma V_{T,sat}$ vary dramatically when Tfin is scaled from 6.4 to 9.6 nm. Recent work [69] has illustrated how the threshold voltage of junctionless nanowire FETs becomes less sensitive to changes in channel doping when the nanowire geometry is miniaturized, which is a consequence arising from improved electrostatic control of the buried channel from the gate. Findings in [68] also agree, although the results suggest that a much larger change in RDF sensitivity should be expected when the body dimensions are scaled to smaller values. Therefore variations in threshold voltage, drive current, leakage current, and drain-induced barrier lowering are heavily impacted by RDF for junctionless FinFETs with sufficiently high channel doping (greater than 10^{19} cm^{-3}).

7.6.4 The Charge-Plasma Concept on Junctionless MOSFET

A new method, by introducing the charge-plasma concept, is proved to have a great effect on suppressing the severe impact of LER on Junctionless transistor [70]. The charge-plasma concept is employed for Junctionless transistor to form S/D N-region in doping-less silicon film by choosing an appropriate metal workfunction electrode [71–74]. The doping-less JLT induces constant electron plasma in S/D region by depositing lateral and top metal over high-k oxide. Figure 7.50 shows the structure of N-channel double-gate conventional JLT and doping-less JLT.

There are two essential requirements in this concept. First, the workfunction of metal used in S/D electrode contact should be less than that of silicon for N-channel MOSFET. Second, the thickness of the silicon body should be less than the Debye length [65].

Figure 7.51 shows 50 groups electrical transfer characteristic curves of conventional JLT and doping-less JLT with the impact of different line edge roughness. We can show that the performance of conventional JLT and doping-less JLT are both affected by LER due to JLT's bulk conduction characteristics. However, it is quite obvious that the LER has more severe impact on conventional JLT. The off-state current of conventional JLT ranges from 3×10^{-14} to 1×10^{-5} A in which

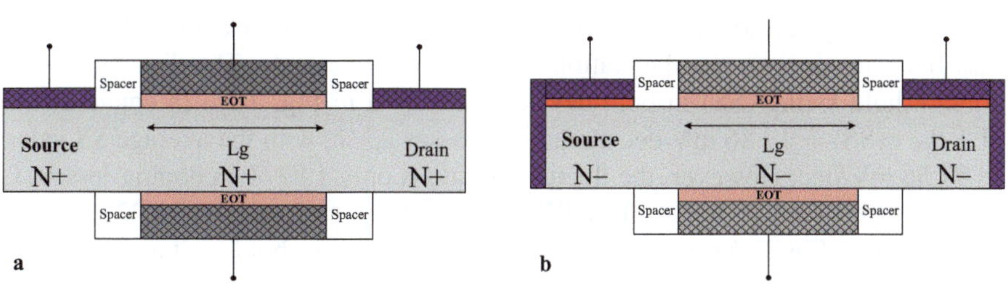

Fig. 7.50 Structure of (**a**) conventional JLT and (**b**) doping-less JLT [70]

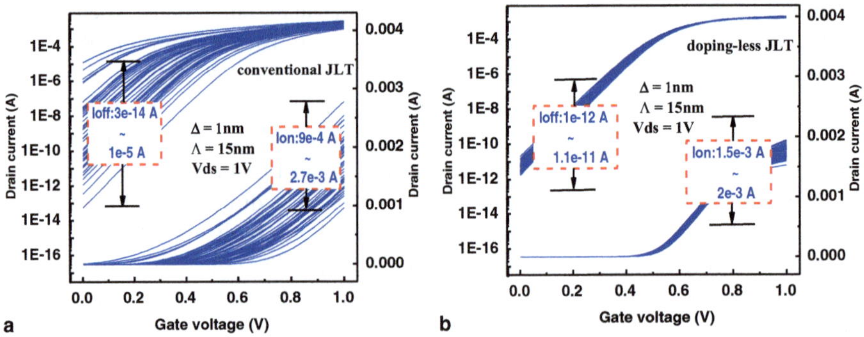

Fig. 7.51 Id-Vg curves of (**a**) conventional JLT and (**b**) doping-less JLT with LER [70]

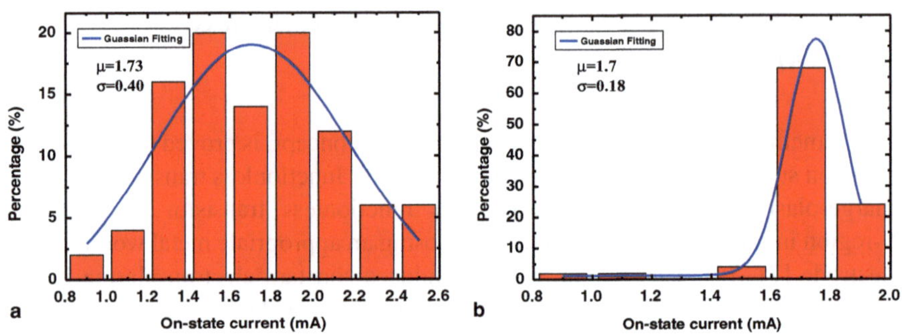

Fig. 7.52 Distributions of On-state current of (**a**) conventional JLT and (**b**)doping-less JLT with LER [70]

the fluctuation surpasses 380 %. Nevertheless, the LER impact of doping-less JLT on the off-state current ranges only from 1×10^{-12} to 1.1×10^{-11} A. The reason is that the dopant concentration is much smaller in doping-less JLT which makes the channel much more easily be fully depleted. The result indicates that the doping-less JLT has a superior immunity to LER.

Figure 7.52 shows the distributions of the on-state current (Ion) in conventional JLT and doping-less JLT. In conventional JLT, the σ(Ion) = 0.4 mA exceeds 23 %, with the average Ion value of 1.73 mA. Whereas the σ(Ion) = 0.18 mA exceeds 10 % in doping-less JLT, with the average Ion value of 1.70 mA. Thus the doping-less JLT can suppress the fluctuations of Ion very effectively. The distributions of subthreshold swing (SS) in both devices are shown in Fig. 7.53. In conventional JLT the σ(SS) = 17.86 mV/dec already surpasses 23 %, with the average SS value of 77.56 mV/dec. However, the fluctuation rate is only 1.82 % in doping-less JLT, with a lower average value of 66.95 mV/dec and lower σ(SS) = 1.22 mV/dec. Furthermore, Fig. 7.54 shows the distributions of threshold voltage (Vth). In conventional JLT the $\sigma(V_{th}) = 0.12$ V, with the average Vth value of 0.32 V, is larger than the $\sigma(V_{th}) = 0.01$ V in doping-less JLT, with the average Vth value of

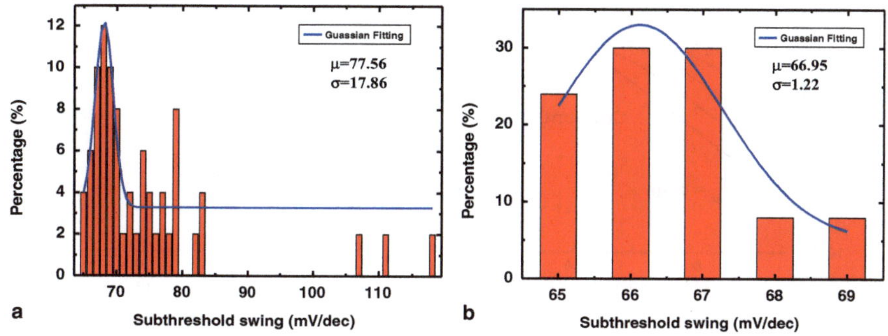

Fig. 7.53 Distributions of Subthreshold swing of (**a**) conventional JLT and (**b**) doping-less JLT with LER [70]

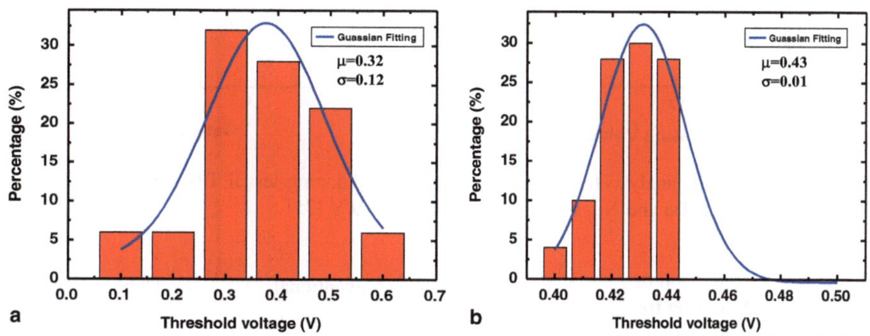

Fig. 7.54 Distributions of Threshold voltage of (**a**) conventional JLT and (**b**)doping-less JLT with LER [70]

0.43 V. The sensitivities of Vth changes by 37.8 % and 2.6 % respectively. The results show that the doping-less Junctionless transistor based on charge-plasma concept has great immunity to LER.

Furthermore, the Junctionless MOSFET based on charge-plasma concept also immunes to RDF [75]. Figure 7.55c, d show the sensitivities of devices for the changes of ±10 % and ±20 % in Nd. The ΔVth/Vth, ΔIon/Ion and ΔSS/SS for +20 % variation in Nd are summarized in Table 7.1. In conventional JLT due to shift in Vth and subthreshold slope, there is significant variation in OFF-state current for +20 % change in Nd that can be seen in Fig. 7.55c, but no variation is observed for doping-less JLT, as shown in Fig. 7.55d. It indicates that the doping-less JLT shows excellent control of short-channel effects in presence of RDF. There is no variation of Vth, Ion and SS being observed for doping-less JLT with the changes of +20 % in Nd, as shown in Table 7.1. It indicates that doping-less JLT shows excellent immunity to the RDF.

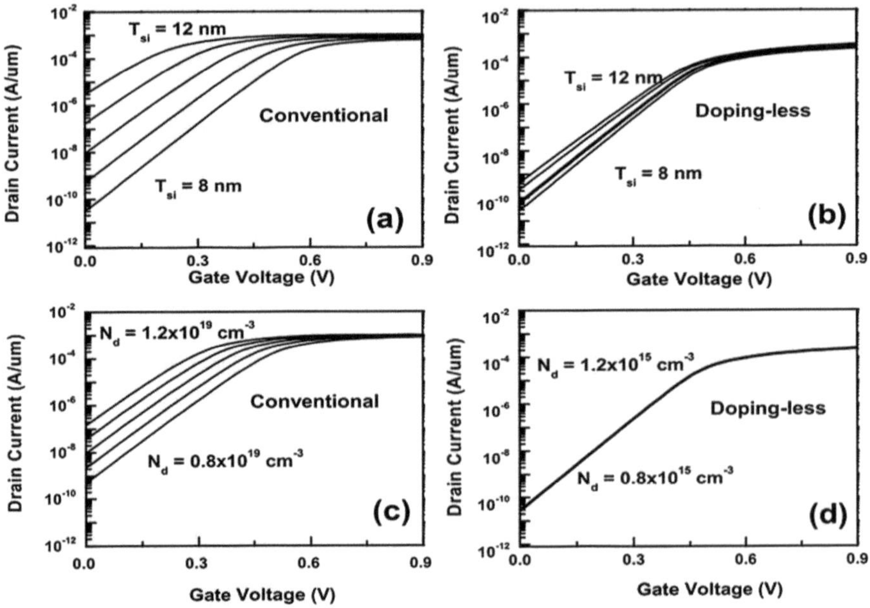

Fig. 7.55 Sensitivity analysis for conventional (**a**), (**c**) and doping-less JLT (**b**), (**d**) for ±10 % and ±20 % variation in Tsi and Nd, respectively at Vds = 0.9 V [75]

Table 7.1 Comparison of Performance Variation for Conventional and Doping-Less JLTs [75]

N-type	Conventional		Doping-less	
Parameters	N_d	T_{si}	N_d	T_{si}
$\Delta V_{th}/V_{th}$ (%)	50.9	102.77	0	15.22
$\Delta I_{on}/I_{on}$ (%)	21.4	43	0	42.63
ΔSS/SS (%)	15.2	22.5	0	10.9

7.7 Conclusion

This chapter introduces the novel transistor, junctionless transistor, from its basic electrical performance and process variation impact to device physics and modeling. The analytical full range drain current models for long and short channel junctionless MOSFETs are proposed, the accuracy and efficiency are ensured as applied in the circuit simulation due the inclusion of source/drain depletion effect and non-iteration. The influences of dual-material gate structure, high-k spacer and strain engineering on junctionless transistor for its current drive, tunneling leakage current and scaling capability are introduced in detail. The process variation impact, including Fin vertical nonuniformity, line edge roughness and random dopant fluctuation, are investigated, then the charge plasma method is successfully utilized to suppress the process variation impact. This work may have a useful guide to understand the physics of junctionless transistor and promote its application in circuit design.

References

1. J.-P. Colinge, C.-W. Lee, A. Afzalian, et al., Nanowire transistors without junctions. Nat. Nanotechnol. **5**(3), 225–229 (2010)
2. C.-W. Lee, I. Ferain, A. Afzalian, et al., Performance estimation of junctionless multigate transistors. Solid-State Electron. **54**(2), 97–103 (2010)
3. J.P. Colinge, C.W. Lee, I. Ferain, et al., Reduced electric field in junctionless transistors. Appl. Phys. Lett. **96**(7), 073510 (2010)
4. R. Rios, A. Cappellani, M. Armstrong, et al., Comparison of junctionless and conventional trigate transistors with down to 26 nm. IEEE Electron Device Lett. **32**(9), 1170–1172 (2011)
5. H. Lou, L. Zhang, Y. Zhu, X. Lin, S. Yang, J. He, M. Chan, A junctionless nanowire transistor with dual-material-gate. IEEE Trans. Electron Devices **59**(7), 1829–1836 (2011)
6. H. Lou, B. Zhang, D. Li, X. Lin, J. He, M. Chan, Suppression of subthreshold characteristics variation for junctionless multigate transistors using high-k spacers. Semicond. Sci. Technol. **30**, 015008 (2015)
7. http://apps.webofknowledge.com.
8. C.H. Park, M.D. Ko, K.H. Kim, R.H. Baek, C.W. Sohn, C.K. Baek, S. Park, M. Deen, Y.H. Jeong, J.S. Lee, Electrical characteristics of 20-nm junctionless Si nanowire transistors. Solid-State Electron. **73**(6), 7–10 (2012)
9. C.H. Park, M.D. Ko, K.H. Kim, S.H. Lee, J.S. Yoon, J.S. Lee, Y.H. Jeong, Investigation of low-frequency noise behavior after hot-carrier stress in an n-channel junctionless nanowire MOSFET. IEEE Electron. Device Lett. **33**(11), 1538–1540 (2012)
10. S. Barraud, M. Berthome, R. Coquand, M. Casse, T. Ernst, M.-P. Samson, P. Perreau, K. Bourdelle, O. Faynot, T. Poiroux, Scaling of trigate junctionless nanowire MOSFET with gate length down to 13 nm. IEEE Electron. Device Lett. **15**(6), 22–27 (2000)
11. S. Migita, Y. Morita, M. Masahara, H. Ota, Electrical performances of junctionless-FETs at the scaling limit (LCH=3 nm). in IEEE International Electron Devices Meeting (IEDM), 2012, pp. 861–864
12. D.D. Zhao, T. Nishimura, C.-H. Lee, R. Ifuku, K. Nagashio, K. Kita, A. Toriumi, Junctionless Ge MOSFETs fabricated on 10 nm-thick GeOI substrate. ECS Trans. **35**(3), 457–464 (2011)
13. R. Yu, I. Ferain, N.D. Akhavan, P. Razavi, R. Duffy, J.P. Colinge, Characterization of a junctionless diode. Appl. Phys. Lett. **99**(99), 013502–013502-3 (2011)
14. Z. Ding, G. Hu, R. Liu, L. Wang, S. Hu, X. Zhou, Analytical models for the electric potential, threshold voltage and drain current of long-channel junctionless double-gate transistors. J. Korean Phys. Soc. **62**(8), 1188–1193 (2013)
15. X. Jin, X. Liu, M. Wu, R. Chuai, J.-H. Lee, J.-H. Lee, A unified analytical continuous current model applicable to accumulation mode (junctionless) and inversion mode MOSFETs with symmetric and asymmetric double-gatestructures. Solid-State Electron. **79**, 206–209 (2013)
16. Z. Chen, Y. Xiao, M. Tang, Y. Xiong, J. Huang, J. Li, X. Gu, Y. Zhou, Surface-potential-based drain current model for long-channel junctionless double-gate MOSFETs. IEEE Trans. Electron. Devices **59**(12), 3292–3298 (2012)
17. X. Lin, B. Zhang, Y. Xiao, H. Lou, L. Zhang, M. Chan, Analytical current model for long-channel junctionless double-gate MOSFETs. IEEE Trans. Electron. Devices **63**(3), 959–965 (2016)
18. Y. Xiao, B. Zhang, H. Lou, L. Zhang, X. Lin, A compact model of subthreshold current with source/drain depletion effect for short channel junctionless sylindrical surrounding-gate MOSFETs. IEEE Trans. Electron. Devices **63**, 2176–2181 (2016)
19. F. Liu, J. He, J. Zhang, Y. Chen, M. Chan, A non-charge-sheet analytic model for symmetric double-gate MOSFETs with smooth transition between partially and fully depleted operation modes. IEEE Trans. Electron. Devices **55**(12), 3494–3502 (2008)
20. A. Ben-Israel, A Newton–Raphson method for the solution of systems of equations. J. Math. Anal. Appl. **15**(2), 243–252 (1966)
21. TCAD, *Sentaurus Device Users Manual* (Synopsys, Mountain View, CA, 2013)

22. H.C. Pao, C.T. Sah, Effects of diffusion current on characteristics of metal-oxide (insulator)-semiconductor transistors. Solid-State Electron. **9**(10), 927–937 (1966)
23. R.F. Pierret, J.A. Shields, Simplified long-channel MOSFET theory. Solid-State Electron. **26**(2), 143–147 (1983)
24. A. Ortiz-Conde, R. Herrera, P.E. Schmidt, F.J.G. Sánchez, J. Andrian, Long-channel silicon-on-insulator MOSFET theory. Solid-State Electron. **35**(9), 1291–1298 (1992)
25. i-MOS. [Online]. Available: http://i-mos.org.
26. A. Gnudi, S. Reggiani, E. Gnani, G. Baccarani, Semianalytical model of the subthreshold current in short-channel junctionless symmetric double-gate field-effect transistors. IEEE Trans. Electron. Devices **60**(4), 1342–1348 (2013)
27. J.-P. Colinge et al., Reduced electric field in junctionless transistors. Appl. Phys. Lett. **96**(7), 073510 (2010)
28. R.T. Doria et al., Junctionless multiple-gate transistors for analog applications. IEEE Trans. Electron. Devices **58**(8), 2511–2519 (2011)
29. T. Holtij, M. Schwarz, A. Kloes, B. Iñíguez, 2D analytical potential modeling of junctionless DG MOSFETs in subthreshold region including proposal for calculating the threshold voltage. in Proc. of the IEEE13th Int. Conf. Ultimate Integr. Silicon, 2012, pp. 81–84.
30. S. Cho, K. Kim, B. Park, I. Kang, RF performance and small-signal parameter extraction of junctionless silicon nanowire MOSFETs. IEEE Trans. Electron Devices **58**(5), 1388–1396 (2011)
31. M. Maglietta, E. Zanazzi, F. Jona, D. Jepsen, P. Marcus, LEED structure analysis of a Co{001} surface. Appl. Phys. A Mater. Sci. Process. **15**(4), 409–412 (1978)
32. R. Lander, J. Hooker, J. Van -Zijl, F. Roozeboom, M. Maas, Y. Tamminga, R. Wolters, A tunable metal gate work function using solid state diffusion of nitrogen. in Proc. of the ESSDERC, 2002, pp. 103–106
33. C. Wu, B. Hung, A. Chin, S. Wang, F. Yen, Y. Hou, Y. Jin, H. Tao, S. Chen, M. Liang, HfSiON n-MOSFETs using low-work function HfSix gate. IEEE Electron. Device Lett. **27**(9), 762–764 (2006)
34. A. Borisov, D. Teillet-Billy, J. Gauyacq, H. Winter, G. Dierkes, Resonant charge transfer in grazing scattering of alkali-metal ions from an Al (111) surface. Phys. Rev. B **54**(23), 17166–17174 (1996)
35. W. Song, M. Yoshitake, A work function study of ultra-thin alumina formation on NiAl (1 1 0) surface. Appl. Surf. Sci. **251**(1–4), 14–18 (2005)
36. A. Chaudhry, M. Kumar, Investigation of the novel attributes of a fully depleted dual-material gate SOI MOSFET. IEEE Trans. Electron. Devices **51**(9), 1463–1467 (2004)
37. Y. Li, K. Lee, C. Yiu, Y. Chiu, R. Chang, Dual-material gate approach to suppression of random-dopant-induced characteristic fluctuation in 16 nm metal-oxide-semiconductor field effect-transistor devices. Jpn. J. Appl. Phys. **50**(4), 04DC07-1–04DC07-7 (2011)
38. G. Leung, C. Chui, Variability of inversion-mode and junctionless FinFETs due to line edge roughness. IEEE Electron. Device Lett. **32**(11), 1489–1491 (2011)
39. S. Potbhare, *Modeling and Characterization of 4H–SIC MOSFETs: HighField, High Temperature, and Transient Effects* (Univ. Maryland, College Park, MD, 2008)
40. C. Chen, J. Lin, M. Chiang, K. Kim, High-performance ultra-low power junctionless nanowire FET on SOI substrate in subthreshold logic application. in Proc. IEEE Int. SOI Conf., 2010, pp. 1–2
41. Y. Tsividis, *Operation and Modeling of the MOS Transistor*, 2nd edn. (Oxford Univ. Press, New York, 1999)
42. K. Bhuwalka, J. Schulze, I. Eisele, Scaling the vertical tunnel FET with tunnel bandgap modulation and gate workfunction engineering. IEEE Trans. Electron. Devices **52**(5), 909–917 (2005)
43. M. Saxena, S. Haldar, M. Gupta, R. Gupta, Design considerations for novel device architecture: Hetero-material double-gate (HEM-DG) MOSFET with sub-100 nm gate length. Solid State Electron. **48**(7), 1169–1174 (2004)

44. A. Khakifirooz, D. Antoniadis, MOSFET performance scaling—Part II: future directions. IEEE Trans. Electron. Devices **55**(6), 1401–1408 (2008)
45. X. Wu, P. Chan, M. Chan, Impacts of nonrectangular FIN cross section on the electrical characteristics of FinFET. IEEE Trans. Electron. Devices **52**(1), 63–68 (2005)
46. International Technology Roadmap for Semiconductors (ITRS)2011www.itrs.net.
47. S.-J. Choi, D.-I. Moon, S. Kim, J.P. Duarte, Y.-K. Choi, Sensitivity of threshold voltage to nanowire width variation in junctionless transistors. IEEE Electron. Device Lett. **32**, 125–127 (2011)
48. C.-H. Park, M.-D. Ko, K.-H. Kim, R.-H. Baek, C.-W. Sohn, C.K. Baek, S. Park, M. Deen, Y.-H. Jeong, J.-S. Lee, Electrical characteristics of 20-nm junctionless Si nanowire transistors. Solid-State Electron. **73**, 7–10 (2012)
49. J.-P. Colinge, C.-W. Lee, N.D. Akhavan, R. Yan, I. Ferain, P. Razavi, A. Kranti, R. Yu, *Junctionless Transistors: Physics and Properties Semiconductor-On-Insulator Materials for Nanoelectronics Applications* (Springer, Berlin, 2011), pp. 187–200
50. J.-P. Colinge, *FinFETs and Other Multi-Gate Transistors* (Springer, New York, 2007)
51. W. Wu, M. Chan, Analysis of geometry-dependent parasitics in multifin double-gate FinFETs. IEEE Trans. Electron. Devices **54**, 692–698 (2007)
52. C.J. Huang, S.C. Lee, Stress effects on self-aligned silicon nanowires junctionless field effect transistors. in 11th IEEE International Conference on Nanotechnology, 2011
53. X. Wu, P. Chan, M. Chan, IEEE Trans. Electron Devices **52**, 63 (2005)
54. D. Zhang, S. Rauf, T. Sparks, IEEE Trans. Plasma Sci. **30**, 114 (2002)
55. Y. Liu, K. Ishii, M. Masahara, T. Tsutsumi, H. Takashima, H. Yamauchi, E. Suzuki, Jpn. J. Appl. Phys. **43**, 2151 (2004)
56. R. Giacomini, J. Martino, J. Electrochem. Soc. **155**, H213 (2008)
57. M. Kushibiki, A. Hara, E. Nishimura, T. Endoh, Jpn. J. Appl. Phys. **50**, 04DA16 (2011)
58. A. Asenov, S. Kaya, A.R. Brown, Intrinsic parameter fluctuations in decananometer MOSFETs induced by gate line edge roughness. IEEE Trans. Electron. Devices **50**(5), 1254–1260 (2003)
59. Y. Liu, K. Ishii, T. Tsutsumi, M. Masahara, E. Suzuki, IEEE Electron. Device Lett. **24**, 484 (2003)
60. H. Lou, D. Li, Y. Dong, X. Lin, S. Yang, J. He, M. Chan, Jpn. J. Appl. Phys. **52**, 104302 (2013)
61. S. Choi, D. Moon, S. Kim, J. Duarte, Y. Choi, IEEE Electron. Device Lett. **32**, 125 (2011)
62. R. Rios, A. Cappellani, M. Armstrong, A. Budrevich, H. Gomez, R. Pai, N. Rahhal-Orabi, K. Kuhn, IEEE Electron. Device Lett. **32**, 1170 (2011)
63. J.P. Colinge, C.W. Lee, A. Afzalian, N.D. Akhavan, R. Yan, I. Ferain, P. Razavi, B. O'Neil, A. Blake, M. White, Nanowire transistor without junctions. Nat. Nanotechnol. **5**, 225–229 (2010)
64. Ying Xiao, Baili Zhang, Haijun Lou, Xiaole Cui, Xinnan Lin, 2015, IEEE International Conference on Electron Devices and Solid-State Circuits (EDSSC)
65. G. Leung, C.O. Chui, Variability impact of random dopant fluctuation on nanoscale junctionless FinFETs. IEEE Electron. Device Lett. **33**(6), 767–769 (2012)
66. A. Gnudi, S. Reggiani, E. Gnani, et al., Analysis of threshold voltage variability due to random dopant fluctuations in junctionless FETs. IEEE Electron. Device Lett. **33**(3), 336–338 (2012)
67. M.J.M. Pelgrom, A.C.J. Duinmaijer, A.P.G. Welbers, Matching properties of MOS transistors. IEEE J. Solid-State Circuits **24**(5), 1433–1439 (1989)
68. M. Aldegunde, A. Martinez, J.R. Barker, Study of discrete dopinginduced variability in junctionless nanowire MOSFETs using dissipative quantum transport simulations. IEEE Electron. Device Lett. **33**(2), 194–196 (2012)
69. R. Trevisoli, R. Doria, M. de Souza, M. Pavanello, Threshold voltage in junctionless nanowire transistors. Semicond. Sci. Technol. **26**(10), 105009 (2011)
70. Wenbo Wan, Haijun Lou, Ying Xiao, Xinnan Lin, 2016 IEEE International Conference on Electron Devices and Solid-State Circuits (EDSSC)

71. B. Rajasekharan, R.J.E. Hueting, C. Salm, T.V. Hemert, R.A.M. Wolters, J. Schmitz, Fabrication and characterization of the charge-plasma diode. IEEE Electron Device Lett. **31**(6), 528–530 (2010)
72. R.J.E. Hueting, B. Rajasekharan, C. Salm, J. Schmitz, The charge plasma P-N diode. IEEE Electron. Device Lett. **29**(12), 1367–1368 (2008)
73. M.J. Kumar, K. Nadda, Bipolar charge-plasma transistor: a novel three terminal device. IEEE Trans. Electron. Devices **59**(4), 962–967 (2012)
74. U. Schwalke, T. Kraussa, F. Wesselya, Dopant-free CMOS on SOI:multi-gate Si-nanowire transistors for logic and memory applications. ECS Trans. **53**(5), 105–114 (2013)
75. C. Sahu, J. Singh, Charge-plasma based process variation immune junctionless transistor. IEEE Electron. Device Lett. **35**(3), 411–413 (2014)

Part II
Sensor

Chapter 8
Effect of Nanoscale Structure on Reliability of Nano Devices and Sensors

Jaya Madan and Rishu Chaujar

8.1 Introduction

CMOS transistors, the core of integrated circuits (ICs), are continuously scaled down to meet the unending ever increasing demand for high speed and dense ICs. Nano-scale CMOS technology has revolutionized the state of art of designing the IC, by providing the stability of operation, low static power dissipation and noise immunity to devices. Aggressive scaling of MOSFET down to nanometer regime is governed by the need to enhance the packing density, low operating power and high switching speed of the device. The continuous scaling of MOSFET increases the switching speed and also the cut-off frequency of MOSFET to gigahertz regime and thus makes it suitable for wireless and high-frequency applications. On the other hand, aggressive scaling down of MOSFET fallouts into a number of critical issues such as short channel effects resulting in performance degradation of MOSFET. The short channel effects of MOSFET increases the subthreshold swing to a value greater than 60 mV/decade and an abrupt increase in the OFF-state current of MOSFET; thereby reducing the switching ratio and puts constraints on the scaling of the supply voltage. Along with various short channel effects, power dissipation (both static and dynamic) is one of the major concerns for the scale downed MOSFET. To overcome the problem of standby power dissipation, the reduction of OFF-state current is essential. To overcome these flaws of MOSFET, novel device and material engineered architectures have been proposed such as multi-gate architectures of MOSFET, new gate dielectric materials, asymmetric channel doping engineering and much more. In digital applications, MOSFET operates like a switch. The prime requirement for switching applications is steeper

J. Madan • R. Chaujar (✉)
Department of Engineering Physics, Delhi Technological University, (Formerly Delhi College of Engineering), Bawana Road, Delhi 110042, India
e-mail: jayamadan.2012@gmail.com

T. Li, Z. Liu (eds.), *Outlook and Challenges of Nano Devices, Sensors, and MEMS*,
DOI 10.1007/978-3-319-50824-5_8

ON/OFF transitions i.e. lower subthreshold swing and higher switching ratio i.e. I_{ON}/I_{OFF} ratio. The steeper subthreshold slope reduces the standby power dissipation and higher I_{ON}/I_{OFF} ratio determines the performance level of the device. The engineering schemes proposed for MOSFET can provide high I_{ON}/I_{OFF} ratio, but the fundamental limit of MOSFET on SS can't go beyond 60 mV/decade. Thus, the basic principle of MOSFET which is governed by the thermionic emission of the carrier over a thermal barrier limits its SS to 60 mV/decade at room temperature, thereby limiting the MOSFET for further scaling. This limit on SS results in challenges for the supply voltage below 1 V due to the leakage current and incremental short channel effects [1, 2]. Thus, a better solution would be to exploit a device which uses a new operating mechanism other than the thermionic emission of the carrier over the barrier. In this regard, Tunnel FET which uses the band to band tunneling mechanism has emerged as a possible replacement candidate. By using the band to band tunneling mechanism, TFET can achieve a subthreshold swing below the diffusion limited value of 60 mV/decade and acts as a promising candidate for further extending the Moore's law [3]. Apart from TFET, there are many other devices whose working principle is not governed by the drift-diffusion mechanism as in the case of MOSFET. These steeper subthreshold devices comprises: impact ionization–MOS or I-MOS [4], nano-electro-mechanical FETs or NEMS FETs [5] and suspended gate MOSFETs [6].

8.2 Tunnel FET: Operating Principle

Tunnel FET is a gated p-i-n structure with gate placed at the entire intrinsic channel region and operates in reverse bias. If compared with MOSFET, the device structure of TFET is analogous to MOSFET, with an exception of asymmetry in source and drain doping. For n (p)-type TFET, the source is p^+ (n^+) doped and drain is n^+ (p^+)-doped. The asymmetry in source and drain doping of TFET results into a staircase like band profile. The basic working principle of Tunnel FET is gate bias controlled band to band tunneling (BTBT) instead of thermionic emission over the potential barrier as in case of a MOSFET [7–12]. For n-TFET, for a reverse biased p-i-n junction and $V_{gs} = 0$ V, i.e. the off state of TFET, the bands alignment is such that there is a large tunneling barrier width present at the source channel (tunneling) junction. This wider barrier width restricts the tunneling of electrons and thus results into a very low off current (generally of the order of femto Amperes). Ideally the tunneling barrier width at $V_{gs} = 0$ V is identical to the channel length.

For a positive gate bias, the electric field at the gate push down the energy bands (both conduction and valence band) in the channel region; such that the overlapping of the valence band of the source with the conduction band of the channel occurs, resulting in lowering of tunneling barrier width. The lowered tunneling barrier width caused by the high electric field thus allows the tunneling of electrons from the valence band of the source to the conduction band of the channel.

After tunneling from source-channel junction via BTBT mechanism, the electrons transport to drain via drift-diffusion, resulting in drain current.

For a negative gate bias, the energy bands in the channel region are lifted upwards. This causes the lowering of barrier width at the drain and channel junction. Lowered barrier width at drain-channel junction becomes favorable for tunneling and thus results into an ambipolar conduction (conduction for both the positive and negative gate bias) of TFET. Thus, TFET do not turn OFF for the negative gate bias as opposed to a n-MOSFET which has a negligible current for negative gate bias.

8.2.1 Advantages of TFET

The built in tunnel barrier at the source-channel junction makes TFET insusceptible from short channel effects, which deteriorates the performance of MOSFET. Furthermore, TFET don't suffer from the fundamental limit of MOSFET on subthreshold swing i.e. 60 mV/decade. In other words, the SS of TFET can be smaller than 60 mV/decade, and thus allows further downscaling of supply voltage. The large barrier width at the source channel junction for lower gate bias blocks the tunneling of electrons from valence band of the source to conduction band of channel and thus offers a very low OFF current (lower than the OFF current of MOSFET) of the order of femto amperes. Moreover, TFET can be fabricated with standard CMOS processing techniques as in the case of MOSFET [13, 14]. These advantages of TFET make it a potential candidate to replace MOSFET, thereby becoming an interesting low power device.

8.2.2 Challenges Faced by Tunnel FET

However, apart from these advantages, there are challenges faced by TFET, which need to be tackled, in adopting TFETs for wide scale applications. The practical problem that TFET suffers is its lower on current and higher ambipolar current. Lower on current of TFET is mainly due to the lower tunneling rate of electrons, and the ambipolar characteristics of TFET are due to the lowering of barrier width at the drain-channel junction for negative gate bias (for the case of n-TFET). A lower on-state current (much smaller than that of MOSFET) limits the device speed, and higher ambipolar current makes the device unsuitable for digital applications by deteriorating the subthreshold swing and enhancing the static power dissipation. To enhance the on current of TFET, the gate dielectric has been introduced with a high-k material [15]. The high-k material used at the gate oxide results in enhancement of the surface potential at a constant gate bias as evident by Eq. 8.1:

$$V_{gs} = -\frac{q_s t_{ox}}{\varepsilon_{ox}} + \phi_s \tag{8.1}$$

where;
qs is the charge density per unit area,
tox is the oxide thickness,
εox is the permittivity of the oxide material and,
ϕs is the surface potential.

The enhanced surface potential enhances the band bending of the silicon surface and thus lowers the tunneling barrier width. Lowered barrier width further allows the electron to tunnel through the source-channel junction (from the valence band of the source to the conduction band of the channel). Furthermore, the tunneling rate and thus the I_{ON} of TFET is higher, for the case of high-k material at the gate dielectric in comparison to SiO_2. However, the challenge faced by high-k material as a gate dielectric in the case of TFET is the increase in ambipolar current of the device. Thus, to overcome these challenges, a high-k material has been locally inserted near the source-channel junction to form the hetero gate dielectric (HD) TFETs [16]. In hetero gate dielectric TFETs, a high-k material is partially located at the source side and thus induces a local minimum of the conduction band at the tunneling junction and the drain side consisting of SiO_2. The presence of high-k material located at the tunneling junction enhances the ON current of the device, and the SiO_2 located near the drain side suppresses the ambipolar behavior of the device. Consequently, a HD TFET simultaneously utilizes the benefits of both SiO_2 dielectric and high-k dielectric, thus lowering I_{OFF} and enhancing I_{ON} respectively. Apart from increasing I_{ON} and reducing I_{AMB}, HD TFET also offers small gate to drain capacitance which is needed for higher switching speed of the device. Other techniques offered for suppressing the ambipolar current is gate-drain underlap or short gate TFETs [17, 18], heterojunction TFET [19–21], and asymmetric source/drain doping [22].

By engineering the electric field at the tunneling junction i.e. the source channel junction, the band to band tunneling rate of electrons can be enhanced. To boost the drain current of TFET, electric field at the tunneling junction must be enhanced. For enhancing the electric field and thus the drain current, many device designs such as high-mobility (i.e. small effective mass) channel [23], hetero-junction source/channel structure [24–26], use of strained silicon [27, 28], use of high-k gate dielectric [29, 30], Nano-wire, Nano-tube structure [31–33], tunnel source i.e. a p-n-p-n [34], multi gate TFET [35] have been proposed.

8.3 Reliability Issues of CMOS Devices

With increase in the complexity and density of VLSI chips, the evaluation of long term reliability of CMOS devices becomes mandatory. The major factors responsible for the device damage problems are: process induced damage [36], radiation induced damage [37], stress induced damage [38] and hot carrier induced damage [39]. In case of MOS devices, the reliability and performance is subjective to the quality of oxide and silicon interface. At the Si-SiO_2 interface, there are four types of charges: (1) fixed oxide charges, (2) mobile oxide charges, (3) oxide trapped charges and (4) interface states. Interface states are mainly attributed to the reduction in device reliability and lifetime of the device. Although by using the better fabrication techniques, the interface states can be reduced, but scaling down of device and enhanced integration density of ICs, results into adverse effect on device performance.

In SiO_2, the atoms of silicon and oxygen are packed in such a manner that each oxygen atom is bonded to two silicon atoms and each silicon atom is bonded to four oxygen atoms. But due to the bond length difference of oxygen and silicon, dangling bonds exists at the interface of Si-SiO_2. Thus, the atoms of silicon at the interface inevitably miss the fourth bond and thus represent the interface defect. The energy level associated with these defects lies in between the energy band gap of silicon i.e. neither in the conduction band nor the valence band. Moreover, the electrons and holes that are present in these states are immobile or are localized. Thus, the electrons from the conduction band and holes from the valence band gets effectively trapped in these interface states. Commonly, there are two types of interface states: the acceptor type and the donor type. The acceptor type interface state is electrically neutral when it is empty, and it becomes negatively charged when filled with an electron; whereas the donor type interface state is electrically neutral when filled with an electron and becomes positively charged when it is empty [40, 41].

8.4 Reliability of TFETs

There is a necessity to study the reliability issues such as the impact of Interface trap charge density and impact of temperature of Tunnel FET, because of its novel working principle in comparison with MOSFET. Moreover, the reliability of TFET is a more prominent problem as compared to MOSFET, because of the change in tunneling field impacted by interface trap charges (ITCs). In case of TFET, the band to band tunneling rate of electrons is highly sensitive to the electric field along the channel direction E_y, being proportional to $\exp(-A/E_y)^6$ [42]. The effect of ITCs on electric field thus strongly impacts the BTBT rate of electrons, and hence the drain current. Furthermore, the high electric field needed at the tunneling junction for enhancing the tunneling rate (or lowering the tunneling barrier width) at the source-channel junction is mainly attributed to the generation of ITCs (both donor and acceptor type). The most important point defects at the interface of Si-SiO_2 are due to the dangling bonds of silicon that are not coordinated with the interfacial oxygen atoms of SiO_2. The concentration of these dangling bonds however, increases significantly with increase in electric field. A lot of work has already been done to study the device reliability of MOSFET owed by the ITC. But there is a need to estimate the performance degradation due to the interface traps present at the $Si\text{-}SiO_2$ interface in TFET as the basic working principle of TFET is entirely different from that of the MOSFET. In this paper, the effect of density of ITCs, polarity of ITCs and temperature affectability on two devices HD-GAA-TFET and GAA-TFET with cylindrical geometry has been compared.

In this chapter, the reliability issues of tunnel FET has been discussed, firstly by studying the effect of interface traps charge density, which are common during the pre and post fabrication process, and secondly by analyzing the effect of temperature on the electrical and analog parameters.

8.4.1 Device Structure and Simulation Methodology

Figure 8.1a, b show the simulation structure and schematic cross sectional view respectively of HD-GAA-TFET. Table 8.1 lists all the device parameters, their physical significance and their values. Silicon is used as source, drain and channel material. To constrain the ambipolarity effect, (tunneling at drain and channel junction) source and drain are doped asymmetrically [43]. The optimum value of gate metal work function $\phi_M = 4.3$ eV has been chosen for better OFF state characteristics in both the devices.

In this work, to study the device reliability issue attributed to the interface charges, the effect of interface traps has been transmuted, into its equivalent positive localized charges (acceptor) and negative (donor) localized charges. The value of N_f is chosen on the basis of various experimental and simulation studies previously published incorporating the process damage, radiation damage and hot carrier damage resulting in trap density of 10^{11}–10^{13} cm^{-2} eV^{-1} [44, 45]. To investigate the impact of interface trap charge density and its polarity on the device performance, both positive and negative charge density as $N_f = \pm$ 1, 2, 3 ($\times 10^{12}$) cm^{-2} has been considered. Further, the effect of interface traps at the $Si\text{-}SiO_2$ interface is considered only near the source side for 10 nm, due to the presence of high transverse electric field at the tunneling junction.

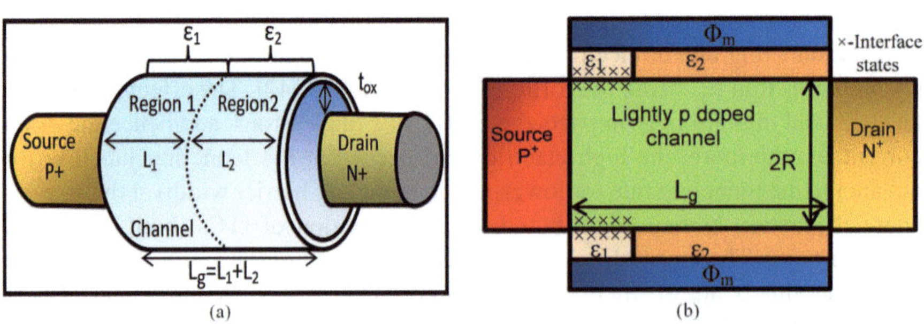

Fig. 8.1 (**a**) 3D view of the simulated structure and (**b**) Schematic 2D cross-sectional view of n-type Heterogeneous Gate Dielectric-GAA-TFET (HD-GAA-TFET)

Table 8.1 Device geometrical parameters and the values used in the analysis

Parameter symbol	Technology parameter	HD GAA TFET	GAA TFET
L_g	Channel length	50 nm ($L_1 = 20$ nm, $L_2 = 30$ nm)	50 nm
R	Channel radius	10 nm	
t_{ox}	Gate oxide thickness	3 nm	
N_s	Source doping (p^+)	10^{20} cm^{-3}	
N_{ch}	Channel Doping lightly p-type	10^{17} cm^{-3}	
N_d	Drain doping (N^+)	5×10^{18} cm^{-3}	
ε_1	Dielectric constant of region 1	21 [HfO_2]	3.9
ε_2	Dielectric constant of region 2	3.9 [SiO_2]	3.9
Φ_m	Metal work function	4.3 eV	

Although the interface traps are also present at the interface of dielectrics in HD GAA TFET and will be absent in the case of GAA TFET, in present work, only the ITCs along the channel length at the silicon and oxide interface has been considered. The main reason for neglecting the ITCs at HfO_2/SiO_2 interface is that in the case of TFETs mainly, the tunneling junction and the drain junction affects the device characteristics [46]. INTERFACE statement is used during simulation to define the density of interface fixed charges and their position present at the interface of Si-SiO_2 [47]. The distribution of interface charges is assumed to be uniform during the entire stimulation.

8.4.1.1 Simulation Methodology

All simulations have been performed using the ATLAS device simulator [47]. The models invoked during simulation are concentration and field dependent mobility model, Shockley-Read-Hall for carrier recombination, non-local band to band tunneling, band gap narrowing and Fermi-Dirac statistics. The most important model for Tunnel FET simulations is the band-to-band tunneling (BTBT) model. In this work the non-local BTBT models, has been used. In local tunneling models (Kane model), the energy barrier is approximated by the triangular barrier and thus the electric field is constant at each point, whereas the non-local model (non-local band to band tunneling models) doesn't make the same assumption, hence making it more real. In non-local tunneling models, the electric field is not uniform but it changes dynamically at each point in the tunneling path depending upon the band edge profile along the tunneling path [48]. Non-local models thus have a more physical basis and don't depend on the electric field at the individual mesh points in the simulated device structure, but rather on band diagrams calculated along cross-sections through the device. Fermi-Dirac statistics are essential to account for certain properties of very highly doped (degenerate) materials such as reduced carrier concentrations in heavily doped regions (statistical approach). As source

Fig. 8.2 Calibration with the published result, I_{ds}-V_{gs} characteristics of 200 nm gate length SiNW-TFET at $V_{ds} = 1.2$ V, $t_{ox} = 4.5$ nm, Diameter = 70 nm

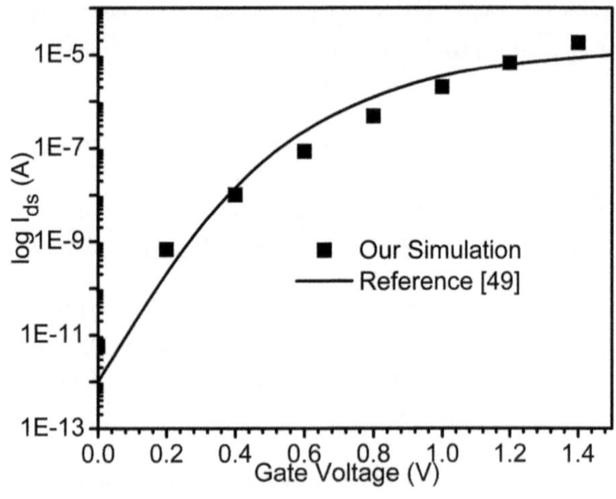

and drain are heavily doped, so Fermi-Dirac statistics is used. For higher doping ($>10^{18}$ cm^{-3}) experimental work has shown that the p-n product in silicon becomes doping dependent. As the doping level increases, a decrease in the bandgap separation occurs. Both the numerical methods Gummel (decoupled) together with Newton's (Fully coupled) has been incorporated to mathematically solve the carrier transport equation.

Calibration

Before simulating the device structure, the simulation models have been calibrated with the already published results [49]. To calibrate the models, the GAA TFET has been designed with the same device parameters as shown in ref [49], as already done in our previously published work [50]. To calibrate the Non-local BTBT model, the tunneling masses were tuned from their default values i.e. me. tunnel = 0.322 m_o and mh.tunnel = 0.549 m_o to make it best fit with the experimental data. Where; m_o is the rest mass of electron. The adjusted value of me. tunnel = 0.22 m_o and mh.tunnel = 0.52 m_o. Fig. 8.2 shows the transfer characteristics of the extracted data from the reference and results obtained from the simulations. The results so obtained are in close proximity, thus validates the model parameters during the simulations.

8.4.2 *Impact of Interface Traps Charge Density*

In this section, the impact of interface trap charge density has been examined. During the analysis of the impact of ITC density, the temperature is kept constant i.e. T = 300 K. The impact of trap charge density has been examined in terms of electric field, BTBT rate of electrons, transfer characteristics, output characteristics, ambipolar current, threshold voltage, subthreshold swing, switching ratio, transconductance, parasitic capacitances, cut off frequency and maximum oscillations frequency.

The presence of ITCs results into additional band bending and hence changes the flat band voltage under the affected region. The amount of change in flat band voltage i.e. ΔV_{fb} is defined by Eq. 8.2:

$$\Delta V_{fb} = \frac{qN_f}{C_{ox}} \tag{8.2}$$

where;
q is the electronic charge,
N_f is the interface charge density,
C_{ox} is the gate oxide capacitance per unit area.

The presence of donor (acceptor) type ITCs enhances (reduces) the band bending at the tunneling junction, thus results in an increase (decrease) in the electric field at the tunneling junction as evident from Fig. 8.3a. The electric field at tunneling junction is much higher for the HD GAA TFET in comparison with GAA TFET, due to the fact that high-k material is present over the source-channel junction that enhances the band bending. Furthermore, the electric field for both the aforementioned devices is almost 30 times higher at tunneling junction than the field at the drain-channel junction. This high electric field at the tunneling junction results into higher tunneling of electrons from the valence band of the source to the conduction band of the channel. Also, the increase in trap charge density of donor/positive (acceptor/negative) type ITCs at the interface of Si-SiO_2 further enhances (decreases) the electric field at both the junctions as evident from Fig. 8.3a, b.

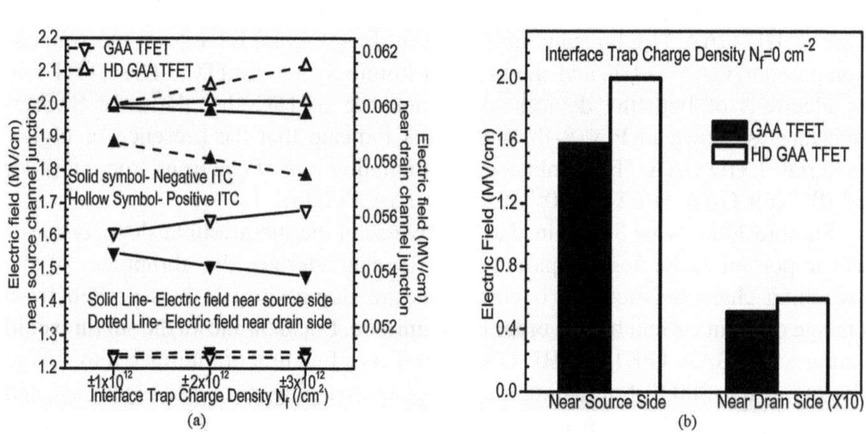

Fig. 8.3 The impact of Interface trap charge densities on Electric field near source-channel and drain-channel junction on GAA TFET and HD GAA TFET at $V_{gs} = 1.2$ V and $V_{ds} = 1.0$ V. (**a**) of donor and acceptor type ITCs (**b**) absence of ITCs

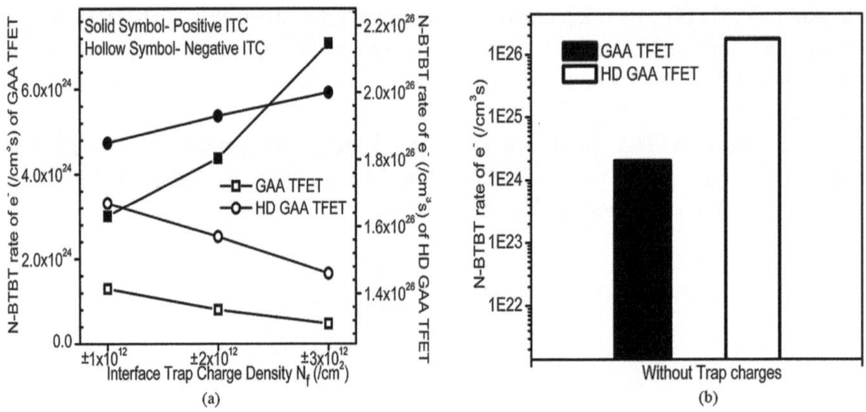

Fig. 8.4 Nonlocal band to band tunneling rate of electrons as a function of density of interface trap charges at the source-channel junction (**a**) for acceptor and donor type trap charges (**b**) in the absence of interface traps at $V_{gs} = 1.2$ V and $V_{ds} = 1.0$ V

As the ΔV_{fb} is inversely proportional to gate oxide capacitance and in the case of HD GAA TFET the high-k dielectric present over the tunneling junction enhances C_{ox} and thus reduces the change in V_{fb}, thereby resulting in a much lesser impact on the electric field in case of HD GAA TFET w.r.t GAA TFET.

The nonlocal band to band tunneling rate of electrons at the source-channel junction for GAA TFET and HD GAA TFET as a function of interface trap charge density for donor/acceptor ITCs and without ITCs are shown in Fig. 8.4a, b respectively. The presence of donor (acceptor) type ITCs increases (decreases) the BTBT rate of electrons because of enhanced (reduced) electric field as shown in Fig. 8.4a. In addition, the increase in donor (acceptor) interface trap charge density further enhances (reduces) the BTBT rate of electrons in both the devices. Moreover, it is found that the presence of high-k material near tunneling junction in case of HD GAA TFET results into a lesser change in BTBT rate of electrons as compared to GAA TFET, and shows better immunity against ITCs. The BTBT rate of electrons of both the devices in the absence of ITCs i.e. the ideal Si-SiO_2 interface is shown in Fig. 8.4b and results indicate that the presence of high-k material in HD GAA TFET enhances the tunneling rate of electrons from an order of 10^{24} (for GAA TFET) to 10^{26} (for HD GAA TFET).

Subthreshold swing SS, defined as the inverse of the subthreshold slope, is one of the important technological parameters. SS characterizes the abruptness of the switching characteristics and is defined as the change in gate bias needed for a change of drain current by one order of magnitude. Fig. 8.5a shows the subthreshold swing SS of GAA TFET and HD GAA TFET as a function of interface trap charge density at T = 300 K. For maintaining a higher switching ratio i.e. a higher I_{ON} and lower I_{OFF}, reduced SS is necessary. Lesser SS is achieved for the HD GAA TFET in comparison to GAA TFET, due to the higher BTBT rate of electrons. The enhanced tunneling rate of electrons enhances the drain current and thus reduces the SS of the device. Moreover, there is an increase (decrease) in SS is found due to

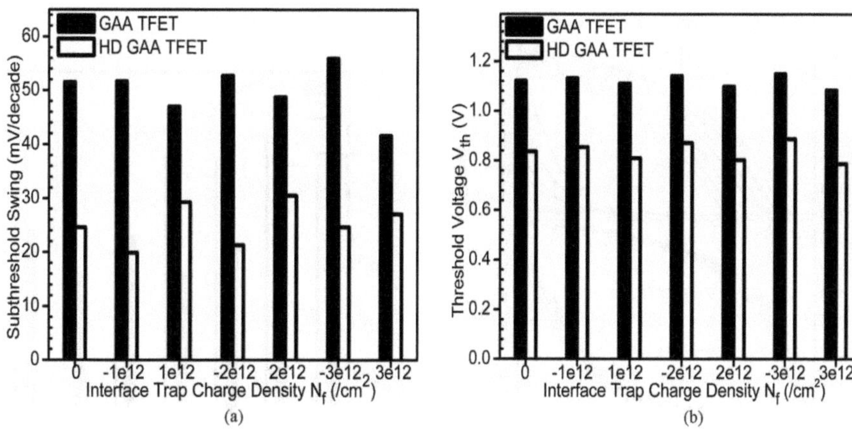

Fig. 8.5 (**a**) Subthreshold swing and (**b**) threshold voltage as a function of trap charge density for GAA TFET and HD GAA TFET

the presence of acceptor (donor) type ITC in the case of GAA TFET. The presence of high-k dielectric near source side reduces (enhances) the SS due to the presence of acceptor (donor) ITCs. With an increase in density of ITCs, the change in SS further increases. For a density of $\pm 3 \times 10^{12}$ cm^{-2}, there is a decrease (increase) in SS of 19.17% (8.65%) for donor (acceptor) type ITCs in GAA TFET. For the case of HD GAA TFET, at $N_f = \pm 2 \times 10^{12}$ cm^{-2}, there is a decrease (increase) in SS of 13.31% (24.28%) in SS for the acceptor (donor) type trap. Further, the effect of interface trap charge density is studied on the threshold voltage as shown in Fig. 8.5b. It is clearly evident from the bar graph that the threshold voltage of HD GAA TFET is 25.35% less than the V_{th} of GAA TFET, because of the presence of high-k material over the tunneling junction which effectively lowers the barrier width as compared to GAA TFET. The presence of positive (negative) ITCs results into decrease (increase) in the V_{th} of both the aforementioned devices. This increase (decrease) in V_{th} is mainly attributed to the reduction (enhancement) in the band bending due to the presence of negative/acceptor (positive/donor) ITCs. Furthermore, the increase in donor (acceptor) trap charge density increases the change in V_{th} for each case. As the trap charge density increases from $+1 \times 10^{12}$ cm^{-2} to $+3 \times 10^{12}$ cm^{-2}, V_{th} decreases by 0.98% to 3.32% for GAA TFET and 3.18% to 5.90% for HD GAA TFET. The increase in trap charge density increases the change in flatband voltage and thus results in an increase in ΔV_{th}.

The impact of trap charge density on drain current of both the devices as a function of gate bias is shown in Fig. 8.6a. Results reveal that donor type fixed charges enhances the drain current, whereas the acceptor type fixed charges leads to decrease in the drain current w.r.t the case when ITCs are not present. As can be seen from Fig. 8.6a that the change in drain current increases with increase in the density of ITCs. At $V_{gs} = 0.5$ V and for a density of $+ (-)\ 1 \times 10^{12}$ cm^{-2}, there is an increase (decrease) of 7.55 (14.15) times and 1.39 (1.46) times in drain current

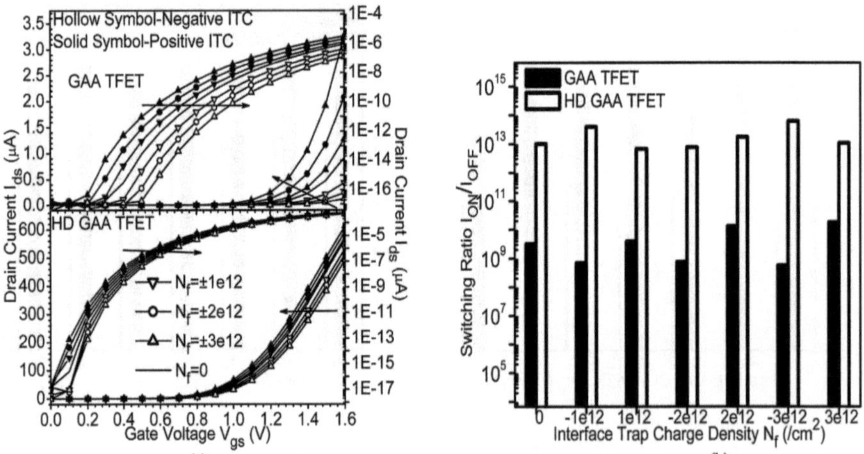

(a) (b)

Fig. 8.6 (**a**) Impact of the density of donor and acceptor interface traps on transfer characteristics of GAA TFET and HD GAA TFET at $V_{ds} = 1.0$ V. (**b**) Switching ratio as a function of trap charge density for GAA TFET and HD GAA TFET at I_{ON} at $V_{gs} = 1.2$ V and I_{OFF} at $V_{gs} = 0.0$ V

w.r.t the case when $N_f = 0$ in the case of GAA TFET and HD GAA TFET respectively. The increase in density of ITCs from 1×10^{12} cm^{-2} to 3×10^{12} cm^{-2} increases the change in drain current. For $N_f = 3 \times 10^{12}$ cm^{-2}, there is an increase (decrease) of 347.24 (1581.03) times and 2.66 (3.21) times in drain current w.r.t the case when $N_f = 0$ in the case of GAA TFET and HD GAA TFET.

The change in drain current is much lesser in the case of HD GAA TFET in comparison to GAA TFET, resulting in the better reliability of HD GAA TFET in terms of ITCs. The switching ratio of both the devices has also been studied w.r.t. variation in density ITCs and is shown in Fig. 8.6b. The switching ratio of HD GAA TFET is significantly larger than GAA TFET. Moreover, the switching ratio increases (decreases) due to the presence of donor (acceptor) trap charges for both the devices. For GAA TFET, I_{ON}/I_{OFF} ratio increases (decreases) by 1.2 (4.65), 4.08 (4.29), 5.61 (5.67) times as N_f increases from + (−) 1×10^{12} cm^{-2}, 2×10^{12} cm^{-2} to 3×10^{12} cm^{-2} respectively. For HD GAA TFET, I_{ON}/I_{OFF} ratio increases (decreases) by 0.66 (0.25), 1.73 (1.31), 1.07 (0.16) times as N_f increases from + (−) 1×10^{12} cm^{-2}, 2×10^{12} cm^{-2} to 3×10^{12} cm^{-2} respectively. Again the change in switching ratio is much lesser in the case of HD GAA TFET as compared to GAA TFET.

One of the important parameters for switching circuits applications is the ambipolar current. The ambipolar behavior in TFET is originated from the tunneling at the drain-channel junction (this is analogous to the phenomenon of gate-induced drain leakage GIDL in MOSFETs). Thus, there is a need to analyze the I_{AMB} of the TFETs. In this work, the ambipolar current has been defined as the current at $V_{gs} = -1.0$ V and $V_{ds} = 1.0$ V. As in this work, the presence of ITCs is

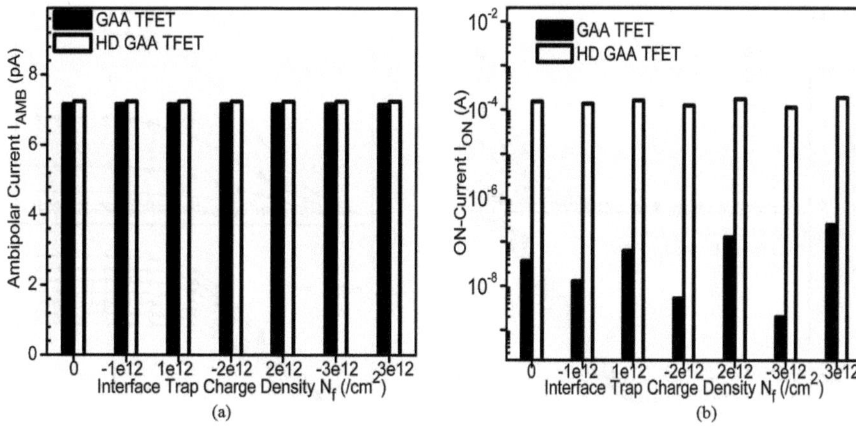

Fig. 8.7 (**a**) Ambipolar current at $V_{gs} = -1.0$ V and $V_{ds} = 1.0$ V and (**b**) On state current of GAA TFET and HD GAA TFET w.r.t. interface trap charge density at $V_{gs} = 1.2$ V and $V_{ds} = 1.0$ V

considered only in the region above and near the source-channel junction, and the ambipolar current is defined by the drain–channel junction and thus there is no effect of ITCs on the I_{AMB} of both the devices as evident from Fig. 8.7a. Moreover, in the case of HD GAA TFET, the high-k dielectric material is present only near the source-channel junction and again the I_{AMB}, which is defined by the drain–channel junction remains unaffected, resulting into almost same I_{AMB} as that of GAA TFET. The major challenge faced by TFET is the lower on current and thus, there is a need to analyze the impact of ITC density on the I_{ON}, which has been plotted in Fig. 8.7b. The presence of donor (acceptor) ITCs increases (decreases) I_{ON}, due to the increase (decrease) in band bending. To add further, the presence of high-k material in case of HD GAA TFET results into higher band bending at the tunneling junction, and thus lowers the tunneling barrier width efficiently resulting into higher I_{ON} as compared to GAA TFET. The change in I_{ON} increases with increase in the density of trap charges. At +(−) 1×10^{12} cm^{-2}, there is an increase (decrease) of 1.72 (0.35) times and 1.06 (0.89) times for GAA TFET and HD GAA TFET respectively.

Device gate transconductance g_m is a critical device parameter for analog circuit design which measures the effectiveness of the control of drain current by the gate voltage. It is defined as the first order derivative of the drain current w.r.t. gate voltage at a constant drain voltage. Figure 8.8a illustrates the magnificent lift in the transconductance g_m of HD GAA TFET as compared to GAA TFET. The increase (decrease) in drain current due to the presence of donor (acceptor) ITCs results into enhancement (degradation) in transconductance of both the devices. At $V_{gs} = 1.2$ V, g_m increases by 1.93 times, 3.56 times and 6.28 times for a density of donor ITCs with $N_f = 1 \times 10^{12}$, 2×10^{12} and 3×10^{12} cm^{-2} respectively in case of GAA TFET. The increase in case of HD GAA TFET is comparatively less due to the presence of high-k material at the tunneling junction i.e. 1.04 times, 1.09 times and 1.14 times for a density of positive fixed ITCs with $N_f = 1 \times 10^{12}$, 2×10^{12} and

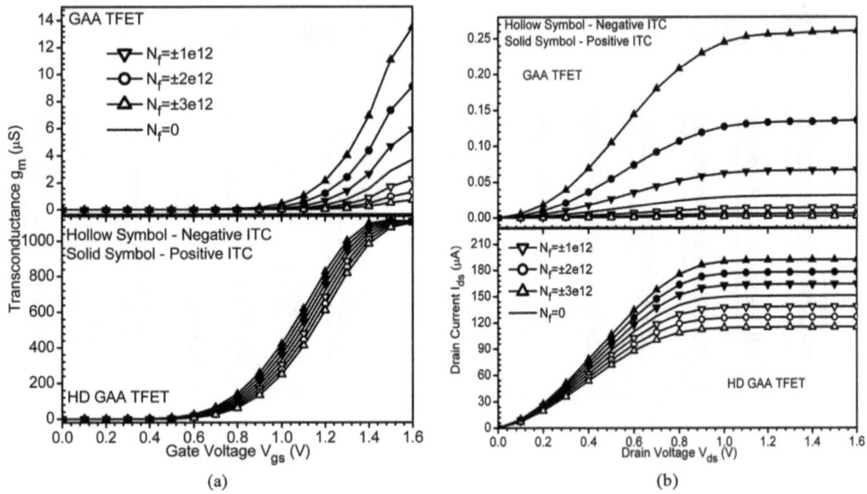

Fig. 8.8 (**a**) Ambipolar current at $V_{gs} = -1.0$ V and $V_{ds} = 1.0$ V and (**b**) On state current of GAA TFET and HD GAA TFET w.r.t. interface trap charge density at $V_{gs} = 1.2$ V and $V_{ds} = 1.0$ V

3×10^{12} cm^{-2} respectively. Figure 8.8b shows the impact of interface trap charge density on output characteristics (I_{ds} –V_{ds}) at a constant gate bias of $V_{gs} = 1.2$ V for GAA TFET and HD GAA TFET. For a constant V_{gs} ($>V_{th}$), the constant amount of electrons will be allowed to tunnel through the source-channel junction (from the valence band of the source to the conduction band of the channel). At constant $V_{gs}>V_{th}$, and at lower V_{ds}, there is negligible small channel resistance and pinning of the conduction band of the channel region by the drain potential, thereby reducing the tunneling barrier width and rapid increase in drain current for lower drain voltage. Further increase in drain bias increases the channel resistance and thus results in saturation of the drain current as is evident from Fig. 8.8b. Again the increase in trap charge density of donor (acceptor) trap charges enhances (reduces) the drain current in both the aforementioned devices.

For digital circuits, propagation delay and power dissipation are the vital performance metrics, and the device parameters affecting these metrics consists of parasitic capacitances. For better digital circuit's designs, Miller capacitance C_{gd} is the major hindrance as it influences response time of the device. Therefore, analysis of interface trap charge density on these parasitic capacitances is essential. As reported in [51, 52], in the case of TFET, the inversion charge distribution is rather different from that of the MOSFET, because of the different transport mechanism in both the devices. In case of MOSFET operating in linear regime, both source and drain are connected to the channel region resulting in equal contribution of gate to drain capacitance (C_{gd}) and gate to source capacitance (C_{gs}) to total gate capacitance (C_{gg}) i.e. $C_{gg}/2 \sim C_{gd} \sim C_{gs}$; for MOSFET operating in saturation region, $C_{gs} \sim 2/3C_{gg}$ and $C_{gd} \sim 0$ [53]. Whereas in the case of TFET, only the drain region is connected to the channel in both linear and saturation regions, resulting in a major contribution of Miller capacitance C_{gd} in total gate

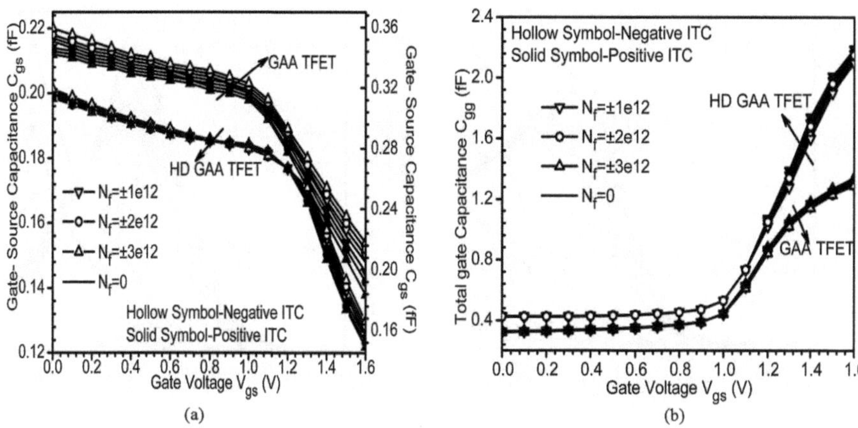

Fig. 8.9 (**a**) Gate to source capacitance C_{gs} (**b**) total gate capacitance as a function of gate bias at a constant drain bias $V_{ds} = 1.0$ V, including the impact of interface trap charge density

capacitance. A plot of C_{gs} as a function of gate bias V_{gs} for both the devices is shown in Fig. 8.9a, which includes the impact of ITC density. These capacitances are extracted by small signal ac device simulation at a constant frequency of 1 x 10^6 Hz. At a constant drain bias, for low gate bias, the inversion layer is formed initially near the drain side, and it extends from drain towards source side with an increase in gate bias. Therefore, this extension of inversion layer decreases the C_{gs} with an increase in gate bias in both the devices. At lower gate bias, the C_{gs} of HD GAA TFET is larger as compared to GAA TFET. The increase in C_{gs} for HD GAA TFET is mainly attributed to the presence of high-k material near the source-channel junction. At higher gate bias, the C_{gs} of HD GAA TFET becomes almost equal to the C_{gs} of GAA TFET. Moreover, there is an increase (decrease) in C_{gs} due to the presence of acceptor (donor) ITCs. The change in C_{gs} is comparatively large in the case of GAA TFET as compared to HD GAA TFET, due to the presence of high-k material at the tunneling junction in HD GAA TFET, thus, revealing the better immunity of HD GAA TFET towards ITCs w.r.t GAA TFET. The total gate capacitance C_{gg} defined as the sum of C_{gs} and C_{gd} is shown in Fig. 8.9b for both the devices including the impact of ITCs. C_{gg} interrelated the active power dissipation ($C_{gg}V_{dd}{}^2f$) as is clearly evident from Fig. 8.9b, there is negligibly small impact of ITC density in C_{gg} in both the devices.

The high-frequency properties are measured by two figures of merit namely the cut-off frequency f_T, and the maximum oscillation frequency f_{MAX}. Cut off frequency f_T also known as the unity current gain frequency is defined as the frequency at which the magnitude of AC current gain falls to 0 dB, whereas, the current gain is defined as the ratio of the amplitude of small-signal drain current to the small-signal gate current. f_T is used as the benchmark to describe the speed of the intrinsic device and is mathematically defined by [46] Eq. (8.3)

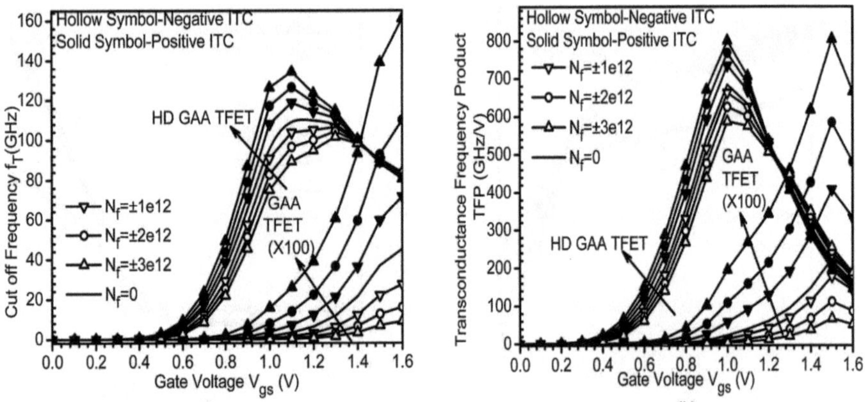

(a) (b)

Fig. 8.10 Comparison of (**a**) Cut off frequency and (**b**) Transconductance frequency product TFP of the GAA TFET and HD GAA TFET as a function of gate bias at a constant drain bias $V_{ds} = 1.0$ V for the variation in interface trap charge density

$$f_T \approx \frac{g_m}{2\pi (C_{gs} + C_{gd})} \tag{8.3}$$

f_{MAX} is defined as the frequency at which the unilateral power gain falls to unity or 0 dB. In order to evaluate the impact of ITCs on RF performance of both the devices, RF figures of merits FOM such as cut-off frequency f_T, transconductance frequency product TFP, and maximum oscillations frequency f_{MAX} has been studied. Figure 8.10a illustrates the variation of f_T as a function of gate bias for various densities of interface trap charges in both the devices. Figure 8.10a reveals that f_T reaches its peak value for a particular gate bias; and after that, further increase in gate bias results into a linear decrease in f_T. Moreover, the cut-off frequency of HD GAA TFET is almost 1000 times greater than the f_T of GAA TFET, indicating that the there is no effect of smaller degradation in the parasitic capacitances in the case of HD GAA TFET in comparison with GAA TFET on high-frequency performance. The presence of donor (acceptor) ITCs results into an increase (decrease) in the f_T in both the devices. There is an increase of 1.91 times, 3.51 times and 6.14 times in f_T of GAA TFET due to the presence of donor type trap density of 1×10^{12}, 2×10^{12} and 3×10^{12} cm^{-2} respectively. Whereas; in the case of HD GAA TFET f_T increases by 1.04 times, 1.04 times and 1.12 times for a donor type trap charge density of 1×10^{12}, 2×10^{12} and 3×10^{12} cm^{-2} respectively. In addition, there is an another important parameter i.e. transconductance frequency product TFP which is defined as the product of device efficiency i.e. g_m/I_{ds} and cut-off frequency f_T. TFP represents the tradeoff between power and bandwidth and is utilized in moderate to high-speed designs [54]. Again the TFP of GAA TFET is found to be 1000 times lesser than the TFP of HD GAA TFET, due to the enhanced f_T of HD GAA TFET. Similar to the case with f_T, the TFP of both the devices

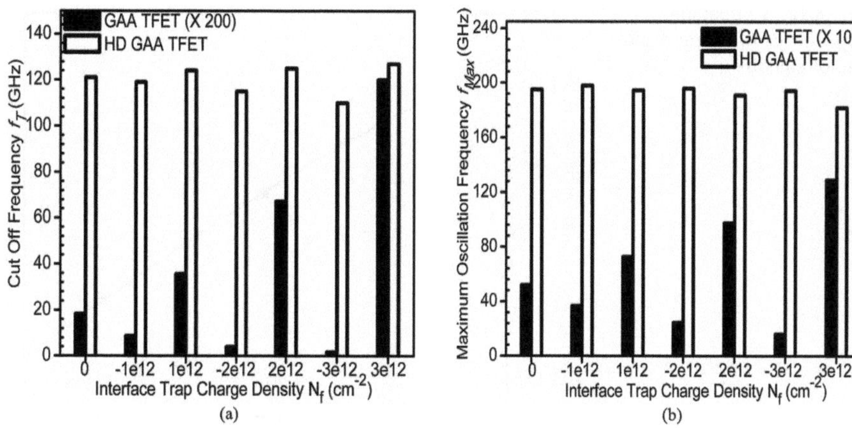

Fig. 8.11 Comparison of (**a**) Cut off frequency and (**b**) Maximum oscillation frequency f_{MAX} of the GAA TFET and HD GAA TFET as a function of interface trap charge density at a constant gate and drain bias $V_{gs} = 1.2$ V and $V_{ds} = 1.0$ V

increases (decreases) due to the presence of positive (negative) ITCs. To add further, the change in TFP increases with the increase in density of trap charges as evident from Fig. 8.10b.

The cut-off frequency extracted from the current gain is shown in Fig. 8.11a. For the case of $N_f = 0\ cm^{-2}$, the f_T of HD GAA TFET is significantly higher (i.e. 1328 times) than the f_T of GAA TFET. Moreover, the presence of donor (acceptor) type trap charges increases (decreases) the f_T, due to the increase (decrease) in transconductance. An increase in f_T has been found in both the devices i.e. 1.94 times, 3.67 times, 6.59 times in case of GAA TFET and for HD GAA TFET the increase is 1.02 times, 1.03 times and 1.04 times for a donor type trap charge density of 1×10^{12}, 2×10^{12} and $3 \times 10^{12}\ cm^{-2}$ respectively. The maximum oscillations frequency f_{MAX} is related to the capability of the device to provide power gain at larger frequencies and is shown in Fig. 8.11b for both the mentioned devices including the impact of ITC density. For the case of $N_f = 0\ cm^{-2}$ i.e. there are no localized charges present at the Si-SiO_2 interface, the f_{MAX} of HD GAA TFET is found to be 37.61 times higher than the f_{MAX} of GAA TFET. For the GAA TFET, the presence of donor (acceptor) type ITCs results into an increase (decrease) in f_{MAX} by 1.39 (1.41) times, 1.87 (2.10) times and 2.48 (3.22) times respectively. In the case of HD GAA TFET, there is very little change in f_{MAX} due to the variations of ITC density.

8.4.3 Impact of Temperature

In this section, the impact of temperature on the device performance has been inspected. While studying the impact of temperature, the interface of Si-SiO_2, has

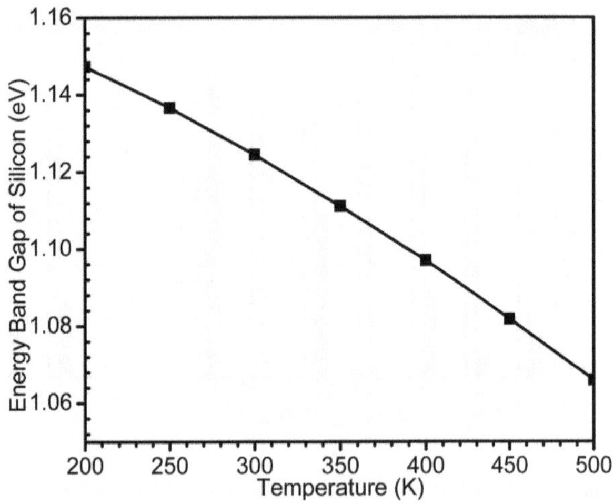

Fig. 8.12 Temperature dependence of energy band gap of silicon

been kept defect free i.e. $N_f = 0\ cm^{-2}$. The impact of temperature has been studied in terms of electrical, analog and RF figures of merits.

Temperature affects the fundamental properties of the semiconductor and hence the performance of the devices. The most fundamental property of semiconductor affected by temperature is the energy band gap Eg, which is dependent on temperature according to Varshni equation [55] is defined as in Eq. (8.4) and is shown in Fig. 8.12.

$$E_g(T) = E_g(0) - \frac{\alpha_E T^2}{T + \beta_E} \tag{8.4}$$

where;

$E_g(0)$ is the band gap energy at absolute zero on the Kelvin scale and is constant for a particular material, for Si, $E_g(0) = 1.170$ eV.

α_E and β_E are the material specific constants for Si $\alpha_E = 4.73 \times 10^{-4}$ eV/K and $\beta_E = 636$ K.

The increase in temperature reduces the energy band gap of semiconductors as evident from Fig. 8.12. Apart from energy band gap of material, carrier density (electron and hole), carrier mobility and diffusion rate of carriers are also dependent on temperature. Thus, it is necessary to investigate the impact of temperature on the device characteristics, as a change in temperature alters the threshold voltage, carrier mobility, tunneling rate of device and thus affects the speed, power and reliability parameters of the device.

The effect of temperature on the electric field at both the junctions i.e. the source channel or the tunneling junction and the drain-channel junction of both GAA TFET and HD GAA TFET is shown in Fig. 8.13a. With an increase in temperature,

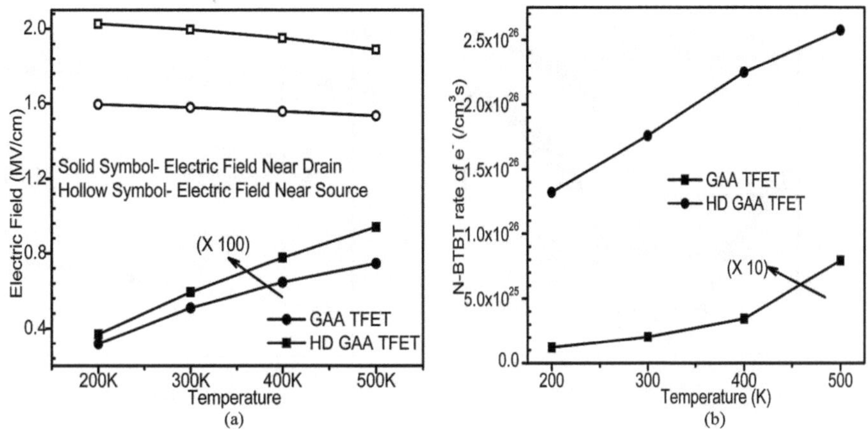

Fig. 8.13 (**a**) Electric field and (**b**) non-local band to band tunneling rate of electrons of GAA TFET and HD GAA TFET at constant $V_{gs} = 1.2$ V and $V_{ds} = 1.0$ V, for increasing operating temperature T

the narrowing of band gap increases, thus, the band gap narrowing results into an increase in the electric field at the drain junction and reduction in the field at the source-channel junction. Moreover, the field at tunneling junction is higher than the field at the drain-channel junction by an order of 30 times at 300 K. The temperature dependence of band to band tunneling rate of electrons is mainly administered by the amount of change (decrease) in the band gap with change (increase) in the temperature and thus increasing the BTBT rate of electrons [56]. Figure 8.13b shows the increase in BTBT rate of electrons as a function of temperature for both the devices. Moreover, the BTBT rate of electrons is approximately one decade higher at T = 300 K in the case of HD GAA TFET as compared to GAA TFET. This higher BTBT rate of electrons is mainly attributed to the high-k dielectric present at the tunneling junction, which lowers the barrier width at the tunneling junction and thus allows more number of electrons to tunnel through the junction. In addition, an increase in temperature from 200 K to 500 K results into an increase of BTBT rate of electrons by 6.52 and 1.94 times for GAA TFET and HD GAA TFET respectively.

Figure 8.14a shows the transfer characteristics of GAA TFET and HD GAA TFET operating in the temperature range of 200–500 K. With an increase in temperature, the bandgap narrowing increases, as drain current in TFET is attributed to the tunneling of electrons from valence band of the source to the conduction band of channel and narrowing of bands favors the tunneling rate of electrons. Therefore, the drain current increases with increase in temperature as evident from Fig. 8.14a. Furthermore, the impact of temperature is more pronounced in the subthreshold region, due to the active participation of Shockley-Read-Hall recombination in this region which has a stronger temperature dependence than the band to band tunneling. Thus, OFF current of the devices increase with temperature and the ON-state current which comes from the BTBT of electrons rises very slightly

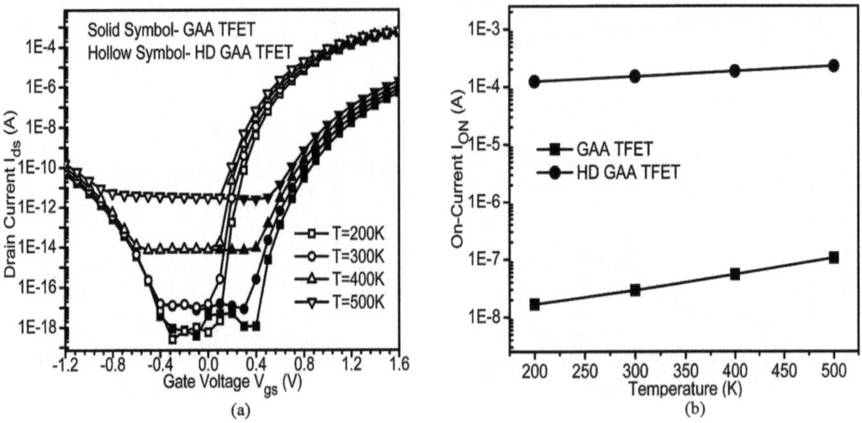

Fig. 8.14 (**a**) Transfer characteristics at $V_{ds} = 1.0$ V and (**b**) On-state current of GAA TFET and HD GAA TFET at constant $V_{gs} = 1.2$ V and $V_{ds} = 1.0$ V, for increasing operating temperature T

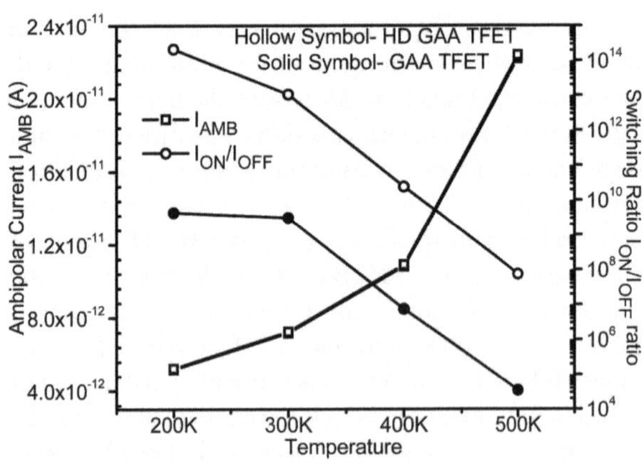

Fig. 8.15 Ambipolar current at $V_{gs} = -1.2$ V and $V_{ds} = 1.0$ V and switching ratio at $V_{gs} = 1.2$ V and $V_{ds} = 1.0$ V of GAA TFET and HD GAA TFET for increasing operating temperature T

with the rise in temperatures. Moreover, the presence of high-k dielectric results into immunity against temperature variations at higher gate bias which can be seen from Fig. 8.14a, that the variations w.r.t temperature is comparatively less in the case of HD GAA TFET. The On-state current of both the devices is shown in Fig. 8.14b. As clearly illustrated that at T = 300 K, the I_{ON} of HD GAA TFET is 5.14×10^3 times greater than the I_{ON} of GAA TFET. There is an increase in tunneling drain current at higher temperatures in both the devices. As the temperature decreases from 500 K to 200 K, the on-state current reduces by 6.29 times and 1.83 times for GAA TFET and HD GAA TFET respectively.

The impact of temperature variation on ambipolar current and switching ratio is shown in Fig. 8.15. As evident from the Fig. 8.15 that the I_{AMB} remains same for both the devices i.e. GAA TFET and HD GAA TFET. Because high-k dielectric is

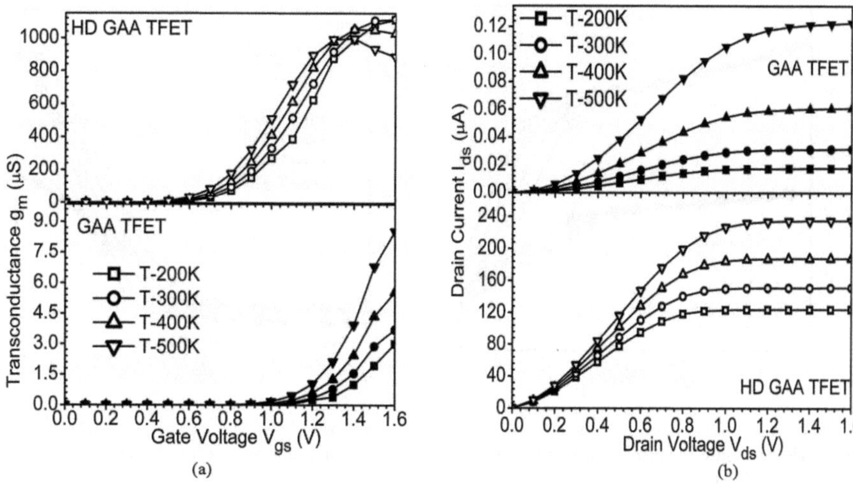

Fig. 8.16 (**a**) Transconductance g_m at $V_{ds} = 1.0$ V (**b**) Output characteristics I_{ds}-V_{ds} at $V_{gs} = 1.2$ V of GAA TFET and HD GAA TFET for increasing operating temperature T

present only at tunneling junction, and the I_{AMB} is reliant on the drain–channel junction, thus the I_{AMB} remains unaltered in HD GAA TFET or is same in both the devices. The switching ratio of both the devices as a function of temperature is plotted on the secondary axis. As clearly illustrated, the switching ratio decreases with an increase in temperature. As switching ratio is inversely dependent on I_{OFF}, and I_{OFF} degrades (increase) at higher temperatures because of the active participation of Shockley-Read-Hall recombination at subthreshold region; this results into degradation in switching ratio at higher temperatures.

Transconductance plays a significant role in the designing of operational amplifiers and operational transconductance amplifiers. Figure 8.16a shows the impact of temperature on the transconductance characteristics of GAA TFET and HD GAA TFET. As clearly observed, there is a significant enhancement in transconductance at elevated temperatures. Moreover, the transconductance of HD GAA TFET is increased by 4.11 times and 1.43 times in case of GAA TFET and HD GAA TFET respectively. The peak of the transconductance characteristics provides the optimum bias point (for the device to be used as an amplifier). Furthermore, the peak of gm shifts towards the lower gate voltages at elevated temperatures and thus causes a serious impact on the device reliability by changing the bias point of the device. Figure 8.16b illustrate the I_{ds}-V_{ds} characteristics of both the aforementioned devices at a constant gate bias, $V_{gs} = 1.2$ V. The drain current is higher at elevated temperatures for both the devices and the drain current for HD GAA TFET is significantly higher in comparison to GAA TFET, which is due to the band gap narrowing at high temperatures.

Among the various factors playing critical roles in static and dynamic power dissipation, parasitic capacitances are of prime importance. Thus, the impact of temperature variations on parasitic capacitances must be taken into consideration.

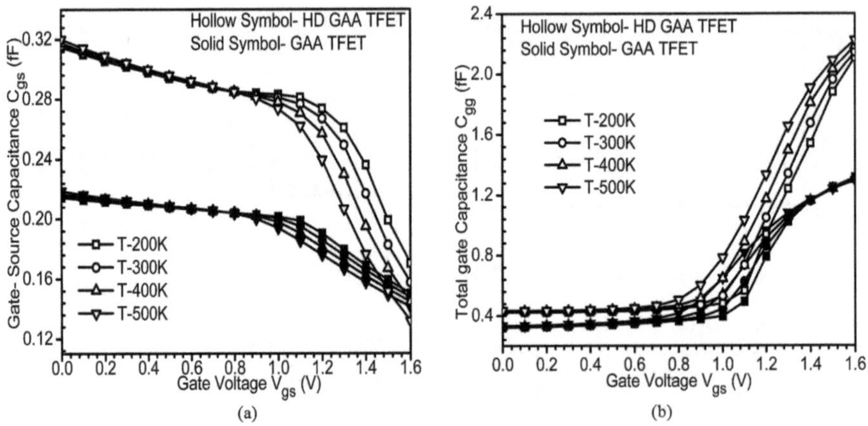

Fig. 8.17 (**a**) Gate to source capacitance C_{gs} (**b**) total gate capacitance as a function of gate bias at a constant drain bias $V_{ds} = 1.0$ V, for increasing operating temperature T

The intrinsic part of the device is the region forming the channel region of device and thus mainly responsible for ON and OFF switching of the device. The charges in the intrinsic region of the device results into intrinsic parasitic capacitances i.e. the gate to source capacitance C_{gs}, gate to drain capacitance C_{gd} and the total gate capacitance C_{gg}. The low parasitic capacitances are needed for faster switching applications. Gate to source capacitance C_{gs} of both the devices as a function of gate bias is plotted in Fig. 8.17a, including the impact of operating temperature. It should be observed that C_{gs} is very less in comparison to C_{gg}, thus, providing an evidence to the fact that only C_{gd} contributes majorly to the total gate capacitance in the case of TFETs. As stated above, C_{gs} is higher for HD GAA TFET due to the presence of high-k material at the tunneling junction, but as $C_{gg} = C_{gs} + C_{gd}$ and the impact of high-k dielectric is concentrated at the C_{gs}, this results into almost same C_{gg} of both the devices at lower gate bias. For HD GAA TFET, the larger variations of C_{gs} at higher gate bias w.r.t. temperature, results into higher C_{gg} at higher gate bias as shown in Fig. 8.17b.

The figures of merits used for the evaluation of the RF performance are the cutoff frequency f_T, maximum oscillation frequency f_{MAX}, and the transconductance frequency product TFP = $(g_m/I_{ds})^*f_T$. Figure 8.18a shows the variations of f_T of GAA TFET and HD GAA TFET as a gate bias for variation in operating temperature. As clearly shown that at elevated operating temperatures, the cutoff frequency f_T is enhanced in both the devices. This enhancement in f_T is mainly attributed to the increase in drain current at elevated temperatures.

Moreover, f_T is much higher for the HD GAA TFET in comparison to GAA TFET, due to the enhanced transconductance of HD GAA TFET. At a gate and drain bias of 1.0 V each, the f_T of HD GAA TFET increases from 90.1 GHz to 103 GHz with a rise of 200–500 K temperature respectively. The comparative plot of TFP of both the devices including the impact of temperature variations has been

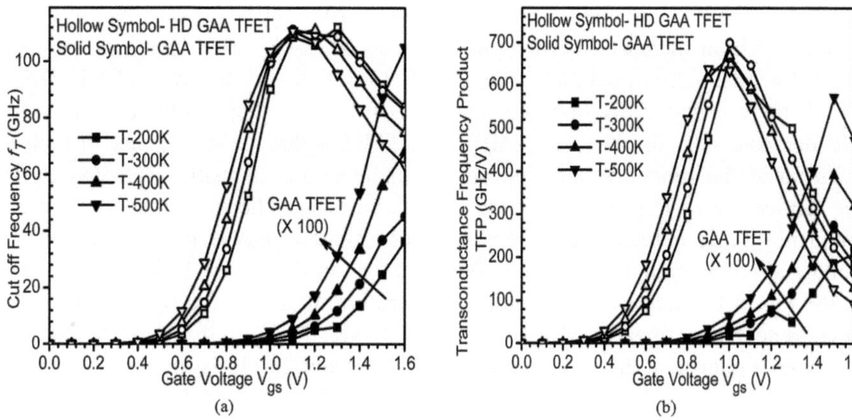

Fig. 8.18 Comparison of (**a**) Cut off frequency f_T and (**b**) Transconductance frequency product TFP of the GAA TFET and HD GAA TFET as a function of gate bias at a constant drain bias $V_{ds} = 1.0$ V for the variation in operating temperature

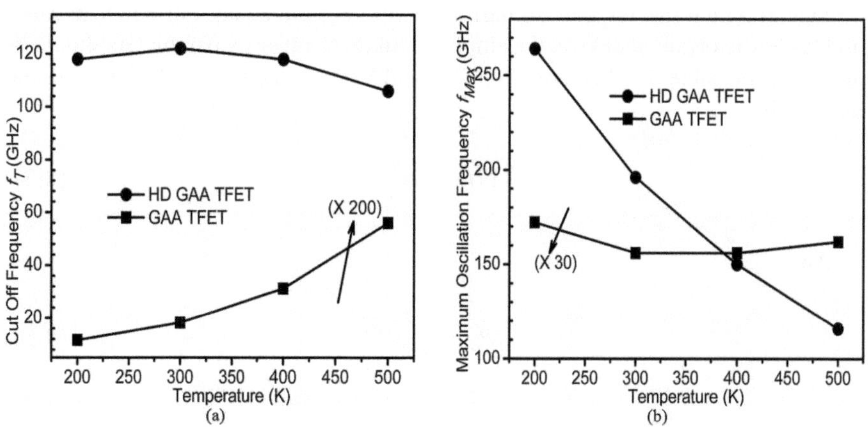

Fig. 8.19 Comparison of (**a**) Cut off frequency and (**b**) Maximum oscillation frequency f_{MAX} of the GAA TFET and HD GAA TFET as a function of temperature at a constant gate and drain bias $V_{gs} = 1.2$ V and $V_{ds} = 1.0$ V

shown in Fig. 8.18b. If compared to GAA TFET, the HD GAA TFET exhibits considerable improvement in TFP at entire gate bias range and each temperature, due to the higher f_T of HD GAA TFET. The peak of TFP curve is shifted to the lower gate bias in the case of HD GAA TFET in comparison with GAA TFET. For the GAA TFET, at $V_{gs} = 1.0$ V and $V_{ds} = 1.0$ V the TFP increases by 3.34 times for a rise in temperature from 200 K to 500 K.

The cut-off frequency f_T of both the devices as a function of temperature, extracted from the plot of current gain at which the current gain drops to 0 dB is shown in Fig. 8.19a. It is clearly illustrated from Fig. 8.19a that at T = 300 K, f_T of

HD GAA TFET is considerable improved i.e. 1340 times larger than that of GAA TFET. With an increase in temperature, the f_T of GAA TFET increases by 4.82 times for a rise of temperature from 200 K to 500 K. For the case of HD GAA TFET, there is negligibly small decrease in f_T for elevated operating temperatures. The decrease in f_T for the case of HD GAA TFET is due to the presence of high-k material at the tunneling junction which enhances the parasitic capacitances of the device (refer Fig. 8.17a, b) and thus reduces f_T. The maximum oscillation frequency f_{MAX} extracted from the curve of unilateral power gain, at which the unilateral power gain drops down to 0dB is shown in Fig. 8.19b. It is clearly shown in the Fig. 8.19b that as the operating temperature increases, the f_{MAX} of both the devices decreases linearly. Moreover, at T = 300 K, the f_{MAX} of HD GAA TFET is 38 times higher than the f_{MAX} of GAA TFET. Taken 200 K as the reference, it is found that the f_{MAX} decreases by 1.06 times and 2.27 times for GAA TFET and HD GAA TFET respectively, as the temperature increases to 500 K.

Further, to avoid the wastage of useful power, the transistors must show the linear behavior. Some important linearity figures of merits are: - VIP_2 (2nd order harmonic) and VIP_3 (3rd order harmonic). VIP_2 shows the input voltage at which the 1st and 2nd harmonic voltages are equal. Likewise, VIP_3 determines the input voltage at which the 1st and 3rd harmonic voltages are equal. For better linearity and lesser distortion of device, the amplitude of these linearity parameters should be as high as possible. Mathematically, VIP_2 and VIP_3 are defined by [57] Eqs. (8.5a) and (8.5b)

$$VIP_2 = 4\frac{g_{m1}}{g_{m2}} \tag{8.5a}$$

And

$$VIP_3 = \sqrt{24\frac{g_{m1}}{g_{m3}}} \tag{8.5b}$$

where;

g_{m1} is the first order transconductance $g_{m1} = \frac{\partial I_{ds}}{\partial V_{gs}}\Big| V_{ds} = const$

g_{m2} is the second order transconductance $g_{m2} = \frac{\partial^2 I_{ds}}{\partial V_{gs}^2}\Big| V_{ds} = const$

g_{m3} is the third order transconductance $g_{m3} = \frac{\partial^3 I_{ds}}{\partial V_{gs}^3}\Big| V_{ds} = const$

Figure 8.20a, b represents the impact of temperature on VIP_2 and VIP_3 on the variation of V_{gs} at a constant V_{ds} = 1.0 V for both the devices mentioned above. Fig. 8.20a clearly reveals a considerable enhancement in VIP_2 in HD GAA TFET in comparison to GAA TFET. The enhanced VIP_2 of HD GAA TFET is mainly attributed to high-k material near tunneling junction, which decreases the tunneling barrier width and thus the tunneling current and the device gain (i.e. transconductance). Result further indicates that at the elevated temperature

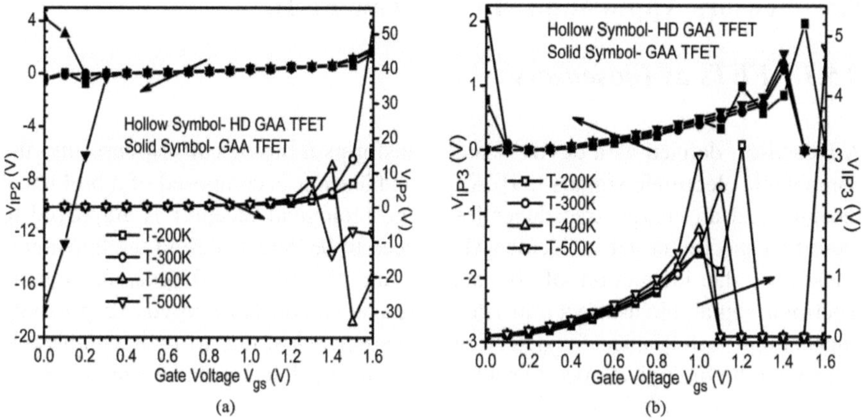

Fig. 8.20 (**a**) VIP_2 and (**b**) VIP_3 of the GAA TFET and HD GAA TFET as a function of gate bias at a constant drain bias $V_{ds} = 1.0$ V for temperature variations

(up to 400 K), the device exhibits significantly enhanced transconductance, thereby improving the linearity and exhibits low-distortions. At T = 500 K, the degradation in transconductance of the device results in a reduction in the peak of VIP_2. Thus, better linearity and lesser distortion of the device is obtained at T = 400 K, in the case of HD GAA TFET. Also, the peak of VIP_2 for HD GAA TFET occurs at lower gate bias if compared with GAA TFET. VIP_3 of both the devices as a function of gate bias has been shown in Fig. 8.20b. The peak of VIP_3 signifies the 2nd order interaction effect and depicts the cancellation of the 3rd order non-linearity by device internal feedback around a 2nd order nonlinearity [57, 58]. Figure 8.20b clearly reveals a considerable enhancement in VIP_3 in HD GAA TFET in comparison to the GAA TFET. The enhanced VIP_3 in HD GAA TFET is mainly attributed to the higher band bending and thus higher tunneling rate of electrons, which increases the drain current and thus the device gain (transconductance). Results further indicate that peak of VIP_3 is higher for the case of HD GAA TFET as compared to the conventional GAA TFET. VIP_3 shows better performance at T = 200 K i.e. to preserve better linearity, device should be operated at lower temperatures. Moreover, the peak of VIP_3 shift towards higher gate bias for an increase of temperature from 200 K to 400 K which implies that higher gate bias is needed for preserving the linearity of the device.

8.5 Sensors Application of Nanoscale FETs

8.5.1 FETs as Biosensors

A biosensor, defined as a device which transforms a biological response into the computable electronic signals [59]. A typical biosenor is composed of a biological receptor and an energy transducer [60]. The biological receptor is employed to identify a specific target molecule (which need to be detected), and the transducer transforms the interaction of the analyte and the biological receptor into an electronic signal. Field-effect transistors based biosensors have posed exceptionally outstanding performance for detection of biomolecules, due to their comparable sizes with the biological species such as proteins, nucleic acids and even microorganism such as bacteria and viruses [61]. The binding of the biomolecule species onto the biosensor alters the electrical properties of the biosensor and thus yields a quantitative determination of the target molecule [62]. Further, FET-based biosensors became popular for electrical detection of biomolecules of both types; charged (such as DNA) and neutral biomolecules such as (biotin-streptavidin). The concept of dielectric modulated FET proposed by [63], have given a new opening for the label-free detection of biomolecules with high sensitivity. The major advantage of dielectric modulated FET is that it overcomes the limitations posed by ion sensitive FET [64], which lacks in the detection of neutral biomolecules. The dielectric modulated FET is used for the label-free DNA detection technique, by considering the dielectric constant as well as the charge possessed by the biomolecule [65]. The basic working principle of dielectric modulated FET-based biosensor is that, if a cavity is carved within the dielectric material for the biomolecules to immobilize inside it, then the presence of biomolecules in the cavity will change the effective coupling between the gate and channel in comparison to the case when there are no biomolecules present. This change in the effective coupling results into change in the electrical parameters (drain current and the threshold voltage) of the biosensor (for the case of presence, absence of biomolecules and also on the properties of biomolecules such as its charges and the dielectric constants) and these changes in the electrical parameters are then used to calibrate for measuring the sensitivity of the biosensor. The large surface to volume ratio of silicon nanowire FET-based biosensors has resulted in outstanding sensitivity and thus has drawn large popularity. Silicon nanowire FET-based biosensors have reported as a powerful nano-electronic biosensor due to its ultrahigh sensitivity, high selectivity, label-free and real-time detection capabilities of DNA and proteins at very low concentration. The high selectivity of these sensors is attributed to their surface functionalization [66]. Moreover, it has been reported that the nano sensors formed by SiNW FETs show such high selectivity that the discrimination of one base mismatched DNA is achievable [67]. The challenges faced by TFET i.e. the ambipolar behavior has also been utilized for the biosensing applications [68]. Highly sensitive biosensors utilizing the advantages of BTBT of TFET has been reported by [69] for sensing biomolecules as well as pH sensor applications. The concept of gate-drain overlapping has also been utilized by [70] for the detection of biomolecules in the biosensor. The ambipolar current of TFET is again utilized for measuring the sensitivity of the device for the case when the biomolecule is present and compared to the case when the biomolecule is absent by quantifying the ambipolar current of TFET.

8.5.2 *FETs as Gas Sensor*

The gas sensor is a subtype of chemical sensors. The main function of the gas sensor is to measure the concentration of the gas present in its vicinity. The gas sensors based on the principle of work function modulation of the metallic gate of a FET is highly appreciable due to its low power consumption and high sensitivity. Palladium Pd has a high selectivity towards H_2 gas [71]. The Tunnel FETs based H_2 gas sensor has been already reported by [72]. The H_2 gas sensor uses palladium metal as the sensing element. When H_2 molecules interact with the gate of the Tunnel FET i.e. the Pd surface, the H_2 molecules dissociate on the surface of the metal and results in the formation of atomic hydrogen. Some of these dissociated H-atoms diffuse through the Pd surface. It is reported earlier that within few nanoseconds, the diffused H-atoms reaches at the interface of metal and oxide layer i.e. at the Pd-SiO_2 interface. The H-atoms reaching the metal-insulator interface gets polarized, and forms surface dipole layer [73, 74]. The dipole layer formed at the Pd-SiO_2 interface in turns modulates the work function, conductivity, etc. and thus affects the electrical properties of the TFET such as the drain current. This change in drain current can thus be calibrated to sense the H_2 gas molecules.

Moreover, Oxygen gas sensor has been proposed using silver as the gate metal for FET type based gas sensor [75]. Silver being an oxygen selective material, acts as a catalyst for oxygen sensing. In addition, the potentiometric sensor was proposed by [76] in which the work function modulation of the iridium oxide (used as the gate contact of the MOSFET), measures the presence of oxygen. The modulation in work function changes the threshold voltage of the MOSFET. The change in threshold voltage has been used as the oxygen sensing parameter. Likewise for sensing a target gas of NO_2, the SnO_2 has been used as the sensing material [77].

8.6 Summary

In this chapter, a theoretical study on the reliability issues of tunnel FET has been discussed. The reliability of TFET has been analyzed in terms of the impact of interface trap charge density, their polarity (positive/negative) and the impact of temperature on the various device performance parameters. Important conclusions drawn from the result are that the presence of positive trap charges present at the

$Si\text{-}SiO_2$ interface, increases the band bending in the channel, and thereby lowers the tunneling barrier width at the source-channel junction. This lowered barrier width enhances the BTBT rate of electrons and thus the drain current, transconductance, and other analog/RF performance metrices. Results also show that increase in density of ITCs enhances the variation in flat band voltage, and thus enhances the change in the device characteristics, in terms of electric field, BTBT rate of electrons, drain current, etc. Decrease (increase) in threshold voltage has been obtained due to the presence of positive (negative) ITCs in the device. In addition, the results show hetero gate dielectric engineering scheme that has been amalgamated to enhance the I_{ON} of TFET also results in better immunity towards ITCs in comparison to conventional GAA TFET. Moreover, while analyzing the impact of temperature, it is found that in the case of TFET, the reliance on temperature is dependent on the operating regime of the device. For the device operating in the subthreshold region, the impact of temperature is prominent. The strong variation with temperature in drain current at lower gate bias is due to the dominance of SRH at low gate bias. However, for the super-threshold operating regime, the BTBT phenomenon overlooks, which has a weak temperature dependence and thus leads to less variation in drain current. Moreover, a positive temperature coefficient has been obtained for the entire range of gate bias in the case of TFET, in contrary to the case of MOSFET where a positive/negative temperature coefficient exists for below/above subthreshold regime respectively. Thus, there is no zero temperature coefficient point in case of TFETs, which is there in case of MOSFET due to the different temperature coefficient of MOSFET. A brief overview of FET based sensors such as biosensors and gas sensors has also been discussed, on the basis of findings of previously reported articles.

Acknowledgment Authors would like to thank to Microelectronics Research Lab, Department of Engineering Physics Delhi Technological University to carry out this work. One of the authors (Jaya Madan) would like to thank University Grants Commission, Govt. of India, for providing the necessary financial assistance during the course of this research work.

References

1. K. Roy, S. Mukhopadhyay, H. Mahmoodi-Meimand, Leakage current mechanisms and leakage reduction techniques in deep-submicrometer CMOS circuits. Proc. IEEE **91**(2), 305–327 (2003)
2. A.M. Ionescu, New functionality and ultra low power: key opportunities for post-CMOS era, in Proc. of the International Symposium on VLSI Technology, Systems and Applications, pp. 72–73, 2008
3. G.E. Moore, L. Fellow, Cramming more components onto integrated circuits. Proc. IEEE **86**(1), 82–85 (1998)
4. M.T. Björk, O. Hayden, H. Schmid, H. Riel, W. Riess, Vertical surround-gated silicon nanowire impact ionization field-effect transistors. Appl. Phys. Lett. **90**, 142110 (2007)

5. H. Kam, D.T. Lee, R.T. Howe, T. King, A New Nano-Electro-Mechanical Field Effect Transistor (NEMFET) Design for low-power electronics, in Proc. of the IEEE IEDM technical digest, pp. 463–466, 2005
6. N. Abele, R. Fritschi, K. Boucart, F. Casset, P. Ancey, A.M. Ionescu, Suspended-Gate MOSFET: bringing new MEMS functionality into solid-state MOS transistor, in Proc. of the IEEE IEDM technical digest, pp. 479–481, 2005
7. J. Madan, R.S. Gupta, R. Chaujar, Analytical drain current formulation for gate dielectric engineered dual material gate-gate all around-tunneling field effect transistor. Jpn. J. Appl. Phys. **54**, 094202 (2015)
8. S.O. Koswatta, M.S. Lundstrom, D.E. Nikonov, Performance comparison between p-i-n tunneling transistors and conventional MOSFETs. IEEE Trans. Electron Devices **56**(3), 456–465 (2009)
9. G. Lee, J. Jang, W.Y. Choi, Dual-dielectric-constant spacer hetero-gate-dielectric tunneling field-effect transistors. Semicond. Sci. Technol. **28**(5), 052001 (2013)
10. T. Nirschl, S. Henzler, J. Fischer, M. Fulde, A. Bargagli-stoffi, M. Sterkel, J. Sedlmeir, C. Weber, R. Heinrich, U. Schaper, J. Einfeld, R. Neubert, U. Feldmann, K. Stahrenberg, E. Ruderer, G. Georgakos, A. Huber, R. Kakoschke, W. Hansch, D. Schmitt-Landsiedel, Scaling properties of the Tunneling Field Effect Transistor (TFET): device and circuit. Solid State Electron. **50**, 44–51 (2006)
11. J. Madan, R.S. Gupta, R. Chaujar, Mathematical modeling insight of hetero gate dielectric-dual material gate-GAA-tunnel FET for VLSI/analog applications. Microsyst. Technol., 1–8 (2016). doi:10.1007/s00542-016-2872-9
12. W.Y. Choi, B. Park, J.D. Lee, T.K. Liu, Tunneling Field-Effect Transistors (TFETs) with Subthreshold Swing (SS) less than 60 mV/dec. IEEE Electron Device Lett. **28**(8), 743–745 (2007)
13. A. Mallik, A. Chattopadhyay, Drain-dependence of tunnel field-effect transistor characteristics: the role of the channel. IEEE Trans. Electron Devices **58**(12), 4250–4257 (2011)
14. J. Wan, C. Le Royer, A. Zaslavsky, S. Cristoloveanu, A tunneling field effect transistor model combining interband tunneling with channel transport. J. Appl. Phys. **110**, 104503 (2011)
15. K. Boucart, A.M. Ionescu, Double-gate tunnel FET with high-k gate dielectric. IEEE Trans. Electron Devices **54**(7), 1725–1733 (2007)
16. W.Y. Choi, W. Lee, Hetero-gate-dielectric tunneling field-effect transistors. IEEE Trans. Electron Devices **57**(9), 2317–2319 (2010)
17. A.S. Verhulst, W.G. Vandenberghe, K. Maex, G. Groeseneken, Tunnel field-effect transistor without gate-drain overlap. Appl. Phys. Lett. **91**, 053102 (2007)
18. J. Zhuge, A.S. Verhulst, W.G. Vandenberghe, W. Dehaene, R. Huang, Y. Wang, G. Groeseneken, Digital-circuit analysis of short-gate tunnel FETs for low-voltage applications. Semicond. Sci. Technol. **26**(8), 085001–085008 (2011)
19. D. Mohata, S. Mookerjea, A. Agrawal, Y. Li, T. Mayer, V. Narayanan, A. Liu, D. Loubychev, J. Fastenau, S. Datta, Experimental staggered-source and N^+ pocket-doped channel III–V tunnel field-effect transistors and their scalabilities. Appl. Phys. Express **4**, 024105 (2011)
20. O.M. Nayfeh, C.N. Chleirigh, J. Hennessy, L. Gomez, J.L. Hoyt, D.A. Antoniadis, Design of tunneling field-effect transistors using strained-silicon/strained-germanium Type-II staggered heterojunctions. IEEE Electron Device Lett. **29**(9), 1074–1077 (2008)
21. L. Wang, E. Yu, Y. Taur, P. Asbeck, Design of tunneling field-effect transistors based on staggered heterojunctions for ultralow-power applications. IEEE Electron Device Lett. **31**(5), 431–433 (2010)
22. K.E. Moselund, M.T. Björk, H. Schmid, H. Ghoneim, S. Karg, E. Lörtscher, W. Riess, H. Riel, Silicon nanowire tunnel FETs: low-temperature operation and influence of high-k gate dielectric. IEEE Trans. Electron Devices **58**(9), 2911–2916 (2011)
23. T. Krishnamohan, D. Kim, S. Raghunathan, K. Saraswat, Double-gate strained-Ge heterostructure tunneling FET (TFET) with record high drive current and <60 mV/dec subthreshold slope, in Proc. of the IEEE International Electron Devices Meeting, pp. 1–3, 2008

24. K.E. Moselund, H. Schmid, C. Bessire, M.T. Björk, H. Ghoneim, H. Riel, InAs–Si nanowire heterojunction tunnel FETs. IEEE Trans. Electron Devices **33**(10), 1453–1455 (2012)
25. Y. Wu, H. Hasegawa, K. Kakushima, K. Ohmori, T. Watanabe, A. Nishiyama, N. Sugii, H. Wakabayashi, K. Tsutsui, Y. Kataoka, K. Natori, K. Yamada, H. Iwai, A novel hetero-junction tunnel-FET using semiconducting silicide-silicon contact and its scalability. Microelectron. Reliab **54**, 899–904 (2014)
26. J. Knoch and J. Appenzeller, Modeling of high-performance p-type III-V heterojunction tunnel FETs, IEEE Electron Device Lett., vol. 31, no. 4, pp-305-307, 2010.
27. S. Saurabh, M.J. Kumar, Impact of strain on drain current and threshold voltage of nanoscale double gate tunnel field effect transistor: theoretical investigation and analysis. Jpn. J. Appl. Phys. **48**, 064503 (2009)
28. N. Patel, A. Ramesha, S. Mahapatra, Drive current boosting of n-type tunnel FET with strained SiGe layer at source. Microelectron. J. **39**, 1671–1677 (2008)
29. K. Boucart, A.M. Ionescu, Length scaling of the double gate tunnel FET with a high-K gate dielectric. Solid State Electron. **51**, 1500–1507 (2007)
30. J. Wan, C. Le Royer, A. Zaslavsky, S. Cristoloveanu, Tunneling FETs on SOI: suppression of ambipolar leakage, low-frequency noise behavior, and modeling. Solid State Electron. **65–66**, 226–233 (2011)
31. H.M. Fahad, M.M. Hussain, High-performance silicon nanotube tunneling FET for ultralow-power logic applications. IEEE Trans. Electron Dev. **60**(3), 1034–1039 (2013)
32. S. Richter, S. Blaeser, L. Knoll, S. Trellenkamp, A. Fox, A. Schäfer, J.M. Hartmann, Q.T. Zhao, S. Mantl, Silicon–germanium nanowire tunnel-FETs with homo- and heterostructure tunnel junctions. Solid State Electron. **98**, 75–80 (2014)
33. J. Appenzeller, Y.-M. Lin, J. Knoch, P. Avouris, Band-to-band tunneling in carbon nanotube field-effect transistors. Phys. Rev. Lett. **93**(19), 196805-1–196805-4 (2004)
34. V. Nagavarapu, R. Jhaveri, J.C.S. Woo, The tunnel source (PNPN) n MOSFET: a novel high performance transistor. IEEE Trans. Electron Devices **55**(4), 1013–1019 (2008)
35. E.-H. Toh, G.H. Wang, L. Chan, D. Sylvester, C.-H. Heng, G.S. Samudra, Y.-C. Yeo, Device design and scalability of a double-gate tunneling field-effect transistor with silicon–germanium source. Jpn. J. Appl. Phys. **47**(4S), 2593–2597 (2008)
36. E.H. Poindexter, MOS interface states: overview and physicochemical perspective. Semicond. Sci. Technol. **4**, 961–969 (1989)
37. Y.H. Lho, K.Y. Kim, Radiation effects on the power MOSFET for space applications. ETRI J. **27**(4), 449–452 (2005)
38. L. Trabzon, O.O. Awadelkarim, Damage to n-MOSFETs from electrical stress relationship to processing damage and impact on device reliability. Microelectron. Reliab. **38**(4), 651–657 (1998)
39. S. Naseh, M.J. Deen, C.-H. Chen, Hot-carrier reliability of submicron NMOSFETs and integrated NMOS low noise amplifiers. Microelectron. Reliab. **46**, 201–212 (2006)
40. D.A. Newmen, *Semiconductor physics and devices-basic principles*, 3 rd edn. (McGraw Hill, New York, NY, 2003)
41. S. Dimitrijev, *Understanding semiconductor devices* (Oxford University Press, New York, 2000)
42. Y. Qiu, R. Wang, Q. Huang, R. Huang, A comparative study on the impacts of interface traps on tunneling FET and MOSFET. IEEE Trans. Electron Dev. **61**(5), 1284–1291 (2014)
43. H. Madan, V. Saripalli, H. Liu, S. Datta, Asymmetric tunnel field-effect transistors as frequency multipliers. IEEE Electron Device Lett. **33**(11), 1547–1549 (2012)
44. T.K. Chiang, A compact model for threshold voltage of surrounding-gate MOSFETS with localized interface trapped charges. IEEE Trans. Electron Devices **58**(2), 567–571 (2011)
45. S. Shabde, A. Bhattacharyya, R.S. Kao, R.S. Muller, Analysis of MOSFET degradation due to hot-electron stress in terms of interface-state and fixed-charge generation. Solid State Electron. **31**(11), 1603–1610 (1988)

46. A. Mallik, A. Chattopadhyay, Tunnel field-effect transistors for analog/mixed-signal system-on-chip applications. IEEE Trans. Electron Devices **59**(4), 888–894 (2012)
47. ATLAS User's guide, SILVACO International, Version 5.14.0.R, 2010
48. A. Biswas, S.S. Dan, C.L. Royer, W. Grabinski, A.M. Ionescu, TCAD simulation of SOI TFETs and calibration of non-local band-to-band tunneling model. Microelectron. Eng. **98**, 334–337 (2012)
49. Z.X. Chen, H.Y. Yu, N. Singh, N.S. Shen, R.D. Sayanthan, G.Q. Lo, D. Kwong, Demonstration of tunneling FETs based on highly scalable vertical silicon nanowires. IEEE Electron Device Lett. **30**(7), 754–756 (2009)
50. J. Madan, R. Chaujar, Interfacial charge analysis of heterogeneous gate dielectric-gate all around-tunnel FET for improved device reliability. IEEE Trans. Device Mater. Reliab, 227–234 (2016). doi:10.1109/TDMR. 2016.2564448
51. J. Madan, R.S. Gupta and R. Chaujar, TCAD analysis of small signal parameters and RF performance of heterogeneous gate dielectric-gate all round tunnel FET. in Proc. of the Advanced Manufacturing, Electronics and Microsystems: Tech Connect Briefs 2015. ISBN 978-1-4987-4730-1, pp. 189–192
52. E. Gnani, E. Baravelli, A. Gnudi, S. Reggiani, G. Baccarani, Capacitance estimation for InAs Tunnel FETs by means of full-quantum k. p simulation. Solid State Electron. **108**, 104–109 (2015)
53. Y. Yang, X. Tong, L. Yang, P. Guo, L. Fan, Y. Yeo, Tunneling field-effect transistor: capacitance components and modeling. IEEE Electron Device Lett. **31**(7), 752–754 (2010)
54. K.P. Pradhan, S.K. Mohapatra, P.K. Sahu, D.K. Behera, Impact of high-k gate dielectric on analog and RF performance of nanoscale DG-MOSFET. Microelectron. J. **45**, 144–151 (2014)
55. K.P.O. Donnell, X. Chen, Temperature dependence of semiconductor band gaps. Appl. Phys. Lett. **58**(25), 2924–2926 (1991)
56. A. Alnuaimi, A. Nayfeh, V. Koldyaev, Electric-field and temperature dependence of the activation energy associated with gate induced drain leakage. J. Appl. Phys. **113**, 044513 (2013)
57. P. Malik, R.S. Gupta, R. Chaujar, M. Gupta, Linearity-distortion analysis of GME-TRC MOSFET for high performance and wireless applications. J. Semicond. Technol. Sci. **11**(3), 169–181 (2011)
58. R. Chaujar, R. Kaur, M. Saxena, M. Gupta, R.S. Gupta, TCAD assessment of Gate Electrode Workfunction Engineered Recessed Channel (GEWE-RC) MOSFET and its multilayered gate architecture — Part I: hot-carrier-reliability evaluation. IEEE Trans. Electron Devices **55**(10), 2602–2613 (2008)
59. J.L. Arlett, E.B. Myers, M.L. Roukes, Comparative advantages of mechanical biosensors. Nat. Nanotechnol. **6**, 203–215 (2011)
60. T. Allsop, R. Neal, M. Dvorak, K. Kalli, A. Rozhin, D.J. Webb, Physical characteristics of localized surface plasmons resulting from nano-scale structured multi-layer thin films deposited on D-shaped optical fiber. Opt. Express **21**(16), 18765–18776 (2013)
61. K. Chen, B. Li, Y. Chen, Silicon nanowire field-effect transistor-based biosensors for biomedical diagnosis and cellular recording investigation. Nano Today **6**(2), 131–154 (2011)
62. N. Elfstro, R. Juhasz, I. Sychugov, T. Engfeldt, A.E. Karlstro, J. Linnros, Surface charge sensitivity of silicon nanowires: size dependence. Nano Lett. **7**(9), 2608–2612 (2007)
63. H. Im, X. Huang, B. Gu, Y. Choi, A dielectric-modulated field-effect transistor for biosensing. Nat. Nanotechnol. **2**(7), 430–434 (2007)
64. P. Bergveld, Development, operation, and application of the ion-sensitive field-effect transistor as a tool for electrophysiology. IEEE Trans. Biomed. Eng. **BME-19**(5), 342–351 (1972)
65. E. Buitrago, M. Fernández-Bolanos, S. Rigante, C.F. Zilch, N.S. Schröter, A.M. Nightingale, A.M. Ionescu, The top-down fabrication of a 3D-integrated, fully CMOS-compatible FET biosensor based on vertically stacked SiNWs and FinFETs. Sens. Actuators B Chem. **193**, 400–412 (2014)

66. C.-H. Kim, C. Jung, H.G. Park, Y.-K. Choi, Novel dielectric modulated field-effect transistor for label-free DNA detection. Biom. J. **2**(2), 127–134 (2008)
67. N. Lu, A. Gao, P. Dai, T. Li, Y. Wang, X. Gao, S. Song, C. Fan, Y. Wang, Ultra-sensitive nucleic acids detection with electrical nanosensors based on CMOS-compatible silicon nanowire field-effect transistors. Methods **63**, 212–218 (2013)
68. A. Gao, N. Lu, Y. Wang, T. Li, Robust ultrasensitive tunneling FET biosensor for point-of-care diagnotics. Sci. Rep. **6**, 22554 (2016)
69. D. Sarkar, K. Banerjee, Proposal for tunnel field effect transistor as ultra sensitive and label free biosensors. Appl. Phys. Lett. **100**, 143108 (2012)
70. D.B. Abdi, M.J. Kumar, Dielectric modulated overlapping gate-on–drain tunnel-FET as a label-free biosensor. Superlattice. Microstruct. **86**, 198–202 (2015)
71. K. Skucha, Z. Fan, K. Jeon, A. Javey, B. Boser, Palladium/silicon nanowire schottky barrier-based hydrogen sensors. Sens. Actuators B Chem. **145**, 232–238 (2010)
72. D. Sarkar, H. Gossner, W. Hansch, K. Banerjee, Tunnel-field-effect-transistor based gas-sensor: introducing gas detection with a quantum-mechanical transducer. Appl. Phys. Lett. **102**, 023110 (2013)
73. L.-G. Ekedahl, M. Eriksson, I. Lundstrom, Hydrogen sensing mechanisms of metal-insulator interfaces. Acc. Chem. Res. **31**(5), 249–256 (1998)
74. J. Fogelberg, M. Eriksson, H. Dannetun, L.G. Petersson, Kinetic modeling of hydrogen adsorption/absorption in thin films on hydrogen sensitive field-effect devices: observation of large hydrogen-induced dipoles at the Pd-SiO_2 interface. J. Appl. Phys. **78**(2), 988–996 (1995)
75. K. Tsukada, D. Kiriake, K. Sakai, T. Kiwa, Silver gate field effect transistor for oxygen gas sensor, in Proc. of the Sensordevices 2011: The Second International Conference on Sensor Device Technologies and Applications, pp. 5–7, 2011
76. J. Hendrikse, W. Olthuis, P. Bergveld, The MOSFET as an oxygen sensor: constant current potentiometry. Sens. Actuators B **59**, 35–41 (1999)
77. C.-H. Kim, I.-T. Cho, J.-M. Shin, K.-B. Choi, J.-K. Lee, J.-H. Lee, A new gas sensor based on MOSFET having a horizontal floating gate. IEEE Electron Device Lett. **35**(2) (2014)

Chapter 9
MEMS/NEMS-Enabled Vibrational Energy Harvesting for Self-Powered and Wearable Electronics

Kai Tao, Jin Wu, Ajay Giri Prakash Kottapalli, Sun Woh Lye, and Jianmin Miao

9.1 Introduction

Recent advances in internet of things (IoT) and low-power electronic devices have given rise to increasing interest in materials and technology that lead to portable and sustainable power sources. Conventional technologies which employ batteries to supply power for wireless sensor networks (WSNs) may not be the choice, since WSNs in use are intended to operate over long period with ease of maintenance. However, as the distributed sensor networks grow and electronic devices become smaller in size, providing sustainable power represents a great challenge. This challenge is further compounded if WSNs are deployed in hazardous and non-accessible areas, where replacement of batteries may not be viable. Secondary power source in addition to or replacement of batteries should therefore be established.

Figure 9.1 depicts the power consumption statistics of various applications in regular daily use around the world. In the future, there will be a great demand for mobile/implantable electronics with extremely low power consumption [1]. To address the constant surge in power consumption, scientists have made significant efforts to develop alternative forms of energy, which usually involves harnessing or harvesting energy from the ambient environment for power. Clearly, harvesting ambient energy to supplement or replace batteries in order to operate low-power wireless electronic devices would be an invaluable alternative.

In the meanwhile, CMOS technology and related micro/nano electromechanical systems (MEMS/NEMS) enabled technologies have been widely employed to

K. Tao • J. Wu • A.G.P. Kottapalli • S.W. Lye • J. Miao (✉)
School of Mechanical and Aerospace Engineering, Nanyang Technological University, 50 Nanyang Avenue, Singapore 639798, Singapore
e-mail: mjmmiao@ntu.edu.sg

T. Li, Z. Liu (eds.), *Outlook and Challenges of Nano Devices, Sensors, and MEMS*,
DOI 10.1007/978-3-319-50824-5_9

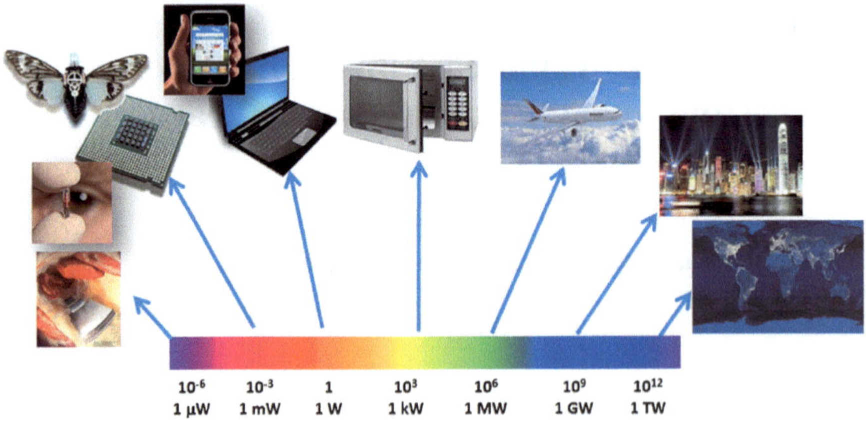

Fig. 9.1 Power requirements for different applications: In the future there will be a huge demand for mobile/implantable electronics with extremely low power consumption [1]

create miniaturized devices and complex circuit modules through batch-fabrication processes. MEMS devices take advantages of integrated circuit fabrication techniques and thus are inherently compatible with microelectronic devices. MEMS technology could facilitate miniaturization, low power consumption, low cost, and ease of integration with electronics, making it an ideal candidate for self-powered and wearable electronics.

There are several existing energy sources in the environment for harvesting, including light, mechanical motion, temperature gradient, acoustic and radiofrequency (RF) energy. Among these, mechanical vibration energy is attractive due to its wide distribution, versatility and ubiquitous availability. Ambient motions can be derived from structures, human body, vehicles, machinery, or air/water flows. Due to their pervasive existence, they are more suitable for small-scale embedded biomedical and wearable applications.

Mechanical energy can be transformed to electricity by exploiting the mechanical strain or relative movement in the transducer. In mechanical strain, the energy from structural deformation can be extracted by employing active materials, such as piezoelectric or magnetostrictive. As for the relative movement, inertial generators with spring-mass inertial structure can be employed to extract the vibration energy. At resonance, the excitation amplitude of external vibration sources can be significantly amplified by relative displacement of the spring-mass frame systems. For this inertial spring-mass structure, there are three commonly used transduction mechanisms, i.e. electromagnetic, electrostatic and triboelectric.

Five transduction mechanisms that convert the mechanical vibration to electricity are summarized in Fig. 9.2, which includes electrostatic, electromagnetic, piezoelectric, triboelectric and magnetostrictive. Piezoelectric harvesters utilize certain piezoelectric materials, such as PZT and PVDF, which have the ability to generate an electrical potential when subjected to mechanical strain. Electromagnetic harvesters are developed on the Faraday's law of induction principle in which electrical power is

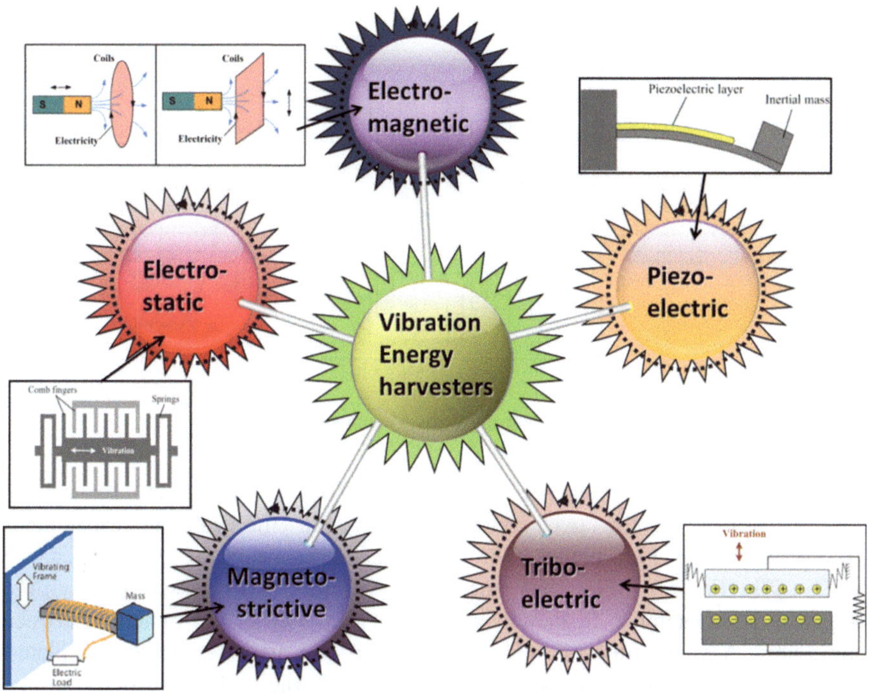

Fig. 9.2 Five reported main vibration-to-electricity transduction mechanisms: electrostatic, electromagnetic, piezoelectric, triboelectric and magnetostrictive

generated via a change of the magnetic flux in the coil. Electrostatic harvesters are based on the capacitance change of the variable capacitors under constant voltage/charge bias, either by an external voltage source or pre-charged electret material. Triboelectric harvesters operate on triboelectric effect where electron migration is created when two dissimilar materials are brought to physical contact. Magnetostrictive harvester is based on Villari effect of certain ferromagnetic materials such as Terfenol-D, which can convert vibration-induced strain to produce a charge in its magnetization. Details of the operating principles and recent advances of these transducers will be compared and presented in the ensuing sections.

9.2 Electrostatic MEMS Vibrational Energy Harvesters

9.2.1 Electret-Free Electrostatic Harvesters

Electrostatic power generators are based on capacitance change of variable capacitors. Electrical power is generated from capacitance change between the two parallel plates, which are electrically isolated from each other by vacuum, air, oil or insulator.

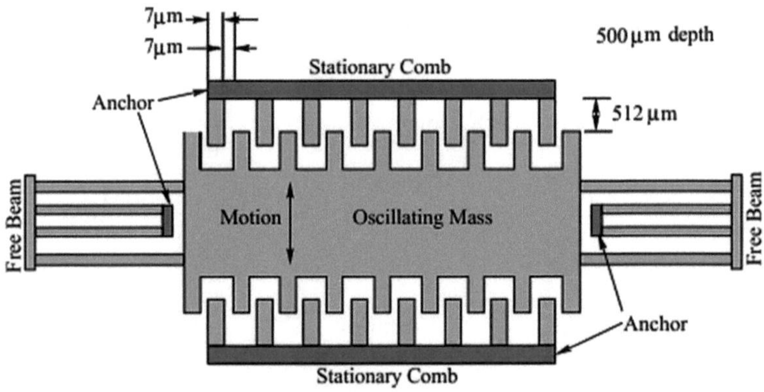

Fig. 9.3 MEMS-based electrostatic transducer by Menninger et al. [2]

The magnitude of the storage of the charge (Q) is determined by voltage (V) difference of the two charging plates and the variable capacitance (C) between the two plates. The fundamental equation of the capacitance is $C = Q/V$. For a parallel capacitor, the capacitance is expressed as $C = \varepsilon A/d$, where C is capacitance in Farads, ε is the permittivity of the material between the plates, A is the area of the plates and d is the distance of the two plates. If the ε_0 is the permittivity of the free space, the voltage difference and energy stored within the capacitor is expressed as $V = Qd/\varepsilon_0 A$ and $E = \frac{1}{2}QV = \frac{1}{2}CV^2 = \frac{1}{2}\frac{Q^2}{C}$, respectively. If the charge of the plates is held constant, the electrostatic force between the variable plates is expressed a

$$E = \frac{1}{2}\frac{Q^2 d}{\varepsilon A} \tag{9.1}$$

If the voltage is kept at constant, the electrostatic force between the variable plates is given by

$$E = \frac{1}{2}\frac{\varepsilon A V^2}{d} \tag{9.2}$$

Harvesting vibration energy by MEMS electrostatic transducer was firstly proposed by Meninger et al. [2, 3] from MIT at 1999. They have successfully designed and fabricated an electrostatic vibration energy harvester based on MEMS technology, as shown in Fig. 9.3. Based on the predicted value of capacitance and simulation, it was estimated that the converter can produce 8.6 μW output power in in-plane overlap varying mode.

Roundy et al. [4] from Berkeley further categorized the electrostatic converters into three conversion types such as in-plane overlap varying, in-plane gap closing and out-of-plane gap varying, as shown in Fig. 9.4. Although variable gap type devices were less sensitive to the parasitic capacitance, these devices exhibited

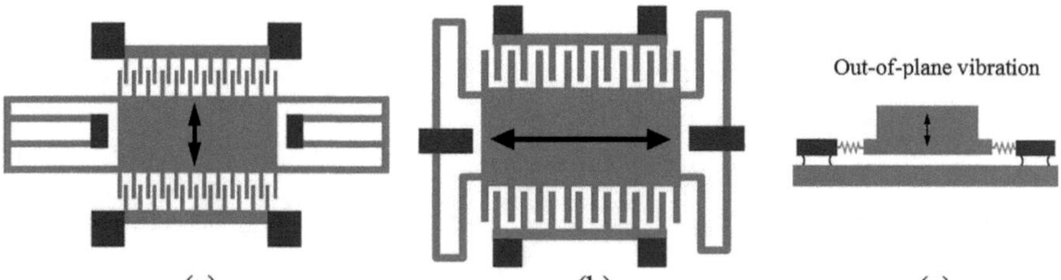

Fig. 9.4 Three different electrostatic conversion modes: (**a**) In-plane overlap varying; (**b**) In-plane gap closing; (**c**) Out-of-plane gap varying [4]

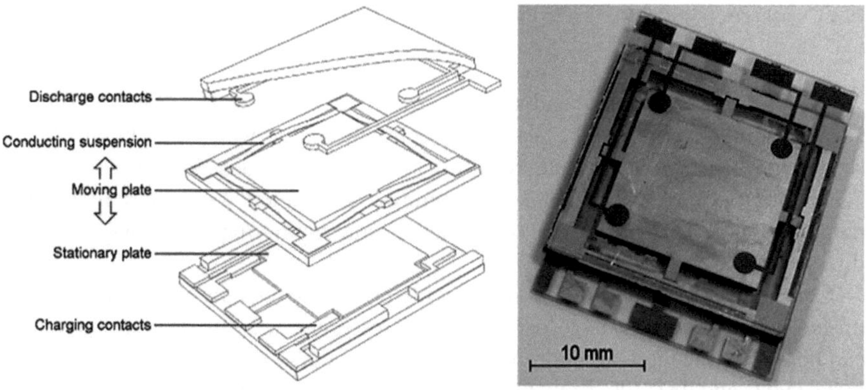

Fig. 9.5 Exploded view (*left*) and prototype (*right*) of power generator [6]. Springer Publishing. Reproduced with permission

higher loss due to air damping and large surface area which might result in electrostatic stability problems. Therefore, an in-plane gap closing type harvester was fabricated by deep reactive ion etching (DRIE) process on the top layer of silicon-on-insulator (SOI) wafers. An optimal output power of 116 $\mu W/cm^3$ was calculated at a minimum gap of 0.25 μm.

Mitcheson et al. [5–9] from Imperial College of London divided vibration-driven MEMS micro generators into three working forms based on different damping conditions, such as velocity-damped resonant generators (VDRGs), Coulomb-damped resonant generators (CDRGs) and Coulomb-force parametric generators (CFPGs). It was concluded that CFPGs could achieve better performance than the other two resonant generators for applications where source frequency was likely to vary. When the source motion amplitude was much larger than internal displacement of the device, a non-resonant generator would be more efficient than a resonant one. For these reasons, a non-resonant device catering to a wide range of input frequencies had been designed and fabricated as shown in Fig. 9.5. The inertial mass was suspended by a low stiffness spring. The bottom plate contained the counter electrode and charging studs. Initially, the moving mass

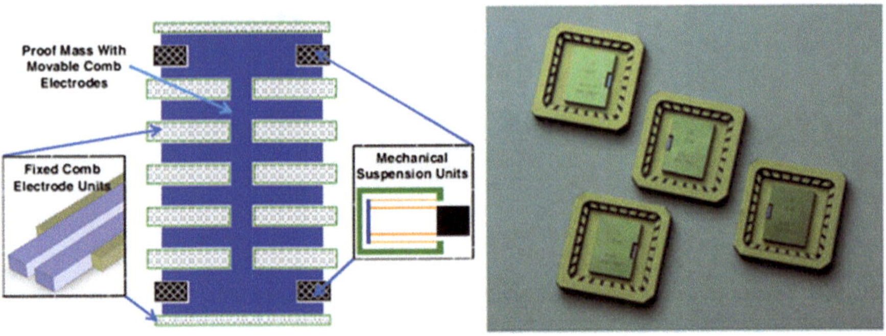

Fig. 9.6 Schematic of the transducer layout (*left*) and micro power generator chips packaged in CLCC packages (*right*) [10].

was at its lowest position and contacted with charging studs. When sufficient acceleration was applied, the plate with constant charge would be pulled up and separated from the bottom plate, resulting in voltage increases. An output power of 120 nJ was obtained by tuning the operating parameters such as reducing the damping, increasing the initial capacitance and modifying the mass vibration direction.

Comb drive designs were most commonly used in developing electrostatic transducers and were adopted by a vast majority of researchers. Comb drive architectures can be realized using silicon-on-insulator (SOI) wafers [10–12]. Researchers from HSG-IMIT and IMTEK from Germany [10] have developed chip size electrostatic generators which can produce a total net energy of 13.38 mJ in a period of 2 h. In their design, an elastic stopper was adopted to limit the mass displacement. The prototype after completely packaged has an area of 5 mm by 6 mm. A schematic of the transducer layout and micro power generator chips packaged in CLCC packages are shown in Fig. 9.6.

However, one of the main problems of the electrostatic energy harvesting is that an external source is required to provide the initial charge or voltage bias to the capacitors. Although the extra charge pump can be designed by the external circuits, this would lead to a substantial increase of the device volume as well as the complexity of the overall system. To address this problem, a new type of electret-based vibration energy harvester (e-VEH) has been proposed recently.

9.2.2 *Electret-Based Electrostatic Harvesters*

In electret-based energy harvesting systems, electrical power is generated from capacitance change between the two parallel plates, which is similar to the conventional electret-free electrostatic energy harvesters. Electret-based energy harvester utilizes an electret as its negative/positive permanent surface voltage source.

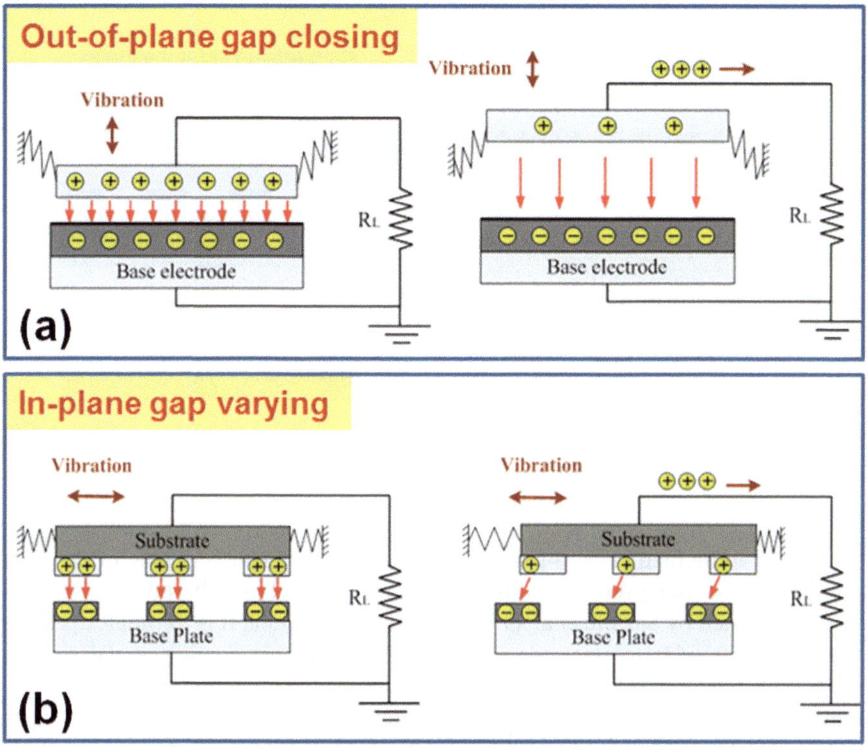

Fig. 9.7 General working principles of e-VEH: (**a**) Charge flow of out-of-plane gap-closing scheme (**d**) Charge flow of In-plane gap varying scheme

Electrets are dielectrics with quasi-permanent electric charge or dipole polarization, which can form a permanent electric field for several years around the structure.

In general, an electret-based generator operates in two schemes: the out-of-plane gap-closing scheme and the in-plane gap varying scheme. The out-of-plane type mainly operates on the change of the dielectric gap, while the in-plane type works on the variation of the overlapping area. Interdigital electrodes and patterned electret layers are usually employed to enhance the rate of change of the overlapping area for in-plane scheme. As depicted in Fig. 9.7, when external excitations give rise to a relative displacement of the two plates either in out-of-plane (Fig. 9.7a) or in in-plane (Fig. 9.7b) direction, an alternative current flow would be generated through external circuit due to capacitance variation.

Generally, electret materials are divided into two categories: SiO2-based inorganic (SiO2, Si3N4) electrets and polymer-based organic (Teflon, Parylene, PVDF, LDPE, CYTOP) electrets. For in-plane overlapping micro energy harvesters, high and stable surface potential on micro-sized electrets is preferred owing to the small change rate in the overlapping area. As the surface charge density of electret could significantly impact the output power of the electrostatic energy harvesters, studies into electret formation and charging are of cardinal significance. Typically, electret

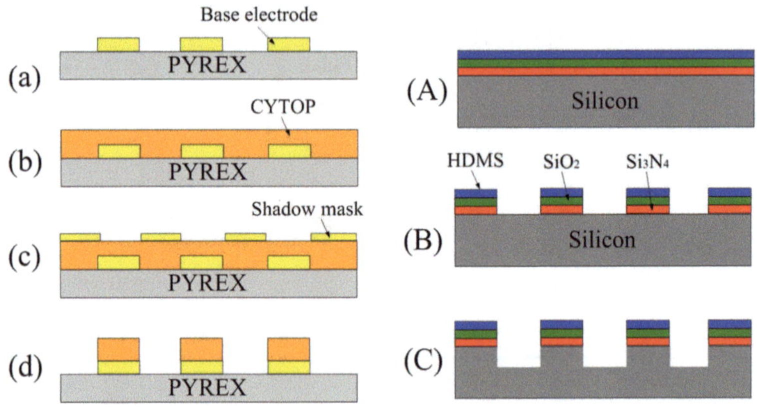

Fig. 9.8 Fabrication process of micro-patterning: (**a**) Polymer-based CYTOP electret [13]; (**b**) SiO2-based SiO2/Si3N4 electret [14]

films need to be firstly micro-patterned etched and then charged. One of the most common fabrication processes of micro-patterning of polymer-based CYTOP electret is presented by Sakane et al. [13], as shown in Fig. 9.8a. It is a complex process that composes of base-electrode sputtering, CYTOP spin coating, metal shadow formation, final CYTOP etching and mask removal. The commonly used micro-patterning of SiO2-based SiO2/Si3N4 electret is presented by Boisseau et al. [14], as shown in Fig. 9.8b. It includes the deposition of SiO2/Si3N4 multilayer structure, patterning the electret and then charging.

One of the challenges in e-VEHs is that most devices reported are only for unidirectional vibration with a narrow bandwidth. Typically, for an in-plane type e-VEH, multi-folded springs with high aspect ratio (HAR) are employed to make the system only flexible along a desired axis while other directions are kept stiff. It is inefficient when the excitation directions do not precisely act along the predefined directions. For out-of-plane type e-VEHs that usually make use of cantilever beam, the response only exhibits a narrow bandwidth. The electrostatic force, as an inherent nonlinear force in the electrostatic energy harvesting system, has not been fully tapped and utilized. Considering the fact that the frequencies and oscillation directions of the ambient vibrations are mostly unpredictable and random, certain broadband and multiple directional energy harvesting techniques are highly demanded.

To make the energy harvester adaptive to multiple excitation directions, two-dimensional (2D) dynamic responses of symmetrical spring-mass resonant systems have been proposed and investigated by Tao et al. [15]. A novel rotational symmetrical resonator was presented that consisted of a movable disk-shaped seismic mass suspended by three sets of spiral-beam springs. The experimental analysis showed that the proposed device could harvest kinetic energy from both out-of-plane and in-plane directions based on its first three vibration modes having resonant frequencies of only several tens of Hz. To enhance the performance and

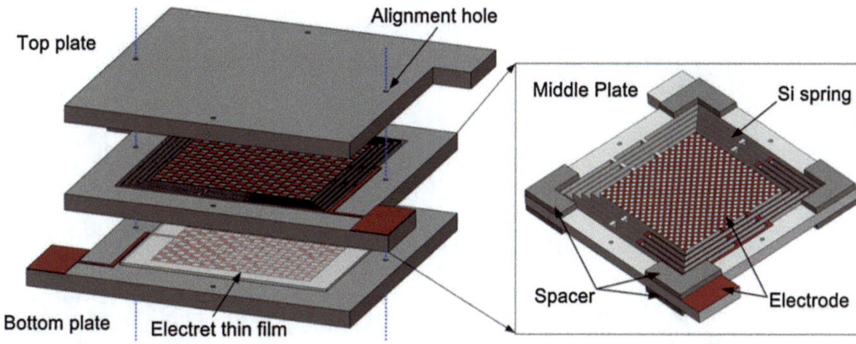

Fig. 9.9 3-D schematic view of the energy harvester [16]. Elsevier Publishing. Reproduced with permission

reduce the parasitic force, a sandwiched structure with a symmetrical resonator was further developed. The structure design composes of a double-sided electrode-patterned seismic mass suspended by four-folded parallel beam flexures [16], as shown in Fig. 9.9. A major advantage of this configuration is that both the vertical pull-in electrostatic force as well as the horizontal damping force can be reduced. A localized corona charging method for patterning charge distribution in electret thin films was applied, where any micro-patterning and chemical etching process could be avoided. The fabricated device and the charge patterns are shown in Fig. 9.10. With the fabricated prototype, an overall output power of 0.12 μW was obtained for the two capacitive parts at an acceleration of 0.2 g at 125 Hz.

To cater to the broadband spectrum exhibited by ambient vibrations, two approaches were intensively investigated that sought to expand frequency spectrum of energy harvesters, namely 'nonlinear technique' and 'multi-frequency energy harvesting'. Nonlinear technique was considered to be an effective way to enlarge the frequency spectrum by utilizing spring softening and hardening effects. Although a number of methods have been proposed in the past to induce non-linearities of energy harvesting system, electrostatic nonlinearity introduced by electret surface potential, as an inherent feature of e-EVHs, was seldom studied. It was also practically more advantageous and readily compatible with MEMS energy harvesting devices. Tao et al. [17] proposed a novel out-of-plane e-VEH device employing gap-closing variable capacitors for low-frequency broadband energy harvesting. Both positive and negative charged electret plates are integrated into a single resonant system to induce strong electrostatic nonlinearity. Figure 9.11 shows the schematic drawing and photograph of fabricated e-VEH MEMS device. At a high excitation level of 0.48 g, the experimental results showed that the 3-dB bandwidth was enlarged by 2.85 times from 1.3 Hz (bandwidth of linear response) to 3.7 Hz. An optimal output power of 0.95 μW was also achieved with a low resonance of 95 Hz.

Compared with other broadband approaches involving extra mechanical components and tuning efforts, multi-frequency technique exploiting multiple vibration

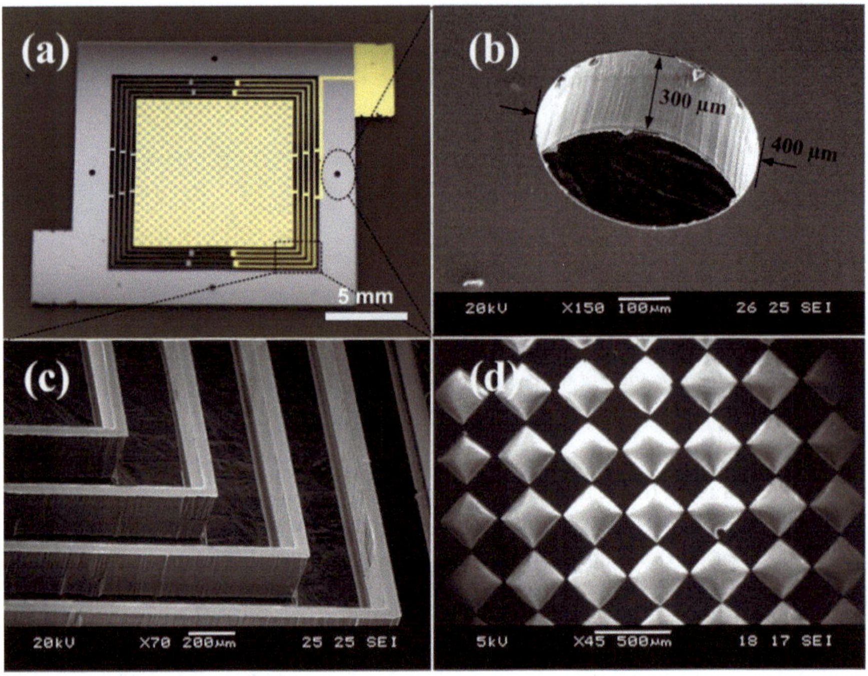

Fig. 9.10 (**a**) Photograph of the fabricated spring-mass structure; (**b, c**) enlarged views of the alignment holes and elastic beam flexures; (**d**) SEM images of the micro-sized charge distribution on LDPE electret thin film (negative charged). Elsevier Publishing. Reproduced with permission

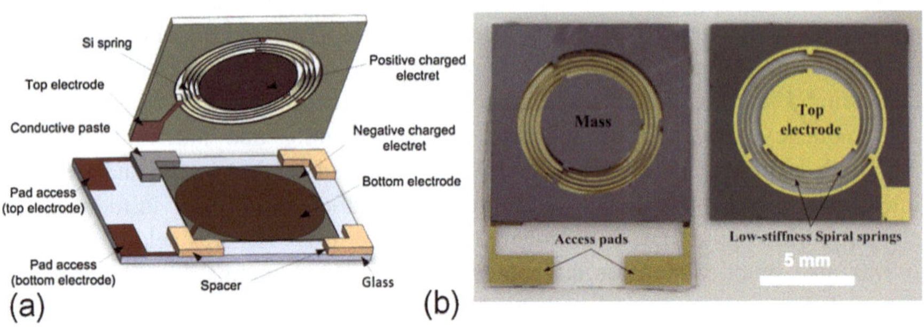

Fig. 9.11 (**a**) Schematic drawing of the proposed e-VEH device; (**b**) Optical images of the assembled MEMS e-VEH device and released spring-mass resonant structure [17]. © IOP Publishing. Reproduced with permission

modes of single two-degree-of-freedom (2DOF) system provides a simple and reliable solution to increase the energy harvesting effectiveness. Therefore, Tao et al. [18, 19] developed and investigated a 2DOF e-VEH system that comprised a primary subsystem for power generation, and an accessory subsystem for frequency tuning. Figure 9.12 shows a schematic and a photograph of the proposed device

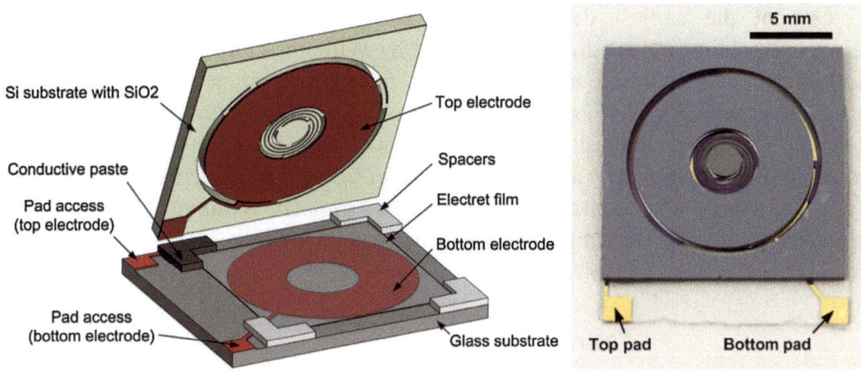

Fig. 9.12 Schematic (left) and fabricated prototype (right) of the proposed electret-based 2DOF MEMS energy harvesting device

structure and the fabricated prototype of the proposed e-VEH device, respectively. By precisely tuning the accessory mass, the first two resonances of primary mass could be tuned close to each other while maintaining comparable magnitudes. With further increased excitation accelerations, it was observed that the 2DOF e-VEH system demonstrated a spring hardening nonlinear effect, where the first peak was capable of being driven towards convergence with the second one to achieve a broadband energy harvesting system. Such novel nonlinear two-degree-of-freedom (2DOF) energy harvesting system combines advantages of both multi-modal energy harvesting and the nonlinear technique, which potentially offers new insight of increasing bandwidth with hybrid broadband mechanisms.

9.3 Electromagnetic MEMS Vibrational Energy Harvesters

The electromagnetic transducer works on the principle of Faraday's law of electromagnetic induction, which states that the change in magnetic flux generates a voltage in the coil. The voltage is proportional to the time rate of change of the magnetic flux linkage, as given below

$$V = -\frac{d\Phi}{dt} = NBl_c\frac{dy}{dt} \tag{9.3}$$

where Φ the flux linkage in the coil, N is the number of turns in the coil, B is the magnetic field strength, l_c is the effective length of the turn, y is the relative distance between the coil and magnet. The electromagnetic induction can be realized by either moving magnet with fixed coil or moving coil with fixed magnet. Owing to the flexibilities of its design, a wide range of possible electromechanical coupling can be implemented.

One of the earliest descriptions of an inertial micro generator utilizing electromagnetic mechanism was presented by Williams et al. [20] in 1996. They formulated and analysed an equation for power generation for linear inertial generators, which served as a fundamental reference for inertial resonator development. The energy harvesting system consists of a moving mass (m) suspended by a spring (k) and damped by an energy transducer (η_e) and parasitic damping element (η_m). For a sinusoidal excitation vibration ($y = Ysin(\omega t)$), Williams and Yates [20] have shown that the output power can be derived in equation as

$$P_{W\&Y} = \frac{m_1 \zeta_t A^2 \omega^3 \left(\frac{\omega}{\omega_n}\right)^3}{\left(1 - \left(\frac{\omega}{\omega_n}\right)^2\right)^2 + \left[2\zeta_t \left(\frac{\omega}{\omega_n}\right)\right]^2} \tag{9.4}$$

where ζ_t is the total damping ratio, ω_n is the resonant frequency, A and ω are the amplitude and frequency of the excitation, respectively. When the excitation frequency equals the natural frequency ($\omega/\omega_n = 1$), the average power can be simplified as

$$p_{av} = \frac{\zeta_e}{4(\zeta_p + \zeta_e)^2} m Y_0{}^2 \omega_n{}^3 \tag{9.5}$$

If further assumed that $\zeta_e = \zeta_p = \zeta/2$, the maximum power output of the energy harvesting system can be obtained and expressed as

$$p_{av} = \frac{m Y_0{}^2 \omega_n{}^3}{16 \zeta_p} \tag{9.6}$$

In this case, the maximum power output can achieve the theoretical conversion efficiency, which is 50% of the total power from the frame vibration. The average power output can be obtained when the natural frequency of the spring-mass system is equal to the excitation vibration and when the electrical damping rate is equal to the mechanical parasitic damping rate. The average power output is found to be proportional to the masses, the square of the excitation amplitude and three times of the natural frequency.

In 1997, Shearwood et al. [21] proposed a preliminary micro power generator based on a 2.4 mg bulk-micromachined SmCo permanent magnet. It could produce a peak power of 0.3 μW at a vibration frequency of 4.4 kHz. This was followed by Yuen et al. [22], where a micro power generators using MEMS laser micromachining technology was developed. The micro generator comprises NdFeB magnets was supported by laser micro machined Cu spring structures as shown in Fig. 9.13. With two micro power transducers connected in series, the electromagnetic micro power generator could gain a maximum power output of 120 μW at the frequency of 80 Hz with input acceleration below 0.5 g. This battery sized generator was able to operate a wireless RF thermometer at a continuous vibration for 32 s.

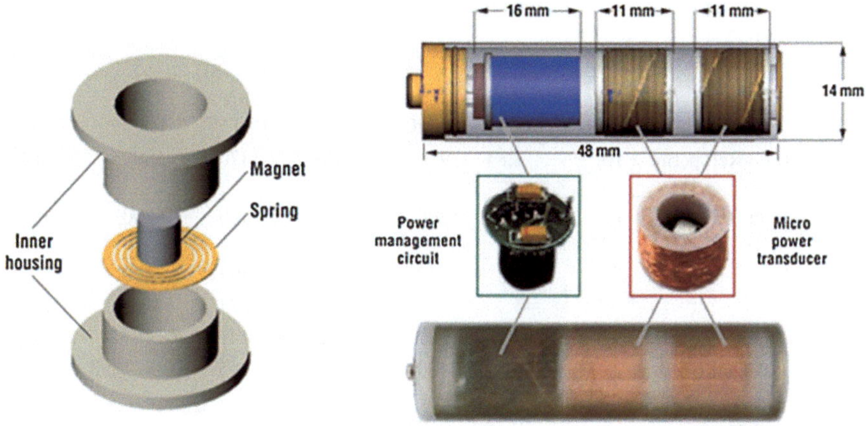

Fig. 9.13 An AA sized vibration based micro generator: the micro power transducer's inner structure (*left*); the power management circuit and transducer, and an assembled AA-sized MPG containing the power manage circuit and two MPTs [22]

In general, electromagnetic energy harvesters can produce relatively high output current level for macro devices. No input voltage sources and no mechanical constraints are needed for electromagnetic devices as compared to electrostatic conversion. However, the power output in electromagnetic energy harvesting strongly relies on their physical size. The performance of electromagnetic generators deteriorates significantly in micro scale [23]. The conversion efficiency of electromagnetic mechanism reduces ten times faster than that of electrostatic one with similar physical size decreases. This was highlighted by a review of electromagnetic vibration-based energy harvester conducted by Scott et al. [24] based on experimentally demonstrated characteristics. Another major concern is the cumbersome integration process. Although various techniques have been developed to integrate the coils and permanent material into micro devices, the system is still complex and bulky and it is difficult to integrate electromagnetic generators with MEMS fabrication process.

9.4 MEMS/NEMS-Enabled Piezoelectric Vibrational Energy Harvesters

9.4.1 MEMS Harvesters with Piezoelectric Thin Films

Piezoelectric effect was first discovered by Jacques Curie and Pierre Curie in 1880 [25]. It is found that certain materials had the ability to generate an electrical potential when subjected to mechanical strain. Conversely, when these materials were subjected to an electrical field, deformations would also take place.

Table 9.1 Key properties of some representative piezoelectric material used in energy harvesting systems [26]

Materials	AlN	ZnO	$BaTiO_3$	PZT-4 'Hard PZT'	PZT-5H 'Soft PZT'	PMN-PT	PVDF
Const. strain Rel. perm. (ε^S_{33})	10	8.84	910	635	1470	680	5–13
Const. strain Rel. perm. (ε^T_{33})	11.9	11.0	1200	1300	3400	8200	7.6
d_{33} pCN^{-1}	5	12.4	149	289	593	2820	−33
d_{31} pCN^{-1}	−2	−5	−58	−123	−274	−1330	21
d_{15} pCN^{-1}	3.6	−8.3	242	495	741	146	−27
Mechanical quality (Q_m)	2800	1770	400	500	65	43–2050	3–10
Electromechanical coupling (k_{33})	–	0.48	0.49	0.7	0.75	0.94	0.19

Electrical power can be generated when the external vibration causes the deflection of the piezoelectric material to undergo compression and tension. Therefore, the performance of piezoelectric power generator is largely dependent on properties of piezoelectric materials. There are commonly several types of piezoelectric materials utilized for piezoelectric power generator in various forms such as hard piezoceramic (PZT-4) and soft piezoceramic (PZT-5H), barium titanate (BaTiO3), thin film (zinc oxide (ZnO) or Alumina nitride (AlN)), thick film based on piezoceramic powder and polymeric materials (polyvinylidenefluoride (PVDF)). Key properties of these representative piezoelectric materials used in energy harvesting systems are given in Table 9.1 [26]. It can be seen that the piezoelectric coefficients d33, d31 and d15 of the ferroelectric material, such as PZT and BaTiO3, are an order of magnitude larger than those of the nonferroelectric materials, such as AlN and ZnO. The relative permittivity ε^S_{33} and ε^T_{33} of ferroelectric materials under constant stress and strain are about two orders larger than those of nonferroelectric materials.

The constitutive equations for piezoelectric material, which describe the behaviour between strain and charge, are given by

$$\delta = \frac{\sigma}{Y} + d_{ij}E \tag{9.7}$$

$$D = \varepsilon E + d_{ij}\sigma \tag{9.8}$$

where δ is the strain, σ is the stress, Y is the Young's modulus, d is the piezoelectric strain coefficient, D is the charge density, E is the electrical field, ε is the material dielectric constant and d_{ij} is the piezoelectric strain coefficient. Because piezoelectric material typically exhibit anisotropic characteristics, the piezoelectric strain coefficient has two subscripts. The first one indicates the direction of the field and the second one denotes the direction of strain. For

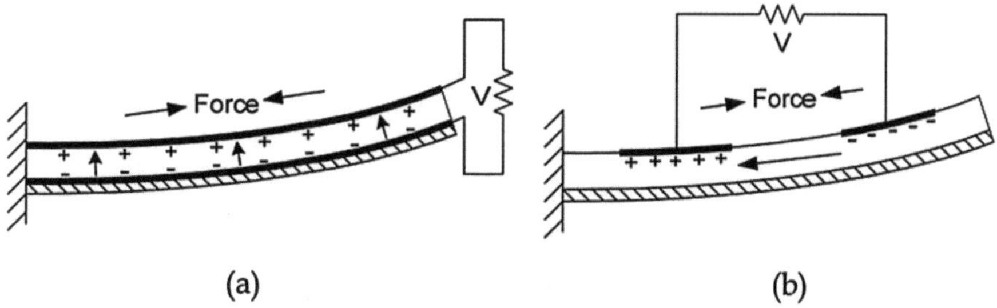

Fig. 9.14 Two types of piezoelectric harvesters (**a**) d31 mode and (**b**) d33 mode [27].

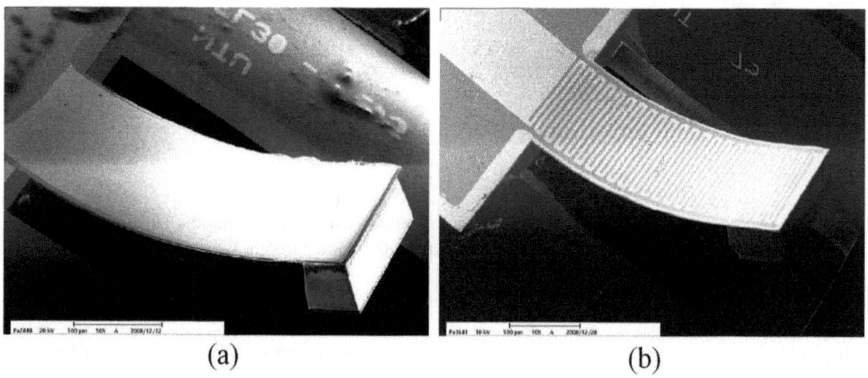

Fig. 9.15 SEM photograph of two types of piezoelectric MEMS generators (**a**) d31 mode and (**b**) d33 mode [29] © IOP Publishing. Reproduced with permission

example, for d33 mode, the applied stress and electric field are in the same direction along the polarization axis; for d31 mode, the stress is the same direction as before, but the electric field is perpendicular to the stress. They are two most common modes used in piezoelectric energy harvesting, as shown in Fig. 9.14 [27].

Both piezoelectric coefficient and coupling constant of d33 mode are higher than that of d31 mode in piezoelectric materials. Although the piezoelectric coefficient of d31 mode is lower than that of d33 mode, it is more flexible, which results in more deformation when subjected to a certain input force [5]. Therefore, d31 mode power generator is commonly used in vibration energy harvesting. When utilized in static force energy harvesting applications, the d33 mode power generators are more efficient [5, 28]. In addition to commonly used parallel electrodes, interdigital electrodes have also been designed to realize a d33 mode piezoelectric coupling. Lee et al. [29] designed and fabricated two piezoelectric energy harvesters with d31 mode and d33 mode electrodes, as shown in Fig. 9.15a, b respectively.

The method of utilizing piezoelectric materials for energy harvesting has been reported by many groups on a wide range of devices and applications.

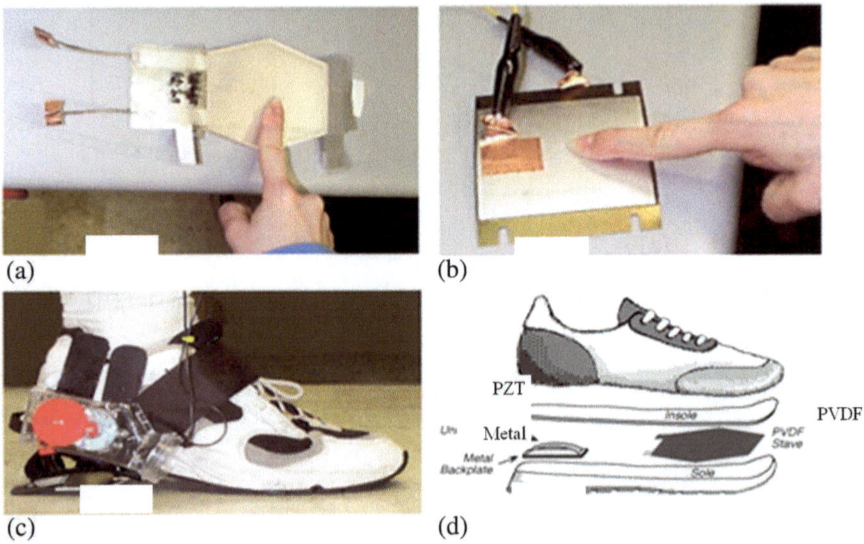

Fig. 9.16 Integrating three devices for parasitic power harvesting in shoes: (**a**) a PVDF insole stave; (**b**) a Thunder PZT unimorph; (**c**) a shoe mounted rotary magnetic generator; (**d**) exploded view of integration of piezos [34]

General reviews of piezoelectric energy harvesting have been published [30–33]. A lot of methods have been exploited in order to enhance the power output such as modifying the piezoelectric materials, altering the poling and stress direction, changing the electrode pattern and increasing the bandwidth or tuning the resonant frequency of the devices. One of the early applications using this power conversion mechanism was performed by Paradiso et al. [34] in 1998, as shown in Fig. 9.16. They successfully implemented these power generators into shoes while maintaining its design and comfort. They integrated three different devices into a shoe: a unimorph strip made from piezoceramic composite material, a stave made from laminated PVDF foil and a shoe mounted rotary magnetic generator. In the self-powered application, a digital RFID signal can be periodically broadcasts while the bearer walks.

AlN is another commonly utilized piezoelectric material for energy harvesting due to its CMOS compatibility and relatively easy deposition process. Elfrink et al. [35] from IMEC/Holst Centre have successfully demonstrated a piezoelectric energy harvester using an uniform AlN micro cantilever. Cantilever beams with length of 1.01–2.01 mm and width of 3–7 mm are encapsulated by two glass wafers, as shown in Fig. 9.17a. The energy harvester prototype mounted on a supportive board is depicted in Fig. 9.17b. The resonant frequencies of the AlN-based MEMS piezoelectric energy harvester can be tuned from 200 to 1200 Hz by adjusting the dimensions of the beam and mass parameters. A maximum power of 60 μW at the resonant frequency of 572 Hz was achieved with an excitation acceleration of 2 g.

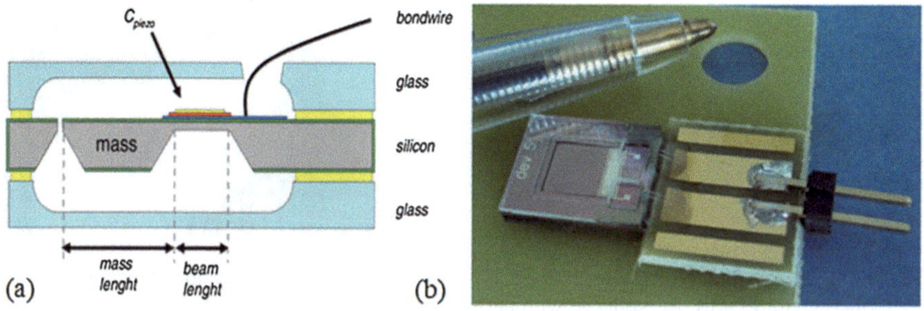

Fig. 9.17 (**a**) AlN-based MEMS piezoelectric energy harvester packaged in between glass substrates; (**b**) the harvester prototype mounted on a supportive board [35]. © IOP Publishing. Reproduced with permission

Generally, most piezoelectric energy harvesters are designed with a cantilever configuration where the piezoelectric beam serves as both mechanical vibration component and energy conversion element. The output power relies on bending and stretching of the piezoelectric beam where fatigue of piezoelectric materials may easily set in after long-term operation. Moreover, such linear cantilever-based structures usually operate in a predefined out-of-plane direction with a narrow bandwidth near its sole resonance. This limits their applicability in ambient environment where the frequency spectrum is wide with various excitation directions. The commonly used brittle and stiff piezoelectric materials, such as PZT and Macro-Fiber Composites (MFC), would also encounter severe challenges in miniaturization and integration.

9.4.2 *Piezoelectric Harvesters with Nanofibers*

To overcome the brittle and stiff nature of piezoelectric ceramics, novel fabrication processes have been proposed to generate PZT in the form of nanofibers, which include epitaxial growth [36], flexible substrate transferring [37, 38] and electrospinning process [39]. Qi et al. [37] proposed a process of transferring PZT ribbons with high crystal concentration onto flexible Polydimethylsiloxane (PDMS) substrate over a large area for flexible energy harvesting purpose. Figure 9.18 shows the schematic and photograph of transferring highly crystalline PZT with flexible PDMS rubble. The PZT nanofibers were sputtered on MgO substrate that was patterned with resist. This was followed by undercut etching to form PZT ribbons. Thereafter, the PZT ribbons were transfer printed by a piece of PDMS. The fabricated PZT ribbons were finally characterized with piezo-force microscopy. The results indicated the possibility to achieve high electromechanical energy conversion coefficient (d33 = 79 pm/V, d33 = 101.0 pm/V).

Based on the transfer printing technology, a highly crystalline PZT with wavy/buckled pattern was also deposited on a prestrained PDMS substrate [38].

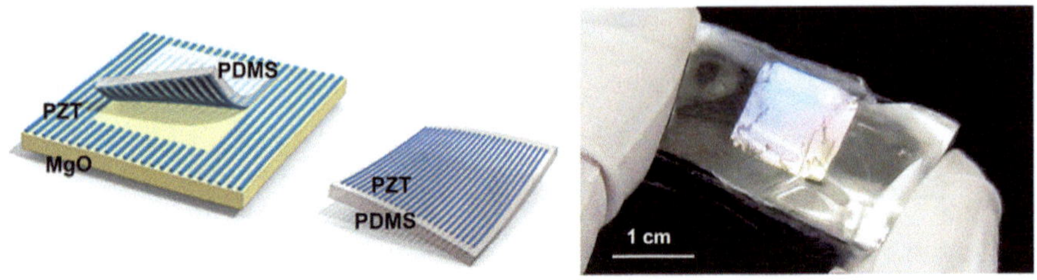

Fig. 9.18 (**a**) Schematic of transferring PZT ribbons on PDMS substrate and (**b**) photograph a PZT ribbon with PDMS. Reproduced with permission [37]. Copyright 2010, American Chemical Society

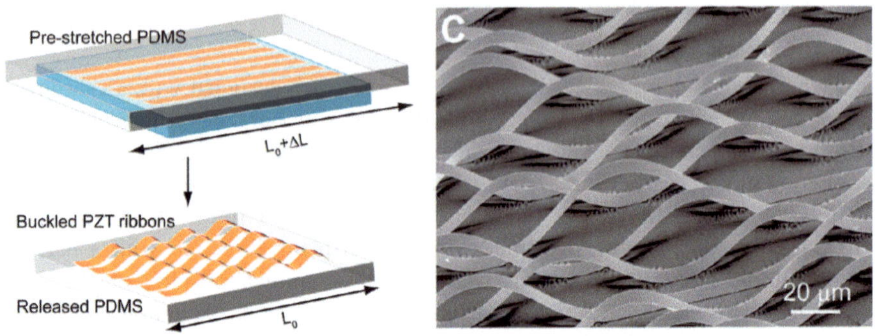

Fig. 9.19 Schematic and SEM image of formation of wavy/buckled PZT fibers through pre-stretch process [38]. Copyright 2011, American Chemical Society

As shown in Fig. 9.19a, a slab of prestrained PDMS was laminated against the ribbons. When released, the PZT ribbons spontaneously buckled to accommodate the release of the prestrain in PDMS. Figure 9.19b shows the SEM image of the wavy/buckled PZT nanofiber ribbons. According to the experimental results, a current output of 60 pA with current density of 2.5 $\mu A/mm^2$ was obtained.

Polymeric polyvinylidine fluoride (PVDF) is a type of semicrystalline polymer with a structure consisting of linear chains with hydrogen and fluoride sequence along with carbon backbone. PVDF nanofibers are naturally flexible and stretchable materials with excellent piezoelectric coefficient, a unique combination which shows high promise for the development of self-powered devices for wearable applications. In general, PVDF is mostly available in α-phase in nature which is typically obtained when the PVDF is cooled and solidified from melt. However, β-phase has been recognized as the only ferroelectric crystalline structure (polar) of PVDF with strong piezoelectric effect. PVDF nanofibers high piezoelectric coefficient and diameters of hundreds of nano-meters could be obtained by mechanical stretching process and the highly electric field (>10 kV) applied during the electrospinning process. Chang et al. [40] proposed PVDF nanogenerators by direct write piezoelectric PVDF nanofiber with near field electrospinning process (Fig. 9.20a). From their characterization as shown in Fig. 9.20c, peak-to-peak current of about 6 nA could be achieved.

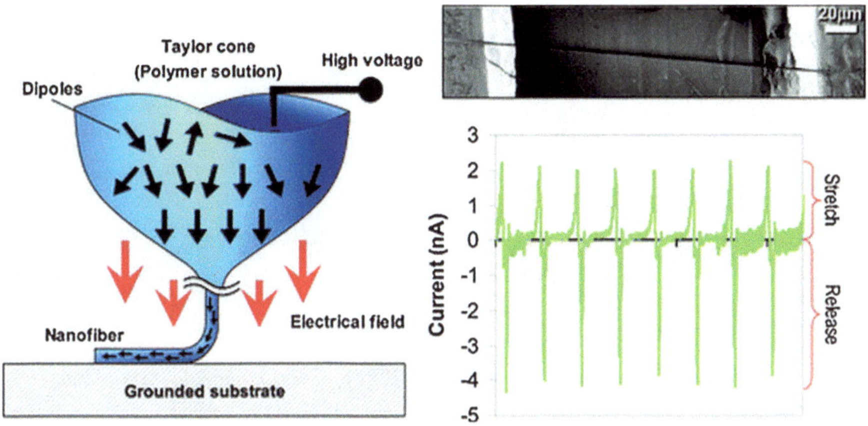

Fig. 9.20 Direct write piezoelectric PVDF nanofiber generator prepared by near field electrospinning. (**a**) Schematic process; (**b**) SEM image of single fiber; (**c**) Output performance at 2 Hz [40]. Copyright 2010, American Chemical Society

9.5 Triboelectric Nanogenerators

Most recently, Wang et al. [41] reported a new method of energy harvesting, namely triboelectric nanogenerators. The fundamental principle involves the contact between the charges created by two dissimilar materials. When the two surfaces separate apart, there would be a potential drop between the two electrodes, leading to an alternative charge flowing through external circuit. Their results demonstrated that an area power density, volume power density and instantaneous conversion efficiency as high as 500 W/m2, 15 MW/m3, ~70% respectively could be achieved [42].

Typically, triboelectric nanogenerators operate mainly on four schemes that are summarized in [43] and shown in Fig. 9.21, including vertical contact-separation scheme, lateral sliding scheme, single-electrode scheme and free-standing electrode scheme. For vertical contact-separation scheme (Fig. 9.21a), the charge is created when the two dielectrics with different electron affinity contact with each other. The charge on the dielectrics would induce opposite charge in the back electrodes. The capacitance change would lead to the induced charge that flows back and forth between the two electrodes. The lateral sliding operation scheme relies on the variance of the overlapping area of two dielectrics (Fig. 9.21b). In the single-electrode scheme, the induced charge in the back electrode can serve as a charge pump, which can extract and release charge from the ground (Fig. 9.21c). Free-standing electrode scheme operates on the unbalanced charge created in two embedded horizontal electrodes, where electrodes on the backside of the dielectric are not necessary (Fig. 9.21d).

Figure 9.22 summaries the work done by Yang et al. [45], showing an example of using triboelectric nanogenerator as acoustic energy harvester and self-powered

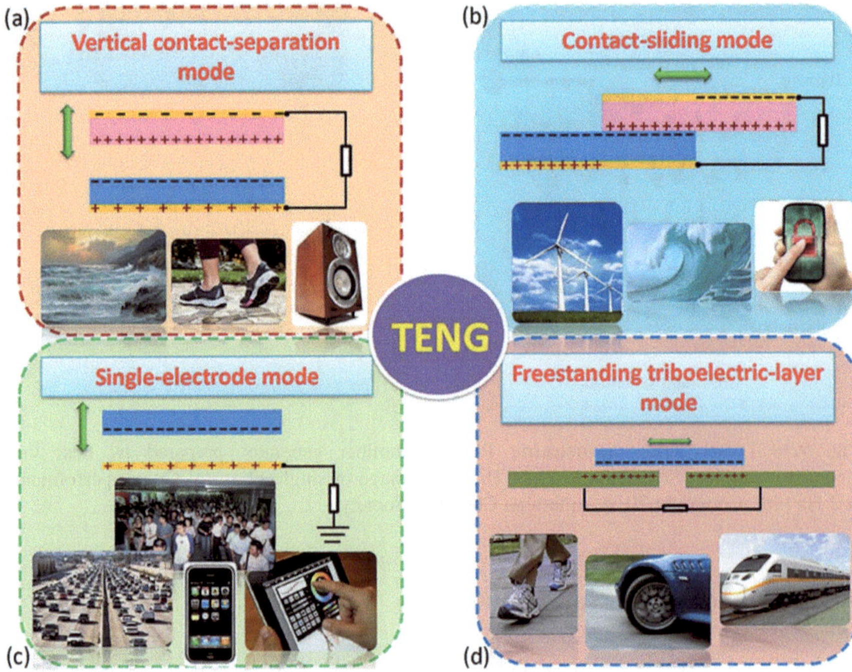

Fig. 9.21 Four operating schemes of triboelectric nanogenerators: (**a**) Vertical contact-separation scheme; (**b**) Lateral sliding scheme; (**c**) Single-electrode scheme; (**d**) Free-standing electrode scheme [44]

sensing. Polytetrafluoroethylene (PTFE) nanowires were created by plasma etching on the organic thin film to increase the contact area of the triboelectric effect. Helmholtz resonator is employed to collect acoustic wave and generate power. The PTFE flexible thin film was attached to the side wall of the cavity as shown in Fig. 9.22a. When the sound wave passed through the narrow neck and reached the cavity, the oscillation of the wave in the cavity would give rise to pressure variation. The relative displacement of the PTFE and Al electrode would induce triboelectrification and thus leads to power generated. As demonstrated in Fig. 9.22b, the generator could produce a maximum electric power density of 60.2 mW/m2 with a pressure from 70 to 110 dB. The generator was also utilized as a self-powered microphone for voice recording (Fig. 9.22c).

9.6 Magnetostrictive Vibrational Energy Harvesters

While piezoelectric materials have the capability to transform mechanical strain to electricity, the magnetostrictive materials demonstrate an ability to convert mechanical strain/stress to magnetic field/induction variation, known as Villari

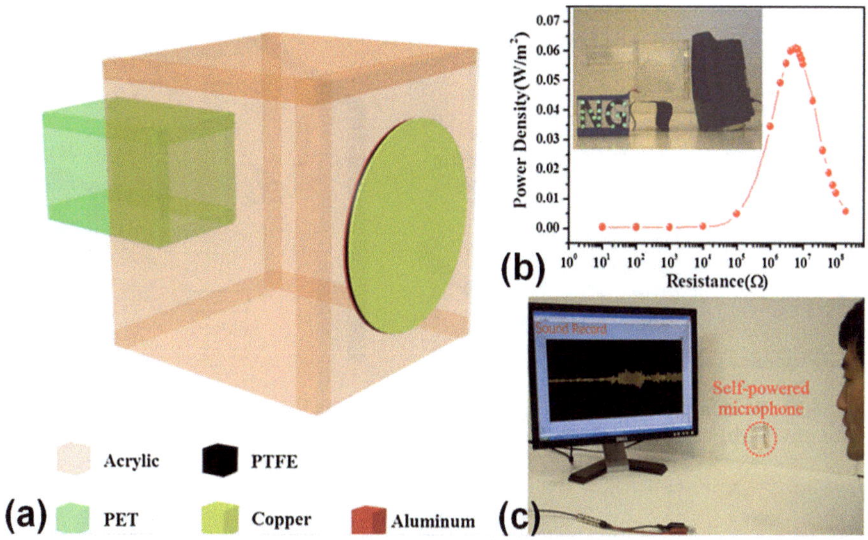

Fig. 9.22 Triboelectric nanogenerator used for acoustic energy harvesting and self-powered Sensing: (**a**) sketch; (**b**) resistance optimization of the generator; (**c**) demonstration of generator as a self-powered microphone for voice recording [45]. Copyright 2014, American Chemical Society

effect. The fluctuation of magnetic flux would give rise to a current flow in the winding coil according to Faraday's law of induction. Several types of magnetostrictive materials have been in the past employed for energy harvesting applications, such as Terfenol (terbiumdisprosium-iron alloy), Galfenol (gallium-iron alloy), and Alfenol (aluminium-iron alloy). Galfenol has the advantage of high tensile strength, ductile nature, and high piezomagnetic constant where more than 1 T change of magnetic induction could be achieved with stress variations. Generally, the magnetostrictive vibrational energy harvester works on two schemes: force driven scheme (Fig. 9.23a) and resonant-based vibration driven scheme (Fig. 9.23b).

Ueno et al. [46] developed a magnetostrictive vibrational energy harvester based on Galfenol alloy wound with coil and York, as shown in Fig. 9.24. A permanent magnet was attached on the back yoke to provide a bias and form a closed-loop of flux, which would give rise to strong magnetic flux change when the yoke and beam contact each other. A current would then be created in the winding coil via the variation of the magnetization in the magnetostrictive alloy due to the bending force on the tip of the device. Based on this configuration, the flux density was varied in the range of 1.16 T at the excitation of 1.2G. This value is higher than the electromagnetic counterparts. As a result, an average power of 3 mW/cm^3 was achieved at 212 Hz and 1.2G with a miniaturized Galfenol alloy rod of $2 \times 0.5 \times 7$ mm^3.

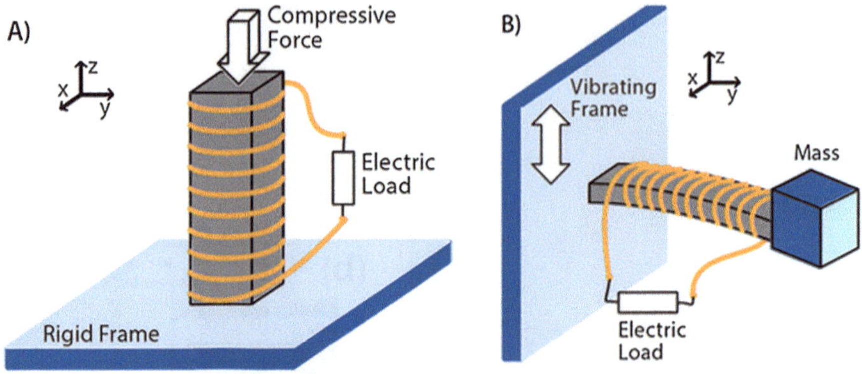

Fig. 9.23 Working principles of magnetostrictive vibrational energy harvesters: (**a**) based on force driven; (**b**) based on resonant-based vibration driven scheme [46]

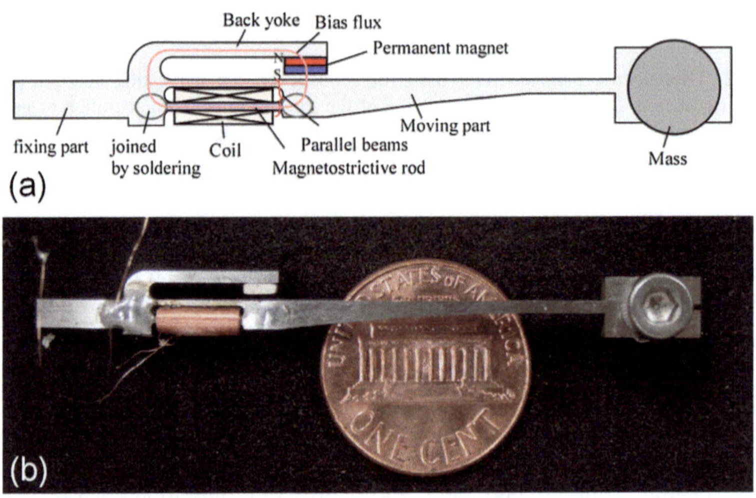

Fig. 9.24 Magnetostrictive vibrational energy harvester with based on Galfenol alloy wound with coil and York structure: (**a**) schematic structure; (**b**) photograph of the device [46]

9.7 Comparative Review of Five Energy Conversion Schemes

This section summarizes a comparative review of five energy conversion schemes that are previously discussed in detail.

Table 9.2 presents a comprehensive and qualitative comparison of the five energy conversion schemes, and lists the advantages and disadvantages of each conversion method. Electrostatic vibration energy harvesters are attractive since they are particularly suited for batch fabrication using Si CMOS compatible processes.

Table 9.2 Comparative review of three energy conversion schemes

Techniques	Advantages	Disadvantages
Electrostatic	1. Easily integrated with MEMS; 2. Flexible with spring-mass design; 3. Suited for low-frequency applications; 4. Good output voltage achieved; 5. Possible large bandwidth with electromechanical coupling; 6. Long-lasting with electrets	1. Initial bias voltage needed or electret pre-charged; 2. Relatively low power density; 3. Large external load impedance required
Electromagnetic	1. Good for low-frequency in micro scale; 2. No external bias source needed; 3. High current and low impedance achieved; 4. Proper power density obtained	1. Not suited with micro device due to magnet and coil manufacturing problems; 2. Large mass displacement required to achieve proper output; 3. Inefficient in microscale
Piezoelectric	1. High output voltage and power density; 2. No external bias source needed; 3. Micro piezoelectric device achievable; 4. Easy implementation method	1. PZT is hard to miniaturize and integrate with micro device; 2. Low coupling coefficient for piezoelectric thin film; 3. Material fatigue problem due to bending of piezoelectric beam
Triboelectric	1. Diverse choice of contact material 2. High instantaneous energy conversion efficiency 3. Easy for wearable and flexible applications 4. Relatively high performance	1. Low performance without modifying the surface 2. Fatigue/wear problems after long-term use 3. Low current output
Magnetostrictive	1. Large magnetic field variation with small movement 2. Robust and easy machining material 3. cost-effective	1. Can not integrate with MEMS/NEMS process 2. External circuitry required to achieve large performance 3. Very few material selectivity

They also offer greater flexibility in device design and conceptualization since mechanical and electrical features can be designed separately. Low-resonant devices that adapt to the vibrations in ambient environment can be easily obtained by incorporating various folded spring designs within a small volume, which is more advantageous than that of piezoelectric and electromagnetic counterparts. Energy density of such devices can be greatly enhanced by decreasing the air gap, increasing the bias voltage, and extending overlapping area. A high bias voltage can be obtained by electret materials which can be utilized as a charge pump to provide a constant bias for years.

Due to the scaling law, electromagnetic conversion mechanisms are more suited for macro energy harvesting applications. As the volume of the devices shrinks to 1% of their original size, the conversion efficiency of the electromagnetic mechanism reduces ten times faster than that offered by electrostatic. Furthermore, although various fabrication methods have been proposed, the miniaturization of coils and magnets is cumbersome and often face major limitations. The high-performance rare earth SmCo or NdFeB bulk magnets are still not compatible with standard MEMS fabrication process.

Piezoelectric conversion mechanisms provide a simple method to convert mechanical vibration to electricity through bending piezoelectric materials. However, this limits the number of design configurations it can have. The performance of the energy harvesters is also highly dependent on the material properties. Although piezoelectric thin films can be deposited through MEMS technologies, high-performance piezoelectric materials, such as PZT and PFC, are hard to be miniaturized and micro-fabricated. Fatigue is another concern due to the bending of piezoelectric beam for long-term operation cycles. Piezoelectric nanofibers may have great potential to address these challenges for flexible and wearable applications.

Triboelectric nanogenerator opens up a new avenue for applying research from various nano-material studies for electrostatic energy conversion. This method has inherent advantages in diversity of choice in selecting contact materials, ease of fabrication and better performance. Therefore, triboelectric generators are highly suitable for wearable applications. Similar to electrostatic transduction, the triboelectric schemes also achieve high energy conversion efficiency. However, the wear/fatigue problem will still be a big challenge for long-term utilization. The surface contamination would also severely deteriorate its overall performance.

Magnetostrictive mechanism with Villari effect shows promise for force driven small-scale energy harvesting applications due to its ductile nature and high piezomagnetic properties. Compared to its counterpart of electromagnetic harvesters, such mechanism enables larger magnetic flux change with small deformation of the material. However, current magnetostrictive materials can only be machined using laser cutting or other traditional micromachining methods, which would restrict its further miniaturization and integration with other electronics.

9.8 Conclusions

Kinetic or mechanical energy is ubiquitous and is a readily available energy source that can be derived from various basic forms of activities like human movement, structural and machinery vibrations. Recent surge in internet of things (IoT) and low-powered electronic devices have further advanced MEMS/NEMS-enabled vibration energy harvesting technologies. This chapter looks into five vibration-to-electricity conversion mechanisms, including electromagnetic, piezoelectric, electrostatic, triboelectric and magnetostrictive. The recent advances and current challenges of each conversion method have been reviewed and summarized.

Even though a significant progress has been made in the last decade in terms of both technological improvement and fundamental understanding, energy harvesting is still an emerging technology and has not been widely adopted by industry. It is believed that energy harvesting technologies will pave the step forward to enabling next-generation of wireless sensing and portable electronic applications. We hope that more people will join in this exciting field and making use of endless power supplied by nature.

References

1. Z.L. Wang, W. Wu, Nanotechnology-enabled energy harvesting for self-powered micro/nanosystems. Angew. Chem. Int. Ed. **51**, 11700–11721 (2012)
2. S. Meninger, J.O. Mur-Miranda, R. Amirtharajah, A.P. Chandrakasan, J.H. Lang, Vibration-to-electric energy conversion. IEEE Trans. Very Large Scale Integr. Syst. **9**, 64–76 (2001)
3. S. Meninger, A low power controller for a mems based energy convertor Ph.D Thesis, MIT, 1999
4. P.K.W. Shad Roundy, Micro-electrostatic vibration-to-electricity converters. ASME Int. Mech. Eng. Congr. Expo., 487–496 (2002)
5. P.D. Mitcheson, E.M. Yeatman, G.K. Rao, A.S. Holmes, T.C. Green, Energy harvesting from human and machine motion for wireless electronic devices. Proc. IEEE **96**, 1457–1486 (2008)
6. P. Miao, P.D. Mitcheson, A.S. Holmes, E.M. Yeatman, T.C. Green, B.H. Stark, MEMS inertial power generators for biomedical applications. Microsyst. Technol. **12**, 1079–1083 (2006)
7. P.D. Mitcheson, Analysis and optimization of energy harvesting micro-generator systems, Ph.D Thesis, Imperial College London, 2005
8. P.D. Mitcheson, P. Miao, B.H. Stark, E.M. Yeatman, A.S. Holmes, T.C. Green, MEMS electrostatic micropower generator for low frequency operation. Sens. Actuators A Phys. **115**, 523–529 (2004)
9. P.D. Mitcheson, T.C. Green, E.M. Yeatman, A.S. Holmes, Architectures for vibration-driven micropower generators. J. Microelectromech. Syst. **13**, 429–440 (2004)
10. D. Hoffmann, B. Folkmer, Y. Manoli, Fabrication, characterization and modelling of electrostatic micro-generators. J. Micromech. Microeng. **19**, 094001 (2009)
11. G.-J. Sheu, S.-M. Yang, T. Lee, Development of a low frequency electrostatic comb-drive energy harvester compatible to SoC design by CMOS process. Sens. Actuators A Phys. **167**, 70–76 (2011)
12. J.A. Yeh, C.N. Chen, Y.S. Liu, Large rotation actuated by in-plane rotary comb-drives with serpentine spring suspension. J. Micromech. Microeng. **15**, 201–206 (2005)
13. Y. Sakane, Y. Suzuki, N. Kasagi, The development of a high-performance perfluorinated polymer electret and its application to micro power generation. J. Micromech. Microeng. **18**, 104011 (2008)
14. S. Boisseau, G. Despesse, and B.A. Seddik, Electrostatic conversion for vibration energy harvesting, arXiv preprint arXiv:1210.5191, 2012.

15. K. Tao, S. Liu, S.W. Lye, J. Miao, X. Hu, A three-dimensional electret-based micro power generator for low-level ambient vibrational energy harvesting. J. Micromech. Microeng. **24**, 065022 (2014)
16. K. Tao, J. Miao, S.W. Lye, X. Hu, Sandwich-structured two-dimensional MEMS electret power generator for low-level ambient vibrational energy harvesting. Sensors Actuators A Phys. **228**, 95–103 (2015)
17. K. Tao, S. Liu, S.W. Lye, J. Miao, X. Hu, Out-of-plane electret-based MEMS energy harvester with the combined nonlinear effect from electrostatic force and a mechanical elastic stopper. J. Micromech. Microeng. **24**, 065022 (2015)
18. K. Tao, J. Wu, L. Tang, X. Xia, S.W. Lye, J. Miao, et al., A novel two-degree-of-freedom MEMS electromagnetic vibration energy harvester. J. Micromech. Microeng. **26**, 035020 (2016)
19. K. Tao, J. Wu, L. Tang, N. Wang, S. Lye, and J. Miao, Broadband energy harvesting using a nonlinear 2DOF MEMS electret power generator, IEEE 29th International Conference on Micro Electro Mechanical Systems (MEMS), 1236–1239, 2016.
20. C.B. Williams, R.B. Yates, Analysis of a micro-electric generator for microsystems. Sens. Actuators A Phys. **52**, 8–11 (1996)
21. C. Shearwood, R.B. Yates, Development of an electromagnetic micro-generator. Electron. Lett. **33**, 1883–1884 (1997)
22. S.C.L. Yuen, J.M.H. Lee, W.J. Li, P.H.W. Leong, An AA-sized vibration-based microgenerator for wireless sensors. IEEE Pervasive Comput. **6**, 64–72 (2007)
23. S.P. Beeby, M.J. Tudor, R.N. Torah, S. Roberts, T. O'Donnell, S. Roy, Experimental comparison of macro and micro scale electromagnetic vibration powered generators. Microsyst. Technol. **13**, 1647–1653 (2007)
24. S.D. Moss, O.R. Payne, G.A. Hart, C. Ung, Scaling and power density metrics of electromagnetic vibration energy harvesting devices. Smart Mater. Struct. **24**, 023001 (2015)
25. W.P. Mason, Piezoelectricity, its history and applications. J. Acoust. Soc. Am. **70**, 1561–1566 (1981)
26. C. Bowen, H. Kim, P. Weaver, S. Dunn, Piezoelectric and ferroelectric materials and structures for energy harvesting applications. Energy Environ. Sci. **7**, 25–44 (2014)
27. Z. Dibin, Vibration energy harvesting: machinery vibration, human movement and flow induced vibration, Intech, 25–54, 2011.
28. W.B. Hobbs, D.L. Hu, Tree-inspired piezoelectric energy harvesting. J. Fluids Struct. **28**, 103–114 (2012)
29. B. Lee, S. Lin, W. Wu, X. Wang, P. Chang, C. Lee, Piezoelectric MEMS generators fabricated with an aerosol deposition PZT thin film. J. Micromech. Microeng. **19**, 065014 (2009)
30. S.W. Ibrahim, W.G. Ali, A review on frequency tuning methods for piezoelectric energy harvesting systems. J. Renew. Sust. Energy **4**, 062703 (2012)
31. H.S. Kim, J.H. Kim, J. Kim, A review of piezoelectric energy harvesting based on vibration. Int. J. Precis. Eng. Manuf. **12**, 1129–1141 (2011)
32. S.R. Anton, H.A. Sodano, A review of power harvesting using piezoelectric materials (2003–2006). Smart Mater. Struct. **16**, R1–R21 (2007)
33. S. Saadon, O. Sidek, A review of vibration-based MEMS piezoelectric energy harvesters. Energy Convers. Manag. **52**, 500–504 (2011)
34. J. Kymissis, C. Kendall, J. Paradiso, N. Gershenfeld, Ieee, Parasitic power harvesting in shoes. Second IEEE International Conference on Wearable Computing, 132–139, 1998.
35. R. Elfrink, T.M. Kamel, M. Goedbloed, S. Matova, D. Hohlfeld, Y.V. Andel, et al., Vibration energy harvesting with aluminum nitride-based piezoelectric devices. J. Micromech. Microeng. **19**, 094005 (2009)
36. S. Xu, B.J. Hansen, Z.L. Wang, Piezoelectric-nanowire-enabled power source for driving wireless microelectronics. Nat. Commun. **1**, 93 (2010)
37. Y. Qi, N.T. Jafferis, K. Lyons Jr., C.M. Lee, H. Ahmad, M.C. McAlpine, Piezoelectric ribbons printed onto rubber for flexible energy conversion. Nano Lett. **10**, 524–528 (2010)

38. Y. Qi, J. Kim, T.D. Nguyen, B. Lisko, P.K. Purohit, M.C. McAlpine, Enhanced piezoelectricity and stretchability in energy harvesting devices fabricated from buckled PZT ribbons. Nano Lett. **11**, 1331–1336 (2011)
39. X. Chen, S. Xu, N. Yao, Y. Shi, 1.6 V nanogenerator for mechanical energy harvesting using PZT nanofibers. Nano Lett. **10**, 2133–2137 (2010)
40. C. Chang, V.H. Tran, J. Wang, Y.-K. Fuh, L. Lin, Direct-write piezoelectric polymeric nanogenerator with high energy conversion efficiency. Nano Lett. **10**, 726–731 (2010)
41. F.-R. Fan, Z.-Q. Tian, Z.L. Wang, Flexible triboelectric generator. Nano Energy **1**, 328–334 (2012)
42. Z.L. Wang, J. Chen, L. Lin, Progress in triboelectric nanogenerators as a new energy technology and self-powered sensors. Energy Environ. Sci. **8**, 2250–2282 (2015)
43. Z.L. Wang, Triboelectric nanogenerators as new energy technology and self-powered sensors–principles, problems and perspectives. Faraday Discuss. **176**, 447–458 (2014)
44. J. Yang, J. Chen, Y. Liu, W. Yang, Y. Su, Z.L. Wang, Triboelectrification-based organic film nanogenerator for acoustic energy harvesting and self-powered active acoustic sensing. ACS Nano **8**, 2649–2657 (2014)
45. D. Davino. Kinetic Energy Harvesting by Magnetostrictive Materials Available: http://www.sigmaaldrich.com/technical-documents/articles/materials-science/kinetic-energy-harvesting.html
46. T. Ueno, Performance of improved magnetostrictive vibrational power generator, simple and high power output for practical applications. J. Appl. Phys. **117**, 17A740 (2015)

Chapter 10
The Application of Graphene in Biosensors

Ting Li, Zebin Li, Jinhao Zhou, Boan Pan, Xiao Xiao, Zhaojia Guo, Lanhui Wu, and Yuanfu Chen

10.1 Synthesis and Properties of Graphene

Graphene, as one of the most attractive materials in the twenty-first century [1], is an archetypal two-dimensional single-atom thick material, which has sp2-bonded carbon atoms faultlessly arranged in a honeycomb lattice. The unique crystalline structure and electronic band, together with controllable synthesis of graphene with various structure, heteroatom doping, thickness, and defect state and density etc., make graphene many potential applications such as radio-frequency electronics [2–7], energy storage and conversion [8–12], field field-emission display [13, 14], optoelectronic devices [15–18], and biomedical sensing [19–22] etc. The excellent electric-field effect, luminescence, photoluminescence, and electrochemical properties make graphene as good candidate for biosensors. The synthesis and properties of graphene for biosensors are summarized as follows.

T. Li (✉) • Z. Li
State Key Lab Elect Thin Film and Integrated Device, University of Electronic Science & Technology of China, Chengdu, China

Shenzhen Geno-Immune Medicine Institute, Shenzhen, Guangdong Province, China
e-mail: liting@uestc.edu.cn

J. Zhou • B. Pan • X. Xiao • Z. Guo • L. Wu • Y. Chen
State Key Lab Elect Thin Film and Integrated Device, University of Electronic Science & Technology of China, Chengdu, China

T. Li, Z. Liu (eds.), *Outlook and Challenges of Nano Devices, Sensors, and MEMS*,
DOI 10.1007/978-3-319-50824-5_10

10.1.1 Synthesis of Graphene

The advancements in graphene fabrication method, are remarkable from the easiest mechanical cleavage to more complex and higher quality construction though it just went through such a short time period. Generally, graphene growth approaches can be classified in two major categories, i.e. top-down and bottom-up approaches (schematically shown in Fig. 10.1).

The "top-down" approach suggests exfoliation of natural or synthetic graphite into mixture of single and few layer graphene nanosheets. Exfoliation covers alternative possibilities including mechanical treatments, ion intercalation, liquidphase and surfactant-assisted exfoliation/dispersion.

Pristine single-layer graphene (SLG) sheets were first obtained by Novoselov and Geim et al. in 2004 [24] through using micromechanical cleavage of highly oriented pyrolytic graphite (HOPG) with scotch tape. The pristine SLG has higher carrier mobility and density [25], room-temperature quantum Hall effect [26],

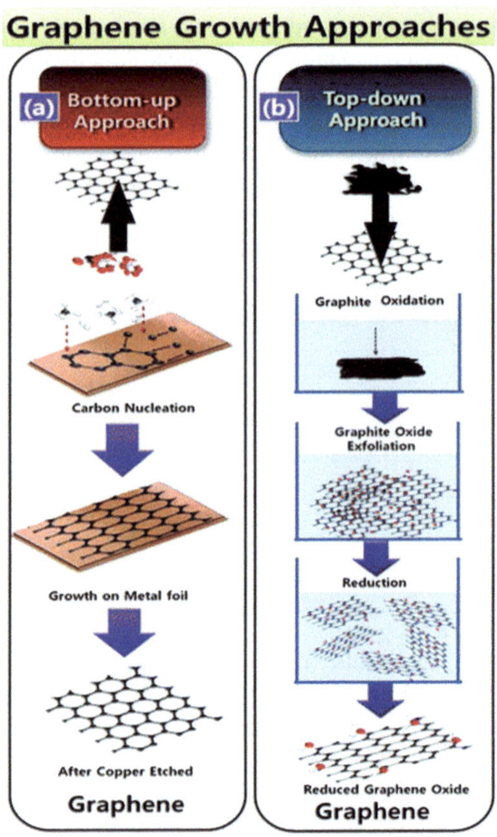

Fig. 10.1 Schematic representation of graphene synthesis processes by the (**a**) "bottom-up" and (**b**) "top-down" approaches [23]

lower intrinsic noise [1, 27, 28], a tunable bandgap [29], ambipolar field effect properties, and so on. However, it limits the practical application of such SLG, due to its very small (micrometer) size, very low yield, and very high cost.

Recently, oxidation-reduction synthesis of graphene nanoplates is widely used, including chemical reduction, thermal reduction and catalytic reduction. To start, graphite need to be oxidized to change graphene lattice from sp2 to sp3 structure, and then the oxygen-containing functional modules (for instance, carboxyl, carbonyl and hydroxyl groups) appear above the surface, which can better improve the hydrophilia through decreasing Van der Waals force between graphite sheets. Next, graphite oxide should be well distributed in deionized water to produce uniform graphene oxide (GO) colloidal solution. Lastly, graphene can be gained after adding dimethyl hydrazine or other strong reducing agents to uniform single layer GO solutions,and heteroatom doped graphene can be very easily obtained. The graphene film can be obtained through the chemical reduction of GO [30]. Although the obvious advantages including the low cost and the high volume production attract many researchers to discuss, the disadvantages of this kind of graphene can never be never be neglected such as a decline in light transmittance, thermal conductivity, and electrical conductivity particularly. In addition, the local structure of graphene prepared by chemical reduction also possess unavoidable defects, which indicate that the reduced graphene oxides (RGO) defect sites are not uniform and cannot be precisely controlled [31].

The bottom-up method is the alternative process to synthesize graphene via atomic level control of their composition and structure. These methods mainly include epitaxial growth on solid substrate and chemical vapor deposition (CVD).

The decomposing silicon carbide (SiC) or silicon (Si) can grow a graphene film with arbitrary size and high quality in a high temperature [32–34]. For instance, by thermal decomposition on the (0001) surface of 6H–SiC, Berger et al. successfully produced ultrathin epitaxial graphite films. In order to remove the oxide, after surface preparation by oxidation or hydrogen etching samples were heated by electron bombardment in ultrahigh vacuum at about 1000 °C, the samples were heated to 1250–1450 °C for 1–20 min to obtain thin graphene film. The layer thickness mainly determined by the temperature [28]. An interesting conclusion has been reported that the connection between graphene and SiC (or Si) result in the graphene bandgap is able to open up to $\approx$0.26 eV [29], which is helpful to the fabrication of FETs. A drawback for this technology for large scale production is the SiC wafers cost and their smaller size (usually no larger than 4") compared to Si wafers. Therefore, this method is not so suitable for fabricate graphene-based biosensors.

CVD is a promising high-yield method in which large size graphene film can be easily produced as a result of chemical decomposition of precursors such as methane, acetylene, methanol and ethanol on catalytic transition metal surfaces [35–37]. Experimental setup commonly used for CVD graphene is presented in Fig. 10.2. It is worth mentioning that doped graphene can be easily obtained by introducing solid, liquid, or gaseous precursors containing desired foreign atoms into the growth furnace together with the carbon sources. The carbon precursor

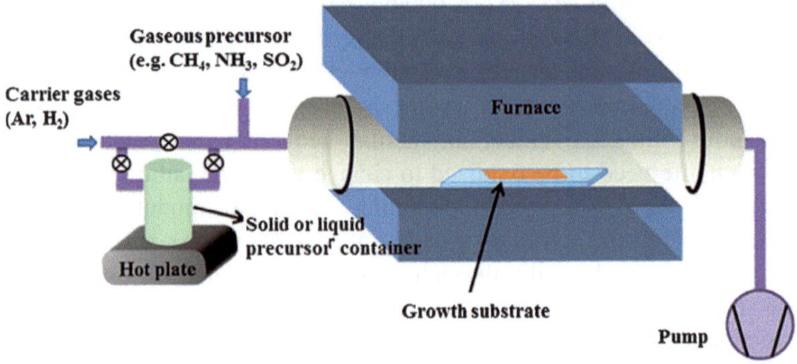

Fig. 10.2 Experimental setup commonly used for CVD graphene growth [38]

decomposes at the high temperature region, and the carbon atoms deposit on the metal substrate gradually to form graphene film. However, the growth mechanisms of graphene on the different metal substrate such as copper and nickel are distinguishing. For instance, the carbon atoms dissolve in nickel to form solid solution under the high temperature firstly, and then the oversaturated carbon separates out on nickel surface to produce graphene. In this process, it is significant for the different layers of graphene growth to control the concentration of dissolved carbon atoms and the cooling rate. However, the mechanisms for copper substrate is different. Because carbon atoms and copper are immiscible between each other, the copper is equal to catalyst for graphene growth. Carbon atoms absorb and crystallize on the surface of the copper to form graphene. The later carbon atoms deposit process is impeded when the single layer graphene form and cover the copper surface, so the single layer graphene can be controlled under certain conditions. It is important to select a suitable carbon source as a precursor because it can influence the quality of graphene sheets greatly. Moreover, impurities and an uncontrollable size of the so-grown graphene film may be affected when transferring the graphene film to an insulating substrate from the metal substrate [31].

In recent years, an increasing attention has been focused on the wide applications of three-dimensional (3D) graphene [39–43]. 3D graphene monolith is seamless, continuous and highly conductive graphene network with free defects and inter-sheet junctions, which is a good way to serve as the scaffold for the fabrication of monolithic composite electrodes. As shown in Fig. 10.3., a novel three dimensional (3D) graphene/Co_3O_4 composite was fabricated by Dong and co-workers in 2012, which can possibly be used as an independent electrode for supercapacitors and biosensors owing to the synergistic effect of the two state-model materials [44]. The 3D graphene/Co_3O_4 electrode offered a very large accessible active area, which may be applied to hydrophobic materials [45, 46], microbial fuel cells [47], and so on.

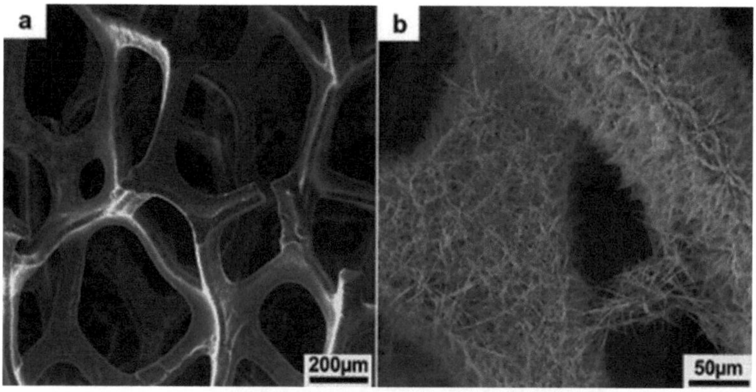

Fig. 10.3. SEM images of (**a**) 3D graphene foam and (**b**) 3D graphene/Co_3O_4 composites [48]

10.1.2 Electronic Structure and Properties of Graphene

Graphene atoms are arranged in a hexagonal crystal lattice as shown in Fig. 10.4a. The hexagonal and symmetrical graphene have four valence electrons, the sp^2-bonded carbon atoms and three carbon atoms adjacent to them form covalent bonds. The structure can be understood as a triangular lattice with a basis of two atoms per unit cell (depicted as blue and yellow in Fig. 10.4a). Since graphene has two atoms per unit cell, the points of particular importance are K and K' points at the corners of the graphene Brillouin zone (Fig. 10.4b). As shown in Fig. 10.4c. The bandgap of graphene is exactly zero, with the conduction band and valence band meeting at the K and K' points, also known as Dirac points or charge neutrality points [49]. Exceptional electrical properties can be shown by graphene, which enables itself to be explained by the velocity of electron movement of around 1000 m/s through interactions among the π-π bonds. The special structural features of single-atom thickness and strong mechanical characters like outstanding robustness and superior flexibility are beneficial for the preparation of sensor devices.

The graphene derivatives such as GO and rGO has an extremely high specific surface area. Besides that, it also exhibits visible and NIR imaging. GO is reactive due to its several chemical groups that can form covalent or noncovalent interactions. Its edges have been shown to be more reactive than its surface because of the additional free reactive oxygen functional groups [50].

Graphene and its derivatives possess outstanding mechanical, electronic, optical, thermal and chemical properties and are also biocompatible. In single layer graphene, the electrons travel as if they carry no mass, which means it has a high electron mobility, vital to field-effect transistor based biosensors and can transfer energy to nearby molecules [50]. Moreover, the properties of graphene can be tuned and modified through physical or chemical doping [51–55]. Additionally, graphene can be easily functionalized with a receptor to detect target substances with a higher accuracy. Owing to these prominent properties, graphene holds immense promise for bio-sensing applications.

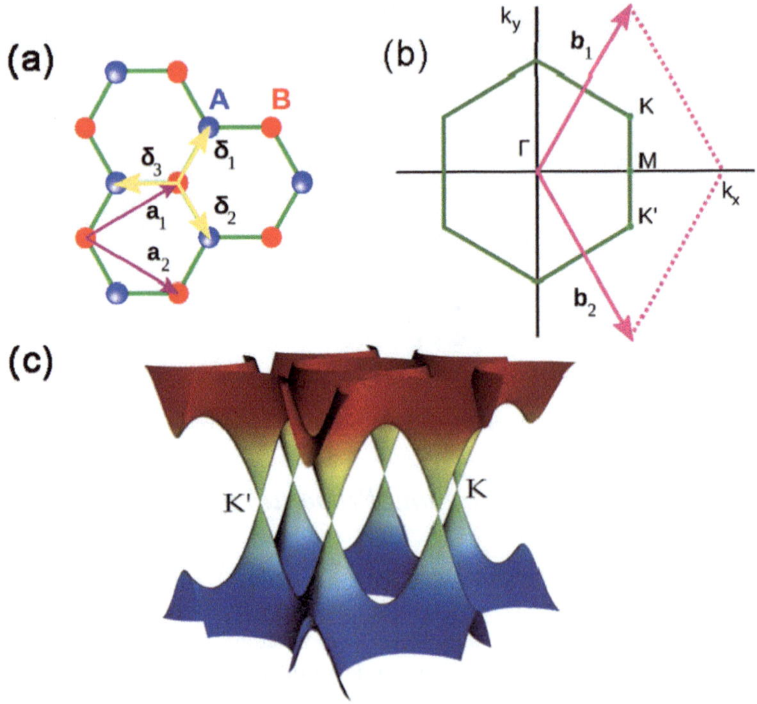

Fig. 10.4 (**a**) Hexagonal crystal lattice of graphene. a1 and a2 are the lattice unit vectors, and δi, i = 1,2,3 are the nearest-neighbour vectors. (**b**) Graphene's Brillouin zone. The Dirac cones are located at the K and K′ points. (**c**) The energy dispersions of graphene crystal lattice. The conductance band touches the valence band at the K and the K′ points [49]

10.2 Bio-Sensing Cancer Biomolecules or Cells

In a near future, biosensors are expected to detect with high sensitivity and selectivity in a wide range of cancer biomarkers exist in body fluids or overexpressed on cancer cells surface in order to diagnosis and monitor the evolution of the disease during the treatment. In this way, bio-sensing systems employing the properties of graphene derivatives for cancer detection have found their way into a wide variety of strategies, which is capable of improving the sensing efficiency of cancer biomarkers. In the following sections, we will describe the application of graphene in optical imaging methods such as photoluminescence and Raman imaging, electrochemical sensors for enzymatic bio-sensing, DNA sensing, and immune-sensing.

The changes that occur in the DNA sequence of the genomes of cancer cells are the cause of cancers [56]. The profound metabolic remodeling of cancer cells, including mitochondrial rearrangement, is an indirect response to cell survival or proliferation, which is controlled by specific cell signaling [57]. The monitoring of molecular and physical events in live cells by visualization is a core access to understand cell biology and it has a profound influence on the process of biomedical sciences. Biomedical sensors that can detect selective signals for biological molecules or physical variables in live cells with spatiotemporal resolution can be regarded as a constant need. Fluorescent spectroscopic techniques offer sensing for intracellular signaling and analysis due to their non-invasiveness and high sensitivity [58]. These optical features are of particular interest in the detection of cancer cells because they allow a sub-cellular resolution and therefore realize the function of identifying tumor phenotypes at their very early stages [59]. Although the organic dyes and fluorescent proteins are powerful molecular probes, they present limitations in making reliable intracellular measurements because of their poor photo bleaching resistance [58]. In addition, their broad emission spectra may hamper practical applications since different fluorophores require multiple excitation wavelengths [60].

The possible chemical interactions of this type of powerful molecular probes or steric hindrance with biomolecules may cause bio-toxicity or perturbation to the systems being investigated which effect is not conceivable in medical applications [61].

10.2.1 Photoluminescence

The advancement of nanoparticles with multifunctional has been a study focus in the past few decade including graphene and its derivatives [62]. The unprecedented characteristics of these nanomaterials lead the researchers to suggest it as a new class of fluorescent probes for biomedical imaging. NG is photos table, non-toxic and easily consumable [62, 63]. NG and NG derivatives can be used as fluorescence probes in photoluminescence imaging. Additionally, photoluminescence studies demonstrate that these carbon fluorescent probes may have two functionalities: specific detection of cancer cells and therapeutic agent (drug/gene vehicle or photodynamic therapy agent), that is present both cancer imaging and therapeutic functionalities [64].

The high contrast bio-imaging formed by NG in the discovery of cancer cells was illustrated in the work by Peng et al. [65] The green NG was applied to enlarge the contrast visualization of the cells compounds. The nucleus of human breast cancer cell lines T47D were stained with DAPI and show blue color under imaging. After staining, cancer cells were treated with green NG during 4 h of incubation. The obtained images clearly showed the phase contrast image of T47D cells, nucleus stained blue with DAPI and high contrast fluorescent image of green NG around each nucleus (Fig. 10.5).

The photoluminescence of GO was first used for cellular imaging by Dai et al. [66] NGO was polyethyleneglycol and conjugated covalently with a B-cell specific antibody Rituxan (anti-CD20) for selective binding to B-cell lymphoma cells. Since NGO has showed luminescence in the visible and near infrared (NIR) regions, the optical identification of cancer cells was possible. Additionally, NGO

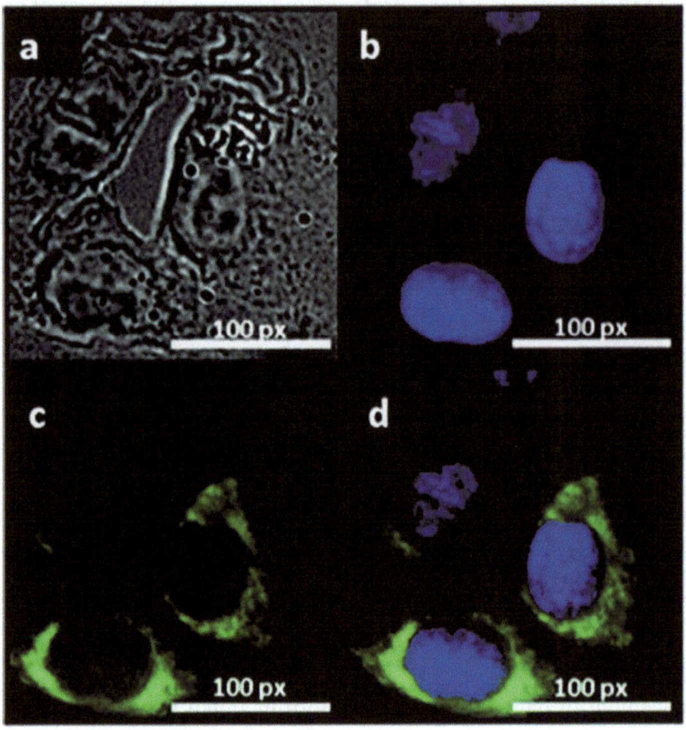

Fig. 10.5 Fluorescent images of human breast cancer cell T47D after incubation with *green* GQDs for 4 h (**a**) phase contrast picture of T47D cells; (**b**) Individual nucleus stained *blue* with DAPI; (**c**) Agglomerated green GQDs surrounding each nucleus; (**d**) The overlay high contrast image of nucleolus stained with *blue* DAPI and GQDs (*green*) staining. Reproduced with permission [65] Copyright 2012, American Chemical Society

presented a dual functionality as drug carrier for Doxorubicin (DOX) and contrast agent for cancer cell imaging.

Using NG prepared by a simple hydrothermal method with polythiophenes in vivo studies, the precursors indicated that NG can act as a multifunctional Nano-platform for the simultaneous imaging and highly efficient in photodynamic therapy (PDT) of cancer [67]. Tumor-targeting ligands such as folic acid (FA) [63] and hyaluronic acid (HAu) [64] could be easily coupled onto NG for tumor targeting. Wang et al. [68] developed a study where NG-FA was incubated with three different cell types that express FA receptor at different levels to demonstrate that FA preserves its binding activity to FA receptor even after its conjugation with NG. It was possible to observe that the fluorescence of NG in the human cervical carcinoma cell line HeLa is considerably stronger than in the adenocarcinoma human alveolar basal epithelial cell line and human embryonic kidney cell line HEK293A. In this study, the authors also proved that the green fluorescence arises from the internalization of NG-FA. Moreover, the results reveal the specific internalization of NG-FA by HeLa cells, consistent with the fact that

HeLa cells overexpress FA receptor while A549 and HEK239A express FA receptor at a low level, indicating that NG-FA is internalized via endocytosis induce by FA receptor.

Under the help of Abdullah-Al-Nahain et al. the hyaluronic acid was conjugated with NG. [64] The Nano-graphene-hyaluronic acid (NG-HA) nanocomposite was tested both on mammalian cell lines MDCK (CD44 negative) and on A549 of lung carcinoma with CD44 overexpressed. The CD44 antigen is expressed in tons of mammalian cells, it is a receptor for hyaluronic acid and participate in a variety of cellular functions, including tumor metastasis. Using findings from the confocal and cellular uptake investigation, it is clearly that a larger amount of NG-HA was taken up by A549 cancer cells. On the contrary, small numbers of non-targeted NG were allowed to enter into both types of cells, in which due to the presence of HA more NG-HA entered the cell cytoplasm via receptor-mediated endocytosis-based target delivery. The imaging fluorescence also studied about vivo biological distribution and tumor specific delivery of NG-HA. To look into the efficient delivery of NG, authors used balb/c mice in 6 weeks of age and then the tumors were grown in 10 days (volume of 100 mm3) on the back of the mice using A549 cells. To tumor tissue and its biological distribution, NG-HA was injected intravenously (10 mg/kg of body weight) through the tail vein. After 2 h, the mice with tumors were sacrificed and their organs dissected. The strong fluorescence from the tumors indicates the targeted accumulation of NG-HA to the tumor site (Fig. 10.6). The biodistribution profiles of NG-HA showed that this nanocomposite was significantly accumulated in the tumor tissue due to the large number of leaky blood vessels surrounding the tumor. NG-HA was also accumulated in great quantities in liver and kidney tissue (Fig. 10.6): blood circulation pass through liver and NG was uptake by the reticuloendothelial system and the accumulation of the kidney indicates a rapid excretion of this nanocomposite. Once again, the fluorescent NG-HA was also used to deliver DOX as treatment to cancer cells. The results showed that 60 % was released within 12 h under the specific conditions in research. Moreover, it was found that almost all drugs were released within 48 h from the NG-HA matrix.

Moreover, NG can not only be helpful to tumor cell detection and recognition, but also to monitor biochemical pathways like apoptosis, the programmed cell death, which can be regarded as a way for multicellular animal disposing of unwanted and damaged cells. The improper regulation of apoptosis has been associated to cancer, neurodegenerative and cardiovascular diseases [69, 70]. Nonetheless, the way that the abnormal cell apoptosis begins and progress in vivo has not yet completely understood. The comprehension of various forms of apoptosis could be the key to unlocking the diagnosis and therapy of the related diseases. Numerous methods have been developed for imaging apoptotic cells in vitro [71, 72], however there are few methods available for imaging apoptotic cells in live animals.

In recent years, Roy et al. found a original method, which uses NG modified with Annexin V antibody (AbA5) to form (AbA5)-modified NG (AbA5-NG) enabling to label apoptotic cells in live zebrafish (Danio-rerio) [73]. Zebrafish showed bright

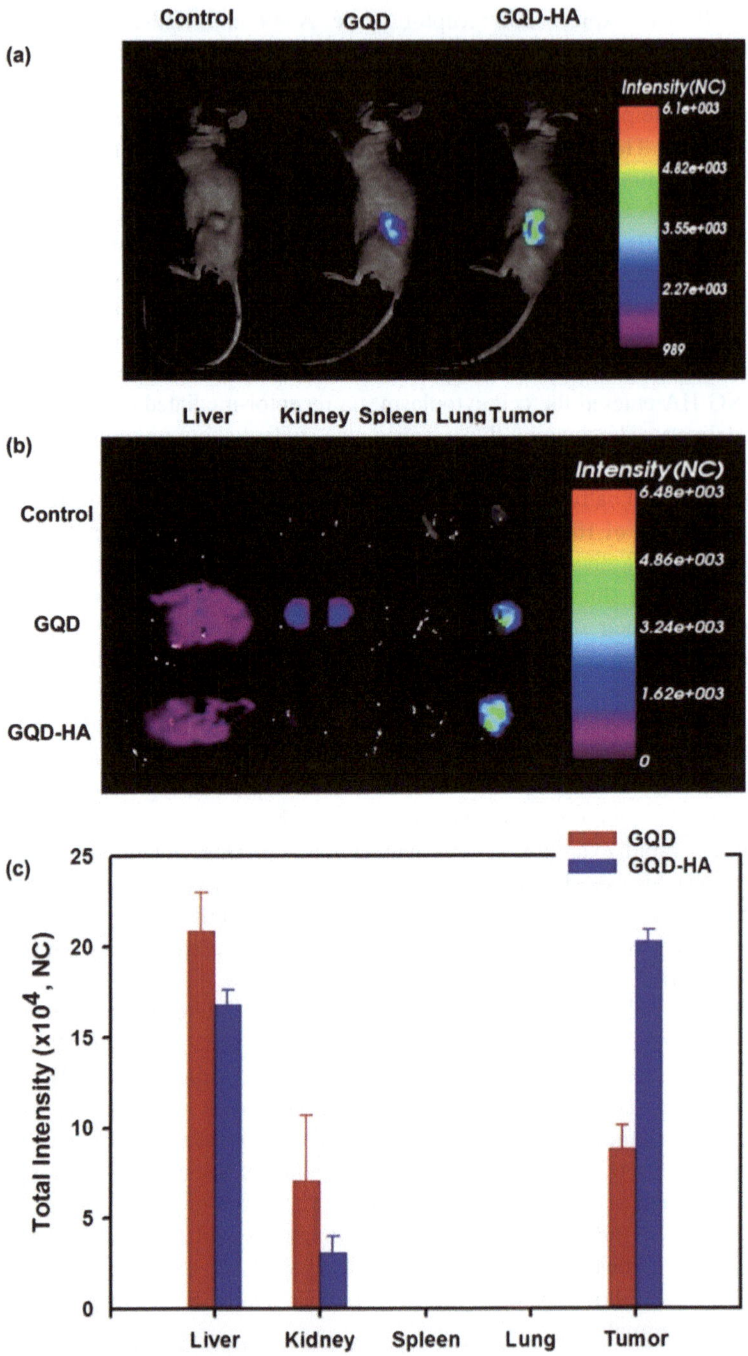

Fig. 10.6 (**a**) In vivo fluorescence images of GQD-HA in mice after tail vein injection; (**b**) Ex vivo images of liver, kidney, spleen, heart, and tumor after dissection; (**c**) Normalized intensity from dissected organs. Reprinted with permission from [64]. Copyright (2013) American Chemical Society.

red photoluminescence in the presence of apoptotic cells. The fluorescent images of normal cells (MCF-10A cells) and cancer cells (HeLa and MCF-7 cells) after treatment with NG (2 mg·mL^{-1}) were separately recorded. It was possible to observe that the photo luminescent spots were distributed around the cell membranes and cytoplasm suggesting that NG penetrated into the cells via receptor-mediated and/or no receptor-mediated endocytosis. This study found another key to visualize the initiation and progression of apoptosis which might help in understanding the apoptotic mechanism better which is of major importance in diagnosis and treatment of cancer.

10.2.2 Surface Enhanced Raman Scattering

In the past decades, the first reports of the feasibility in vivo imaging of cancer with biocompatible surface-enhanced Raman scattering (SERS) probes have emerged although its application is traditionally in vitro. Fluorescence imaging is the most common biological imaging technique [74, 75] for the reasons mentioned in the previous sub-section but various intrinsic limitations including photo bleaching due to unstable fluorescent dyes, auto fluorescence from the biological samples, or limited multispectral detection owing to the spectral overlap between broadband fluorescence spectra hinders the further applications of fluorescence microscopy imaging [76, 77]. In contrast, Raman scattering provides narrow spectral bandwidth and is resistant to photo bleaching and auto fluorescence and suitable for long-term monitoring of cellular processes [78, 79]. The low Raman scattering of an analytic, especially of biological nature, can be strongly enhanced by locating it near the surface of a noble metal nanostructure. This process known as surface-enhanced Raman scattering (SERS) makes that the amplification of the received signal overcomes the limitation of the intrinsically low Raman scattering efficiency of Raman spectroscopy [75]. SERS is capable of enhancing the Raman signals of analyses located near the surfaces of noble metal nanostructures by up to 10 orders of magnitude [74, 80]. This allows the probing and imaging of cells with high sensitivity using SERS labels [81]

10.3 Electrochemical Sensors

10.3.1 Immuno-Sensors

Immunoassay has already become one of the most significant tools for clinical diagnosis of cancer. The detection of cancer biomarkers related to certain tumors in early stage of the disease becomes one important contribution to successful treatment. Graphene sensors promise to be an innovative technology in next generation electronics on cancer diagnostic applications, due to exceptional electronic properties and extreme surface to volume ratio able to offer greatly enhanced sensitivity for the detection of cancer biomarkers. New graphene-based technology allows the development of portable and rapid diagnostic sensors whit potential to overcome the features of conventional enzyme-linked immune absorbents assays (ELISAs). After that, we will describe the potential use of graphene and its derivatives on the development of new electrochemical sensors for detection of specific cancer biomarkers.

10.3.2 Carcinoembryonic Antigen

Carcinoembryonic antigen (CEA) has become one of the popular tumor markers in early detection of lung, ovarian carcinoma, breast cancers and cystadenocarcinoma. A novel electrochemical immune sensor based on *carbon ionic liquid electrode* (CILE) modified with AuNPs, rGO and poly (l-Arginine) nanocomposite material as sensitive platform for detection of CEA [82]. The synergistic effects of nanocomposite resultant from the individual properties of each material allow an enhanced capacity of adsorption of CEA antibody and a consequently improvement of the electrochemical responses. Under optimized conditions, differential pulse voltammetry responses were proportional to CEA concentration in the range from 0.5 to 200 ng·mL^{-1} with the detection limit as 0.03 ng·mL^{-1}.

Additionally, in another work it reported the advancement of a novel electrochemical immune sensor on the basis of the deposition of ultrathin Au-Pt nanowire-decorated Lauth's violet/reduced graphene oxide (AuPtNWs/THI/rGO) on the surface of glass carbon electrode (GCE) [83]. The differential pulse voltammetry responses were proportional to CEA concentration which ranges from 50 FG·mL^{-1} to 100 ng·mL^{-1} with a detection of 6 FG·mL^{-1}.

One simple method has been successfully employed for the fabrication of ultrasensitive electrochemical immune sensor for CEA, which uses GO-Thi-Au nanocomposites as sensing platform for the surface modification of GCE [84]. The immune sensor demonstrates a linear calibration plot in a wide concentration range from 0.1 FG·mL^{-1} to 1×10^9 FG·mL^{-1} and ultralow detection limit of 0.05 FG·mL^{-1}.

Electrochemical immune sensors *on the basis of Au electrodes modified* with graphene nanocomposites as a sensing platform showed to be also a viable approach for the detention of CEA. Samanman et al. reported the Au electrode surface modification by deposition of a thin film of silver and then with an AuNPs-rGO-chitosan nanocomposite prepared by cryogen- functionalized with anti-CEA [85]. At optimal conditions, the decrease of the cyclic voltammetry silver peak current was proportional to the CEA concentration over a range of from 1.0 FG to 1.0 ng·mL^{-1} with a detection limit of 0.1 FG·mL^{-1}.

In another work Au electrode surface was modified with rGO-SWCNTs-Chitosan nanocomposite (Chi-rGO-CNTs) at the beginning and then followed by electrodeposition with Au and Pt NPs. Later, it was functionalized with anti-CEA. The proposed immune sensor demonstrated that CEA determination of range 0.1 pg·mL^{-1} to 40 ng·mL^{-1} with a limit of detection down to 30 FG·mL^{-1} [86].

Recently Kumar et al. reported for the first time a new graphene paper electrode based on poly(3, 4-ethylenedioxythiophene):poly(styrenesulfonate)(PEDOT:PSS)/ rGO composite as a promising alternative over the expensive conventional electrodes (ITO, gold and GCE) [87]. RGO PEDOT: PSS film was treated with ethylene glycol in order to enhance conductivity, electrochemical activity and charge transfer kinetics and functionalized through the immobilization of the anti-CEA for electrochemical detection of CEA cancer biomarker. This simple, low-cost, flexible PEDOT: PSS/rGO based biosensor shows the sensitivity of 25.8 μA·ng^{-1}·mL·cm^{-2} in the detection linear range of 2–18 ng·mL^{-1}.

Graphene biosensors *based on sandwich-type immune sensors methods* were also be extensively explored in the detection of CEA. High-affinity antibodies and appropriate labels are usually employed for the amplification of detectable signal [88]. Recent research has set eye on the development of innovative and powerful novel graphene labels or graphene sensing platforms, which can control and tailor their properties in a very predictable manner to meet the requirements for increasing the detection level and specificity for cancer biomarker.

Using ZnO nanoparticles (ZnONPs) and glucose oxidase (GOD), a sandwich luminal electro-chemical-luminescence (ECL) immune sensor decorated rGO as labels and in situ generated hydrogen peroxide as co-reactant (ZnONPs@rGO-GOD-Ab2) [88]. The sensing platform at GCE electrode surface was obtained by the assembly with a hybrid architecture of AuNPs/rGO able to provide high immobilization sites for antibody (Ab1). After the biological recognition event of CEA the ZnONPs@rGO-GOD-Ab2 labels were able to be immobilized on the electrode surface (GCE/rGO-AuNPs/Ab1) via sandwich immunoreactions. Enhanced sensitivity was obtained in situ generating hydrogen peroxide with glucose oxidase and the catalysis of ZnONPs to the ECL reaction of luminol-$1H_2O_2$ system. The rGO-based immune sensor exhibited a detection range of CEA from 10 pg·mL^{-1} to 80 ng·mL^{-1} and a detection limit of 0.0033 ng·mL^{-1}.

In recent days, another work depicts the use of rGO on both sensor sensing platform and construction of the "sandwich-type" immune sensor with ultrahigh efficiency for CEA detection as a signal amplifying factor [89]. Graphene sensing platform was prepared by π-π stacking with 1, 5-diaminonaphthalene (DN) an then coated with AuNPs conjugated with antigen CEA (Ab1). Graphene tracer label was also based on graphene/DN composite coated with Ag/Au nanoparticles and conjugated with antigen (Ab2). The biological recognition event occurs after the addition of CEA, by immune reactions of antigens and antibodies, which results in the great current signal owing to the existence of electroactive element Ag/AuNPs. The cyclic voltammetry results showed a linear increased of current with the increasing antigen CEA concentrations in the range of 10–1.2 × 10^5 pg·mL^{-1} and a limit of detection 8 pg·mL^{-1}.

Graphene produced by CVD was also developed as electrochemical sensing platform, in the form of high surface area conducting electrodes surfaces for the high sensitive detection of CEA. The graphene sensing surface was modified with magnetic beads (MBs) functionalized with CEA detection antibody (Gr/MBs-Ab1), by applying an external magnetic field. In order to enhance the sensitivity of the sensor, AuNPs were modified with horse radish peroxidase (HRP) and detection antibody (Ab2), to form the conjugate Ab2-1AuNPs-1HRP [90]. The electrochemical immune sensor analysis was based on the biological recognition event with CEA, using GR/MBs-1Ab1/CEA/Ab2-AuNPs-1HRP system as a tracer and H_2O_2 as an enzyme substrate, in a sandwich-type immunoassay format. Experimental results showed linear relationship between peak current and CEA concentration for the range of 5–160 ng·mL^{-1}. The limit of detection was 5 ng·mL^{-1}.

10.3.3 Prostate Specific Antigen.

Prostate specific antigen (PSA) is a typical tumor marker of **prostate cancer** applied in prostate cancer diagnosis and screening. The first attempt to use graphene for sensitive detection of PSA was through the fabrication of a label-free electrochemical immune sensor. The sensor technology was based on the surface functionalization of GCE electrode with a thin film of the nanocomposite, rGO, 1-pyrenebutanoic acid, succinimidyl ester (PBSE) and CoNPs, as a sensing platform [91]. The PBSE was conjugated with Anti-PSA antibody in order to increase the selectivity for PSA. The measurements suggested that electronic activity of CoNPs was greatly enhanced in the presence of rGO due to its great electron-transfer ability. The specific antibody-antigen immune complex formed on the electrode resulted in the decrease of ampere-metrical signal, linearly dependent with PSA concentration in the range of 0.02–12 ng·mL^{-1} with a low detection limit of 0.01 ng·mL^{-1}.

Recently a new approach based on highly conductive crumpled like structure rGO/AuNPs nanocomposite as a sensing platform for detection of PSA was reported [92]. The 3D immune sensor showed a good linear relationship between the current change and different concentrations of PSA from 0 to 110 ng·mL^{-1} and the detection limit 0.59 ng·mL^{-1}.

Recently a novel label electrochemical graphene-based immune sensor for PSA was reported. The sensing surface of the GCE was constructed based on the deposition of new amino-functionalized graphene sheet ferrocene pyrrole aldehyde composite material (NH2-rGO@FCA) [93]. The sensor label was based on silver hybridized mesoporous silica NPs (Ag@NH2-MCM48), which is capable of improving the electron transfer ability of the immune sensor system. The workability of immune sensors system Ag@NH2-MCM48/Ab2 and NH2-GS@FCA was based on the linear catalytic current increase with the PSA concentration. The immune sensor showed a good relationship between the current responses toward PSA concentration on the range from 0.01 to 10.0 ng·mL^{-1} similar to the non-labeled ones, however the detection limit was as much lower 0.002 ng·mL^{-1}.

10.3.4 Carbohydrate Antigen 19-9 and 15-3

As a screening test for cancer, *carbohydrate Antigen 19-9 (CA 19-9)* is not sensitive or specific to use but also can be *applied as a tumor marker* if is produced in abnormal amounts. When CA 19-9 level is elevated to high concentrations it is often correlated to with gastrointestinal malignancies such as cholangiocarcinoma, pancreatic cancer, or colon cancer.

Yang et al. proposed a novel sandwich-type graphene-based electrochemical immune sensor for the detection of CA19-9 [94]. The sensing surface of GCE was built through the assembly Au-rGO nanocomposite and posterior immobilization of anti-CA19-9, forming the structure Ab1/Au-rGO/GCE. The signal enhancers were designed though the GO surface modification with core shell Au and Pd NPs, Lauth's violet (Thi), anti-CA19-9 (Ab2) and Horseradish peroxidase (HRP) forming signal probes Au and Pd-Gra/Thi-Ab2/HRP. During the biological recognition event, the target protein CA19-9 was sandwiched between the primary antibody Ab1/Au-rGO/GCE and the prepared biological conjugates Au and Pd-Gra/Thi-Ab2/HRP, resulting in enhancement of the detectable signal. The results showed that peak currents of the electrochemical immunoassay increased with the increase of CA19-9 concentrations, and exhibited a linear relationship range from 0.015 to 150 $U{\cdot}mL^{-1}$, with low detection limit of 0.006 $U{\cdot}mL^{-1}$.

Carbohydrate Antigen 15-3 (CA 15-3) is a tumor marker applied to monitor certain cancers, *particularly breast cancer*. The first technological method for the sensitive detection of CA 15-3 uses graphene based on materials which are composed of in the development of label-free electrochemical immune sensor [95]. The immune sensor, was designed by the surface modification of GCE with a highly conductive N-doped graphene sheets, that allows a significantly increase of electron transfer and high sensitivity toward CA 15-3. The immune sensor exhibited a low detection limit at 0.012 U/mL and a broad linear response in the range of 0.1–120 U/ml.

Recently, it was reported that a novel electrochemical graphene based on immune sensor for sensitive detection of CA15-3 based on dual signal amplification strategy [96]. For trace tag signal amplifier was synthesized a nanocomposite material based on Nano-porous TiO2 functionalized with Cd^{2+} for increasing the electron transfer rate, followed by the immobilization of the antibody CA15-3 (Ab2), resulting in the following sequence (Ab(2)-f-TiO2-Cd^{2+}). The GCE sensing platform was developed by immobilization of rGO functionalized with ionic liquid and primary CA15-3 antibody (Ab (1)) resulting in the sequence GS-Ab (1). Through biological recognition reaction, the target protein CA15-3 was sandwiched between the primary antibody GS-Ab1 and the prepared (Ab2-f-TiO2-Cd^{2+}) biological conjugates, resulting in an enhanced detectable signal. The resultant immune sensor displayed a wide range of linear response (0.02–160 U/mL), ultra-low detection limit (0.008 U/mL), good reproducibility, selectivity and stability towards CA15-3.

10.3.5 Protein p53

In the control of cell growth and modulation of DNA repair processes, Protein p53 plays a role as a popular tumor suppressor and a transcription factor fundamental [97]. The loss of p53 function caused by the conformational changes in p53 protein structure has given rise to an induction of tumors and gene mutation [98]. Some clinical evidences showed the implication of p53 phosphorylation in human cancers, particularly expression of p53 protein phosphorylated, serine 15 (Ser15) serine 20 (Ser20) and serine 392 (Ser392). In that sense the quantitative analysis of phosphorylated p53 is crucial for early cancer diagnosis. Most of the electrochemical graphene-based immune sensors engineered for sensitive detection of phosphorylated p53, were based on sandwich type strategy. The fist graphene-based immune sensor reported was dedicated to the specific detection of phosphorylated p53 on serine 392. For immune sensor fabrication, it was used in screen-printed carbon electrode (SPCE) modified with AuNPs to self-assemble a layer of N-hydroxysuccinimideactivated hex (ethylene glycol) undecane thiol (NHS) for primary attachment of phospho-p53392 capture antibody (Ab1/NHS/AuNPs-SPCE) [99]. The HRP-p53392Ab2-GO conjugate was synthetized by the functionalization of carboxylate GO with HRP and the immobilization of capture antibody p53392 (Ab2) by amidation reaction. After sandwich immunoreaction, the HRP-p53392Ab2-GO captured onto the electrode surface produced an amplified electro catalytic response by the reduction of enzymatically oxidized Lauth's violet in the presence of H_2O_2 (Fig. 10.7). The increase of response current was proportional to the phospho-p53392 concentration in the range of 0.02–12 NM with the detection limit of 0.01 NM.

Another work reported the development of graphene-based immune sensor for specific detection of phosphorylated p53 on serine 15. In that case graphene was

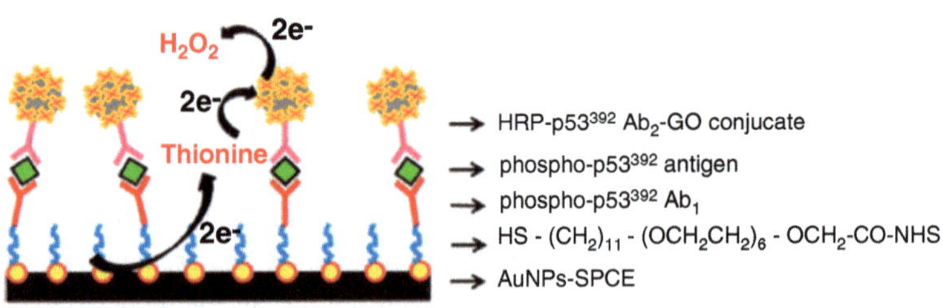

$$H_2O_2 + HRP_{(Red)} \longrightarrow HRP_{(Ox)} + H_2O_{(Red)}$$

$$HRP_{(Ox)} + Thionine_{(Red)} \longrightarrow HRP_{(Red)} + Thionine_{(Ox)}$$

$$Thionine_{(Ox)} + 2H^{+} + 2e^{-} \longrightarrow Thionine_{(Red)}$$

Fig. 10.7 Schematic Illustration of the Multi-enzyme Labeling Amplification Strategy Using HRP-p53392Ab2-GO Conjugate. Reproduced with permission [99]. Copyright 2011, American Chemical Society

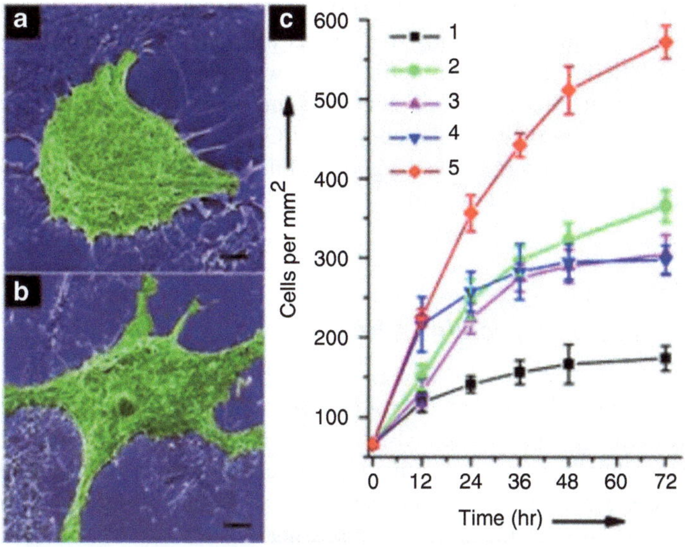

Fig. 10.8 SEM images of MCF-7 cells cultured (12 h) on (**a**) ITO/ (AP) 10 and (**b**) ITO/ (graphene-AP) $_{10}$. The scale bar is 5 μm for (**a**) and (**b**). (**c**) Proliferation curves of cells cultured on films with different compositions: 1. ITO/ (AP) $_{10}$, 2. ITO/ (graphene) $_{10}$, 3. ITO/ (graphene-AP) $_{10}$, 4. ITO/ (graphene-AP) $_{10}$-laminin (only one layer of laminin on top of the structure), and 5. ITO/ (graphene-AP-laminin) $_{10}$. Reproduced with permission [112]. Copyright 2010, Wiley-VCH Velar GmbH & Co

integrated on the sensing platform instead of the sandwich conjugate platform. For the fabrication of the immune sensor, the SPCE was functionalized with graphene and chitosan for primary antibody attachment phospho-p5315 (Ab1/graphene-1CHI/SPCE) [100]. The sandwich immune complex was formed among phospho-p5315 capture antibody, phosphop5315 antigen, biotinylated phospho-p5315 detection antibody and HRP-labeled streptavidin, resulting in the following system: HRP-1 streptavidin-1biotinAb2/phospho-p5315/Ab1/graphene-1CHI/SPCE. The linear response can be observed for concentration range of phospho-p5315, which changes from 0.2 to 10 ng·mL^{-1} with the detection limit of 0.1 ng·mL^{-1}.

10.3.6 Vascular Endothelial Growth Factor (VEGF)

Vascular endothelial growth factor (VEGF) is an important growth factor that regulates the progression of angiogenesis. VEGF overexpression appears in many human cancers, such as brain tumors, lung cancer, breast cancer, gastrointestinal cancer and urinary tract tumors [101].

Lin et al. originated an easy and reusable strategy to detect the VEGF concentration in human serum via using a graphene based biosensor [102]. The sensing platform deposited at Au electrode surface consisted of a nanocomposite of GO-F_3O_4NPs functionalized with Avastin, as the specific biological recognition element for VEGF. The ampere-metric signal was measured by differential pulse voltammetry to quantify the VEGF concentration, being the linear response of the biological sensor observed for the range of VEGF concentrations from 0.03125 to 2 ng·mL^{-1}. The biosensors showed a good stability (RSD = 2.36 %, n = 50) and a limit of detection of 0.03125 ng·mL^{-1}.

10.3.7 Alpha-Fetoprotein (AFP)

Alpha-fetoprotein (AFP) is a glycoprotein of fetal component produced during the embryonic period. However, high AFP serum levels can be found mainly in patients with hepatocellular carcinoma or other types of cancer such as, lung cancer, biliary cancer, gastric cancer, pancreatic cancer, teratocarcinoma of the testis. Graphene-based materials have been explored as a sensing

Platform for the advancement of noble electrochemical immunoassay can amplify the sensitivity in the detection of AFP. The pioneer work reported a graphene-based ultrasensitive multiplexed electronic chemiluminescent (ECL) immunoassay method for the detection AFP tumor marker [103]. The graphene probe was synthesized by the assembly with poly (methyl chloride) (PDDA) and luminal-capped gold nanoparticles (PDDA-G and Lu-Au). The obtained PDDA-G and Lu-Au composite particles were used as substrate for antibody and HRP immobilization with high luminal capacity load efficiency. The second probe consisted on multi-able-antibody functionalized core-shell Fe_3O_4@Au composites (GMPs). The combination of AFP with the two labels results in sandwich-type immune complexes PDDA-G and Lu-Au~HRP-Ab2/AFP/Ab1~GMP. The magnetic sandwich-type immune complexes were immobilized on the working screen-printed carbon electrode (SPCEs) by applying a magnetic field. That approach avoids some adverse effects obtained through the surface functionalization of the electrodes. Under the optimized conditions, the ECL method shows a linear range of AFP detection from 0.002 to 20 ng·mL^{-1} and a low detection limit of 0.2 pg·mL^{-1}.

Zhuo et AL recently proposed the advancement of a Nobel electronical chemiluminescent immune sensor based on a new improvement of per-oxydisulfate system for the detection of AFP [104]. The sensor probe design was based on rGO/AuNPs nanocomposites functionalized by supramolecular assembly with arginine covalently bond to 3, 4, 9, 10-perylenetetracarboxylic acid (PTC-Arg/Au and Gra). The PTC-Arg/Au and Gra biological conjugate was used as multifunctional Nano-carriers for the absorption of the second antibody of alpha-fetoprotein (Ab2), which formed sandwich-type immunoassay format through association with AFP and the first antibody (Ab1) previously assembled at Au electrode surface. With PTC-Arg/Au and Gra as enhancer of per-oxydisulfate system, the immune sensor exhibited a wide dynamic range of 0.001–110 ng·mL^{-1} and detection of limit of 0.3 pg·mL^{-1}.

10.3.8 Matrix Metalloproteinases (MMPs)

Chen et al. put forward a novel approach for sensitive and rapid detection of magnetic-activated cell sorting (Matrix Metalloproteinases, MMP-7, a biomarker involved in the degradation of various macromolecules) based on a polypeptide (JR2EC) functionalized reduced graphene oxide (rGO) field effect transistor (FET) [105]. MMP-7 specifically digests negatively charged JR2EC immobilized on rGO, thereby modulating the conductance of rGO-FET. The proposed assay enabled detection of MMP-7 at clinically relevant concentrations with a limit of detection (LOD) of 10 ng/mL (400 p.m.), attributed to the significant reduction of the net charge of JR2EC upon digestion by MMP-7. Quantitative detection of MMP-7 in human plasma was further demonstrated with a LOD of 40 ng/mL, illustrating the potential for the proposed methodology for tumor detection and carcinoma diagnostic (e.g., lung cancer and salivary gland cancer). Additionally, excellent specificity of the proposed assay was demonstrated by using matrix metallopeptidase 1 (MMP-1), a protease of the same family. With appropriate selection and modification of polypeptides, the proposed assay could be extended for detection of other enzymes with polypeptide digestion capability.

10.3.9 Multiplex Cancer Tumor Markers Detection

Graphene is first to be explored as an effective platform for electrochemical multiplex detection of actively targeted biomarkers, which has clinical relevance. Recently it was reported the development of a suspended single crystalline graphene (SCG) biosensor for multiplex lung cancer tumor markers detection [106]. After the immobilization process of the antibodies, the label-free suspended SCG sensor was ready to detect the important lung cancer biomarkers: ANXA2, VEGF, and ENO1. SCG label-free lung cancer biosensor showed a resistive response shift due to the absorption of biomolecules by the bio-receptors on the surface of graphene that cause the change in density and mobility of charge carriers of the film according to different concentrations of the three types of lung cancer tumor markers, ANXA2, VEGF, and ENO1. The biosensor sensitivity studies performed with presence of the three tumor markers showed that the response to the specific antigen is one order of magnitude larger than the non-specific ones. The high performance of the SCG biosensor results from synergistic effects obtained from the combination of two important bio-events on its surface: strong interaction between antigen and its bio-receptors at graphene surface and the blocking of the non-specific binding sites on graphene surface by BSA, which further avoids the absorption of non-specific molecules. Furthermore, the suspended structure of SCG biosensor, which has better carrier mobility, sensing area, and lower frequency noise is responsible for the enhanced sensitivity with detection limit of 0.1 $pg \cdot mL^{-1}$, good specificity and large linear detection range from 1 pg·mL^{-1} to 1 μg·mL^{-1}. This SCG label-free lung cancer biosensor showed to have the lower detection limits ever reported for the three different tumors markers ANXA2, VEGF, and ENO1.

10.3.10 DNA Biosensors

DNA sensors are common used approach for early cancer diagnosis and mutation detection. Electrochemical sensors are one of the most used approaches in DNA technology advancement, because DNA bases are electroactive which allows direct electrochemical signals recording in the electronical oxidation process. Graphene illustrates superior electrochemical performance for oxidation of DNA bases because of its outstanding electrochemical properties and large surface-to-volume ratio, which play an integral part in the improvement of new electrodes with increased sensitivity and simultaneous detection [107]. Electrochemical DNA graphene based sensor for low-concentration detection of Breast Cancer 1 (BRCA1) DNA sequences is an emerging technology recently [108]. It was reported that mutations on this gene were related to an 80% increase in the risk of breast and ovarian cancer [109]. The biosensors construct was based on the functionalization of GCE surface with rGO as a sensing platform for the immobilization of DNA capture probe (DNA-c with amino group at 51 end). In this system, one-half of the DNA target (DNA-t) is allowed to hybridize with the immobilized capture probe (DNA-c), while the other half interacts with the reporter probe (DNA-r) conjugated with gold nanoparticles. Electrochemical oxidation of gold nanoparticle was monitored using cyclic voltammetry to detect the concentration of DNA-t. The results showed that the sensor was stable, reproducible and sensitive and it could detect up to 6 FG·mL^{-1} of DNA target.

In another work, Wang et al. proposed a novel sensitive and selective electrochemical DNA graphene-based sensor for the detection of BCR/ABL fusion gene in chronic myelogenous leukemia (CML) [110]. CML is a clonal neoplastic disorder of hematopoietic stem cells caused by expression of the chimeric BCR/ABL fusion oncogene (abnormality occurs in more than 95 % patients) [111]. The proposed sensor structure consist on the surface functionalization of GCE trough the self-assembly of graphene/chitosan nanocomposite, followed by the electropolymerization of polyaniline (PANI) and electrodeposition of Au nanoparticles and finally decorated with a functional hairpin structure probe for the detection of BCR/ABL fusion gene of CML. The hybridization event was sensitively transduced to the enzymatically amplified electrochemical current signals, which presents a low detection limit 2.11 p.m. and a high DNA detection sensitivity.

10.3.11 Cell Sensors

Graphene based sensors were also explored for modern diagnosis of cancer by the detection and quantification of specific cancer cells. Graphene based electrochemical sensors has been revealed as the most promising method. Graphene based materials on cells sensors can provide an increase of electrochemical properties in the electrodes and furthermore provides very rich anchor sites to bind recognition species in order to create an interface with high selectivity.

Guo et al. reported a smart multilayer ampere-metric sensor based on deposition of rGO onto an ITO electrode, followed by the functionalization with Prussian blue as H_2O_2 catalyst and laminin matrix protein to promote cell adhesion. On this system it was observed that rGO can provide a compatible interface for the growth of human cells and excellent electrical conductivity for electrical detection, allowing for the first time, the in situ selective and quantitative extracellular H_2O_2 detection. Breast cancer cells were grown on the electrode surface and the sensor was able to detect H_2O_2 with a LOD of 0.1 μM upon stimulation [112].

Recently scientists began to focus on the advancement of new graphene ultrasensitive sensors which are able to quantify cells that can directly grow on its sensing surface or specifically detect cancer cells from a very complex mixture. Wu et al. develop a high sensitive graphene based electrochemical sensor through the surface modification of GCE with: chitosan/electrochemically rGO film covalently modified with anti-Ep-CAM antibodies [113]. Then this biological interface was applied to detect Hep3B cancer cells. After that, a sandwich system was generated using CdTe- and ZnSe coated silica nanoparticles that could easily serve as tracing tags to label anti-Ep-CAM and anti-GPC3. Each biological recognition event yields a distinct voltammetry peak, in which position and size represents the corresponding identity and amount of the respective antigen. The combination of ultrasensitive and highly specific electrochemical and fluorescent immune sensing methods showed great potential for application on the detection of circulating tumor cells. The immune sensor developed exhibited high sensitivity and specificity with excellent stability, reproducibility, and accuracy, with a LOD low then 10 cells·mL^{-1}. In another work was reported the preparation of a graphene electrode modified with a new conjugate of peptide nanotubes and folic acid for the selective detection of human cervical cancer cells (HeLa cells) over-expressing folate receptors. The increases number of HeLa cells immobilized at electrode surface promotes a formation of insulating layer inducing the reduction of the recorded intensity of the electrochemical signals. This electrode provides a new sensing alternative for cancer cells detection, with high simplicity related with the synthetic process when compared with the previous one, however it showed higher LOD of 250 cells·mL^{-1} [114].

Feng et al. discovered a reusable electrochemical graphene-based sensor via using aptamer clinical trial II AS1411 for label-free cancer cell detection [115]. Functionalized with 3, 4, 9, 10-perylene tetra-carboxylic acid (PTCA) through π-π stacking and hydrophobic interactions graphene surface was able to introduce more

-COOH groups. NH2-modified aptamer strand was linked to PTCA/rGO via covalent bond as the recognition element that can bind to the overexpressed nucleoli on the plasma membrane of cancer cells. The binding of cancer cells decreased the access of the redox probe to the electrode interface. The electrochemical apt-sensor was able to differentiate three types of cancer cells (HeLa cells (human cervical carcinoma cell), MDA-MB-231 (human breast cancer cell), K562 cells (leukemia line)) and normal ones (NIH3T3 cells).

Another electrochemical graphene-based sensor was developed for specific detection of tumor marker Her2 overexpressed on surface of SKOV-3 tumor cells [116]. The GO modified electrode was covalently functionalized with Anti-Her2 for the capture of SKOV-3 cells, and the unbound HER2 was available to connect with the anti-HER2 graft dsDNA, integrating the advantages of graphene electrochemical properties and DNA marker. This approach provided a very low LOD 5.2 cells·mL^{-1} for SKOV-3 cells.

Recently, graphene/inorganic nanoparticles hybrids started to be used as a sensing platform at electrodes surface in order to increase the sensitivity of the sensor. Liu et al. developed a new ZnO/graphene composite modified with S6 aptamer as sensitive photo electrochemical (PEC) strategy for the specific detection of SK-BR-3 cancer cells [117]. This hybrid structure was used as sensing platform on ITO working electrode, in order improve its PEC performance. The increase of photocurrent intensity results from the increase in steric hindrances with S6 aptamer with SK-BR-3 cancer cells receptors. The results showed high selectivity with analogous cells and a LOD of 58 cells·mL^{-1}.

Yan et al. reported a novel graphene immune sensor for detection of cancer cells MCF-7 though the surface modification of the working electrode with GO/AuNPs nanocomposite [118]. The composite was functionalized with aptamer to recognize and bind the tumor marker on the surface of cancer cells MCF-7. Additionally, Lauth's violet functionalized Nano-porous Pt-Fe alloy was employed as signal amplifier for cells detection. The results gained demonstrates the increase of the electrochemical signal measured proportional to the concentration of MCF-7 cells. The biosensor showed a good selectivity, acceptable stability and reproducibility with a LOD of 38 cell·mL^{-1}. Another work reported the use of small bifunctional composite quantum dot (Fe_3O_4/Cd-Se QD), with intense electronic chemiluminescence (ECL) and excellent magnetic property, as reported as enhancer signal probe for sensitive detection of cancer cells via DNA cyclic amplification technique using GO sensing platform to immobilize DNA capture probes (c-DNA1) at the surface of Au electrode [119]. The increased concentration of target cells results in scission of more QDs/DNA signal probe, thus more decrease of ECL signal was obtained. The relationship between the changes of ECL signal and the concentrations of target cells was for the range from 300 to 9000 cells·mL^{-1} with a detection limit of 98 cells·mL^{-1}.

10.4 Luminescence Sensors

Graphene as we discussed before was characterized to process very interesting properties, in particular high conductivity and luminescence quenching ability. In fact, the ultra-high quenching efficiency is attributed to three possible mechanisms: Fortier resonance energy transfer (FRET), surface energy transfer (SET) and photo-induced electron transfer [120]. As a quencher for diverse luminescence energy donors, graphene materials have been used in combination with several probes.

Graphene oxide (GO) which is based on fluorescent biosensor was ready for real-time in situ detection of integrin αvβ3. Integrin plays an integral role in cancer proliferation, cell adhesion, migration and metastasis [121]. The biosensor concept is based on the capacity of the sp^2 aromatic domains of GO to establish noncovalent interactions with pyrene molecules resulting in the quenching effect via fluorescence resonance energy transfer or dipole-dipole coupling effects. In this work GO based biosensor system was initially at a quenching state due to the proximity of RGD-pyrene to GO upon π-π stacking interactions [122]. However, the competitive binding of an RGD receptor, integrin avb3, to the RGD ligand disturbs the adsorption of RGD-pyrene onto the GO surface, resulting in the recovery of pyrene fluorescence with the increase of concentrations of integrin as illustrated in Fig. 10.9. The biosensors developed showed the capacity for detection of purified integrin protein in buffer and the effectiveness for in situ detection of integrin overexpression on cancer cells surface (MDA-MB-435 cell line).

A multiplex microfluidic chip integrated with the GO-based FRET strategy to create a screening assay for in situ detection of tumor cells was also reported [123]. For this purpose, a 33-channel microfluidic chip integrated with the GO-based FRET apt-sensor was designed and employed to detect target CCRF-CEM cells with a confocal fluorescence scanning microscope (Fig. 10.10). Initially the FRET probe of GO/FAM-Sgc8 exhibited a quenched fluorescence because of π-π stacking interactions between FAM-Sgc8 and GO. The strong interaction between FAM-Sgc8 and CCRF-CEM cells promotes the release FAM-Sgc8 from GO surface resulting in the recovery of the FAM-Sgc8 fluorescence. The changes observed in fluorescence intensity measurement allowed the quantification of the target cancer cells CCRF-CEM, with the linear response in a concentration range from 2.5×10^1 to 2.5×10^4 cells·mL^{-1} and a detection limit about 25 cells·mL^{-1}. Another work suggested a new graphene based fluorescent biosensor system based on a development of a novel molecular aptamer beacon (MA) as fluorescently labelled and GO as the acceptor for target detection by employing long range resonance energy transfer (Lr-RET) of cellular prion protein (Pr-PC) [124]. Initially the fluorescence of the designed MAB was completely quenched by GO, however after addition of Pr-PC the quenched fluorescence was recovered significantly. The results obtained showed that Pr-PC can be detected over a wide range of 10.2–78.8 μg·mL^{-1} with a detection limit of 0.309 μg·mL^{-1}.

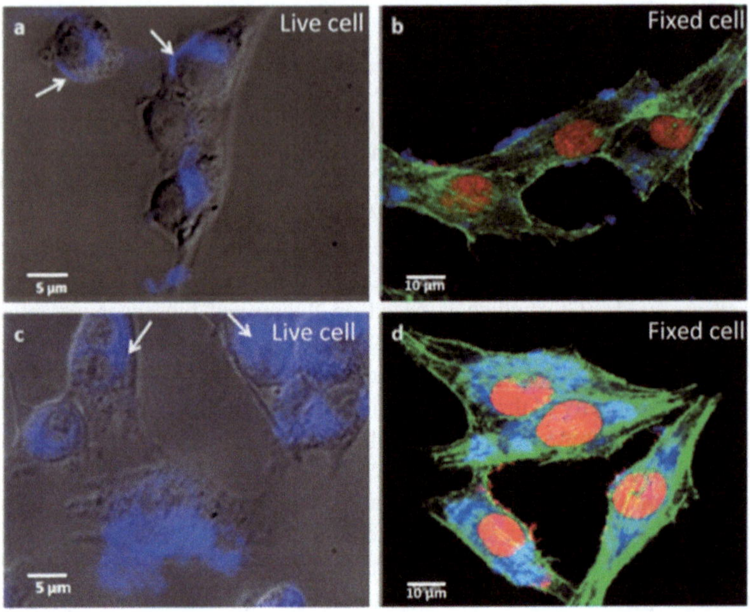

Fig. 10.9 Real-time in situ detection of breast cancer cell surface integrin expression by the RGD-pyrene-GO probe: (**a**) probe fluorescence recovery by live MDA-MB-435 cancer cells which overexpress integrin avb3 on the cell surface. The recovered fluorescence is mainly detected on the cell membrane as indicated by white arrows; (**b**) probe fluorescence recovery by MDA-MB-435 cancer cells followed by 4% formalin fixing; (**c**) equivalent concentration of free RGD-pyrene incubated with liveMDA-MB-435 demonstrates significant endocytosis as indicated by white arrows; (**d**) equivalent concentration of RGD-pyrene incubated with MDA-MB-435 followed by 4% formalin fixing. Reproduced with permission [122]. Copyright 2012, Royal Society of Chemistry

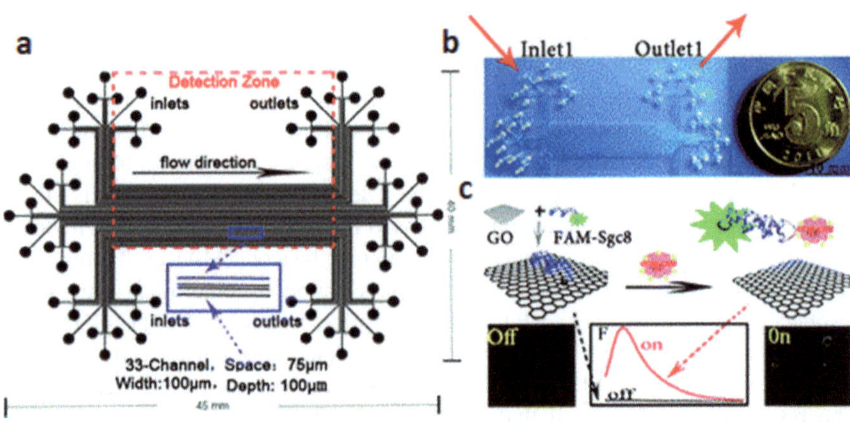

Fig. 10.10 (**a**) Schematic setup and (**b**) the photograph of the GO-based FRET apt-sensing microfluidic chip. Scale bar: 10 mm. (**c**) The principle of a 'signal-on' apt-sensor for detecting CCRF-CEM cells by assaying the cell-induced fluorescence recovery of GO/FAM-Sgc8. Reproduced with permission [123] Copyright 2012, Royal Society of chemistry

obtained electrode when compared with CFE and glassy carbon electrode (GCE) for the oxidation of AA, DA and UA. Also, the linear calibration plots for AA, DA and UA were observed, respectively, in the ranges of 45.4–1489.23 μM, 0.7–45.21 μM and 3.78–183.87 μM in the individual detection of each component. By simultaneously changing the concentrations of AA, DA and UA, their oxidation peaks appeared at 0.05 V, 0.16 V and 2.6 V, and the good linear responses ranges were 73.52–2305.53 μm, 1.36–125.69 μM and 3.98–371.49 μm, respectively.

10.5 Outlook

Although the great potentialities of graphene in bio sensing technologies, its use could present some health concern issues since a large number of toxicological studies presented contradictory conclusions. Some studies have reported that the cytotoxicity of graphene and its derivatives depend on many factors, including preparation protocol, purification processes and final physicochemical properties. Thus, there is a long way to go until these graphene based sensing-systems become optimize and subsequently available for real-world clinical applications. In fact, more detailed studies concerning, for example, the safety of graphene and its derivatives should be addressed with the purpose of monitoring and controlling their metabolic pathway, cellular-uptake mechanism and long term toxicity, which are key-points for nanomaterials applications in bio-imaging, drug delivery and cancer therapy.

The scalable, controllable and reproducible methods of synthesis of graphene-based nanomaterials are other relevant challenges that need to be improved in order to obtain consistent results.

Overall, cancer is a disease with high innate heterogeneity, very difficult to detect and diagnose via a single biomarker with high specificity and sensitivity. The detection of cancer biomarker in real biological samples such as blood plasma, blood serum cerebrospinal fluid, saliva or urine is a real challenge since these samples are very complex systems with different kinds of proteins, ions and other chemical species able to promote false positive responses in label free assays. Possible ways to overcome this issue are the passivation of the sensors surface with an antifouling agent or, more particularly, the previous filtration/purification of the biological sample in order to concentrate the target entity.

Thus, as it was discussed, the ability of graphene based materials to conjugate their outstanding chemical and physical features into remarkable sensing properties appears to be a solid milestone on the long pathway to achieve an early and optimal cancer diagnostics. Indeed, researchers around the world are continuing to explore the wide range of sensing strategies opened by graphene in the last few years due to its proven adaptability and functionality in different cancer microenvironments. For the near future, the ultimate goal is to develop graphene based devices capable of simultaneously detect multiple cancer biomarkers.

Graphene-based sensors for biomedical detection actually show better performances than their corresponding graphene-less ones and that surely the best results in terms of sensor sensitivity and detection limit are obtained when metal nanoparticles are co-immobilized with graphene, hydrogen peroxide, L-Cysteine, dopamine, sometimes in presence of a proper conductive polymer. In comparison with other nanostructured materials, graphene-based sensors often provide the best performances.

Acknowledgement This work is supported by the National Natural Science Foundation of China (Grant No. 61308114), the national natural science fund projects (Grant No. 61675039), and Shenzhen Peacock team project (Grant No. KQTD20140630143254).

References

1. F. Schedin, A.K. Geim, S.V. Morozov, E.W. Hill, P. Blake, M.I. Katsnelson, K.S. Novoselov, Nat. Mater. **6**, 652 (2007)
2. Y.Q. Wu, Y.M. Lin, A.A. Bol, K.A. Jenkins, F.N. Xia, D.B. Farmer, Y. Zhu, P. Avouris, Nature **472**, 74 (2011)
3. F.S.F. Schwierz, Nat. Nano-technol. **5**, 487 (2010)
4. J.S. Wu, W. Pisula, K. Mullen, Chem. Rev. **107**, 718 (2007)
5. Q.Y. He, S.X. Wu, Z.Y. Yin, H. Zhang, Chem. Sci. **3**, 1764 (2012)
6. G.S. Kulkarni, Z.H. Zhong, Nano Lett. **12**, 719 (2012)
7. W. Fu, M. El Abbassi, T. Hasler, M. Jung, M. Steinacher, M. Calame, C. Schönenberger, G. Puebla-Hellmann, S. Hellmüller, T. Ihn, A. Wallraff, Appl. Phys. Lett. **104**, 013102 (2014)
8. Y.Q. Sun, Q.O. Wu, G.Q. Shi, Energy Environ. Sci. **4**, 1113 (2011)
9. D.A.C. Brownson, D.K. Kampouris, C.E. Banks, J. Power, Sources **196**, 4873 (2011)
10. X.H. Cao, Z.Y. Yin, H. Zhang, Energy Environ. Sci. **7**, 1850 (2014)
11. J.X. Zhu, D. Yang, Z.Y. Yin, Q.Y. Yan, H. Zhang, Small **10**, 3480–3498 (2014). doi:10.1002/smll.201303202
12. Z.Y. Yin, J.X. Zhu, Q.Y. He, X.H. Cao, C.L. Tan, H.Y. Chen, Q.Y. Yan, H. Zhang, Adv. Energy Mater. **4**, 1300574 (2014)
13. C.K. Huang, Y.X. Ou, Y.Q. Bie, Q. Zhao, D.P. Yu, Appl. Phys. Lett. **98**, 263104 (2011)
14. Z.S. Wu, S.F. Pei, W.C. Ren, D.M. Tang, L.B. Gao, B.L. Liu, F. Li, C. Liu, H.M. Cheng, Adv. Mater. **21**, 1756 (2009)
15. J.K. Wassei, R.B. Kaner, Mater. Today **13**, 52 (2010)
16. X. Huang, X.Y. Qi, F. Boeya, H. Zhang, Chem. Soc. Rev. **41**, 666–686 (2012)
17. X. Huang, Z.Y. Yin, S.X. Wu, X.Y. Qi, Q.Y. He, Q.C. Zhang, Q.Y. Yan, F. Boey, H. Zhang, Small **7**, 1876 (2011)
18. X. Huang, Z.Y. Zeng, Z.X. Fan, J.Q. Liu, H. Zhang, Adv. Mater. **24**, 5979 (2012)
19. J.H. Jung, D.S. Cheon, F. Liu, K.B. Lee, T.S. Seo, Angew. Chem. Int. Ed. **49**, 5708 (2010)
20. Y. Bo, H. Yang, Y. Hu, T. Yao, S. Huang, Electrochim. Acta **56**, 2676 (2011)
21. A. Kakatkar, T.S. Abhilash, R. De Alba, J.M. Parpia, H.G. Craighead, Nanotechnology **26**, 125502 (2015)
22. H.Y. Yue, H. Zhang, J. Chang, X. Gao, S. Huang, L.H. Yao, X.Y. Lin, E.J. Guo, Anal. Biochem. **488**, 22 (2015)
23. U. Patil, S.C. Lee, S. Kulkarni, J.S. Sohn, M.S. Nam, S. Han, S.C. Jun, Nanoscale **7**, 6999 (2015)
24. K.S. Novoselov, A.K. Geim, S.V. Morozov, D. Jiang, Y. Zhang, S.V. Dubonos, I.V. Grigorieva, A.A. Firsov, Science **306**, 666 (2004)

25. A.K. Geim, K.S. Novoselov, Nat. Mater. **6**, 183 (2007)
26. Y. Zhang, Y.-W. Tan, H.L. Stormer, P. Kim, Nature **438**, 201 (2005)
27. Y.M. Lin, P. Avouris, Nano Lett. **8**, 2119 (2008)
28. A.N. Pal, A. Ghosh, Appl. Phys. Lett. **95**, 082105 (2009)
29. M. Han, B. Ozyilmaz, Y. Zhang, P. Jarillo-Herero, P. Kim, Phys. Status Solidi B **244**, 4134 (2007)
30. H.P. Boehm, A. Clauss, G.O. Fischer, U. Hofmann, Z. Naturforsch. **17**, 150 (1962)
31. C.M.C. Mattevi, H. Kim, M. Chowilla, J. Mater. Chem. **21**, 3324 (2011)
32. W.A. de Heer, C. Berger, X.S. Wu, P.N. First, E.H. Conrad, X.B. Li, T.B. Li, M. Sprinkle, J. Hass, M.L. Sadowski, M. Potemski, G. Martinez, Solid State Commun. **143**, 92 (2007)
33. P.W. Sutter, J.I. Flege, E.A. Sutter, Nat. Mater. **7**, 406 (2008)
34. S. Shivaraman, R.A. Barton, X. Yu, J. Alden, L. Herman, M.V.S. Chandrashekhar, J. Park, P.L. McEuen, J.M. Parpia, H.G. Craighead, M.G. Spencer, Nano Lett. **9**, 3100 (2009)
35. W.C. Ren, L.B. Gao, L.P. Ma, H.M. Cheng, New Carbon Mater. **26**, 71 (2011)
36. X.C. Dong, P. Wang, W.J. Fang, C.Y. Su, Y.H. Chen, L.J. Li, W. Huang, P. Chen, Carbon **49**, 3672 (2011)
37. A. Reina, X.T. Jia, J. Ho, D. Nezich, H.B. Son, V. Bulovic, M.S. Dresselhaus, J. Kong, Nano Lett. **9**, 30 (2009)
38. X. Wang, G. Sun, P. Routh, D. Kim, W. Huang, P. Chen, Chem. Soc. Rev **43**, 7067 (2014)
39. X.H. Cao, Y.M. Shi, W.H. Shi, G. Lu, X. Huang, Q.Y. Yan, Q.C. Zhang, H. Zhang, Small **7**, 3163 (2011)
40. X. Cao, Y. Shi, W. Shi, X. Rui, Q. Yan, J. Kong, H. Zhang, Small **9**, 3433–3488 (2013)
41. X.H. Cao, Z.Y. Zeng, W.H. Shi, P. Yep, Q.Y. Yan, H. Zhang, Small **9**, 1703 (2013)
42. X.H. Cao, B. Zheng, X.H. Rui, W.H. Shi, Q.Y. Yan, H. Zhang, Angew. Chem. Int. Ed. **53**, 1404 (2014)
43. W.J. Zhou, X.H. Cao, Z.Y. Zeng, W.H. Shi, Y.Y. Zhu, Q.Y. Yan, H. Liu, J.Y. Wang, H. Zhang, Energy. Environ. Sci. **6**, 2216 (2013)
44. X.C. Dong, H. Xu, X.W. Wang, Y.X. Huang, M.B. Chan-Park, H. Zhang, L.H. Wang, W. Huang, P. Chen, ACS Nano. **6**, 3206 (2012)
45. X.C. Dong, J. Chen, Y.W. Ma, J. Wang, M.B. Chan-Park, X.M. Liu, L.H. Wang, W. Huang, P. Chen, Chem. Comum. **48**, 10660 (2012)
46. Z.P. Chen, F. Houshmand, W.C. Ren, Y. Peles, H.M. Cheng, N. Koratkar, Small **9**, 75 (2013)
47. Y.C. Yong, X.C. Dong, M.B. Chan-Park, H. Song, P. Chen, ACS Nano. **6**, 2394 (2012)
48. B. Zhan, C. Li, J. Yang, G. Jenkins, W. Huang, X. Dong, Graphene Sens. **10**(20), 4042–4065 (2014)
49. S. Lee, Z. Zhong, Nanoscale **6**, 13283 (2014)
50. F.M.P. Tonelli, V.A.M. Goulart, K.N. Gomes, M.S. Ladeira, A.K. Santos, E. Lorencon, L.O. Ladeira, R.R. Resende, Nanomedicine-UK **10**, 2423 (2015)
51. Z. Wang, P. Li, Y. Chen, J. He, J. Liu, W. Zhang, Y. Li, J. Power, Sources **263**, 246 (2014)
52. Z. Wang, P. Li, Y. Chen, J. Liu, H. Tian, J. Zhou, W. Zhang, Y. Li, J. Mater. Chem. C. **2**, 7396 (2014)
53. Z. Wang, P. Li, Y. Chen, J. He, W. Zhang, O.G. Schmidt, Y. Li, Nanoscale **6**, 7281 (2014)
54. J. He, Y. Chen, P. Li, F. Fu, Z. Wang, W. Zhang, J. Mater. Chem. A **3**, 18605 (2015)
55. J. He, Y. Chen, W. Lv, K. Wen, Z. Wang, W. Zhang, Y. Li, W. Qin, W. He, ACS Nano **10**, 8837 (2016)
56. M.R. Stratton, P.J. Campbell, P.A. Futreal, The cancer genome. Nature **458**, 719–724 (2009)
57. T.L. Serafim, P.J. Oliveira, in *Tumor Metabolome Targeting and Drug Development*, ed by S. Kanner. Regulating mitochondrial respiration in cancer (Springer New York, New York, NY, 2014), pp. 29–73
58. Q. Li, L. Liu, J.-W. Liu, J.-H. Jiang, R.-Q. Yu, X. Chu, Nanomaterial-based fluorescent probes for live-cell imaging. TrAC Trends Anal. Chem. **58**, 130–144 (2014)
59. S.W. Perry, R.M. Burke, E.B. Brown, Two-photon and second harmonic microscopy in clinical and translational cancer research. Ann. Biomed. Eng. **40**, 277–291 (2012)

60. H. Yuan, J.K. Register, H.N. Wang, A.M. Fales, Y. Liu, T. Vo-Dinh, Plasmatic Nano-probes for intracellular sensing and imaging. Anal. Biological Anal. Chem. **405**, 6165–6180 (2013)
61. L. Baù, P. Tecilla, F. Mancin, Sensing with fluorescent nanoparticles. Nanoscale **3**, 121–133 (2011)
62. J.L. Li, B. Tang, B. Yuan, L. Sun, X.G. Wang, A review of optical imaging and therapy using Nano-sized graphene and graphene oxide. Biomaterials **34**, 9519–9534 (2013)
63. J.M. Yoo, J.H. Kang, B.H. Hong, Graphene-based nanomaterials for versatile imaging studies. Chem. Soc. Rev. **44**, 4835–4852 (2015)
64. N. Abdullah Al, J.E. Lee, I. In, H. Lee, K.D. Lee, J.H. Jeong, S.Y. Park, Target delivery and cell imaging using hyaluronic acid-functionalized graphene quantum dots. Mol. Pharm. **10**, 3736–3744 (2013)
65. J. Peng, W. Gao, B.K. Gupta, Z. Liu, R. Romero-Aburto, L. Ge, L. Song, L.B. Alemany, X. Zhan, G. Gao, et al., Graphene quantum dots derived from carbon fibers. Nano Lett. **12**, 844–849 (2012)
66. X.M. Sun, Z. Liu, K. Welsher, J.T. Robinson, A. Goodwin, S. Zaric, H.J. Dai, Nano-graphene oxide for cellular imaging and drug delivery. Nano Res. **1**, 203–212 (2008)
67. J. Ge, M. Lan, B. Zhou, W. Liu, L. Guo, H. Wang, Q. Jia, G. Niu, X. Huang, H. Zhou, et al., A graphene quantum dot photodynamic therapy agent with high singlet oxygen generation. Nat. Commun. **5** (2014)
68. X. Wang, X. Sun, J. Lao, H. He, T. Cheng, M. Wang, S. Wang, F. Huang, Multifunctional graphene quantum dots for simultaneous targeted cellular imaging and drug delivery. Coll. Surf. B Bio-inters. **122**, 638–644 (2014)
69. M.P. Mattson, Apoptosis in neurodegenerative disorders. Nat. Rev. Mol. Cell Biol. **1**, 120–129 (2000)
70. M. Vila, S. Przedborski, Targeting programmed cell death in neurodegenerative diseases. Nat. Rev. Neuroscience. **4**, 365–375 (2003)
71. H. Lecoeur, Nuclear apoptosis detection by flow cytometry: influence of endogenous endonucleases. Exp.Cell Res. **277**, 1–14 (2002)
72. Y. Gavrieli, Y. Sherman, S.A. Ben-Sasson, Identification of programmed cell death in situ via specific label-ingot nuclear DNA fragmentation. J. Cell Biol. **119**, 493–501 (1992)
73. P. Roy, A.P. Periasamy, C.Y. Lin, G.M. Her, W.J. Chiu, C.L. Li, C.L. Shu, C.C. Huang, C.T. Liang, H.T. Chang, Photo-luminescent graphene quantum dots for in vivo imaging of apoptotic cells. Nanoscale **7**, 2504–2510 (2015)
74. Z.M. Liu, Z.Y. Guo, H.Q. Zhong, X.C. Qin, M.M. Wan, B.W. Yang, Graphene oxide based surface-enhanced Raman scattering probes for cancer cell imaging. Phys. Chem. Chem. Phys. **15**, 2961–2966 (2013)
75. B. Kann, H.L. Offerhaus, M. Windbergs, C. Otto, Raman microscopy for cellular investigations—from single cell imaging to drug carrier uptake visualization. Adv. Drug Deliv. Rev. **89**, 71–90 (2015)
76. J.E. Aubin, Auto-fluorescence of viable cultured mammalian cells. J. History-chem. Cystochem. **27**, 36–43 (1979)
77. L. Song, E.J. Hennink, I.T. Young, H.J. Tanke, Photo-bleaching kinetics of fluorescein in quantitative fluorescence microscopy. Biophys. J. **68**, 2588–2600 (1995)
78. C. Zavaleta, A. de la Zerda, Z. Liu, S. Keren, Z. Cheng, M. Schipper, X. Chen, H. Dai, S.S. Gambhir, Noninvasive Raman spectroscopy in living mice for evaluation of tumor targeting with carbon nanotubes. Nano Lett. **8**, 2800–2805 (2008)
79. K. Ock, W.I. Jeon, E.O. Ganbold, M. Kim, J. Park, J.H. Seo, K. Cho, S.W. Joo, S.Y. Lee, Real-time monitoring of glutathione-triggered thiol-purine anticancer drug release in live cells investigated by surface-enhanced Raman scattering. Anal. Chem. **84**, 2172–2178 (2012)
80. E.C. Le Ru, E. Blackie, M. Meyer, P.G. Etchegoin, Surface enhanced Raman scattering enhancement factors: a comprehensive study. J. Phys. Chem. C **111**, 13794–13803 (2007)

81. X. Guo, Z. Guo, Y. Jin, Z. Liu, W. Zhang, D. Huang, Silver-gold core-shell nanoparticles containing methylene blue as seers labels for probing and imaging of live cells. Micro-chem. Acta **178**, 229–236 (2012)
82. S. Yu, X. Cao, M. Yu, Electrochemical immunoassay based on gold nanoparticles and reduced graphene oxide functionalized carbon ionic liquid electrode. Micro-chem. J. **103**, 125–130 (2012)
83. W. Lu, J. Ge, L. Tao, X. Cao, J. Dong, W. Qian, Large-scale synthesis of ultrathin au-pt nanowires assembled on thionine/graphene with high conductivity and sensitivity for electrochemical immune-sensor. Electro-chemistry. Acta **130**, 335–343 (2014)
84. J.M. Han, J. Ma, Z.F. Ma, One-step synthesis of graphene oxide-thionine-au nanocomposites and its application for electrochemical immune-sensing. Biosens. Bio-electron. **47**, 243–247 (2013)
85. S. Samanman, A. Numnuam, W. Limbut, P. Kanatharana, P. Thavarungkul, Highly-sensitive label-free electrochemical carcinoembryonic antigen immune-sensor based on a novel au nanoparticles-graphene-chitosan nanocomposite cryogen electrode. Anal. Chem. Acta **853**, 521–532 (2015)
86. X. Jiang, Y. Chai, R. Yuan, Y. Cao, Y. Chen, H. Wang, X. Gan, An ultrasensitive luminal cathodic electro-chemiluminescence immune-sensor based on glucose oxidase and nanocomposites: Graphene-carbon nanotubes and gold-platinum alloy. Anal. Chem. Acta **783**, 49–55 (2013)
87. S. Kumar, S. Kumar, S. Srivastava, B.K. Yadav, S.H. Lee, J.G. Sharma, D.C. Doval, B.D. Malhotra, Reduced graphene oxide modified smart conducting paper for cancer biosensor. Bio-sens. Bio-electron. **73**, 114–122 (2015)
88. X. Pei, B. Zhang, J. Tang, B. Liu, W. Lai, D. Tang, Sandwich-type immune-sensors and immunoassays exploiting nanostructure labels: a review. Anal. Chem. Acta **758**, 1–18 (2013)
89. J. Huang, J. Tian, Y. Zhao, S. Zhao, Ag/au nanoparticles coated graphene electrochemical sensor for ultrasensitive analysis of carcinoembryonic antigen in clinical immunoassay. Sens. Actuators B-Chem. **206**, 570–576 (2015)
90. B. Jin, P. Wang, H. Mao, B. Hu, H. Zhang, Z. Cheng, Z. Wu, X. Bian, C. Jia, F. Jing, et al., Multi-nanomaterial electrochemical biosensor based on label-free graphene for detecting cancer biomarkers. Bio-sens. Bio-electron. **55**, 464–469 (2014)
91. T. Li, M. Yang, H. Li, Label-free electrochemical detection of cancer marker based on graphene-cobalt hexa-cyanoferrate nanocomposite. J. Electro-Anal. Chem. **655**, 50–55 (2011)
92. H.D. Jang, S.K. Kim, H. Chang, J.-W. Choi, 3D label-free prostate specific antigen (PSA) immune-sensor based on graphene-gold composites. Bio-sens. Bio-electron. **63**, 546–551 (2015)
93. Y. Li, J. Han, R. Chen, X. Ren, Q. Wei, Label electrochemical immune-sensor for prostate-specific antigen based on graphene and silver hybridized mesoporous silica. Anal. Bio-chem. **469**, 76–82 (2015)
94. F. Yang, Z. Yang, Y. Zhuo, Y. Chai, R. Yuan, Ultrasensitive electrochemical immune-sensor for carbohydrate antigen 19-9 using au/porous graphene nanocomposites as platform and au and pd core/shell bimetallic functionalized graphene nanocomposites as signal enhancers. Bio-sens. Bio-electron. **66**, 356–362 (2015)
95. A. Lipatov, A. Varezhnikov, P. Wilson, V. Sysoev, A. Kolmakov, A. Sinitskii, Highly selective gas sensor arrays based on thermally reduced graphene oxide. Nanoscale **5**, 5426–5434 (2013)
96. L. Zhao, Q. Wei, H. Wu, J. Dou, H. Li, Ionic liquid functionalized graphene based immune-sensor for sensitive detection of carbohydrate antigen 15-3 integrated with cd2+−functionalized Nano-porous TiO2 as labels. Bio-sens. Bio-electron. **59**, 75–80 (2014)
97. F. Toledo, G.M. Wahl, Regulating the p 53 pathway: In vitro hypotheses, in vivo VERITAS. Nat. Rev. Cancer **6**, 909–923 (2006)
98. A.M. Bode, Z.G. Dong, Post-translational modification of p 53 in tumorigenesis. Nat. Rev. Cancer **4**, 793–805 (2004)

99. D. Du, L.M. Wang, Y.Y. Shao, J. Wang, M.H. Engelhard, Y.H. Lin, Functionalized graphene oxide as a Nano-carrier in a multi-enzyme labeling amplification strategy for ultrasensitive electrochemical immunoassay of phosphorylated p 53 (s392). Anal. Chem. **83**, 746–752 (2011)
100. Y.Y. Xie, A.G. Chen, D. Du, Y.H. Lin, Graphene-based immune sensor for electrochemical quantification of phosphorylated p 53 (s15). Anal. Chem. Acta **699**, 44–48 (2011)
101. N. Ferrara, H.P. Gerber, J. Le Counter, The biology of verge and its receptors. Nat. Med. **9**, 669–676 (2003)
102. C.-W. Lin, K.-C. Wei, S.-S. Liao, C.-Y. Huang, C.-L. Sun, P.-J. Wu, Y.-J. Lu, H.-W. Yang, C.-C.M. Ma, A reusable magnetic graphene oxide-modified biosensor for vascular endothelial growth factor detection in cancer diagnosis. Bio-sens. Bio-electron. **67**, 431–437 (2015)
103. N. Gan, J.G. Hou, F.T. Hu, Y.T. Cao, T.H. Li, L. Zheng, J. Wang, Sandwich-type electrochemiluminescent immune-sensor based on PDDA-g and lu-au composite for alpha-fetoprotein detection. Int. J. Electro-chem. Sci. **6**, 5146–5160 (2011)
104. Y. Zhuo, M. Zhao, W.J. Qiu, G.F. Gui, Y.Q. Chai, R. Yuan, Supramolecular assembly of beryline derivatives on au functionalized graphene for sensitivity enhancement of electrochemiluminescent immune-sensor. J. Electro-anal. Chem. **709**, 106–110 (2013)
105. D.T. Puerta, J.A. Lewis, S.M.J. Cohen, Am. Chem. Soc. **126**, 8388–8389 (2004)
106. P. Li, B. Zhang, T. Cui, Towards intrinsic graphene biosensor: a label-free, suspended single crystalline graphene sensor for multiplex lung cancer tumor markers detection. Bio-sens. Bio-electron. **72**, 168–174 (2015)
107. A. Ambrosi, M. Pumera, Stacked graphene nanofibers for electrochemical oxidation of DNA bases. Phys. Chem. Chem. Phys. **12**, 8944–8948 (2010)
108. P.A. Rasheed, N. Sand-Haran, Graphene-dna electrochemical sensor for the sensitive detection of brca1 gene. Sens. Actuators B-Chem. **204**, 777–782 (2014)
109. R. Sifri, S. Gang-ad-Harappa, L.S. Acheson, Identifying and testing for hereditary susceptibility to common cancers. Ca-a Cancer J. Clin. **54**, 309–326 (2004)
110. L. Wang, E. Hua, M. Liang, C. Ma, Z. Liu, S. Sheng, M. Liu, G. Xie, W. Feng, Graphene sheets, polyaniline and aunts based DNA sensor for electrochemical determination of BCR/ABL fusion gene with functional hairpin probe. Bio-sens. Bio-electron. **51**, 201–207 (2014)
111. R.E. Champlin, D.W. Golde, Chronic myelogenous leukemia: recent advances. Blood **65**, 1039–1047 (1985)
112. C.X. Guo, X.T. Zheng, Z.S. Lu, X.W. Lou, C.M. Li, Bio-interface by cell growth on layered graphene-artificial peroxidase-protein nanostructure for in situ quantitative molecular detection. Adv. Mater. **22**, 5164–5167 (2010)
113. Y.F. Wu, P. Xue, Y.J. Kang, K.M. Hui, Highly specific and ultrasensitive graphene-enhanced electrochemical detection of low-abundance tumor cells using silica nanoparticles coated with antibody-conjugated quantum dots. Anal. Chem. **85**, 3166–3173 (2013)
114. J.J. Castillo, W.E. Svendsen, N. Rozlosnik, P. Escobar, F. Martineza, J. Castillo-Leon, Detection of cancer cells using a peptide nanotube-folic acid modified graphene electrode. Analyst **138**, 1026–1031 (2013)
115. L.Y. Feng, Y. Chen, J.S. Ren, X.G. Qu, A graphene functionalized electrochemical apt sensor for selective label-free detection of cancer cells. Biomaterials **32**, 2930–2937 (2011)
116. Y.F. Xia, P.Y. Gao, Y. Bo, W.Q. Wang, S.S. Huang, Immunoassay for skov-3 human ovarian carcinoma cells using a graphene oxide-modified electrode. Micro-Chem. Acta **179**, 201–207 (2012)
117. F. Liu, Y. Zhang, J. Yu, S. Wang, S. Ge, X. Song, Application of ZNO/graphene and s6 aptamers for sensitive photo-electrochemical detection of sk-br-3 breast cancer cells based on a disposable indium tin oxide device. Bio-sens. Bio-electron. **51**, 413–420 (2014)
118. M. Yan, G.Q. Sun, F. Liu, J.J. Lu, J.H. Yu, X.R. Song, An apt-sensor for sensitive detection of human breast cancer cells by using porous go/au composites and porous PTFE alloy as effective sensing platform and signal amplification labels. Anal. Chem. Acta **798**, 33–39 (2013)

119. G.F. Jie, Y.B. Zhao, S.Y. Niu, Amplified electro-chemiluminescence detection of cancer cells using a new bifunctional quantum dot as signal probe. Bio-sens. Bio-electron. **50**, 368–372 (2013)
120. D. Rodrigo, O. Limaj, D. Janner, D. Etezadi, F. Javier Garcia de Abajo, V. Pruneri, H. Altug, Mid-infrared plasmatic biological sensing with graphene. Science **349**, 165–168 (2015)
121. G. Niu, X. Chen, Why integrin as a primary target for imaging and therapy. Thera-Notices **1**, 30–47 (2011)
122. Z. Wang, P. Huang, A. Bhirde, A. Jin, Y. Ma, G. Niu, N. Neamati, X.Y. Chen, A nanoscale graphene oxide-peptide biosensor for real-time specific biomarker detection on the cell surface. Chem. Commun. **48**, 9768–9770 (2012)
123. L. Cao, L. Cheng, Z. Zhang, Y. Wang, X. Zhang, H. Chen, B. Liu, S. Zhang, J. Kong, Visual and high-throughput detection of cancer cells using a graphene oxide-based fret apt sensing microfluidic chip. Lab Chip **12**, 4864–4869 (2012)
124. H.L. Zhuang, S.J. Zhen, J. Wang, C.Z. Huang, Sensitive detection of prion protein through long range resonance energy transfer between graphene oxide and molecular aptamer beacon. Anal. Methods **5**, 208–212 (2013)

Chapter 11
Modelling and Optimization of Inertial Sensor-Accelerometer

Zakriya Mohammed, Waqas Amin Gill, and Mahmoud Rasras

11.1 Introduction

Inertial Sensors are the collective term given to accelerometer and gyroscope sensors. The realization of these sensors using MEMS technology had drastically reduced their size and lowered their cost. Starting from pressure sensor in 1960 till today numerous sensors are designed and fabricated using MEMS technology. After pressure sensors, the most commercially successful and mass produced Micromachined sensor is the accelerometer. It is highly anticipated that the gyroscopes will also increase in the mass production at a similar rate. The reason that accelerometers have high sales is driven by their applications to the automotive industry [1, 2]. Accelerometers are used to activate safety systems in vehicles such as airbags, automatic stability system and suspension system. Accelerometer has also gained attention in various other sectors such as biomedical, consumer electronics, heavy industry, oil & gas and defense and aerospace, owing to low cost and small size [3, 4]. In the biomedical field, they are used for activity monitoring. Image stabilization in cameras, virtual reality and orientation detection in smartphones, are some of the applications relating to consumer electronics. Accelerometers are used in condition monitoring systems (CMS) to prevent sudden failure of large mechanical rotating machines. Additionally, highly sensitive accelerometers are used in oil exploration. In defense and aerospace sector, accelerometers are used in vehicle stabilization, missile guidance systems and void detection.

Gyroscopes are used for measuring the angular rate of rotation. These sensors can also be used like accelerometers in applications such as vehicle stabilization in automobiles, image stabilization in digital cameras and smart phones. Other

Z. Mohammed (✉) • W.A. Gill • M. Rasras
Department of Electrical and Computer Engineering, Masdar Institute of Science and Technology, Abu Dhabi, UAE
e-mail: zmohammed1@masdar.ac.ae

T. Li, Z. Liu (eds.), *Outlook and Challenges of Nano Devices, Sensors, and MEMS*, DOI 10.1007/978-3-319-50824-5_11

applications include a wide variety of virtual reality systems, robotics, industrial and military uses. Conventional gyroscopes are very bulky and expensive, making them weak contenders for commercialization. Micromachining can reduce the size as well as cost of the sensors, thus enabling these sensors to be employed in all the above mentioned applications. Miniaturization and large volume production of these sensors on silicon wafer, reduces the cost per device.

The combination of accelerometers and gyroscopes in one platform can be used for accurate position tracking. This combination is generally used for inertial navigation [5]. These two sensors are used in Inertial Measurement Units (IMUs), to track the motion of footslogger, automotive and aerospace vehicles [6, 7]. An IMU consists of 3-axis accelerometer, 3-axis gyroscope and 3-axis magnetometer and provides 9 degrees of freedom. But such complex integration increases the overall cost of the device. For land navigation, the cheapest configuration consists of 2-axis accelerometer and 1-axis gyroscope [8]. This chapter focuses on the dual-axis accelerometer for relatively cheap and cost-effective integration.

The main focus of the work is placed towards the modelling and optimization of a dual axis accelerometer. This book chapter is organized as follows: a brief description of various application of inertial sensors is given, it is followed by discussing different types of accelerometer sensors and their limitations. As the designs sensing is based on capacitive change detection, an overview of working principle of the capacitive accelerometers is given in detail. In the next section, applications and major challenges in designing dual-axis accelerometers are explained. Finally, modelling and simulation of a dual-axis accelerometer is presented. The performance is benched marked to the ones available in literature.

11.2 Types of Accelerometer Sensors

An accelerometer can be modelled as a second order spring mass damper system, see Fig. 11.1. The basic model is shown in Fig. 11.1. It consists of a suspended proof mass having mass ‘m’, anchored to a substrate through beams of spring constant ‘k’. There is also damping of ‘γ’ in the system effecting the dynamics of the mass.

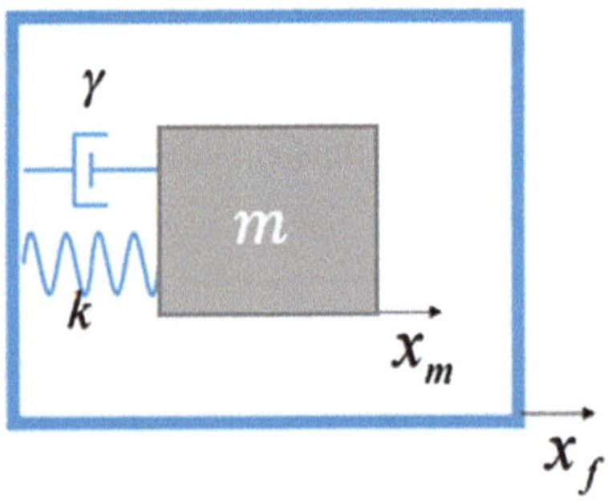

Fig. 11.1 Basic model of accelerometer

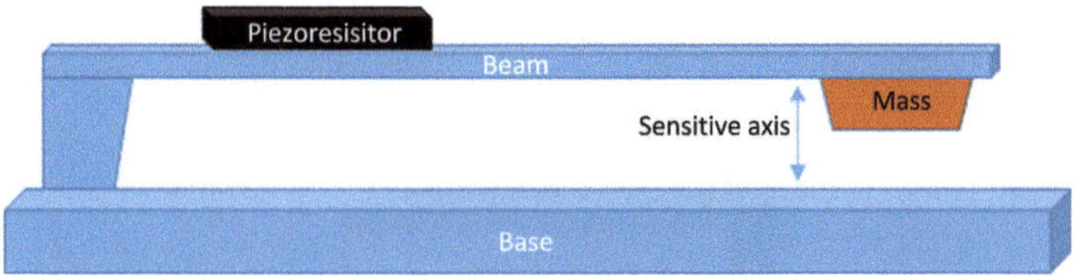

Fig. 11.2 Cross section of a piezo resistive accelerometer

Different transduction mechanism can be used to detect acceleration. External acceleration moves the proof mass and this displacement can be sensed with different transduction mechanism. Depending on the transduction mechanism used, acceleration sensing mechanism can be piezo resistive, tunneling, resonant, thermal, piezoelectric, optical and capacitive.

Piezoresistive accelerometers are widely used and they are the first micro machined accelerometer sensors to be commercialized [9]. They are easy to fabricate and they need relatively simple readout circuits. They operate according to simple principle of change in electrical resistance due to mechanical stress. However, they suffer from poor sensitivity and high temperature dependency compared to capacitive counter parts [10]. In order to sense acceleration, piezo resistive material is embedded in springs, the regions where stress produced is high due to acceleration (see Fig. 11.2). By measuring the relative change in resistance, the acceleration is determined. Piezoresistive accelerometers are fabricated using bulk micromachining techniques [11].

Tunneling accelerometers have been also proposed [12]. They provide high resolution, large bandwidth and high sensitivity. They detect acceleration by measuring tunneling current that occurs between two electrode plates separated by a small distance (in angstroms Å). Due to acceleration, one of the electrode plate moves, changing the tunneling current [13]. By measuring this change, acceleration is detected. Low noise levels are achieved with this type of sensor. However, high fabrication cost and high drift values limit it from commercialization.

Resonant accelerometers directly detect the acceleration. Proof mass has the natural vibration frequency, which changes due to inertial force of the acceleration. The shift in the resonant frequency is used to measure the acceleration [14]. Initial resonant accelerometers were fabricated using quartz material [15]. To achieve high sensitivity and resolution using silicon, many designs uses thick proof mass [16]. Still these devices suffer from low bandwidth.

In thermal accelerometers, a proof mass is analogous to a hot bubble placed between two conductive plates. Under normal condition, the temperature difference between the two plates is fixed. However, in presence of external acceleration, the air bubble moves changing the temperature difference between the electrodes [17].

Piezoelectric accelerometers are also available where stress in the piezoelectric material is activated by the applied acceleration similar to piezoresistive ones [18]. Acceleration causes stress in the piezoelectric material that is embedded in

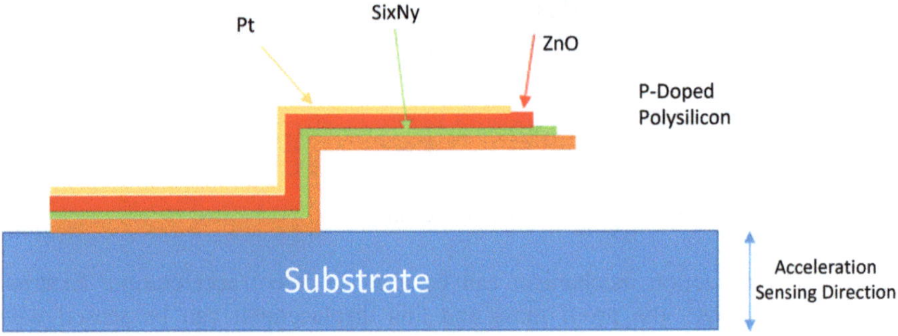

Fig. 11.3 Cross section of a piezoelectric accelerometer

the device which generates electric charges proportional to the applied stress. A cross section of a piezoelectric accelerometer with embedded piezoelectric film (ZnO) is shown in Fig. 11.3 [19]. The piezoelectric accelerometers are active devices and require no power supply. They are extensively used in Condition Monitoring Systems (CMS) to monitor the rotating parts of wind turbine for wear and tear. Low cost fabrication makes them very attractive. The main drawback faced by piezoelectric accelerometers is they do not have a dc response, under constant stress charge always leaks away. Therefore they are un-operational for low frequency. Thus piezoelectric accelerometers cannot be used in many applications.

Optical Accelerometers are high performance accelerometers. They are immune to electromagnetic interference and have high resistance to temperature. They detect acceleration by measuring change in the intensity/or refractive index of the material in which light is propagating through due to applied acceleration [20]. However, the complexity and cost of the fabrication make them suitable for only high end applications.

Compared to all different accelerometers, capacitive accelerometer is most researched and widely used sensor in the MEMS industry [21]. Capacitive accelerometers provide high performance and can be realized in a small foot-print. Moreover, they are produced at relatively low cost and have good thermal stability [22]. Low noise, good DC response, high sensitivity, low power dissipation and low drift are some of the important characteristics of capacitive accelerometer. Their main disadvantage is that they are susceptible to electromagnetic interference. Therefore they need high quality packaging for mechanical as well as for electronic readout circuit.

11.3 Capacitive Accelerometers

The capacitive accelerometers produce capacitance change with applied acceleration. Figures 11.4 and 11.5 show block diagrams of simple lateral capacitive accelerometers. In these structures, capacitance is formed by overlapping of proof

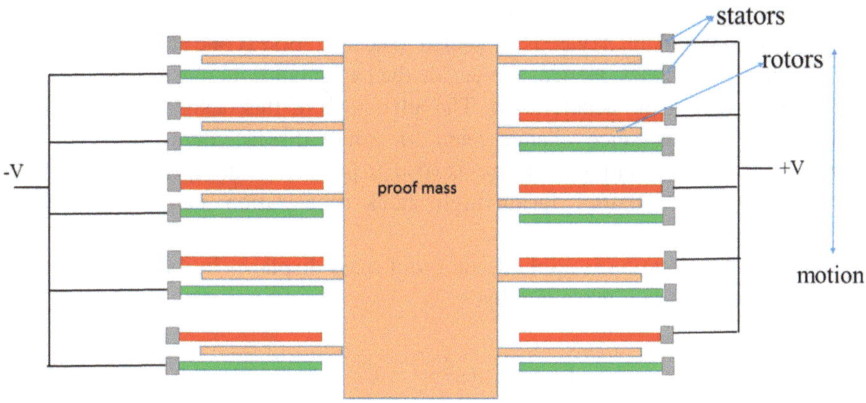

Fig. 11.4 Block and connection diagram of variable gap capacitive accelerometer

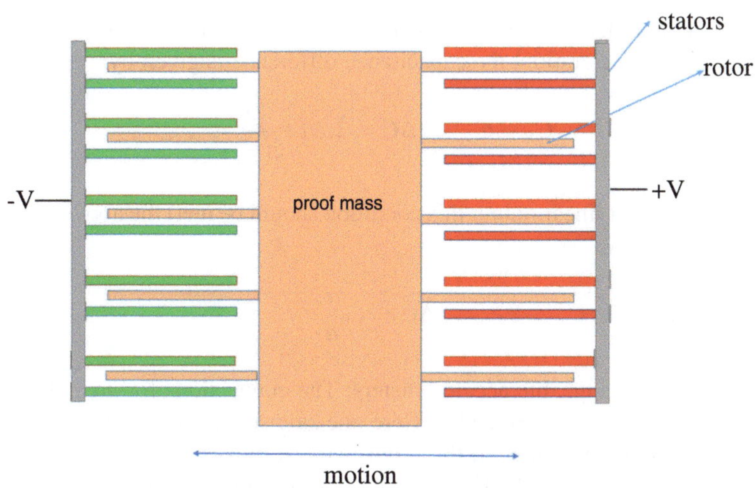

Fig. 11.5 Block and connection diagram of area changeable accelerometer

mass fingers (rotors) and fixed fingers (stators). When the device is subjected to acceleration, the proof mass deflects along with its fingers, causing a capacitance change which is directly proportional to the acceleration [23, 24]. A read out electronic circuit is used to convert this capacitance change into voltage. These accelerometers can be either gap-varying, in which finger gap spacing changes with acceleration or area-varying, where the overlap length changes with the acceleration.

In-plane gap varying accelerometers: the connections for a varying gap capacitive accelerometer is shown in Fig. 11.4. In order to enhance the capacitance change the connection is made in the following manner. All the top stator plates are wired together and all the bottom plates are connected together. There is an isolation between top and bottom plates. This type of connection is essential for distance changeable accelerometer, otherwise capacitance change obtained will not be very large. When the device is subjected to acceleration at one side gap increases and at other it decreases. Hence, at one side the capacitance increases while at other it decreases with same acceleration. The advantage of this design is its ability to produce high capacitance change even for small displacement. However, the response is highly nonlinear. There is also a problem with the rotors sticking permanently to the stators during high shock acceleration, causing permanent failure [25].

When an acceleration is applied, the proof mass displaces by 'x'. The resulting capacitances C_1 and C_2 are given by:

$$C_1 = N\epsilon\, t\frac{L}{d-x} = C_0 + \Delta C \tag{11.1}$$

$$C_2 = N\epsilon\, t\frac{L}{d+x} = C_0 - \Delta C \tag{11.2}$$

If no acceleration is applied, then there is zero displacement and $C_1 - C_2 = 0$. In the presence of acceleration, the capacitance difference is given by:

$$C_1 - C_2 = 2\Delta C = 2\epsilon NLt\frac{x}{g_1^2 - x^2} \tag{11.3}$$

When x ≪ d (for linear output this condition should be met) the above model can be approximated by

$$\Delta C = C_0\frac{x}{d} \tag{11.4}$$

In-plane area changeable accelerometers: The connections for area changeable accelerometer is shown in Fig. 11.5. Here one side stators are connected together. When an acceleration is applied the proof mass, it moves parallel to stator plates causing change in the overall effective area. The advantage of this design is that its output is highly linear. The drawback is, the design has relatively poor sensitivity. The fabrication is simpler as there is no isolation between the top and bottom stators and all the stators can be anchored together.

When an acceleration is applied, the proof mass displaces by 'x'. The resulting capacitances C_1 and C_2 are given by:

$$C_1 = N\epsilon\, t\frac{L+x}{d} = C_0 + \Delta C \tag{11.5}$$

$$C_2 = N\epsilon\, t\frac{L-x}{d} = C_0 - \Delta C \tag{11.6}$$

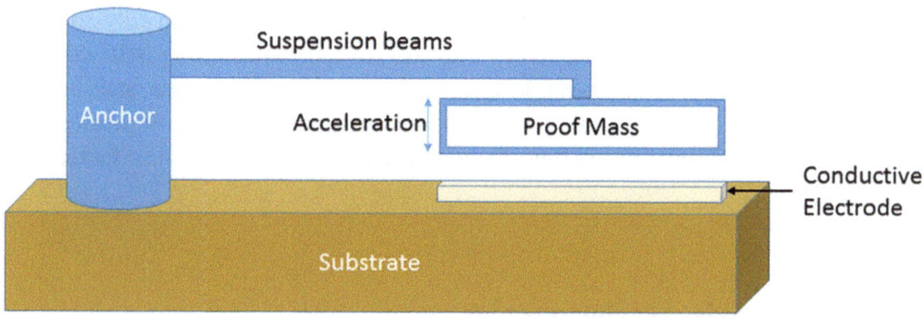

Fig. 11.6 Vertical accelerometer structure for Z-acceleration sensing

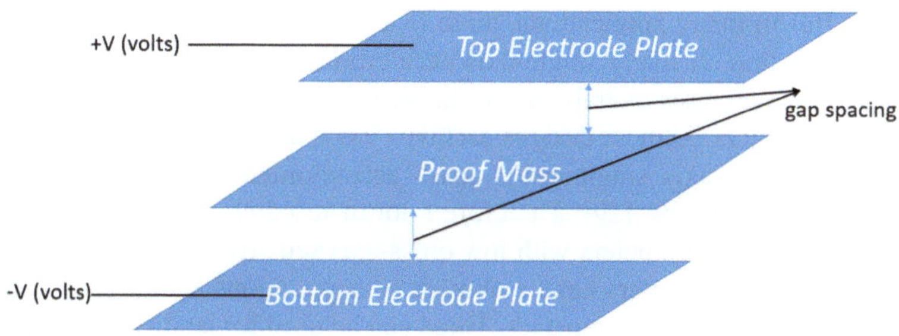

Fig. 11.7 Differential sensing scheme for vertical accelerometer

In the presence of acceleration, the capacitance difference is given by:

$$C_1 - C_2 = 2\Delta C = 2\epsilon Nt\frac{x}{d} \tag{11.7}$$

The differential capacitance output is linear and there is no need of approximation

$$\Delta C \, \alpha \frac{x}{d} \tag{11.8}$$

Z-axis accelerometers: The out-of-plane Z-axis acceleration sensor generally uses the vertical structure shown in Fig. 11.6. In these designs, one large electrode plate beneath the proof mass forms a parallel plate capacitor [26]. Acceleration along Z-direction moves the proof mass resulting in changing the gap spacing between electrode plate and proof mass. If an electrode is placed on top of the proof mass, differential sensing can be achieved similar to lateral accelerometers (Fig. 11.7).

The methods used to fabricate capacitive accelerometers are surface micromachining and bulk micromachining. In surface micromachining, sensor is fabricated layer by layer on top of silicon substrate. While, in bulk micromachining, the sensor is fabricated by etching the substrate. The surface micromachining technology is most commonly used to fabricate accelerometers for industrial and automotive applications [27, 28]. The most successful among them is Analog

Devices' ADXL series. This series uses surface micromachining with polysilicon as the structural layer (proof mass layer). The surface micromachining is IC compatible, devices have small foot print, low cost. However, it suffers noise, stability and flexibility issues. On the other hand bulk micromachining has low noise, high stability and flexibility, but suffers from bulky designs and complex fabrication process.

11.4 Dual-axis Accelerometers

Single axis accelerometers are most extensively used and are simpler to design and fabricate. But in many applications, there is a need to sense acceleration in more than one direction. If many single-axis accelerometers oriented orthogonally are used, then this increases overall size of the sensor. The only advantage with single axis accelerometers is, most designs archives cross axis sensitivity less than 1%. However multiple axis single proof mass accelerometers still have cross axis sensitivity more than 2% [29]. Therefore a lot of research is focused on compact size Multi-axis accelerometers with low cross-axis sensitivity.

Dual-axis accelerometers have a number of applications and are cost effective compared to tri-axial accelerometers. These can be used for land navigation, where acceleration is confined to only two directions. The use of 3-axis accelerometer for land navigation increases the cost and complexity [30]. Apart from inertial sensing, dual axis accelerometers have many specific applications such as enhancing the gaming experience in game pad controller, and enhancing user experience in personal digital assistant (PDA). They can also be used in anti-theft devices, robotic and automotive electronic devices [31].

In Multi-axis accelerometers that use single proof mass, the proof mass along with rotors have multiple degree of freedom. This causes cross-axis sensitivity between different outputs: undesired response in sensing axis, while acceleration is applied on different axis. Moreover, sensing modal frequency also gets coupled with cross-axis mode [32]. To minimize cross coupling, different mechanical designs are proposed in the literature, leading to complex fabrication process, increasing the overall cost of the sensor [33]. Another study reports very low cross-axis coupling, utilizing many proof masses. However, it increased the overall die area [34].

11.4.1 Modelling of a Dual-Axis Accelerometer

The proposed accelerometer consist of a single square shaped proof mass anchored to the substrate by four crab-leg springs, as shown in Fig. 11.8. The proof mass fingers (rotors) are placed on all four sides of the square proof mass. The fixed fingers (stators) are placed strategically forming two gap spacing (gap and anti-gap) [35]. The device is

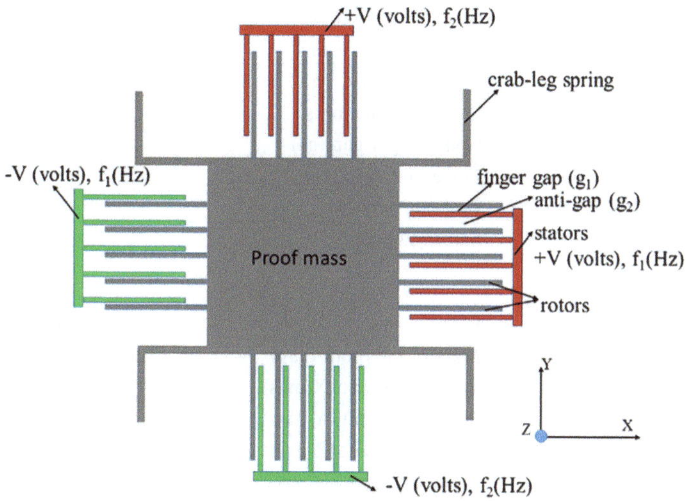

Fig. 11.8 Block diagram of proposed design

protected from shocks by adding extra mask in the regions near the springs with a spacing slightly less than the finger gap (g1). These blocks act as mechanical stoppers, preventing damage of the device. The aspect ratio of the springs are determined by simulation, such that it maximizes the sensing displacement and minimizes the cross-axis one. To make the sense capacitance and sensitivity identical for both the axes, the number of fingers are set equal on all the sides.

In the absence of acceleration the gap spacing on both the sides remain the same. When the device is subjected to acceleration, the proof mass displaces by a small amount. Due to this displacement, the finger gap spacing will vary. At one side it increases and at other it decreases. Acceleration is detected by measuring the difference between these two capacitances.

The design of this accelerometer followed design rules of GlobalFoundries Inertial Measurement Unit (IMU) platform. The device layer is made up of low resistance single crystalline silicon. The thickness of the structural layer is 30 μm, which increases the capacitive area, and hence enhances sensitivity. The finger gap (g1) is 1.25 μm. This space plays extremely important role in the sensitivity. Designers tend to use smallest gap possible, to increase capacitance, and hence the sensitivity. But this space is process dependent, and depends on fabrication platform. Experimental results confirm that the utilized process achieves the minimum gap spacing of 1.25 μm. In order to enhance the change in capacitance, finger anti-gap spacing (g2) is determined by simulation, which depends upon spring constant and comb length. It has been found that, at a critical value of ‘g2’, the sensitivity is high. If the selected value is above or below that critical value, sensitivity drops [36]. The critical value for this design is found to be 3.75 μm. The complete device parameters are summarized in Table 11.1.

Table 11.1 Summary of design parameters

Design parameters	Values (μm)
Overall sensor size	1500 × 1500
Proof mass size	1040 × 1040
Structural thickness	30
Comb finger gap (g_1)	1.25
Finger anti-gap (g_2)	3.75
Crab-leg springs	220 × 4
Sensing fingers	150 × 5

Table. 11.2 Summary of performance parameters

Parameters	Values
Mass of the proof mass	9.0778×10^{-8} kg
Spring constant (X-axis)	60 N/m
Spring constant (Y-axis)	63 N/m
Spring constant (Z-axis)	704 N/m
Sense capacitance/axis	8.7 pF

The proof mass moves only in the X and Y directions. When the sensor experiences acceleration in the Y-axis, the lateral displacement causes capacitance change in right and the left side combs (Y-electrodes). But due to inherent coupling nature of this device it also causes capacitance change in top and bottom combs (X-electrodes). When the device is under acceleration in the X-axis, top and bottom combs will endure the capacitance change and causes cross-axis sensitivity in Y-electrodes. The device is modelled in such a fashion to reduce the cross-axis sensitivities between X and Y directions.

There are various proposed readout circuits in the literature which can be used to detect capacitance change in the 2-axis accelerometer [37]. As the design is made symmetric in the X and Y direction, a frequency modulation scheme is used to separate the signals. Carrier signals with different frequencies f1 and f2 are used to modulate the complementary set of electrodes. This way detection of the simultaneous accelerations, in both in-plane directions can be done. The output signal is later converted into voltage by trans-impedance amplifier. Finally these signal are demodulated and filtered by a low pass filter.

As the sensing is done differentially, the device is not affected by the Z-axis acceleration. But poor stiffness in the Z-direction, causes some rotation, which leads to cross-axis capacitance change, even when differential sensing is employed. Low spring constant in Z also causes excessive sag of the proof mass. Also, during shock conditions, the excessive displacement along Z-axis causes the proof-mass to touch top cap, leading to permanent failure of the accelerometer. All these problems can be solved by improving the stiffness in Z-direction. If we make the springs stiffer in Z-axis, then their stiffness also increases in X- and Y-directions. Therefore, the present design improves the Z-direction stiffness with minimum reduction in sensitivity. The performance parameters are summarized in Table 11.2.

11.4.2 Simulation Results–Accelerometer

The design is modelled in MEMS+ and detailed analysis was done using Finite Element Analysis (FEA) tool CoventorWare. The modal analysis indicates that, the first two vibration modes are in X and Y directions with frequencies of 4.0 and 4.2 kHz, respectively. The third mode is in Z-direction recording a frequency of 14.0 kHz. Figure 11.9 shows the exaggerated view of various mode shapes.

The displacement sensitivity in the cross-axis Z-direction is 1.2644 nm/g. Therefore during shock conditions, where acceleration is close to 1000 g, the displacement is only 1.2644 µm/g.

Thus, the device is well protected against shocks in Z-directions. The displacement sensitivity in the Y-direction is calculated to be 0.013 µm/g. Figure 11.10 shows the displacement of proof mass for 1 g of acceleration (Y-axis). The open-

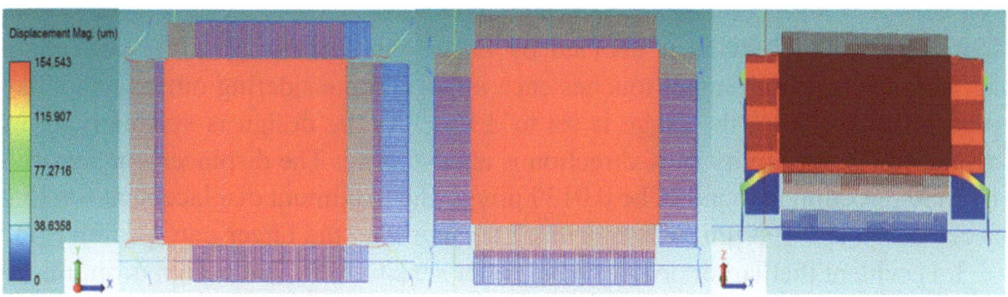

Fig. 11.9 Mode shapes X-direction (4.0 kHz), Y-direction (4.2 kHz) and Z-direction (14.0 kHz) [left to right]

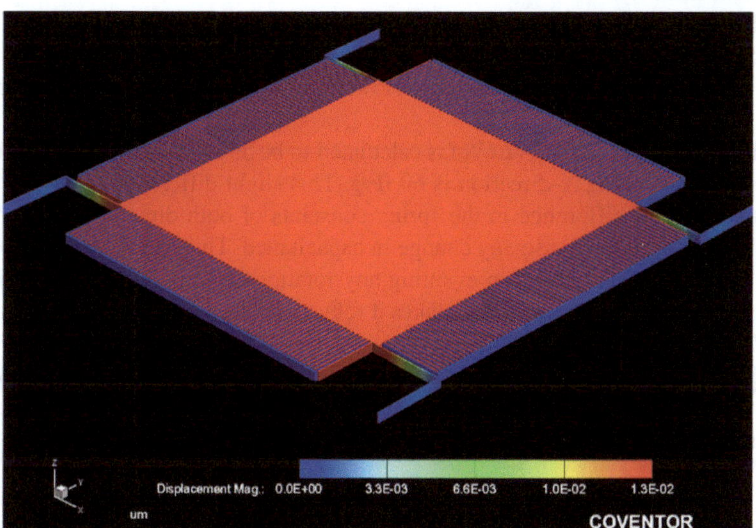

Fig. 11.10 Displacement of proof mass for 1 g of acceleration (Y-direction)

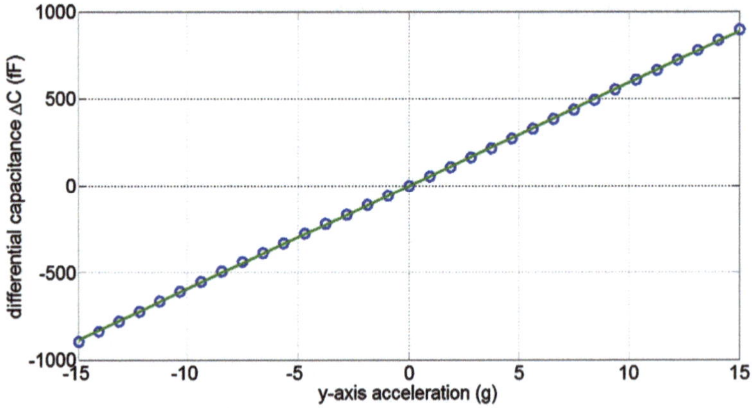

Fig. 11.11 Sensitivity plot of accelerometer (Y-direction)

loop range of the device is decided by finger gap spacing (g1). The maximum acceleration at which finger touches each other. But considering other factors like pull in and linearity, the range is set to ±15 g. As the design is symmetrical, its displacement sensitivity in X-direction is almost same. The displacement sensitivity in X-direction is found to be 0.0139 μm/g. The maximum displacement for 15 g acceleration is 0.195 μm, which is much smaller than the finger gap spacing (g1).

It is evident that larger devices have better sensitivities. The design is optimized to give higher sensitivity in the small overall dimension of 1.5 mm^2. The relation between the input acceleration and the resulting differential capacitance is shown in Fig. 11.11. It is observed that, within the operating range, the device exhibits excellent linearity.

This linearity is obtained because the variation in gap of the fingers remain small compared to nominal finger gap spacing of 1.5 μm. Furthermore, non-linearity is very low under operating conditions. However, its starts to increase significantly after 15 g acceleration.

The in-plane sensitivity ($\Delta C/g$) is calculated to be 59 fF/g along the Y-direction. The sensitivity along X-direction is 60 fF/g. This slight difference in sensitivity is due to the slight difference in the spring constants of both directions. The Z-axis acceleration does not cause any change in capacitance. This happens as stiffness in the Z-direction is very high, preventing any rotation motion. Additionally, differential sensing is employed which makes it robust. Thus cross-axis sensitivity due to the Z-direction acceleration is zero.

The capacitance in the output of X-electrodes due to acceleration along Y-axis is shown in Fig. 11.12. The output error due to off-axis input is very low. The cross-axis sensitivity effect is calculated to be 1.33%.

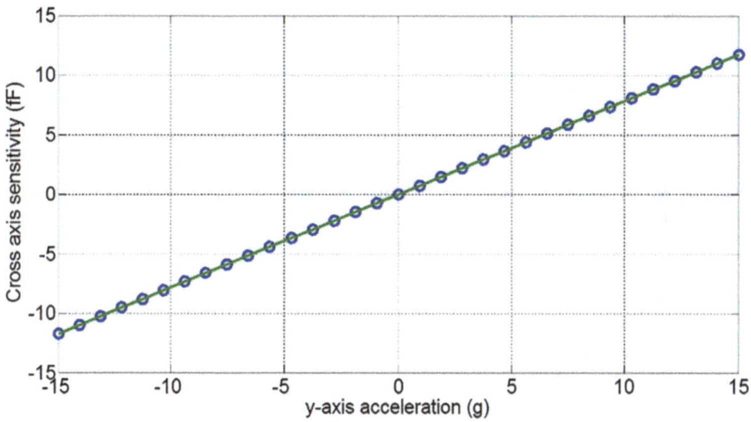

Fig. 11.12 Cross-axis sensitivity plot of accelerometer

11.5 Conclusion

In this chapter, after providing a brief overview of the applications, operation principle and limitations of MEMS accelerometers, the chapter presents the modelling and simulation details of a dual axis accelerometer with low cross axis sensitivity. The design uses crab-leg spring suspension. The sensor is designed on a foot print of 1.5 mm^2. As the design employs a single proof mass, its die size can be reduced while achieving high sensitivity. In the target input range of ± 15 g, the capacitance change is almost linear. Furthermore, the sensor is designed with modal frequency separation of 10 kHz between in-plane and out-of-plane. The displacement sensitivity in X and Y is 0.013 μm/g. The Z-axis displacement sensitivity is 0.00126 μm/g. The differential capacitance sensitivity is calculated to be 59 fF/g. The average XY cross-axis sensitivity is 1.33%. There is no cross-axis sensitivity due to Z-direction acceleration. The device was designed using GlobalFoundries Inertial Measurement Unit platform.

Acknowledgement This work was funded by Mubadala Development Company–Abu Dhabi, Economic Development Board–Singapore and GLOBALFOUNDRIES–Singapore under the framework of 'Twinlab' project with participation of A*STAR Institute of Microelectronics–Singapore (IME), Masdar Institute of Science and Technology–Abu Dhabi and GLOBALFOUNDRIES–Singapore. Further, we would like to acknowledge Mr. Aveek Chatterjee of GlobalFoundries for design evaluations, and IME's Dr. Peter Kee and Dr. Ilker Ocak for all the support they have provided us with standard cells, design reviews, and process information.

References

1. I. Frosio, F. Pedersini, N.A. Borghese, Autocalibration of MEMS Accelerometers. IEEE Trans. Instrum. Meas. **58**(6), 2034–2041 (2009)
2. M. Ferraresi, S. Pozzi, in *Advanced Microsystems for Automotive Applications.* MEMS Sensors for Non-Safety Automotive (Springer, Berlin Heidelberg, 2009), pp. 355–367
3. S. Luczak, W. Oleksiuk, M. Bodnicki, Sensing tilt with MEMS accelerometers. IEEE Sensors J. **6**(6), 1669–1675 (2006)
4. R. Perez, Ú. Costa, M. Torrent, J. Solana, E. Opisso, C. Caceres, J.M. Tormos, J. Medina, E.J. Gómez, Upper limb portable motion analysis system based on inertial technology for neurorehabilitation purposes. Sensors **10**(12), 10733–10751 (2010)
5. J. Söderkvist, Micromachined gyroscopes. Sens. Actuators, A **43**(1–3), 65–71 (1994)
6. C.-W. Tan, S. Park, Design of accelerometer-based inertial navigation systems. IEEE Trans. Instrum. Meas. **54**(6), 2520–2530 (2005)
7. Z.F. Syed, P. Aggarwal, C. Goodall, X. Niu, N. El-Sheimy, A new multi-position calibration method for MEMS inertial navigation systems. Meas. Sci. Technol. **18**(7), 1897–1907 (2007)
8. P. Aggarwal, Z. Syed, A. Noureldin, N. El-Sheimy, *MEMS-Based Integrated Navigation (GNSS Technology and Applications)* (Artech House, Norwood, Massachusetts 2010)
9. L.M. Roylance, J.B. Angell, A batch-fabricated silicon accelerometer. IEEE Trans. Electron Devices **26**(12), 1911–1917 (1979)
10. N. Yazdi, F. Ayazi, K. Najafi, Micromachined inertial sensors. Proc. IEEE **86**(8), 1640–1659 (1998)
11. H.V. Allen, S.C. Terry, D.W. Bruin, Accelerometer systems with self-testable features. Sens. Actuators **20**(1–2), 153–161 (1989)
12. C. Yeh, K. Najafi, A low-voltage tunneling-based silicon microaccelerometer. IEEE Trans. Electron Devices **44**(11), 1875–1882 (1997)
13. C.-H. Liu, T.W. Kenny, A high-precision, wide-bandwidth micromachined tunneling accelerometer. J. Microelectromech. Syst. **10**(3), 425–433 (2001)
14. A.A. Seshia, M. Palaniapan, T.A. Roessing, R.T. Howe, R.W. Gooch, T.R. Schimert, S. Montague, A vacuum packaged surface micromachined resonant accelerometer. J. Microelectromech. Syst. **11**(6), 784–793 (2002)
15. J. Danel, F. Michel, G. Delapierre, Micromachining of quartz and its application to an acceleration sensor. Sens. Actuators, A **23**(1–3), 971–977 (1990)
16. E.W. Bums, R.D. Homing, W.R. Herb, J.D. Zook, H.Guckel, Resonant microibeam accelerometers. The 8th International Conference on Solid-State Sensors and Actuators, Stockholm, Sweden, 1995
17. A.M. Leung, J. Jones, E. Czyzewska, J. Chen, B. Woods, Micromachined accelerometer based on convection heat transfer. The Eleventh Annual International Workshop on Micro Electro Mechanical Systems, Heidelberg, 1998
18. P.-L. Chen, R.S. Muller, R.D. Jolly, G.L. Halac, R.M. White, A.P. Andrews, T.C. Lim, M.E. Motamedi, Integrated silicon microbeam PI-FET accelerometer. IEEE Trans. Electron Devices **29**(1), 27–33 (1982)
19. D.L. DeVoe, A.P. Pisano, Surface micromachined piezoelectric accelerometers (PiXLs). J. Microelectromech. Syst. **10**(2), 180–186 (2001)
20. A. Llobera, V. Seidemann, J.A. Plaza, V.J. Cadarso, S. Buttgenbach, SU-8 optical accelerometers. J. Microelectromech. Syst. **16**(1), 111–121 (2007)
21. T. Tsuchiya, H. Funabashi, A z-axis differential capacitive SOI accelerometer with vertical comb electrodes. Sens. Actuators, A **116**(3), 378–383 (2004)
22. B.E. Boser, Electronics for micromachined inertial sensors. International Conference on Solid State Sensors and Actuators, Chicago, 1997
23. H. Luo, G.K. Fedder, L.R. Carley, A 1 mG lateral CMOS-MEMS accelerometer. The Thirteenth Annual International Conference on Micro Electro Mechanical Systems, Miyazaki, 2000

24. J. Chae, H. Kulah, K. Najafi, A hybrid silicon-on-glass (SOG) lateral micro-accelerometer with CMOS readout circuitry. The Fifteenth IEEE International Conference on Micro Electro Mechanical Systems, Las Vegas, NV, USA, 2002
25. I.Y. Park, C.W. Lee, H.S. Jang, Y.S. Oh, B.J. Ha, Capacitive sensing type surface micromachined silicon accelerometer with a stiffness tuning capability. The Eleventh Annual International Workshop on Micro Electro Mechanical Systems, Heidelberg, 1998
26. Y. Matsumoto, M. Iwakiri, H. Tanaka, M. Ishida, T. Nakamura, A capacitive accelerometer using SDB-SOI structure. Sens. Actuators, A **53**(1–3), 267–272 (1996)
27. H. Xie, G.K. Fedder, A CMOS z-axis capacitive accelerometer with comb-finger sensing. The Thirteenth Annual International Conference on Micro Electro Mechanical Systems, Miyazaki, 2000
28. K.-Y. Park, C.-W. Leeb, H.-S. Janga, Y. Ohc, B. Hac, Capacitive type surface-micromachined silicon accelerometer with stiffness tuning capability. Sens. Actuators, A **73**(1–2), 109–116 (1999)
29. Y.-W. Hsu, J.-Y. Chen, H.-T. Chien, S. Chen, S.-T. Lin, L.-P. Liao, New capacitive low-g triaxial accelerometer with low cross-axis sensitivity. J. Micromech. Microeng. **20**(5), 1–10 (2010)
30. A. Alfaifi, F. Nabki, M.N. El-Gamal, A dual-axis bulk micromachined accelerometer with low cross-sensitivity. 19th IEEE International Conference on Electronics, Circuits and Systems (ICECS), Seville, 2012
31. A. Baschirotto, A. Gola, E. Chiesa, E. Lasalandra, F. Pasolini, M. Tronconi, T. Ungaretti, A ±1-g dual-axis linear accelerometer in a standard 0.5-μm CMOS technology for high-sensitivity applications. IEEE J. Solid State Circuits **38**(7), 1292–1297 (2003)
32. Z. Mohammed, G. Dushaq, A. Chatterjee, M. Rasras, Bi-axial highly sensitive ±5 g polysilicon based differential capacitive accelerometer. 17th IEEE International Conference on Thermal, Mechanical and Multi-Physics Simulation and Experiments in Microelectronics and Microsystems, Montpellier, France, 2016
33. C.-M. Sun, M.-H. Tsai, Y.-C. Liu, W. Fang, Implementation of a monolithic single proof-mass tri-axis accelerometer using CMOS-MEMS technique. IEEE Trans. Electron Devices **57**(7), 1670–1679 (2010)
34. J. M. Darmanin, I. Grech, J. Micallef, E. Gatt, O. Casha, A. Briffa, Design consideration for three-axis MEMS accelerometers using an asymmetric proof mass. IEEE EUROCON, Zagreb, 2013
35. Z. Mohammed, O.T. Waheed, I. Elfadel, A. Chatterjee, M. Rasras, Design, analysis and system level modelling of single axis MEMS capacitive accelerometer. ASME 2016 International Mechanical Engineering Congress and Exposition, Phoenix, Arizona, USA, 2016
36. Z. Mohammed, M. Rasras, Optimization of finger spacing and spring constant in comb type capacitive accelerometer. 7th IEEE International Nanoelectronics Conference (INEC), Chengdu, China, 2016
37. Y. Terzioglu, S.E. Alper, K. Azgin, T. Akin, A capacitive MEMS accelerometer readout with concurrent detection and feedback using discrete components. IEEE/ION Position, Location and Navigation Symposium–PLANS 2014, Monterey, CA, 2014

Chapter 12
Graphene for Future High-Performance Gas Sensing

Jin Wu, Kai Tao, Jianmin Miao, and Leslie K. Norford

12.1 Introduction

With the development of industry and agriculture, an increasing amount of toxic, volatile and flammable gases and chemicals has been generated [1]. Among them, NOx, SOx, COx and ozone (O_3) are the most important and harmful air pollutants that threaten people's health [1, 2]. To monitor these gaseous chemicals, reliable gas sensors that enable the detection of trace amounts of chemicals are highly demanded. A number of materials have been employed as the sensing materials, including semiconducting metal oxide, conductive polymers, optical fibers and carbon-based materials [3]. Metal oxides based gas sensors display large response to various gases. However, the requirement of external heater to maintain the high operation temperature leads to high energy consumption and thermal safety problems [4]. In comparison, graphene (Gr) gains the advantage of high sensitivity, low limit of detection (LOD) and facile operation at low temperature. The high performance gas sensing achieved by Gr is mainly attributed to the unique structures and outstanding electronic properties, such as the atom-thick two-dimensional structures, high surface to volume ratio, small size, high electronic mobility and high sensitivity to perturbations from gas molecules [3]. As each atom on the single-layer Gr can be regarded as the surface atom, the surface to volume ratio of Gr reaches the largest. The large surface

J. Wu • K. Tao • J. Miao (✉)
School of Mechanical and Aerospace Engineering, Nanyang Technological University, Singapore 639798, Singapore
e-mail: mjmmiao@ntu.edu.sg

L.K. Norford
Center for Environmental Sensing and Modeling (CENSAM), Singapore-MIT Alliance for Research and Technology (SMART) Centre, Singapore 117543, Singapore

Department of Architecture, Massachusetts Institute of Technology, Cambridge, MA 02139, USA

T. Li, Z. Liu (eds.), *Outlook and Challenges of Nano Devices, Sensors, and MEMS*,
DOI 10.1007/978-3-319-50824-5_12

areas of Gr facilitate the adsorption of gas molecules, raising the sensitivity. Furthermore, Gr displays the extremely high carrier mobility of $2 \times 105\ cm^2\ V^{-1}\ s^{-1}$ and the high carrier density of $1012\ cm^{-2}$ at room temperature, making Gr have extremely low resistivity (10–6 Ω) [3]. Since Gr has very low electronic noise, the adsorption of a trace amount of gas molecules on Gr surface induces a noticeable resistance variation of Gr, making Gr even sensitive to a single gas molecule [3, 5]. In this chapter, we will systematically discuss Gr-based gas sensors from the fundamental principles to high-performance gas sensing applications.

12.2 Principles of Gas Sensors Based on Gr

The principle of Gr-based sensors is the resistance or conductance change of Gr upon exposure to test gases. The interaction between gas molecules and Gr can be weak van der Waals forces, electrostatic force, hydrogen bond or strong covalent bonding [3]. All these interactions cause charge transfer between Gr and gas molecules, leading to a resistance/conductance change [3]. Different gas molecules interact with Gr in different fashions. Oxidizing gases such as NO_2 and CO_2 withdraw electrons from the Gr upon adsorption [3, 6]. As the Gr behaves as a p-type semiconducting behavior at room temperature, the electron transfer from Gr to NO_2 increases the carrier (hole) concentration in Gr, reducing its resistance [3, 6]. In contrast, NH_3 is an electron-donating molecule; the adsorption of NH_3 on Gr surface leads to decreased (hole) concentration in Gr, increasing its resistance [3]. Different from NO_2 and NH_3 in causing charge transfer between Gr and gas molecules, H_2O does not give rise to detectable localized impurity states, but redistributes electrons within the Gr sheets or between substrate and the Gr sheets.

12.3 Configurations of Gr-based Gas Sensors

The most widely utilized configurations of Gr-based sensors are chemiresistor and field effect transistors (FETs). The chemiresistor has been widely employed as the platform for gas sensing based on various materials due to its simple fabrication and facile measurement. Normally, interdigital electrodes (IEs) consisting of conductive metal lines are fabricated by micromachine technologies, nanofabrication or shadow mask methods. The Gr sheets can be deposited on the IEs by various methods such as drop casting, dip coating, printing or spin coating. After evaporation of solvent, Gr sheets bridge the gaps on the IEs and thus can be utilized for gas sensing. Figure 12.1a illustrates the configuration of gas sensor based on a two-point interdigitated chemiresistor, which is contributed by our group [2]. The IEs are bridged by three-dimensional (3D) reduced graphene oxide hydrogel (RGOH). Compared with 2D counterparts, the enlarged surface areas of 3D RGOH considerably increase the sensitivity of gas sensing. Importantly, a

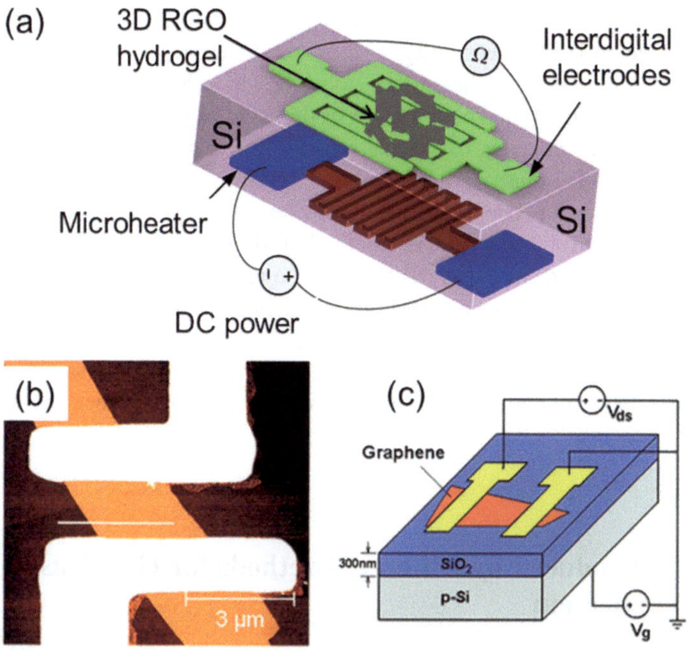

Fig. 12.1 (**a**) Scheme illustrating a 3D RGOH-based chemiresistor with an imbedded microheater. 3D RGOH bridged the gaps on the IEs on the top side of the Si/SiO_2 substrate, while a microheater was fabricated on the bottom side [2]. Reproduced with permission from [2]. Copyright American Chemical Society. (**b**) AFM image of a Gr-based FET gas sensing device. A Gr sheet bridges the gap between IEs [7]. (**c**) Schematic illustrating Gr-based gas sensor in (**b**) with an FET platform. Reproduced with permission from [7]. Copyright American Chemical Society

Pt-based microheater is fabricated on the bottom side of sensor to raise the substrate temperature. It is found that the elevated temperature can improve both the selectivity and recovery of NO_2 sensing.

FETs have attracted wide attention in chemical sensors due to their high sensitivity [8, 9]. In this configuration, the gate voltage can be used to adjust the source-drain current [3]. The presence and concentration of a gas can be determined by monitoring the change of source-drain current. Usually, a larger on/off ratio of source-drain current leads to a higher sensitivity [7]. Hence, many efforts have been made to increase the on/off ratio of FETs, for example, preparing Gr nanoribbon and tiny Gr sheets by nanopatterning or other methods [7]. Figure 12.1b, c displays the structure of a FET-based Gr sensor, in which a Gr sheet bridges the source and drain electrodes and serves as the conducting channel. The local charge carrier concentration changes upon the adsorption of NH_3 molecules on Gr surface, leading to the changes of source-drain current. It was found that a polymer resist residue left by nanolithography on Gr surface can affect the sensitivity [7]. The contamination layer chemically dopes the Gr and acts as an adsorption layer to concentration gas molecules, increasing the sensitivity.

Other strategies such as the surface acoustic wave (SAW) are also utilized for Gr-based gas sensors [10]. The principle of this kind of sensor is based on conductance or resistance changes of the Gr sensing layers and the frequency changes brought about by the mass variations upon exposure to test gases. For example, a SAW-based sensor has been reported to display the response of 1.7 or 7.0 Hz to 10,000 ppm H_2 or 1000 ppm CO, respectively [10]. In addition to SAW, transistors based on surface work function (SWF) are also employed for gas sensors [11]. In this method, as Gr displays a p-type semiconducting property, the adsorption of gas molecules changes the electron affinity and dipole moment, leading to the change of SWF on Gr. SWF-based gas sensors have the advantage of rapid response (less than 10 s), which is mainly attributed to the change of the surface electronic properties of Gr.

12.4 Gr Produced by Different Methods for Gas Sensing Application

Since Novoselov et al. reported the detection of single NO_2 molecules using mechanically exfoliated Gr in 2007 [5], Gr-based materials prepared by different methodologies, including mechanical exfoliation [5], chemically [6] or thermal reduction [12], chemical vapor deposition (CVD) [13], and epitaxial growth [14, 15], have been exploited for gas sensing applications [1, 3, 16]. Mechanical exfoliation can be employed to separate single-layer or multiple-layer Gr sheets from bulk graphite with near-perfect and defect-free crystalline structure [3]. Based on Gr obtained by mechanical exfoliation, Novoselov et al. utilized the Hall geometry to achieve the highest response to the charge carrier density change near the Dirac point of Gr [3]. Figure 12.2a reveals that the charge carrier concentration in single-layer Gr increased linearly with the concentration of NO_2 gas. Figure 12.2b shows the step-like Hall resistivity changes upon exposure of the sensor to extremely diluted 1 ppm NO_2 gas. The change of Hall resistivity is caused by adding or reducing one electron charge, which is marked by the grid lines [3]. After analyzing the data, the authors claim that the sensor could detect single NO_2 molecules. The high sensitivity is attributed to the exceptionally low electronic noise of Gr material electronically [5]. Following the pioneering work described above, the pristine mechanically exfoliated Gr has been employed to detect various gases e.g. NO_2, NH_3, CO_2 and so on, both experimentally and theoretically [3]. Although mechanical exfoliation can be employed to fabricate Gr with high quality and minimal defects, the low yield of this methodology prevents its mass-production for gas sensor applications.

The low-yield shortcoming of mechanical exfoliation can be overcome by chemical vapor deposition (CVD) synthesis of Gr and chemical production of

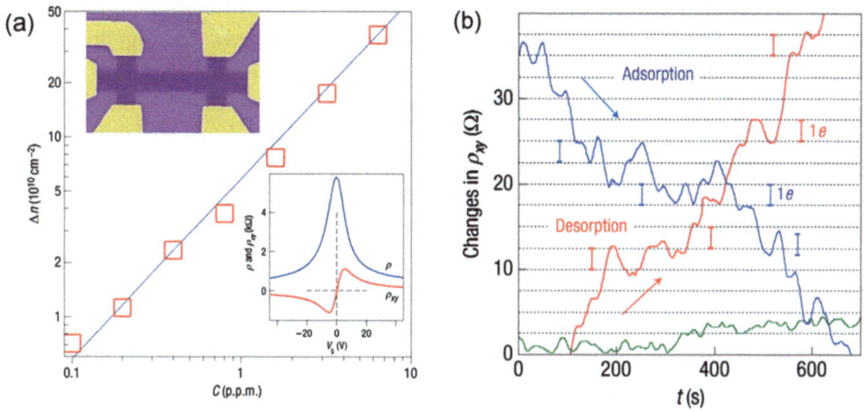

Fig. 12.2 (**a**) The concentration change of charge carriers in single-layer Gr with NO_2 concentration. *Upper inset*: Scanning electron micrograph (SEM) image of this sensor. *Lower inset*: Electric-field effect is utilized to characterize the Gr device [5]. (**b**) Hall resistivity changes observed near the neutrality point during the adsorption of highly diluted NO_2 (*blue curve*) and the desorption in vacuum at 50 °C (*red line*) [5]. The *green curve* is a reference when the sensor is exposed to pure He. Reproduced with permission from [5]. Copyright Nature Publishing Group (2007)

reduced graphene oxide (RGO). The CVD technique is utilized to synthesize Gr on various metal surfaces e.g. Ni, Cu, Co. The hydrocarbons are decomposed at moderate temperatures (<1000 °C) [3, 12]. The CVD synthesized Gr could be transferred to other substrates to fabricate gas sensors. The CVD-synthesized Gr has been utilized to detect O_2, NO_2 and NH_3 [13, 14]. O_2 and NO_2 behave as p-type dopants to withdraw electrons from Gr materials upon adsorption, while NH_3 acts as an n-type dopant to donate electrons to Gr. Plasma-enhanced CVD has been employed to fabricate patterned Gr on the surface of gold or SiO_2 surfaces for NO_2 and NH_3 sensing [13]. The sensor displayed increased voltage upon exposure to 1% NH_3, while exhibited reduced voltage to 100 ppm NO_2 [13]. It is found that the sensitivity is influenced by both temperature and humidity. For example, the highest response is achieved at temperatures ranging from 150–200 °C. The adsorption of water molecules leads to the decrease of resistance of Gr. Thus, Gr or graphene oxide can also be utilized for humidity sensing [15, 16].

Among various methods of Gr synthesis, chemically derived RGO has attracted widespread attention for gas sensing due to the low-cost and high-yield production of RGO. RGO is usually reduced from graphene oxide (GO), which can be synthesized by oxidizing graphite using $KMnO_4$ in acidic environment [17]. In comparison with pristine Gr, GO is heavily functionalized with oxygenated groups, such as carboxyl groups at sheet edges and the hydroxyl and epoxy groups on at the basal plane [3]. These oxygenated groups can act as the gas adsorption sites. Nevertheless, these oxygen-rich functional groups make the conductivity of GO very low, which is not suitable for gas sensing application [3]. Chemical or thermal treatment of GO can form RGO, leading to increased conductivity. Although most of the oxygenated groups have been removed, some oxygen functional groups still existed on RGO, which can be exploited to tailor the gas sensing properties. Furthermore, the reduction can introduce some structural vacancies and defects, which can act as the active sites for gas adsorption [18]. It was demonstrated that the defects adsorption play the dominant role in changing the resistance of defective Gr [3]. Nevertheless, the gas desorption on the defects is very slow. Hence, a balance should be made between the sensitivity and recovery speed by optimizing the density of defects.

Jesse D. Fowler etc. reported practical chemical sensor by chemically derived RGO [6]. Dispersions of RGO in anhydrous hydrazine are formed from GO. The RGO is sensitive to NO_2, NH_3, and 2,4-dinitrotoluene. Furthermore, a micro hot plate sensor substrate is employed to increase the recovery speed at the cost of sensitivity for NO_2 sensing [6]. In addition to hydrazine, other weak reductants such as p-phenylenediamine (PPD) and ascorbic acid were also employed to obtain RGO from GO for gas sensing application [3, 19]. As PPD and ascorbic acid are weaker reductants than hydrazine, the obtained RGO has more oxygenated groups and defects. The ascorbic acid reduced RGO was deposited on a flexible substrate such as poly-ethylene terephthalate (PET) by inkjet printing [19]. The resultant sensors can detect a variety of gases such as NO_2 and Cl_2 with the LOD of 100 ppm and 500 ppb, respectively. However, the recovery process is slow due to the strong interaction between gas molecules and the oxygenated groups on RGO. In this case, UV illumination can be utilized to speed up the signal recovery and achieve the full signal recovery. In addition to chemical reduction, thermal reduction can also be employed to reduce GO into RGO [20]. It is reported that thermal annealing at low temperature such as 200 °C can be deployed to partially reduce GO [20]. The generated RGO can be employed for a high performance gas sensor. The sensitivity of 1.41 was obtained and the complete recovery was achieved after air purge for 0.5 h. The high sensitivity is attributed to the formation of vacancies and defects during the thermal annealing process. These defects favor the gas adsorption, increasing the response.

12.5 Gas Sensor Based on Chemically Functionalized Gr

Both theoretical and experimental studies demonstrate that chemical functional groups can be exploited to increase the sensitivity significantly. For example, a first-principles study indicates that the defective Gr displays the highest adsorption energy with NO_2, NO and CO molecules [21]. Furthermore, B-doped Gr shows the tightest binding with NH_3. The interaction between Gr and gas molecules depends on the electronic properties of Gr [21]. A density functional theory study also demonstrates that Al-doped Gr shows strong response to CO because of the formation of Al—CO bonds [3]. In contrast, pristine Gr shows weak adsorption with CO molecules because of the long distance and the small binding energy between Gr and CO molecules [3]. S-doping is an effective means to enhance the NO_2

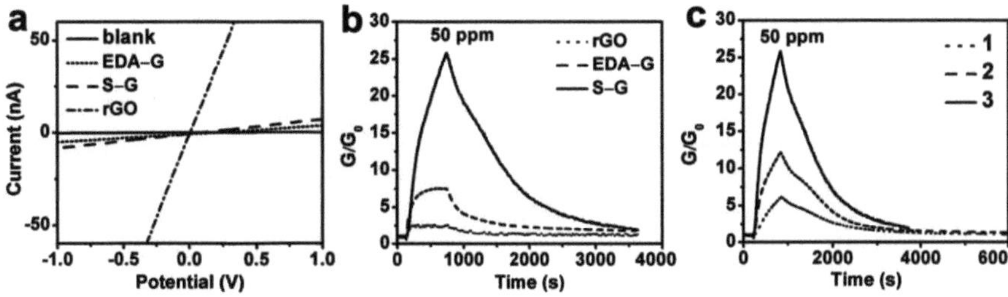

Fig. 12.3 Chemically modified Gr for NO_2 sensing. (**a**) I–V curves of blank-IE-, RGO, EDA-G and S-G-based sensors [23]. (**b**) Comparison of the responses of RGO, EDA-G and S-G-based sensors to 50 ppm NO_2. (**c**) Responses of the S-G-based sensors to 50 ppm NO_2 with the sensing layers thicknesses of 50 nm (1), 10 nm (2), and a few layers (1–6 nm) (3) of S-G sheets

sensing of RGO [16, 22]. Yuan et al. reported sulfonated RGO and ethylenediamine (EDA)-modified RGO display a 4–16 times stronger response toward NO_2 gas than those of the unmodified RGO counterpart, demonstrating the advantages of utilizing S and N doping to enhance the performance of NO_2 sensing [16]. Furthermore, the sulfonated RGO shows good reversibility and selectivity (Fig. 12.3). The better performance of the sulfonated RGO gas sensor is attributed to the chemical functionalization of RGO with functional SO_3-groups. RGO exhibits an electron-withdrawing property at room temperature. NO_2 molecules have an electron-withdrawing property, and favor adsorption on electron-rich sites such as the lone-pair electrons of O or S atoms on SO_3-groups [16]. After NO_2 adsorption, the electron-withdrawing property of RGO is enhanced. Thus, the resistance of sulfonated RGO decreases significantly. Since the electrostatic interaction between S-RGO and NO_2 is weak, NO_2 molecules can be detached from the sensing materials easily by purging the device with N_2, leading to a good reversibility. The authors also find that higher response can be achieved by using a thinner RGO layer. In addition, sulfonated RGO modified with Ag nanoparticles (NPs) (Ag–S-RGO) has also been reported to display high sensitivity, rapid response and recovery to NO_2 [22].

In addition to S and N-doping, B-doping is also employed to boost the performance of Gr-based NO_2 sensor significantly [24]. For example, Lv et al. reported ultrasensitive detection of NO_2 by large-area boron-doped Gr [24]. They confirmed that B-doped Gr displayed p-type semiconducting behavior. A unique croissant-like feature can be observed within the lattice of B-doped Gr, which is attributed to the presence of B–C trimers embedded in the Gr lattice [24]. The sensor can detect NO_2 with the concentration as low as 1 ppb. Two element co-doping can also be utilized to boost the gas sensing performance. Based on density functional theory (DFT) calculations, Choudhuri et al., reported that B–N@Graphene can be exploited for highly sensitive and selective gas sensor [25].

The oxygenated groups on RGO exert an important effect in inducing the response of the RGO-based sensor upon exposure to gases. Hence, various methods have been explored to introduce oxygen functional groups in RGO to improve the

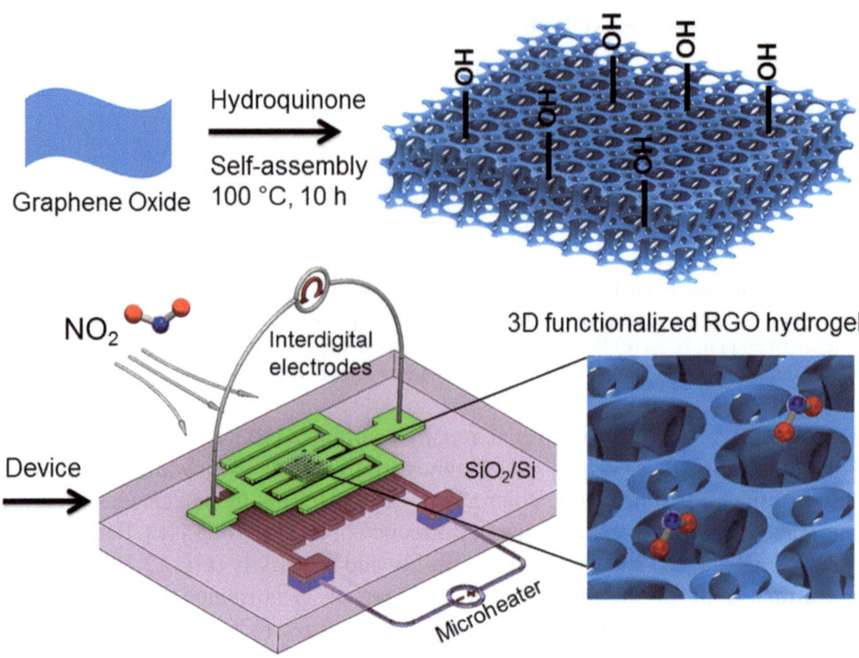

Fig. 12.4 Scheme illustrating one-step hydrothermal synthesis of 3D RGOH and simultaneous functionalization of it with hydroquinone molecules. The synthesized 3D RGOH is employed for a high-performance gas sensor. A microheater is imbedded at the bottom side of sensor to improve the selectivity of NO_2 detection [21]

sensitivity of gas sensing. For instance, we have utilized one-step hydrothermal synthesis to prepare chemically modified 3D reduced graphene oxide hydrogel (FRGOH) with the aid of hydroquinone molecules (Fig. 12.4) [21]. In the one-step synthesis process, the hydroquinone molecules act both as the reductant and the functionalizing molecules. In comparison with the unmodified RGOH-based sensor, the FRGOH sensor not only exhibits two times higher sensitivity to both NO_2 and CO_2 sensing, but also displays increased recovery percentage and lower LOD, demonstrating the advantages of chemical functionalization. The sensor can experimentally detect NO_2 with the low concentration of 200 ppb. Furthermore, the theoretical LOD is calculated to be as low as 57 ppb NO_2. Each hydroquinone molecule contains two hydroxyl groups. The superior sensitivity achieved by the FRGOH is attributed to the oxygenated hydroxyl groups contained in hydroquinone molecules. The hydroxyl groups significantly enhance the sensitivity of NO_2 sensing. First, the hydroxyl groups function as the dominant adsorption sites for NO_2 molecules. Second, the simulation indicates the distance between the NO_2 molecule and the carbon basal plane decreases [21]. The calculated binding energy between NO_2 molecules and hydroxylated Gr (218.5 meV) is much larger than that for unmodified Gr (87 meV). The increased binding energy may be attributed to the formation of hydrogen bonds. Third, a previous report indicates the relaxation and rotation of hydroxyl groups in the adsorption and desorption processes of NO_2 molecules result in high sensitivity [22]. Significantly, an imbedded microheater is exploited to enhance the selectivity of NO_2 sensing, as well as to speed up the response and recovery processes.

Oxygenated functional groups on Gr can also be introduced by other routes, such as ozone treatment [26]. The ozone treatment introduced a large amount of oxygen-rich functional groups, such as C═O and C─O on Gr surface, which acted as favorable gas adsorption sites with high binding energy. Hence, the ozone-treated Gr (OTG) sensor displayed remarkable enhanced sensing performances such as response, LOD and response time. For instance, the OTG exhibits twofold higher response than that of pristine Gr sensor when exposed to 200 ppm NO_2, demonstrating that ozone treatment is an effective way to improve the performance of Gr-based sensor due to its simplicity, effectiveness and low-cost. In addition to RGO, pristine GO can also be reported as the active sensing material with reversible and high response to NO_2 at room temperature because it contains a large amount of oxygen-rich functional groups [22]. GO-based sensors overcome the drawback of sluggish and irreversible recovery in RGO-based sensors. The first-principles calculations together with experimental results demonstrate the critical role of hydroxyl groups rather than epoxy groups in increasing the sensitivity of NO_2 detection. It also demonstrates that the rotation of these groups for the adsorption of NO_2 molecules and relaxation to the original states during the desorption of NO_2 molecules, are responsible for the fast and reversible NO_2 sensing.

12.6 Defective Gr for Improved Gas Sensor

The defects on Gr/RGO play an important role in enhancing the gas sensing performance, which was collaborated by both theoretical calculation and experimental studies [3]. The first-principles study indicates that defective Gr has strong interaction with NO_2, CO and NO molecules, but displays weak interaction with NH_3 molecules [27]. Because of the high binding energy at the defect sites, gas molecules favor adsorption on the defect regions. Masel and co-workers made a comparison between response of polycrystalline Gr with line defects and that of pristine Gr to 1,2-dichlorobenzene and toluene vapors, and found that the Gr with line defects displayed significantly enhanced sensitivity compared with the pristine counterpart (Fig. 12.5) [28]. This is because conduction paths do not exist around the line defects easily. In contrast, pristine Gr sensors are less sensitive to gas molecules since gas adsorbates tend to bind to point defects. But point defects have low resistance pathways around them. It demonstrates the line defects have strong impact on the sensitivity, while point defects have a small effect on the response [28].

The defect sites on Gr can also be created by many other methods, such as lithography and laser reduction [3, 15]. A laser can be employed to simultaneously pattern and reduce GO film. For instance, two-beam laser interference was

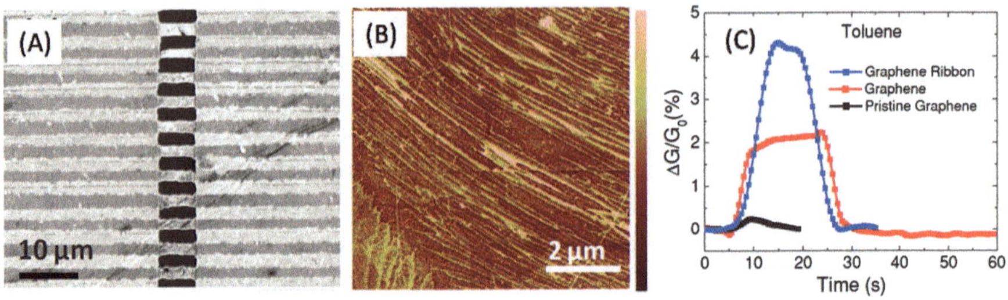

Fig. 12.5 Defective Gr ribbons as a chemiresistor for gas sensor. (**a**) SEM image of CVD Gr ribbons. (**b**) Atomic-force microscopy (AFM) image of Gr ribbons utilized for gas sensor. (**c**) Responses of the Gr microribbon, defective Gr and pristine Gr to 1014 toluene molecules

deployed to simultaneously pattern and reduce GO nanostructures on flexible polyethylene terephthalate substrates [15]. The generated RGO patterns were used for humidity sensing with high performance. It was found that the hierarchical RGO nanostructures hold great promise for increasing the sensing performance by enlarging the interaction areas with gas molecules. Furthermore, the content of oxygen functional groups can be tuned by adjusting the laser power. The response/recovery time can be programmed by controlling the magnitude of laser power since the oxygen functional groups interact strongly with water molecules. Besides the laser patterning, various lithographic methods such as nanosphere lithography, soft lithography and block copolymer lithography, can also be used to create Gr patterns with defects for the enhancement of gas sensing [3]. Nanosphere lithography combined with reactive ion etching was utilized to create Gr nanomesh with a large number of edge defects and unsaturated grain boundary [29]. The increased defect density on Gr nanomesh provides more adsorption sites for ethanol detection.

12.7 Gr/Nanoparticles and Gr/Polymer Composites for Improved Gas Sensor

Metal oxide nanostructures such as SnO_2, ZnO and Cu_2O nanowires or nanorods show high sensitivity to various gases mainly because of their high surface to volume ratio [3]. However, their gas sensing performance is limited by their large resistance. Meanwhile, pristine Gr suffers from poor selectivity and long response time [30]. Blending Gr with metal oxide nanostructures can combine the merits of both Gr and metal oxide nanostructures. For example, SnO_2/Gr composites have been designed to achieve high-performance NO_2 sensing in many works [31–34]. It was found that 3D SnO_2/RGO-4 composites prepared from Sn^{4+} precursor displayed much higher response than SnO_2/RGO-2 composites obtained from Sn^{2+} precursor in NO_2 detection (Fig. 12.6) [31]. The promising sensing performance of 3D SnO_2/RGO-4 is attributed to the extremely large surface area, good electrical

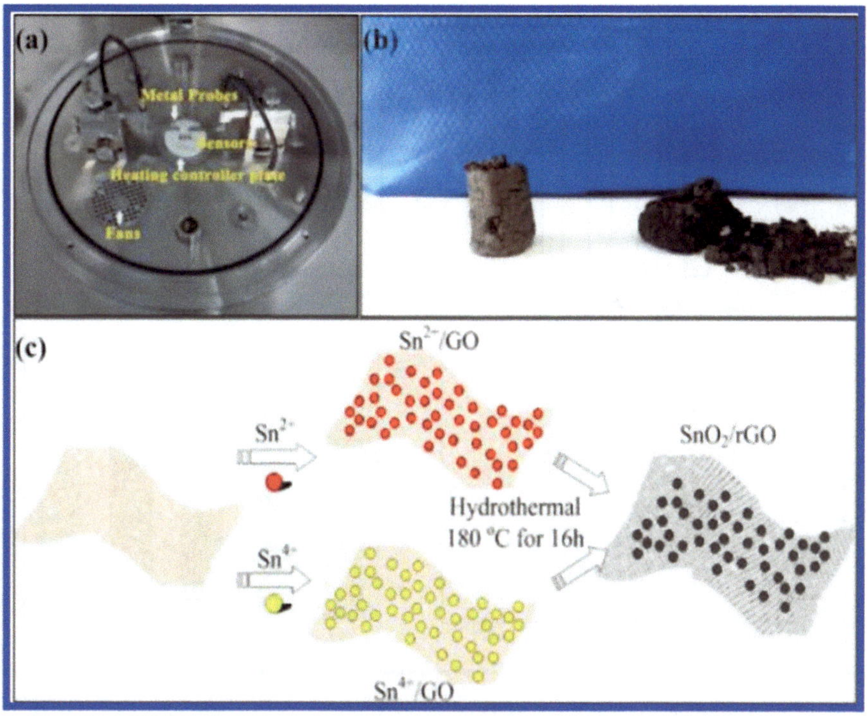

Fig. 12.6 (**a**) and (**b**) Photographs of CGS-1TP gas sensing system and 3D SnO_2/RGO aerogel structures, respectively [31]. (**c**) Schematic illustrating the synthesis of SnO_2/RGO composites. Reproduced with permission from [34]. Copyright American Chemical Society

conductivity and increased adsorption sites. In another work carried out by Mao etc. indicated that decoration of RGO with SnO_2 enhanced NO_2 but weakened NH_3 sensing compared with unmodified RGO, leading to an increased selectivity of NO_2 detection [32]. Furthermore, the hybrid SnO_2/RGO sensor displays low LOD and good repeatability to target gases at room temperature, demonstrating the effect of semiconducting materials on tuning the selectivity and sensitivity of RGO sensors [32].

In addition to metal oxide nanostructures, decoration of noble metal nanoparticles such as Ag, Au and Pt on RGO is also an effective strategy to boost the sensing performance of RGO [35–38]. For instance, the decoration of Ag and Au nanoparticles on RGO/CNT surfaces has been utilized to increase the response of NO_2 detection [36]. It is found that the charge transfer between RGO and metal nanoparticles causes a potential drop at the interface of RGO-metal nanoparticles [36]. The metal nanoparticles induce a p-type doping to RGO. Furthermore, the Au/Ag nanoparticles can act as gas adsorption sites, leading to improved sensitivity to H_2S and NO_2. The interaction between oxidizing gases and RGO leads to increased hole mobility, while the interaction between H_2S and RGO results in decreased hole mobility. In another work performed by Li etc., CVD grown Gr

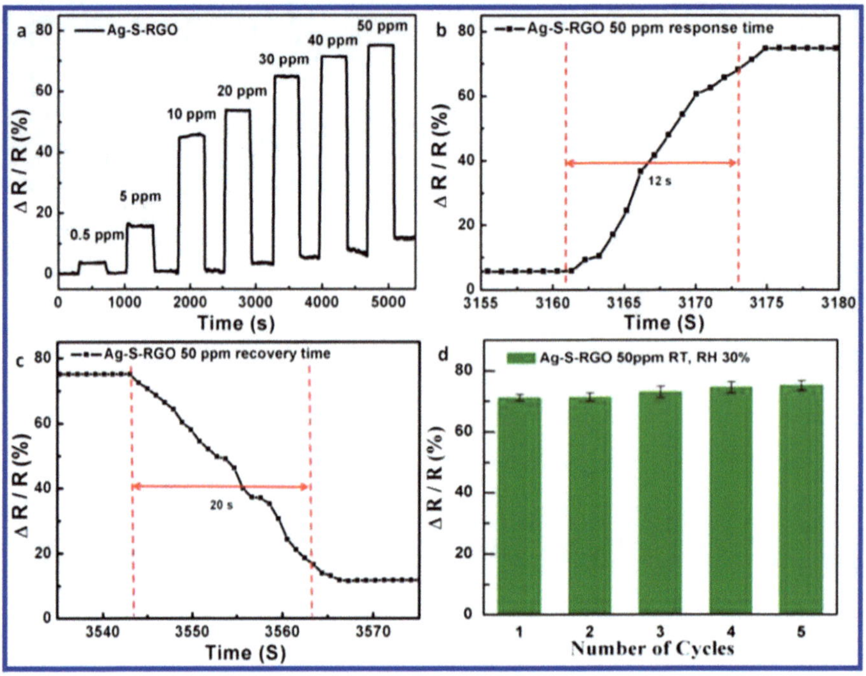

Fig. 12.7 (**a**) Response of the Ag–S-RGO sensor to NO_2 with various concentrations [35]. (**b**, **c**) Fast response and rapid recovery of this sensor to 50 ppm NO^2, respectively. (**d**) Repeated responses of the sensor to 50 ppm NO_2 in five cycles. Reproduced with permission from [22]. Copyright American Chemical Society

contacts of Pd-RGO are exploited to improve the stability and sensitivity of the gas sensing devices because of the matching of work function between CVD-grown RGO and Gr [38]. The sensor exhibits good sensitivity, recovery and reliability to NO gas with the concentration ranging from 2 to 420 ppb. In addition to boosted response, Ag nanoparticle functionalization is also deployed to accelerate the response and recovery. Lei Huang etc. reported NO_2 sensors with rapid response based on sulfonated RGO modified with Ag nanoparticles (Ag–S-RGO) (Fig. 12.7) [35]. The sensor displays a short recovery time of 20 s and a response time of 12 s. Au nanoparticle-RGO has also been utilized to detect chemical vapor based on micro-gravimetric platform [29]. First, Au NPs are synthesized on GO sheets surface in situ with the reduction of oleylamine. Since GO sheets are spaced by the Au NPs, 3D stacked porous nanostructure is formed. Second, the hydrophilic GO sheets are transferred to hydrophobic RGO by a gas-phase reducing method. Third, 11-mercaptoundecanoic acid (11-MUA) is modified on Au NPs, forming a self-assembled monolayer (SAM). The sensors displayed a high sensitivity, good selectivity and rapid response to trimethylamine (TMA) vapor with ppm-level concentration. In addition to Gr-NPs composites, the Gr-polymer hybrid structures have also been widely employed to enhance gas sensing [3, 39]. Especially, it is found that some photoresist polymer such as polymethyl methacrylate (PMMA) can act as chemical dopant to dope Gr and enhance the carrier scattering [7]. Furthermore, this polymer can function as an adsorbent layer to concentrate gas molecules, increasing the sensitivity. Without the functional polymer residue, the intrinsic response of Gr to ammonia vapor is very small. In addition to photoresist polymer, conductive polymers have been widely utilized to boost the sensing performance of Gr/RGO. For example, Gaoquan Shi's group reported GO/polypyrene (PPr) composite displays a fast, linear and reversible response to toluene [40]. The same group also reported ultrasensitive and selective NO_2 sensor based on self-assembled graphene/nanofibers composite [40]. After polymer nanofibers were deposited on IEs by electrospinning, RGO sheets were self-assembled onto the nanofibers surface, forming an ultrathin coating layer. This sensor displayed a high sensitivity of 1.03 ppm^{-1}, good selectivity and reversibility to NO_2 at room temperature. Furthermore, the sensor exhibited a low LOD of 150 ppb at room temperature. This concentration is lower than the threshold exposure limit of 200 ppb that proposed by American Conference of Governmental Industrial Hygienists, demonstrating the capability of the sensor to detect low concentration NO_2.

12.8 3D Gr/Porous Structured Gr for Gas Sensing

In addition to chemical functionalization, structural modification can also be exploited to significantly boost the performance of Gr/RGO sensing [2, 41]. In comparison with its 2D counterpart, 3D porous RGO provides increased surface areas and reactive sites to facilitate the adsorption of gas molecules. The large space between porous structures also facilitates the diffusion and transportation of electrons, ions and gas molecules. Consequently, 3D Gr/RGO structures prepared by various methods have been employed for enhanced gas sensing. For example, our group reported facile, one-step and hydrothermally synthesized 3D RGO hydrogel for high performance gas sensing [2]. The 3D RGOH devices exhibit more than ten times increased response compared to that of 2D RGO devices. Importantly, a microheater is integrated on the substrate of sensor to improve both the sensitivity and the signal recovery process. The sensor shows the capability to detect NO_2 and NH_3 at low concentrations of 200 ppb and 20 ppm, respectively, at room temperature. In another work of our group, 3D chemically functionalized reduced graphene oxide hydrogel (FRGOH) is employed for high-performance NO_2 and CO_2 detection for the first time [21]. Hydroquinone molecules are chemically modified on RGO surface in a facile, one-step hydrothermal synthesis process of RGO. The FRGOH sensor not only exhibits doubled sensitivity to both NO_2 and CO_2, but shows much lower theoretical LOD to NO_2 compared with unmodified reduced graphene oxide hydrogel (RGOH) counterpart. Furthermore, an imbedded microheater is deployed to improve the selectivity of NO_2 detection. Specifically, the response of the FRGOH sensor to NO_2 is not obviously deteriorated at elevated

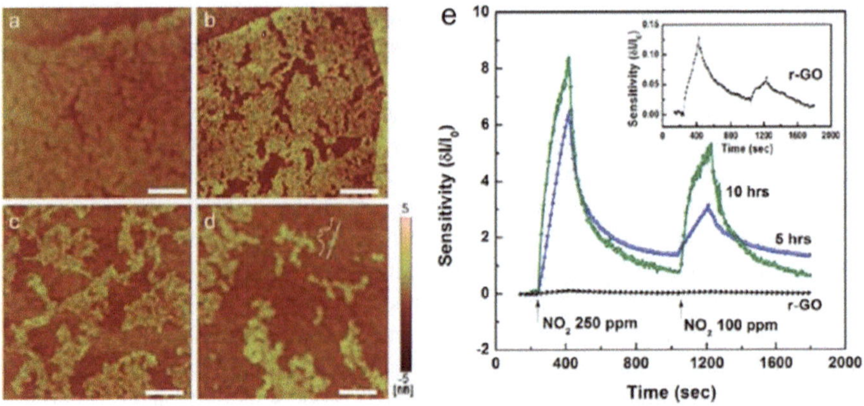

Fig. 12.8 (**a–d**) AFM images showing the morphology of GO sheets after streaming for 5, 10, 20, and 30 h, respectively [42]. (**e**) Comparison between thermally reduced RGO and the steamed GO in NO_2 vapor sensing. Reproduced with permission from [44]. Copyright American Chemical Society

temperature. However, the response to CO_2 can be suppressed effectively by elevated temperature, leading to improved selectivity to NO_2. Hydrothermal steam etching has been employed to create porous GO for high-performance chemical sensing [42]. The degree of etching and the formed porosity can be tuned conveniently by controlling the etching time. Compared with nonporous RGO without steam etching treatment, the nanoporous GO displayed two order of magnitude higher response to NO_2 (Fig. 12.8). This work demonstrates the efficacy of nanoporous structures and defects in the enhancement of gas sensing. Besides naturally formed 3D porous RGO, external 3D templates such as SU-8 templates are also employed for self-assembly of RGO on 3D structures to enhance gas sensing performance via increased interaction cross-section [1]. In addition to RGO, 3D CVD grown Gr is also exploited to fabricate high performance NO_2 and NH_3 sensors with good reliability [43].

12.9 Conclusion

In summary, Gr materials hold great promise in high-performance practical gas sensing because of their high response to a variety of gases, such as NO_2, NO, NH_3, CO_2, organic vapors and so on. Compared with conventional metal oxide based gas sensors, Gr-based gas sensors achieve improved sensitivity, limit of detection and reversibility. Furthermore, Gr-based gas sensors gain advantage in operation at room temperature, bypassing the requirement of external heaters that are utilized in metal oxide based gas sensors. Conducting polymers are promising for room-temperature gas sensors due to their facile synthesis, high conductivity and easy processability [30]. However, the poor recovery and stability limits the practical applicability of conducting polymers based gas sensors. Importantly, the facile chemical functionalization and structural modification can be used to tune the gas sensing properties of Gr, leading to the emergence of a wide variety of novel materials, such as graphene/metal nanocomposites, graphene/polymers and so on. This provides new insights in optimizing the sensing properties of Gr materials.

However, there are still several challenges associated with Gr-based gas sensors that prevent their commercialization. First, the selectivity of Gr-based gas sensors needs to be improved. Different kinds of gas molecules tend to adsorb on the same Gr surfaces and cause similar resistance change, leading to a poor selectivity. Functionalization of Gr with capture molecules that can specifically bind to target gas molecules can be exploited to improve the selectivity [32]. Temperature modulation can also be employed to increase the selectivity for certain gas detection. Second, good stability and repeatability need to be ensured. During long-term application, the gas molecules or moisture water may adsorb on Gr surface, changing the resistance of Gr sensor. Third, controllable technologies are required to fabricate sensors massively in large scale. Currently, techniques such as drop casting, spin coating and inkjet printing are employed to deposit sensing materials on electrodes. Nevertheless, it is hard to control the amount of Gr participating in gas sensing and ensure the same amount of Gr on different sensors. Last but not the least, the cost of sensor needs to be considered. With the development of Gr synthesis strategies, the cost of materials production will be affordable, since the amounts of Gr materials for each sensor is very small. However, the sensor fabrication and assembly cost may be an issue. After addressing above issues, Gr-based gas sensors have great potential to be commercialized and replace current electrochemical and metal oxide based gas sensors in near future.

References

1. L.T. Duy, D.-J. Kim, T.Q. Trung, V.Q. Dang, B.-Y. Kim, H.K. Moon, N.-E. Lee, Adv. Funct. Mater. **25**, 883–890 (2015)
2. J. Wu, K. Tao, J. Miao, L.K. Norford, ACS Appl. Mater. Interfaces **7**, 27502 (2015)
3. W. Yuan, G. Shi, J. Mater. Chem. A **1**, 10078 (2013)
4. Y.H. Kim, S.J. Kim, Y.-J. Kim, Y.-S. Shim, S.Y. Kim, B.H. Hong, H.W. Jang, ACS Nano **9**, 10453–10460 (2015)
5. F. Schedin, A.K. Geim, S.V. Morozov, E.W. Hill, P. Blake, M.I. Katsnelson, K.S. Novoselov, Nat. Mater. **6**, 652–655 (2007)
6. J.D. Fowler, M.J. Allen, V.C. Tung, Y. Yang, R.B. Kaner, B.H. Weiller, ACS Nano **3**, 301–306 (2009)
7. Y.P. Dan, Y. Lu, N.J. Kybert, Z.T. Luo, A.T.C. Johnson, Nano Lett. **9**, 1472–1475 (2009)
8. Q. He, S. Wu, Z. Yin, H. Zhang, Chem. Sci. **3**, 1764 (2012)
9. G.H. Lu, S. Park, K.H. Yu, R.S. Ruoff, L.E. Ocola, D. Rosenmann, J.H. Chen, ACS Nano **5**, 1154–1164 (2011)
10. R. Arsat, M. Breedon, M. Shafiei, P.G. Spizziri, S. Gilje, R.B. Kaner, K. Kalantar-zadeh, W. Wlodarski, Chem. Phys. Lett. **467**, 344–347 (2009)

11. M. Qazi, T. Vogt, G. Koley, Appl. Phys. Lett. **91**, 233101 (2007)
12. G. Lu, L.E. Ocola, J. Chen, Nanotechnology **20**, 445502 (2009)
13. A. Salehi-Khojin, D. Estrada, K.Y. Lin, M.H. Bae, F. Xiong, E. Pop, R.I. Masel, Adv. Mater. **24**, 53–57 (2012)
14. M.W.K. Nomani, R. Shishir, M. Qazi, D. Diwan, V.B. Shields, M.G. Spencer, G.S. Tompa, N.M. Sbrockey, G. Koley, Sens. Actuators, B **150**, 301–307 (2010)
15. R. Pearce, T. Iakimov, M. Andersson, L. Hultman, A.L. Spetz, R. Yakimova, Sens. Actuators, B **155**, 451–455 (2011)
16. W. Yuan, A. Liu, L. Huang, C. Li, G. Shi, Adv. Mater. **25**, 766–771 (2013)
17. X.S. Li, W.W. Cai, J.H. An, S. Kim, J. Nah, D.X. Yang, R. Piner, A. Velamakanni, I. Jung, E. Tutuc, S.K. Banerjee, L. Colombo, R.S. Ruoff, Science **324**, 1312–1314 (2009)
18. K. Yu, P. Wang, G. Lu, K.-H. Chen, Z. Bo, J. Chen, J. Phys. Chem. Lett. **2**, 537–542 (2011)
19. C.W. Chen, S.C. Hung, M.D. Yang, C.W. Yeh, C.H. Wu, G.C. Chi, F. Ren, S.J. Pearton, Appl. Phys. Lett. **99**, 243502 (2011)
20. L. Guo, H.-B. Jiang, R.-Q. Shao, Y.-L. Zhang, S.-Y. Xie, J.-N. Wang, X.-B. Li, F. Jiang, Q.-D. Chen, T. Zhang, H.-B. Sun, Carbon **50**, 1667–1673 (2012)
21. Y.H. Zhang, Y.B. Chen, K.G. Zhou, C.H. Liu, J. Zeng, H.L. Zhang, Y. Peng, Nanotechnology **20**, 185504 (2009)
22. L. Huang, Z. Wang, J. Zhang, J. Pu, Y. Lin, S. Xu, L. Shen, Q. Chen, W. Shi, ACS Appl. Mater. Interfaces **6**, 7426–7433 (2014)
23. F. Yavari, N. Koratkar, J. Phys. Chem. Lett. **3**, 1746–1753 (2012)
24. V. Dua, S.P. Surwade, S. Ammu, S.R. Agnihotra, S. Jain, K.E. Roberts, S. Park, R.S. Ruoff, S.K. Manohar, Angew. Chem., Int. Ed. **49**, 2154–2157 (2010)
25. G. Lu, L.E. Ocola, J. Chen, Appl. Phys. Lett. **94**, 083111 (2009)
26. R.T. Lv, G.G. Chen, Q. Li, A. McCreary, A. Botello-Mendez, S.V. Morozov, L.B. Liang, X. Declerck, N. Perea-Lopez, D.A. Culleni, S.M. Feng, A.L. Elias, R. Cruz-Silva, K. Fujisawa, M. Endo, F.Y. Kang, J.C. Charlier, V. Meunier, M.H. Pan, A.R. Harutyunyan, K.S. Novoselov, M. Terrones, Proc. Natl. Acad. Sci. U. S. A. **112**, 14527–14532 (2015)
27. S. Borini, R. White, D. Wei, M. Astley, S. Haque, E. Spigone, N. Harris, J. Kivioja, T. Ryhanen, ACS Nano **7**, 11166–11173 (2013)
28. I. Choudhuri, N. Patra, A. Mahata, R. Ahuja, B. Pathak, J. Phys. Chem. C **119**, 24827–24836 (2015)
29. J. Wu, K. Tao, J. Zhang, Y. Guo, J. Miao, L.K. Norford, J. Mater. Chem. A **4**, 8130–8140 (2016)
30. Y.R. Choi, Y.-G. Yoon, K.S. Choi, J.H. Kang, Y.-S. Shim, Y.H. Kim, H.J. Chang, J.-H. Lee, C.R. Park, S.Y. Kim, H.W. Jang, Carbon **91**, 178–187 (2015)
31. M.G. Chung, D.H. Kim, H.M. Lee, T. Kim, J.H. Choi, D.K. Seo, J.-B. Yoo, S.-H. Hong, T.J. Kang, Y.H. Kim, Sens. Actuators, B **166–167**, 172–176 (2012)
32. H. Yu, P. Xu, D.W. Lee, X. Li, J. Mater. Chem. A **1**, 4444 (2013)
33. J. Zhang, X. Liu, G. Neri, N. Pinna, Adv. Mater. **28**, 795–831 (2016)
34. L. Li, S. He, M. Liu, C. Zhang, W. Chen, Anal. Chem. **87**, 1638–1645 (2015)
35. D.C. Marcano, D.V. Kosynkin, J.M. Berlin, A. Sinitskii, Z.Z. Sun, A. Slesarev, L.B. Alemany, W. Lu, J.M. Tour, ACS Nano **4**, 4806–4814 (2010)
36. S. Mao, S. Cui, G. Lu, K. Yu, Z. Wen, J. Chen, J. Mater. Chem. **22**, 11009 (2012)
37. G. Neri, S.G. Leonardi, M. Latino, N. Donato, S. Baek, D.E. Conte, P.A. Russo, N. Pinna, Sens. Actuators, B **179**, 61–68 (2013)
38. Q. Lin, Y. Li, M. Yang, Sens. Actuators, B **173**, 139–147 (2012)
39. V. Tjoa, W. Jun, V. Dravid, S. Mhaisalkar, N. Mathews, J. Mater. Chem. **21**, 15593 (2011)
40. B. Cho, J. Yoon, M.G. Hahm, D.-H. Kim, A.R. Kim, Y.H. Kahng, S.-W. Park, Y.-J. Lee, S.-G. Park, J.-D. Kwon, C.S. Kim, M. Song, Y. Jeong, K.-S. Nam, H.C. Ko, J. Mater. Chem. C **2**, 5280 (2014)
41. W.W. Li, X.M. Geng, Y.F. Guo, J.Z. Rong, Y.P. Gong, L.Q. Wu, X.M. Zhang, P. Li, J.B. Xu, G.S. Cheng, M.T. Sun, L.W. Liu, ACS Nano **5**, 6955–6961 (2011)

42. W. Yuan, L. Huang, Q. Zhou, G. Shi, ACS Appl. Mater. Interfaces **6**, 17003–17008 (2014)
43. L. Zhang, C. Li, A. Liu, G. Shi, J. Mater. Chem. **22**, 8438 (2012)
44. T.H. Han, Y.K. Huang, A.T. Tan, V.P. Dravid, J. Huang, J. Am. Chem. Soc. **133**, 15264–15267 (2011)
45. F. Yavari, Z. Chen, A.V. Thomas, W. Ren, H.M. Cheng, N. Koratkar, Sci. Rep. **1**, 166 (2011)

Chapter 13
Nanoporous Palladium Films Based Resistive Hydrogen Sensors

Shuanghong Wu, Han Zhou, Mengmeng Hao, and Zhi Chen

13.1 Introduction

Hydrogen is an extremely clean energy source for use in fuel cells and internal combustion engines. As a colourless, odourless and tasteless flammable gas, hydrogen cannot be detected by human senses, and other means are therefore required to detect its presence and quantify the concentration. The development of effective hydrogen sensors to quickly and sensitively response to hydrogen gas leak and to monitor manufacturing and distribution is crucial for the safe deployment of all hydrogen-based applications [1–12]. Especially for the detection of hydrogen leaks below the lower explosive limit of 4% (by volume ratio of hydrogen to air), hydrogen sensor must response in a short time. There are many kinds of hydrogen sensors based on different working mechanism, such as catalytic [8, 13–15], electrochemical [16–18], resistance based [19–24], work function based [25–27], optical [28–30] and so on. As we all know, there is no one sensor type which can demonstrate optimum performance overall, and which technology is most suited to a given application depends on the operating requirements for that application.

Resistance based hydrogen sensors have an attraction due to their low cost, good stability, easy to construct and so on [31–35]. Two important resistance based hydrogen sensors are semiconducting metal-oxide sensor and metallic resistors,

S. Wu (✉) • H. Zhou • M. Hao
School of Optoelectronic Information, University of Electronic Science and Technology of China, Chengdu 610054, China
e-mail: shuanghongwu1983@gmail.com

Z. Chen
School of Optoelectronic Information, University of Electronic Science and Technology of China, Chengdu 610054, China

Department of Electrical and Computer Engineering and Center for Nanoscale Science and Engineering, University of Kentucky, Lexington, KY 40506, USA

T. Li, Z. Liu (eds.), *Outlook and Challenges of Nano Devices, Sensors, and MEMS*,
DOI 10.1007/978-3-319-50824-5_13

Table 13.1 Characteristics of resistive hydrogen sensors [7]

Sensor type	Measurand	Advantages	Disadvantages
Semiconducting metal-oxide	Resistance	High sensitivity, fast response, acceptable lifetime, wide operating temperature range, low cost, modest power consumption	Poor selectivity, interference from humidity and temperature, high operation temperature, contamination, susceptible to aging and memory effects, requires O_2 to operate
Metallic resistor	Resistance	Very wide detection range, rapid response, selective, long term stability, can operate in the absence of O_2	Dependence on temperature, affected by total gas pressure, poisoned by SO_2 and H_2S, susceptible to aging effects

respectively. Table 13.1 presents the characteristics comparison of semiconducting metal-oxide sensor and metallic resistors. The working mechanism is that the electrical properties of metal oxides and metal change when they are exposed to reducing gases.

In 1962, Taguchi et al. [36] patented and subsequently marketed a SnO_2 based semiconductor device capable of detecting low concentrations of combustible and reducing gases using a simple electrical circuit. Semiconducting oxides, including SnO_2, ZnO, TiO_2, FeO, Fe_2O_3, NiO, Ga_2O_3, In_2O_3, Sb_2O_5, MoO_3 and WO_3, show a large change in their electrical resistance following adsorption of hydrogen gas. In recent years microelectro-mechanical system (MEMS) technology has been applied to the fabrication of a small-sized, low power consumption hydrogen sensor based on a nano-crystalline, fluorine-doped SnO_2 hydrogen sensor [37–39]. Typically metallic resistor sensors are fabricated by deposition of a film of metal or alloy on a substrate (e.g. silicon) between two electrical contacts, the electrical resistivity of which changes markedly on the absorption of hydrogen. Hydrogen has a high solubility in palladium (Pd) in particular and this interaction is also selective making Pd the metal of choice in this type of sensor. Pd lattice can transit to α phase after absorbing hydrogen, which leads to lattice expansion and resistance increase. The resistance change in the presence of hydrogen can be detected as hydrogen concentration signal.

In 2007, Prakash et al. investigated the effect of thin film morphology, carbon monoxide (CO) and resistor geometry on the response of hydrogen sensitive thin film palladium resistors [22]. Pd thin film morphology was found to strongly influence sensor response in terms of hydrogen sensitivity and rate of response. Figure 13.1 describes the response of 100 nm Pd films with different deposition conditions to 1% H_2 and gas mixture containing 5% CO and 1% H_2 in Ar at 100 Torr pressure and 75°C substrate temperature. Sensor response, response time and resistance to CO poisoning have been shown to depend on the palladium film morphology which is dependent on the method and conditions of film fabrication. Film-1 sensors displayed a fast response to H_2 that was strongly influenced by the presence of CO. The H_2 response was significantly retarded when CO was

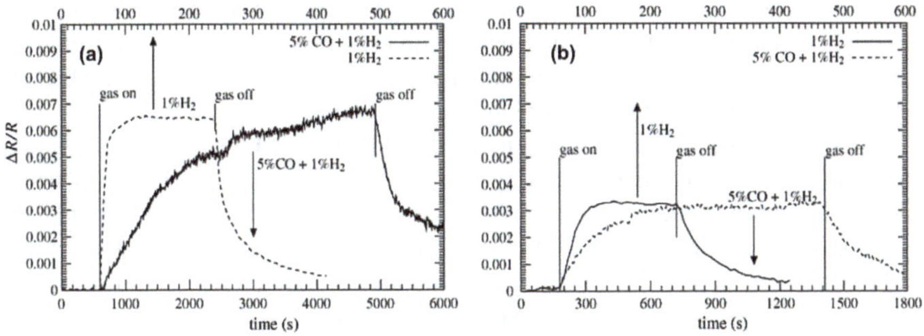

Fig. 13.1 Response of 100 nm Pd films with different deposition conditions to 1% H_2 and gas mixture containing 5% CO and 1% H_2 in Ar at 100 Torr pressure and 75°C substrate temperature. (**a**) Film-1 deposited at 5 mTorr, 20 W power and a substrate temperature of 40°C. (**b**) Film-2 deposited at 25 mTorr pressure, 20 W power and substrate temperature of around 40°C [22]

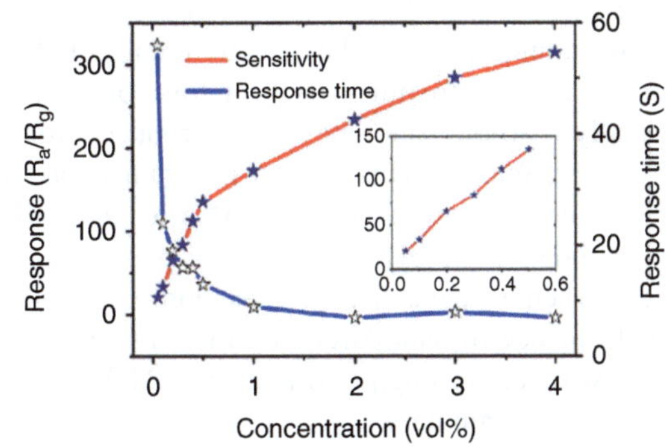

Fig. 13.2 The plots of response and response time of the sensor made from directly formed Pd NCs loaded WO_3 NPs versus hydrogen concentration. The *inset* is the response of the sensor as the concentration is below 0.5 vol% [41]

added to the gas mixture. The lower sensitivity of film-2 sensor is due to the increased resistivity of the film as compared to film-1. In the presence of CO and H_2, film-2 sensors are less influenced by the poisoning effects of CO. While the time to equilibrium response is slowed compared to the H_2 only case, the film-2 resistors achieve steady state significantly faster than film-1 sensors in CO/H_2 mixtures.

Pd alloy has also been reported for hydrogen sensing. Pd-yttrium based hydrogen thin film sensor using silicon oxide substrate was demonstrated by Jamshidi et al. in 2013, which shows better sensing performance than pure Pd film [40]. A thin palladium coated Mg-Ni alloy film has also been reported for hydrogen detection in the range 10 ppm-10% at room temperature [35]. A hydrogen sensor comprising a thin film Pd-Ni alloy has been demonstrated and claims a wide measuring range of 0.1–100% [31]. Liu et al. studied room-temperature hydrogen sensing performance of mechanically mixed Pd nanocubes (NCs) and WO_3 nanoplates (NPs) and directly formed Pd NCs loaded WO_3 NPs [41]. Figure 13.2 depicts response and response time of the sensor made from directly formed Pd NCs loaded WO_3 NPs versus hydrogen concentration. The response time of the sensor decreases sharply from 54 s at 0.05 vol% to 13 s at 0.5 vol% and further slides gently to 7 s at 2 vol%. The greatly enhanced hydrogen sensing performance of the directly formed composite indicates that the direct formation of shape-controlled Pd nanocrystals on metal oxides may be an effective and workable strategy to fabricate high-performance room-temperature hydrogen sensors.

Progress in nano research has led to the development of H_2 sensors based on Pd nanomaterials. In this chapter, a way of fabricating nanoporous Pd films using anodic aluminum oxide (AAO) template as substrated has been demonstrated for hydrogen sensors. Nanoporous Pd films based on AAOs were found to have a quick and reversible response due to their enhanced absorption and disorption of hydrogen compared with dense Pd films. We have investigated post-deposition annealing temperatures of Pd films and pore-widening treatment of AAOs on the performance of hydrogen sensors. A response time as short as 30 s at 1% hydrogen concentration with an anneal temperature of 200°C has been obtained. The sensing performance of hydrogen sensor based on nanoporous Pd supported by AAOs was enhanced by pore-widening treatment using phosphoric acid (H_3PO_4) as etching solution. The optimized hydrogen sensor shows a fast response time of 19 s at hydrogen concentration of 1% and a detection range of H_2 concentration from 0.1% to 2% by pore-widening treatment with a time of 30 min and 5% H_3PO_4 concentration. Finally, carbon nanotubes are introduced for novel Pd nanostructure resistive hydrogen sensors. This chapter is organized as follows. Firstly, we have fabricated nanoporous Pd films using AAOs as the substrates and studied the effect of annealing on the hydrogen sensing performance of nanoporous Pd films in Sect. 13.2. Then we have discussed the pore-widening treatment of AAOs on the sensing performance of nanoporous Pd films in Sect. 13.3. Effect of carbon nanotubes on the hydrogen sensing performance of Pd film has been discussed in Sect. 13.4. Final section is a summary.

13.2 Annealing on the Performance of Nanoporous Pd Films Based Hydrogen Sensors

Pd is one of the most important hydrogen sensitive materials, leading to lattice expansion and resistance increase after absorbing hydrogen. However, slow response and limited H_2 concentration measured hinder their widespread application of Pd based hydrogen sensors. Improvement of the sensing performance of Pd materials using nanotechnology has been done. Pd nanomaterials like nanowires [31, 42–44], nanoclusters [45, 46] and nanopores [47–52] show fast response owing to huge surface area and short diffusion time. For example, Yang et al. have demonstrated that Pd nanowires have the capacity for detecting H_2 much more rapidly than Pd film [53]. Furthermore, the limit of detection for hydrogen is below 10 ppm.

Favier et al. fabricated hydrogen sensors and hydrogen-activated switches from arrays of mesoscopic palladium wires, which were prepared by electrodeposition

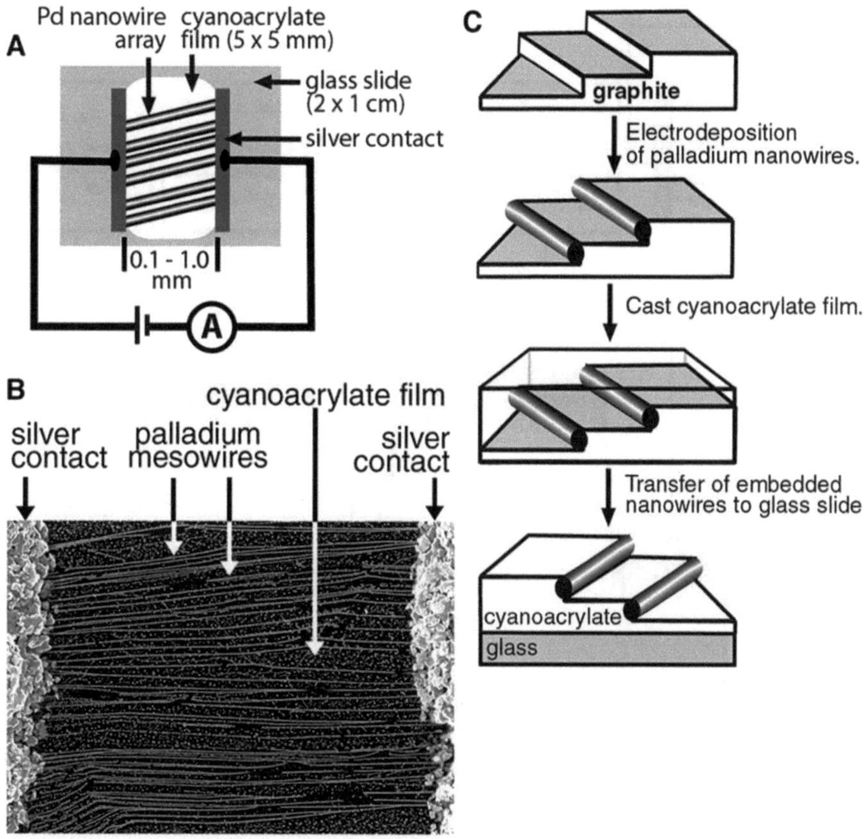

Fig. 13.3 (**a**) Schematic diagram of a Pd mesowire array based hydrogen sensor or switch. (**b**) SEM image [400 mm (h) by 600 mm (w)] of the active area of a Pd mesowire array based hydrogen sensor. (**c**) Pd mesowire arrays were prepared by electrochemical step edge decoration at graphite surfaces and transferred to a cyanoacrylate film [44]

onto graphite surfaces and were transferred onto a cyanoacrylate film [44]. Figure 13.3a shows the schematic diagram of a Pd mesowire array based hydrogen sensor or switch. The making process of Pd mesowire array is described in Fig. 13.3c. Exposure to hydrogen gas caused a rapid (less than 75 ms) reversible decrease in the resistance of the array that correlated with the hydrogen concentration over a range from 2 to 10%. Two different modes of sensor operation were observed. Mode I sensors remained conductive in the absence of H_2. The resistance of a mode I sensor decreased in the presence of H_2, with the decrease related to the H_2 concentration. Mode II sensors were hydrogen-activated switches. In the absence of H_2, the resistance of a mode II sensor became large. Above a threshold of approximately 2% H_2, the switch closed and a device resistivity became measurable. The mechanism proposed is shown in Fig. 13.4. The sensor response appears to involve the closing of nanoscopic gaps or "break junctions" in wires

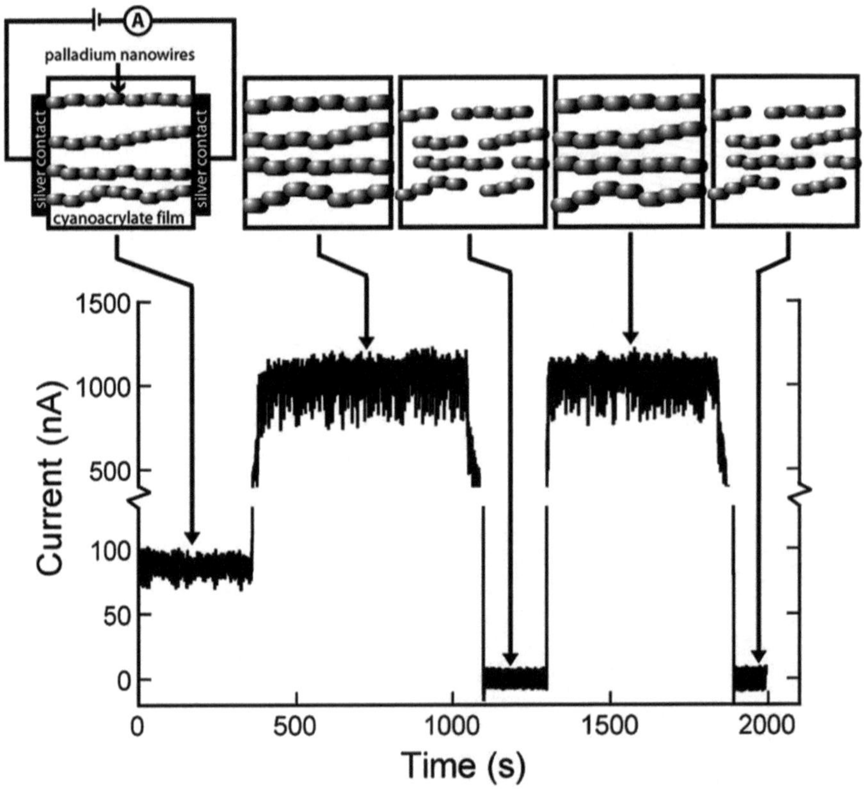

Fig. 13.4 The first exposure of a new sensor to hydrogen. In this case, an irreversible transition from mode I to mode II operation was observed (*bottom*). Shown at top is the mechanism proposed for mode II sensor operation. Mode I sensors operate by an identical mechanism, except that some mesowires remain conductive in the absence of hydrogen [44]

caused by the dilation of palladium grains undergoing hydrogen absorption. Wire arrays in which all wires possessed nanoscopic gaps reverted to open circuits in the absence of hydrogen gas. However, the hydrogen sensors based on Pd nanowires have problems of complex fabrication process and performance stability.

Xu et al. proposed a sensor comprising a discontinuous ultrathin palladium film on a siloxane treated glass substrate in 2005 [55]. Conductance responses of a 3.3 nm thick Pd film on siloxane coated glass (SCG) to various concentrations of H_2 are shown in Fig. 13.5. This type of nanocluster palladium film detects hydrogen based on the expansion of isolated nano-sized palladium particles following hydrogenation. In the process of expansion some particles touch forming new electrical connections, and creating new conducting pathways, which results in an overall net decrease in the films electrical resistance. The resulting Pd nanocluster film can detect 2% H_2 with a rapid response time of 70 ms and is sensitive to 25 ppm hydrogen, detectable by a 2% increase in conductance due to the hydrogen-induced palladium lattice expansion.

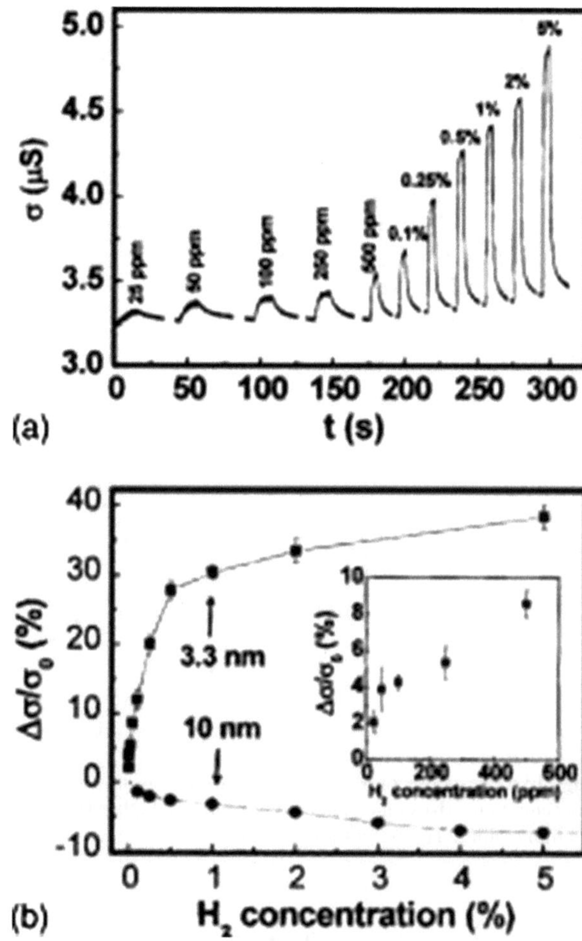

Fig. 13.5 (**a**) Conductance responses of a 3.3 nm thick Pd film on SCG to various concentrations of H_2. (**b**) conductance change vs H_2 concentration for the same sample (*solid square*). The inset is a magnified view of the data in the low concentration region. Data obtained for a continuous Pd film (10 nm thick) on SCG is also given (*solid dot*) for comparison [55]

Our group have demonstrated a series of resistive hydrogen sensors using AAO template as the substrate [49]. Figure 13.6 shows the response of hydrogen sensors based on different Pd films at different H_2 concentrations. Nanoporous Pd film has a faster response time of about 5 min at higher H_2 concentrations compared with dense Pd film, which has a response time of about 10 min. However, the response time of several minutes is still too long for hydrogen leak detection. Meanwhile, there is no response for dense Pd film at dilute H_2 concentrations. However, nanoporous Pd film can detect H_2 concentration less than 250 ppm with low sensitivity. In this section, we have improved the response speed of the nanoporous Pd film supported by AAO template through anneal. A response time as short as 30 s was obtained at 1% hydrogen concentration with the deposited Pd film annealed at 200°C. The sensor exhibited good performance under hydrogen concentrations from 0.1% to 1%. This information is critical for both sensing material optimization and understanding of thermal-treatment effect.

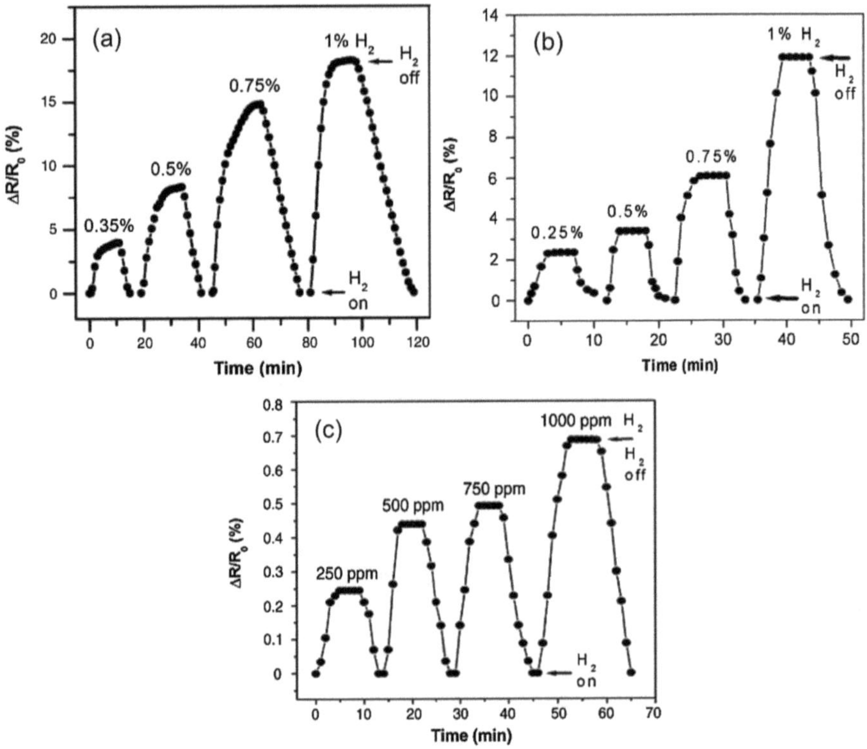

Fig. 13.6 Response of Pd based hydrogen sensors. (**a**) 45 nm dense Pd film at higher H_2 concentrations. (**b**) 45 nm nanoporous Pd film at higher H_2 concentrations. (**c**) 45 nm nanoporous Pd film at dilute H_2 concentrations [49]

13.2.1 Materials and Methods

Figure 13.7 demonstrates the schematic illustration of fabrication process of hydrogen sensors based on nanoporous Pd films supported by AAO templates. After clean of aluminum sheet, it was anodized to form AAO nanostructure by a two-step anodic oxidation method using platinum as a counter electrode. The first anodizing was conducted at 40 V in 0.3 mol/L oxalic acid with platinum as a counter electrode. After a period of anodization (~9 h), the anodic oxide film was chemically removed using 1.8 wt% chromic acid (H_2CrO_4) and 6 wt% H_3PO_4 in 60°C bath for 2 h. The second anodizing was conducted under the same condition as the first step for 1 h. Morphology of the nanoporous alumina was examined by a scanning electron microscope (SEM) (FEI-Inspect F50, Holland). Then Pd film was deposited on the AAO template by DC magnetron sputtering with a thickness of 45 nm. After deposition, Pd film was annealed using the heater wire in the chamber at temperatures of 150°C, 200°C, 300°C, 400°C, respectively. All

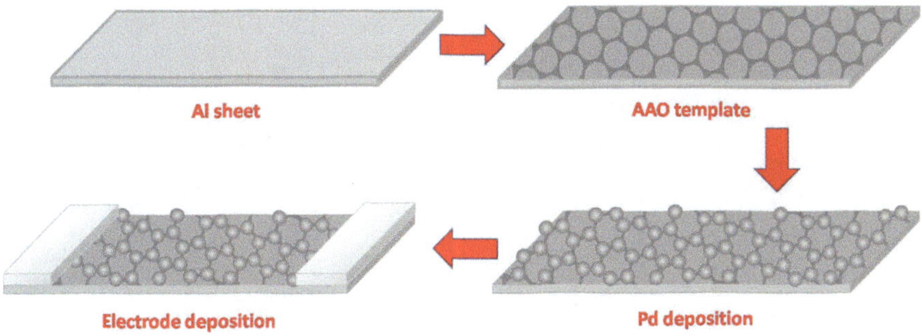

Fig. 13.7 The schematic illustration of the fabrication process of nanoporous Pd films based hydrogen sensors

annealing treatments were done in the vacuum chamber. Temperature was measured by an thermocouple and maintained for half an hour. The microstructure of the films was studied by X-ray diffraction (XRD, Bede D1) using a Cu Kα (1.54 Å) radiation. Gold was deposited on the Pd films by thermal evaporation to fabricate the electrodes (size of 10 × 3 mm). Then the sensors were connected by copper wire to a Keithley 2700 multimeter (Tektronix (China) Co., Ltd., Chengdu, China) for resistance measurement in a gas flow cell made of teflon at room temperature. The gas flow cell has a gas inlet and an outlet. Hydrogen at various concentrations using nitrogen as the carrier gas was introduced with a constant flow rate of 200 sccm.

13.2.2 Results and Discussion

Figure 13.8. presents the top and cross-sectional view SEM images of the bare AAO template and nanoporous Pd film on AAO template. The morphology of Pd layer (Fig. 13.8.b) simply duplicates that of bare AAO template (Fig. 13.8.a). After breaking off the samples, the capping layer of Pd is observed (Fig. 13.8.d). It can be concluded that the Pd film is well deposited on the AAO substrate. The pore diameters of AAO template are around 70-80 nm. The H_2 response of this developed nanoporous Pd film strongly depends on the pores. The nanostructure of Pd film can provide large specific area, which is important for rapid absorption/desorption of hydrogen gas. Hydrogen absorption on the surface of Pd film can be described by Langmuir absorption model [54]. Langmuir suggested that absorption takes place through this mechanism: $H_2 + 2S \rightleftharpoons 2H\text{-}S$, where S is an absorption site on the surface of Pd film. So absorption depends strongly on the specific surface and active sites. Hydrogen can be absorbed on pore ends and walls of the developed Pd film, which leads to fast absorption rate. Compared with the response time of dense Pd film, much shorter response time was found for nanoporous Pd film.

The hydrogen sensing properties were measured by exposing nanoporous Pd films to various hydrogen concentrations: from 250 ppm to 1%. Figure 13.9 shows

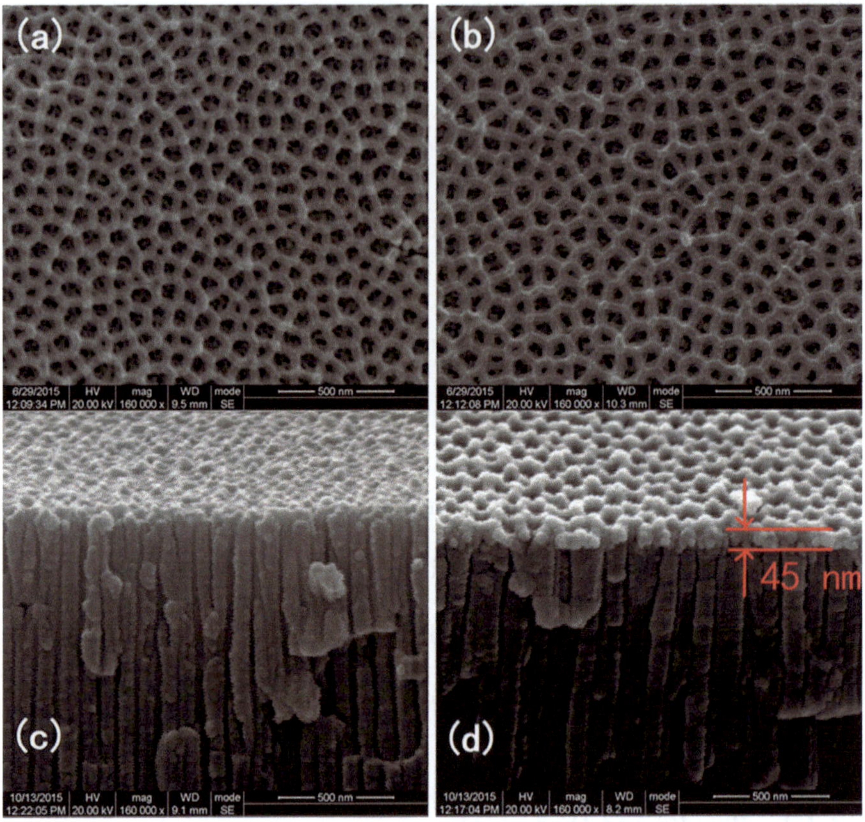

Fig. 13.8. SEM images of an AAO template and a nanoporous Pd film on AAO template: (**a**) *top view* of the AAO substrate, (**b**) *top view* of the 45-nm nanoporous Pd film on AAO substrate (**c**) *cross-sectional view* of the AAO substrate, (**d**) *Cross-sectional view* of the 45-nm nanoprous Pd film deposited on the AAO substrate

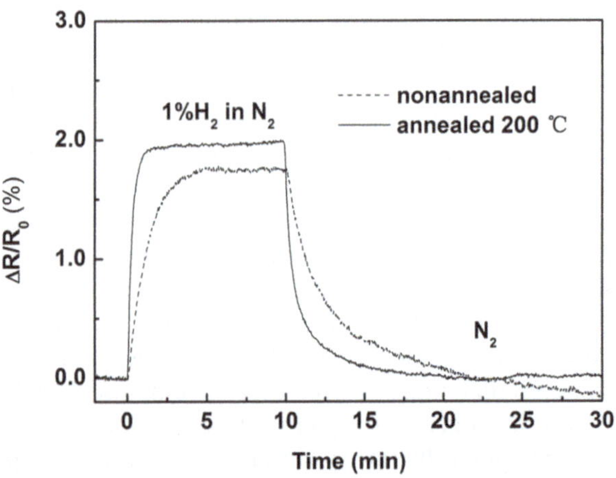

Fig. 13.9 Resistance responses of the 45-nm nanoprous Pd films exposed to 1% H_2 concentration. The dash line refers to the nonannealed nanporous Pd film. The solid line refers to the nanoporous Pd film annealed at 200°C

the steady-state response (defined as the relative resistance change, $\Delta R/R_0$) of hydrogen sensors based on the nanoporous Pd film (45 nm) with anneal at 200°C and that without anneal. The nonannealed sensor shows a relatively slow absorption and desorption. A response time (time delay to reach 90% of resistance change) of about 2.5 min has been obtained for the nonannealed sensor at H_2 concentration of 1%, which is a little shorter than that reported in our previous work (within 3 or 4 min) [49]. Deposited gold electrods have contributed to the fast response for good contact between Pd films and copper wires. Compared with sensors without anneal, the sensor with anneal responses faster. The response time of the sensor with anneal is 30 s at H_2 concentration of 1%, which is much shorter than that (about 2.5 min) of the sensor without anneal. Annealing treatment demonstrates significant improvement of the response time. Furthermore, we do not observe a drift in baseline for the annealed sensor, which is widely displayed by the nonannealed film and leads to instability issues.

As shown in the Fig. 13.10a, the flat response peaks suggest that the response of the annealed nanoporous Pd film can quickly reach a steady state. The steady-state response increases with hydrogen concentrations. Figure 13.10b shows the repeated

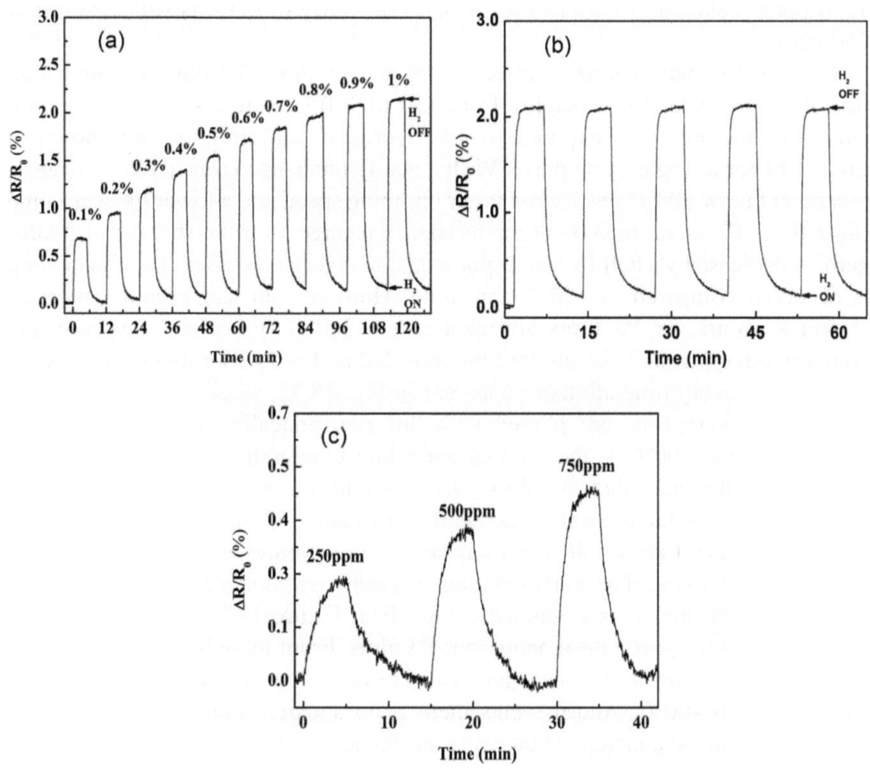

Fig. 13.10 Response of the 45-nm nanoporous Pd film annealed at 200°C. (**a**) at various hydrogen concentrations ranged from 0.1% to 1%. (**b**) at a repeated hydrogen concentration of 1%. (**c**) at low hydrogen concentrations

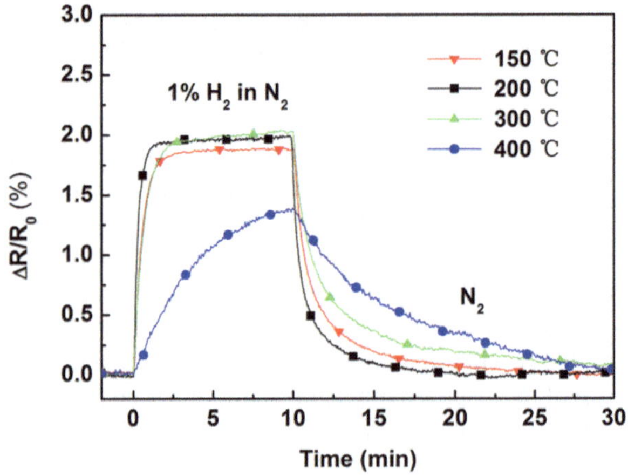

Fig. 13.11 Comparison of resistance responses of the annealed 45-nm nanoporous Pd films exposed to 1% hydrogen concentration. The *lines with solid inverted triangle*, *solid box*, *solid triangle* and *solid circle* refer to the nanoporous Pd films annealed at 150°C, 200°C, 300°C and 400°C for half an hour, respectively

response of the annealed nanoporous Pd film exposed to 1% hydrogen concentration at room temperature. It is observed that the sensitivity is about 2.1% at 1% H_2 concentration. Response depending on low hydrogen concentrations below 1000 ppm is shown in Fig. 13.10c. A low sensitivity of 0.3% is still obtained at 250 ppm.

In order to study annealing effect, a set of nanoprous Pd films via annealing treatment at temperatures ranging from 150 to 400°C were tested. The response curves of the Pd films exposed to 1% hydrogen concentration are shown in Fig. 13.11 for a single-cycle pulse. With a nearly unchanged sensitivity, annealing treatment below 300°C results in a faster climbing speed and a lower recovery time (time delay to retreat to 90% of the resistance change after the shut-off of hydrogen). A decrease by a half for the response and recovery time after 200°C annealing is observed compared to 150°C annealing. However, an unexpected prolonged response occurs for Pd films annealed above 300°C. Especially response and recovery are quite slow for the Pd film annealed at 400°C. The trends in response time and recovery time are more apparent in Fig. 13.12, where the response time and the recovery time are plotted with different annealing temperatures. It is determined that 200°C is the optimal annealing temperature. The response time for the Pd film annealed at 200°C decreases to about 30 s at 1% hydrogen concentration and the recovery time decreases to about 2 min.

As is discussed above, different annealing temperatures have been studied for nanoprous Pd films. The corresponding surface morphologies of the annealed nanoporous Pd films was examined using SEM. Figure 13.13 presents the SEM images of various post-anneal naoporous Pd films. From these images, we can find that the porosity remains unchanged with the annealing temperatures increasing from 150°C to 400°C. And it seems there is no apparent morphology change in visual. So it can be concluded that annealing in the vacuum chamber below 400°C maybe lead to less apparent morphology change for nanoporous Pd films on AAO. Since no apparent morphology change occurs, the microstructure of the annealed nanoporous Pd films was investigated using XRD.

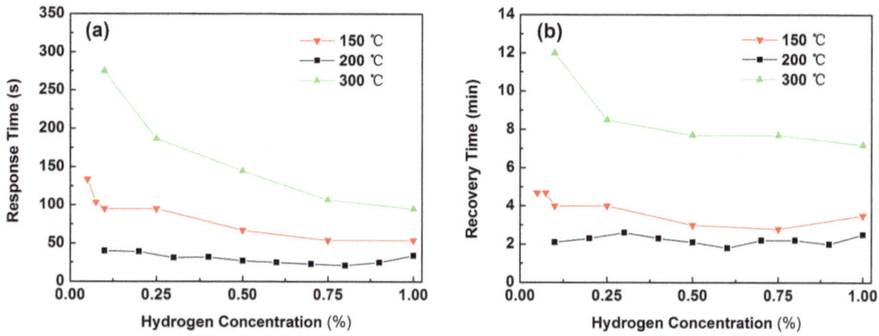

Fig. 13.12 Comparison of the response and recovery time of the annealed 45-nm nanoporous Pd films depending on hydrogen concentrations. (**a**) Response time, (**b**) Recovery time. The lines with solid inverted triangle, *solid box* and *solid triangle* refer to the nanoporous Pd films annealed at 150°C, 200°C and 300°C, respectively

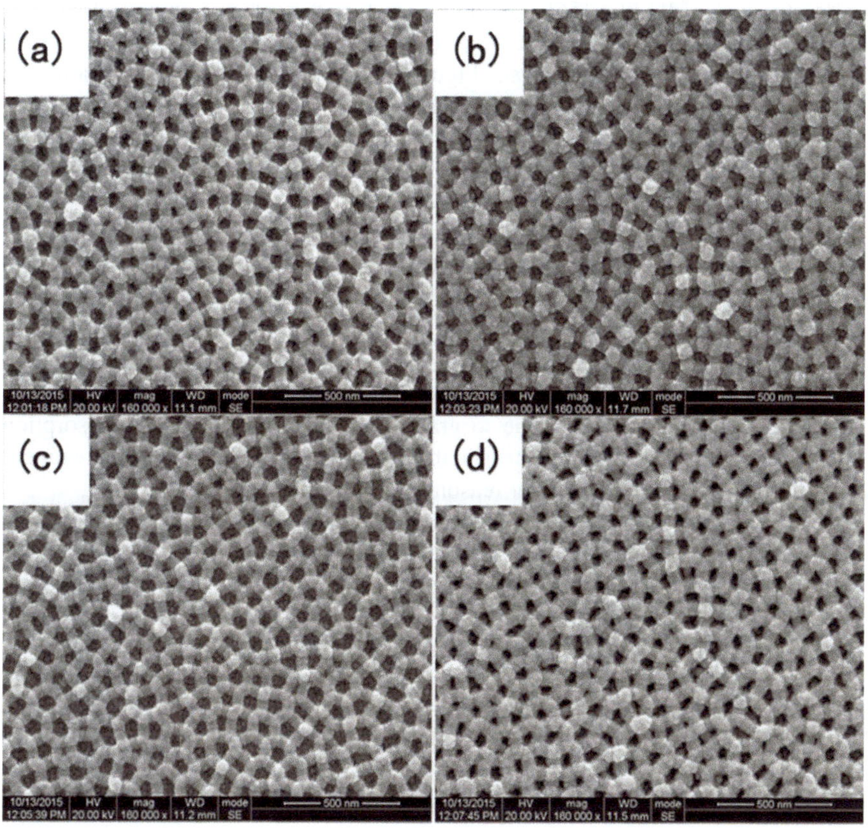

Fig. 13.13 SEM-revealed surface morphologies of the 45-nm nanoporous Pd films annealed at different temperatures: (**a**) 150°C, (**b**) 200°C, (**c**) 300°C, (**d**) 400°C

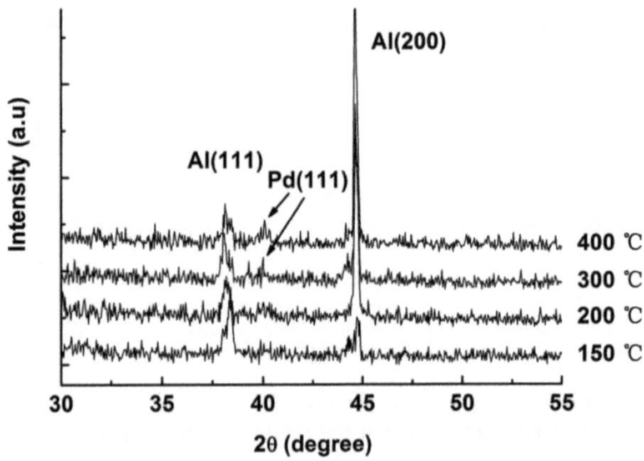

Fig. 13.14 Comparison of the normal θ–2θ scan XRD patterns of the annealed 45-nm nanoporous Pd films. The annealing temperatures are denoted on the right side of the spectra

It was reported that annealing pretreatment can change the corresponding microstructure of Pd film [56]. The microstructure dependence on annealing temperatures and its correlation with the resulting hydrogen absorption characteristics were examined using XRD analysis. Figure 13.14 presents the XRD spectra for the nanoporous Pd films annealed at temperatures of 150°C, 200°C, 300°C, 400°C, respectively. There are two significant peaks at about 38° and 44° in all the four samples. The two peaks are corresponding to the Al (111) and Al (200) planes (because the porous AAO layer is based on Al foil). The Pd (111) peak is at about 40°. There is no obvious Pd (111) peak for the nanoporous Pd film annealed at 150°C. The Pd peak is very weak since the thickness of the Pd film is only 45 nm. After annealing temperature increases to 400°C, the Pd peak is apparently stronger, which is due to the grain size growth after annealing treatment. Grain growth observed may be one of the factors causing the hydrogen absorption enhancement for the film annealed up to 200°C. As discussed in the literature [56], the reduction in disordered grain boundary due to grain growth may increases the absorption sites. Furthermore, the internal stress in the as-deposited film after anneal treatment is reduced, which may be another reason for the fast response time.

An unexpected slow response appears for nanoporous Pd films annealed above 300°C. The interaction between Pd and AAO support may explain this effect. It has been reported that a thin layer of aluminate phase is formed after calcination at 450°C on Pd/Al_2O_3 [57]. It can be speculated that the aluminate layer formed during annealing treatment blocks the hydrogen absorption on Pd surface. This effect strongly affects the sensing characteristics of Pd film.

In conclusion, we have shown a hydrogen sensor with fast response based on annealed nanoporous Pd films shaped with AAO template. After annealed at 200°C, the sensor showed a response time of 30 s and a recovery time of ~2 min upon exposure of 1% H_2 concentration. In addition, the baseline shift was not observed. Improved sensing performance may be attributed to internal stress relaxation and grain size growth, which was studied via XRD analyses. A post-deposition annealing proved to be an efficient way to improve the performance of this hydrogen sensor.

13.3 Pore-widening Treatment on the Performance of Nanoporous Pd Films Based Hydrogen Sensors

To develop a low cost and efficient Pd based hydrogen sensor, AAO are widely used as the substrate due to its honeycomb nanostructure, which can significantly increase the specific surface area of sensitive layer [58]. However, there is rare work studying the influence of pore-diameter of AAO on the performance of hydrogen sensor. In this section, we have researched the sensing performance of Pd based hydrogen sensors using AAO as substrates by pore-widening treatment with phosphoric acid as corrosive liquid based on Sect. 13.2 [53]. AAO templates can provide uniform nanoporous structures with controllable diameters as well as enhance the stability of hydrogen sensors [47–49]. It is demonstrated that different concentrations of H_3PO_4 and different pore-widening time lead to different pore-diameters of AAO, resulting in different performance of hydrogen sensors. The optimized hydrogen sensor shows a fast response time of 19 s at hydrogen concentration of 1% by pore-widening treatment of AAO with a time of 30 min and 5% H_3PO_4 concentration.

13.3.1 Materials and Methods

AAO templates were prepared using the same way in Sect. 13.2. Different concentrations of H_3PO_4 were used to make pore diameter wider. Conditions of samples with different concentrations of H_3PO_4 and different pore-widening times are shown in Table 13.2, named as a-series, b-series, and c-series, respectively. A 30 nm Pd thin film was deposited onto the upper surface of AAO templates by DC magnetron sputtering with the sputtering rate of 1.5 Å/s. The thickness of Pd film decreases due to that 30 nm thick Pd film has better sensing performance than that with a thickness of 45 nm in our study. So we chose 30 nm as the thickness of Pd film. We didn't use any annealing treatment when preparing Pd film. An aluminum layer with a thickness of 150 nm was deposited on both sides of the surface of nanoporous Pd thin film through E-beam evaporation to fabricate a couple of electrodes (size of 10 × 3 mm), due to its lowcost and good performance similar to that with Au electrode. The test of hydrogen sensors is done in the same way with Sect. 13.2. A mass flow controlled by the gas calibration system was used for mixing high purity nitrogen and 2% H_2 gas balanced in nitrogen for different

Table 13.2 Pore-widening conditions of different samples

Sample	a1	a2	a3	b1	b2	b3	c1	c2	c3	d
Concentration of H_3PO_4	2.5%	2.5%	2.5%	5%	5%	5%	6.5%	6.5%	6.5%	0
Time (min)	10	30	60	10	30	60	10	30	60	0

concentrations, and purging them into the chamber at a constant gas flow of 200 sccm. The difference is that we reduced the disdance between the chamber and gas calibration system for lightening the influence of gas transfer on the response speed.

13.3.2 Results and Discussion

Figure 13.15 presents the steady-state response $\Delta R/R_0$ of different samples at various H_2 concentrations. It is shown that Sample a2, b2, b3, and c2 have higher responses compared to other samples at various H_2 concentrations, which declares their corresponding pore-widening conditions may be appropriate. The steady-state response of Sample c2 is found to be the best (nearly 7% at 2% H_2 concentration) compared with Sample a2 (nearly 3% at 2% H_2 concentration), b2 (nearly 4% at 2% H_2 concentration) and b3 (nearly 4.5% at 2% H_2 concentration). The response of Sample b3 is higher than Sample b2 probably due to the larger pore diameter of AAO by pore-widening treatment. However, the response of sample c3 is relatively worst, which may be the result of damage from the hexagonal cells of AAO template treated with high concentration of H_3PO_4 and too long pore-widening time. The response of hydrogen sensor based on AAO without (W/O) pore-widening treatment is shown too (sample d).

The dynamic resistances of hydrogen sensors with different pore-widening treatments depending on time at RT are presented in Fig. 13.16. These devices are exposed to different concentrations of H_2 gas at 20 min intervals under measurement. With the increasing of pore-widening time and the concentration of H_3PO_4 solution, the base resistances of hydrogen sensors increase with the pore width of AAOs by pore-widening treatment. However, sample c3 only has a resistance of 1.42 Ω, which indicates that the AAO's surface may be destroyed by H_3PO_4 totally. Considering the same concentration of H_3PO_4, the results clearly indicate that hydrogen sensors with 30 min pore widening time demonstrate the best stable dynamic response.

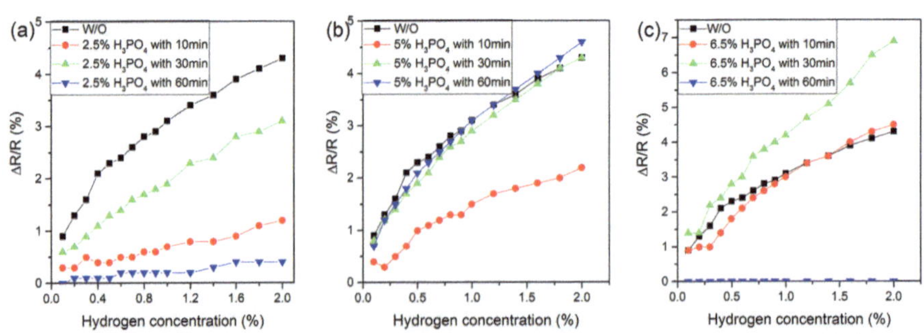

Fig. 13.15 Resistance change vs H_2 concentration. (**a**) a-series, (**b**) b-series, (**c**) c-series

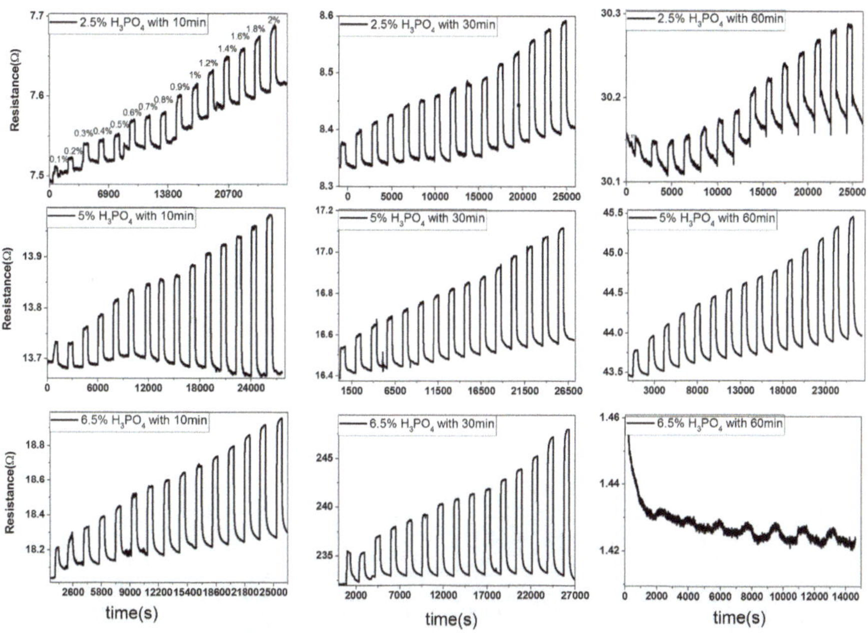

Fig. 13.16 Resistance response of a-series, b-series and c-series. The concentration of the hydrogen gas ranges from 0.1% to 2%. The step length is 0.1% from 0.1% to 1% and 0.2% from 1% to 2%, respectively

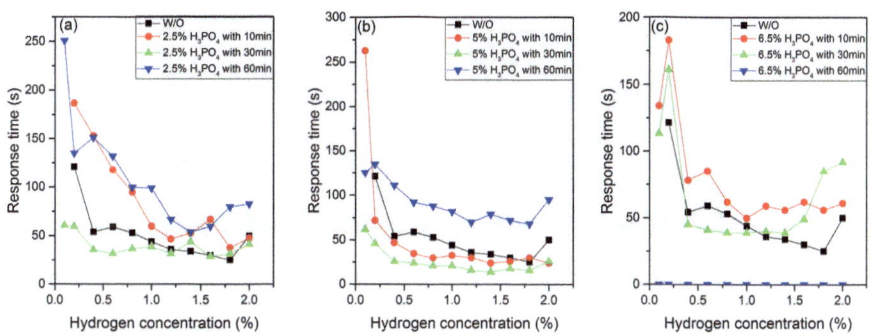

Fig. 13.17 Response time vs H_2 concentration. (**a**) a-series, (**b**) b-series, (**c**) c-series. The response time of sample d is shown too

In order to achieve the best pore widening condition, Fig. 13.17 and Fig. 13.18 describe the response time and recovery time of three series and sample d, respectively. The response time of hydrogen sensor based on AAO without pore-widening treatment is about 40 s at 1% H_2 concentration. There are obvious fluctuations of response times and recovery times for the devices treated with 2.5% and 6.5% concentration of H_3PO_4. It is shown that b-series have relatively stable response time and recovery time compared to a-series and c-series. From Fig. 13.17b, it is

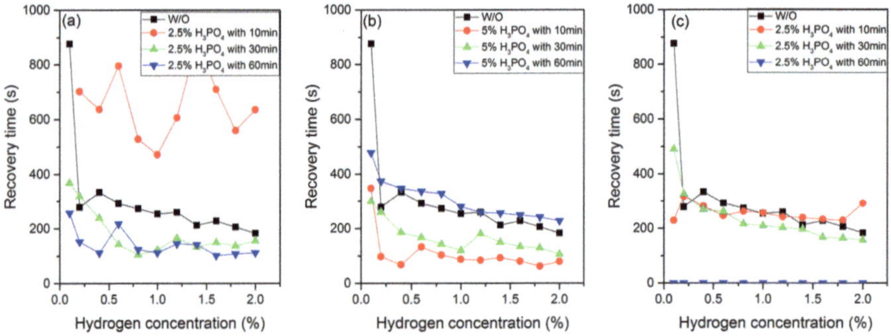

Fig. 13.18 Recovery time vs H_2 concentration. (**a**) a-series, (**b**) b-series, (**c**) c-series. The recovery time of sample d is shown too

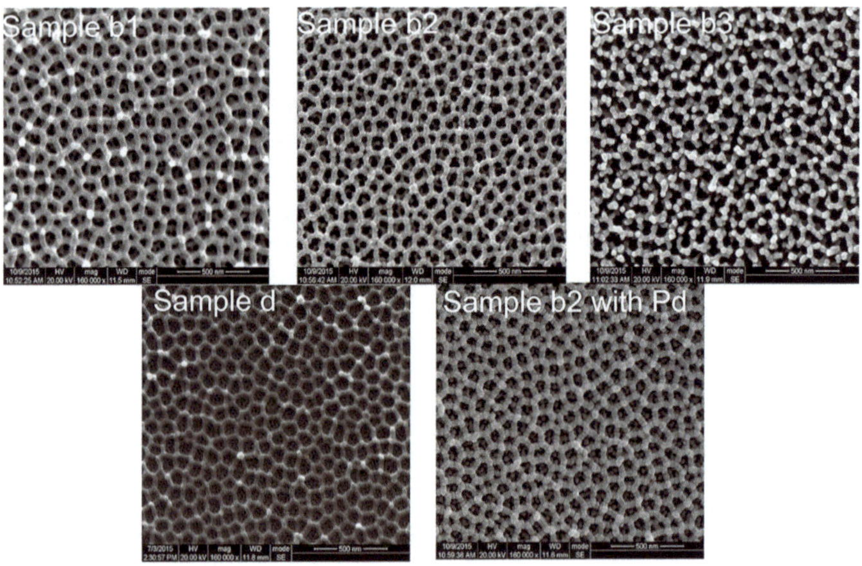

Fig. 13.19 SEMs of AAOs of Sample b1, b2, b3 and d without Pd film and SEM of AAO of Sample b2 with 30 nm Pd film

observed that sample b2 shows a fast response time of 19 s at H_2 concentration of 1%, which responses much faster than Sample b1, b3 and d. Obviously, the pore-widening treatment of AAO substrate through H_3PO_4 solution can affect the sensing performance of hydrogen sensor. The hydrogen sensor has the fastest response time based on AAO with pore-widening treatment in 5% H_3PO_4 solution for 30min, which should be the best condition of pore-widening treatment.

In order to study the mechanism of enhanced response time by pore-widening treatment of AAO, we researched SEMs of AAOs with different conditions of pore-widening treatment of b-series in detail. The top view SEM images of AAOs which were pore widened through 5% H_3PO_4 with different times are shown in Fig. 13.19. It is observed that pore diameters of AAOs become bigger than that of AAO without pore-widening treatment. In addition, longer pore-widening time results in bigger pore diameters. From the results of response time and recover time in Figs. 13.17 and 13.18, we find the best pore-widening treatment condition is 5% H_3PO_4 solution with 30 min. As is known to all, the large surface area of Pd film is beneficial for rapid absorption/desorption of hydrogen and its diffusion in Pd film [59]. The number of surface absorption/desorption sites of Pd film has influence on the response time of hydrogen sensor, that is larger surface absorption areas leads to fast and efficient response. Adams et al. [60] have found that the resistance of Pd thin film increases when exposing to hydrogen due to the formation of PdHx while the resistance decreases with the effect of break junction [44, 49]. In our work, the response time of hydrogen sensor is improved due to its larger surface area of Pd film changed by pore-widening treatment as well as the suitable break junction appears. However, it is found that Sample b3 has the slowest response, which may be induced by the damage of the hexagonal cells of AAO and introduction of more break junctions than other samples clear shown in Fig. 13.19.

In summary, we researched the sensing performance of hydrogen sensor through widening pore diameter of AAO templates using different concentration of H_3PO_4 solution with different pore-widening times. We find the best pore-widening treatment condition is 5% H_3PO_4 solution with a pore-widening time of 30min. A response time of 19 s at 1% H_2 concentration and a sensitivity of 3% have been achieved for the hydrogen sensor based on AAO with pore-widening treatment in 5% H_3PO_4 solution with a time of 30min. This kind of hydrogen sensor can detect a range concentration of H_2 from 0.1% to 2%.

13.4 Carbon Nanotube for High Performance of Nano Pd Films Based Hydrogen Sensors

To improve H_2 sensing performance, nanostructured Pd materials have been developed, and they have demonstrated a much quicker response to H_2 gas. Break-junctioned Pd nanowires reported by Favier et al. have shown the fastest response because of a volume-expansion-induced resistance decrease upon the absorption of H_2 [44]. However, these Pd nanowires are only sensitive to high concentrations of H_2 and have a bad stability. Our group reported a simple but effective method to fabricate a nanoporous Pd film sensor that has a remarkable sensing performance and is able to detect both dilute and high concentrations of H_2 gas (ranging from 250 ppm to 10% H_2) with quick response times at room temperature, through the use of pyrolytic carbon as a transition layer between the Pd film and the AAO template/substrate [47]. The device structure of resistive H_2 sensors based on nanoporous Pd film supported by pyrolytic carbon layer is shown in Fig. 13.20. As a robust and insulating substrate, the AAOs are a uniform nanotemplate for shaping nanoporous carbon and the subsequent nanoporous Pd film. Figure 13.21 shows the steady-state response of a nanoporous carbon-supported sensor at dilute

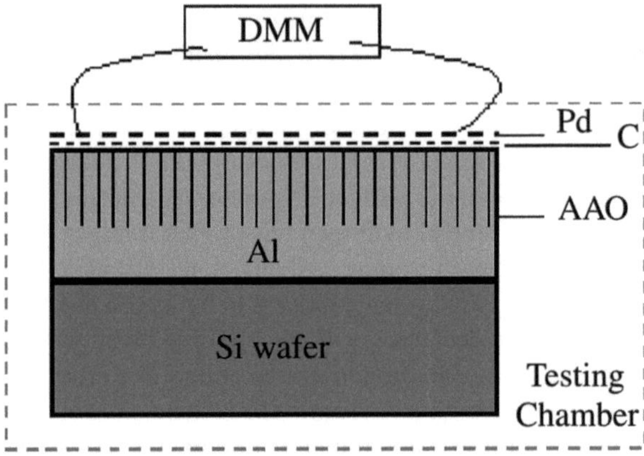

Fig. 13.20 Resistive H_2 sensors based on a pyrolytic, carbon-supported, nanoporous Pd film

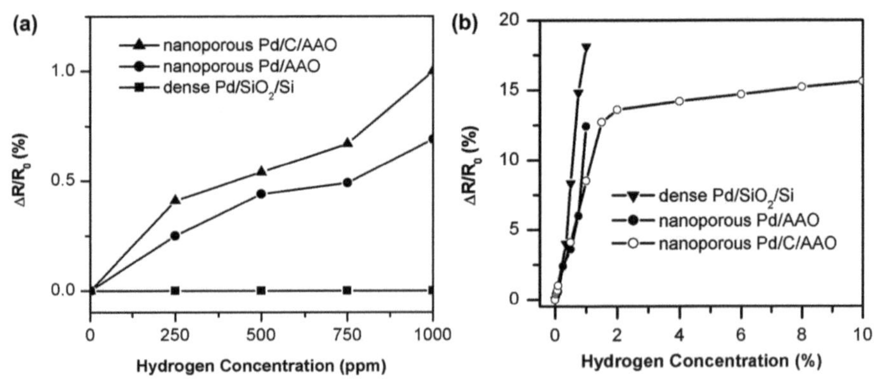

Fig. 13.21 Steady-state response of a nanoporous carbon-supported sensor at dilute (**a**) and at various (**b**) H_2 concentrations. Response curves of an AAO-supported sensor and a dense film sensor supported by an oxidized wafer are also presented

and at various H_2 concentrations. With the aid of pyrolytic carbon, a great improvement in the response to high H_2 concentrations can be achieved for nanoporous Pd film. As shown in Fig. 13.22, a typical response time for the carbon-supported nanoporous Pd film sensor is less than 4 min for dilute H_2 gas. At H_2 concentrations of 10%, the response time is less than 30 s.

Carbon nanotubes (CNTs) have the possibility to be used to prepare gas sensors because of their large surface area, excellent mechanical and electron properties [61, 62]. Kong et al. [63] first proposed gas sensors by using CNTs as gas absorption material. But researches have shown that pristine single-walled carbon nanotubes (SWCNTs) can hardly response to hydrogen, because there is lack of very strong interaction between hydrogen molecule and SWCNTs [64]. Our group reported a CNT-supported Pd film sensor, which is sensitive to hydrogen gas at both dilute and medium concentrations ranging from 100 ppm to 1.5% H_2 [48]. The structure diagram of CNT-supported Pd film sensor is shown in Fig. 13.23. Aligned CNTs

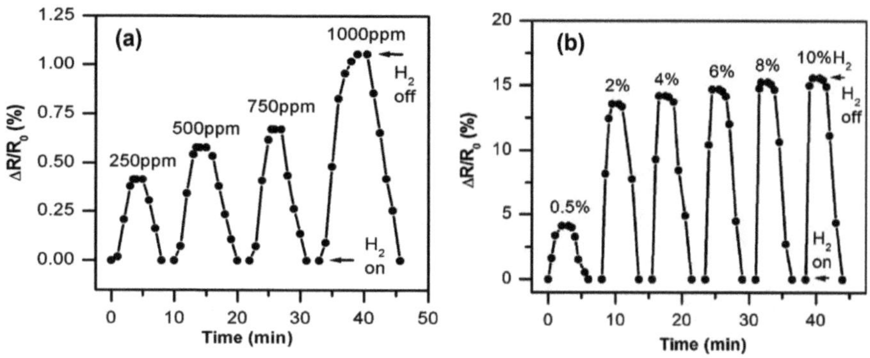

Fig. 13.22 Response of a nanoporous carbon-supported Pd sensor at dilute (**a**) and high (**b**) H_2 concentrations

Fig. 13.23 The structure diagram of CNT-supported Pd film sensor

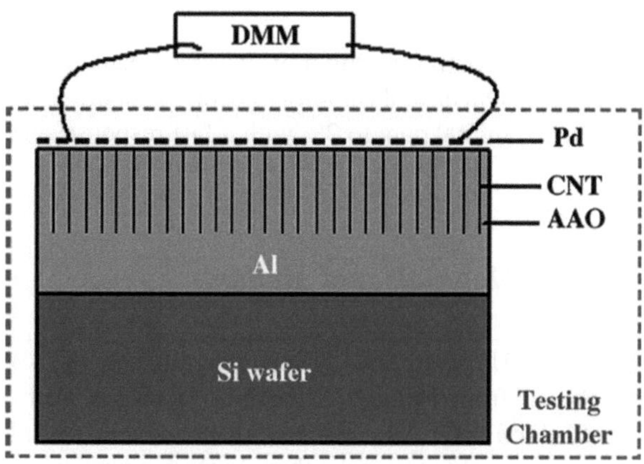

with partially crystallized structures were grown in the AAO template through pyrolysis of 10% acetylene in argon at 650°C. As shown in Fig. 13.24, typical response time of the CNT-supported Pd film sensor is less than 7 min for dilute H_2 gas, and less than 4 min for medium concentration hydrogen.

Palladium, which has remarkable ability to absorb abundant H_2 and highly selective response to H_2, has been used as a catalyst to improve the performance of SWCNTs based H_2 sensors. SWCNTs grafted with Pd nanoparticles (Pd-NPs) have been developed, and they showed a much faster response to H_2 gas. Sayago et al. [65] used the DC magnetron sputtering to deposit the Pd nanoparticles on the SWCNTs in order to improve the speed of response. Based on the same sensor structure, Jennifer et al. [66] used thermal evaporaion to prepare SWCNTs and Pd nanoparticle based H_2 sensors. However, these CNTs-Pd-NPs H_2 sensors don't response fast enough. Ju et al. [67] found that the SWCNTs modified with dendrimers before grafted with Pd-NPs had a much fast response time (3 s) to H_2 gas. A novel simple and effective way to fabricate H_2 sensors based on SWCNTs

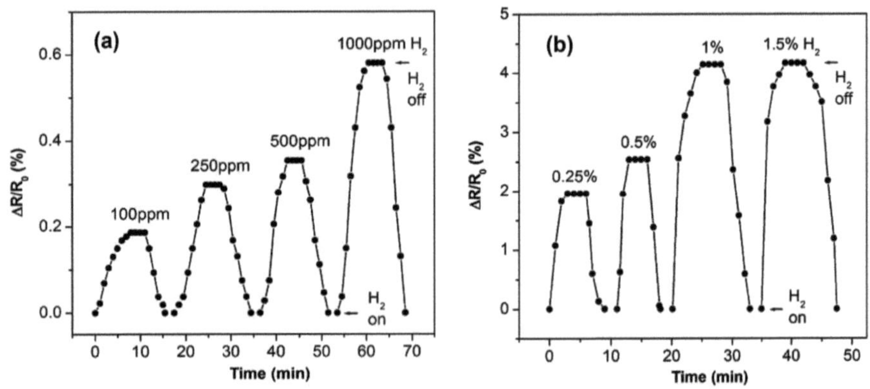

Fig. 13.24 Response of the CNT-supported Pd film sensor at dilute (**a**) and medium (**b**) H_2 concentrations

and Pd nanocomposite thin films has been demonstrated in our work. It was found that H_2 sensors have the ability to detect both dilute and high-concentration H_2 gas ranging from 500 ppm to 2% with a fast response speed at room temperature.

13.4.1 Materials and Methods

SWCNTs (Ultrapure SWCNTs aqueous dispersion, SWCNTs content 1 wt %) were purchased from Nanjing XFNANO Materials Tech Co., Ltd. (XFNANO). We mixed 1 ml SWCNTs aqueous dispersion in 49 ml deionized water and used ultrasonic machine to make them homogeneous mixing (That is SWCNTs content is changed to 0.02 wt%). Figure 13.25a shows the preparation process flow diagram of the hydrogen sensors. First, a p-type silicon wafer (20 × 20 mm) was rinsed using the RCA cleaning procedure. Subsequently, Sample I was coated with intersecting parallel SWCNTs by a two-step spray coating. Using 0.02 wt Ultrapure SWCNTs aqueous dispersion with a volume of 1 ml, we spray coated the column SWCNTs on the p-type silicon masked by an aluminum comb mask. The structure of the mask was made of 22 fingers of 250 μm width, and the distance between fingers is 250 μm. Then we spray coated the row SWCNTs by rotating the mask with 90 degrees clockwise. The amount of Ultrapure SWCNTs aqueous dispersion was the same as in the first step. For comparative study, Sample II was spray coated with SWCNTs on the surface of the silicon substrate without the aluminum comb mask. Then a 15 nm thick Pd thin film was deposited on the surface of the SWCNTs in an ultra-high vacuum DC magnetron sputtering system with a base pressure of 4×10^{-4} Pa. The rate of deposition was 0.15 nm/s. To fabricate hydrogen sensors, aluminum counter electrode (an area of 20 × 1 mm) with a thickness of 150 nm was deposited on Samples I and Samples II via electron beam evaporation. Figure 13.25b shows the side view of Sample I.

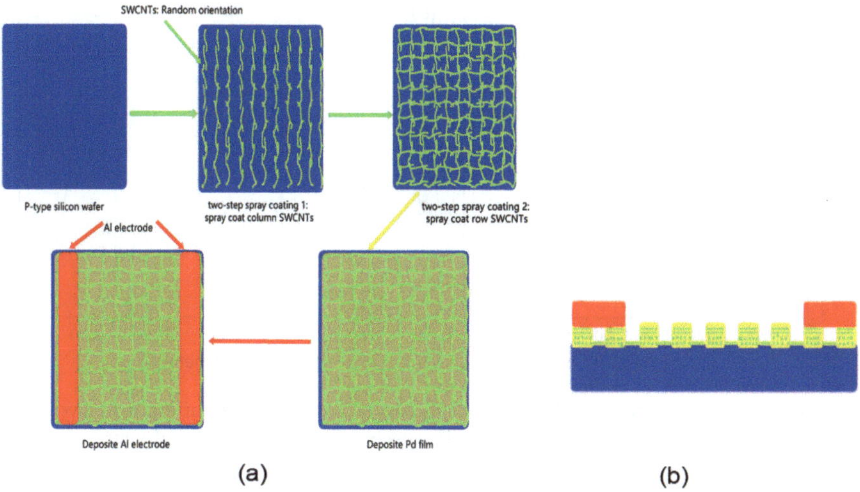

Fig. 13.25 (**a**) Preparation process flow diagram of the hydrogen sensor, (**b**) Side view of Sample I

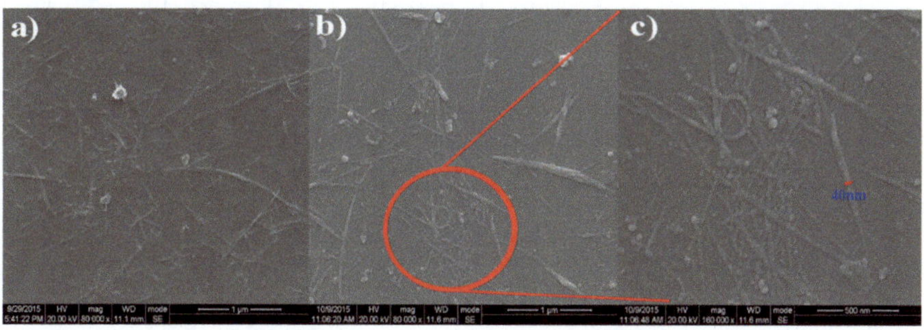

Fig. 13.26 SEM of the intersecting parallels SWCNTs and Pd nanocomposite thin film. (**a**) SWCNTs without Pd, (**b**) SWCNTs with Pd, (**c**) lager magnification factor of SWCNTs with Pd

13.4.2 Results and Discussion

Figure 13.26 shows the SEM image of the intersecting parallels SWCNTs and Pd nanocomposite thin film grown on the Sample I. Figure 13.26a, b are the images of the SWCNTs without Pd and the SWCNTs with Pd, respectively. As shown in Fig. 13.26b, the deposition of 15 nm thick Pd thin film did not change the surface morphology of SWCNTs. We investigated the H_2 gas sensing performance of Samples I and Samples II by introducing different H_2 concentrations (from 500 ppm to 2%) to the test chamber using the dry air as the carrier gas. Figure 13.27a shows that response of Sample I at different H_2 concentration levels exhibits an increased relationship between the response and the concentration of H_2, and a high

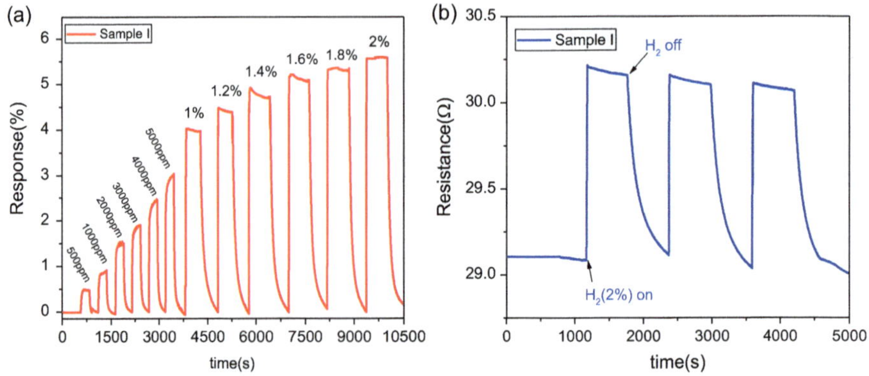

Fig. 13.27 Response of Samples I at different H_2 concentrations and the corresponding veriation in the resistance under repeated testing at 2% H_2 concentration balanced with dry air

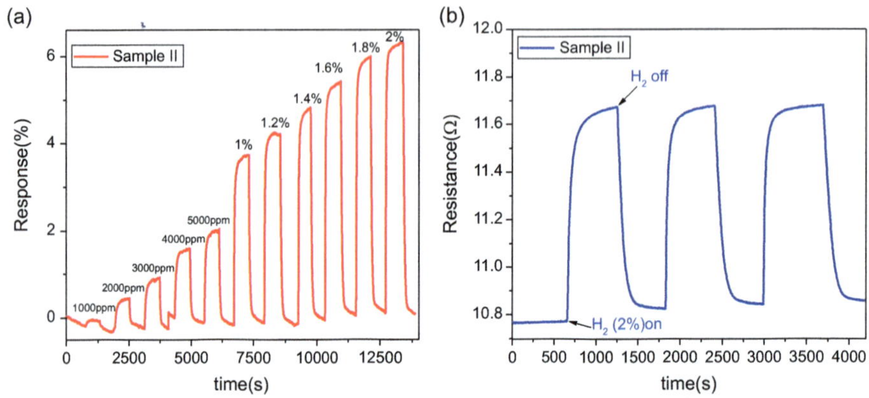

Fig. 13.28 Response of Samples II at different H_2 concentrations and the corresponding veriation in the resistance under repeated testing at 2% H_2 concentration balanced with dry air

response of 5% is achieved at 2% H_2 concentration. A typical response time of Sample I at different H_2 concentration levels is less than 1 min as shown in Fig. 13.27a. At the H_2 concentration of 500 ppm, the response time is less than 40s. At the H_2 concentration of 2%, the response time is about 8 s, which is faster than that of CNT-based hydrogen sensors reported by Yong et al. [68]. Meanwhile, Fig. 13.27b shows a typical veriation in the resistance under repeated testing at 2% H_2 concentration. We did not observe any shift in base line for the intersecting parallel SWCNTs-Pd (Sample I) hydrogen sensors. Figure 13.28 shows the response and repeated tests of Sample II, which shows a slower response of 54 s at 2% H_2 concentration than that of Sample I (8 s at 2% H_2 concentration).

Figure 13.29 shows the response time and recovery time of Sample I and Sample II. Obviously, through the use of SWCNTs, the improvement for detecting H_2 in a

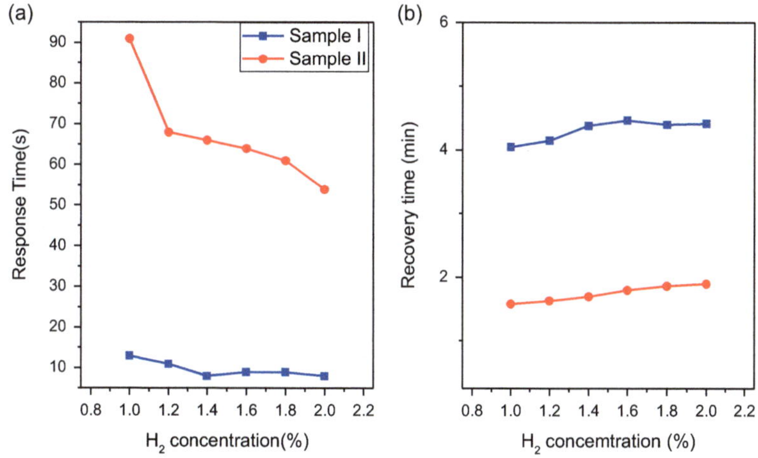

Fig. 13.29 (**a**) Response time and (**b**) Recovery time of Sample I and Sample II

wide range of concentration is quite clear in comparison to the dense Pd thin film [49]. There is a difference in the response time of Sample I and Sample II because of the intersecting parallel SWCNTs using a two-step spray coating. With the aid of intersecting parallel SWCNTs, a great improvement of response speed at different H_2 concentrations can be achieved. It is known that nanoporous Pd films can provide rapid absorption/desorption of H_2 due to its porous structure compared to dense Pd films [48, 69]. Kong et al. [63] showed that Pd functionalized SWCNTs have remarkably enhanced the H_2 detection for the underlying interactions between H_2, Pd and the nanotube. As reported by Kong et al., the DC magnetron sputtering deposition led Pd decorating the sidewall of the SWCNTs, which is similar to our SEM images in Fig. 13.26, suggesting that the Pd nanoparticles adhere to SWCNTs. Sample I and Sample II have the same fundamental mechanism due to the interaction of SWCNTs, Pd and H_2. When exposed to H_2 at room temperature, hydrogen molecules dissociate into H atom on Pd surface. This process results in lowering the work function of Pd. Then electrons in Pd transfer to SWCNTs, which causes decreasing of hole-carriers in the p-type SWCNTs and thus increasing of hydrogen sensor's resistance. After we turn off hydrogen, oxygen in dry air plays a key role in the reverse process. Oxygen combines two H atoms on the Pd surface, which generates two hydroxyls. Then a hydroxyl and a H atom combine to form H_2O.

In conclusion, we have demonstrated a new way for fast response hydrogen sensor using the intersecting parallels SWCNTs by a two-step spray coating with Pd thin film decoration. Sensors based on the intersecting parallels SWCNTs displayed a very fast response time of nearly 8 s at 2% hydrogen. Our simple fabrication method can optimize the performance of hydrogen sensors for high sensitivity, large detection range of H_2 and fast response.

13.5 Summary

Hydrogen sensors can play an important role in ensuring the safety of people and property wherever hydrogen is produced, stored, transported or used. There are a number of technologies that have been developed and demonstrated for the detection of hydrogen gas. Pd is considered a good hydrogen sensing material, especially in nano structures. Pd based resistive H_2 sensors have attracted much attention due to their simple device structure and fabrication process. In this chapter, AAO substrates have been used for fabricating nanoporous Pd films and the resistive hydrogen sensors based on nanoporous Pd films have been investigated. This kind of sensor can be simplely fabricated especially for fuel cells and transportation applications owing to their small size, low temperature operation and fast response. We enhanced the response speed of nanoporous Pd films based hydrogen sensors by post-deposition annealing treatment of Pd films and pore-widening treatment of AAOs. A fast response time of 19 s at hydrogen concentration of 1% was obtained by pore-widening treatment of AAO with a time of 30 min and 5% H_3PO_4 concentration. SWCNTs and Pd nanocomposite thin films were investigated for fast response hydrogen sensors, which shows a fast response time of nearly 8 s at 2% hydrogen concentration. In order to meet the current and emerging technical applications, further work and new technology are required for the ongoing development of hydrogen sensing technologies, in terms of basic research into new materials and sensor principles as well as applied research.

Acknowledgments This work was supported by National Natural Science Foundation of China under Grant Nos. 61405026, 61371046, 61474016 and 61421002, National Higher-education Institution General Research and Development Funding (ZYGX2014J049), and China Postdoctoral Science Foundation (Grant No. 2014 M552330).

References

1. C. Schwandt, D.J. Frey, Hydrogen sensing in molten aluminum using a commercial electrochemical sensor. Ionics **6**, 3–4 (2000)
2. W.J. Buttner et al., An overview of hydrogen safety sensors and requirements. Int. J. Hydrog. Energy **36**, 2462–2470 (2010)
3. L. Boon-Brett, J. Bousek, G. Black, P. Moretto, P. Castello, T. Hübert, U. Banach, Identifying performance gaps in hydrogen safety sensor technology for automotive and stationary applications. Int. J. Hydrog. Energy **35**, 373–384 (2010)
4. A. Katsuki et al., H_2 selective gas sensor based on SnO_2. Sensors Actuators B Chem. **52**, 30–37 (1998)
5. B.S. Kang et al., AlGaN/GaN-based metal-oxide-semiconductor diode-based hydrogen gas sensor. J. Appl. Phy. Lett. **84**, 1123 (2004)
6. Y. Sakamoto et al., Hysteresis behaviour of electrical resistance of the Pd-H system measured by a gas-phase method. J. Phys. Condens. Matter **8**, 10511–10520 (1996)
7. T. Hübert et al., Hydrogen sensors-A review. Sensors Actuators B Chem. **157**(2), 329–352 (2011)

8. E. Lee, I. Hwang, J. Cha, H. Lee, W. Lee, J. Pak, J. Lee, B. Ju, Micromachined catalytic combustible hydrogen gas sensors. Sens. Actuators B Chem **153**(2), 392–397 (2010)
9. R.C. Thomas, R.C. Hughes, Sensors for detecting molecular hydrogen based on Pd metal alloys. J. Electrochem. Soc. **144**(9), 3245–3249 (1997)
10. G. Ertl, J. Koch, Adsorption studies with a Pd(111) surface in Adsorption-Desorption Phenomena: Proceedings of the II International Conference, April 1971, 1971, 345-357.
11. M.Z. Jacobson, W.G. Colella, D.M. Golden, Cleaning the air and improving health with hydrogen fuel-cell vehicles[J]. Science **308**(5730), 1901–1905 (2005)
12. J. Lee, J.S. Noh, S.H. Lee, B. Song, H. Jung, W. Kim, W. Lee, Cracked palladium films on an elastomeric substrate for use as hydrogen sensors. Int. J. Hydrog. Energy **37**, 7934–7939 (2012)
13. J.G. Firth, A. Jones, T.A. Jones, The principles of the detection of flammable atmospheres by catalytic devices. Combust. Flame **20**, 303–311 (1973)
14. V.R. Katti, A.K. Debnath, S.C. Gadkari, S.K. Gupta, V.C. Sahni, Passivated thick film catalytic type H_2 sensor operating at low temperature. Sens. Actuators B Chem. **84**, 219–225 (2002)
15. E. Jones, in *Solid State Gas Sensors*, ed by P. Moseley, B. C. Tofield. The pellistor catalytic gas detection (Adam Hilger, Bristol, 1987), pp. 17–31
16. J.R. Stetter, J. Li, Amperometric gas sensors-a review. Chem. Rev. **108**, 352–366 (2008)
17. X. Lu et al., Solid-state amperometric hydrogen sensor based on polymer electrolyte membrane fuel cell. Sens. Actuator B **107**, 812–817 (2005)
18. L.P. Martin, A.-Q. Pham, R.S. Glass, Electrochemical hydrogen sensor for safety monitoring. Solid State Ionics **175**, 527–530 (2004)
19. F.C. Lin, Y. Takao, Y. Shimizu, M. Egashira, Zinc oxide varistor gas sensors: effect of the Bi2O3 content on the H_2-sensing properties. J. Am. Ceram. Soc. **78**, 2301–2306 (1995)
20. S.J. Ippolito et al., Hydrogen sensing characteristics of WO_3 thin film conductometric sensors activated by Pt and Au catalysts. Sens. Actuators B Chem. **108**, 154–158 (2005)
21. A. Lee, B. Reedy, Temperature modulation ins emiconductor gas sensing. Sens. Actuators B Chem. **60**, 35–42 (1999)
22. J. Ravi Prakash, A.H. McDaniel, M. Horn, L. Pilione, P. Sunal, R. Messier, R.T. McGrath, F.K. Schweighardt, Hydrogen sensors: role of palladium thin film morphology. Sens. Actuators B Chem. **120**, 439–446 (2007)
23. E. Sennik et al., Sputtered platinum thin films for resistive hydrogen sensor application. Mater. Lett. **177**, 104–107 (2016)
24. S. Öztürka et al., Pd thin films on flexible substrate for hydrogen sensor. J. Alloys Compd. **674**, 179–184 (2016)
25. K. Potje-Kamloth, Semiconductor junction gas sensors. Chem. Rev. **108**, 367–399 (2008)
26. S. Basu, S. Roy, C. Jacob, Ruthenium as Schottky metal for SiC-based high temperature hydrogen sensors. Mater. Technol. Hydrog. Econ. **801**, 193–198 (2004)
27. C. Pandis, N. Brilis, E. Bourithis, D. Tsamakis, H. Ali, S. Krishnamoorthy, A.A. Iliadis, M. Kompitsas, Low-temperature hydrogen sensors based on Au nanoclusters and Schottky contacts on ZnO films deposited by pulsed laser deposition on Si and SiO_2 substrates. IEEE Sens. J. **7**, 448–454 (2007)
28. M.A. Butler, Optical fibre hydrogen sensor. Appl. Phys. Lett. **45**, 1007–1009 (1984)
29. A. Trouillet, E. Marin, C. Veillas, Fibre gratings for hydrogen sensing. Meas. Sci. Technol. **17**, 1124–1128 (2006)
30. B. Chadwick, J. Tann, M. Brungs, M. Gal, A hydrogen sensor based on the optical generation of surface plasmons in a palladium alloy. Sens. Actuators B Chem. **17**, 215–220 (1994)
31. R.C. Hughes, W.K. Schubert, Thin films of Pd/Ni alloys for detection of high hydrogen concentrations. J. Appl. Phys. **71**, 542–544 (1992)
32. M.K. Kumar, M.S. Ramachandra Rao, S. Ramaprabhu, Structural, morphological and hydrogen sensing studies on pulsed laser deposited nanostructured palladium thin films. J. Phys. D. Appl. Phys. **39**, 2791–2795 (2006)

33. P. Kumar, L.K. Malhotra, Palladium capped samarium thin films as potential hydrogen sensors. Mater. Chem. Phys. **88**, 106–109 (2004)
34. R.K. Joshi, S. Krishnan, M. Yoshimura, A. Kumar, Pd nanoparticles and thin films for room temperature hydrogen sensor. Nanoscale Res. Lett. **4**, 1191–1196 (2009)
35. K. Yoshimura et al., New hydrogen sensor based on sputtered Mg-Ni alloy thin film. Vacuum **83**, 699–702 (2009)
36. N. Taguchi, Japan. Pat. 45-38200 (1962), 4738840 (1963) 50-23317
37. S. Shukla et al., Hydrogen-discriminating nanocrystalline doped-tin-oxide room-temperature microsensor. J. Appl. Phys. **98**, 104306 (2005)
38. S. Shukla et al., Room temperature hydrogen response kinectics of nanomicro-integrated doped tin oxide sensor. Sens. Actuators B Chem. **120**, 573–583 (2007)
39. C.-H. Han, S.-D. Han, I. Singh, T. Toupance, Micro-bead of nano-crystalline F doped SnO_2 as a sensitive hydrogen gas sensor. Sens. Actuators B Chem. **109**, 264–269 (2005)
40. M.M. Jamshidi et al., Development of palladium-based hydrogen thin film sensor using silicon oxide substrate. Indian J. Phys. **87**, 511–515 (2013)
41. B. Liu et al., Improved room-temperature hydrogen sensing performance of directly formed Pd/WO_3 nanocomposite. Sens. Actuators B **193**, 28–34 (2014)
42. X.Q. Zeng, Y.L. Wang, H. Deng, M.L. Latimer, Z.L. Xiao, J. Pearson, W.K. Kwok, Networks of ultrasmall Pd/Cr nanowires as high performance hydrogen sensors. ACS Nano **4**, 7443–7452 (2011)
43. J.Y. Wang, E.H. Yue, G. Yu, Y.K. Xiao, Z.Z. Chen, Preparation of Pd-Ni alloy nanowires by AAO template. Rare Metal Mater. Eng. **36**, 126–129 (2007)
44. F. Favier, E.C. Walter, M.P. Zach, T. Benter, R.M. Penner, Hydrogen sensors and switches from electrodeposited palladium mesowire arrays. Science **293**, 2227–2231 (2001)
45. N. Liu, M.L. Tang, M. Hentschel, H. Giessen, A.P. Alivisatos, Nanoantenna-enhanced gas sensing in a single tailored nanofocus. Nat. Mater. **10**, 631–636 (2011)
46. B. Xie, S. Zhang, F. Liu, X. Peng, F. Song, G. Wang, M. Han, Response behavior of a palladium nanoparticle array based hydrogen sensor in hydrogen–nitrogen mixture. Sens. Actuators A Phys. **181**, 20–24 (2012)
47. D. Ding, Z. Chen, A pyrolytic, carbon-stabilized, nanoporous Pd film for wide-range H_2 sensing. Adv. Mater. **19**, 1996–1999 (2007)
48. D. Ding, Z. Chen, S. Rajaputra, V. Singh, Hydrogen sensors based on aligned carbon nanotubes in an anodic aluminum oxide template with palladium as a top electrode. Sens. Actuators B Chem. **124**, 12–17 (2007)
49. D. Ding, Z. Chen, C. Lu, Hydrogen sensing of nanoporous palladium films supported by anodic aluminum oxides. Sens. Actuators B Chem. **120**, 182–186 (2006)
50. X. Yang, X. Wei, S. Wu, T. Wu, Z. Chen, S. Li, High-quality self-ordered TiO_2 nanotubes on fluorine-doped tin oxide glass. J. Mater. Sci. Mater. Electron. **26**, 7081 (2015)
51. M. Hao, S. Wu, H. Zhou, et al., Room-temperature and fast response hydrogen sensor based on annealed nanoporous palladium film[J]. J. Mater. Sci. **51**(5), 2420–2426 (2016)
52. C. Lu, Z. Chen, K. Saito, Hydrogen sensors based on Ni/SiO_2/Si MOS capacitors. Sens. Actuators B Chem. **122**, 556–559 (2007)
53. F. Yang, S.C. Kung, M. Cheng, J.C. Hemminger, R.M. Penner, Smaller is faster and more sensitive: the effect of wire size on the detection of hydrogen by single palladium nanowires. ACS Nano **4**, 5233–5244 (2010)
54. K.J. Laidler, *Chemical Kinetics* (McGraw-Hill Book Company, New York, 1965)
55. T. Xu, M.P. Zach, Z.L. Xiao, D. Rosenmann, U. Welp, W.K. Kwok, G.W. Crabtree, Self-assembled monolayer-enhanced hydrogen sensing with ultrathin palladium films. Appl. Phys. Lett. **86**, 203104 (2005)
56. Z. Zhao, M.A. Carpenter, Annealing enhanced hydrogen absorption in nanocrystalline Pd Au sensing films. J. Appl. Phys. **97**, 124301–124307 (2005)

57. A.S. Ivanova, E.M. Slavinskaya, R.V. Gulyaev, V.I. Zaikovskii, O.A. Stonkus, I.G. Danilova, L.M. Plyasova, I.A. Polukhina, A.I. Boronin, Metal–support interactions in Pt/Al_2O_3 and Pd/Al_2O_3 catalysts for CO oxidation. Appl. Catal. B Environ. **97**, 57–71 (2010)
58. C. Lu, Z. Chen. Anodic Aluminum Oxide-Based Nanostructures and Devices [M] Encyclopedia of Nanoscience and Nanotechnology. American Scientific Publishers, 2011, 11(259): 235-259.
59. T. Kiefer, L.G. Villanueva, F. Fargier, J. Brugger, The transition in hydrogen sensing behavior in noncontinuous palladium films [J]. Appl. Phys. Lett. **97**(12), 121911 (2010)
60. B.D. Adams, A. Chen, The role of palladium in a hydrogen economy [J]. Mater. Today **14**(6), 282–289 (2011)
61. L.Q. Rong, C. Yang, Q.Y. Qian, X.H. Xia, Study of the nonenzymatic glucose sensor based on highly dispersed Pt nanoparticles supported on carbon nanotubes. Talanta **72**, 819–824 (2007)
62. C. Thelander, M.H. Magnusson, K. Deppert, L. Samuelson, P.R. Poulsen, J. Nygård, J. Borggreen, Gold nanoparticle single-electron transistor with carbon nanotube leads. Appl. Phys. Lett. **79**, 2106–2108 (2001)
63. J. Kong, M.G. Chapline, H. Dai, Functionalized carbon nanotubes for molecular hydrogen sensors [J]. Adv. Mater. **13**, 1384 (2001)
64. A. Fediai, D.A. Ryndyk, G. Cuniberti, Electron transport in extended carbon-nanotube/metal contacts: Ab initio based Green function method. Phys. Rev. B **91**, 165404 (2015)
65. I. Sayago, E. Terrado, E. Lafuente, M.C. Horrillo, W.K. Maser, A.M. Benito, J. Gutierrez, Hydrogen sensors based on carbon nanotubes thin films. Synth. Met. **148**, 15–19 (2005)
66. J. Sippel-Oakley, H.T. Wang, B.S. Kang, Z. Wu, F. Ren, A.G. Rinzler, S.J. Pearton, Carbon nanotube films for room temperature hydrogen sensing. Nanotechnology **16**, 2218 (2005)
67. S. Ju, J.M. Lee, Y. Jung, E. Lee, W. Lee, S.J. Kim, Highly sensitive hydrogen gas sensors using single-walled carbon nanotubes grafted with Pd nanoparticles. Sens. Actuators B Chem. **146**, 122–128 (2010)
68. Y. Du, Q. Xue, Z. Zhang, F. Xia, Z. Liu, W. Xing, Enhanced hydrogen gas response of Pd nanoparticles-decorated single walled carbon nanotube film/SiO_2/Si heterostructure. AIP Adv. **5**, 027136 (2015)
69. J. Kong, N.R. Franklin, C. Zhou, M.G. Chapline, S. Peng, K. Cho, H. Dai, Nanotube molecular wires as chemical sensors. Science **287**, 622–625 (2000)

Part III
Nanostructured Oxide

Chapter 14
Microstructure of the Nanostructured Oxide Composite Thin Films and Its Functional Properties

Xingkun Ning

14.1 Introduction

Nanocomposite oxide film is one of the most important research topics in nanoscale devices and sensors since availability of appropriate nanomaterial represents the foundation for exploring novel physical properties and applications associated with the nanometer length scale. The coexistence of multi-physical properties of the two phases engenders the material with the "product" property (i.e., the composite exhibits responses that are not available in the individual component phases), thus allowing an additional degree of freedom in the design of actuators, transducers, and storage devices [1]. For example, nanocomposite films of FE material with a FM have been synthesized toward future devices based on the control of magnetization with an electric field [2, 3], nanocomposite films of binary oxide with the manganites are particularly interesting as designing the memory-readers of digital devices and magnetic sensor due to the large values of low field magnetoresistance (LFMR) effect [4–6], and nanocomposite films combined of nanoparticles of noble metal and the CD oxides or DE oxides could potentially be applied for the electrode or optical devices as the enhanced conductivity or optical properties [7–10]. In addition, the phases with various well-controlled sizes and morphologies in the self-assembled nanocomposite films have attracted tremendous attention for creating a range of nanoscale structure types, leading to novel functionalities by interface coupling between the constituents [5, 11].

In this chapter, a basic characterization of the microstructure of the self-assembled nanocomposite oxide films is given together with general description of the forming mechanism. Then the typical microstructure and the corresponding

X. Ning (✉)
Hebei Key Lab of Optic-electronic Information and Materials, The College of Physics Science and Technology, Hebei University, Baoding 071002, People's Republic of China
e-mail: xkning@alum.imr.ac.cn

T. Li, Z. Liu (eds.), *Outlook and Challenges of Nano Devices, Sensors, and MEMS*,
DOI 10.1007/978-3-319-50824-5_14

physical properties in some simple oxide nanocomposite, multiferroics based, manganites based, and CD or DE based nanocomposite films, are described, followed by some discussions on the potential applications. In this chapter we confine ourselves to phenomenological descriptions where the microstructure and the interface coupling are mainly taken into account with simple structure models and circuit models.

14.2 Growth Mechanisms

Research in oxide nanocomposites-artificially created various microstructures involving transition metal oxide compounds-has flourished over the last decade [12, 13]. With the developments during the past decade in our ability to make extremely high quality nanocomposite oxide films, and therefore the corresponding microstructure, considerable interest has now grown in the science of microstructure-related and microstructure-tuned functionalities [14, 15]. It is then noted that designing the microstructure of the nanocomposite films is most important to tune the physical properties. In order to be able to modulate the microstructures, one of the key issues of research focus in this context is the mechanism of nanocomposite film growth and the corresponding microstructure.

14.2.1 Nucleation and Growth

For the nanocomposite oxide films with nucleation and growth mechanism, the themodynamics of the two phase, ratios of the two phase, epitaxial growth relation with the substrates (Fig. 14.1) and the surface/interface energies play important role in determining the microstructure of the nanocomposite films. On the one hand, there is no doubt that the volume ratio of the two phases plays critical role in determining the mircostructure. In general, when the ratios are much larger than 1:1, nanogranular microstructure would formed with the majority phase as the parent phase and the minority phase as the second phase. For the system with nearly equally content, the form of the microstructure depend on the different surface and interficial energies of the two phase. Under these circumstance, the microstructure of the nanocomposite films could be modulated by different substrate or using different crystal orientations. For example, as shown in Fig. 14.1, in a model $BiFeO_3/CoFe_2O_4$ (BFO/CFO)system, a (001) substrate results in rectangular-shaped CFO nanopillars in a BFO matrix; in contrast, a (111) substrate leads to triangular-shaped BFO nanopillars in a CFO matrix, dependent on the surface and interface energy [16]. On the other hand, he thermodynamics of the two phases strongly affect the microstructure of the nanocomposite films. With lower growth rate and higher insitu deposition temperature, the second phase completely wets the parent phase, forming the nanocolumnar structure embed in the parent

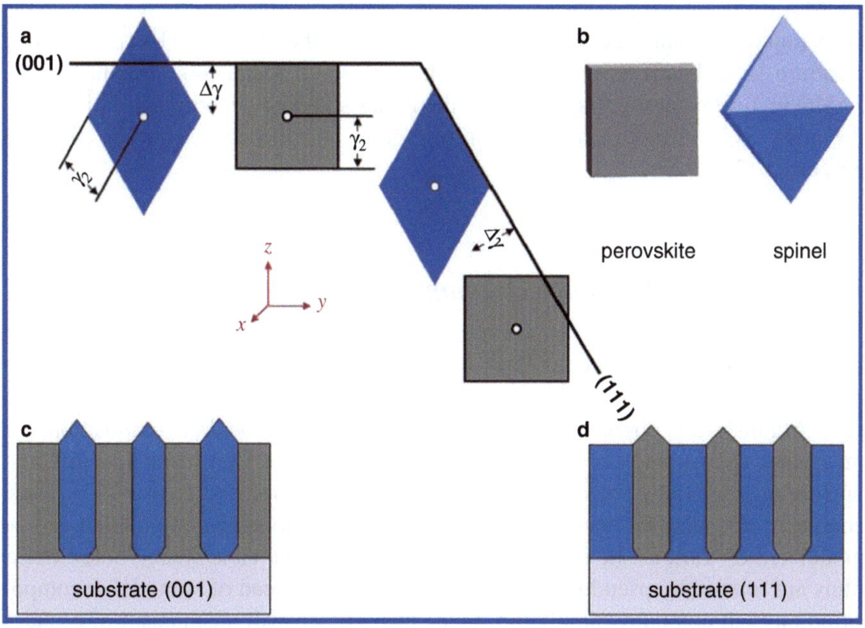

Fig. 14.1 Schematics of perovskite-spinel nanostructures on the (100) and (111) surfaces of $SrTiO_3$[16]

phase. In contrast, with higher growth rate and lower in-situ deposition temperature, the random nanoparticles structure forms instead of nanocolumnar structure because the insufficient kinetic energy to nucleate and grow on the surface of the parent phase, forming the nanogranular structure.

14.2.2 *Spinodal Decomposition*

Spinodal decomposition is a mechanism for the rapid unmixing of a mixture of liquids or solids from one thermodynamic phase, to form two coexisting phases. Spinodal decomposition can be constrasted with nucleation and growth where it occurs when the mixture is such that there is essentially no barrier to nucleation of the new phase. Recently, spinodal decomposition is of interest for the oxide materials which are mainly depend on the temperature, oxygen vacancy and distortion [17]. As have been reported in the $BaTiO_3$ (BTO)/STO self-assembled nanocomposite films, it is only possible that the spinodal decomposition happen around 300°C [18]. The microstructure after the spinodal decomposition is determined by the oxygen content and the distortion. Research in distortion-induced spinodal decomposition at the nanocomposite has recently. For example, the ordered nanocheckerboard-like structure have been reported in the $Mg(Mn,Fe)_2O_4$ spinel structure [19]. In addition to this, Guion et al. also have been found the nanocheckerboard-like structure in the $(Nd_{1-x}Li_x)TiO_3$ perovskite structure [20]. In the year of 2008, Park et al. have found this structure in the $ZnMnGaO_4$ films at the first time, and then after spinodal decomposition is one of the main tools to modulated the microstructure and the corresponding physical properties [21].

14.2.3 Pseudo-Spinodal Decomposition

Pseudo-spinodal growth contains of both spinodal decomposition and nucleation and growth and was first observed in metallic alloy systems more than 20 years ago [22]. Although this phase transition is in line with the spinodal decomposition, the interfacial energy and the lattice distortion energy should be taken into account when using the Gibbs free energy to describe the nucleation and growth [23]. Thus, compared the spinodal decomposition, the pseudo-spinodal decomposition can be modulated by various. In 2009, Ni et al. showed that the nanocheckerboard oxide films are formed by pseudo-spinodal decomposition instead of spinodal decomposition through computer simulation. Since then, the growth mechanism of pseudo-spinodal decomposition becomes an important research area in nanocomposite films [24].

14.3 Synthesis and Microstructure

In the nanocomposite films, microstructures with nanogranular, nanomultilayer, nanocolumnar and nanocheckerboard characteristics can be divided according to the size, shape, distribution, arrangement and other characteristics of the second phase. Figure 14.2 shows the cross-sectional schematic view of the structure of nanogranular, nanomultilayer, nanocolumnar. Nanogranular (0-3 structure) configuration means that there are two phases in the composite, one consisting of zero-dimension particulates with the same shape and size, and the other is three-dimensional bulk [25]. Nanomultilayer (2-2 structure) refers to that the second phase and the parent phase are layered with each other. At last, the nanocolumnar (1-3 structure) refers to that the second phase formed columnar structure which uniformly embedded in the parent phase.

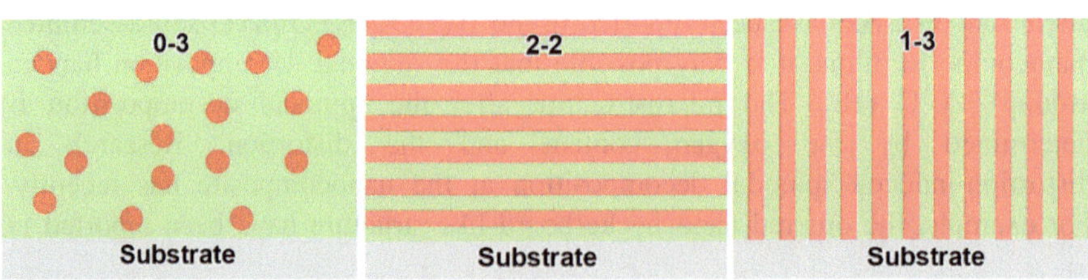

Fig. 14.2 Schematic of the 0-3, 2-2 and 1-3 structures

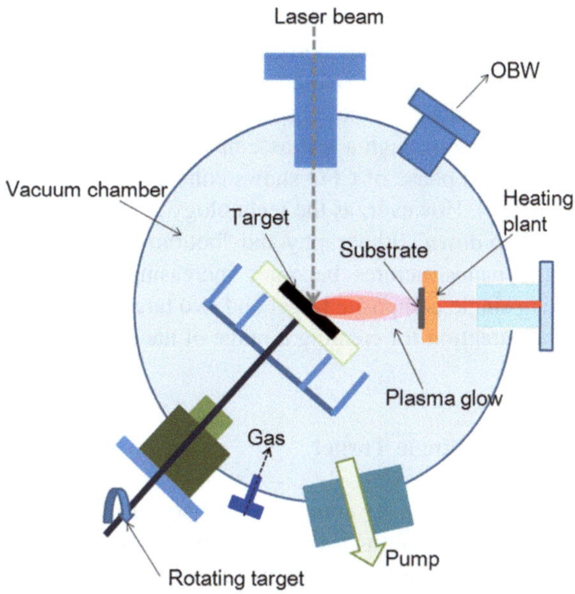

Fig. 14.3 A sketch of pulsed laser deposition (PLD) system

14.3.1 Synthesis Method

Researchers have synthesized FE/ferromagnetic, manganite/metal oxide, superconductor/metal oxide, DE/metal oxide, and CD oxide/noble metal nanocomposite films with nanogranular, nanolayer, nanocolumnar and nanocheckerboard configurations via physical deposition techniques (e.g., pulsed laser deposition (PLD), magnetron sputtering, molecular beam epitaxy (MBE)) and chemical method (e.g., sol-gel method). Among these method, PLD is a widespread technique for well-controlled epitaxial growth of complex oxide films (Fig. 14.3). Here, we just describe three important methods to synthesize nanocomposite films that are based on the PLD technique. The first, templated self-assembly of functional oxide nanocomposites films have been applied to synthesize nanocompostie films with nanocolumnar structures. The second, a single composite targets is used to deposit nanocomposite oxide film with different microstructure. The third, two targets with the single phase are used to deposit nanocomposite oxide film with different microstructure.

14.3.1.1 Template

Templating of the locations and geometry of the self-assembled columnar would dramatically increase the utility of the nanocomposite oxide films. Seeding the nucleation of the nanocolumnar was first demonstrated for 100–200 nm period, 20–30 nm thick BFO/CFO nanocomposites grown by pulsed electron deposition on a substrate patterned with CFO seed particles made from a CFO film by electron beam lithography and ion milling [26]. Aimon et al. using a new method for templating BFO/CFO nanocomposites based on selective nucleation in topographc feature produced in the substrate by either top-down focused ion beam irradiation or wet etching through a self-assembled triblock terpolymer mask. In this procedure, the second phase of CFO shows enhanced aspect ratio and large magnetic anisotropy [27]. However, as the technology is pushed to ever-smaller length scale, both the "top-down" lithography and "bottom-up" self-assembly methods used to fabricated nanostructures becomes increasingly challenging. Self-assembly process using single composite target and two target with single phase have attract tremendous attention for creating a range of nanoscale structure types [28, 29].

14.3.1.2 Single Target

For the one single composite targets, the self-assembly mechanisms of oxide nanocomposite films, as well as the advantageous of Self-organization are important to modulate the microstructure of the films. In order to predict new compositions one need to select systems that have the potential to undergo these spinodal processes. The kinetic and thermodynamic parameters are a guide to this selection [15]. For example, Bi_2FeMnO_6 with supercell structure have been prepared using the composite target in which compounds Bi_2O_3, Fe_2O_3 and MnO_2 were mixed in stoichiometric ratio [30]. Chen et al. also synthesized $La_{0.7}Sr_{0.3}MnO_3$/ZnO (LSMO/ZnO) nanoculomnar structures using the $(LSMO)_{0.5}/(ZnO)_{0.5}$ composite target [31]. The main challenge for single composite target method is to create structures of the desired compositions. Since the compositions rely on phase separation from a single starting composition, achieving the desired compositions is harder than for standard multilayer growth where each individual composition is grown from a separate target.

14.3.1.3 Two Targets

For the two targets with single phase, deposition is always carried out from two targets sequentially, producing submonolayer films from each target to ensure intermixing of the film [32–34]. The target was rotated sufficiently fast to ensure that much less than one unit cell was deposited per one rotation so that complete intermixing of the two constituents could take place [35]. For example, Imai et al. have epitaxially grown $Bi_4Ti_3O_{12}$ (BTO):CFO nanocomposite films using BTO and CFO single targets which were either simultaneously or alternatively deposited on 0.5 wt.% Nb-doped STO (001) (Nb:STO) by PLD [36]. Zhao et al. Have successfully synthesized composite films with nanocolumnar configurations in the LSMO/V_2O_3 (VO) system. The average deposition fluxes for LSMO and VO at this condition are 2 Å/min and 1.3 Å/min, respectively. To grow LSMO/VO nanocomposite, LSMO and VO targets were alternately switched in situ to the deposition position for certain amount of time for the deposition with LSMO grown first on the substrate [37]. A dual target technique [38] has also been employed to synthesize CFO/$SrRuO_3$ nanostructured thin films [39]. Using the two targets systems, the nanocomposite thin films with nanogranular, nanomultilayer, nanocolumnar and nanocheckerboard configuration have been successfully in the LSMO:NiO and $La_{0.7}Ca_{0.3}MnO_3$:NiO (LCMO:NiO) [4, 5]. For example, as shown in Fig. 14.4. to produce the LCMO:NiO composite films, two separated targets of

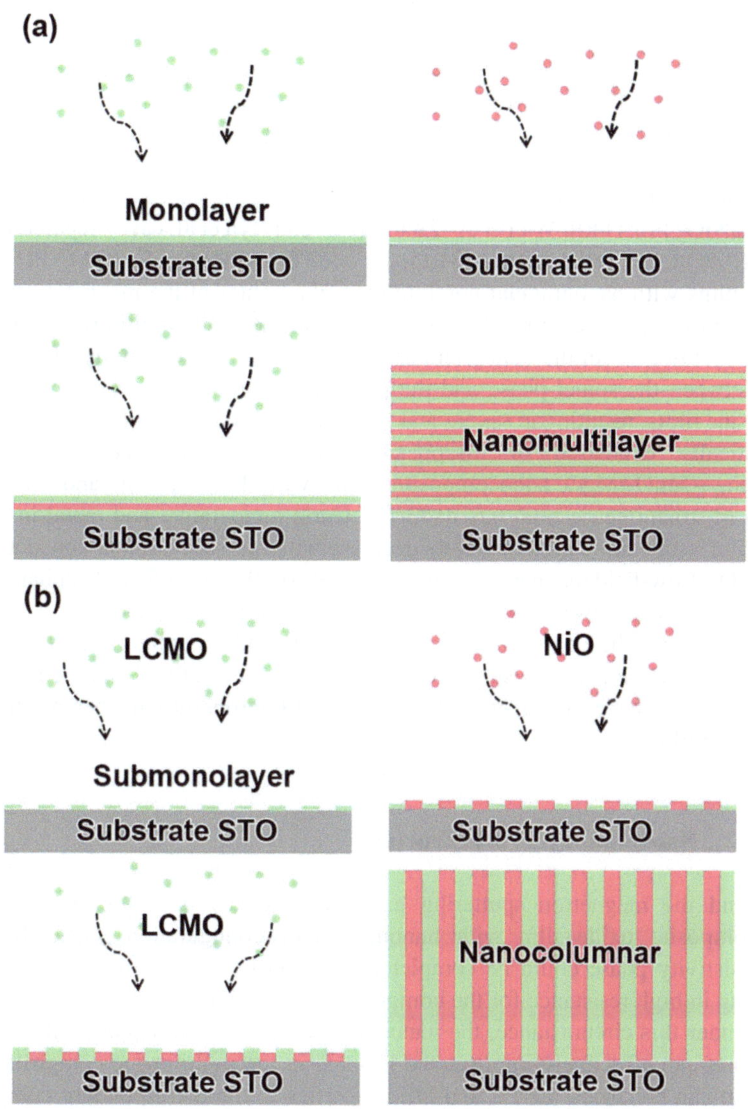

Fig. 14.4 Schematic illustrations of the growth modes for the LCMO:NiO composite films with equal volume fractions of LCMO and NiO but with different microstructures: (**a**) nanomultilayer configuration and (**b**) nanocolumnar configuration [5]

LCMO and NiO were alternately used to deposit LCMO and NiO on the STO substrate. By controlling the deposition time for LCMO and NiO, LCMO:NiO nanocomposite films with the configurations of nanocolumnar, nanogranular and nanomultilayer have been formed [5].

14.3.2 Microstructure

14.3.2.1 Nanogranular (0-3 Structure)

In this kind of nanocomposite structure, The physical properties such as the magnetoelectric coupling (MEC), DE, magnetoresistance and electrical conductivity have been effectively improved. For example, Bi et al. fabricated a CD composite film of $LaNiO_3$/Pt with the nanogranular configuration by radiofrequency co-sputtering. The conductivity of the composite films enhanced compared with the pure $LaNiO_3$ films [40]. Wan et al., Zhang et al. and Ryu et al. have prepared Pb(Zr, Ti)O_3(PZT)/CFO, $(x)Bi_{3.15}Nd_{0.85}Ti_3O_{12}/(1\text{-}x)$CFO and PZT/NiFe2O4 (NFO) composite films with nanogranular configuration, the values of the magnetic coupling coefficient increased compared with the single phase of the multiferroics material [41–43]. However, in the nanogranular structures, the low resistivity of the interface between the spinel phase and the perovskite parent phase form a CD paths across the films, resulting in severe leakage of the films and the FE properties were dramatically decreased or disappeared. Staruch et al. have synthesized $La_{0.67}Sr_{0.33}MnO_3$:MgO nanocomposite films with La, Sr, Mn, and Mg were mixed in stoichimetric ratio 0.67/0.33/1/x. Curie temperature and metal-insulator transition temperature systematically decrease with increasing molar concentration of Mg(O). Low-field magnetoresistance (LFMR) of films significantly enhanced by Mg addition, values were ~35.5% and ~83.2% respectively [44, 45].

In summary, the composite films with nanogranular configuration can effectively improve the magnetoelectric coupling effect, conductivity and the magnetoresistance, but have an adverse effect on the FE properties in the multiferroic composite films.

14.3.2.2 Nanomultilayer (2-2 Structure)

PLD and the magnetron sputtering have been widely used for preparing the nanocomposite oxide film with nanomultilayer configuration. Generally, the low-resistivity phase could be completely isolated by the high-resistance phase, resulting in high resistance for the composite films with nanomultilayer configuration. Under this circumstance, the nanocomposite films with nanomultilayer configuration can effectively improve the leakage in the multiferroic composite film and also can enhanced the FE and DE properties in the corresponding composite films. In addition to this, the LFMR in the manganites based composites films also improved in this microstructure [46]. Ma et al. and Deng et al. synthesis BTO/CFO and BTO/NFO layered structure and measured the response of the FE and ferromagnetic coupling coefficient, showing an enhanced magnetoelectric coupling properties compared with the nanocomposite with the nanogranular configuration [47, 48]. Recently, researchers have found that by minimizing the thickness of each layer in the nanomultilayer structure one can increase the areas of the interface, leading to enhanced coupling properties and also explore novel physical properties. For example, Dawber et al. synthesised a FE/DE superlattices and showed that the FE properties is dramatically improved. Novel physical properties also have been found in this system such as the novel polarization reversal mode [49, 50]. In FE superlattices, modulating the periodic thickness is a flexible and efficient method to regulate the physical properties [51, 52]. Hesse et al. showed that a remarkable enhancement in polarization and DE constants, and unusual ferroelectricity have been observed in ferroelectric superlattices which are expected to be used in the preparation of high-performance microelectronic devices [53, 54].

In summary, compared with the nanogranular configuration, nanocomposite oxide films with the nanomultilayer configuration can effectively improve the physical properties such as magnetoelectric coupling in multiferroic system and LFMR in manganites composite film.

14.3.2.3 Nanocolumnar (1-3 Structure)

Recently, there has been significant interest in the design and development of the the nanocomposite oxide films with nanoculomnar configuration due to the exciting prospects for next generation devices in data processing and storage [55]. PLD and template methods have been widely used to synthesis the composites films with nanocolumnar configuration. Compared with the lateral interface of, the effect of vertical interface on the physical properties of metal oxide films is profound [56]. The dominant strain mechanism for the two phase in the nanocolumnar structure arises from the vertical interface, rather than from the lateral interface which are much larger than the pure films of the two phase. It is noted that the vertical interface is more efficient to control the out-of-plane lattices constant than the lateral interface [57], particular in the relatively thick composites films. The long-range strain field across the films can effectively tune the physical properties which could be used to modulated the functionalities of the composite films. For example, Zhao et al. have precisely tuned the microstructure of the $(YBa_2Cu_3O_{7-\delta})_{1-x}/(BaZrO_3)_x$ (YBCO/BZO), and the Curie temperature of the YBCO have been widely tuned for the microstructures with the nanocolumnar configuration [11]. Yang et al. have investigated the vertical interface effect on the physical properties of epitaxial BFO/Sm_2O_3 (BFO/SmO) nanocomposite films. The nanocomposite shows a lower than expected DE loss, due to the presence of a larger vertical interfacial area in the nanocomposite, causing a reduction of the oxygen depletion in the BFO phase relative to a pure BFO film [56, 58]. In addition, MacManus-Driscoll et al. and Ning et al. have prepared LSMO:ZnO and LSMO:NiO composite films with nanocolumnar configuration.

The value of the LFMR have been largely enhanced due to the strong electron scattering at the ferromagnetic/insulator interfaces and magnetic tunnel junctions (MTJs) of LSMO/insulator/LSMO at the nanometer scale [4, 6, 31].

In summary, compared with the nanocomposite films with the nanogranular and nanomultilayer configuration, nanocolumnar structure shows large advantage in multiferrotic based and manganite base composite films by tuning the DE and LFMR properties.

14.3.2.4 Nanocheckerboard (0-0 Structure)

One major challenge for further enhancing the LFMR in a high temperature range in the manganites based composite films is how to tune the crystalline grain size of the second phase. It is also noted that in the manganites based composite thin films, if manganites phase contacts with an insulator phase and then other manganites phase forming a manganites/insulator/manganites structure, a magnetic tunneling junction (MTJ) can be formed. The MTJs of manganites/insulator/manganites are important to enhance the LFMR properties of composite films. To achieve this object, the major challenge is how to control the grain size of the insulator phase within 1–2 nm, which is the optimized thickness of a middle layer in a MTJ structure [59, 60]. As have been mentioned in the method section, the nanocheckerboard oxide films which are formed by pseudo-spinodal decomposition shows controllable grain size within 10 nm. MacManus-Driscoll et al. have reported the in-plane nanocheckerboard-like structure in the LSMO:ZnO nanocolumnar structure (nanocheckerboard configuration along the in-plane direction, but nanocolumnar configuration along the out-of-plane direction.) [15]. Although, the values of the LFMR are largely enhanced in the low-temperature range, but rapidly decreased near the Curie temperature. In 2014, nanocheckerboard configuration along the out-of-plane direction have been successfully synthesized at the first time by Ning et al. [4], as shown in Fig. 14.5. The composite films with the checkerboard-like structure exhibit a large LFMR in the high temperature range from 200 to 300 K with the highest value of 17% at 250 K under a magnetic field of 1 T.

14.4 Nanocomposite Films and the Potential Application

14.4.1 Multiferroic Nanomposite Film

For the multiferroic composites, which incorporate both ferri-/ferromagnetic (FE/FM) phases, typically yield giant magnetoelectric coupling (ME) response above room temperature, makes them ready for technological applications. The availability of high-quality nanostructured composites makes it easier to tailor their properties through epitaxial strain, atomic-level engineering of chemistry, and

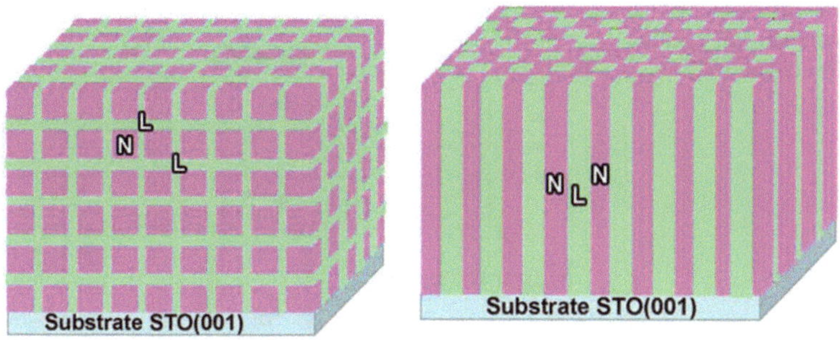

Fig. 14.5 Schematic illustrations of (**a**) the ideal checkerboard-like structure and (**b**) the nanocolumnar structure in the nanocomposite thin films. "L" stands for LSMO phase and "N" stands for NiO phase

interfacial coupling. Of interest, motivated by on-chip integration in microelectronic devices, the coupling interaction between nanosized FE and magnetic oxides nanostructured composites have recently been deposited in a film-on substrate geometry. Thus the ME composites are ready for technological applications. Promising applications include magnetic field sensors thus complementing Hall sensors and current measurement probes, transducer, filters, oscillators, phase shifters, memory devices, and so on [61]. Figure 14.6 shows a sketch of the ME gyrators and the equivalent circuit which have important applications as voltage gain devices, current sensors, and other power conversion devices. An extremely high voltage gain effect under resonance drive has been reported in long-type ME laminates consisting of Terfenol-D and PZT layers [62].

Recently, perpendicular recording technology has been introduced in hard disk drives, as the traditional longitudinal recording technology is approaching its storage limit due to superparamagnetic effects. Large and well controlled magnetic anisotropy in nanocolumnar FE/FM compostite films is thus necessary to guarantee bit stability, while preserving high density storage capability [63]. Integration of nanocomposites on a Si platform would provide a path towards large scale and low cost devices such as multiferroic memory and logic. Integration of self-assembled epitaxial BFO/CFO multiferroic nanocomposites on silicon substrates have now been successfully synthesised by Kim et al. [64].

In 2004, Zheng et al. have firstly synthesis BTO/CFO naocompsoites films with nanocolumnar structure by PLD, in which CFO formed prismatic columns which extended from the substrate to the top surface, embedded in a BTO matrix [1]. Since this pioneering work, the nanocomposite films in this systems with the structure of nanoculumnar configuration have been widely reported due to the higher magnetoelectric coupling than their laminate counterparts and the nanogranular structures [65–68]. Figure 14.7 shows the typical SEM images of the BFO/CFO nanocomposite films with nano columnar structures reported by Dix et al. [69].

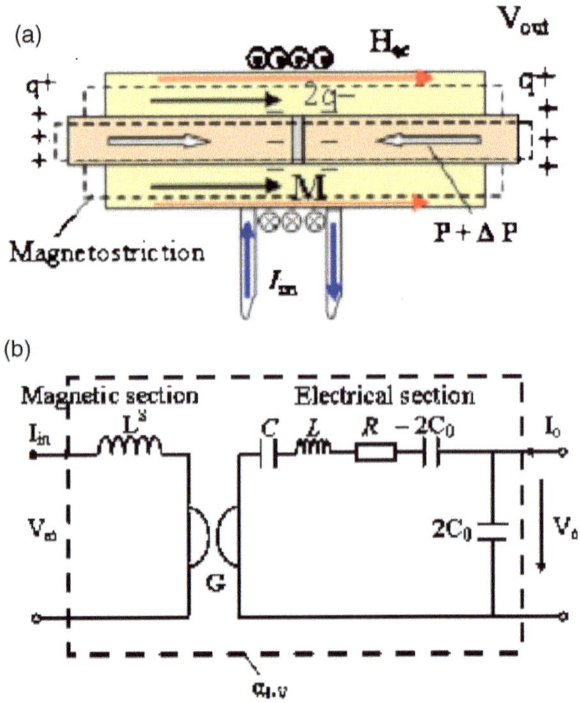

Fig. 14.6 (**a**) a sketch of the ME gyrator (**b**) ME gyration equivalent circuit [62]

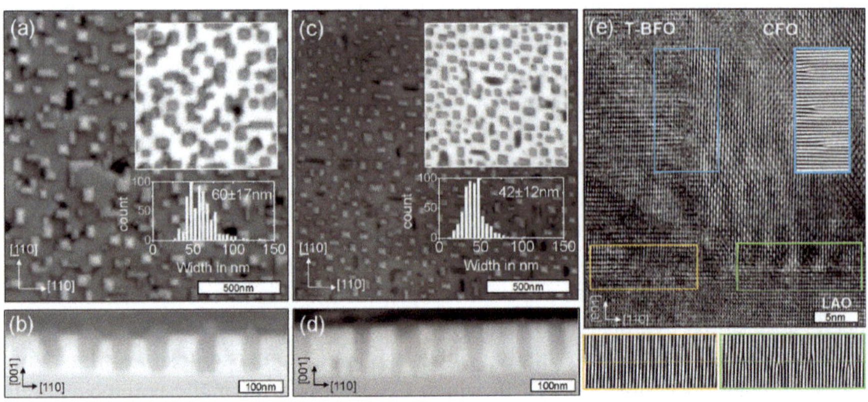

Fig. 14.7 SEM images (secondary electrons) of BFO/CFO on $LaNiO_3$ (LNO)/$LaAlO_3$ (001). The area in the top right inset was imaged with backscattered electrons[69]

In this systems that have nanoculomnar configuration thus far, the ferromagnetic material is chosen from the AB_2O_4 spinel family (such as CFO or NFO) and the FE materials is chosen from ABO_3 perovskite (such as BFO, BTO, and $PbTiO_3$), respectively [36]. In this nanoculmnar configuration, whereas the resistive CFO pillars exhibited single-domain magnetic contrast with high anisotropy due to the magnetoelasticity of the spinel phase. Magnetoelectric coupling was observed in which an applied voltage led to reversal of the magnetic pillars [3].

However, in nanocomposites on STO (001), the spontaneous polarization of BFO is at around 55° with respect to the normal direction, whereas in the CFO nanopillars the out-of-plane is the magnetic easy axis. Clearly, the fixed directions of the FE and FM easy axes in a columnar nanocomposite restrict its development for optimal exploitation of both ferroic properties and magnetoelectric response [36]. Dix et al. demonstrate that epitaxial strain engineering is an efficient method to manipulate the ferromagnetic and FE properties in BFO/CFO columnar nanocomposites [36]. On the one hand, indeed, electric-field induced switching of magnetization has been observed, at room-temperature, in CFO nanopillars in BFO matrix deposited on STO (001) [2, 70, 71] and it is believed that elastic interaction mediates the magnetoelectric coupling [61]. On the other hand, Wang et al. report growth of various phase architectures of self-assembled BFO/CFO thin films on differently oriented STO substrates. CFO forms segregated square, stripe, and triangular nanopillars embedded in a coherent BFO matrix on (001)-, (110)-, and (111)-oriented STO substrates, respectively. Nanostructures with an aspect ratio of up to 5:1 with a prominent magnetic anisotropy were obtained on both (001) and (110) STO along out-of-plane and in-plane directions. Magnetic easy axis rotation from in-plane to out-of-plane directions was realized through aspect ratio control [72]. This paper will pave the way on the perpendicular recording technology as the well controlled magnetic anisotropy in this nanocolumnar FE/FM compostite films.

The recent progress in the multiferroic nanocomposites films not only provide a strategy for producing a new class of delicately functional materials, but also shed promising light on fabricating the ME part of the devices. In the future, the possibility of further tuning the electric polarization or magnetization of the multiferroic nanocomposites films should be evidenced by the coupling and charge transfer of the 3*d* ions at the phase interfaces.

14.4.2 Manganites Nanocomposite Film

Lanthanum manganite perovskites $La_{1-x}A_xMnO_3$ (A is a divalent cation) with a Curie temperature (T_C) above room temperature (LSMO) or with the T_C below room temperature (LCMO) have attracted great attention due to their complex magnetic properties and the colossal magnetoresistance (CMR) occurring around the T_C [73, 74]. These CMR materials are particularly interesting as designing the memory-readers of digital device, magnetic sensor, p-n junctions or spintronic applications involving GMR effect with the high-T_C superconductivity (Fig. 14.8) [75–77]. However, on the one hand, the intrinsic CMR value in this system can only be triggered at a high magnetic field of several Teslas and within a narrow temperature range [78], which obstructed applications such as magnetic head sensors or low-cost magnetic sensors that should be operated at a low magnetic field and stable temperature-sensitive environment [79, 80]. On the other hand,

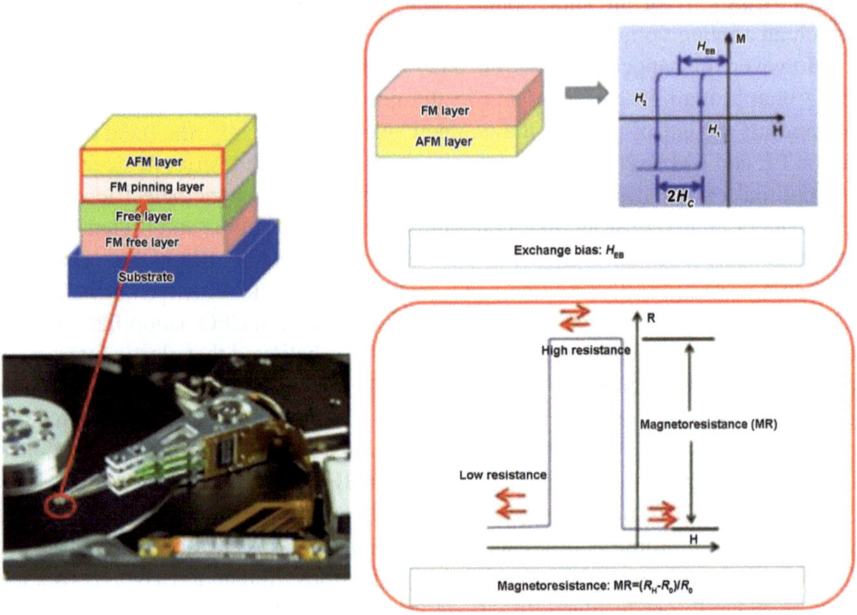

Fig. 14.8 A sketch of the magnetic head sensor and the magnetoresistance effect

lower temperatures of the metalinsulator transition and the magnetoresistance enhancement at low temperatures were obtained by reducing the manganite film thickness or by creating manganiteinsulator superlattices with periods of $n \leq 10$ [81]. However, few observations of unusual CMR behavior in these superlattices were reported below 90 K, the range of interest for combination with high-temperature superconductivity (HTS) materials [46].

14.4.2.1 LSMO Nanocomposites

Recently, due to the deepening cognition of the intrinsic and extrinsic CMR effect, attention has been paid to polycrystalline manganites to obtain the extrinsic low-field magnetoresistance (LFMR) by structuring grain boundaries, nanosized inclusions, interface phase, and artificial grain boundaries. What's more, the LFMR in composites or films can be further improved by introducing a secondary phase (usually non-magnetic or antiferromagnetic insulators) [37, 82–84]. For example, the LFMR values of −6% at 5 K and 0.16 T in LSMO/STO nanomultilayer structure [73], −17.5% at 30 K and 1 T in LSMO:ZnO nanocolumnar structure [31], −8% at 10–150 K and 1 T in LSMO/MgO nanorod arrays structure [85], −12% at 77 K and 0.4 T in LSMO/Ag nanogranular structure [86] and −17% at 250 K and 1 T in LSMO/NiO nanocheckerboard structure [4] have been reported. Figure 14.9 shows the LFMR properties for the LSMO based composites.

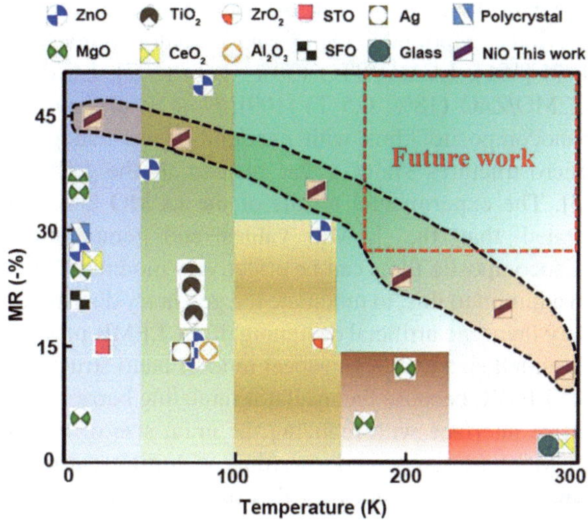

Fig. 14.9 Temperature range of the LFMR for the LSMO composites.NiO: refs [4, 5], ZnO: refs [87, 88], MgO: refs [85, 89, 90] , TiO_2: refs [91–93] , CeO_2: ref [94] , ZrO_2: ref [95] , Al_2O_3: ref [88] , STO: ref [96] , SFO: ref [97] , Ag: ref [88] and glass: ref [98]

However, neither the early results obtained from LSMO polycrystalline films or recent results obtained from LSMO/ZnO composite films with a nanocolumnar structure [31], enhanced LFMR values are always displayed in a low temperature range (below room temperature). Systems featuring a large LFMR at temperatures close to or even higher than room temperature are more important owing to their potential applications in magnetic field sensing and data storage [99]. However, no LFMR properties above room temperature have been reported until now. As shown in Fig. 14.7, compared with other microstructures, more research has been carried out on nanomultilayer and nanocolumnar configurations as the significant enhancement of the LFMR at high temperatures range, for example, in the LSMO/MO system (where MO is a binary metal oxide, such as NiO, CeO_2, ZnO, and VO, etc.) [44, 57, 87, 97, 99–104]. Dey et al. have found that the large LFMR will keep steady in a higher temperature range when the grain size is in nanoscale and gets with the decrease in particle size [100]. It is that the spin pinned effect occurs at the nanosized grain surface defect sites or spin coupling at the boundaries. In addition, the short range FM coupling order may add the opportunity of spin scattering even near the Curie temperature. Thus it is speculated that the high-temperature LFMR in LSMO composite films can be got through tuning the grain size and the FM coupling at the phase boundary. Under these circumstances, the future work should be done for the nanocompsoite films with interfacial FM coupling.

14.4.2.2 LCMO Composites

As have been mentioned above, LFMR behavior in the LCMO based composite films were also important below 90 K, the range of interest for combination with HTS materials [46]. Recently studies have shown that the MR% values (>80% at a magnetic field of 1 T) in the nanomultilayer films are much higher than those previously reported in the LCMO-insulator composites such as LCMO/MgO (20%, 1 T) [106], LCMO/BTO (30%, 1 T) [107], LCMO/STO (70%, 2 T) [108] and LCMO/ZnO (18%, 0.5 T) [109]. Ning et al. have reported that the LCMO/NiO nanocomposite films with nanomultilayer, nanogranular and the nanocolumnar microstructures show a flat plateau in the LFMR versus temperature curves [5]. The experimental results of the LCMO based nanocomposite films demonstrated that the LFMR values and temperature range of self-assembled nanocomposite films can be effectively modulated by tuning their microstructures. In addition to this, to modulate the geometry distribution of these nano-sized phases may favor the artificial designing of the LFMR properties. For the composite films, the small size and large aspect ratio of nano-structure insulator phase would favor the LFMR, because the insulator tunneling barriers and the scattering centers at the phase interface with reducing the grain size of the insulator phase. Actually, the LCMO base nanocomposite films with different microstructure can be explained within the framework of a Resistance–capacitance (R_C) and series circuit model [4, 46], arising as a result of the well-ordered 3D geometric shapes and arrangements of the LCMO and the second phases in the composite films, as shown in Fig. 14.10 [4].

However, the unusual plateau in the LFMR versus temperature curves in LCMO based composites films makes the high LFMR values remain up to a sufficiently higher temperature of approximately 100 K. More importantly, the unusual flat plateau region in the LFMR versus temperature curves implies a broadened temperature range for potential applications in field sensor devices, especially for the

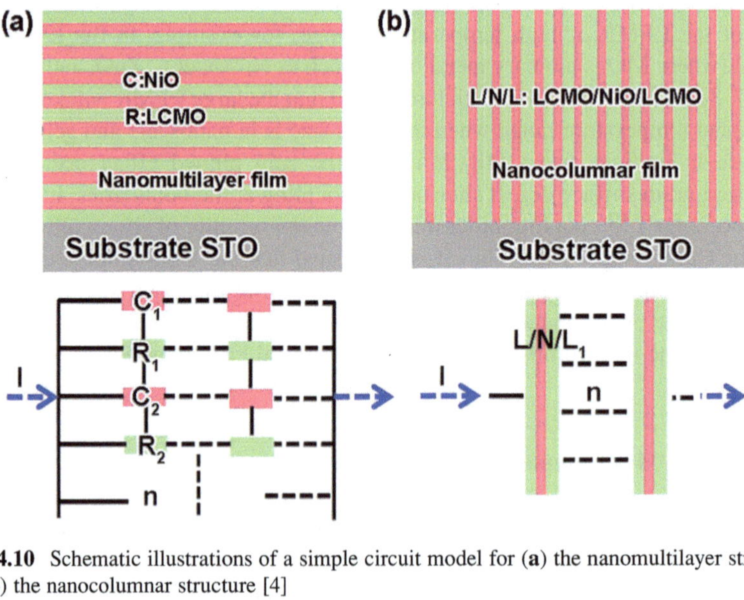

Fig. 14.10 Schematic illustrations of a simple circuit model for (**a**) the nanomultilayer structure and (**b**) the nanocolumnar structure [4]

heterostructures composed of cuprate and manganite components that are used for spintronic applications employing LFMR and HTS.

14.4.2.3 Conclusion

To sum up, the LFMR of the manganites based nanocomposites films can be effectively modulated by tuning the microstructure. On the one hand, the LFMR on the LSMO based nanocomposite thin films with the T_C above room temperature can be used as the memory-readers of digital device. On the other hand, the LFMR on the LCMO based nanocomposite thin films with the T_C below room temperature can be used for spintronic applications employing LFMR and HTS. However, significant challenges still need to be overcomed before manganites nanocomposite films based devices become a reality.

14.4.3 Superconductor Nanocomposite Film

Recently, the superconductor based nanocomposites oxide films have been studied extensively. On the one hand, the coupling between the antagonistic superconducting (SC) and FM orders in thin film multilayers is the subject of intensive research [110]. The proximity effect at superconductor-ferromagnet interfaces produces damped oscillatory behavior of the Cooper pair wave function within the ferromagnetic medium. In 2003, Ustinov et al. proposed using the π junction as a phase shifter in rapid single flux quantum circuits. The π junctions can scale the dimension of superconducting logic circuits down to the submicron size [111]. The structures consisting of 0 and π Josephson junctions are also very important to design phase based devices. The exchange interaction between the SC and FM could strongly affects Andreev reflection at the F/S interface which could providing a powerful tool to probe ferromagnets and measure their spin polarization [112]. On the other hand, special attention is paid to the striking nonmonotonic dependence of the critical temperature and the magnetic-flux pinning based device in the SC nanocomposite films also have been well studied.

14.4.3.1 YBCO:LCMO Nanomposites

The pioneering work on spin-polarized transport in conventional superconductors has been recently extended to high transparency ferromagnet-superconductor heterostructures where the two-particle process of Andreev reflection plays an important role [113–115]. The role of Andreev bound states across a YBCO/LSMO interfaces have been reported by Chen et al. [116]. The suppression of Andreev bound states in the a-b plane by the spin-polarized transport across the interface consistent with the reported photoemission data. In 2004, Holden et al.

have studied the proximity induced metal-insulator transition in YBCO/LCMO nanomultilayers and suggest that either a long-range charge transfer from the YBCO to the LCMO layers or alternatively a strong coupling of the charge carriers to the different and competitive kind of magnetic correlations in the LCMO and YBCO layers is at the heart of the observed metal-insulator transition [117]. In 2005, Hoffmann et al. studied the magnetic properties of YBCO/LCMO nanomultilayer using the polarized neutron reflectometry and shown that the magnetization is reduced due to an inhomogenous magnetization depth profile arising from the suppression of magnetization near the YBCO/LCMO interface [118]. After then, the magnetic proximity effect at the YBCO/LCMO interface of the composite films is becoming an increasingly important issue [119–121]. They have demonstrated that in such a heterostructure the inhomogeneous exchange field enhances the proximity effect. Satapathy et al. shows that the magnetic proximity effect strongly depends on the electronic state of the manganite layers, being pronounced for the ferromagnetic-metallic LCMO and almost absent for ferromagnetic-insulating $LaMnO_3$ [110]. The charge transfer effect from Mn to Cu and the orbital reconstruction across the interface are important factors in determining the proximity effect of this heterostructure [122]. However, De Luca et al. show that, even in the absence of direct Cu-O-Mn covalent bonding, the interfacial CuO_2 planes of superconducting $La_{1.85}Sr_{0.15}CuO_4$ thin films develop weak ferromagnetism associated to the charge transfer of spin-polarised electrons from the $La_{0.66}Sr_{0.33}MnO_3$ ferromagnet [123]. Chien et al. studied the buried interfaces between cuprate and manganite layers using cross-sectional scanning tunnelling microscopy and spectroscopy together with atomic-resolution electron microscopy. The results show that the fundamental length scale of the electronic evolution between YBCO and LCMO is confined to the subnanometre range [124].

14.4.3.2 YBCO:BaZrO3 Nanomposites

In 2005, MacManus-Driscoll et al. have found a strongly enhanced current densities in superconducting coated conductors of YBCO/$BaZrO_3$ (BZO) [125]. They also have pointed out that by changing different types of heteroepitaxial addition and concentration, it is likely that other concentrations and different heteroepitaxial second-phase additions will lead to yet greater enhancements in pinning. Then after, the YBCO based nanocomposite films with different microstructure have been widely explored. Kang et al. have found an enhancements of the critical current in self-field as well as excellent retention of this current in high applied magnetic fields in the thick films via incorporation of a periodic array of extended columnar defects, composed of self-aligned nanodots of nonsuperconducting material extending through the entire thickness of the film. These columnar defects are highly effective in pinning the superconducting vortices or flux lines, thereby resulting in the substantially enhanced performance of this wire [126]. Depending on the angle-dependent critical-current measurements, Gutiérrez et al. demonstrates that a strong and isotropic flux-pinning mechanism is extremely effective at high temperatures and high magnetic fields leading to high-temperature superconductors with record values of pinning force (65 K, 78 GN m^{-3}) [127]. However, retaining a dissipation-free state while carrying large electrical currents is a challenge that needs to be solved to enable commercial applications of high-temperature superconductivity. In 2009, Maiorov et al. show that the controlled microstructures is possible and effective which is used to produce thick films with remarkable IC (H) and nearly isotropic angle dependence [128]. Recent studies are then focus on the microstructure of the YBCO based composites films and the corresponding functional properties, especially for the critical temperature [129, 130]. In 2014, Zhao et al., have realized the precise tuning of the microstructure from vertical nanocolumnar structure to horizontal nanomultilayered in the YBCO/BZO systems [11], as shown in Fig. 14.11.

14.4.3.3 Conclusion

The concept of combining materials with superconductors in the form of artificially grown nanocolumnar and nanomultilayer can be useful in advancing our fundamental understanding of superconductive properties and the way they change with the addition of nanosized correlated defects and second phases. The flux pinning and critical current can be readily tuned with the shape, order, and size of embedded nanoparticles. The future challenge is explore the subtle changes in interfacial composition and coupling effect at the phase interfaces.

14.4.4 Ferroelectric, Dielectric and Conductive Nanocomposite Film

FE, DE and CD oxides incorporating 3*d* transition metals, in particular ABO_3 perovskites, have been fertile ground for the discovery and application in the electronic industry. For example, BTO, $Pb(Zr_xTi_{1-x})O_3$ (PZT), (1-*x*)Pb(Mg, Nb) $O_{3-x}PbTiO_3$ (PMN-PT), LNO etc. which show excellent piezoelectric, DE, FE, pyroelectric and CD properties have been widely used in the sensors, capacitors, nonvolatile memory, surface acoustic wave devices and electrode for the microelectronic devices (Fig. 14.12). These three nanocomposite material are the key materials for the FE random access memory (FRAM) and microelectro-mechanical systems (MEMS).

Recently, remarkable improvement in oxide thin-film techniques have allowed for the growth and characterization of complex oxide heterostructures with (near) atomic precision, opening an avenue for the fabrication of the composite films with different microstructures which shows high polarization performance FE composite

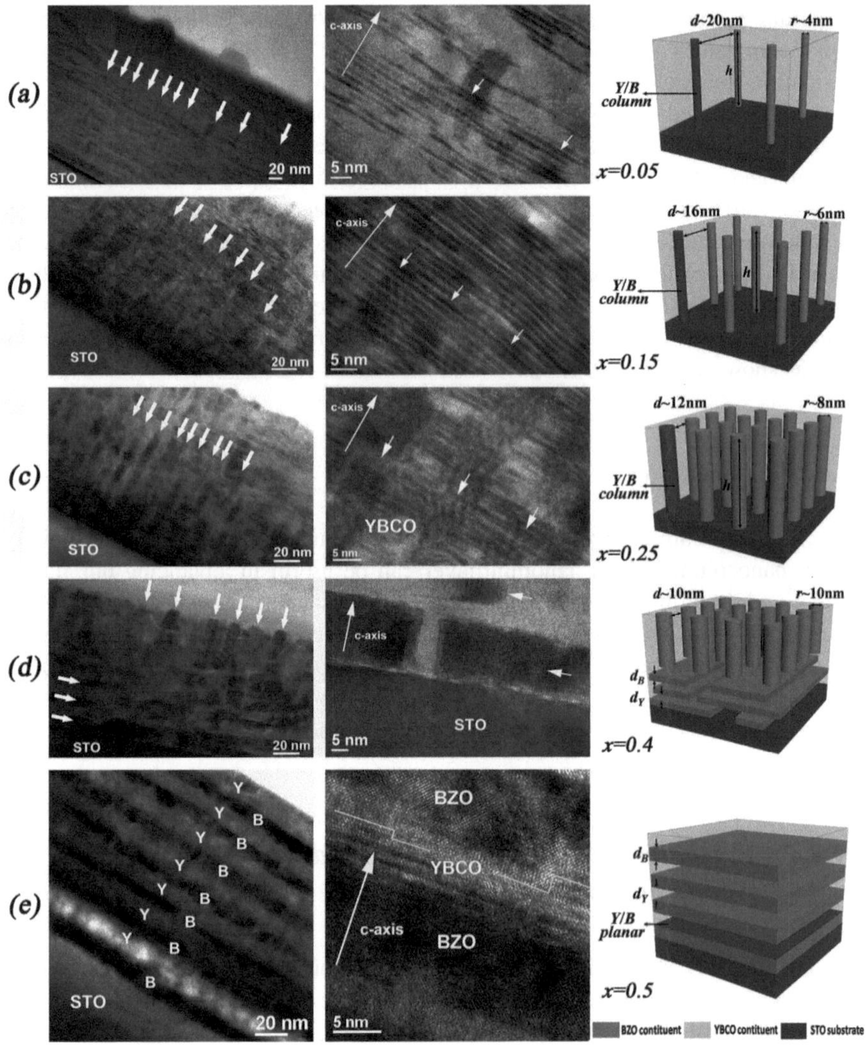

Fig. 14.11 Cross-sectional TEM images of YBCO/BZO thin films with different microstructure [11]

films, high DE properties for the DE composite films and low resistivity for the CD composite films, leading to the rapidly development of the research in the field of FRAM devices.

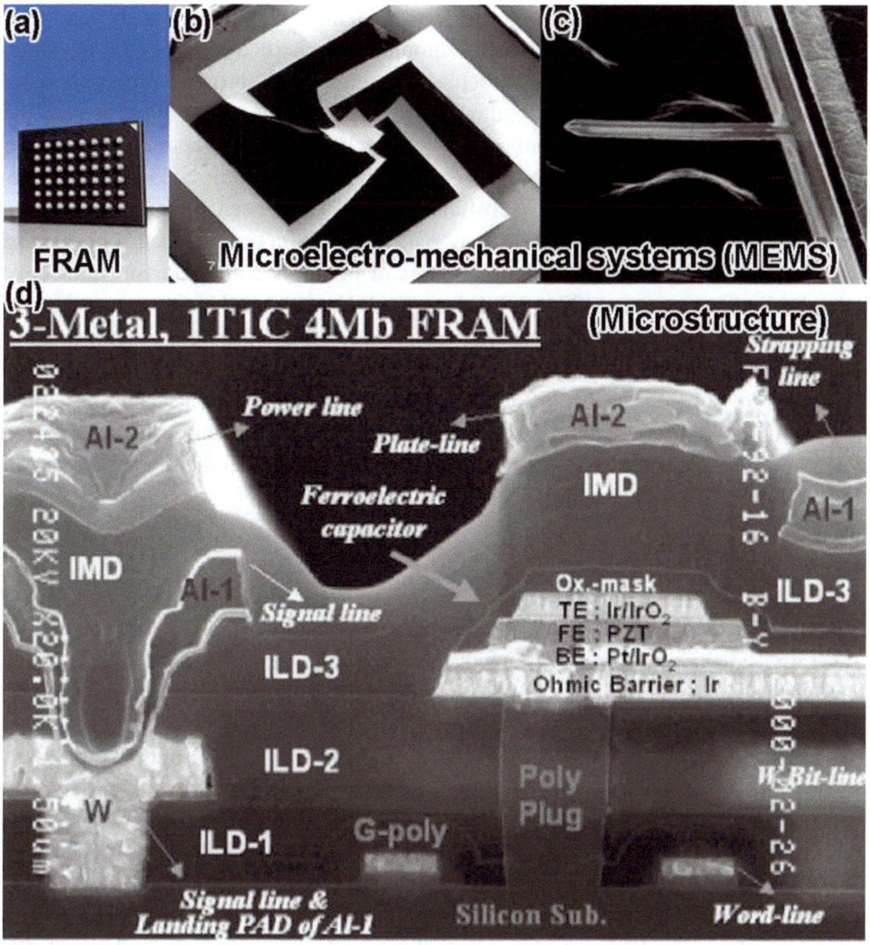

Fig. 14.12. (**a**) the FE random access memory (FRAM) (**b**) and (**c**) Microelectro-mechanical systems (MEMS) (**d**) the corresponding microstructure of the FRAM

14.4.4.1 FE Nanomposites

In this part, we focus on the FE nanocomposite films with nanomultilayer configuration. As have been mentioned above, a remarkable enhancement in polarization and resistivity, and unusual ferroelectricity have been observed in FE superlattices which are expected to be used in the preparation of high-performance microelectronic devices [131–133]. Recently, artificial FE composites films with the nanomultilayered configuration have drawn considerable attention. [134, 135]. The FE nanocomposite films with nanomultilayer configuration have the periodic structures containing different kinds of FE materials. The FE nanomultilayer films can be simply divided into symmetric and asymmetric superlattices according to the thickness of the two component. In 1994, Tabata et al. have firstly synthesis BTO/STO naocompsoites films with nanomultilayer structure by PLD, in which BTO and STO formed layered structure and epitaxially grown with each other [136]. Since this pioneering work, the nanocomposite films in this systems with the structure of nanomultilayer configuration have been widely reported due to the excellent polarization performance. As compared to the their corresponding counterparts epitaxially grown on the substrates, which have been already extensively studied, the strain is one of the most important issues to modulated the FE properties. The theoretical dependence of the properties of FE films with different states have been studied extensively [137, 138]. Ederer et al. Have show that the epitaxial strain dependence of the polarization varies considerably for the different systems, and in some cases is, in fact, very small [139]. The strain state can be simply controlled by selecting the substrates with different lattice parameters. There are some scientific papers report on strain dependency of the nanomultilayers on various substrates and shows that the FE and piezoelectric behavior is consistent with the theoretical model for the films with different strain states [140–143].

14.4.4.2 DE Nanomposites

Recently, on the one hand, composite thin films of metal-perovskite oxides have produced a tremendous flurry of research interest because of their unique physical and chemical properties. Ge et al. Have shown that the Co and Ni nanocrystals in the face-center cubic structure dispersed well in the single BTO matrix monitored by in situ reflection high energy electron diffraction. Their results provide an efficient way to engineer BTO/Ni or BTO/Co nanocomposite films with desired qualities [144, 145]. Metal-DE nanocomposite films, such as BTO/Co, BTO/Au and $PbTiO_3$/Pb (PTO/Pb) composite films, have showed enhanced optical and electrical properties [8, 9, 146] On the other hand, composite thin films of DE:metal-oxides have also been studied extensively as the low leakage and the enhanced DE properties. Lee et al. have studied the $Ba_{0.6}Sr_{0.4}TiO_3/Sm_2O_3$ (BSTO:SmO) nanocomposite films with nanocolumnar structure. They created a nanoscaffold composite which has a very high tunability which scales inversely with loss. Furthermore, low DE loss values of <0.01 significantly lower than reference pure films (Fig. 14.13) [147]

14.4.4.3 CD Nanomposites

Adding metal nanoparticles into the CD oxide could also reduce the resistivity of the composite films. For example, Tin doped indium oxide films (ITO) with embedded silver nanoparticles were prepared by Boen et al. [148]. Delafossite structure of tin doped $AgInO_2$ was found at lower annealing temperatures in all compositions. The interconnected network of Ag phases provides effective

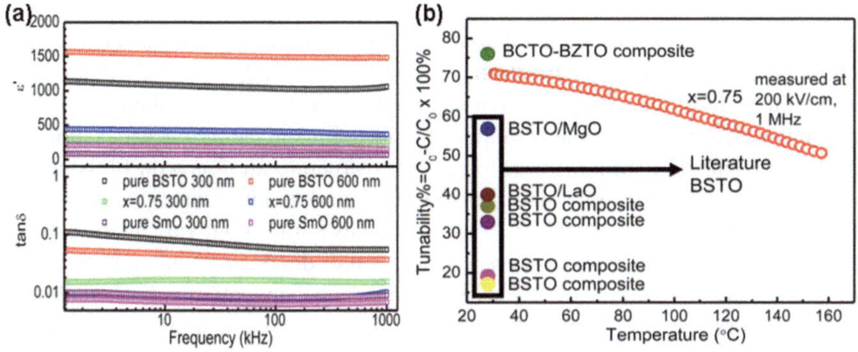

Fig. 14.13 (**a**) Relative permittivity, ε′, versus frequency (*upper*) and loss tangent, tan δ, versus frequency (*lower*) for 300 nm and 600 nm pure BSTO, SmO, and nanocomposite films; (**b**) tunabilities for 600 nm, $x = 0.75$ nanocomposite films as a function of temperature (25–160°C) in comparison to similar thickness BSTO and doped BSTO composite films on different substrates, as well as BCTO-BZTO composite film from the literature

conducting paths that reduced the electrical resistivity. Qiao et al. have synthesis a CD composite film of LNO containing 36.1 at.% Pt by radio frequency co-sputtering. The film consists of two individual phases of separated LNO and Pt. Temperature-dependent resistivity measurement indicates a weak electron–phonon interaction and strong electron-electron scattering in the film [40]. Wang et al. have reported that LNO/Au nanocomposite thin films with an excellent electrical conductivity can be prepared by one-step chemical solution deposition [7]. Metal-CD oxide nanocomposite thin films are regarded as promising electrodes for FE devices to overcome the intrinsic drawbacks of conventional metal or CD oxides electrodes. Wang et al. studied the FE PZT thin films with a thickness of 240 nm were deposited on metal CD oxide nanocomposite LNO/Au electrodes by a sol–gel method. It was observed that the PZT thin films fabricated on the LNO/Au bottom electrode exhibited enhanced ferroelectricity [149].

14.5 Conclusions

This review has mainly discussed the nanocomposite oxide films with different microstructure in the field of multiferroics, manganites, supperconductor, FE, DE and CD. New developments are occurring at a rapid pace, throwing further light onto the intricacies of these materials. The dramatic progress in thin film heterostructure and nanostructure growth has been a key enabler fueling these discoveries [12]. However, until now, there are still significant challenges to be overcome before the nanocomposite oxide films base devices become a reality. For example, it should be noted that PLD method as the main method used in this review does not allow large area fabrication of such composite films and the nanocomposite structure (nanogranular, nanomultilayer, nanocolumnar and nanocheckerboard) is not maintained after a certain thickness that imposes certain limitations. In addition, the functional properties are degraded when epitaxially grown on Si substrate using the chemical method. Although the enhanced performance have been found on nanocomposite oxides films in the past decades, most of these nanocomposite films with different microstructure are focus on films grown on single-crystal perovskite and perovskite-related oxide substrates, including STO, $LaAlO_3$, $DyScO_3$, $NdGaO_3$, etc., to achieve high epitaxial quality. Such single-crystal substrates are expensive and in small dimension, thus not desirable for large-scale integration of oxide thin films with conventional semiconductor devices. Therefore, it is highly attractive to grow these nanocomposite oxide thin films on traditional semiconductor substrates such as silicon (Si) while maintaining a satisfactory performance in the future.

Another problems in these nanocomposite films have have to be taken into considerations. As a class of emerging nanocomposite, most nanocomposite oxide films are still limited to take the advantage of their bimodal physical properties inherited from their single component counterparts. New functions such as the ME coupling effect, charge transfer and FM interactions due to the strong coupling at the interfaces between the two different phases in individual nanocomposite films are barely explored although performance enhancement of the nanocomposite films compared to their corresponding single-component nanoparticles has been observed in several systems. Under this circumstance, precise characterization of the two component interfaces in the nanocomposites films and comprehensive understanding of the interfacial behaviors are critical to reveal the coupling mechanisms that are responsible for improving the functional properties as well as discovering new properties.

Acknowledgements The authors thank Shufang Wang, Mingjing Chen for collaborations. Xingkun thank the National Natural Science Foundation of China with Grant No. 11604073 and 51372064. The Nature Science Foundation for Distinguished Young Scholars of Hebei University.

References

1. H. Zheng, J. Wang, S.E. Lofland, et al., Multiferroic $BaTiO_3$-$CoFe_2O_4$ nanostructures. Science **303**(5658), 661–663 (2004)
2. F. Zavaliche, D. Hong, H.Q. Chiang, et al., Electric field-induced magnetization switching in epitaxial columnar nanostructures. Nano Lett. **5**(9), 1793–1796 (2005)
3. N.M. Aimon, D.H. Kim, X.Y. Sun, C.A. Ross, Multiferroic behavior of templated $BiFeO_3$-$CoFe_2O_4$ self-assembled nanocomposites. ACS Appl. Mater. Interfaces **7**(4), 2263–2268 (2015)
4. X.K. Ning, Z.J. Wang, Z.D. Zhang, Large, temperature-tunable low-field magnetoresistance in $La_{0.7}Sr_{0.3}MnO_3$:NiO nanocomposite films modulated by microstructures. Adv. Funct. Mater. **24**(34), 5393–5401 (2014)

5. X.K. Ning, Z.J. Wang, Z.D. Zhang, Controllable self-assembled microstructures of $La_{0.7}Ca_{0.3}MnO_3$:NiO nanocomposite thin films and their tunable functional properties. Adv. Mater. Interfaces **2**(15), 1500302–1500311 (2015)
6. B.S. Kang, H. Wang, J.L. MacManus-Driscoll, Y. Li, Q.X. Jia, Low field magnetotransport properties of $(La_{0.7}Sr_{0.3}MnO_3)_{(0.5)}$:$(ZnO)_{(0.5)}$ nanocomposite films. Appl. Phys. Lett. **88**(19), 192514–192517 (2006)
7. H.L. Wang, X.K. Ning, Z.J. Wang, Enhanced electrical conductivity of Au-LaNiO3 nanocomposite thin films by chemical solution deposition. RSC Adv. **5**(94), 76783–76787 (2015)
8. Z.R. Wang, T. Hu, L.W. Tang, et al., Ag nanoparticle dispersed $PbTiO_3$ percolative composite thin film with high permittivity. Appl. Phys. Lett. **93**(22), 222901–222904 (2008)
9. W.D. Wu, Y.J. He, F. Wang, et al., Preparation and characterization of Co-$BaTiO_3$ nanocomposite films by the pulsed laser deposition. J. Cryst. Growth **289**(1), 408–413 (2006)
10. M.L. Hammock, A. Chortos, B.C.K. Tee, et al., 25th Anniversary article: the evolution of electronic skin (e-skin): a brief history, design considerations, and recent progress. Adv. Mater. **25**(42), 5997–6038 (2013)
11. R. Zhao, W.W. Li, J.H. Lee, et al., Precise tuning of $(YBa_2Cu_3O_{7-\delta})_{1-x}$:$(BaZrO_3)_x$ thin film nanocomposite structures. Adv. Funct. Mater. **24**(33), 5240–5245 (2014)
12. L.W. Martin, Y.-H. Chu, R. Ramesh, Advances in the growth and characterization of magnetic, ferroelectric, and multiferroic oxide thin films. Mater. Sci. Eng. R Rep. **68**(4), 89–133 (2010)
13. J.S. Andrew, J.D. Starr, M.A. Budi, Prospects for nanostructured multiferroic composite materials. Scr. Mater. **74**, 38–43 (2014)
14. A. Llordés, A. Palau, J. Gázquez, et al., Nanoscale strain-induced pair suppression as a vortex-pinning mechanism in high-temperature superconductors. Nat. Mater. **11**(4), 329–336 (2012)
15. J.L. MacManuas-Driscoll, Self-assembled heteroepitaxial oxide nanocomposite thin film structures: designing interface-induced functionality in electronic materials. Adv. Funct. Mater. **20**(13), 2035–2045 (2010)
16. H. Zheng, Q. Zhan, F. Zavaliche, et al., Controlling self-assembled perovskite-spinel nanostructures. Nano Lett. **6**(7), 1401–1407 (2006)
17. N.N. Padurow, Mischbarkeit im system rutil-zinnstein. Naturwissenschaften **43**(17), 395–396 (1956)
18. D. Fuks, S. Dorfman, S. Piskunov, E.A. Kotomin, Ab initio thermodynamics of $Ba_cr_{(1-c)}TiO_3$ solid solutions. Phys. Rev. B **71**(1), 014111–014120 (2005)
19. C.L. Zhang, S. Yeo, Y. Horibe, Y.J. Choi, et al., Coercivity and nanostructure in magnetic spinel Mg $(Mn, Fe)_2O_4$. Appl. Phys. Lett. **90**(13), 3123–3126 (2007)
20. B.S. Guiton, P.K. Davies, Nano-chessboard superlattices formed by spontaneous phase separation in oxides. Nat. Mater. **6**(8), 586–591 (2007)
21. S. Park, Y. Horibe, T. Asada, et al., Highly aligned epitaxial nanorods with a checkerboard pattern in oxide films. Nano Lett. **8**(2), 720–724 (2008)
22. B. Zhang, M. Lelovic, W.A. Soffa, The formation of polytwinned structures in Fe Pt and Fe Pd alloys. Scr. Metall. Mater. **25**(7), 1577–1582 (1991)
23. Y.L. Bouar, A. Loiseau, A.G. Khachaturyan, Origin of chessboard-like structures in decomposing alloys. Theoretical model and computer simulation. Acta Mater. **46**(8), 2777–2788 (1998)
24. Y. Ni, W. Rao, A.G. Khachaturyan, Pseudospinodal mode of decomposition in films and formation of chessboard-like nanostructure. Nano Lett. **9**(9), 3275–3281 (2009)
25. R.E. Newnham, D.P. Skinner, L.E. Cross, Connectivity and piezoelectric-pyroelectric composite. Mater. Res. Bull. **13**(5), 525–536 (1978)
26. R. Comes, H. Liu, M. Khokhlov, et al., Directed self-assembly of epitaxial $CoFe_2O4$-$BiFeO_3$ multiferroic nanocomposites. Nano Lett. **12**(5), 2367–2373 (2012)

27. N.M. Aimon, K.C. Hong, X.Y. Sun, et al., Templated self-assembly of functional oxide nanocomposites. Adv. Mater. **26**(19), 3063–3067 (2014)
28. G.M. Whitesides, B. Grzybowski, Self-assembly at all scales. Science **295**(5564), 2418–2421 (2002)
29. J.M. Lehn, Toward self-organization and complex matter. Science **295**(5564), 2400–2403 (2002)
30. L.G. Li, W. Zhang, F. Khatkhatay, et al., Strain and interface effects in a novel bismuth-based self-assembled supercell structure. ACS Appl. Mater. Interfaces **7**(21), 11631–11636 (2015)
31. A. Chen, Z. Bi, C.F. Tsai, et al., Tunable low-field magnetoresistance in $(La_{0.7}Sr_{0.3}MnO_3)_{0.5}$: $(ZnO)_{0.5}$, self-assembled vertically aligned nanocomposite thin films. Adv. Funct. Mater. **21** (13), 2423–2429 (2011)
32. H.K. Dong, B. Lei, N.M. Aimon, et al., Combinatorial pulsed laser deposition of Fe, Cr, Mn, and Ni-substituted $SrTiO_3$ films on Si substrates. ACS Comb. Sci. **14**(3), 179–190 (2012)
33. D.H. Kim, N.M. Aimon, C.A. Ross, Self-assembled growth and magnetic properties of a $BiFeO_3$-$MgFe_2O_4$ nanocomposite prepared by pulsed laser deposition. J. Appl. Phys. **113** (17), 17B510 (2010)
34. N.M. Aimon, H.K. Dong, K.C. Hong, et al., Deposition of epitaxial $BiFeO_3/CoFe_2O_4$ nanocomposites on (001) $SrTiO_3$ by combinatorial pulsed laser deposition. Appl. Phys. Lett. **100**(9), 092901–092905 (2012)
35. H.M. Christen, D.P. Norton, L.A. Géa, et al., Pulsed laser deposition of solid-solution films using segmented targets. Thin Solid Films **312**(312), 156–159 (1998)
36. A. Imai, X. Cheng, H.L. Xin, et al., Epitaxial $Bi_5Ti_3FeO_{15}$-$CoFe_2O_4$ pillar-matrix multiferroic nanostructures. ACS Nano **7**(12), 11079–11086 (2013)
37. Z. Liao, P. Gao, S. Stadler, et al., Tuning properties of columnar nanocomposite oxides. Appl. Phys. Lett. **103**(103), 043112–043116 (2013)
38. C.H. Yang, F. Yildiz, S.H. Lee, et al., Synthesis of nanoscale composites of exchange biased $MnFe_2O_4$ and Mn-doped $BiFeO_3$. Appl. Phys. Lett. **90**(16), 163116–163119 (2007)
39. H.J. Liu, L.Y. Chen, Q. He, et al., Epitaxial photostriction-magnetostriction coupled self-assembled nanostructures. ACS Nano **6**(8), 6952–6959 (2012)
40. Q. Liang, X. Bi, Nanostructure and performance of Pt–$LaNiO_3$, composite film for ferroelectric film devices. Acta Mater. **57**(14), 4109–4114 (2009)
41. J.G. Wan, X.W. Wang, Y.J. Wu, et al., Magnetoelectric $CoFe_2O_4$-$Pb(Zr,Ti)O_3$ composite thin films derived by a sol-gel process. Appl. Phys. Lett. **86**(12), 122501–122504 (2005)
42. X.L. Zhong, J.B. Wang, M. Liao, et al., Multiferroic nanoparticulate $Bi_{3.15}Nd_{0.85}Ti_3O_{12}$ oFe_2O_4 composite thin films prepared by a chemical solution deposition technique. Appl. Phys. Lett. **90**(15), 152903–152906 (2007)
43. H. Ryu, P. Murugavel, J. H. Lee, and S. C. Chae, Magnetoelectric coupling of multilayered $Pb(Zr_{0.52}Ti_{0.48})O_3$-$CoFe_2O_4$ film by piezoresponse force microscopy under magnetic field, Appl. Phys. Lett., vol. 89, no. 10, 102907, Jun. 2006.
44. M. Staruch, D. Hires, A. Chen, Z. Bi, Enhanced low-field magnetoresistance in $La_{0.67}Sr_{0.33}MnO_3$:MgO composite films. J. Appl. Phys. **110**(11), 3913–3918 (2011)
45. M. Staruch, C. Cantoni, M. Jain, Systematic study of magnetotransport properties and enhanced low-field magnetoresistance in thin films of $La_{0.67}Sr_{0.33}MnO_3 + Mg(O)$. Appl. Phys. Lett. **102**(6), 6436–6444 (2013)
46. S.A. Fedoseev, A.V. Pan, S. Rubanov, et al., Large, Controllable Spikes of Magnetoresistance in $La_{2/3}Ca_{1/3}MnO_3/SrTiO_3$ Superlattices. ACS Nano **7**(1), 286–293 (2013)
47. Y.G. Ma, W.N. Cheng, M. Ning, C.K. Ong, Magnetoelectric effect in epitaxial Pb $(Zr_{0.52}Ti_{0.48})O_3/La_{0.7}Sr_{0.3}MnO_3$ composite thin film. Appl. Phys. Lett. **90**(15), 15152911–15152913 (2007)
48. C. Deng, Y. Zhang, J. Ma, Y. Lin, Magnetic-electric properties of epitaxial multiferroic $NiFe_2O_4$-$BaTiO_3$, heterostructure. J. Appl. Phys. **102**(7), 7074114–7074118 (2007)
49. M. Dawber, C. Lichtensteiger, M. Cantoni, et al., Unusual behavior of the ferroelectric polarization in $PbTiO_3/SrTiO_3$ superlattices. Physics **95**(17), 177601–177604 (2005)

50. S.J. Callori, J. Gabel, D. Su, et al., Ferroelectric $PbTiO_3/SrRuO_3$ superlattices with broken inversion symmetry. Phys. Rev. Lett. **109**(6), 067601–067606 (2012)
51. M. Dawber, N. Stucki, C. Lichtensteiger, et al., Tailoring the properties of artificially layered ferroelectric superlattices. J. Adv. Mater. **19**(23), 4153–4159 (2007)
52. S.S.A. Seo, J.H. Lee, H.N. Lee, et al., Ferroelectricity in artificial bicolor oxide superlattices. J. Adv. Mater. **19**(18), 2460–2464 (2007)
53. K. Boldyreva, L. Pintilie, A. Lotnyk, et al., Thickness-driven antiferroelectric-to-ferroelectric phase transition of thin $PbZrO_3$ layers in epitaxial $PbZrO_3/Pb(Zr_{0.8}Ti_{0.2})O_3$ multilayers. Appl. Phys. Lett. **91**(12), 122915 (2007)
54. I. Vrejoiu, Y.L. Zhu, G.L. Rhun, et al., Structure and properties of epitaxial ferroelectric $PbZr_{0.4}Ti_{0.6}O_3/PbZr_{0.6}Ti_{0.4}O_3$ superlattices grown on $SrTiO_3$ (001) by pulsed laser deposition. Appl. Phys. Lett. **90**(7), 072909–072912 (2007)
55. R. Ramesh, N.A. Spaldin, et al., Multiferroics: progress and prospects in thin films. Nat. Mater. **6**(1), 21–29 (2007)
56. D. Tian, Q. Chen, F.Q. Nie, et al., Patterned wettability transition by photoelectric cooperative and anisotropic wetting for liquid reprography. Adv. Mater. **21**(37), 3744–3749 (2009)
57. J.L. MacManus-Driscoll, P. Zerrer, H. Wang, et al., Strain control and spontaneous phase ordering in vertical nanocomposite heteroepitaxial thin films. Nat. Mater. **7**(4), 314–320 (2008)
58. H. Yang, H. Wang, G.F. Zou, et al., Leakage mechanisms of self-assembled $(BiFeO_3)_{0.5}$:$(Sm_2O_3)_{0.5}$ nanocomposite films. Appl. Phys. Lett. **93**(14), 142904–142907 (2008)
59. X.F. Liu, J. Shi, Magnetic tunnel junctions with Al_2O_3 tunnel barriers prepared by atomic layer deposition. Appl. Phys. Lett. **102**(20), 202401–202405 (2013)
60. B. Huang, Y. Liu, R. Zhang, et al., Low-field MR behaviour in La0.67Ca0.33MnO3/ZrO2 composite system. J. Phys. D Appl. Phys. **36**(16), 1923–1927 (2003)
61. C.W. Nan, M.I. Bichurin, S. Dong, et al., Multiferroic magnetoelectric composites: historical perspective, status, and future directions. J. Appl. Phys. **103**(3), 031101–031135 (2008)
62. S. Dong, J. Zhai, J.F. Li, D. Viehland, et al., Magnetoelectric gyration effect in $Tb_{1-x}Dy_{yx}Fe_{2-y}$/$Pb(Zr,Ti)O_3$ laminated composites at the electromechanical resonance. Appl. Phys. Lett. **89** (24), 243512–243515 (2006)
63. Z. Wang, Y. Yang, R. Viswan, J. Li, et al., Giant electric field controlled magnetic anisotropy in epitaxial $BiFeO_3$-$CoFe_2O_4$ thin film heterostructures on single crystal $Pb(Mg_{1/3}Nb_{2/3})_{0.7}Ti_{0.3}O_3$ substrate. Appl. Phys. Lett. **99**(4), 043110–043112 (2011)
64. H.K. Dong, N.M. Aimon, Y.S. Xue, et al., Integration of self-assembled epitaxial $BiFeO_3$-$CoFe_2O_4$, multiferroic nanocomposites on silicon substrates. Adv. Funct. Mater. **24**(37), 5889–5896 (2014)
65. I. Levin, J. Li, J. Slutsker, et al., Design of self-assembled multiferroic nanostructures in epitaxial films. Adv. Mater. **18**(15), 2044–2047 (2006)
66. H. Zheng, Q. Zhan, F. Zavaliche, et al., Controlling self-assembled perovskite-spinel nanostructures. Adv. Mater. **6**(7), 1401–1407 (2006)
67. H. Zheng, F. Straub, Q. Zhan, et al., Self-assembled growth of $BiFeO_3$-$CoFe_2O_4$ nanostructures. Adv. Mater. **18**(20), 2747–2752 (2006)
68. Q. Zhan, R. Yu, S.P. Crane, et al., Structure and interface chemistry of perovskite-spinel nanocomposite thin films. Appl. Phys. Lett. **89**(17), 172902–172904 (2006)
69. N. Dix, R. Muralidharan, J.M. Rebled, et al., Selectable spontaneous polarization direction and magnetic anisotropy in $BiFeO_3$-$CoFe_2O_4$ epitaxial nanostructures. ACS Nano **4**(8), 4955–4961 (2010)
70. N. Dix, R. Muralidharan, J. Guyonnet, et al., On the strain coupling across vertical interfaces of switchable $BiFeO_3$-$CoFe_2O_4$ multiferroic nanostructures. Appl. Phys. Lett. **95**(95), 062907–062910 (2009)
71. L. Yan, Z.P. Xing, Z.G. Wang, T. Wang, et al., Direct measurement of magnetoelectric exchange in self-assembled epitaxial $BiFeO_3$-$CoFe_2O_4$ nanocomposite thin films. Appl. Phys. Lett. **94**(19), 192902 (2009)

72. Z.G. Wang, Y.X. Li, R. Viswan, B.L. Hu, et al., Engineered magnetic shape anisotropy in $BiFeO_3$-$CoFe_2O_4$ self-assembled thin films. ACS Nano **7**(4), 3447–3456 (2013)
73. S.J. Zhu, J. Yuan, B.Y. Zhu, et al., Exchange bias effect and enhanced magnetoresistance in $La_{0.67}Sr_{0.33}MnO_3/SrTiO_3$ superlattices. Appl. Phys. Lett. **90**(11), 112502–112505 (2007)
74. Y.-M. Kang, H.-J. Kim, S.-I. Yoo, Excellent low field magnetoresistance properties of the $La_{0.7}Sr_{0.3}Mn_{1+d}O_3$-manganese oxide composites. Appl. Phys. Lett. **95**(5), 052510–052513 (2009)
75. S. Ornes, Giant magnetoresistance, Proceedings of the National Academy of Sciences, vol. 110, no.10, pp. 3710, 2013
76. B.B. Nelsoncheeseman, F.J. Wong, R.V. Chopdekar, et al., Room temperature magnetic barrier layers in magnetic tunnel junctions. Phys. Rev. B **81**(21), 214421–214428 (2010)
77. P. P. Deen, F. Yokaichiya, A. de Santis, et al., Ferromagnetic clusters and superconducting order in $La_{0.7}Ca_{0.3}MnO_3/YBa_2Cu_3O_{7-\delta}$ heterostructures, Phys. Rev. B, vol. 74, no. 22, pp. 224414, 2006
78. P.P. Deen, F. Yokaichiya, A. de Santis, F. Bobba, Spin-polarized intergrain tunneling in $La_{2/3}Sr_{1/3}MnO_3$. Phys. Rev. Lett. **77**(10), 2041–2044 (1996)
79. H. Li, J.R. Sun, H.K. Wong, Enhanced low-field magnetoresistance in $La_{2/3}Ca_{1/3}MnO_3/Pr_{2/3}Ca_{1/3}MnO_3$ superlattices. Appl. Phys. Lett. **80**(4), 628–630 (2002)
80. S. Jin, T.H. Tiefel, M. Mccormack, et al., Thousandfold change in resistivity in magnetoresistive la-ca-mn-o films. Science **264**(5157), 413 (1994)
81. M. H. Jo, N. D. Mathur, J. E. Evetts, et al., Magnetotransport and interface magnetism in manganite heterostructures: implications for spin polarized tunneling, MRS Proceedings. Cambridge University Press, vol. 602, pp. 3, 1999
82. V. Moshnyaga, B. Damaschke, O. Shapoval, et al., Corrigendum: structural phase transition at the percolation threshold in epitaxial $(La_{0.7}Ca_{0.3}MnO_3)_{1-x}$:$(MgO)_x$ nanocomposite films. Nat. Mater. **2**(4), 247–252 (2003)
83. D. Bhadra, M.G. Masud, S.K. De, et al., Observation of large magnetodielectric and direct magnetoelectric behavior in LCMO/PVDF 0-3 nanocomposites. Appl. Phys. Lett. **102**(7), 072902–072907 (2013)
84. Y.K. Tang, X.F. Ge, X.F. Si, et al., Influence of magnetic correlations on low-field magnetoresistance in $La_{2/3}Sr_{1/3}MnO_3/SrTiO_3$ composites. Phys. Status Solidi A **210**(6), 1195 (2013)
85. L. Li, X. Zhang, L. Li, et al., Magnetoresistance of single-crystalline $La_{0.67}Sr_{0.33}MnO_3$/MgO nanorod arrays. Solid State Commun. **171**(10), 46 (2013)
86. J. Li, Q. Huang, Z.W. Li, et al., Enhanced magnetoresistance in Ag-doped granular $La_{2/3}Sr_{1/3}MnO_3$ thin films prepared by dual-beam pulsed-laser deposition. J. Appl. Phys. Lett. **89**(11), 7428–7430 (2001)
87. M. Staruch, H. Gao, P. Gao, et al., Low-field magnetoresistance in $La_{0.67}Sr_{0.33}MnO_3$:ZnO composite film. Adv. Funct. Mater. **22**(17), 3591 (2012)
88. L. Yan, L.B. Kong, T. Yang, et al., Magnetoresistance and current-voltage characteristics in $La_{2/3}Sr_{1/3}MnO_3$/ZnO composites. J. Appl. Phys. **96**(3), 1568–1571 (2004)
89. S.A. Köster, V. Moshnyaga, K. Samwer, et al., Doping of interfaces in $(La_{0.7}Sr_{0.3}MnO_3)_{1-x}$:$(MgO)_x$ composite films. Appl. Phys. Lett. **81**(9), 1648–1650 (2002)
90. Z. Zhang, R. Ranjith, B.T. Xie, et al., Enhanced low field magnetoresistance in nanocrystalline $La_{0.7}Sr_{0.3}MnO_3$ synthesized on MgO nanowires. Appl. Phys. Lett. **96**(22), 222501 (2010)
91. P. Kameli, H. Salamati, M. Eshraghi, et al., The effect of TiO_2 doping on the structure and magnetic and magnetotransport properties of $La_{0.75}Sr_{0.25}MnO_3$ composite. J. Appl. Phys. **98**(4), 043908–043912 (2005)
92. A. Gaur, G.D. Varma, H.K. Singh, Enhanced low field magnetoresistance in $La_{0.7}Sr_{0.3}MnO_3/TiO_2$ composite. J. Phys. D Appl. Phys. **39**(16), 3531–3535 (2006)
93. A. Gaur, G.D. Varma, Electrical and magnetotransport properties of $La_{0.7}Sr_{0.3}MnO_3/TiO_2$ composites. Cryst. Res. Technol. **42**(2), 164–168 (2007)

94. S. Valencia, O. Castaño, J. Fontcuberta, et al., Enhanced low field magnetoresistive response in $(La_{2/3}Sr_{1/3}MnO_3)_x/(CeO_2)_{1-x}$ composite thick films prepared by screen printing. J. Appl. Phys. **94**(4), 2524–2528 (2003)
95. W.J. Lu, Y.P. Sun, X.B. Zhu, et al., Low-field magnetoresistance in $La_{0.8}Sr_{0.2}MnO_3/ZrO_2$ composite system. Mater. Lett. **60**(27), 3207–3211 (2006)
96. K. Dörr, T. Walter, M. Sahana, et al., Magnetotransport of $La_{0.7}Sr_{0.3}MnO_3/SrTiO_3$ multilayers with ultrathin manganite layers. J. Appl. Phys. **89**(11), 6973–6975 (2001)
97. Z. Zi, Y. Fu, Q. Liu, et al., Enhanced low-field magnetoresistance in LSMO/SFO composite system. J. Magn. Magn. Mater. **324**(6), 1117–1121 (2012)
98. S. Gupta, R. Ranjit, C. Mitra, et al., Enhanced room-temperature magnetoresistance in $La_{0.7}Sr_{0.3}MnO_3$-glass composites. Appl. Phys. Lett. **78**(3), 362–364 (2001)
99. R.N. Mahato, H. Lülf, M.H. Siekman, et al., Ultrahigh magnetoresistance at room temperature in molecular wires. Science **341**(6143), 257–260 (2013)
100. P. Dey, T. K. Nath, Effect of grain size modulation on the magneto-and electronic-transport properties of $La_{0.7}Ca_{0.3}MnO_3$ nanoparticles: the role of spin-polarized tunneling at the enhanced grain surface, Phys. Rev. B, vol. 73, no. 21, pp. 214425, Jun. 2006.
101. A. Sadhu, S. Bhattacharyya, Enhanced low-field magnetoresistance in $La_{0.71}Sr_{0.29}MnO_3$ nanoparticles synthesized by the nonaqueous Sol-Gel route. Chem. Mater. **26**(4), 1702–1710 (2014)
102. T.H. Kim, M. Uehara, S.W. Cheong, et al., Large room-temperature intergrain magnetoresistance in double perovskite $SrFe_{1-x}(Mo\ or\ Re)_xO_3$. Appl. Phys. Lett. **74**(12), 1737–1739 (1999)
103. H.J. Liu, V.T. Tra, Y.J. Chen, et al., Large magnetoresistance in magnetically coupled $SrRuO_3$–$CoFe_2O_4$ self-assembled nanostructures. Adv. Mater. **25**(34), 4753–4759 (2013)
104. W. Zhang, A. Chen, F. Khatkhatay, et al., Integration of self-assembled vertically aligned nanocomposite $(La_{0.7}Sr_{0.3}MnO_3)_{1-x}:(ZnO)_x$ thin films on silicon substrates. ACS Appl. Mater. Interfaces **5**(10), 3995–3999 (2013)
105. A.P. Chen, W. Zhang, J. Jian, et al., Role of boundaries on low-field magnetotransport properties of $La_{0.7}Sr_{0.3}MnO_3$-based nanocomposite thin films. J. Mater. Res. **28**(13), 1707–1714 (2013)
106. Z. Sheng, Y. Sun, X. Zhu, et al., Enhanced low-field magnetization and magnetoresistance in nano-MgO added $La_{2/3}Ca_{1/3}MnO_3$ composites. J. Phys. D Appl. Phys. **40**(11), 3300–3305 (2007)
107. L. Joshi, S. Keshri, Extrinsic behavior in $La_{0.67}Ca_{0.33}MnO_3$-$BaTiO_3$ composites. Ceram. Int. **38**(7), 5889–5896 (2012)
108. L. Fei, L. Zhu, X. Cheng, et al., Structure and magnetotransport properties of epitaxial nanocomposite $La_{0.67}Ca_{0.33}MnO_3$: $SrTiO_3$ thin films grown by a chemical solution approach. Appl. Phys. Lett. **100**(8), 082403–082408 (2012)
109. P.K. Siwach, P. Srivastava, J. Singh, et al., Broad temperature range low field magnetoresistance in $La_{0.7}Ca_{0.3}MnO_3$: nano-ZnO composites. J. Alloys Compd. **481**(1), 17–21 (2009)
110. D.K. Satapathy, M.A. Uribe-Laverde, I. Marozau, et al., Magnetic Proximity Effect in $YBa_2Cu_3O_7/La_{2/3}Ca_{1/3}MnO_3$ and $YBa_2Cu_3O_7/LaMnO_{3+\sigma}$Superlattices. Phys. Rev. Lett. **108**(19), 197201–197206 (2012)
111. A.V. Ustinov, V.K. Kaplunenko, Rapid single-flux quantum logic using π-shifters. J. Appl. Phys. **94**(8), 5405–5407 (2003)
112. A.I. Buzdin, Proximity effects in superconductor-ferromagnet heterostructures. Rev. Mod. Phys. **77**(3), 935–976 (2005)
113. M.J.M. De Jong, C.W.J. Beenakker, Andreev reflection in ferromagnet-superconductor junctions. Phys. Rev. Lett. **74**(9), 1657–1660 (1995)
114. R.J. Soulen, J.M. Byers, M.S. Osofsky, et al., Measuring the spin polarization of a metal with a superconducting point contact. Science **282**(5386), 85–88 (1998)
115. S. K. Upadhyay, A. Palanisami, R. N. Louie, et al., Probing ferromagnets with Andreev reflection, Phys. Rev. Lett., vol. 81, no. 15, pp. 3247, Oct. 1998.

116. Z. Y. Chen, A. Biswas, I. Žutić, et al, Spin-polarized transport across a $La_{0.7}Sr_{0.3}MnO_3$/ $YBa_2Cu_3O_{7-x}$ interface: role of Andreev bound states, Phys. Rev. B, vol. 63, no. 21, pp. 212508, May 2001.
117. T. Holden, H.U. Habermeier, G. Cristiani, et al., Proximity induced metal-insulator transition in $YBa_2Cu_3O_7/La_{2/3}Ca_{1/3}MnO_3$ superlattices. Phys. Rev. B **69**(6), 064505–064512 (2004)
118. A. Hoffmann, S.G.E. Te Velthuis, Z. Sefrioui, et al., Suppressed magnetization in $La_{0.7}Ca_{0.3}MnO_3/YBa_2Cu_3O_{7-\delta}$ superlattices. Phys. Rev. B **72**(14), 140407–140411 (2005)
119. J. Chakhalian, J.W. Freeland, G. Srajer, et al., Magnetism at the interface between ferromagnetic and superconducting oxides. Nat. Phys. **2**(4), 244–248 (2006)
120. A.F. Volkov, K.B. Efetov, Proximity effect and its enhancement by ferromagnetism in high-temperature superconductor-ferromagnet structures. Phys. Rev. Lett. **102**(7), 077002–077006 (2009)
121. R. Werner, C. Raisch, A. Ruosi, et al., $YBa_2Cu_3O_7/La_{0.7}Ca_{0.3}MnO_3$ bilayers: interface coupling and electric transport properties. Phys. Rev. B **82**(22), 224509–224516 (2010)
122. J. Chakhalian, J.W. Freeland, H.-U. Habermeier, et al., Orbital reconstruction and covalent bonding at an oxide interface. Science **318**(5853), 1114–1117 (2007)
123. G. M. De Luca, G, Ghiringhelli, C. A. Perroni, et al., "Ubiquitous long-range antiferromagnetic coupling across the interface between superconducting and ferromagnetic oxides," Nat. Commun., vol. 5, no. 6, pp. 5626-5633, 2014.
124. L.F. Kourkoutis, J. Chakhalian, B. Gray, et al., Visualizing short-range charge transfer at the interfaces between ferromagnetic and superconducting oxides. Nat. Commun. **4**(4), 2336–2343 (2013)
125. J.L. MacManus-Driscoll, S.R. Foltyn, Q.X. Jia, et al., Strongly enhanced current densities in superconducting coated conductors of $YBa_2Cu_3O_{7-x}$+ $BaZrO_3$. Nat. Mater. **3**(7), 439–443 (2004)
126. S. Kang, A. Goyal, J. Li, et al., High-performance high-T*c* superconducting wires. Science **311**(5769), 1911–1914 (2006)
127. J. Gutiérrez, A. Llordes, J. Gazquez, et al., Strong isotropic flux pinning in solution-derived $YBa_2Cu_3O_{7-x}$ nanocomposite superconductor films. Nat. Mater. **6**(5), 367–373 (2007)
128. B. Maiorov, S. A. Baily, H, Zhou, et al., "Synergetic combination of different types of defect to optimize pinning landscape using $BaZrO_3$-doped $YBa_2Cu_3O_7$," Nat. Mater., vol. 8, no. 5, pp. 398-404, 2009.
129. C. Cantoni, Y. Gao, S.H. Wee, et al., Strain-driven oxygen deficiency in self-assembled, nanostructured, composite oxide films. ACS Nano **5**(6), 4783–4789 (2011)
130. S.H. Wee, Y. Gao, Y.L. Zuev, et al., Self-assembly of nanostructured, complex, multication films via spontaneous phase separation and strain-driven ordering. Adv. Funct. Mater. **23**(15), 1912–1918 (2013)
131. M. Dawber, E. Bousquet, New developments in artificially layered ferroelectric oxide superlattices. MRS Bull. **38**(12), 1048–1055 (2013)
132. H.N. Lee, H.M. Christen, M.F. Chisholm, et al., Strong polarization enhancement in asymmetric three-component ferroelectric superlattices. Nature **433**(7024), 395–399 (2005)
133. E. Bousquet, M. Dawber, N. Stucki, et al., Improper ferroelectricity in perovskite oxide artificial superlattices. Nature **452**(7188), 732–736 (2008)
134. C.H. Ahn, K.M. Rabe, J.M. Triscone, Ferroelectricity at the nanoscale: local polarization in oxide thin films and heterostructures. Science **303**(5657), 488–491 (2004)
135. P. Zubko, S. Gariglio, M. Gabay, et al., Interface physics in complex oxide heterostructures. Annu. Rev. Condens. Matter Phys. **2**(1), 141–165 (2011)
136. H. Tabata, H. Tanaka, T. Kawai, Formation of artificial $BaTiO_3/SrTiO_3$ superlattices using pulsed laser deposition and their dielectric properties. Appl. Phys. Lett. **65**(15), 1970–1972 (1994)
137. N. A. Pertsev, V. G. Kukhar, H. Kohlstedt, et al., Phase diagrams and physical properties of single-domain epitaxial $Pb(Zr_{1-x}Ti_x)O_3$ thin films, Phys. Rev. B, vol. 67, no. 5, pp. 054107, Feb. 2003.

138. J. X. Zhang, D. G. Schlom, L. Q. Chen, et al., Tuning the remanent polarization of epitaxial ferroelectric thin films with strain, Appl. Phys. Lett., vol. 95, no. 12, pp. 122904, Sep 2009.
139. C. Ederer, N. A. Spaldin, Effect of epitaxial strain on the spontaneous polarization of thin film ferroelectrics, Phys. Rev. Lett., vol. 95, no. 25, pp. 257601, Dec. 2005.
140. M. D. Nguyen, M. Dekkers, E. Houwman, et al., Misfit strain dependence of ferroelectric and piezoelectric properties of clamped (001) epitaxial $Pb(Zr_{0.52}, Ti_{0.48})O_3$ thin films, Appl. Phys. Lett., vol. 99, no. 25, pp. 252904, Dec. 2011.
141. K.J. Choi, M. Biegalski, Y.L. Li, et al., Enhancement of ferroelectricity in strained $BaTiO_3$ thin films. Science **306**(5698), 1005–1009 (2004)
142. J.H. Haeni, P. Irvin, W. Chang, et al., Room-temperature ferroelectricity in strained $SrTiO_3$. Nature **430**(7001), 758–761 (2004)
143. D. G. Schlom, L. Q. Chen, C. B. Eom, et al., Strain tuning of ferroelectric thin films, Annu. Rev. Mater. Res. vol. 37, no. 126, pp. 589-626, May. 2007.
144. F.F. Ge, L. Bai, W.D. Wu, et al., The controllable growth of Co-$BaTiO_3$ nanocomposite epitaxial film by laser molecular beam epitaxy. J. Cryst. Growth **312**(16), 2489–2493 (2015)
145. F.F. Ge, X.M. Wang, L.H. Cao, et al., Self-organized Ni nanocrystal embedded in $BaTiO_3$ epitaxial film. Nanoscale Res. Lett. **5**(5), 834–838 (2010)
146. Y.J. Wu, H.L. Wang, X.K. Ning, et al., Dielectric properties of Au-$BaTiO_3$ nanocomposite films prepared by Sol-Gel method. J. Inorg. Mater. **30**(2), 207–213 (2015)
147. O.J. Lee, S.A. Harrington, A. Kursumovic, et al., Extremely high tunability and low loss in nanoscaffold ferroelectric films. Nano Lett. **12**(8), 4311–4317 (2012)
148. B. Houng, Tin doped indium oxide transparent conducting thin films containing silver nanoparticles by sol-gel technique. Appl. Phys. Lett. **87**, 251922 (2005)
149. H.L. Wang, Y. Bai, X.K. Ning, Z.J. Wang, Enhanced electrical properties in ferroelectric thin films on conductive Au-$LaNiO_3$ nanocomposite electrodes via modulation of Schottky potential barrier. RSC Adv. **5**(126), 104203–104209 (2015)

Chapter 15
Fabrication and Characterizations of Bi_2Te_3 Based Topological Insulator Nanomaterials

Z.H. Wang, Xuan P.A. Gao, and Z.D. Zhang

15.1 Introduction

Recent theoretical work has predicted a new class of quantum matter with an insulating bulk gap and gapless edge or surface states, that is the topological insulators in two (2D) and three dimensions (3D) [1–3]. 3D topological insulators have a gapped bulk band structure, but a band inversion induced by spin-orbit interaction leads to a metallic topological surface state. The topological surface state can exhibit a Dirac-like dispersion which is protected by time-reversal symmetry, resulting in unique properties such as spin momentum locking and prohibited backscattering [1]. Experimentally, the 2D topological insulator has been first observed by transport measurements on the edges of thin buried layers of HgTe in quantum wells. And the 3D topological insulator phase is reported in $Bi_{1-x}Sb_x$ alloy with complicated topological surface states [4, 5]. On the other hand, 3D topological insulators with the simplest possible surface states consisting of a single Dirac cone have been proposed theoretically in stoichiometric compounds Bi_2Se_3, Bi_2Te_3 and Sb_2Te_3 [1]. Bi_2Se_3, Bi_2Te_3, and Sb_2Te_3 share the same rhombohedral crystal structure with the space group D3d5 (R3m) with five atoms in the trigonal unit cell. As an example of Bi_2Te_3 crystal structure, each quintuple layer with a thickness of ~1 nm consists of five atomic planes arranged in the sequence of Te(1)-Bi-Te(2)-Bi-Te(1). The coupling is strong within one quintuple layer but weak between any two quintuples, which are bonded by the van der Waals force. Bi_2Te_3 based nanomaterials such as nanoplates, nanowires, and thin films have

Z.H. Wang (✉) • Z.D. Zhang
Shenyang National Laboratory for Materials Science, Institute of Metal Research, Chinese Academy of Sciences, 72 Wenhua Road, Shenyang 110016, People's Republic of China
e-mail: zhwang@imr.ac.cn

X.P.A. Gao
Department of Physics, Case Western Reserve University, Cleveland, OH 44106, USA

T. Li, Z. Liu (eds.), *Outlook and Challenges of Nano Devices, Sensors, and MEMS*,
DOI 10.1007/978-3-319-50824-5_15

been widely studied as topological insulator materials. Because of the weak van der Waals force between epilayer and substrate, the lattice mismatch between these films and the substrate is not crucial such as the high quality single crystalline and twin domain free Bi_2Te_3 thin films realized on Si(111) substrates [6]. It is easy to grow selenide or telluride topological insulators on many different substrates, such as CdTe, Al_2O_3, BaF, InP, GaAs (111), InP (111), InP(001), and $SrTiO_3$ (111) [7–14].

The surface properties of the 3D topological insulators have been mainly investigated on the cleaved surface of bulk crystals and single crystalline nanostructure by angle resolved photoemission spectroscopy (ARPES), and scanning tunneling microscopy (STM), which confirmed the presence of Dirac states at the surface [15–18]. These surface states can also be probed by magneto-transport measurements. Numerous works on magneto-transport measurements have proved the existence of surface states by means of weak anti-localization (WAL) and observation of Shubnikovde Has (SdH) oscillations in the single crystals and thin films of Bi_2Te_3, Bi_2Se_3 [19, 20]. However, due to the highly doped and metallic bulk, the transport properties presented originate dominantly from the bulk carriers and the contribution of surface state carriers is negligible [21]. It is possible to control the Fermi surface by synthesized ternary/quaternary compounds, gating or doping with Cu, Sb and Sn to push it into the bulk band gap [22]. However, these doping in turn create defects, it is a great if not impossible challenge to realize an intrinsic topological insulator [23]. Looking forward to future device applications, it is desirable to develop growth procedures for intrinsic undoped thin films of topological insulators.

In this manuscript, we follow the recent experimental research progresses in topological insulators Bi_2Se_3 and Bi_2Te_3 based nanostructures (e.g. nanoflakes, nanoplats, nanosheets, nanowires, and thin films), and discuss various nanomaterial synthesis, characterization methods, and notable transport properties.

15.2 Fabrication of Bi_2Te_3 Based Topological Insulator Nanomaterials

The contribution of the bulk carriers can be suppressed by reducing sample size. As a new member of topological insulator nanomaterials, ultrathin topological insulator nanomaterials have an extremely large surface-to-volume ratio and can be electrically gated more effectively than the bulk form, potentially enhancing surface state effects in transport measurements. Single crystalline Bi_2Te_3, Bi_2Se_3, and various elements doping Bi_2Se_3 and Bi_2Te_3 nanoflakes, nanoplates, nanowires, nanoribbons can be synthesized or prepared by chemical vapor deposition (CVD) [22, 24, 25], chemical method, [26] vapor–solid growth, van der Waals epitaxy [27, 28], mechanically exfoliation methods, and polyol method [29–31]. All the above methods are useful for growing single-crystal nanomaterials of Bi_2Se_3 and

Bi_2Te_3 with well aligned orientation, controlled thickness, and specific positional placement. Furthermore, the high-quality, large-area nanoplates grown on transparent insulating substrates, such as oxidized silicon, mica, can serve as a straightforward platform for spectroscopy, device fabrication, and electronic measurements. Optical images reveal thickness dependent color and contrast for Bi_2Se_3 and Bi_2Te_3 nanoplates grown on oxidized silicon (300 nm SiO_2/Si) [32]. It has been demonstrated that the molecular beam epitaxy (MBE) method grown on various substrates can be used to grow high-quality thin films of Bi_2Te_3, Bi_2Se_3 [33–35], and various element doped Bi_2Se_3 and Bi_2Te_3 thin films [36, 37]. On the other hand, it has also been demonstrated that this family of thin films can be obtained by pulsed laser deposition (PLD), CVD method, vapor–solid growth and mechanically exfoliation methods [38–42].

Figure 15.1 illustrates a schematic diagram of the experimental setup used to synthesize nanomaterials (nanosheets, nanoplates, films) of binary (Bi_2Te_3, Bi_2Se_3) and ternary (such as $(Bi_{1-x}Se_x)_2Te_3$, $(Bi_{1-x}Sb_x)_2Se_3$) topological insulators on mica/silicon/glass substrates by the CVD method [22, 43–46]. In a typical experiment, solid powder sources of one or few kinds are loaded into the quartz tube reactor inside a tube furnace by quartz transfer rods. If the binary topological insulators are prepared, only a single powder source is needed and the corresponding source is placed in the quartz tube reactor. If the ternary topological insulators are synthesized, one of the two powder sources is placed at the center of quartz tube; the other powder source is placed upstream of the tube reactor in order to keep the concentration low both in the vapor and in the as-deposited nanomaterials/thin film. A cleaned silicon/freshly-cleaved mica/glass substrate is placed at 13–18 cm downstream from the center of the Lindberg/Blue M single zone furnace to set the growth temperature lower than the temperature at the center of the furnace. The furnace is initially pumped down to 0.5 Pa, and then a carrier gas of Ar with 10 % H_2 at a typical flow rate such as 50 sccm is used to transport the vapor from the powder sources to the substrate. The furnace is heated to the temperature which the experiment used within short time (e.g. a few minutes) with a stated pressure for example 30 Pa. The flow rate and pressure can be controlled in order to get the aimed samples (nanoplates, nanowires and films). After the growth is carried out for right amount of time, the furnace is then allowed to cool down naturally to room temperature with the carrier gas flowing.

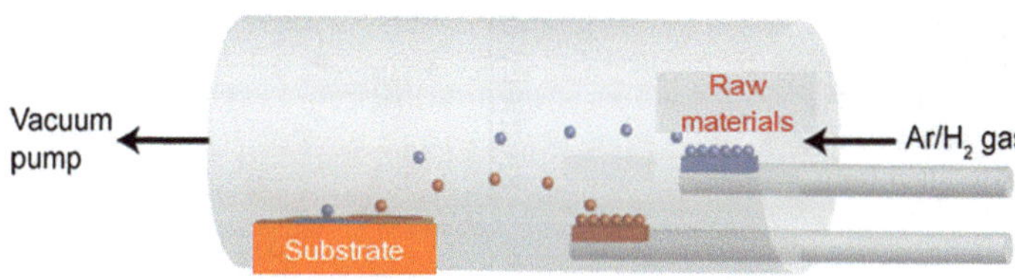

Fig. 15.1 Schematic diagram of the vapor transport setup used for the growth of topological insulator nanomaterials/films. Modified from Lee et al. [43]

15.3 Characterizations of Bi_2Te_3 Based Topological Insulator Nanomaterials and Films

15.3.1 Characterizations of Bi_2Te_3 Based Nanomaterials (e.g. Nanoplates, Nanowires and Nanoribbons)

15.3.1.1 Characterizations of Binary Bi_2Te_3 Nanomaterials

The ability to transfer and manipulate nanomaterials is central to their practical use in electronics [47]. Many experimental investigations involving mechanical processing and transfer of the sample generally assume such processes have no impact on the sample properties, particularly for the topological insulators research. However, if mechanical transfer can change the surface character of topological insulator nanostructures, then it is important to carefully investigate the sample's fundamental physical properties in different contexts. The comparison is performed between un-manipulated topological insulator nanoplates on an original substrate ("as-grown") and the characteristics of such as-grown nanoplates after they have been transferred to a new substrate [43]. Two infrared (IR)-active phonon modes are observed in Raman scattering from as-grown thin (~10 nm) nanoplates of a representative topological insulator Bi_2Te_3. As shown in Fig. 15.2, the two IR-modes centered at 93 and 113 cm^{-1} are observed in nanoplates of Bi_2Te_3 with thickness about 8 nm, which are attributed to the out-of-plane vibrations $A^1{}_{1u}$ and $A^2{}_{1u}$, respectively. This emergence of IR-active modes in Raman spectra reveals a breakdown of inversion symmetry in two-dimensional Bi_2Te_3. Both the IR and Raman features remain unchanged after spin-coating with a layer of polymer and after device fabrication of the nanoplates, suggesting that the phonon characteristics of the nanoplates are stable, as shown in Fig. 15.3. The Raman features also are affected by the thickness of Bi_2Te_3 nanoplates. As shown in Fig. 15.4, the SEM image shows the as-grown nanoplates of Bi_2Te_3 with the thickness of 320 nm and 11 nm. The IR-active phonon modes are absent in Bi_2Te_3 with the thickness of 320 nm. The nanoplates of Bi_2Te_3 with the same thicknesses (about 10 nm) but

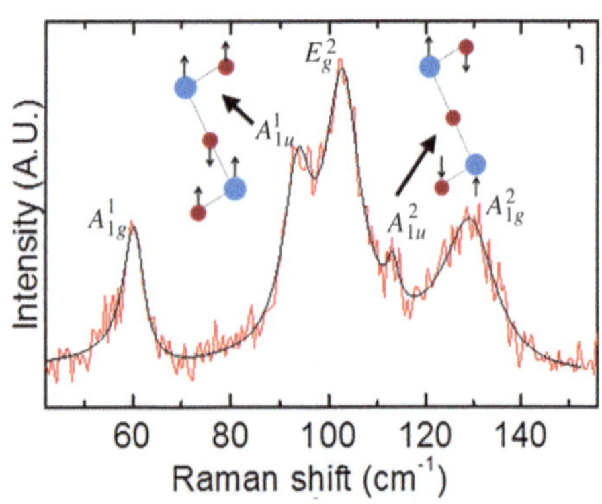

Fig. 15.2 Raman spectra of as-grown Bi_2Te_3 nanoplates with the thickness of 8 nm. The black line shows a fit of the Raman peaks with Lorentzian or Fano line shapes. The $A^1{}_{1g}$ and $A^2{}_{1g}$ modes are fit with a Fano line shape. Modified from He et al. [43]

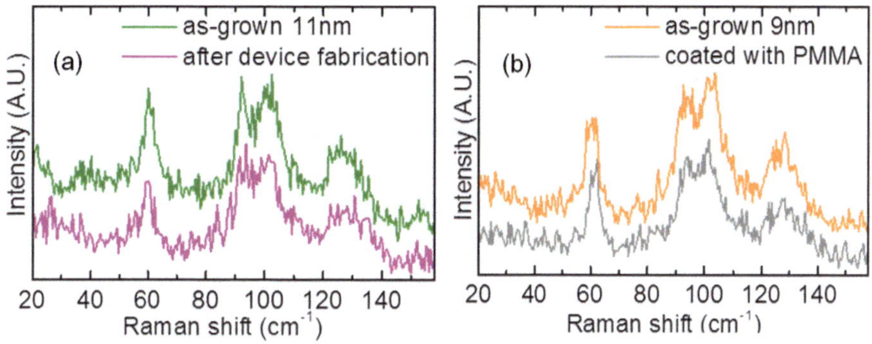

Fig. 15.3 (**a**) Comparison of Raman spectra from a nanoplate before and after device fabrication. (**b**) Raman spectra from a nanoplate before and after spin-coating with a PMMA layer of 100 nm thick. Modified from He et al. [43]

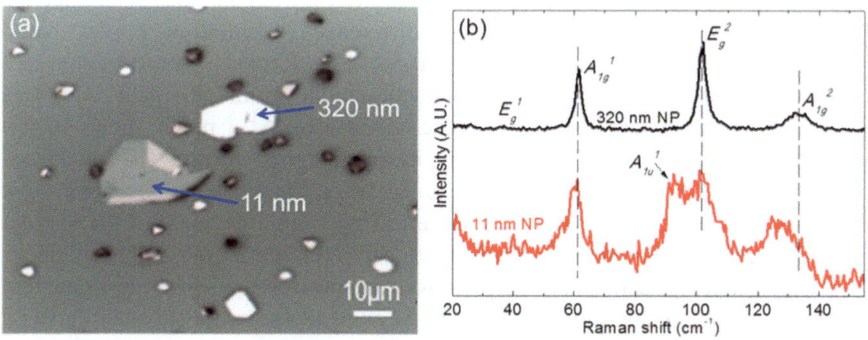

Fig. 15.4 (**a**) SEM images and (**b**) Raman spectra of as-grown Bi_2Te_3 nanoplates with thickness of 320 nm and 11 nm

transferred to another SiO_2 substrate are studied, which show that the IR-active phonon modes are absent and that the Raman spectra are largely similar to those of bulk Bi_2Te_3, as shown in Fig. 15.5. These results reveal that not only thickness induces a breakdown of inversion symmetry and a concomitant appearance of IR-active modes in Raman scattering, but also the absence of inversion symmetry breaking in transferred Bi_2Te_3 nanoplates could be ascribed to weak interactions between the nanoplates and the substrate. Thicknesses less than 20 nm thin nanoplates with stoichiometry different from Bi_2Te_3 are also studied [48]. Raman lines from non-stoichiometric nanoplates display characteristic changes with the increase of Bi concentration, which are different from those of stoichiometric ones. The optical absorption coefficient in thin Bi-Te nanoplates strongly depends on their stoichiometry, because thin nanoplates with the same thickness but different stoichiometries show different color contrast compared to the substrate in the optical image. Two IR-active vibrational modes are observed in thin Bi2Te3nanoplates by Raman scattering, which show the evidence of an inversion

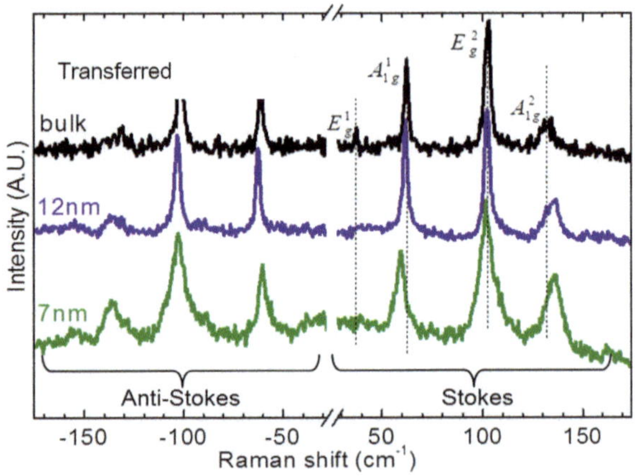

Fig. 15.5 Raman spectra of transferred Bi_2Te_3nanoplateswith different thickness and bulk Bi_2Te_3. Modified from He et al. [43]

symmetry breakdown in thin as-grown nanoplates [43]. Raman map is an effective method to characterize the thickness difference of ultrathin Bi_2Te_3 based nanoplates. Both Stokes and anti-Stokes Raman spectroscopy of Bi_2Se_3nanoplatelets resolve four optical phonon modes from individual nanoplates down to 4 nm, where the out-of-plane vibrational $A^1{}_{1g}$ mode shows a few wavenumbers red shift as the thickness decreases below 15 nm. This thickness-dependent red shift is tentatively explained by a phonon softening due to the decreasing of the effective restoring force arising from a decrease of the van der Waals forces between adjacent layers [49]. Strain effects in the Bi_2Se_3/graphene nanoplates heterostructures is studied by Raman spectroscopy, showing 1 quantum layer (QL) and 2 quantum layers(QLs) Bi_2Se_3 nanoplates experience tensile stress [27]. The ultrathin Bi_2Te_3 nanoplates with the thickness of about six QLs are obtained by the modified hot wall epitaxy technique [50]. The Raman vibration mode $A^1{}_{1g}$ from the nanoplates exhibits an obviously red shift with decreasing thickness. The thickness variation of one nanoplate is obtained by the Raman map derived from the vibration frequency of $A^1{}_{1g}$ mode.

The hexagonal Bi_2Te_3 nanoplates can exhibit the mapped multiple surface plasmon modes with energies that cover the entire visible range, which are identified using transmission electron microscopy (TEM)-based electron energyloss spectroscopy (EELS) and cathodoluminescence (CL) spectroscopy [51]. At the center and edge of the single Bi_2Te_3nanoplate, different plasmon modes are observed. And a breathing mode is discovered in a nonnoble metal. Theoretical calculation showed that the interplay between spin–orbit couplings induced metallic surface states, which gives rise to surface plasmons in the visible range. Owing to the high modulation depth and the large saturation intensity, the Bi_2Te_3 nanoplates as a saturated absorber is utilized in stable passively Q-switched laser pulses, which are successfully conducted in the 1.5 μm wavelength region [52]. The electric and dielectric properties of the Bi_2Te_3 based nanoplates could be significantly changed by the outside environments [53, 54]. The high, fast and reversible response indicates a high possibility and potential of this material for chemical sensing applications. Topological insulators are ideal materials for an electronic chemical sensor, because an ideal topological insulator only has conducting channels on the surface, which are directly exposed to outside environment.

As-grown Bi_2Te_3 usually displays a metallic resistivity due to the Fermi energy (E_F) lies in the valence band (VB). In experiments, synthesized Bi_2Se_3 is often heavily n-type doped due to selenium vacancies. Furthermore, Bi_2Se_3 single crystals gets additional n-type doping after exposure to the atmosphere, which reduces the relative contribution of surface state in total conductivity. Transport measurements on Bi_2Se_3 nanoribbons provide evidence of environmental doping process. The topological surface contribution in transport is hindered by residual bulk carriers from environmental doping or crystal defects [53]. Therefore, to exploit the surface transport properties of topological insulators, it is crucial to achieve a bulk-insulating state in a topological insulator material. The nonmetallic Bi_2Te_3 is obtained by selective cleaving of Bi_2Te_3 grown with a weak compositional gradient [55]. The SdH oscillations arising from the surface states are observed in the nonmetallic crystals of Bi_2Te_3. Very large surface-to-volume ratios have been achieved in Bi_2Se_3 nanoplates and nanoribbons, which found to be all n-type doped [24, 56]. The drastic difference in sheet resistance between 1–2 QLs and 4–5 QLs is showed in the local conductivity of Bi_2Se_3 nanoribbons. The nonmetallic temperature dependence of resistance in Bi_2Se_3(<5 QLs) device is observed [56]. The linear magnetic resistance (LMR) of Bi_2Se_3 nanoribbons under perpendicular magnetic fields is observed [57, 58]. The angle-dependent measurements of MR of Bi_2Se_3 nanoribbons at 2 K from the vertical magnetic field configuration to the parallel magnetic field are performed. With fixed magnetic fields, the MR is only sensitive to the vertical component of magnetic field, which directly demonstrates a 2D LMR. As a prototypical 3D topological insulator, the nontrivial surface conduction of Bi_2Se_3 is likely to be responsible for this 2D LMR. Furthermore, the close relationship between MR and mobility is investigated by comparing the Hall mobility and longitudinal resistance under different temperatures [57]. The consistent behavior of MR and Hall mobility throughout the temperature range gives direct evidence for a mobility fluctuation induced LMR model by Parish and Littlewood (P-L model) [57]. Concomitantly, the resistance-versus-temperature curves of thin Bi_2Se_3 sheets evolve from metallic to semiconductor-like, and increasingly strong WAL behavior is manifested. Analysis of the MR data reveals two contributions, namely from the bulk conduction band and from a state inside the bulk gap. The latter is responsible for the linear MR and likely represents the topologically protected surface state [27].

Well-distinguished SdH oscillations are observed in Bi_2Se_3 nanoplates and nanobeams. Careful analysis of the SdH oscillations suggests the existence of Berry's phase, which confirms the quantum transport of the surface Dirac fermions [59]. On Bi_2Se_3 nanoplates, three types of carriers, one of 3D and two of 2D character, are identified by analyzing the angular dependence of SdH oscillations up to 45 T, which confirmed the coexistence of bulk carriers and band bending induced 2D electron gas [60]. The distribution of the Aharonov-Bohm (AB) periods provides clear evidence to distinguish the surface conductance from the bulk in

Bi_2Se_3 nanoribbons [61]. The surfaces of thin Bi_2Se_3 are strongly electrostatically coupled, and a gate electrode can completely remove bulk charge carriers and bring both surfaces through the Dirac point simultaneously. The ultrathin Bi_2Se_3 (approximately three quintuple layer) are n-type and have a clear OFF state at negative gate voltage [62–64]. For thin Bi_2Se_3, when the gate-tuned chemical potential is near or below the Dirac point, the carrier density is strongly temperature-dependent, reflecting thermal activation from the nearby bulk valence band. When above the Dirac point, n-type surface conduction is observed with negligible thermal activation of bulk carriers [62]. The modulation of the surface states by using a gate voltage to control quantum oscillations is found in Bi_2Te_3 nanoribbons. Both the voltage-controlled SdH and the Aharonov–Bohm oscillations suggest that by applying an external gate voltage, the Fermi level can be tuned away from the valence band, leading to the suppressed bulk conduction and the enhanced surface states in thin Bi_2Te_3 nanoribbons. The possibility of surface state control in topological insulator nanostructures is demonstrate in thin Bi_2Te_3 nanoribbons as conducting channels in field-effect transistor (FET) structures for the first time [65, 66]. Environmental doping significantly affects the field emission properties of the Bi_2Se_3 nanoflake arrays. Transport measurements on Bi_2Se_3 nanoribbons provide additional evidence of such environmental doping process. Such surface oxidation is likely the material origin of the degradation of topological surface state [54, 67].

15.3.1.2 Characterizations of Ternary and Quaternary Bi_2Te_3 Based Nanomaterials

Topological surface states were initially observed by ARPES experiments. Then many papers reported the bulk insulating phase achieved in Bi_2Te_3 based topological insulators as observed by both ARPES and STM [68–70]. Ideally, stoichiometric Bi_2Se_3 and Bi_2Te_3 are in the intrinsic topological insulator regime. However, owing to the predominance of bulk carriers from environmental doping, crystal defects or thermal excitations, the transport measurements have been limited intensively by the bulk defects. The response to magnetic and non-magnetic bulk dopants is studied in Bi_2Te_3 and Bi_2Se_3. Far from the Dirac point, helicity provides remarkable resilience to backscattering even in the presence of ferromagnetism. However, approaching the Dirac point, where the surface states' wavelength diverges, bulk doping results in pronounced nanoscale spatial fluctuations of energy, momentum and helicity. Although backscattering and localization are absent for Dirac topological surface states, reducing charge defects is required for both tuning the chemical potential to the Dirac point and achieving high electrical mobility for novel states [71]. Synthesis of ternary and quaternary topological insulator compounds is a useful method to decrease crystal defects. Individual Bi_2Te_2Se nanoplates presence a 2D-WAL effect, originating from the topologically protected surface states, which is confirmed by gate-dependent charge transport performed at different temperatures. The Bi_2Te_2Se nanoplatelets on hBN enables

Fig. 15.6. SEM images of $Bi_2(Te_{1\text{-}x}Se_x)_3$ nanoribbons with different length and width. Modified from Wang et al. [22]

the observation of well-developed SdH oscillations and are sufficiently thin to permit tuning the Fermi level position over a wide range by applying a back gate voltage [27, 72]. In $Bi_2(Te_{1\text{-}x}Se_x)_3$ nanoribbons, the bulk carrier transport can be effectively suppressed, inducing semiconducting behavior in the temperature-dependent resistance [19, 22, 73]. The atomic ratio of Se/Te in $Bi_2(Se_xTe_{1-x})_3$ single-crystalline nanostructures can be controlled by the molecular ratio of Bi_2Se_3 to Bi_2Te_3 in the source powder mixture. Magneto-transport measurements on $Bi_2(Se_xTe_{1-x})_3$ nanoribbons show 2D-WAL based on angle-dependent studies [73]. Gate voltage enables ambipolar modulation of resistance with thicknesses around or larger than 100 nm, indicating significantly enhanced contribution in transport from the gapless surface states [22]. Few typical images of $Bi_2(Te_{1\text{-}x}Se_x)_3$ nanoribbons prepared by CVD method is shown in Fig. 15.6.. The length of the ribbons ranges from several μm to tens of μm and the width ranges from a few hundred nm to a few μm. For $Bi_2(Te_{1\text{-}x}Se_x)_3$ nanoribbons with thickness ~100–300 nm, the material can be tuned gradually from metallic to semiconducting by increasing the Se concentration. The concentrations of Bi, Te and Se in $Bi_2(Te_{1\text{-}x}Se_x)_3$ nanoribbons are estimated by scanning energy-dispersive X-ray spectroscopy (EDS) on individual nanoribbons. To investigate the transport properties and explore the topological surface conduction in $Bi_2(Te_{1\text{-}x}Se_x)_3$ nanoribbons, back-gated field effect transistor (FET) devices are fabricated on doped Si substrate with 300 nm thick SiO_2 on surface and measured. The AFM image of a typical $Bi_2(Te_{1\text{-}x}Se_x)_3$ nanoribbon device and the thickness of the nanoribbon are shown in Fig. 15.7. The thickness and width of the nanoribbon are about 200 nm and 600 nm, respectively. Figure 15.8 shows the resistance versus temperature of a series $Bi_2(Te_{1\text{-}x}Se_x)_3$ nanoribbons with different selenium concentration x. The resistances of different samples are normalized by the values at 300 K in order to compare the temperature dependence directly. A clear trend shows that the material turns from metallic into semiconducting when x is more than ~10 %. When Se concentration is (20 ± 5) %, resistance with temperature shows an insulating behavior, but it saturates below about 75 K due to the metallic surface transport. Temperature dependent resistance measurements of $Bi_2(Te_{1\text{-}x}Se_x)_3$ nanoribbons with different x show that substituting Te with Se can indeed suppress the residual bulk carriers and tune the Fermi level into bulk bandgap. Moreover, ambipolar field effect from surface state conduction is achieved by applying aback-gate voltage for semiconducting $Bi_2(Te_{1-x}Se_x)_3$nanoribbons with x ~20 % with a thickness of ~100 nm or even larger. Figure 15.9a, b shows the resistance as a function of the back gate voltage (V_g) at various temperatures without magnetic field for $Bi_2(Te_{1-x}Se_x)_3$ nanoribbon with x ~20 %, whose thickness is ~105 nm. It is interesting to note that the gate modulated resistance showed distinct behaviors at different temperature. At 300 K, the resistance increased with decreasing the gate voltage, exhibiting a clear n-type feature. With the temperature decreasing and down to 200 K, the device kept a weak n-type characteristic with high gate voltage, and the resistance started to decrease when the applied gate voltage is below about +20 V,

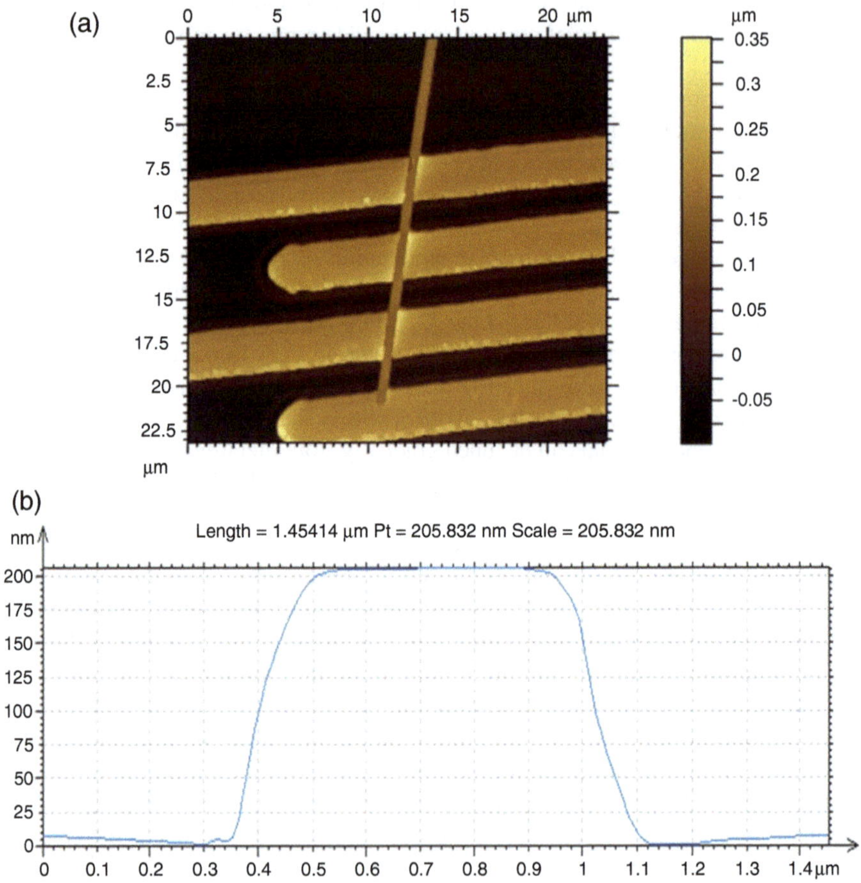

Fig. 15.7 (**a**) AFM image of a typical $Bi_2(Te_{1-x}Se_x)_3$ nanoribbon device and (**b**) the thickness and width of the nanoribbon can be estimated by AFM

Fig. 15.8 The dependence of resistance with temperature showing the transport property changing from metallic into semiconducting as the concentration of Se increases in $Bi_2(Te_{1-x}Se_x)_3$ nanoribbons. Modified from Wang et al. [22]

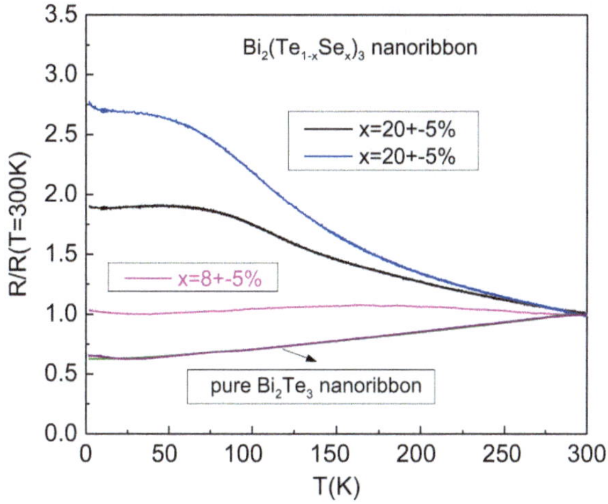

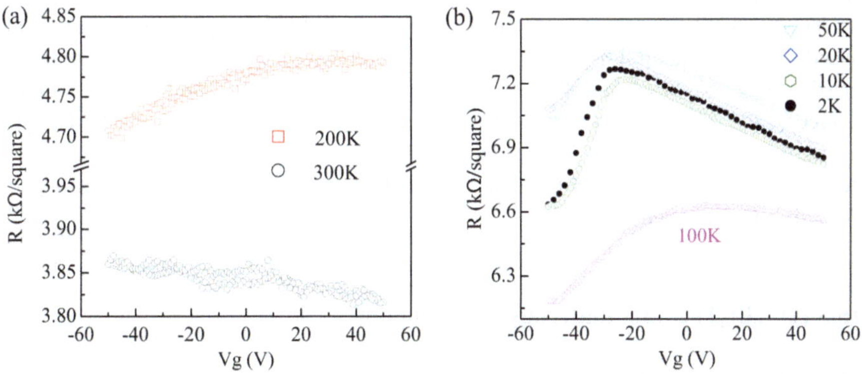

Fig. 15.9 (**a**, **b**) The dependence of resistance per square vs. back-gate voltage at different temperatures for a 105 nm thick $Bi_2(Te_{0.8}Se_{0.2})_3$ nanoribbon. Modified from Wang et al. [22]

which indicating a transition from n-type to p-type conduction. As temperature decreases down to 2 K, an ambipolar characteristic become evidently in the $R(V_g)$ curve and the turning point moved to ~ −20 V, as shown in Fig. 15.9b. The thickness of $Bi_2(Te_{1-x}Se_x)_3$nanoribbons is also found to be important for observing ambipolar field effect [22]. As shown in Fig. 15.10, for thicker nanoribbon devices with thickness ~205 nm and ~320 nm, the dependence of resistance on V_g showed a weaker tunability. The ambipolar effect disappeared at 20 and 10 K for devices with thickness ~205 nm and ~320 nm, respectively. By tuning the ratio of bismuth to antimony in $(Bi_{1-x}Sb_x)_2Se_3$ nanostructure, the bulk carrier density is reduced, while maintaining the topological insulator properties. The introduction of Sb in Bi_2Se_3 effectively suppresses the electron density at room temperature in $(Bi_{1-x}Sb_x)_2Se_3$ at x~0.15, while maintaining the metallic transport behavior. A metal–insulator transition (MIT) is observed in $(Bi_{1-x}Sb_x)_2Se_3$ nanosheets with x >0.20, indicating that

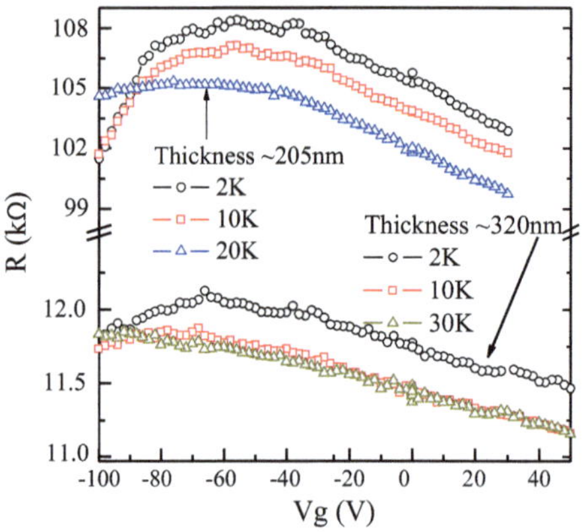

Fig. 15.10 Gate voltage tuned ambipolar conduction at various temperatures in $Bi_2(Te_{1-x}Se_x)_3$ nanoribbons with thickness about 205 nm and 320 nm. Modified from Wang et al. [22]

the system has transformed into an insulator in which the metallic surface conduction is blocked. The appearance of vibrational modes arising from Sb-Sb and Sb-Se bonds at high Sb concentrations is revealed by Raman spectroscopy, confirming the emergence of Sb_2Se_3 crystal structure in the sample [43]. In quaternary materials, based on a two-channel model, 99 % surface transport contribution can be realized in $Bi_{1.5}Sb_{0.5}Te_{1.8}Se_{1.2}$ nanoflakes by analyzing the resistivity and Hall resistance devices [74, 75]. Ambipolar electric field effect by back gate and WAL with magnetic field is observed. The surface channel is estimated based on the thickness dependence of electrical conductance and the result of the SdH oscillations in $Bi_{1.5}Sb_{0.5}Te_{1.7}Se_{1.3}$ flakes [74].

15.3.1.3 Bi_2Te_3 Nanomaterials With Doping

Elements doping is also a useful method to get semiconducting topological insulators by suppressing the bulk resistance. The gate tunable surface states are realized in Na-doped Bi_2Te_3 nanoplates, and the topological insulators can be tuned from p-type to n-type by applying a back gate voltage. The gate-controlled SdH oscillations in the Na-doped Bi_2Te_3nanoplates are found at low temperatures down to 1.9 K with an out-of-plane magnetic field applied to the nanoplates' surface [65]. Varying the Ca concentration in high-quality Bi_2Se_3, the insulating transport behavior is obtained. From the optical and electrical transport measurements, the band gap and the reduced effective mass of the electrons and holes are determined [76]. And the sufficiently low bulk carrier density can change the sign of the Hall density with the gate voltage Vg. The conductance, WAL and metallic resistance-temperature profile identify the protected surface state [29]. According to the structural characterization and compositional analysis of Cu-doped Bi_2Te_3 hexagonal nanoplates, the Cu2p ions are found to substitute Bi3p ions in the lattice. A paramagnetic state is observed in these nanoplates from 2 to 295 K, which is a significant difference from their diamagnetic un-doped Bi_2Te_3 [77]. The surface states of topological insulators are robustly protected by time-reversal symmetry.

Breaking time-reversal symmetry of topological insulators by doping magnetic impurities is predicted to open a gap in the otherwise gapless surface states [78, 79]. Very few studies on magnetic doped Bi_2Te_3 nanomaterials have been reported. The Mn-doped Bi_2Te_3 nanoplates are analyzed from the structural and compositional characterization, which find that the Mn^{2+} and Mn^{3+} ions substitute Bi^{3+} ions in the lattice. Ferromagnetic ordering at the high Curie temperature (up to 45 K) is reported in Mn2p/Mn3p substituted Bi_2Te_3nanoplates [78]. The Kondo effect, a saturating resistance upturn at low temperatures, is observed in ferrocene-doped Bi_2Se_3 nanoribbons, indicating presence of localized impurity spins. Magneto-conductance of the ferrocene-doped ribbons displays both WL and WAL [80, 81]. Although the presence of the magnetic dopants in the Bi_2Se_3 nanoribbons is confirmed by the observation of the Kondo effect [80], bulk carrier concentration is still too high in these magnetically doped Bi_2Se_3 nanoribbons to test whether a small gap opens in the surface states. Future work includes reducing the bulk carrier concentration in these Bi_2Se_3 nanoribbons and correlating the amount of Fe atoms in the ribbons with the carrier density [80]. To further study the interface between a conventional superconductor and a topological insulator, Pb-Bi_2Te_3-Pb lateral and sandwiched junctions are fabricated, and electron transport measurements are performed down to low temperatures. The strong superconducting proximity effect between Bi_2Te_3 and Pb are shown at a temperature very close to the superconducting T_C of Pb. Moreover, a Josephson current can be established over several microns in the lateral direction between two Pb electrodes on the Bi_2Te_3 surface. The critical current of the devices exhibits s-wave-like interference and Fraunhofer diffraction patterns. With improved designs, Josephson devices of this type would provide a test-bed for exploring novel phenomena such as Majorana fermions in the future [82].

15.3.2 *Bi_2Te_3 Based Films*

15.3.2.1 Binary Bi_2Te_3 Films Without Doping

Atomically smooth, single crystalline Bi_2Se_3 thin films are prepared by MBE. The films grow by layer-by-layer mode, with controllable thickness down to one quintuple layer [38, 83, 84]. ARPES measurement results on Bi_2Se_3 and Bi_2Te_3 thin films show that the as-grown films without doping exhibit a low defect density, and become a bulk insulator at a thickness of ten quintuple layers. The bulk band structure and the topological surface states are observed in these thin films by ARPES [83]. The electronic properties of the Bi_2Te_3 thin films can be regulated by altering the growth conditions without extrinsic dopants. As revealed by the ARPES, the Fermi energy of the Bi_2Te_3 films can be tuned with the Si substrate temperature when it is increased above a critical temperature. A conversion from n- to p-type is observed and correlated with a change in the growth mode from layer-by-layer to step-flow [85–88]. The interfacial and internal structure of Bi_2Te_3 films are studied, the results revealed a Te-dominated buffer layer, and a large interfacial spacing properly stacked bulk like Bi_2Te_3 film [85]. Raman scattering spectra and atomic force microscope (AFM) measurements indicate that single crystalline films of Bi_2Te_3 grow in a layer-by-layer mode under Te-rich conditions. Transport measurements prove the insulating behavior of the films grown in this way [85]. The evolution of band structure of Bi_2Te_3 films with a thickness from 1 QL to 5 QLs shows that the coupling between top and bottom surfaces is strong enough to open up a whole insulating gap for the 1 QL film. Starting from the 2 QLs film, the inter-surface coupling becomes progressively weaker, and the topological features are recovered. With increasing thickness, the Fermi level moves toward the bulk gap [33]. These results demonstrate that both quality and stoichiometry of the film become better as the film grows thick. The micro-Raman spectroscopy investigation of single-crystal Bi_2Te_3 films with thickness ranging from few-nanometers to bulk limit [38, 89, 90]. It is found that the optical phonon mode appears in the atomically-thin films due to crystal-symmetry breaking. The crystal symmetry breaking in few-quintuple films results in appearance of A_{1u}-symmetry Raman peaks.

There was a standing fundamental issue of topological insulators that their theoretically predicted nanoscale metallic surface state has to be demonstrated substantially by experiments. An experiment was designed to clarify this issue by measuring the surface-state and bulk-state resistances of Bi_2Te_3 and Sb_2Te_3 thin films [91]. The surface-state resistivity is lower than that of the bulk-state by 5 orders of magnitude, which indicated that the nanoscale surface state is metallic. Further, the surface-state resistances decrease monotonically with the decreasing temperature, which clearly showed that the metallic temperature dependence is consistent with the presence of the topologically protected Dirac surface state. The bulk state resistances firstly decrease, and reach a rock bottom near 125 K, then increase with decreasing temperature and achieve the saturated state near 50 K. These experiments have provided the direct evidence of the theoretically predicted nanoscale metallic surface state of topological insulators [91]. Through gate dependent SdH oscillations, two surface conductions (top/bottom) in Bi_2Te_3 thin films could be separated [92]. By sweeping the gate voltage, only the Fermi level of the top surface is tuned while that of the bottom surface remains unchanged due to strong electric field screening effects arising from the high dielectric constant of Bi_2Te_3. In addition, the bulk conduction can be modulated from n- to p-type with a varying gate bias [92]. In situ scanning tunneling microscopy (STM) shows that Bi growth mode changes from quasi bilayer-by-bilayer to step-flow with increasing substrate temperature [93]. The bilayer Bi exhibits an electron donor behavior, causing an downshift of the Dirac point of Bi_2Te_3.The film thinning to few-quintuples and tuning of the Fermi level can help achieving the topological-insulator surface transport regime with an extraordinary thermoelectric efficiency [89].

Few years ago, the rapid expansion of the emerging field of topological insulators has again stimulated intensive research on the magnetic-resistance (MR) in these topological materials with non-trivial zero-gap Dirac-like surface states.

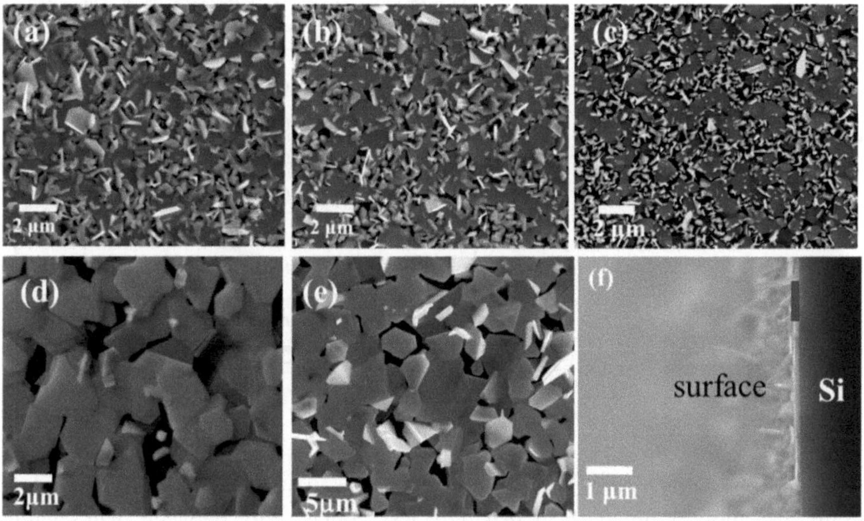

Fig. 15.11 (**a–e**) The SEM images of Bi_2Te_3 films with different sizes of nanoflakes, (**e**) the side view of a film. Modified from Wang et al. [43]

Especially, the initial discovery of a LMR together with the 2D quantum oscillations of MR from the surface states [57], and then many further experiments are performed to confirm the existence of LMR in topological insulator materials. These investigations on high quality Bi_2Se_3, Bi_2Te_3 films have enabled observation of extremely large effect, e.g. up to 60 Tesla in Bi_2Te_3 films, and thickness or gate tunable LMR [94–96]. The giant linear and non-saturating MR in topological insulator Bi_2Te_3 can be tuned by adjusting the degree of granularity of interconnected nanoplates in films prepared by CVD method [43]. Typical SEM images of Bi_2Te_3 films are shown in Fig. 15.11a–e. As showing in SEM images, the films are formed by interconnected grains of Bi_2Te_3nanoplateswith different size. The thickness of films can be estimated from the SEM images of sample taken from the side view (shown in Fig. 15.11e). The layer thickness of films is typically about 100 nm.

The temperature dependent resistances for sample (a)–(d) at zero magnetic fields are shown in Fig. 15.12.They are represented the low resistivity, medium resistivity, nearly semiconducting and semiconducting samples, respectively. For the first three samples (sample (a)–(c)), the resistance vs. temperature curve, R(T) is metallic from $T = 2$ K to about 250 K, presumably because that the unintentionally formed defects during the growth induce high carrier concentration and give rise to metallic bulk conduction, similar to other chalcogenide topological insulator materials without intentional doping. However, the R(T) of sample (d) decreases with increasing temperature showing semiconductor behavior in most of the temperature range, due to the much stronger scattering and porosity in this type of samples. It is insightful to compare between the sample's R(T) and granularity. Sample (a)–(d) consists of more and more loosely packed nanoplate networks, which correlated

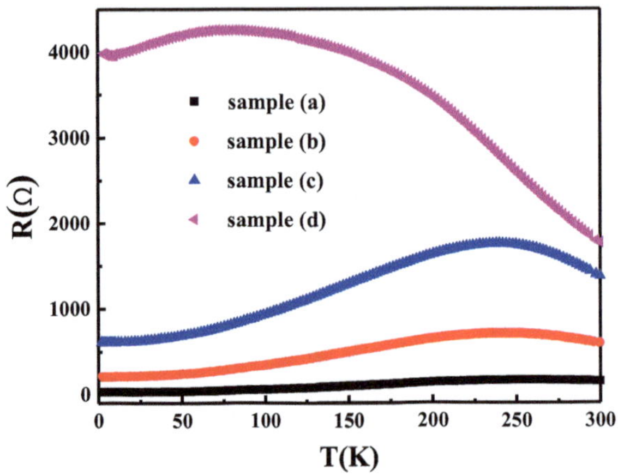

Fig. 15.12 The resistance plotted as a function of temperature for representative Bi_2Te_3 samples (**a**–**d**). Modified from Wang et al. [43]

with the increasing resistance. Moreover, sample (a)–(c) have similar nanoplate grain size about 1 μm with increasingly less connectivity between plates while sample (d) is made of larger size grains. It can be seen that the films with loosely packed nanoplate network and larger size grains are inclined to get semiconducting samples.

The MR of sample (a)–(d) are measured in perpendicular magnetic field. ΔR(B)/R(0) (defined as [R(B)-R(0)]/R(0)) of these samples are presented in Fig. 15.13. All the granular Bi_2Te_3 nanoplate film samples show LMR effect which increases with decreasing temperature. It is shown that the ΔR(B)/R(0) first shows a quadratic growth below certain threshold field, and then transforms into a linearly rising behavior with increasing magnetic field without any sign of saturation. There is a clear difference in the magnitude of LMR between these samples. A large LMR (~450% at 14 T) is obtained at low temperatures in sample (a) with most densely packed nanoplates. However, as the film became less densely packed, ΔR(B)/R(0) decreases until the maximal ΔR(B)/R(0) in sample (d) is merely one tenth of that in sample (a). Interestingly, in sample (d) which contains the largest nanoplate grains and weakest classical LMR effect, a sharp resistance dip in the ΔR(B)/R(0) curve near zero magnetic field appears at low temperature, manifesting the WAL effect. The typical Hall resistance R_{xy} as functions of magnetic field at various temperatures is shown in Fig. 15.14. The sign of R_{xy} indicates n-type conduction in all the samples. The corresponding shape of R_{xy} with magnetic field changes from linear to non-linear with decreasing temperature, suggesting a change from one-band band transport to two (or multiple) band transport due to coexisting surface and bulk channels of carriers, and/or electron and hole puddles.

The proportional relation between mobility, sample's non-uniformity and LMR shows clear evidence for the phenomenological model of Parish-Littlewood, which attributes the LMR to large spatial fluctuations in the conductivity of the material [43]. In chalcogenide topological insulator materials, the positive MR dip around zero magnetic fields associated with the WAL effect and the LMR effect generally persists to high fields and high temperatures. The MR of topological insulator

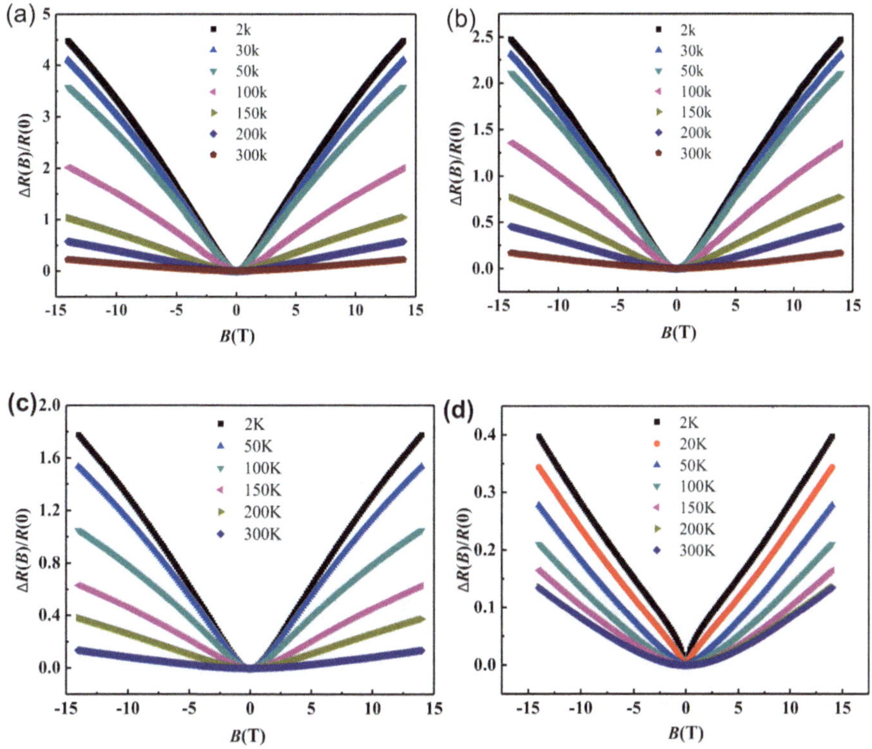

Fig. 15.13 (**a–d**) Magneto-resistance ΔR(B)/R(0) as a function of magnetic field at different temperatures for Bi_2Te_3 film sample (**a–d**). Modified from Wang et al. [43]

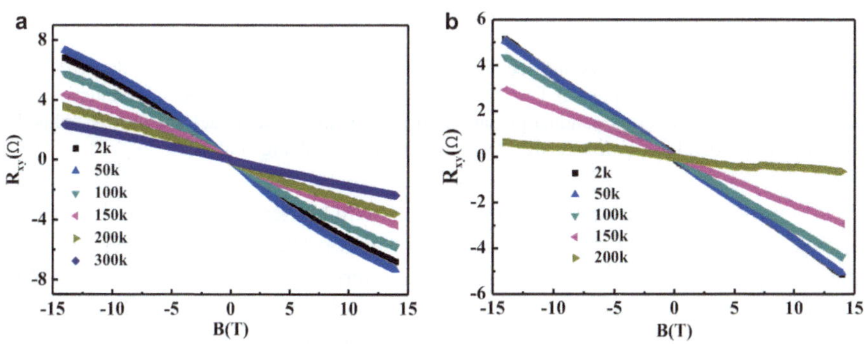

Fig. 15.14 The Hall resistance data of Bi_2Te_3 (**a**) sample (**a**), and (**b**) sample (**d**). Modified from Wang et al. [43]

Bi_2Te_3 films from the metallic to semiconducting transport regime has been studied [43]. In metallic samples, the WAL is difficult to identify due to the smallness of the WAL compared to the samples' conductivity, the sharp WAL dip in the MR is

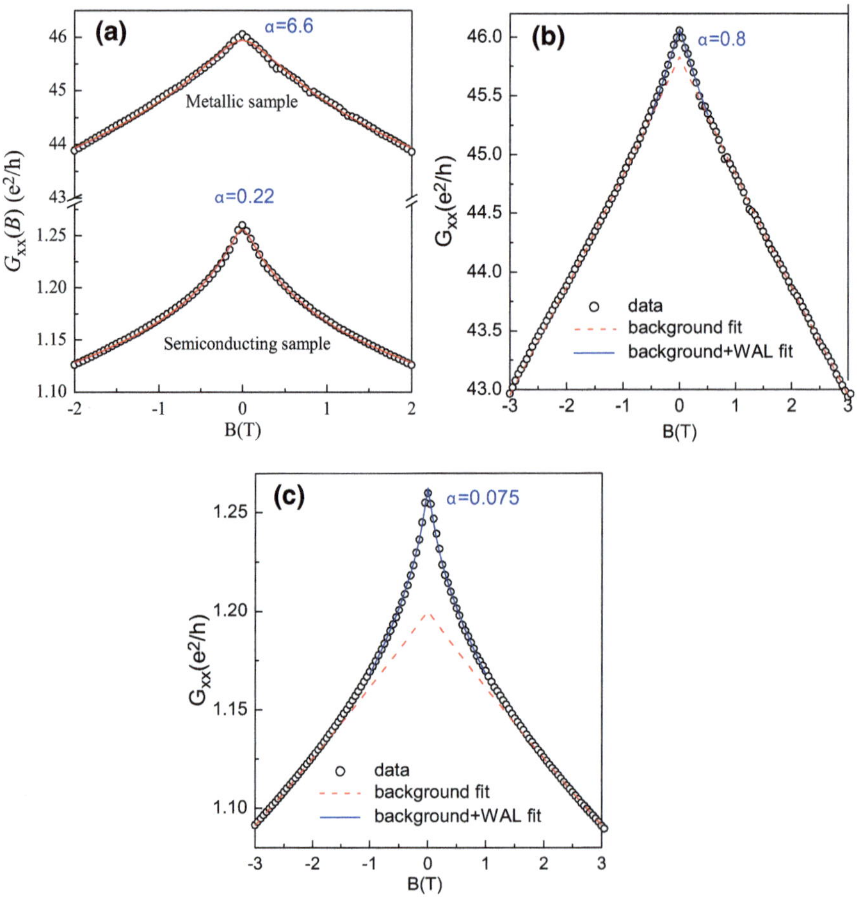

Fig. 15.15 (**a**) The direct weak anti-localization (WAL) fitting (*solid line*) of raw data (*dots*) for typical two Bi_2Te_3 film samples at T = 2 K. Fitting (*blue solid line*) of magneto-conductivity for metallic sample (**b**) and semiconducting sample (**c**) at 2 K using the model of 2D WAL plus the classical background due to linear MR effect. Modified from Wang et al. [43]

clearly present in the semiconducting samples with higher resistivity. In the various analysis of 2D WAL effect in topological insulators, the prefactor α, representing the strength of WAL is a key factor in the interpretation of the origin of WAL. Typical experimental values of α-cover the range from 0.1 to 1 in Bi_2Te_3 systems, depending on the Fermi level position and thickness of the sample. The α-value of 1 is interpreted as the topological insulator film having two independent layers of surface states each contributing $\alpha = 1/2$, while $\alpha < 1$ indicates interlayer or surface-bulk coupling. Two Bi_2Te_3 films showing metallic and semiconducting are selected to investigate the values of α. The magneto-conductivity data is fitted directly according to the 2D WAL according to the Hikami-Larkin-Nagaoka (HLN) model [97]. A reasonably good fit is obtained as shown in Fig. 15.15a. However, the fitted α is unrealistically large (α > 6) for the metallic sample. With the LMR effect considered as a background contribution in the fitting of magneto-conductance $G_{xx}(B)$ data, the fitting parameter α became significantly smaller and more reasonable, especial for the sample with higher conductivity value and larger classical magneto-conductance change, as shown in Fig. 15.15b, c. To correctly account for the low field MR by the quantitative theory of WAL according to HLN model, the classical LMR effect should be taken into account together with the WAL quantum correction. Otherwise the WAL fitting alone yields an unrealistically large coefficient α in the HLN analysis [43].

15.3.2.2 Bi_2Te_3 Films Doped With Nonmagnetic Elements

A full range of Sb-Bi compositions of $(Bi_{1-x}Sb_x)_2Te_3$ films have been studied by researchers in order to obtain the lowest possible bulk conductivity. With optimized Sb compositions, the carrier type can be tuned from n-type to p-type across the whole thickness with the help of a back-gate. SdH oscillations associated with the top and bottom surface states can also be observed and distinguished by the top gate biases in $(Bi_{1-x}Sb_x)_2Te_3$. The surface gap is opened in the films of $(Bi_{0.57}Sb_{0.43})_2Te_3$ below six quintuple layers. Competition between WL and WAL at low magnetic fields can be tuned by effective tuning the Fermi level via gate-voltage control [98–100]. By systematically studying the structural and electronic properties of Cu-doped Bi_2Te_3 films by different doping methods, it was found that Cu acts as electron donors when deposited onto the surface while behaving as acceptors when doped in the growth process [101–103]. Superconductivity was found in the Cu-intercalated Bi_2Se_3 with a transition temperature of Tc = 3.8 K. The superconducting phase $Cu_xBi_2Se_3$ occurs only in a small composition range. A combined study is performed to clarify the doping nature of Cu atoms in Bi_2Se_3 films, which find that Cu atoms behave as donors at intercalated and interstitial sites in Bi_2Se_3 films. Only the interstitial defect density plays an important role in the observation of Landau quantization of the topological surface states in Bi_2Se_3 [101]. The band structure of superconducting electrochemically intercalated $Cu_xBi_2Se_3$ is studied by ARPES, which find the band gap at the point is much larger than in pristine Bi_2Se_3. Compared with the results of band structure calculations, which indicates the origin of the large gap is come from the internal stress which created by the Cu intercalation [101]. In order to confirm that Bi_2Te_3 polycrystalline film has a robust topological surface state, and the surface Dirac fermions can be protected from localization in transport, a systematic transport measurement is investigated based on a Sn-doped Bi_2Te_3polycrystalline film with Hall configuration. The strong electron–electron interaction effect is observed, which can help realize an insulating ground state. Although the film has many structural defects and doped with non-magnetic, the surface state of the film exhibits stable WAL features and the Hall resistance can present a significant nonlinear dependence on magnetic fields. These two characteristics indicate that the film has a robust topological surface state, and the surface electrons cannot be localized. Owing to the lack of topological protection, the weak anti-localization transport of bulk electrons cannot be guaranteed, and weak localization behavior may appear in the bulk channels [104–106]. Then the Bi_2Te_3films with appropriate Sn doping exhibit a novel MR switching effect when a parallel magnetic field is applied. The WL effect from the bulk state is observed in Sn-doping Bi_2Te_3 films. The experimental data shows a transition from WAL to WL in the parallel MR curve of the low-doped film with increasing the magnetic field, and the inversion phenomenon from a positive to a negative MR cusp at low fields driven by the temperature is observed in the parallel MR in the high-doped sample [104]. High performance current-controlled magnetic field detectors can be fabricated using the Sn doping Bi_2Te_3films. When a parallel magnetic field is applied, the as-fabricated device exhibits a stable and reproducible MR switching behavior [104]. A collapse in the transport lifetime is observed in $(Bi_{1-x}In_x)_2Se_3$ with certain In substitution, that indicates the destruction of the topological phase. The thickness dependent collapse of the transport lifetime is observed, showing the unusual role that finite-size effects play in this topological quantum phase transition. The transformation from a topologically nontrivial metal into a topologically trivial band insulator through three quantum phase transitions is reported in $(Bi_{1-x}In_x)_2Se_3$ thin film [107, 108]. The transition from a topologically nontrivial metal to a trivial metal is happened at x about 3–7 %, and the metal becomes a variable-range-hopping insulator when x $\approx$15 %. Finally, the system becomes a true band insulator with its resistance immeasurably large even at room temperature when x above 25 %.

15.3.2.3 Bi_2Te_3 Films Doped With Magnetic Elements

The breaking of time-reversal symmetry by ferromagnetism is predicted to yield profound changes to the electronic surface states of a topological insulator. The two-dimensional WAL effect of Bi_2Te_3 thin films associated with surface carriers is revealed in the tilted magnetic field dependence of magneto-conductance. The WAL is observed when deposition of nonmagnetic Au impurities on the surface of Bi_2Te_3 thin films, but it disappeared by the deposition of magnetic Fe impurities which destroy the Berry phase of the topological surface states [11]. The magneto-conductance of Bi_2Te_3 film with thickness of 5 nm shows a crossover from symplectic to unitary classes with the deposition of Fe impurities. For Cr content up to 5 % the Bi_2Se_3 films are good crystalline quality. The Curie temperature reached a maximum T_C of 20 K for 5.2 % Cr. Well-defined ferromagnetic hysteresis in the magnetization and in the MR is also observed. The magneto-transport behavior of magnetically Cr doped Bi_2Se_3 exhibits a systematic crossover between WAL and WL with the change of magnetic impurity concentration, temperature, and magnetic field [85, 109]. The surface ferromagnetism in the temperature range $15\ K < T < 100\ K$ is observed in Mn-Bi_2Se_3 thin films, accompanied by a suppressed density of surface states at the Dirac point [110]. An inhomogeneous distribution of Mn atoms, with a tendency to segregate towards the sample surface is revealed by Secondary-ion mass spectroscopy and scanning tunneling microscopy. Concerning the magnetic properties, $Bi_{2-x}Mn_xTe_3$ does display long range ferromagnetism, but the Curie temperature is too low to need for applications. The long-range ferromagnetism at ambient temperature is induced in $Bi_{2-x}Mn_xTe_3$ by the magnetic proximity effect through deposited Fe overlayer [111]. STM studies of Fe on the Bi_2Te_3 and Bi_2Se_3 films surface are performed. For Fe-doped Bi_2Se_3 films, most Fe atoms substitute for Bi sites in the Fe3+state. The Fe3+ substitution ion can be switched into Fe2+ with a STM tip in a controlled manner. Fe on the Bi_2Te_3 surface show distinct impurity structures which exhibit temperature-dependent characteristics. In low-temperature deposition, Fe exists in metastable states of which only the transition-metal Bi split interstitial is STM visible. After room-temperature annealing, iron forms substitutional FeBi. Using density functional theory with spin-orbit coupling, it is showed that Fe prefers highly coordinated subsurface configurations [112, 113]. Magnetic quantum phase transition is observed in Cr-doped $Bi_2(Se_xTe_{1-x})_3$ topological insulator films. It is found that the bulk band topology is the fundamental driving force for the magnetic quantum phase transition [114]. Both the hole-mediated Ruderman Kittel-Kasuya-Yosida (RKKY) coupling and carrier-independent van Vleck magnetism are demonstrated in Cr-doped $(Bi_xSb_{1-x})_2Te_3$ thin films with effective top-gate modulations [115, 116]. Most importantly, the interplay between the two magnetic orders is found by varying the Cr doping concentrations from 2 % to 20 %,and the valid approach to either enhance or suppress each individual contribution is established. Robust ferromagnetism in thin films of Cr-doped $(Bi_xSb_{1-x})_2Te_3$ is observed against the variation in both carrier density and type, suggesting that the van Vleck mechanism due to the valence electrons with inverted band structure is responsible for the ferromagnetism [115]. Furthermore, an unusually large anomalous Hall (AH) effect has been observed in these films when the carrier density is reduced to a very low level by doping or gating, representing a step towards to the realization of the quantum anomalous Hall (QAH) effect. The magnetic responses are studied from the magnetically modulation-doped $(Bi_zSb_{1-z})_2Te_3$/ $Cr_x(Bi_ySb_{1-y})_2Te_3$ bilayer films. It is showed that the top topological insulator surface carriers can effectively mediate the magnetic impurities and generate robust ferromagnetic order by electrically tuning the Fermi level across the Dirac point [36].

In a QAH insulator, spontaneous magnetic moments and spin-orbit coupling combine to give rise to a topologically nontrivial electronic structure, leading to the quantized Hall Effect without an external magnetic field. First-principles calculations predicted the topological insulators Bi_2Te_3 and Bi_2Se_3 to form magnetically ordered insulators when doped with transition metal elements (Cr or Fe). In 2D thin films, the magnetic order gives rise to a topological electronic structure characterized by a finite Chern number, with the Hall conductance quantized in units of e^2/h, where e is the charge of an electron and h is Planck 's constant. A detailed theoretical introduction to the QAH effect based on magnetic topological insulators in 2D and 3D is presented [117, 118]. In 2D topological insulators, magnetic order breaks the symmetry between the counter-propagating helical edge states, changing the quantum spin Hall Effect into the QAH effect. In 3D topological insulators, magnetic order opens a gap for the topological surface states, and chiral edge state has been predicted to exist on the magnetic domain walls. Most recently, the QAH effect has been observed in the thin films of magnetic topological insulator Cr-doped $(Bi,Sb)_2Te_3$ [36, 119]. Sb_2Te_3 and Bi_2Te_3 have the same crystal structure and close lattice constants, moreover the former is usually p-doped and the latter is usually n-doped, so it is possible to tune the chemical potential by mixing the two compounds with proper ratio. The MBE growth of Cr-doped $(Bi_xSb_{1-x})_2Te_3$ films is carried out on sapphire (0001) or $SrTiO_3$ (111) substrates. The ferromagnetism is confirmed by direct magnetization measurements with SQUID magnetometer. The QAH effect is discovered at 30 mK in optimized samples of 5QLs $Cr_{0.15}(Bi_{0.1}Sb_{0.9})_{1.85}Te_3$films grown on 0.25-mm $SrTiO_3$(111) substrates by gating. At zero magnetic fields, the gate-tuned anomalous Hall resistance reaches the predicted quantized value of h/e^2 accompanied by a considerable drop in the longitudinal resistance. Under a strong magnetic field, the longitudinal resistance vanishes, whereas the Hall resistance remains at the quantized value. Thin films of $Cr_x(Bi_{0.2}Sb_{0.8})_{2-x}Te_3$ are grown on semi-insulating InP(111) substrates by MBE method [120], in which the Dirac point of the surface states is isolated within the 3D bulk band gap. When the films have a thickness of approximately 8 nm, and $x = 0.22$, the QAH effect is observed at temperatures as low as 50 mK. The quantum criticality behavior near the QAH phase transition is further studied, and shown that it can be described by the renormalization group theory for the integer QH states. In another experiment, thin films of $(Cr_{0.12}Bi_{0.26}Sb_{0.62})_2Te_3$ are grown by MBE, and the QAH effect is observed in 10 QLs and 6 QLs thin films at temperature $T = 85$ mK [121]. The Hall resistance reaches the quantized value h/e^2 while a non-zero residual longitudinal resistance remains. The magnitude of the non-local resistances is changed by changing the direction of magnetization, which provides strong evidence for the coexistence of non-chiral edge states with the chiral edge state in the QAH system.

15.4 Conclusions

In summary, the investigation of Bi_2Te_3 based topological insulator materials is developing quickly due to their unique surface state properties, especially in the nanostructures due to the large surface-to-volume ratio. The review here focuses on the nanoflakes, nanoplats, nanosheets, nanowires, and thin film. The chemical vapor deposition (CVD) method is descripted in detail as an example growth method for topological insulator nanomaterials. The characterizations of Bi_2Te_3 nanoplates by Raman spectroscopy reveal that the thickness and weak interactions between the nanoplates and the substrate induce a breakdown of inversion symmetry and a concomitant appearance of IR-active modes in Raman scattering. Topological surface states have been observed in the bulk insulating Bi_2Te_3 based topological insulators. However, owing to the predominance of bulk carriers from environmental doping, crystal defects or thermal excitations, the transport measurements of topological surface states remain challenging. Synthesis of ternary/quaternary topological insulator compounds, doping topological insulators with elements or gating are useful means to suppress the bulk carriers. Both the SdH and the AB oscillations suggest the existence of Berry's phase for surface conduction, which confirms the quantum transport of the surface Dirac fermions. Furthermore, the breaking of time-reversal symmetry by ferromagnetism is predicted to yield profound changes to the electronic surface states of a topological insulator. The QAH effect is observed in the thin films of magnetic topological insulator Cr-doped (Bi, $Sb)_2Te_3$.

Acknowledgements The authors thank Rui He, CheeHuei Lee, Dong Liang, Richard L.J. Qiu, Mohan Sankaran, Hao Tang for collaborations. Z.D. Zhang and Z.H. Wang thank the National Natural Science Foundation of China with Grant No. 51331006 and 51522104. X.P.A. Gao thanks the National Science Foundation (DMR-1151534) of U.S. for funding support.

References

1. M.Z. Hasan et al., Colloquium: topological insulators. Rev. Mod. Phys. **82**(4), 3045–3067 (2010)
2. J.E. Moore, The birth of topological insulators. Nature **464**(7286), 194–196 (2010)
3. H. Zhang et al., Topological insulators in Bi_2Te_3, Bi_2Te_3 and Sb_2Te_3 with a single Dirac cone on the surface. Nat. Phys. **5**(6), 438–442 (2009)
4. B.A. Bernevig et al., Quantum spin Hall effect and topological phase transition in HgTe quantum wells. Science **314**(5806), 1757–1761 (2006)
5. D. Hsieh et al., A topological Dirac insulator in a quantum spin Hall phase. Nature **452**(7190), 970–9U5 (2008)
6. J. Kampmeier et al., Suppressing twin domains in molecular beam epitaxy grown Bi_2Te_3 topological insulator thin films. Cryst. Growth Des. **15**(1), 390–394 (2015)
7. S. Cho et al., Antisite defects of Bi_2Te_3 thin films. Appl. Phys. Lett. **75**(10), 1401–1403 (1999)
8. J. Zhang et al., Band structure engineering in $(Bi_{1-x}Sb_x)_2Te_3$ ternary topological insulators. Nat. Commun. **2**, 574 (2011)
9. Z. Aabdin et al., Sb_2Te_3 and Bi_2Te_3 thin films grown by room-temperature MBE. J. Electron. Mater. **41**(6), 1493 (2012)
10. S. Shimizu et al., Gate control of surface transport in MBE-grown topological insulator $(Bi_{1-x}Sb_x)_2Te_3$ thin films. Phys. Rev. B **86**(4), 045319 (2012)
11. H.T. He et al., Impurity effect on weak antilocalization in the topological insulator Bi_2Te_3. Phys. Rev. Lett. **106**(16), 166805 (2011)
12. N.V. Tarakina et al., Comparative study of the microstructure of Bi_2Te_3 Thin Films Grown on Si(111) and InP(111) Substrates. Cryst. Growth Des. **12**(4), 2012 (1913-1918)
13. Z. Xu et al., Anisotropic topological surface states on high-index Bi_2Se_3 films. Adv. Mater. **25** (11), 1557–1562 (2013)
14. G. Zhang et al., Growth of topological insulator Bi_2Se_3 thin films on $SrTiO_3$ with large tunability in chemical potential. Adv. Funct. Mater. **21**(12), 2351–2355 (2011)
15. L. Plucinski et al., Robust surface electronic properties of topological insulators: Bi_2Te_3 films grown by molecular beam epitaxy. Appl. Phys. Lett. **98**(22), 222503 (2011)
16. R.K. Gopal et al., Weak-antilocalization and surface dominated transport in topological insulator Bi_2Se_2Te. AIP Adv. **5**(4), 047111 (2015)
17. P. Roushan et al., Topological surface states protected from backscattering by chiral spin texture. Nature **460**(7259), 1106–1U64 (2009)

18. Z. Alpichshev et al., STM imaging of electronic waves on the surface of Bi_2Te_3: topologically protected surface states and hexagonal warping effects. Phys. Rev. Lett. **104**(1), 016401 (2010)
19. J.J. Cha et al., Weak antilocalization in $Bi_2(Se_xTe_{1-x})_3$ nanoribbons and nanoplates. Nano Lett. **12**(2), 1107–1111 (2012)
20. L. Bao et al., Weak Anti-localization and quantum oscillations of surface states in topological insulator Bi_2Se_2Te. Sci. Rep. **2**, 726 (2012)
21. H.L. Cao, R. Venkatasubramanian, et al., Topological insulator Bi_2Te_3 films synthesized by metal organic chemical vapor deposition. Appl. Phys. Lett. **101**(16), 162104 (2012)
22. Z.H. Wang, L. Richard, J. Qiu, et al., Ambipolar surface conduction in ternary topological insulator $Bi_2(Te_{1-x}Se_x)_3$ nanoribbons. ACS Nano **7**(3), 2126–2131 (2013)
23. D. Hsieh, Y. Xia, et al., A tunable topological insulator in the spin helical Dirac transport regime. Nature **460**(7259), 1101–1U59 (2009)
24. D.S. Kong et al., Topological insulator nanowires and nanoribbons. Nano Lett. **10**(1), 329–333 (2010)
25. B. Yu et al., Chemical assembly and electrical characteristics of surface-rich topological insulator Bi_2Te_3 nanoplates and nanoribbons. Appl. Phys. Lett. **101**(14), 143103 (2012)
26. G. Ramanath et al., Microsphere bouquets of bismuth telluride nanoplates: room-temperature synthesis and thermoelectric properties. J. Phys. Chem. C **114**(4), 1796–1799 (2010)
27. K. Kern et al., Two-dimensional magnetotransport in Bi_2Te_2Se nanoplatelets. Appl. Phys. Lett. **101**(2), 023116 (2012)
28. Z.F. Liu et al., Epitaxial heterostructures of ultrathin topological insulator nanoplate and graphene. Nano Lett. **10**(8), 2870–2876 (2010)
29. J.G. Checkelsky et al., Bulk band gap and surface state conduction observed in voltage-tuned crystals of the topological insulator Bi_2Se_3. Phys. Rev. Lett. **106**(19), 196801 (2011)
30. L. Lu et al., Proximity effect at superconducting Sn-Bi_2Te_3 interface. Phys. Rev. B **85**(10), 104508 (2012)
31. Q.H. Xiong et al., Enhanced thermoelectric properties of solution grown $Bi_2Te_{3-x}Se_x$ nanoplatelet composites. Nano Lett. **12**(3), 1203–1209 (2012)
32. D.S. Kong et al., Few-layer nanoplates of Bi_2Se_3 and Bi_2Te_3 with highly tunable chemical potential. Nano Lett. **10**(6), 2245–2250 (2010)
33. Y.Y. Li et al., Intrinsic topological insulator Bi_2Te_3 thin films on si and their thickness limit. Adv. Mater. **22**(36), 4002–4007 (2010)
34. A. Kapitulnik et al., Weak localization effects as evidence for bulk quantization in Bi_2Te_3 thin films. Phys. Rev. B **88**(12), 121103 (2013)
35. M.R. Lang et al., Revelation of topological surface states in Bi_2Se_3 thin films by in situ al passivation. ACS Nano **6**(1), 295–302 (2012)
36. K. He et al., From magnetically doped topological insulator to the quantum anomalous hall effect. Chin. Phys. B **22**(6), 067305 (2013)
37. X.F. Kou et al., Manipulating surface-related ferromagnetism in modulation-doped topological insulators. Nano Lett. **13**(10), 4587–4593 (2013)
38. H.B. Zhang et al., High-performance Bi_2Te_3-based topological insulator film magnetic field detector. ACS Appl. Mater. Interfaces **5**(22), 11503–11508 (2013)
39. H.B. Zhang et al., Magnetoresistance switch effect of a sn-doped Bi_2Te_3 topological insulator. Adv. Mater. **24**(1), 132 (2012)
40. R.J. Cava et al., A ferromagnetic insulating substrate for the epitaxial growth of topological insulators. J. Appl. Phys. **114**(11), 114907 (2013)
41. Y.L. Chen et al., Observing electronic structures on ex-situ grown topological insulator thin films. Phys. Status Solidi-Rapid. Res. Lett. **7**(1–2), 130–132 (2013)
42. A.A. Balandin, Crystal symmetry breaking in few-quintuple Bi_2Te_3 films: applications in nanometrology of topological insulators. Appl. Phys. Lett. **96**(15), 153103 (2010)
43. C.H. Lee et al., Metal-insulator transition in variably doped $(Bi_{1-x}Sb_x)_2Se_3$ nanosheets. Nanoscale **5**(10), 4337–4343 (2013)

44. R. He et al., Observation of infrared-active modes in Raman scattering from topological insulator nanoplates. Nanotechnology **23**(45), 455703 (2012)
45. Z.H. Wang et al., Linear magnetoresistance versus weak antilocalization effects in Bi_2Te_3. Nano Res. **8**(9), 2963–2969 (2015)
46. Z.H. Wang et al., Granularity controlled nonsaturating linear magnetoresistance in topological insulator Bi_2Te_3 films. Nano Lett. **14**(11), 6510–6514 (2014)
47. Z.Y. Fan et al., Toward the development of printable nanowire electronics and sensors. Adv. Mater. **21**(37), 3730–3743 (2009)
48. R. He et al., Laser induced oxidation and optical properties of stoichiometric and non-stoichiometric Bi_2Te_3 nanoplates. Nano Res. **8**(3), 851–859 (2015)
49. J. Zhang et al., Raman spectroscopy of few-quintuple layer topological insulator Bi_2Se_3 nanoplatelets. Nano Lett. **11**(6), 2407–2414 (2011)
50. J.H. Guo et al., Growth and microstructures of ultrathin Bi_2Te_3 nanoplates by modified hot wall epitaxy. Nano **9**(6), 1450056 (2014)
51. M. Zhao, M. Bosman, M. Danesh, et al., visible surface plasmon modes in single Bi_2Te_3 nanoplate. Nano Lett. **15**(12), 8331–8335 (2015)
52. X. He, H. Zhang, W. Lin, et al., PVP-assisted solvothermal synthesis of high-yielded Bi_2Te_3 hexagonal nanoplates: application in passively Q-switched fiber laser. Sci. Rep. **5**, 15868 (2015)
53. B. Liu, W.Y. Xie, H. Li, et al., Surrounding sensitive electronic properties of Bi_2Te_3 nanoplates-potential sensing applications of topological insulators. Sci. Rep. **4**, 4639 (2014)
54. D.S. Kong et al., Rapid surface oxidation as a source of surface degradation factor for Bi_2Se_3. ACS Nano **5**(6), 4698–4703 (2011)
55. D.X. Qu et al., Quantum oscillations and hall anomaly of surface states in the topological insulator Bi_2Te_3. Science **329**(5993), 821–824 (2010)
56. S.S. Hong et al., Ultrathin topological insulator Bi_2Se_3 nanoribbons exfoliated by atomic force microscopy. Nano Lett. **10**(8), 3118–3122 (2010)
57. H. Tang et al., Two-dimensional transport—induced linear magneto-resistance in topological insulator Bi_2Se_3 nanoribbons. ACS Nano **5**(9), 7510–7516 (2011)
58. Y. Yan et al., Large magnetoresistance in high mobility topological insulator Bi_2Te_3. Appl. Phys. Lett. **103**(3), 033106 (2013)
59. D.P. Yu et al., Synthesis and quantum transport properties of Bi_2Se_3 topological insulator nanostructures. Sci. Rep. **3**, 1264 (2013)
60. L. Lu et al., Coexistence of bulk and surface shubnikov-de haas oscillations in Bi_2Se_3. J. Low Temp. Phys. **170**(5–6), 397–402 (2013)
61. H.L. Peng et al., Aharonov-Bohm interference in topological insulator nanoribbons. Nat. Mater. **9**(3), 225–229 (2010)
62. S. Cho et al., Insulating behavior in ultrathin bismuth selenide field effect transistors. Nano Lett. **11**(5), 1925–1927 (2011)
63. D. Kim et al., Surface conduction of topological Dirac electrons in bulk insulating Bi_2Se_3. Nat. Phys. **8**(6), 459–463 (2012)
64. D. Kim et al., Intrinsic electron-phonon resistivity of Bi_2Te_3 in the topological regime. Phys. Rev. Lett. **109**(15), 166801 (2012)
65. F.X. Xiu et al., Manipulating surface states in topological insulator nanoribbons. Nat. Nanotechnol. **6**(4), 216–221 (2011)
66. Y. Wang et al., Gate-controlled surface conduction in na-doped Bi_2Te_3 topological insulator nanoplates. Nano Lett. **12**(3), 1170–1175 (2012)
67. Y. Yan et al., Synthesis and field emission properties of topological insulator Bi_2Se_3 nanoflake arrays. Nanotechnology **23**(30), 305704 (2012)
68. Y. Xia et al., Observation of a large-gap topological-insulator class with a single Dirac cone on the surface. Nat. Phys. **5**(6), 398–402 (2009)
69. M.Z. Hasan et al., A topological insulator surface under strong Coulomb, magnetic and disorder perturbations. Nat. Phys. **7**(1), 32–37 (2011)

70. D.S. Dessau et al., Mapping the orbital wavefunction of the surface states in three-dimensional topological insulators. Nat. Phys. **9**(8), 499–504 (2013)
71. A. Yazdani et al., Spatial fluctuations of helical Dirac fermions on the surface of topological insulators. Nat. Phys. **7**(12), 939–943 (2011)
72. K. Kern et al., Growth of high-mobility Bi_2Te_2Se nanoplatelets on hBN sheets by van der waals epitaxy. Nano Lett. **12**(10), 5137–5142 (2012)
73. N. Samarth et al., Superconducting proximity effect and possible evidence for Pearl vortices in a candidate topological insulator. Phys. Rev. B **84**(16), 165120 (2011)
74. B. Xia et al., Indications of surface-dominated transport in single crystalline nanoflake devices of topological insulator $Bi_{1.5}Sb_{0.5}Te_{1.8}Se_{1.2}$. Phys. Rev. B **87**(8), 085442 (2013)
75. T.C. Hsiung et al., Enhanced surface mobility and quantum oscillations in topological insulator $Bi_{1.5}Sb_{0.5}Te_{1.7}Se_{1.3}$ nanoflakes. Appl. Phys. Lett. **103**(16), 163111 (2013)
76. Z.Y. Wang et al., Tuning carrier type and density in Bi_2Se_3 by Ca-doping. Appl. Phys. Lett. **97**(4), 042112 (2010)
77. Z.G. Chen et al., Paramagnetic Cu-doped Bi_2Te_3 nanoplates. Appl. Phys. Lett. **104**(5), 053105 (2014)
78. Q. Liu et al., Magnetic impurities on the surface of a topological insulator. Phys. Rev. Lett. **102**(15), 156603 (2009)
79. L. Cheng et al., High curie temperature $Bi_{1.85}Mn_{0.15}Te_3$ nanoplates. J. Am. Chem. Soc. **134** (46), 18920–18923 (2012)
80. J.J. Cha et al., Effects of magnetic doping on weak antilocalization in narrow Bi_2Se_3 nanoribbons. Nano Lett. **12**(8), 4355–4359 (2012)
81. J.J. Cha et al., Magnetic doping and kondo effect in Bi_2Se_3 nanoribbons. Nano Lett. **10**(3), 1076–1081 (2010)
82. F. Qu et al., Strong superconducting proximity effect in $PbBi_2Te_3$ hybrid structures. Sci. Rep. **2**, 339 (2012)
83. G.H. Zhang et al., Quintuple-layer epitaxy of thin films of topological insulator Bi_2Te_3. Appl. Phys. Lett. **95**(5), 053114 (2009)
84. C.L. Song et al., Topological insulator Bi_2Se_3 thin films grown on double-layer graphene by molecular beam epitaxy. Appl. Phys. Lett. **97**(14), 143118 (2010)
85. G. Wang et al., Topological insulator thin films of Bi_2Te_3 with controlled electronic structure. Adv. Mater. **23**(26), 2929–2932 (2011)
86. Y. Liu et al., Interfacial bonding and structure of Bi_2Te_3 topological insulator films on Si(111) determined by surface x-ray scattering. Phys. Rev. Lett. **110**(22), 226103 (2013)
87. P.P.J. Haazen et al., Ferromagnetism in thin-film Cr-doped topological insulator Bi_2Se_3. Appl. Phys. Lett. **100**(8), 082404 (2012)
88. H.W. Liu et al., Growth of topological insulator Bi_2Te_3 ultrathin films on Si(111) investigated by low-energy electron microscopy. Cryst. Growth Des. **10**(10), 4491–4493 (2010)
89. V. Goyal et al., Mechanically-exfoliated stacks of thin films of Bi_2Te_3 topological insulators with enhanced thermoelectric performance. Appl. Phys. Lett. **97**(13), 133117 (2010)
90. K.M.F. Shahil et al., Micro-Raman spectroscopy of mechanically exfoliated few-quintuple layers of Bi_2Te_3, Bi_2Se_3, and Sb_2Te_3 materials. J. Appl. Phys. **111**(5), 054305 (2012)
91. H.B. Zhang et al., Experimental evidence of the nanoscaled topological metallic surface state of Bi_2Te_3 and Sb_2Te_3 films. Europhys. Lett. **95**(5), 56002 (2011)
92. X.X. Yu et al., Separation of top and bottom surface conduction in Bi_2Te_3 thin films. Nanotechnology **24**(1), 015705 (2013)
93. M. Chen et al., Molecular beam epitaxy of bilayer Bi(111) films on topological insulator Bi_2Te_3: a scanning tunneling microscopy study. Appl. Phys. Lett. **101**(8), 081603 (2012)
94. H.T. He et al., High-field linear magneto-resistance in topological insulator Bi_2Te_3 thin films. Appl. Phys. Lett. **100**(3), 032105 (2012)
95. S.X. Zhang et al., Magneto-resistance up to 60 Tesla in topological insulator Bi_2Te_3 thin films. Appl. Phys. Lett. **101**(20), 202403 (2012)

96. B.F. Gao et al., Gate-controlled linear magnetoresistance in thin Bi_2Se_3 sheets. Appl. Phys. Lett. **100**(21), 212402 (2012)
97. S. Hikami et al., Spin-orbit interaction and magnetoresistance in the two dimensional random system. Prog. Theor. Phys. **63**(2), 707–710 (1980)
98. X. He et al., Highly tunable electron transport in epitaxial topological insulator $(Bi_{1-x}Sb_x)_2Te_3$ thin films. Appl. Phys. Lett. **101**(12), 123111 (2012)
99. L. He et al., Evidence of the two surface states of $(Bi_{0.53}Sb_{0.47})_2Te_3$ films grown by van der Waals epitaxy. Sci. Rep. **3**, 3406 (2013)
100. M. Lang et al., Competing weak localization and weak antilocalization in ultrathin topological insulators. Nano Lett. **13**(1), 48–53 (2013)
101. X.G. Zhu et al., Doping nature of Cu in epitaxial topological insulator Bi_2Te_3 thin films. Surf. Sci. **617**, 156–161 (2013)
102. Y.L. Wang et al., Structural defects and electronic properties of the Cu-doped topological insulator Bi_2Se_3. Phys. Rev. B **84**(7), 075335 (2011)
103. A. Ribak et al., Internal pressure in superconducting Cu-intercalated Bi_2Se_3. Phys. Rev. B **93** (6), 064505 (2016)
104. H.B. Zhang et al., Robust topological surface transport with weak localization bulk channels in polycrystalline Bi_2Te_3 films. J. Phys. D Appl. Phys. **49**(9), 095003 (2016)
105. H.B. Zhang et al., Weak localization bulk state in a topological insulator Bi_2Te_3 film. Phys. Rev. B **86**(7), 075102 (2012)
106. Q. Yang et al., Emerging weak localization effects on a topological insulator–insulating ferromagnet (Bi_2Se_3-EuS) interface. Phys. Rev. B **88**(8), 081407 (2013)
107. L. Wu et al., A sudden collapse in the transport lifetime across the topological phase transition in $(Bi_{1-x}In_x)_2Se_3$. Nat. Phys. **9**(7), 410–414 (2013)
108. M. Brahlek et al., Topological-metal to band-insulator transition in $(Bi_{1-x}In_x)_2Se_3$ thin films. Phys. Rev. Lett. **109**(18), 186403 (2012)
109. M.H. Liu et al., Crossover between weak antilocalization and weak localization in a magnetically doped topological insulator. Phys. Rev. Lett. **108**(3), 036805 (2012)
110. D.M. Zhang et al., Interplay between ferromagnetism, surface states, and quantum corrections in a magnetically doped topological insulator. Phys. Rev. B **86**(20), 205127 (2012)
111. I. Vobornik et al., Magnetic proximity effect as a pathway to spintronic applications of topological insulators. Nano Lett. **11**(10), 4079–4082 (2011)
112. C.L. Song et al., Gating the charge state of single Fe dopants in the topological insulator Bi_2Te_3 with a scanning tunneling microscope. Phys. Rev. B **86**(4), 045441 (2012)
113. D. West et al., Identification of magnetic dopants on the surfaces of topological insulators: experiment and theory for Fe on Bi_2Te_3 (111). Phys. Rev. B **85**(8), 081305 (2012)
114. J.S. Zhang et al., Topology-driven magnetic quantum phase transition in topological insulators. Science **339**(6127), 1582–1586 (2013)
115. X.F. Kou et al., Interplay between different magnetisms in Cr-doped topological insulators. ACS Nano **7**(10), 9205–9212 (2013)
116. C.Z. Chang et al., Thin films of magnetically doped topological insulator with carrier-independent long-range ferromagnetic order. Adv. Mater. **25**(7), 1065–1070 (2013)
117. J. Wang et al., Quantum anomalous Hall effect in magnetic topological insulators. Phys. Scr. **T164**, 014003 (2015)
118. R. Yu et al., Quantized anomalous Hall effect in magnetic topological insulators. Science **329** (5987), 61–64 (2010)
119. C.Z. Chang et al., Experimental observation of the quantum anomalous hall effect in a magnetic topological insulator. Science **340**(6129), 167–170 (2013)
120. J.G. Checkelsky et al., Trajectory of the anomalous Hall effect towards the quantized state in a ferromagnetic topological insulator. Nat. Phys. **10**(10), 731–736 (2014)
121. X. Kou et al., Scale-invariant quantum anomalous Hall effect in magnetic topological insulators beyond the two-dimensional limit. Phys. Rev. Lett. **113**(13), 137201 (2014)

Part IV
Other Nanoscale Devices

Chapter 16
Micro/Nanoscale Optical Devices for Hyperspectral Imaging System

Li Li, Chengjun Huang, and Haiying Zhang

16.1 Introduction

The concept of hyperspectral imaging system was originally developed as a remote sensing technique in 1970s, and has been an area of active research ever since. As a powerful tool, it extends the functionality of optical inspection system by combining the spatial information and patterns of 2D images with full spectral intensity variation of each pixel. Theoretically, the hyperspectral imaging system contains rich and detailed spectral information with tens or hundreds of narrow and contiguous bands for each pixel in an image. With a high resolution sampling over spectral data, potentially it can be used to detect and analyze various of substances. Since its successful proof of concept as an airborne geographic imaging system to identify minerals on earth's surface, its application has been extended into vegetation mapping, agricultural production monitoring for nutrient shortage and insect infestation, environmental monitoring for pollution detection and control, mineralogy, resource management. With the same concept, the airborne system has also been brought down to the ground and widely implemented in food and pharmaceutical manufacturing industry for chemical and substance determination, quality control and monitoring in agricultural commodities, meats, and medicine. And right now, further work has extended the hyperspectral imaging system to medical and clinical field, such as for cancer detection (benign and cancerous tumor differentiation, skin cancer, cervical cancer and breast cancer diagnosis), surgical guidance and health monitoring (oxygenation status of the brains, eye balls). And right now people are still keeping expanding new areas where hyperspectral imaging techniques can be helpfully utilized. The system of hyperspectral imaging camera is a combination of various optics, spectral modulation instruments,

L. Li • C. Huang (✉) • H. Zhang
Institute of Microelectronics of Chinese Academy of Sciences, Beijing, China
e-mail: huangchengjun@ime.ac.cn

T. Li, Z. Liu (eds.), *Outlook and Challenges of Nano Devices, Sensors, and MEMS*,
DOI 10.1007/978-3-319-50824-5_16

photodetectors, image reconstruction algorithm, computational devices. However, due to the functionality of extracting both 2D spatial and 1D spectral intensity information, the system has to include a spectrometry mechanism and even mechanics to tuning the device, this will in turn increase the size, complexity, alignment challenges and maintenance cost. Therefore, hyperspectral imaging system has been mostly only available to academic researchers, highly demanded industrial manufacturers and military. With the development of data processing algorithm, computational devices, miniature of the instrument components, improved detectors and new applications, many novel solutions for the spectral imaging system have been proposed and resulted in diverse optics and system design. It is now possible to build cheap and robust hyperspectral imaging system and even bring the application into the daily household appliance, such as car imaging system for ice detection on the road and horticultural management for detecting failed plants. A more detailed system level description is a much extended topic and is out of the scope of this chapter, readers who has intended to learn more can refer to the related materials.

The core functionality of a hyperspectral imaging system is to extract the spectral intensity variation over a desired range for every single pixel within the same scenery. It is can be achieved either by capturing images at sequentially scanned spectral band using monochromatic photodetector(s) or acquiring full spectrum intensities of only a small area subsequentially scanned over a scenery. Therefore, it is has to include the spectral tuning mechanism. Conventionally, spectral modulation can be realized through various tunable or non-tunable spectral filters, such as filter wheel, liquid crystal tunable filters (LCTF), acoustic optical tunable filters (AOTF), Fourier transform spectrometers, linear variable filter, various forms of Fabry-Perot etalon interferometers, optical dispersive devices such as prism and diffractive gratings. Usually, the bench top tunable optical filter containing complex system of lenses and moving parts, thus are bulky, fragile and expensive. With development of microfabrication technology, it is possible to achieve integration between microoptics, microelectronics and micromechanics; and the small scale integration of those filters onto chip. With everything is trying to shrink down nowadays, it might not be an optimum option for optical system due to the optical efficiency consideration for trade-offs between optical performance and device size. However, those microfabricated spectrometers can offer significant advantages over the existing instruments, including reduced size, system integration, compactness, low cost, fast data acquisition, robust, high reliability, simplified alignment and assembly. For many applications, these advantages are of higher importance than reduced resolution.

A hyperspectral imaging system requires integration among electronics, detectors, optics, spectral tuning and spatial tuning components. And various novel architectures for hyperspectral imaging system has been proposed, and many of which involving implementing microfabricated optics, such as micro tunable filter, dispersive optics and microscanners. With the development of microfabrication techniques and materials, a microfabricated optical devices can potentially integrate micro mechanical structures, microoptics and IC circuits in a single chip. In this chapter, the novel structure designs, fabrication solutions for microspectrometers are focused, readers interesting in the detailed performance and system architecture can refer to the quoted academic papers. First, the general principle and design consideration for each types of microspectrometers are described in the followed section. Then, the existing examples of microfabricated spectrometers are reviewed in terms of state-of-the-art research status, novel structure design, modeling methodology, fabrication techniques and optimization methods. At last, the challenges and future tendency are summarized.

16.2 Tunable Microspectrometers

One of the most popular forms of microfabricated tunable optical filter is Fabry-Perot filter, which is composed of two paralleled micromirror separated by an optical cavity. When a broadband light beam is incident onto the filter, the transmitted optical light is resonantly oscillating between two semi-reflective micromirror, for the spectrum experiencing a travelling distance of exactly N times of its half wavelength is then transmitted through the filter due to constructive interference of multiple in-phase reflected beams; for those light beams not satisfying this condition are not transmitted due to the deconstructive interference among out-of-phase reflected light beams. Thus, the filtered wavelengths of a Fabry-Perot interferometer can be tuned by varying the cavity length between two reflective layers.

The reflective micromirror structure and its actuating mechanism of Fabry-Perot interferometers can be well adapted in MEMS devices using microfabrication technology. Because the displacement actuation of planar structures and fabricating reflective micromirrors have been intensively researched for numerous MEMS devices with various forms. And the microfabricated Fabry-Perot filters has been widely studied and implemented in fiber optics telecommunication, spectrometry and spectral imaging applications. However, the design of tunable micro Fabry-Perot filter for a telecommunication application is very different from the one used for spectroscopy and imaging system. This is because that the optical filters for telecommunication application usually require a narrow bandwidth, high transmittance, relatively small tuning range, while the ones designed for spectroscopy and spectral imaging require wide tunable spectral range and large aperture size [1]. In this section, only the Fabry-Perot filters design for imaging applications are reviewed.

Micro Fabry-Perot filter can be bulk [2] or surface micromachined. Generally, the design for a microfabricated Fabry-Perot filter need to consider the material selections, geometry, fabrication techniques and its results optical performance, such as the aperture size, actuation mechanism, cavity lengths, reflective and material of reflectors, transmittance, bandwidth, spectral range and etc.

First of all, reflectivity (R) of two micromirror structures determines spectral resolution of a Fabry-Perot filter which is quantified by FWHM (full-width and half

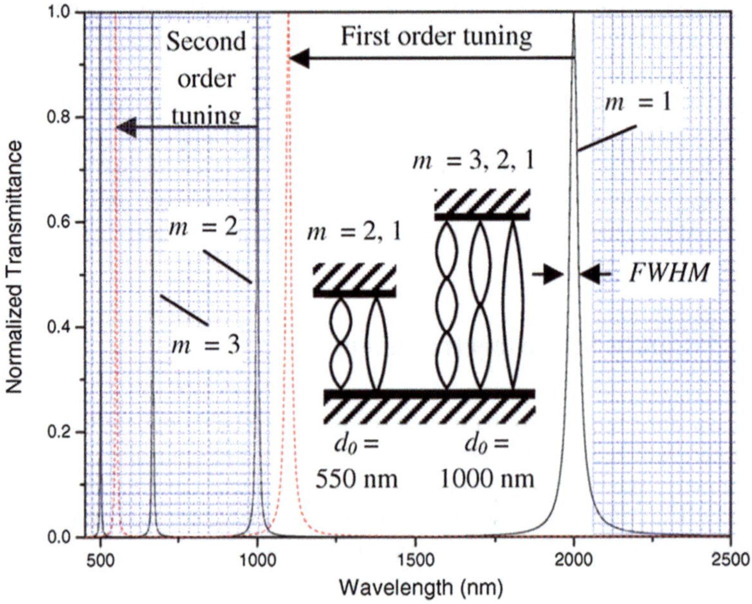

Fig. 16.1 FWHM of filtered wavelength of a Fabry-Perot filter with 95% reflectivity for both micromirrors and cavity length of 550–1000 nm respectively [1]

maximum) of filtered wavelength (λ). Based on Eq. (16.1), the higher the micromirrors' reflectivity are (i.e. >98%), the narrower or finer the output wavelength as shown in Fig. 16.1.

$$\text{FWHM} = \frac{\lambda(1-R)}{m\pi\sqrt{R}} \tag{16.1}$$

There are two types of microfabricated reflective structures. The reflective layer can be either deposited in metal layers of aluminum (Al), silver (Ag) or gold (Au) through electron beam (E-beam) evaporation, sputtering and patterned through lift-off, or alternatively, it can be designed in the form of dielectric Bragg mirror and fabricated by LPCVD or PECVD depositing alternating dielectric layers of different refractive index. However, the optical performances for these two types reflective material are quite opposite to each other. For example, the metallic mirror has good reflectivity over a broad spectrum range. The reflectivity of metallic layer is dependent on its thickness, however, the metallic mirror has high optical absorption through its thickness. Therefore, the metallic mirror cannot provide both high spectral finesse and high transmittance simultaneously [3]. In the contrary, Bragg dielectric mirror which is multiple stacks of two alternating arranged dielectric layers with different refractive index, is featured by a high reflectivity (>90%) over a narrow spectrum range. The reflectivity of a Bragg mirror is dependent on the contrast of two repeated dielectric layers as well as the number of stacks deposited. Therefore, in order to achieve a high reflectivity (>90%) for two particular materials, multiple repeated layer pairs need to deposited (i.e. 6–7 pairs), which in turn challenges the complexity fabrication processes and yield rate. The thickness of each dielectric layer is usually designed to be a quarter of wavelength for which the peak transmission is centered at, thus results in a quart-wave Bragg dielectric micromirror design for high reflective in a narrow wavelength range. It is also possible to design a broadband dielectric mirror by optimizing all the layers in the stack, such as designing groups of several quarter-wave Bragg micromirror of different center wavelengths together or even having all layers' thicknesses optimized. But to achieve both the reflectivity and broad spectrum range, more numbers of stacks will be required compared to the conventional quarter-wave dielectric mirror (i.e. 54 layers [4]). This will in turn increase the totally thickness of reflector (i.e. >3 mm [4]). Because of these limitations, the selection between the metallic and dielectric reflect layers need to take consideration of both available fabrication processes and the minimum required optical performance, such as FWHM and spectrum range. Despite of all that, quarter-wave dielectric Bragg micromirror is still more preferred by Fabry-Perot filter designers for spectral imaging application for only a limited spectral tuning range. This is preferable for those spectral imaging systems detecting only a limited number of or a particular substance(s) where spectrum of interest is fixed rather than analyzing a wide range of substances. Among microfabrication materials, silicon is a popular candidate for Fabry-Perot reflectors. Because, silicon is highly absorptive for all the wavelength less than 1.1 μm but transparent for entire NIR, SWIR spectral regions and beyond [5], which makes it a good candidate material for IR wavelengths. With the development of infrared photodetector, it is now possible to study the monolithic integration of Fabry-Perot filter with a single or even an array of IR photodetectors [6].

Apart from the material selection for micromirrors, the fabrication processes and aperture size also require carefully designing and evaluating for structure stress and curvature control. In order to ensure a good optical transmittance uniformity over the entire filter, flat, smooth and homogenous reflectors are required. And especially, when a large aperture size is required, the stress and curvature could be an issue. Firstly, residual stress within a layer usually arises from the temperature difference between fabrication and the actually operating condition. And a suspended plate microstructure with multiple layers, the mismatched stress between layers causes the structure to bend, or to have a surface curvature. Typical aperture size of Fabry-Perot filter or of reflective layers can range from several hundreds of micrometer to several millimeters in diameter, while the thickness of the reflective layer is much less which can range from several tens of nanometers to several micrometer depended on reflector designed. With such an aspect ratio, the multilayer could bend under either residual stress within a single layer or stress mismatch between multiple stacked layers. Therefore, the fabrication parameters are required to be specific tailored for a controlled curvature over the micromirror area; and sometimes an additional post-process treatment is required for a further cure for a flatter micromirror plate. Stress of a material layer is dependent on the deposited thickness (stress ICPCVD deposited Si and PECVD oxide) [7]. For multilayer deposition, the thickness and fabrication conditions need to be carefully tailored in order to cancel out the overall bending effect by matching the interlayer stress. Common post-process technique used is thermal annealing at a critical stress conversion temperature (i.e. >350°C) in a N_2 environment [5]. The curvature of the plate could also be cured by oxygen plasma treatment. For example, a 600-s low power oxygen plasma treatment was applied to improve the flatness of dielectric micromirror [8, 9].

Also, anchor effect should be considered for the suspended reflector. There are generally two options for support the movable reflector. One is to conformably support at the periphery of micromirror. This configuration can bring the best support and stability. However, because the micromirror is constrained fully at the rim, there could be unavoidably non-uniform flatness around anchor area and sever bowing effect during actuation which effectively reduce the reflector optical area and actuation range. The other anchor design is using flexible supporting microstructures or springs to connect movable micromirror to the substrate at symmetric points around mirror rim. This type of connection design can provide a flexible connection and hence potentially gives a larger deflection range with a lower actuation voltage. During actuation, because the maximum stress is centered at the flexible connectors, the micromirror has less bowing effect and remain relatively flat during actuation. But at the same time, due to the flexibility of these structures, it is less stable at rest position as well as during actuation compared to the micromirror conformably anchored at the rim. This instable displacement can also be brought up geometric inconsistency among microsprings arranged around the micromirror due to the fabrication variation. Apart from that, any suspended structures need to avoid the low frequency vibration [5].

In order to tune cavity length between two reflective micromirrors, a movable structure configuration and actuation mechanism are required. Usually, tunable Fabry-Perot filter is designed to have one movable micromirror while the other one is fixed to either the substrate or on top of the integrated photodetector. To create a free space for movable micromirror, a sacrifice layer is predefined at the maximum tunable cavity length and is removed later during the fabrication as well as releasing the movable micromirror. And actuation of the micromirror plates are commonly achieved by using electrostatic actuators, due to a fast switch speed of typically no less than several tens of micron seconds. A fast switching actuation mechanism is essential for spectral imaging applications. In order to obtain a hyperspectral imaging data cube, a sequential scanning over a spectral range within the same scenery is applied when using a tunable Fabry-Perot filter. The duration of a full range spectral scan requires to be as short as possible, where the targeted sample or object have negligible movement and considered to be steady. In the case of any shift of objector or scenery, the spectral smudge will exhibited where spectral information is mismatched between pixels. This could be corrected through a mapping algorithm when this misalignment is limited and predictable. However, for very severe spectral smudge, the mismatch cannot be fixed and spectral data cube will be useless. During actuation, when applying a voltage between two reflective layers, the attractive electrostatic force will pull the reflective layers closer. Since, the hyperspectral imaging system and some spectroscopy require a wide tunable spectral range over a narrow width of filtered spectral (FWHM). This in turn demands the actuator to prove a large displacement for the movable micromirror. In many designs, the movable electrode of electrostatic actuator either deposited to all over or partly over a micromirror area or on its rim, and to create a parallel electrode pair with the bottom fixed electrode. However, this parallel-plate actuation by using electrostatic actuators exhibit drawbacks of limited tunable range and instability. That is, the parallel plate electrostatic actuators can only actuated to 33% of initial gap [9]. When the two electrodes are too close to each other, the two actuators will snap into each other and stick together, and results in a short circuit between two electrodes. This snap-in phenomena is also commonly referred as the pull-in effect for electrostatic actuation mechanism. One solution is to design the electrodes gap differ from the micromirror gap to extend the tunable range of Fabry-Perot mirrors [9], or introduce a feedback control [10], however both of these methods increase the fabrication complexity. Apart from the limited tuning range and instability, the electrostatically tuned Fabry-Perot filter also requires high actuating voltage, for example 10–27 V.

16.3 Non-tunable Microspectrometers

A tunable Fabry-Perot filter requires mechanical actuation mechanism to change the cavity length between two reflector mirrors and therefore is capable of filtering spectral wavelengths continuously. The acquiring of series of spectra can also be achieved by using an array of fixed Fabry-Perot filters with different cavity lengths. In this way, when the same input light beam is incident to filter array, different spectrum wavelengths are filtered at equivalent Fabry-Perot filter simultaneously. Therefore, similar to a prism or a grating, the array of different Fabry-Perot filters spatially disperse a series of spectrum simultaneously but in the discrete format. Because of no mechanical structures presented, an array of fixed Fabry-Perot filters is more reliable, and the constraints limited by the actuation instability, deflection worsened stress and bowing effect can be eliminated. Moreover, less power is required because of actuation free (Fig. 16.2).

However, the design and fabrication of this type of devices is not necessarily simpler. Because, the filtered wavelengths by fixed Fabry-Perot filters array are geometrically tuned simultaneously. Therefore, one of the main challenges for this type of geometrically tuned filters is a precisely controlled fabrication process to create the accurately varying geometrical dimensions as desired.

First of all, the number of output wavelengths, or the resolving capability, is depended on the number of filter elements can be achieved. Because, only a limited number of filters can be designed within a limited planar area for cost and application facts. In the meanwhile a sufficient area for each Fabry-Perot filter should be arranged for optical efficiency guarantee, and the filter elements need to be arranged

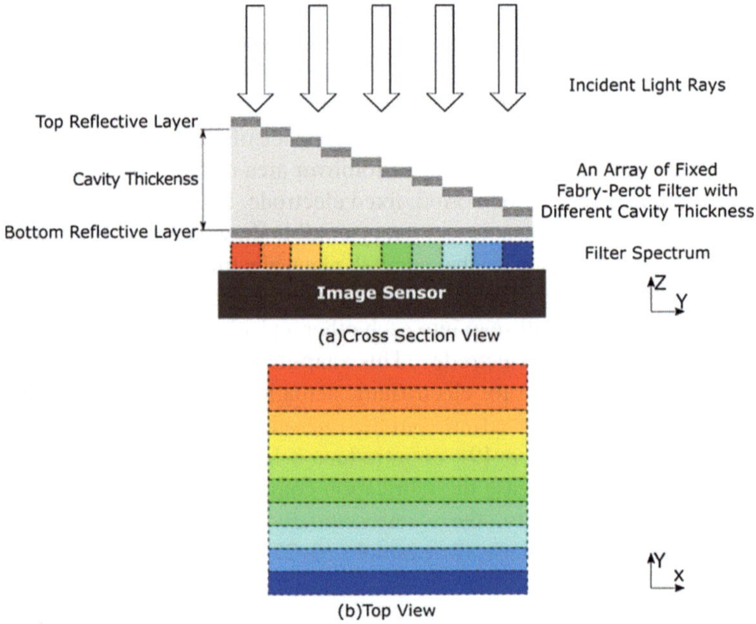

Fig. 16.2 Illustration of a light beam filtered by an array of Fabry-Perot filters with different cavity lengths

next to each other with very small gap or no gap. Secondly, in order to guarantee the output of the desired wavelengths, high accuracy is required for cavity thicknesses. The cavities thickness can be formed within the same material layer at multiple etching steps. Because multiple etching steps within the same layer are required, the commonly used etching stop layer technique cannot be applied, which is usually a thin layer of oxide underneath the targeted material layer used to adapt the etching tolerance. And the cavities' thicknesses are calculated from wavelengths to be filtered with precision of only a few nanometers. Therefore, the timing and flow rate of etchant need to be carefully tested and controlled. Also, with more filters designed, more etching steps are needed. For filter with the same thickness increment, a combinatorial etching technique has been developed which allowing two deposits and N times etching steps for 2^N filter elements [11]. This technique has greatly improved the fabricating efficiency for a large number of different Fabry-Perot filter elements in tradeoff in the number of available cavity thicknesses. Despite of this, the fabrication error can be built up after multiple etching steps and results in a larger etching error for the thinner cavities. And the optical performance consistency will be sensitive to fabrication variation among wafers and between fabrication batches.

For both the electrostatically tunable Fabry-Perot filters and filter array of various fixed cavity lengths, the two reflectors of which require to be ideally parallel to each other. Any curvature, tilt and surface roughness is considered to be imperfection degrading the finesse of Fabry-Perot filter. However, a linear variable

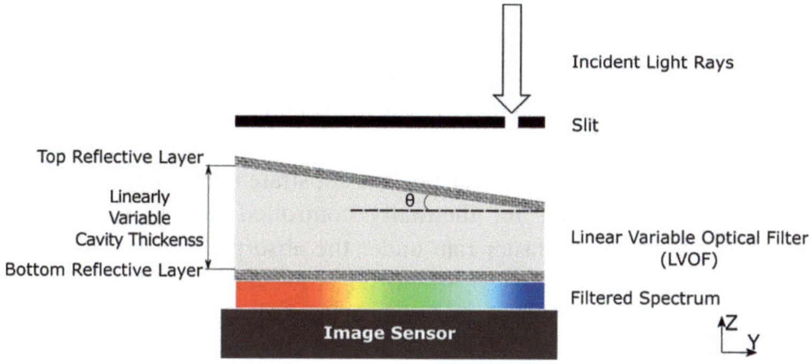

Fig. 16.3 Illustration of a light beam is filtered and dispersed by a linear variable optical filter based on a Fabry-Perot filter with unparalleled reflectors

optical filter (LVOF) has implemented the tilting angle between two reflectors to achieve the wavelengths filtering as well as dispersing the filter spectral spatially along the varying cavity length. This can be explained by the cavity length between two unparalleled micromirrors continuously varies and can be represented by an unlimited number of Fabry-Perot filters. Therefore, the spectral wavelengths transmitted through the filter continuously vary along the filter and can be readout using an array of detectors. LVOF can filter continuously varied wavelengths with ranges defined by thickness variation of cavity just like an electrostatically tuned Fabry-Perot filter but is more stable and robust. In this way, a LVOF based microspectrometer can provide a high resolution spectral analysis (Fig. 16.3).

The key point of a LVOF is to fabricate an accurately controlled angle between two micromirrors. There are several methods to fabricate a layer with a wedged cross-section, such as, gray scale lithography [12], inclined/rotated UV lithography [13], laser direct writing [14], shadow mask [15], polishing, triangle-shaped mask [16] and resist reflow [17]. The gray scale lithography method requires high cost especially with high resolution requirement, the alternation moving of the lithography equipment. The resist reflow technique is intensively studied by Professor Wolffenbuttel's research group [17]. This technique performed first by spin-coating a layer of photoresist on top of a layer of dielectric resonant cavity material. The process utilizes the reflow characteristics of photoresist to reshape it into the tapered geometry with unparalleled surface to the substrate. The conventional dry etch is then performed, and the photoresist's profile is transferred onto the bottom dielectric layer due to the equal etching rate for difference thickness in photoresist layer. Therefore, LVOF cavity can be fabricated through one deposition step, resist reflow and then one lithography etching step, where accurate fabrication of the angle for tapered dielectric layer is depended on the photoresist reshape. The photoresist reflows when the temperature is increased above its glass transition temperature or exposed to a solvent where the photoresist experiences a decreasing viscosity and become soft to flow along the substrate under surface tension. The transition process can be controlled by resist and wetting solvent material selection, reflow temperature as well as defining the initial rectangular outer contour. This resist reflow techniques is very attractive due to its low cost and IC compatibility. Because the resist reflow process can be operated above the glass transition temperature of photoresist which is typically around 120–140°C. But the whole process of the thermal controlled reflow is reported to require a long period of time to ensure a smooth surface. Also, resist on the substrate hardly flows when the two adjacent flowing resist joint for thermally controlled reflow. In the contrary, photoresist flows at a much faster rate under the absorption of vaporized solvent at a slightly elevate temperature (i.e. 40–60°C). However, due to fast transition rate of this chemical reflow technique, it is much harder to control than the thermal reflow technique. Therefore, a combined method was proposed by the research team where the thermal and chemical reflow techniques are applied alternatively. In this way, the cross-linking between resist can be avoided and the transition processes is much more efficient. After intensive experiments, analysis and theoretical modelling of resist reflow process, the research group has successfully fabricated a LVOF with controllable taper angle ranging from 0.001° to 0.1° [17].

The design of the LVOF based spectrometer requires the tradeoff between the spectral range and available FSR. Therefore, the optical performance of this LVOF also has a limited FSR. High dielectric Bragg mirror is the most utilized reflectors for LVOF. Thus, the LVOF based spectrometers are widely used for spectral analysis in a narrow spectral range. The lithographic pattern mask and the resist reflow process can be modeled by Finite Element Modeling software. And the optical performance can be simulated using TFCalc software (Software Spectra, Portland, OR, USA).

It appears that with the stable structure, low cost fabrication, continuously wavelengths output, a LVOF are more advantageous over both the electrostatically tuned Fabry-Perot filter and filter array of fixed cavity lengths. However, there are many limitations on optical performance and practical operations for LVOFs. First of all, multiple reflections within LVOF cavity give rise to variation of wavefronts of different spectrum. Transmitted wavelengths laterally shift along the cavity which results in various problem. Theoretically, the transmittance of Fabry-Perot filters which can reach 100% for all the filtered wavelengths and reflectively, however, the peak transmittance decreases at a higher mirror reflectivity and operational mode [18]. This resulted in an asymmetrical shape of the transmitted peaks of filtered wavelengths. Moreover, the phases of various filtered wavelengths both along the cavity width as well as towards the transmitted direction. The incident angle and position should be optimized for each LVOF-detector separation. A positive incident angle (where the transmitted beam walks away from LVOF apex) should be avoided, due to the enhance the walk-off effect and thus deteriorating the optical performance [18]. Secondly, the width of a single LVOF where the cavity thick varies along with as to be extended to several millimeters (typically 3–5 mm) to stretch the spatial distribution of spectrum peaks to ensure a spectral readout. Therefore, thus existing structures are utilized as a single element for single point spectral analysis. Thirdly, although, all Fabry-Perot based optical filters are an incident angle dependent optical devices. However, since a LVOF simultaneously filters several spectrum which are spatially distributed, a different incident angle can shift the distribution pattern of filtered wavelengths and complex the calibration between peaks transmittance and wavelengths during readout on the photodetectors.

16.4 Scanning Micromirrors

MEMS scanning mirror is another type of MEMS devices that have been investigated for hyperspectral imaging applications. Usually, micromirrors are designed to reflect light beam either for optical path modulation or beam wavefront modulation. MEMS microscanner is suitable to provide fast optical path tuning for imaging system with tight design volume, such as the OCT (optical coherence tomography), confocal microscopy and endoscope. Different from previously mentioned HSI microoptics which was implemented right before the photodetector(s), a microscanner is usually arranged between optics in an imaging system as a spatial light modulator and has to occupy necessary volume. However, due to integration of mechanical actuation parts and micromirror into a single tunable microoptics, it is still capable of benefitting hyperspectral imaging application with a compact system design. Besides, due to the micro range size, the micromirror can be operated at its structural resonant frequencies to provide a two dimensional scanning range of fast scanning range (typically in thousands of Hertz) as well as a large scanning angle.

A small micromirror could have a diameter in micrometer range in order to match with diameter of a laser spot in laser excited fluorescence microscope. But typical diameter of a micromirror for imaging application are usually in range of millimeters for efficient light throughput. The implementation of scanning micromirror in an imaging application is typical taken advantage for its optical path tuning ability at either a fast scanning rate or at stable tilt motion. For hyperspectral imaging system, when spatial scanning data acquisition method is adopted, a fast frame rate is essential to provide the system a spectral-smudge free data cube.

Both two orthogonal scanning direction of a microscanner can be chosen to be electrostatically, electrothermally and electromagnetically actuated. For resonantly actuated microscanner, electrostatic and electromagnetic actuation are the most adopted. The scanning micromirrors implemented in this fashion are usually actuated in its nature resonant frequencies for each individual axis to perform a Lissajous scanning pattern to achieve fast scanning. However, electrostatic actuated micromirror is limited by high actuation voltage due to highly enforced safety restriction. And the electromagnetic actuated one is limited by its volume due to actuation through electric coil and permanent magnets.

Digital micromirror (DMD) array provides various spatial and spectral analyses on the scene data. The DMD has been invented by TI with a dedicated fabrication process, it has been successfully implemented in various optical system, such as, commercial DLP, adaptive compressive sampling system. Due to its wide applications, the detailed fabrication and structures of TI DMD have been detailed described by many other literatures, therefore, only its application in hyperspectral imaging system will be summarized in this chapter. And it is even been investigated for its application in hyperspectral imaging system due to its powerful and flexible spatial light modulation ability. A DMD contains thousands digital micromirrors where it can be individually switched between two angles, which is usually described as the 'on' position towards the optical path and the 'off' position away from the optical path. Therefore, it can be used hyperspectral imaging or analysis system as a scanner for spatial modulating of illumination, or as a spectrum band switcher.

16.5 Current Research Status

One of popular tendency for microspectrometers is to integrate or even monolithic integrate the spectral modulating filters with a single or even an array of photodetector sensors. This work has been consistently focused by several research groups. The monolithically integration between an array of 16 non-tunable micro Fabry-Perot filters was demonstrated by Professor Wolffenbuttel's research group from Delft University of Technology as early as 1999 [3]. Each element of the filter array area is 500 × 500 μm and has difference resonance cavity lengths ranging from 225 to 300 nm with a 5-nm increment. The filters was post fabricated at a low temperature onto a CMOS chip containing photodetector array and readout circuit. The design utilizes metal layer as reflective layer for simplification. It evaporates a layer of 45 nm Al as the bottom reflective layer and 20 nm Ag as the top layer which is implemented at the last step. The PEVCD silicon dioxide is deposited for the cavity and was etched to 16 different thickness through 4 etching steps using 4 masks. However, at that time the device was reported to have poor signal to noise ratio due to the high stray light, beam divergence and surface roughness (Figs. 16.4 and 16.5).

This FP filter array based hyperspectral imaging sensor design has then been further implemented and commercialized by IMEC (Interuniversity Microelectronics Centre) [19, 20]. In their devices, the FP filters of different cavities are arranged on a single CMOS photodetector chip for spatially line-scan hyperspectral imaging system, or repeated groups of FP filters are arrange on a chip for snapshot hyperspectral imaging system.

The research work on electrostatically actuated Fabry-Perot filters has been intensively focused by the research team in the University of Western Australia. In 2005, they reported a monolithically integrated Fabry-Perot based microspectrometers by using a low temperature (<125°C) and low stress fabrication processes compatible with their HgCdTe SWIR (short wavelength infrared) photodetector techniques [6, 21, 22]. The fabricated device consists of a 90 × 90 μm FP filter on top of and a single HgCdTe photodetector and is

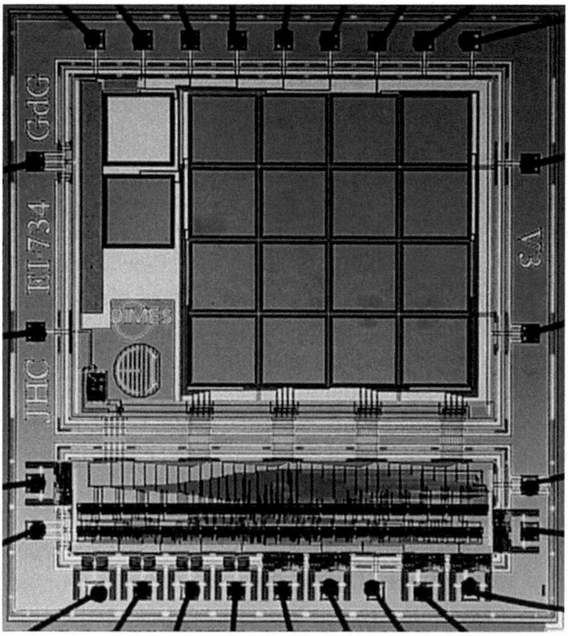

Fig. 16.4 Illustration of the monolithically integrated Faby-Perot filter array [3]

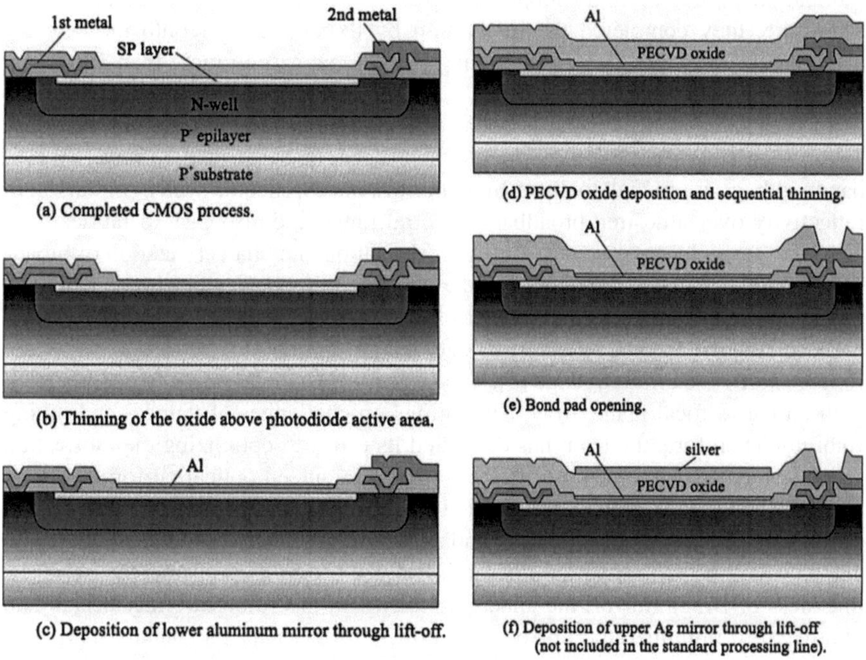

Fig. 16.5 A CMOS post-process fabrication sequence for a monolithically integrated spectrometric sensor chip [3]

characterized to have transmittance of over 60% covering a range from 2.2 to 1.85 μm with roughly 100 nm FWHM when actuated up to 7.5 V. After this proof-of-concept work, they have been keep investigating methods to improve the deposition method to obtain a controlled curvature and stability.

Realizing limitations of the monolithically integrated tunable microspectrometer have affected the optical performance, the team has chosen to focus on optimizing the MEMS filter and detector fabrication processes separately to provide more environmentally stable and controlled curvature for the micromirror plates, and then using die bonding to integrated filter and detector. In 2009, they studied the structure and lithography mask design of electrostatic actuated Fabry-Perot SWIR filter using their existing processes to introduce a fixed-fixed beam actuators and micromirror connecting structure to improve the tunable range and optical performance (FWHM of filtered wave) [9]. Due to this strain-stiffen effect [23, 24] during the actuation state of this fixed-fixed beam microactuator, the stiffness of this double-end clamped beam increases as being bent by electrostatic force and is less likely to snap down to its attracting electrode counter-part compared to the more flexible circular or square plate design [25, 26]. As a results, they have improve the tuning range by 10%, and has achieve the best tuning range at that time [9]. However, in order to improve the environmental stability of Si_xN_y support layer, they slightly modify the fabrication parameters, such as increasing the deposition temperature for Si_xN_y to 200°C rather than 125°C, as a results this fabrication processes is no longer compatible with the HgCdTe photodetector. Therefore, they completed the integration by fixing the filter onto the detector using an epoxy, which as a result having a non-ideal case of more than 1 mm gap between the two.

In their later work, they have been focusing on optimizing filter and photodetector separately. To have a wide tunable spectral range and narrow FWHM bandwidth, it is crucial to have high performance reflector which is flat, high reflectivity over a desired broadband spectral range and also easy to fabricate. In 2015, the team demonstrated the design, fabrication and characterization of broadband Bragg dielectric reflectors for the spectral imaging application. First, they proposed a bulk micromachined electrostatic Fabry-Perot filter designed with Si-SiOx-Si Bragg dielectric reflectors of size ranging from 500 × 500 μm to 3 × 3 mm for the shortwave infrared range [5]. The large mirror size is chosen to be integrated with an entire focal plane array. This brings up challenges of flatness and tuning stability. Therefore, the team has dedicated its effort in optimizing the fabrication processes to control the overall flatness for a guaranteed optical performance. As reported, ICPCVD (inductively coupled plasma chemical vapour deposition) low stress thin film deposition process and post-process of annealing treatment upto 420°C was applied to both the filter and Bragg Mirrors. However, Because of the thickness of Bragg mirror, the annealing treatment has to be carefully design and optimized to minimize the mirror from bowing effect near the anchor area. As a results, the total flatness of its Si-SiOx-Si Bragg mirror was measured to be around 30 nm within approx. 75–85% designed area; and its reflectance was measured to be almost 90% across 1560–2050 nm; and the transmission of rested filter structure was measured to be ~94% at 1940 nm with FWHM of 250 nm.

Later in the same year, in order to increase reflectivity of Bragg dielectric reflector by increasing the refractive index contrast of two alternating layers and eliminated curvature by film stress mismatch, the team proposed and surface micromachined a symmetric Si-Air-Si based Bragg dielectric reflector for electrostatically tuned Fabry-Perot filter for SWIR (1500–3000 nm) and MWIR (3000–6000 nm) range respectively [7]. In order to release this air gap within Bragg mirror, a polyimide sacrifice layer was used to define the gap, and array of etching holes of around 1% reflector area was created on the top silicon membrane. To reduce the stray light effect, they deposited gold pad under each etching hole at the bottom layer of Bragg reflector. The whole membrane was conformably supported at the membrane periphery. As a result, the 20–30 nm flatness was measured within 70–75% of the Bragg dielectric reflector area. And the overall reflectance was measured to be more than 85% over 3750–6000 nm MWIR and 1950–2700 nm SWIR wavelength range. And this high reflectance range can be further increased by designing a layer of quarter wavelength thick silicon oxide on the substrate before the filter. The structural design, fabrication and optical performance evaluation of both Bragg mirror designs detailed described, and an electrostatically tuned Fabry-Perot filter design based on Si-Air-Si Bragg mirrors was proposed (Fig. 16.6).

Finally, in early 2016, they reported and demonstrated the design and actuation characteristic of a Fabry-Perot filter with an asymmetric reflectors' configuration for thermal imaging band in long-wave infrared [27]. In this design, the top reflector is a single layer of Ge membrane while the bottom reflector is a Ge-ZnS-Ge-ZnS Bragg dielectric reflector. And the top Ge membrane is electron-beam evaporated at room temperature and designed to extend beyond the filter optical area overall the whole chip which resulted in a flat deformable membrane at rest position. Therefore, this design does not require additional post-process stress release technology. For the LWIR spectral range transparency, a low oxygen float-zone silicon wafer was used as the filter substrate. As a result, the 200 × 200 μm filter shown a continuous wavelength tuning over a range of 8.5–11.5 μm with peak transmission in 80–87% approximately 0.5 μm FWHM and less than 1.2% peak variation of transmitted wavelengths. However, as a proof-of-concept design, the filter was electrostatically driven up to 160 V during actuation. A series of large size filters with varying optical area have been fabricated using the same process. But due to the conformably anchored at the membrane periphery, the micromirrors has demonstrated a reduced tuning range and a worsen mirror surface bowing during actuation which result in a cavity variation at the actuated state. Because of the mirror is anchored conformably to the support structure, a severe surface bowing has observed which is the height difference between the edge and the center, for example, for optical area of 1 × 1 μm, a surface bowing of 550 nm was measured at 85 V and a poor peak variation of around 8.6% for filtered wavelengths. Their research on the hyperspectral imaging system featuring an electrostatically tuned microspectromtere has been pushed forward by this work and it will be interesting

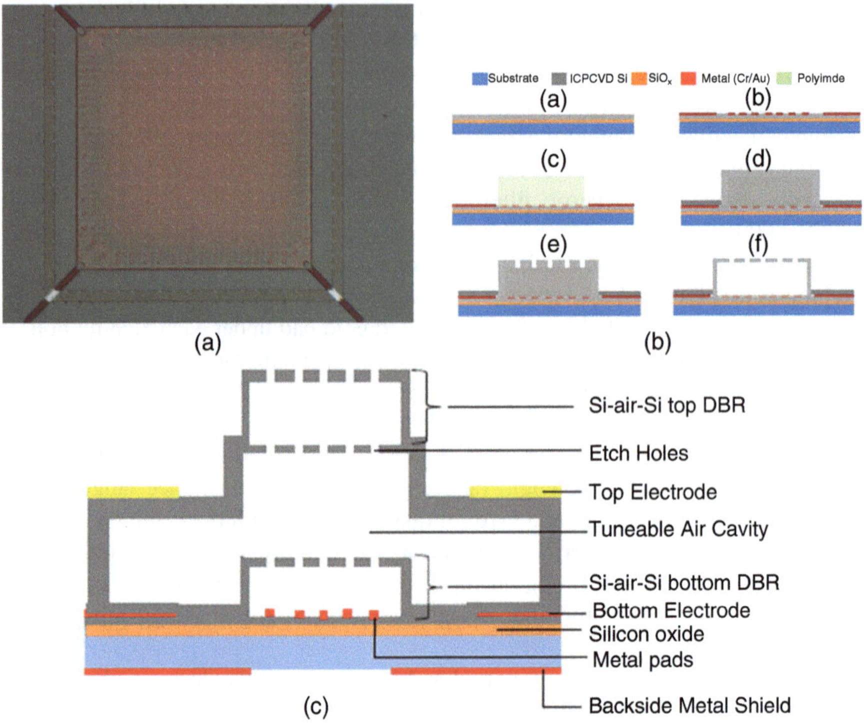

Fig. 16.6 (**a**) A photograph of a 2 × 2 mm fabricated DBR; (**b**) the fabrication processes of a Si-air-Si BDR; (**c**) a cross-section schematic illustration of electrostatic tuned Fabry-Perot filter concept based on the Si-air-Si DBR [7]

to see the filter design to be integrated with the LWIR photodetectors and operated at a cryogenic cooled temperature in the future. From their dedicated work on the electrostatically tunable Fabry-Perot infrared microspectroeters, it can be seen that the best optical performance can be obtained by optimizing the reflector, tunable filter, integrated with substrate and photodetectors separately (Fig. 16.7).

Apart from a single Fabry-Perot filter designed to integrate with an array of photodetectors. An array of Fabry-Perot microfilters with each element aligned with a single detector can also be an option for hyperspectral imager design. This type of work can be referred to a device proposed by a research team in Wuhan National Laboratory for Optoelectronics in Huazhong University of Science and Technology. In 2015, the team proposed concept of a 128-by-128 Fabry-Perot filter array which can be electrostatically actuated individually [28]. The team have concentrated on designing an Al based movable platform for supporting Bragg dielectric reflector. Due to the implementation of symmetrically arranged flexible connectors to connect mirror platform to substrate, the maximum deflection is measured at the connecters rather than on the platform. With this flexible connections, an in-actuation flatness can be guaranteed for the movable micromirror. The optical

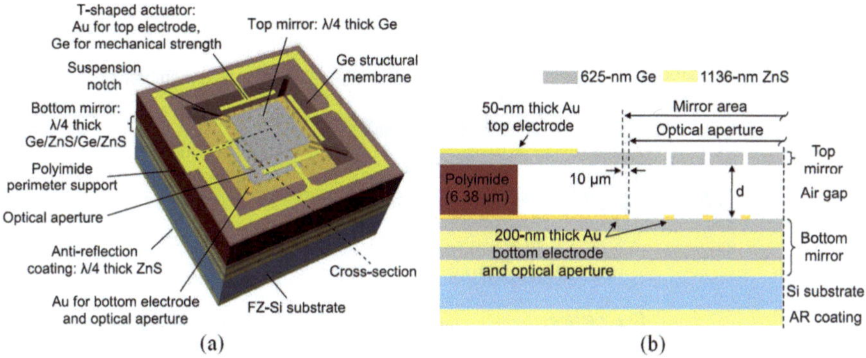

Fig. 16.7 (**a**) Graphical illustration of electrostatically tuned Fabry-Perot filter scheme; (**b**) Cross-section view [27]

performance of proposed structures with Ge-Al_2O_3-Ge Bragg dielectric reflectors has been simulated using TFcalc to expect a tunable range over 3–5 mm range and a spectral resolution of 91 nm when driving upto 25.5 V. The proposed microfilter array with individually actuated element could be used for spectral analysis only for regions of interest within the scenery. However, due to the 120 × 120 μm filter element and an optical aperture of 50 × 50 μm, current proposed array format give rise in a low fill factor of around 8–15% (ratio between the effective photon sensitive area and the totally chip area). The proposed design was only reported with fabricated sample and simulated results, and it will interesting to investigate the actuation consistence among all the 128 × 128 elements as well as the results with integrated photodetector array. This configuration requires extra design constraint on alignment between detector and filter array element as well as complicated electric route design for actuating each filter (Fig. 16.8).

For LVOF based microspectrometers, the research works have widely investigated by Professor Wolffenbuttel's research group from Delft University of Technology using IC-compatible resist reflow techniques. They have successfully fabricated and implemented a wide range of LVOFs for narrowband spectrometry using resist reflow technique, such as, a UV LVOFs for 300–400 nm [29–31],visible LVOFs for 580–720 nm [32], 610–680 nm [33], IR LOVFs for 1400–2500 nm [34], 3000–5000 nm [35]. A microfabricated LVOF provides significant advantages over gratings and Fabry-Perot resonator especially for UV spectral range (230–280 nm) where rather small feature are required for gratings and cavity length less than 140 nm are required in optical design [29]. The spectral resolving power of a LVOF based microspectrometer is the filtered spectral range divided by the number of photodetector along the dispersed wavelengths. It is claimed that the spectral resolution of microfabricated LVOFs could be further improved to 0.5 nm by an optimized algorithm [36]. However, the readout of those fine spectral peaks is limited by existing pixel size of a detector, which is typically 3 μm or larger (Fig. 16.9).

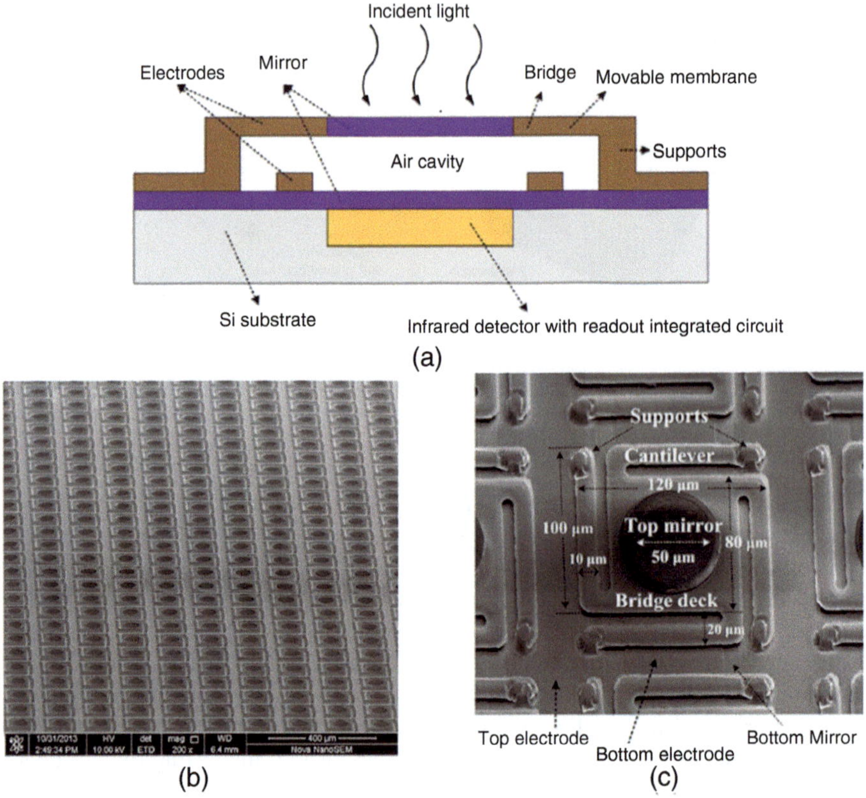

Fig. 16.8 (**a**) cross-section structure design of a Fabry-Perot filter; (**b**) SEM photograph of the fabricated Fabry-Perot filter array; (**c**) a detailed close-up view of a single fabricated Fabry-Perot filter element [28]

Fig. 16.9 (**a**) Photograph UV LVOFs fabricated on a glass substrate [37]; (**b**) LVOF mounted on a CMOS camera [33]

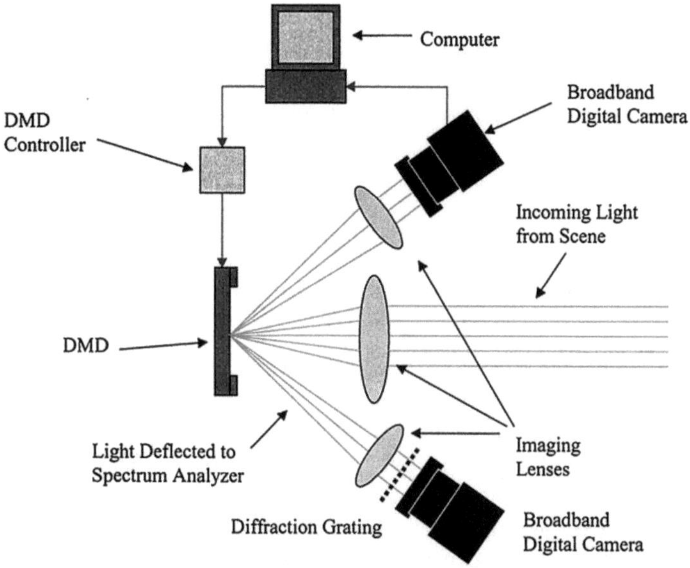

Fig. 16.10 Schematic depiction of an ACTIVE-EYES real-time pixel spectral analyzing integrated imaging system [38]

The Texas Instrument DMD (Digital Micromirror Device) is usually implemented in a hyperspectral imaging system as optical path switches for switching between spectral bands dispersed by a previous optics. As demonstrated in Fig. 16.10, a DMD is implemented in color imaging system, where part of the digital micromirrors to switch to one direction for normal color imaging, while use the other part of digital micromirrors to switch to the other direction for spectral analysis of only region of interest. In this way, it can achieve a pixel scan spectral analysis while perform a real-time imaging photography or even video [38]. Or, DMD can perform as a spectral scanning device to sequentially switch spectral band either towards the object as a illuminator [39] or towards the imaging sensor [40, 41] as tunable spectral filter. Such hyperspectral imaging system first requires light to be dispersed by either a prism, grating or a LVOF and then is projected on the DMD where rows of digital micromirrors of the same spectral band is switched in-line with the optical path towards either the sample to be illuminated or towards the CCD/CMOS image sensor to be recoded.

A single point spectral diagnosis is very commonly used in biomolecule and chemical analysis. It typically requires a single or an array of fast response and highly sensitive photodetector(s), such as PMT to provide precise photon detection at low light intensity. With the implementation of the scanning micromirror, the system is capable of scanning excitation light source over a 2D planar area and in turn to convert a single point diagnosis system to a 2D scanning imaging system. Since 2010, a research team from University of Texas, US has a designed a spectral imaging system for medical sample analysis utilizing the 2D MEMS scanning micromirrors. The system implements electrostatically or electromagnetically actuated MEMS scanner to scan an excitation laser beam in a Lissajous 2D scanning pattern over the sample area to achieve large field of view fluorescence microscope [42–44]. The system has shown that with implementation of a MEMS scanner, the fluorescence hyperspectral imaging system can be bring the conventional tabletop setup down to an handheld portable device.

16.6 Summary

Hyperspectral imaging system has been an active researcher area over decades ever since its first proof of concept. As a complex system, it requires obtain two-dimensional images of scenery as well as one dimensional spectrum intensity for every single pixel as fine as over 100 spectral bands. This has unavoidable resulted in a bulky, complicated and fragile optical imaging system. Comparably, microfabrication technologies have been matured and diverse during the same period of time, and are still expanded with new micro/nano devices and materials. The resulted MEMS devices are characterized as cheap, sensitive, small and robust. Therefore, a microspectrometer can be more compact, reliable, fast, low-cost and high sensitive compared to the conventional stand-alone spectral modulating optics. Moreover, with the development of both the fabrication technology, improved sensitivity of photodetector, signal processing algorithms and fast computing devices, the advantages brought by a microspectrometer is preferred over its lowered resolution.

The most popular microfabricated spectral tuning devices is a Fabry-Perot filter. One popular research tendency on microspectrometers is the integration or even monolithic integration of these optical filters with photodetector or detector array. Monolithic integration between electrical circuit, photodetector and optical filter is critically important, because this can reduce the assembly cost, and provide a good alignment between the optical filter and photodetector. However, the monolithic fabricated microspectrometers are limited by the optical performance. Therefore, more effort has been focus on optimizing material fabrication and structure stability to provide a desired optical performance in specific spectral range for spectral imaging application. And the integration of detectors and microoptics can be solved by dedicatedly designed assembly mechanism. Currently, the main challenges of designing such a microspectrometer for hyperspectral imaging application are a high reflective reflectors over a wide spectral range, a flat and smooth reflector surfaces, stability, reliability and simple and low cost fabrication processes.

For a reflector, a Bragg dielectric stacks are widely adapted in micro Fabry-Perot design due to the fact that it less absorptive compared to the metallic reflective layer. Flatness of the micromirror is deteriorated by the stress introduced during the fabrication processes at a rise temperature different to the operating temperature. During the fabrication of the micromirror plate, the micromirror will usually exhibit tilting, curvature and roughness. All of these can affect FWHM of the filtered spectrum. Especially for the dielectric Bragg mirror where multiple layers are fabricated and the mismatch of the intrinsic stress of each layer can result in a curvature of the micromirror. Therefore, fabrication processes have to be carefully designed for produce a dielectric stacks with desired flatness as well as consideration of fabrication simplicity. At the same time, the fabrication technology is also critical in terms of producing an accurate thickness for each alternating layer of dielectric mirror. Besides, a fabrication variation can affect the ideal yield rate of the microspectrometers which is affected by the resulted variability between devices on the same wafer and wafers from different fabrication batches.

Fabry-Perot filter can be designed in various forms. One popular realization is electrostatically actuated Fabry-Perot filter with a movable reflector and movable electrodes. This type of tunable microspectrometer has been designed as a single device with large optical area and an array form. The electrostatic actuation can provide a fast switching rate and is widely adapted in various MEMS devices. However, because of combining microoptics and movable actuation, the design requires trade-off consideration between effective optical area, suspension microstructure, stability and optical performance. First of all, a wide spectral range requirement for hyperspectral imaging application challenges the tunable range of the paralleled electrodes actuation which is limited by pull-in effect. Secondly, stability of a suspended reflector at rest or even during actuation is critical for the optical performance. At the same time, the drift along time can also challenge the reliability of overtime optical performance. Besides, electrostatic actuators require high actuating voltage.

Another Fabry-Perot filter design is implementing an array of fixed Fabry-Perot with different cavity thicknesses. In this way, the devices is more reliable and power consumption because of no suspended structures and actuation mechanism required. However, this doesn't not necessarily means a simplified fabrication processes. The more spectrum bands are required, the more discrete Fabry-Perot filters will be designed, which challenges the accuracy of multiple etching steps on the same cavity layer. On the other hand, a LVOF is capable of filtering a continuously varying spectrum wavelengths over a wide spectral range theoretically by utilizing a slightly tilted angle between two reflectors. However, this devices is limited by the available free spectral range and incident angle sensitivity.

As reviewed in this chapter, although there are many novel structures has been proposed for microspectrometers designs, however a limited number of cases are demonstrated as implementation in a hyperspectral imaging system. The ones that attempted to do so has reported with an optical performance to be improved. In the future, material combination and fabrication for a broadband Bragg reflector designs and an improved optical reflectivity over desired spectral range are still essential and highly attractive research area for microspectrometers using MEMS techniques. Because IR spectral range can be used for detecting combustible gases, the microspectrometers designs for this spectral range will still be active. With the development of IR photodetector on the sensitivity and processes, the integration between the IR photodetector or photodetector array and tunable spectral filters using microfabrication techniques will be constantly explored.

References

1. L.P. Schuler, J.S. Milne, J.M. Dell, L. Faraone, MEMS-based microspectrometer technologies for NIR and MIR wavelengths. J. Phys. D. Appl. Phys. **42**(13), 33001 (2009)
2. J.H. Correia, M. Bartek, R.F. Wolffenbuttel, Bulk-micromachined tunable Fabry–Perot microinterferometer for the visible spectral range. Sensors Actuators A Phys. **76**(1–3), 191–196 (1999)
3. J.H. Correia, G. de Graaf, S.H. Kong, M. Bartek, R.F. Wolffenbuttel, Single-chip CMOS optical microspectrometer. Sensors Actuators A Phys. **82**(1–3), 191–197 (2000)
4. N.M. Lin, S.C. Shei, S.J. Chang, Design and Fabrication of a TiO_2/SiO_2 Dielectric broadband and wide-angle reflector and its application to GaN-based blue LEDs. IEEE J. Quantum Electron. **51**(7), 1–5 (2015)
5. D.K. Tripathi, F. Jiang, R. Rafiei, K.K.M.B.D. Silva, J. Antoszewski, M. Martyniuk, J.M. Dell, L. Faraone, S. Member, Suspended large-area MEMS-based optical filters for multispectral shortwave infrared imaging applications. J. Microelectron. Syst. **24**(4), 1102–1110 (2015)
6. C.A. Musca, J. Antoszewski, K.J. Winchester, A.J. Keating, T. Nguyen, K.K.M.B.D. Silva, J.M. Dell, L. Faraone, P. Mitra, J.D. Beck, M.R. Skokan, J.E. Robinson, Monolithic integration of an infrared photon detector with a MEMS-based tunable filter. IEEE Electron. Device Lett. **26**(12), 888–890 (2005)
7. D.K. Tripathi, H. Mao, K.K.M.B.D. Silva, J.W. Bumgarner, M. Martyniuk, J.M. Dell, L. Faraone, Large-area MEMS-based distributed bragg reflectors for short-wave and mid-wave infrared hyperspectral imaging applications. J. Microelectromech. Syst. **24**(6), 2136–2144 (2015)
8. T.G. Bifano, H.T. Johnson, P. Bierden, R.K. Mali, Elimination of stress-induced curvature in thin-film structures. J. Microelectromech. Syst. **11**(5), 592–597 (2002)
9. J.S. Milne, J.M. Dell, A.J. Keating, L. Faraone, Widely tunable MEMS-based Fabry-Perot filter. J. Microelectromech. Syst. **18**(4), 905–913 (2009)
10. M.S.C. Lu, G.K. Fedder, Position control of parallel-plate microactuators for probe-based data storage. J. Microelectromech. Syst. **13**(5), 759–769 (2004)
11. S.-W. Wang, M. Li, C.-S. Xia, H.-Q. Wang, X.-S. Chen, W. Lu, 128 channels of integrated filter array rapidly fabricated by using the combinatorial deposition technique. Appl. Phys. B Lasers Opt. **88**(2), 281–284 (2007)
12. C.M. Waits, A. Modafe, R. Ghodssi, Investigation of gray-scale technology for large area 3D silicon MEMS structures. J. Micromech. Microeng. **13**(2), 170 (2003)
13. M. Han, W. Lee, S.-K. Lee, S.S. Lee, 3D microfabrication with inclined/rotated UV lithography. Sensors Actuators A Phys. **111**(1), 14–20 (2004)
14. W.X. Yu, X.-C. Yuan, N.Q. Ngo, W.X. Que, W.C. Cheong, V. Koudriachov, Single-step fabrication of continuous surface relief micro-optical elements in hybrid sol-gel glass by laser direct writing. Opt. Express **10**(10), 443–448 (2002)
15. D. Y. Choi, Shadow mask and method of fabricating vertically tapered structure using the shadow mask, US7425275 B2, 2008
16. B. Sheng, P. Chen, C. Tao, R. Hong, Y. Huang, D. Zhang, Linear variable filters fabricated by ion beam etching with triangle-shaped mask and normal film coating technique. Chin. Opt. Lett. **13**(12), 122301 (2015)
17. A. Emadi, H. Wu, S. Grabarnik, G. de Graaf, R.F. Wolffenbuttel, Vertically tapered layers for optical applications fabricated using resist reflow. J. Micromech. Microeng. **19**(7), 074014 (2009)
18. N.P. Ayerden, G. de Graaf, R.F. Wolffenbuttel, Compact gas cell integrated with a linear variable optical filter. Opt. Express **24**(3), 2981–3002 (2016)
19. N. Tack, A. Lambrechts, P. Soussan, L. Haspeslagh, A compact, high-speed, and low-cost hyperspectral imager, Proc. SPIE 8266, Silicon Photonics VII, 82660Q (2012)

20. A. Lambrechts, P. Gonzalez, B. Geelen, P. Soussan, K. Tack, M. Jayapala, A CMOS-compatible, integrated approach to hyper- and multispectral imaging, *2014 I.E. International Electron Devices Meeting*, 2014 pp. 10.5.1–10.5.4
21. J. Antoszewski, K. J. Winchester, A. J. Keating, T. Nguyen, K. Silva, H. Huang, C. A. Musca, J. M. Dell, L. Faraone, P. Mitra, J. D. Beck, M. R. Skokan, J. E. Robinson, A monolithically integrated HgCdTe SWIR photodetector and tunable MEMS-based optical filter, in *Infrared Technology and Applications XXXI, Pts 1 and 2*, 5783(1–2), 2005, pp. 719–727
22. A.J. Keating, K.K.M.B.D. Silva, J.M. Dell, C.A. Musca, L. Faraone, Optical characterization of Fabry-Perot MEMS filters integrated on tunable short-wave IR detectors. IEEE Photon. Technol. Lett. **18**(9), 1079–1081 (2006)
23. R.C. Batra, M. Porfiri, D. Spinello, Electromechanical model of electrically actuated narrow microbeams. J. Microelectromech. Syst. **15**(5), 1175–1189 (2006)
24. H. Rong, Q.-A. Huang, M. Nie, W. Li, An analytical model for pull-in voltage of clamped–clamped multilayer beams. Sensors Actuators A Phys. **116**(1), 15–21 (2004)
25. J.S. Milne, A.J. Keating, J. Antoszewski, J.M. Dell, C.A. Musca, L. Faraone, Extending the tuning range of SWIR microspectrometers. Proc. SPIE **6542**, 65420M-1–65420M-11 (2007)
26. J. Milne, J. Dell, A. Keating, L. Faraone, Extended tuning range Fabry-Perot etalon with doubly-supported beam actuators. in *IEEE/LEOS International Conference on Optical MEMS and Their Applications Conference, 2006,* 2006, pp. 134–135
27. H. Mao, K.K.M.B.D. Silva, M. Martyniuk, J. Antoszewski, J. Bumgarner, B.D. Nener, J.M. Dell, L. Faraone, MEMS-based tunable Fabry-Perot filters for adaptive multispectral thermal imaging. J. Microelectromech. Syst. **25**(1), 227–235 (2016)
28. Q. Meng, S. Chen, J. Lai, Y. Huang, Z. Sun, Multi-physics simulation and fabrication of a compact 128x128 micro-electro-mechanical system Fabry-Perot cavity tunable filter array for infrared hyperspectral imager. Appl. Opt. **54**(22), 6850–6856 (2015)
29. A. Emadi, H. Wu, G. de Graaf, K. Hedsten, P. Enoksson, J.H. Correia, R.F. Wolffenbuttel, An UV linear variable optical filter-based micro-spectrometer. Proc. Eng. **5**, 416–419 (2010)
30. A. Emadi, Spectral measurement with a UV linear-variable optical filter microspectrometer, *IEEE Sensors*, 2011, pp. 420–423
31. A. Emadi, H. Wu, G. de Graaf, P. Enoksson, J.H. Correia, R. Wolffenbuttel, Linear variable optical filter-based ultraviolet microspectrometer. Appl. Opt. **51**(19), 4308–4315 (2012)
32. A. Emadi, H. Wu, S. Grabarnik, G. De Graaf, K. Hedsten, P. Enoksson, J.H. Correia, R.F. Wolffenbuttel, Fabrication and characterization of IC-compatible linear variable optical filters with application in a micro-spectrometer. Sensors Actuators A Phys. **162**(2), 400–405 (2010)
33. A. Emadi, H. Wu, G. de Graaf, R. Wolffenbuttel, Design and implementation of a sub-nm resolution microspectrometer based on a Linear-Variable Optical Filter. Opt. Express **20**(1), 489–507 (2012)
34. N. P. Ayerden, M. Ghaderi, M. F. Silva, A. Emadi, P. Enoksson, J. H. Correia, G. de Graaf, R. F. Wolffenbuttel, Design, fabrication and characterization of LVOF-based IR microspectrometers, in *Proc. SPIE 9130, Micro-Optics 2014*, vol 9130, p. 91300 T, 16–19 April 2014
35. M. Ghaderi, N.P. Ayerden, A. Emadi, P. Enoksson, J.H. Correia, G. de Graaf, R.F. Wolffenbuttel, Design, fabrication and characterization of infrared LVOFs for measuring gas composition. J. Micromech. Microeng. **24**(8), 84001 (2014)
36. A. Emadi, S. Grabarnik, H. Wu, G. de Graaf, K. Hedsten, P. Enoksson, J. H. Correia, R. F. Wolffenbuttel, Spectral measurement using IC-compatible linear variable optical filter, in *Micro-optics 2010*, 2010, vol 7716
37. A. Emadi, H. Wu, G. de Graaf, P. Enoksson, J. H. Correia, R. Wolffenbuttel, Design, fabrication and measurements with a UV linear-variable optical filter microspectrometer, in *Optical Sensing and Detection II*, 2012, vol. 8439
38. M.P. Christensen, G.W. Euliss, M.J. McFadden, K.M. Coyle, P. Milojkovic, M.W. Haney, J. van der Gracht, R.A. Athale, ACTIVE-EYES: an adaptive pixel-by-pixel image-

segmentation sensor architecture for high-dynamic-range hyperspectral imaging. Appl. Opt. **41**(29), 6093–6103 (2002)
39. A. Bednarkiewicz, M. Bouhifd, M.P. Whelan, Digital micromirror device as a spatial illuminator for fluorescence lifetime and hyperspectral imaging. Appl. Opt. **47**(9), 1193–1199 (2008)
40. M.L. Mangum, E. Livingston, K.J. Zuzak, Visible to NIR DLP hyperspectral imaging system for surgical utility using inherent chromophores and fluorescent probes. Proc. SPIE **7932**, 793203–793211 (2011)
41. E. Livingston, K. Zuzak, NIR DLP hyperspectral imaging system for medical applications. Proc. SPIE—Int. Soc. Opt. Eng. **7932**, 793204–793209 (2011)
42. Y. Wang, Y.D. Gokdel, N. Triesault, L. Wang, Y.-Y. Huang, X. Zhang, Magnetic-actuated stainless steel scanner for two-photon hyperspectral fluorescence microscope. J. Microelectron. Syst. **23**(5), 1208–1218 (2014)
43. Y. Wang, S. Bish, J.W. Tunnell, X. Zhang, MEMS scanner based handheld fluorescence hyperspectral imaging system. Sensors Actuators A Phys. **188**, 450–455 (2012)
44. Y. Wang, S. Bish, A. Gopal, J. W. Tunnell, X. Zhang, MEMS scanner enabled real-time depth sensitive hyperspectral imaging, in *Optical MEMS and Nanophotonics (OPT MEMS), 2010 International Conference on*, 2010, pp. 2–4

Chapter 17
Fabrication and Physical Properties of Nanoscale Spin Devices Based on Organic Semiconductors

Xianmin Zhang

17.1 Introduction

Organic spintronics devices have attracted extensive research interest as they combine the advantages of spintronics and organic semiconductors [1–14]. These devices can be used to manipulate the properties of spin and charge inherent in electrons, and the organic nature of such microelectronic devices could provide high structural flexibility, low production cost, and large area processing. Importantly, the strength of the spin–orbit interaction in organic semiconductors is considerably weaker when compared with that in inorganic materials as most organic semiconductors comprise elements with low atomic numbers [1, 2, 14]. This makes the spin relaxation time in an organic semiconductors longer than that in inorganic counterparts. Moreover, the interface coupling between molecular layer and ferromagnetic metal plays important roles in determining spin injection/detection efficiency of traditionally organic spin devices and designing molecular spin memory devices. Therefore, organic device is ideal to fully modulate and utilize the spin of the electron to construct novel organic functional devices for next-generation applications.

Organic materials can be divided into small molecules and polymers. Both materials have been studied in spin based devices [9, 15, 16]. Limited by the length of this chapter, we will mainly discuss the magnetoresistance effects in vertical spin devices using molecules, polymer devices [11] and spin-organic light-emitting diode [17] could be referred elsewhere. In the following parts, we will separately report the investigation of organic spinterface and spin transport behaviors in organic spin valves.

X. Zhang (✉)
Key Laboratory for Anisotropy and Texture of Materials from Ministry of Education, Northeastern University of China, Shenyang 110819, P. R. China
e-mail: zhangxm@atm.neu.edu.cn

T. Li, Z. Liu (eds.), *Outlook and Challenges of Nano Devices, Sensors, and MEMS*,
DOI 10.1007/978-3-319-50824-5_17

17.2 Organic Spinterface

Interface between molecular layer and ferromagnetic metal plays an important role in determining spin injection and detection efficiency of organic spin valve devices [1, 2], and constructing molecular spin memory devices [5] as well as developing novel magnetic materials [18]. Investigation and understanding of the interface interactions are essential for fundamental science and organic device fabrications [19–23]. A complex by Co reacting with 8-hydroxyquinoline-aluminum (Alq_3) at their interface [24] and a magnetically-coupled interface layer of Alq_3 with Co or Fe were observed [25], respectively. A significant magnetic polarization was observed in C_{60} π^*-derived orbitals due to the hybridization between the frontier orbitals of C_{60} and continuum states of Fe [26]. Recently, charge transfers at molecular interfaces for metal/C_{60} [27] or oxide/C_{60} [28] could modify spin polarization or magnetization. Mari et al., reported that the electronic states of non-ferromagnetic materials, copper and manganese, could be altered by connecting with C_{60}, and then make them ferromagnetic at room temperature [18]

The traditional optical pump probe method used for studying spin dynamics and transport in inorganic semiconductors is not suitable for OSs because of the weak spin–orbit interaction and the absence of a crystalline structure with inversion asymmetry in organic semiconductors [29, 30]. The use of an organic spin valve device is still the most popular technique for investigating spin dependent transport (SDT) in organic semiconductors.

It is known that very thin Co or Fe films (~1 nm) show strong perpendicular magnetic anisotropy (PMA), depending on buffer and capping layer materials. Most of the PMA originates from the electron orbital anisotropy mainly inducing at the interfaces. Thus, this could be used as a probe to study the interface interaction at the spinterface. The six types of distribution for d electron oribital are shown in Fig. 17.1. The orbital moment and the resultant magnetic anisotropy originate from the spin orbital (SO) interaction given by the Hamiltonian

$$\mathrm{Hso} = \xi_{3d}\mathrm{l} \cdot \mathrm{s} \quad (17.1)$$

where ξ_{3d} is the SO coupling constant of Co or Fe. The matrix elements with the same spin $<dxy|Hso|dx^2\text{-}y^2>$and $<dxz|Hso|dyz>$ produce the perpendicular magnetic moment, which dominantly determine the PMA. The absolute value of $<dxy|Hso|dx^2\text{-}y^2>$ is the largest among nonzero matrix elements. When the crystal field locates the dxy and $dx^2\text{-}y^2$ states near to E_F with one being below E_F and the other above E_F, the perpendicular magnetic moment was significantly enhanced [31]. For a very thin film with only several atomic monolayers, the crystal field could be modified by the linked materials through interface interaction.

The on-top interface effect on PMA of very thin Co layer in the Pt (111)/Co/molecule system using five different molecular layers: fullerene (C_{60}), 8-hydroxyquinoline-aluminum (Alq_3), 5,6,11,12-tetraphenylnaphthacene (rubrene), pentacene and phthalocyanine (CuPc).

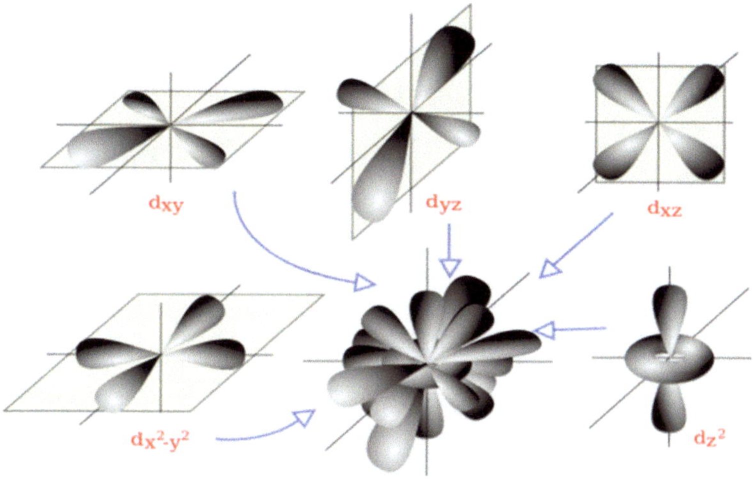

Fig. 17.1 The orbital distribution of *d* electron

Figure 17.2 show the M–H curves for the samples with the 0.7-nm-thick Co capped respectively by above five types of molecules. It is definitely observed that all the samples have an easy magnetization axis parallel to the film normal. This means the PMA properties predominate in the ultra-thin Co layer. However, compare the results in Fig. 17.2a–c with Fig. 17.2d, e, it is observed that the out-of-plane coercivity of both pentacene and CuPc-capped samples is larger than that of other three samples, as summarized in Table 17.1. To systemically study this phenomenon, films with Pt/Co/molecular layer structure were fabricated by varying the thickness of Co from 0.5 nm to 1.8 nm. The dependence of the out-of-plane coercivity on Co thickness is plotted in Fig. 17.3. The optimized thickness of the ultra-thin Co layer for maximum out-of-plane coercive field was around 0.7 nm for all samples. Above and below this thickness, the coercivity of the samples begins to decrease and PMA disappears when increasing Co thickness to 1.8 nm. Figure 17.4 is the saturation magnetization (Ms) for the samples with the thickness of Co around 0.7 nm, showing the Ms is roughly independent of sorts of capping molecules (see Table 17.1).

In order to compare the PMA properties for all series of films quantitatively, we estimated effective magnetic anisotropy energy (K_u^{eff}) for the Co layer of 0.7 nm using the following equation, $K_u^{eff} = M_s H_k^{eff}/2$, the effective anisotropy field (H_k^{eff}) was defined as the extrapolated intersection of the in-plane M-H loop with the saturation magnetization value of the out-of-plane M-H curve. A higher magnetic field was tried when measuring the M-H curves to get an exact value for H_k^{eff}, but the signal-to-noise ratios at higher magnetic field are too low to obtain them. Herein, the value had to be estimated from the slope of M-H curves in Fig. 17.2 and the errors are also added. The estimated K_u^{eff} values are around 3×10^6 erg/cc

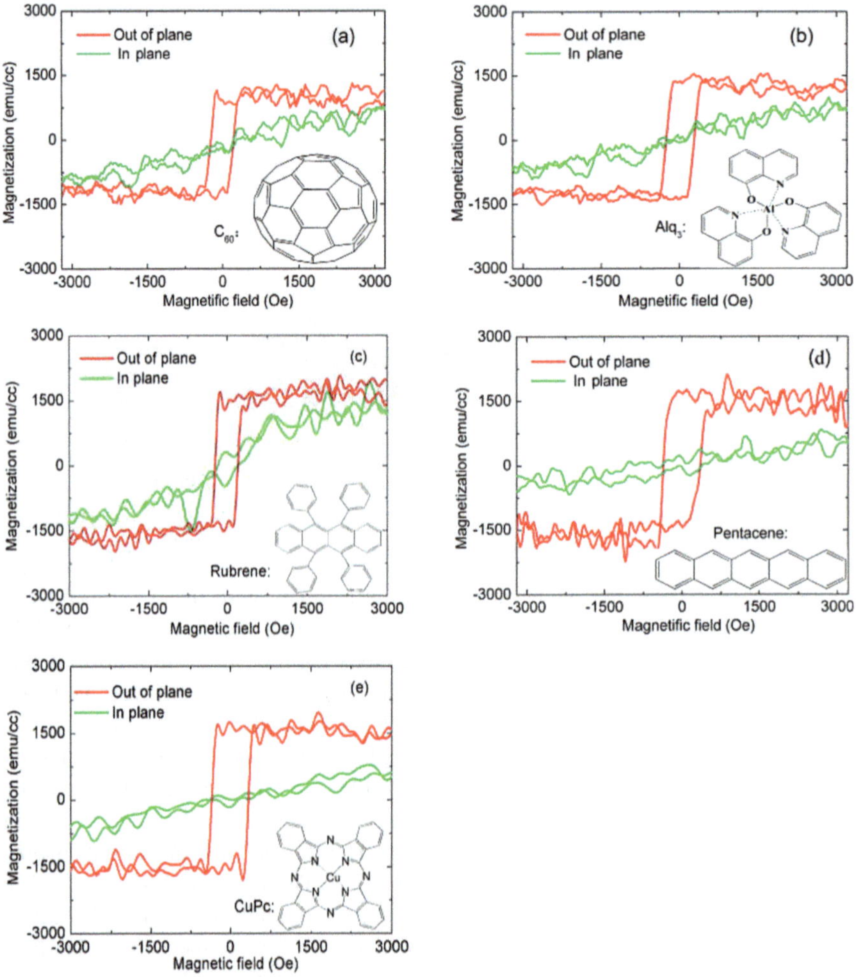

Fig. 17.2 Magenetization curves with 0.7 nm Co by different capping layers. The capping layers are C_{60} (**a**), Alq_3 (**b**), rubrene (**c**), pentacene (**d**) and CuPc (**e**), respectively. The insert was the molecule structure (from Zhang et al. Appl. Phys. Lett. 99, 162,509 (2011). Zhang et al. J. Appl. Phys. 111, 07B320 (2012))

Table 17.1 The calculated K_u^{eff} for samples with 0.7 nm Co

Capping layer	M_s (emu/cc)	H_k (Oe)	K_u^{eff}($\times 10^6$ erg/cc)
C_{60}	1410 ± 70	4600 ± 200	3.0 ± 0.3
Alq_3	1440 ± 60	4700 ± 200	3.1 ± 0.3
Rubrene	1440 ± 80	4400 ± 200	2.9 ± 0.3
Pentacene	1440 ± 90	7100 ± 350	5.1 ± 0.4
CuPc	1460 ± 60	6600 ± 200	4.9 ± 0.3

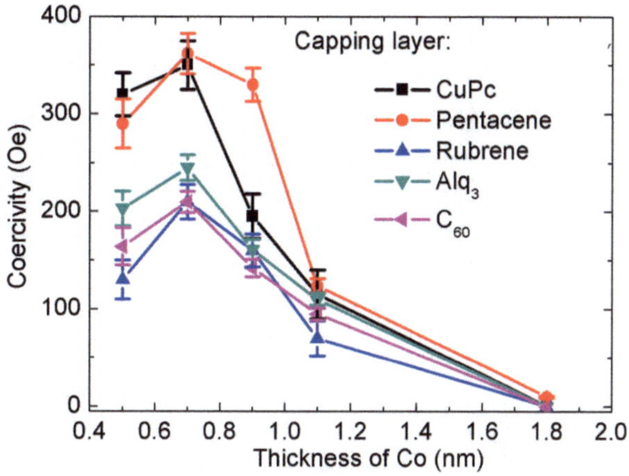

Fig. 17.3 Out-of-plane coercivity dependence on Co thickness

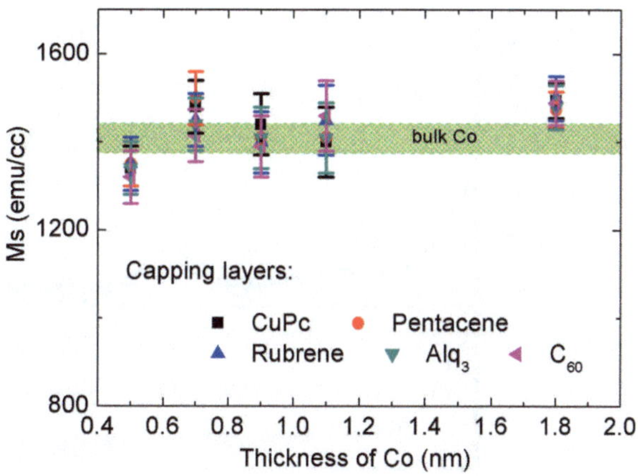

Fig. 17.4 Ms dependence on Co thickness

and 5×10^6 erg/cc for both C_{60}, rubrene, Alq_3 and pentacene, CuPc- capped samples, respectively. This means that PMA properties could be modified by different molecular types.

Figure 17.5 shows the respective Θ–2Θ XRD patterns for the samples with and without capped layers. The diffraction peak of Si (004) was observed at around 69° in the Si/SiO_2 substrate. After Pt was deposited on the substrate, the strongest peak associated with Pt (111) can be clearly seen. No additional peaks in the films capped by C_{60}, Alq_3 and rubrene were found, even though the organic layer thickness is 70 nm. This indicates that both these three organic layers are amorphous. Notably, with both Pc and CuPc-capped layers, several additional peaks are observed (labeled by a square with dashed lines). These peaks are confirmed to originate from (001), (002), (003) and (004) planes of Pc layer. This indicates that the Pc layer has a triclinic structure [32] with the following lattice parameters: $a = 0.79$ nm, $b = 0.606$ nm, $c = 1.601$ nm, $\alpha = 101.9°$, $\beta = 112.6°$, and $\gamma = 85.8°$. These Bragg reflection peaks indexed as (002), (004), and (008) of

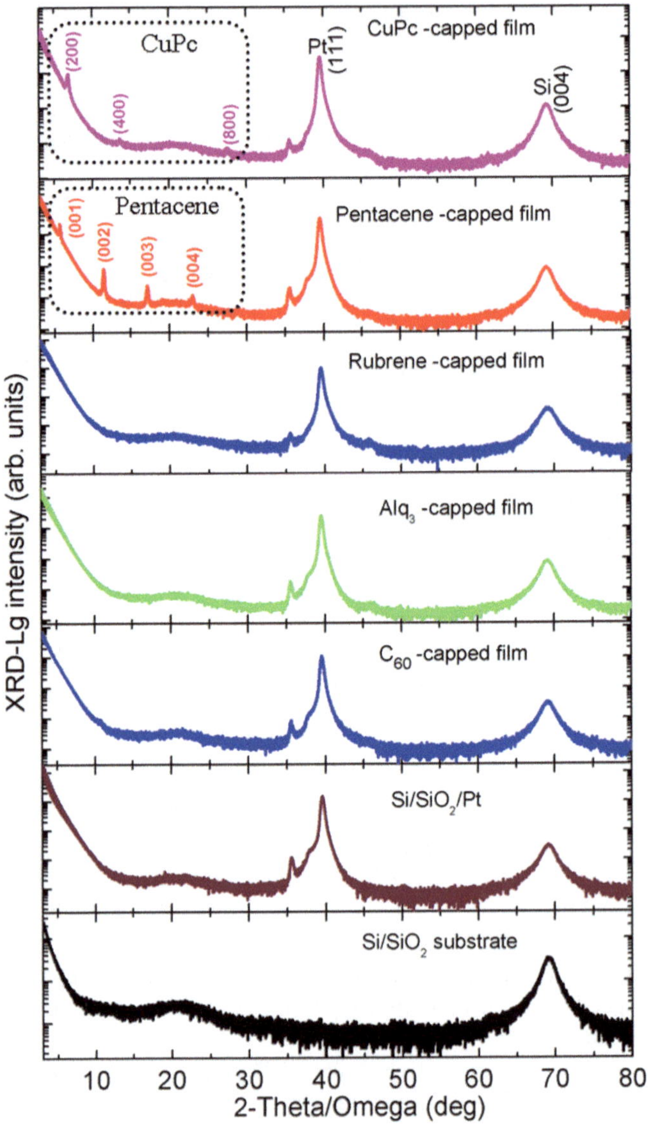

Fig. 17.5 XRD of the films

CuPc (monoclinic, *C*2/*c*, $a = 2.592$ nm, $b = 0.379$ nm, $c = 2.392$ nm, $\beta = 90.4°$) were consistent with the observation in the previous literature [33]. In addition, we note that the Co layer is too thin to show any diffraction peaks given the measurement limitations.

The above results imply the change of PMA may have relations with the crystal structure of the molecular layer. Stacking structure in molecular layer might reflect the interaction of Co/C_{60}, Co/Alq_3, Co/rubrene, Co/CuPc and Co/pentacene are because interface plays a dominant role for original growths of molecular stacking, probably owing to their molecule structures.

To explore the reason for the change of PMA, it is necessary to consider the spin-orbit (SO) interaction because it determines the orbital moment and the resultant magnetic anisotropy, as shown in the formula 1. Different molecules likely have different bonding by different interaction with Co layer. The interface effect by physisorption is very weak because the dominant force holding both parts together is the van der Waals force. Thus, the SO interaction for the Co surface layer will not be altered observably. Comparatively, chemisorption of various molecules can significantly alter the SO interaction of surface atomic layer, by states being removed from the vicinity of the Fermi level during the formation of covalent bonds [34]. Pc adsorbed on top of surface Co layer by physisorption [35, 36] and around 5% decrease of in-plane coercivity was observed in comparison to that without capping layer [35]. It was also reported that with an aromatic 6π-electron system, the benzene, showed the weaker hybridization of the out-of-plane π (p_z) atomic-type orbital with the *d* states (d_{z^2}, d_{xz}, d_{yz} of Fe), compared to the cyclopentadienyl radical (C_5H_5) with a 5π-electron system [37]. Pc is composed of benzene, whereas C_{60} contains both 6π-electron and 5π-electron systems, shown in the inset of Fig. 17.2. Herein, it is considered that the interaction of Pc with *d* electrons is weaker than that of C_{60}. XRD results show that all the rubrene, Alq_3 and C_{60} -capped layers are amorphous when deposited on Co layer. By contrast, the CuPc and Pc-capped layer is of crystal structure. It was approved that the chemical bond existed at the interface of Co/Alq_3 and Co/C_{60} via the 3*d* states hybridizing with the π orbits [38–42]. In particular, a strong bond was formed by Co interacting with either the nitrogen heteroatom [39] or the oxygen heteroatom [40] of Alq_3. Cummings et al. reported that the strength of C_{60}-Co bond interaction was stronger than for the Co-Co bond [42]. Therefore, we suggest that the strong interface couplings of Co/C_{60} and Co/Alq_3 significantly alter the orbitals of 3*d* electrons near the interface and the weak interaction of Co/Pc has limited effect on 3*d* states. This results in a larger mismatch between the E_F and the energies associated with the orbitals d_{xy} and $d_{x^2-y^2}$, and for C_{60}, rubrene and Alq_3 capped films, compared to CuPc and Pc-capped samples. Consequently, a lower PMA in rubrene, Alq_3- and C_{60} -capped samples and a relative larger one in CuPc and Pc-capped films were observed.

It was reported that such interface interaction can result in the molecular-level engineering of metal/organic interfaces not only to overcome the conductivity mismatch issue but also to customize spin injection as well for bringing new electrical functionalities to the spintronics devices. Raman et al., reported that the interface-engineered templates for molecular spin memory devices [5]. A neutral planar phenalenyl-based molecule, zinc methyl phenalenyl (ZMP, $C_{14}H_{10}O_2Zn$), were grown on a ferromagnetic surface. The interface spin transfer causes a hybridized organometallic supramolecular magnetic layer, which shows a large magnetic anisotropy and spin-filter properties. This interface layer creates a spin

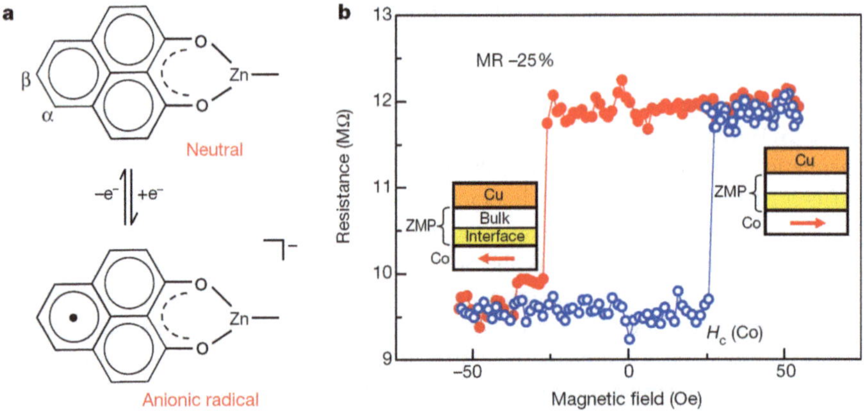

Fig. 17.6 Molecular structure of zinc methyl phenalenyl (ZMP) (**a**) and Interface magnetoresistance effect (**b**). ZMP was in a neutral state with no net spin (*top*) and the chemical state of the phenalenyl moiety changed from neutral to an anionic radical (*bottom*) with net moment by the hybridization on the ferromagnetic surface. b, Magnetoresistance response (data points) of a device with a structure of (Co (8 nm)/ZMP(40 nm)/Cu(12 nm)), measured at 4.2 K in a magnetic field. *Blue* and *red data points* refer to positive and negative field sweeps respectively [5]. (from Raman et al. Nature. 2013, 493, 509–513)

dependent resistance and gives rise to an interface magnetoresistance (IMR) effect. Molecular structure of ZMP is shown in Fig. 17.6a. ZMP was normally in a neutral state with no net spin (top). When ZMP was grown on ferromagnetic film, such as Co, the chemical state of the phenalenyl moiety changed from neutral to an anionic radical (bottom) with net moment by the hybridization on the ferromagnetic surface. Thus, an interface layer was created. Figure 17.6b shows a magnetoresistance measurement (data points) of a device with a structure of (Co (8 nm)/ZMP (40 nm)/Cu(12 nm)), measured at 4.2 K in a magnetic field.

Chemical modification of the ferromagnet/ZMP molecule interface originated from the charge transfer and hybridization of the molecular orbitals with the d-orbitals of the ferromagnet, creating new hybrid, metal–organic interface states that directly affect the electronic and magnetic properties of the adsorbed molecule. These interactions strongly depend on the specific molecular structure and morphology, influencing the efficiency of spin polarization and spin injection at the interface. In the case of ZMP, the planar geometry of the molecule provides an ideal situation for strong orbital overlap leading to d–p hybridization effects. Spin-polarized first-principles calculations using density-functional theory (DFT) were carried out in zero magnetic field on such a hybrid metal–organic interface, comprising five Co(111) layers and either one or two molecular layers of ZMP on top.

Very recently, Mari et al., reported that the electronic states of non-ferromagnetic materials, copper and manganese, could be altered by connecting with C_{60}, and then made them ferromagnetic at room temperature [18]. The Cu/C_{60} charge transfer and interface reconstruction results in substantial

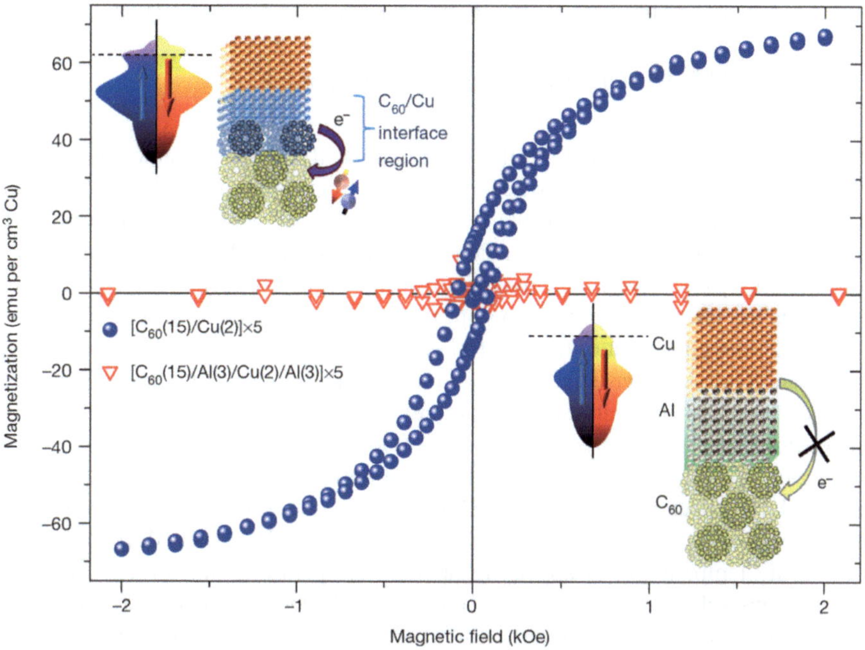

Fig. 17.7 Effect of molecular interfaces. Schematics and room-temperature magnetization for a Ta(5 nm)/[C_{60}(15 nm)/Cu(2 nm)]$_5$/Al(5) and a Ta(5 nm)/[C_{60}(15 nm)/Al(3 nm)/Cu(2 nm)/Al (3 nm)]$_5$ sample [18] (from Ma'Mari et al. Nature. 2015, 524, 69–73)

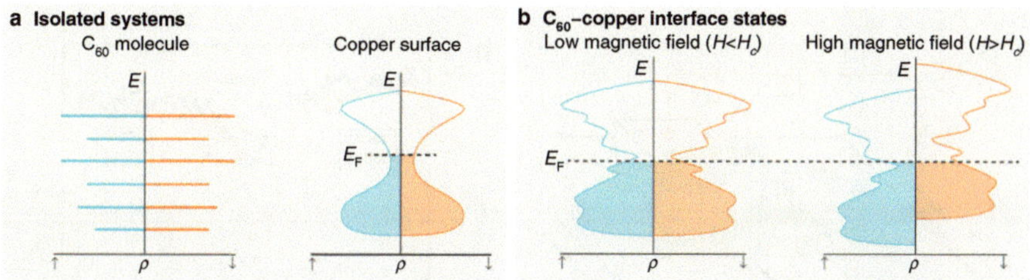

Fig. 17.8 Magnetic copper–buckyball interfaces. (**a**). The C_{60} molecule (*left*) and copper (*right*) are non-magnetic in their isolated form. E_F is the Fermi energy. (**b**). In materials containing C_{60}–copper interfaces. Above a critical value Hc, the interface becomes spontaneously magnetized (*right*) [43]. (from Raman et al. Nature. 524, 42–43 (2015))

changes in the density of states (DOS) of the metallic film and a band splitting that leads to magnetic ordering. On the other hand, an Al spacer between both materials screens the charge transfer from the 3d transition metal and stops the band splitting, as shown in Fig. 17.7. The phenomenon could be found in Fig. 17.8. Both C_{60} and Cu are non-magnetic in their isolated form. C_{60}/Cu interface states are still non-magnetic with a small magnetic field strength. When the field was increased above the critical valve, the C_{60}/Cu interface became spontaneously magnetized.

17.3 Spin Transport Behaviors in Organic Spin Valves

17.3.1 GMR Effect in Organic Spin Valve

Organic spin valves (OSVs) are the most popular devices in organic spintronics. Typically, OSVs can be classified as lateral structure devices [1] and vertical structure devices [2] based on the device structure. Both types of structures are composed of two ferromagnetic (FM) electrodes separated by an organic semiconductors spacer layer. The magnetoresistance (MR) effect in the spin valve device is the phenomenon of change in the electrical resistance depending on the magnetization alignment of the two FM electrodes that are modified by an external magnetic field. A larger (smaller) resistance is usually observed when the magnetization directions are anti-parallel (parallel). This phenomenon can be utilized to investigate spin transport properties in OSV devices.

Figure 17.9 shows the structure and transport properties of the fabricated organic spin-valve devices with a typical LSMO/Alq_3/Co structure. Giant Magnetoresistance (GMR) ratio is around −40% obtained at 11 K, which decreases with increasing temperature, disappearing at room temperature. Since then, Alq_3 has been extensively studied in organic spintronics. At room temperature, MR ratios around 6% and 10% were reported by Moodera's and Koopmans's groups using ultra-thin Alq_3 layers that were 1.6 nm and 1 nm thick, respectively [23, 44]. Dediu et al. successfully observed an inverse MR effect at room temperature using a 100-nm-thick Alq_3 layer, which showed a MR ratio of 0.15% [13]. However, achieving a large MR effect at room temperature in organic spintronics is still

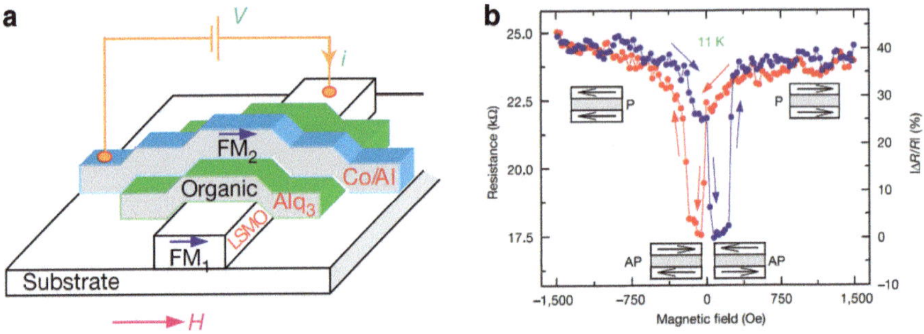

Fig. 17.9 Schematic representation of a typical device that consists of two FM electrodes (FM1 and FM2) and an organic semiconductor spacer. Spin-polarized electrical current I flows from FM1 (LSMO), through the organic semiconductor spacer (Alq_3), to FM2 (Co) when a positive bias V is applied. An in-plane magnetic field, H, is swept to switch the magnetization directions of the two FM electrodes separately. (**b**) GMR loop of a LSMO (100 nm)/Alq3 (130 nm)/Co (3.5 nm) spin-valve device measured at 11 K. The *blue* (*red*) curve denotes GMR measurements made while increasing (decreasing) H. The anti-parallel (AP) and parallel (P) configurations of the FM magnetization orientations are shown in the insets at low and high H, respectively. The electrical resistance of the device is higher when the magnetization directions in FM1 and FM2 films are parallel to each other (from Xiong et al. Nature. 2004,427,821–824)

challenging because most reports on the spin-dependent transport behavior in Alq_3 films are generally reported at low temperatures [21, 45–50]. Possible reasons for this are the low Curie temperature and small spin polarization efficiency of the LSMO electrodes when operated at room temperature, which limits the spin transport efficiency in these devices. To obtain a large MR ratio at room temperature in organic spin valve devices, FM metals have been used as the electrode because they have a high Curie temperature (Tc ~ 1000 K) when compared with LSMO (Tc ~ 300 K). Unfortunately, these attempts were not very effective. This means that the spin injection efficiency from the FM metal to the Alq_3 film is likely not high enough to induce room temperature spin transport because of the small spin polarization (~35%) in the FM metal. Herein, a search for a new material with a large spin polarization for use as an electrode may be helpful. Density-functional calculations showed that the electrons in magnetite (Fe_3O_4) at the Fermi level were 100% spin-polarized [51, 52]. Experimental measurements showed that the spin polarization of Fe_3O_4 was around 80% at room temperature (Tc ~ 850 K) [53]. Thus, Fe_3O_4 has potential for use as an efficient electrode material in spintronics.

In this part, we mainly report on a functional organic SV device with a Fe_3O_4 and Co as electrodes, respectively. Organic spacer will include C_{60}, C_{70}, Alq_3, rubrene, etc. Spin dependent transport will be discussed to understand the criteria for organic semiconductors selection and design to improve the performance in spin based devices.

The hysteresis curves for the Fe_3O_4 and Co films were measured at 300 K, which were deposited on MgO and Si/SiO_2 substrates with an Alq_3 buffer layer, respectively. Figure 17.10a showed that the saturation magnetization of the Fe_3O_4 film was 370 emu/cc and the coercivity was around 140 Oe. In Fig. 17.10b, the hysteresis curve for a Co film showed a nearly rectangular shape with a saturation magnetization of 1400 emu/cc and a coercivity of about 30 Oe. The clear difference between the coercivity for the Fe_3O_4 and Co films provided a precondition for magnetic anti-parallel alignment. The anisotropic magnetoresistance (AMR) effects on both of the individual films were also evaluated at 300 K, as shown in Fig. 17.11.. It was noted that the AMR ratio was around −0.04% for the Fe_3O_4 film and the value was nearly zero for the Co film. The Co film deposited on a Si/SiO_2 substrate with a C_{60} buffer layer showed similar results. The AMR effect was measured in the film plane and the external magnetic fields were applied parallel and perpendicular to the direction of the electric current in the Fe_3O_4 and Co films, respectively. Meanwhile, the Verwey transition temperature was observed to be 117 K for a Fe_3O_4 film (see Fig. 17.12), confirming the high quality of the film.

17.3.1.1 Fabrication of Organic Spin Valve

An etching technique used to fabricate inorganic spin valve is not suitable for the fabrication of organic spin valves because the chemical solvent for cleaning would pollute organic semiconductor layers. Thus, shadow masks have to be used to form

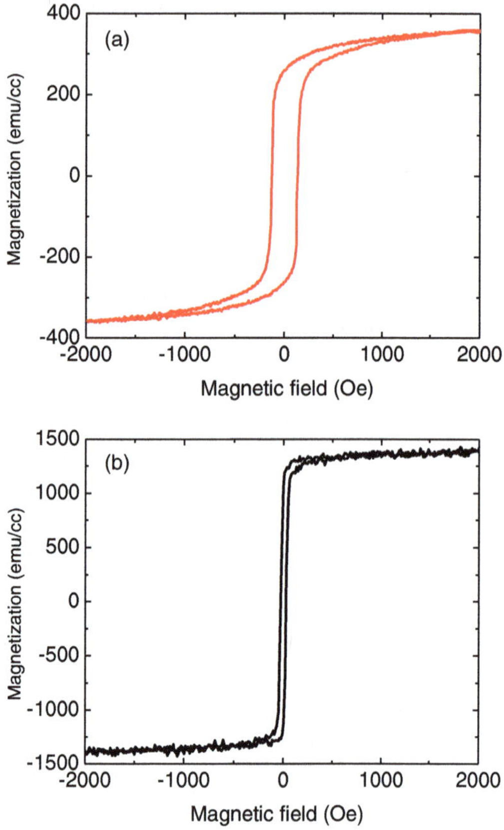

Fig. 17.10 Hysteresis curves for the (**a**) Fe_3O_4 film and (**b**) the Co film measured at 300 K. (from Zhang, et al. J. Appl. Phys. 115, 172,608 (2014))

a cross junction in organic spin valve. Figure 17.13 shows a fabrication process for a typical organic spin valve with stacking structure device. Mask 1 is used to grow the bottom Fe_3O_4 strip. A circle shadow mask (mask 2) was used to fabricate organic semiconductor layer. Finally, mask 3 is used to prepare the top electrode strip.

Figure 17.14a shows a schematic diagram of C_{60}-based OSV device with an MgO substrate/Fe_3O_4/AlO/C_{60}/Co/Al stacking structure. A stable Al-O layer was used to protect the surface of the Fe_3O_4 layer for mask exchange and increased spin injection efficiency. A C_{60} molecule was used as the organic layer, and Co was used as the top FM electrode. C_{60} is stable and easily forms a high-quality molecular film via thermal evaporation. In particular, the hyperfine interaction in C_{60} is considerably weak due to the absence of polarised hydrogen nuclei and the low natural abundance of ^{13}C nuclear spins in the C_{60} molecule. The weak hyperfine interaction has been predicted to increase the SDT length [54], which has been demonstrated at low temperatures using a deuterated polymer as a spacer in organic SV devices [55]. Furthermore, the C_{60} molecule is nearly isotropic due to its high symmetry and rotational freedom. Therefore, a polarised electron can easily move from one

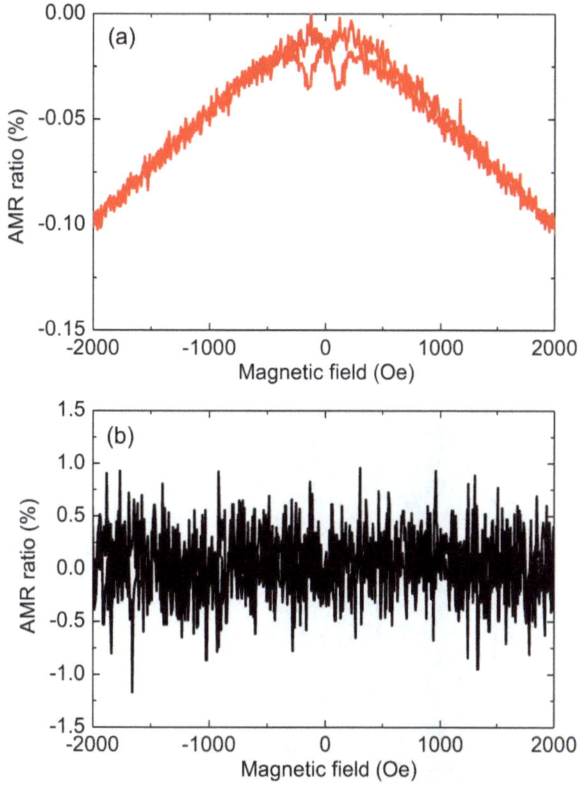

Fig. 17.11. Anisotropic magnetoresistance effect for the (**a**) Fe_3O_4 film and (**b**) the Co film measured at 300 K. (from Zhang, et al. J. Appl. Phys. 115, 172,608 (2014))

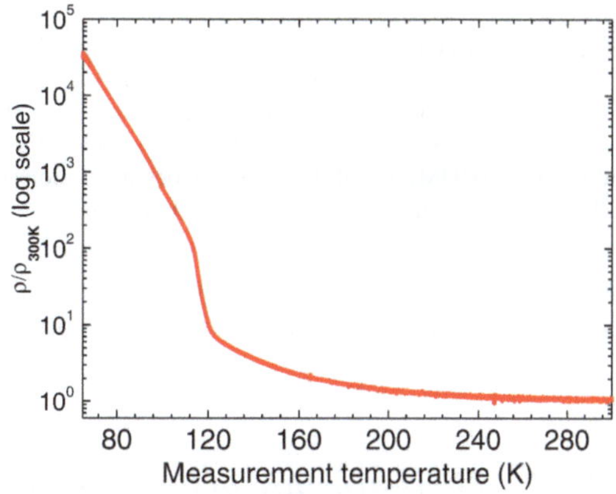

Fig. 17.12 The Verwey transition temperature was observed to be 117 K for a Fe_3O_4 film

molecule to the next with a smaller spin flip because the energy loss of the carrier is small between hopping states [56] compared to other strongly anisotropic OSEs, such as Alq_3. These characteristics of C_{60} can potentially increase the SDT length in spin-based organic devices. Furthermore, the energy levels of Fe_3O_4, C_{60}, and Co

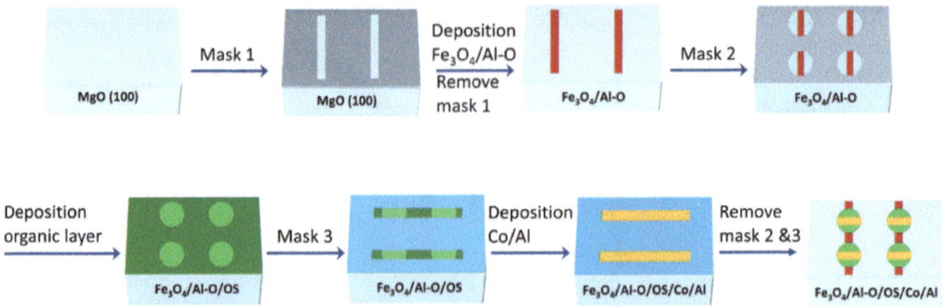

Fig. 17.13 Schematic fabrication process (from Zhang, et al. J. Appl. Phys. 115, 172,608 (2014))

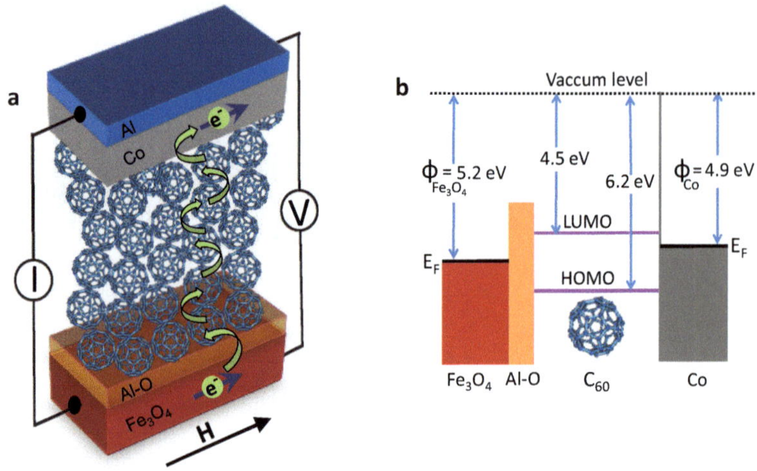

Fig. 17.14 Structure of designed organic spin valve devices. (**a**). Schematic diagram of the device structure with an MgO-substrate/Fe_3O_4/Al-O/C_{60}/Co/Al stacking structure (**b**). Schematic band diagram of the organic device in the rigid band approximation shows the Fermi levels (E_F), the work functions of the two ferromagnetic electrodes (Fe_3O_4 and Co), the highest occupied molecular orbital (HOMO), and the lowest unoccupied molecular orbital (LUMO) levels of C_{60}.(from Zhang, et al. Nature Communications.4,1392 (2013))

are well matched, as shown in Fig. 17.14b. Thus, in principle, electron spin injection, transport, and detection can be realised in the present organic SV device.

17.3.2 Microstructure of an Organic Device

To further explore the microstructure of the organic SV devices, scanning transmission electron microscopy and elementary mapping by energy dispersive X-ray spectroscopy were performed for an organic SV device. The microscopy images of Fig. 17.15 show that each layer had a sharp interface and that the C_{60} layer was

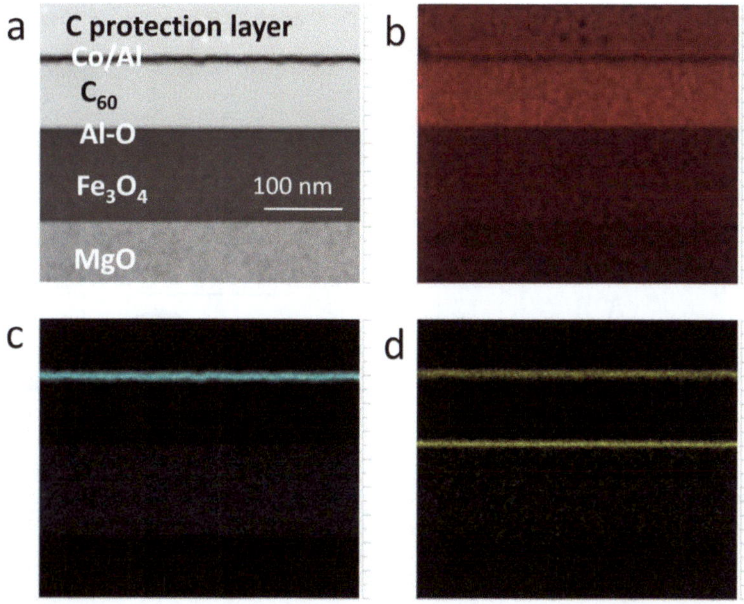

Fig. 17.15 Microstructure of an organic spin valve device. (**a**) Scanning transmission electron microscopy of a device with an MgO-substrate/Fe_3O_4/Al-O/C_{60}/Co/Al stacking structure; the top C layer was deposited to protect the device layers during microscopy measurements; Scale bar 100 nm. (**b–d**) are the elementary mappings of C (*red*), Co (*cyan*), and Al (*yellow*), respectively, for K-shell electrons performed by energy dispersive X-ray spectroscopy (from Zhang, et al. Nature Communications.4,1392 (2013))

homogeneous. The elementary mappings of C, Co, and Al showed a clear separation for each layer, as shown in Fig. 17.15b–d, respectively. These studies further demonstrate that there was no appreciable diffusion of Co into the C_{60} layer. Thus, it is very likely that the "ill-defined" layer due to metal inclusions that is assumed for the organic layer in Alq_3-based SV devices [2] does not appear in our devices. One possible reason for the absence of this layer is the difference between the spatial structures of the C_{60} and Alq_3 molecules. The spherical structure of C_{60} makes the sustenance of the upper metallic layer easy.

The MR curves for the device with an 80-nm thick C_{60} layer measured at 300 K, 250 K, 200 K, and 150 K are shown in Fig. 17.16a. The MR ratio was calculated using the expression $\Delta R/R_p = (R_{ap}/R_p - 1) \times 100\%$, where R_{ap} and R_p denote the resistance in the anti-parallel and parallel states, respectively. The R_p value was obtained at a magnetic field of 3000 Oe. The calculation shows an MR ratio of 5.3% at 300 K, which is one of the highest MR ratios that have been reported at room temperature. Note that both the MR ratios and device resistance increased with decreasing measurement temperatures. The MR ratios were 6.1% at 250 K, 6.7% at 200 K, and 6.9% at 150 K. The bias-voltage dependence of the MR ratio at 300 K and 150 K is shown in Fig. 17.16b. The clearly nonlinear nature of the current-voltage curves can be observed in Fig. 17.16c. Similar conductance-bias curves

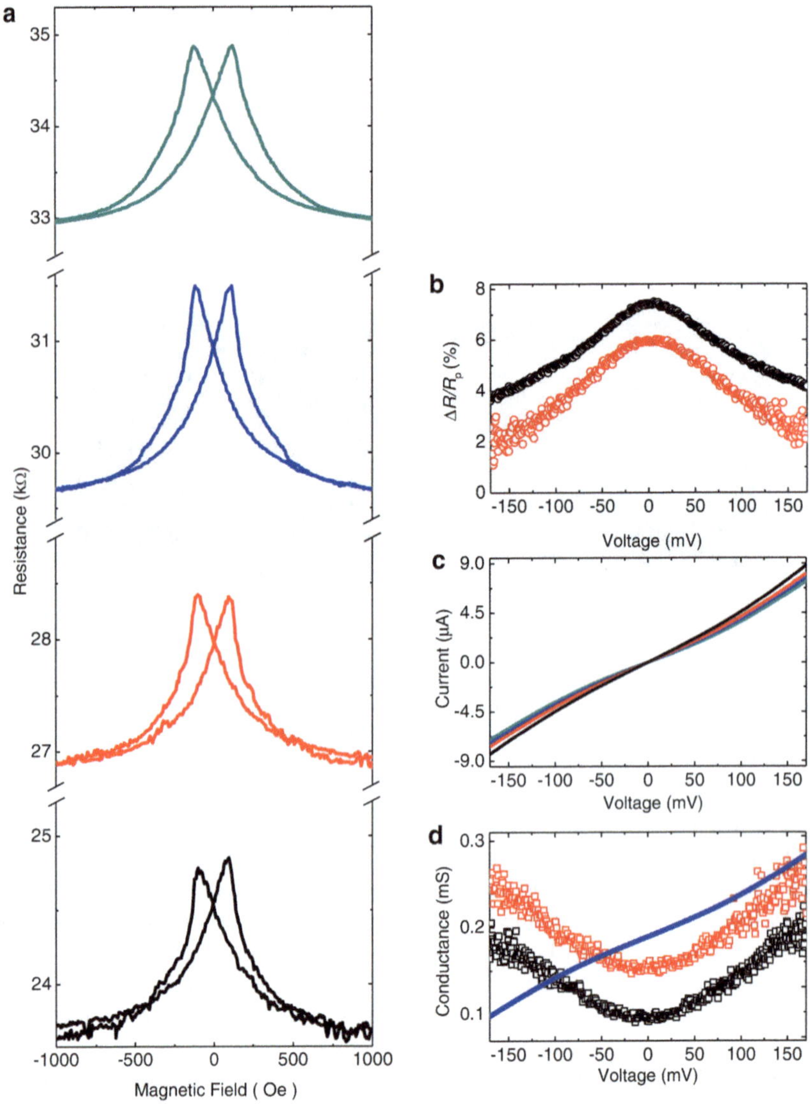

Fig. 17.16 Transport properties of C_{60} spin valve devices. (**a**) Magnetoresistance curves of the organic device composed of a 80-nm-thick C_{60} layer measured at different temperatures, showing magnetoresistance ratios of 5.3% at 300 K (*black curve*), 6.1% at 250 K (*red curve*), 6.7% at 200 K (*blue curve*), and 6.9% at 150 K (*dark cyan curve*). The bias voltage was fixed at 30 mV to measure the magnetoresistance effect. (**b**) Bias-voltage dependence of $\Delta R/R_p$ at 300 K (*red circles*) and 150 K (*black circles*) for a device with of a 80-nm-thick C_{60} layer. (**c**) Nonlinear current-voltage curves for the device with a 80-nm-thick C_{60} layer at 300 K (*black curve*), 250 K (*red curve*), 200 K (*blue curve*), and 150 K (*dark cyan curve*). (**d**) Conductance versus bias curves for a device with a 10-nm-thick C_{60} layer measured at 300 K (*red squares*) and 150 K (*black squares*) and the corresponding nonlinear current-voltage curve (*blue circles*) at 300 K (from Zhang, et al. Nature Communications.4,1392 (2013))

were observed at 300 K and 150 K for the device with a 10-nm-thick C_{60} layer, as shown in Fig. 17.16d. The primary difference was the down-shift of the curve at 150 K due to the increase in the resistance, indicating that the C_{60} layer may have been free of magnetic inclusions.

17.3.3 Dependence of Magnetoresistance Ratio on C_{60} Thickness

A set of organic SV devices consisting of different C_{60} layer thicknesses (10–110 nm) were fabricated. The measured MR curves were similar to that shown in Fig. 17.16 for all devices. The dependence of the MR ratio on the C_{60} layer thickness at 300 K and 150 K is plotted in Fig. 17.17a. The MR ratios at 150 K were larger than those at 300 K for each corresponding C_{60} layer thickness due to the decrease in spin scattering (such as magneto-excitation) at lower temperatures. Notably, an unusual and interesting dependence of the MR ratio on the C_{60} layer thickness was observed, which differs from previous reports. The MR ratios sharply increased upon increasing the C_{60} layer thickness from 10 nm to 40 nm and then

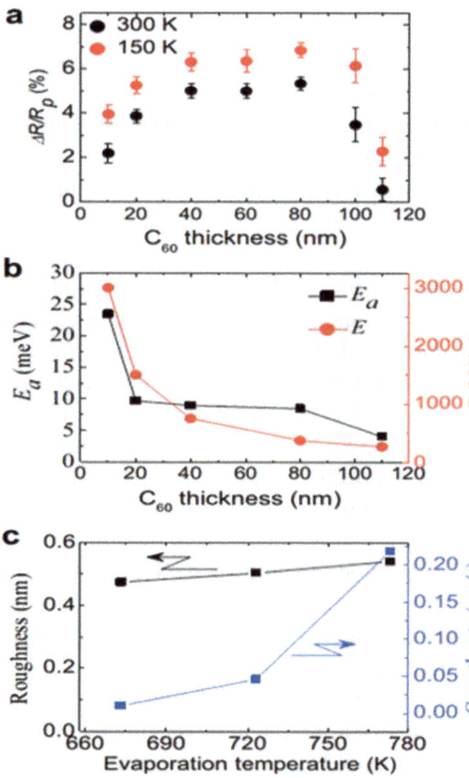

Fig. 17.17 Thickness and temperature dependence. (**a**). $\Delta R/R_p$ as a function of the C_{60} layer thickness measured at 300 K (*black circles*) and 150 K (*red circles*). The bias voltage was fixed at 30 mV to measure magnetoresistance effect, and the error bars were estimated for four devices at each thickness (**b**). Activation energy (*black squares*) and electric field magnitude (*red circles*) as a function of the C_{60} layer thickness (**c**). Dependence of the roughness (*black squares*) and growth rate (*blue squares*) on the evaporation temperature for the 80-nm-thick C_{60} films; the solid curves in **b** and **c** are only visual guides (from Zhang, et al. Nature Communications.4,1392 (2013))

slowly increased for increases in the C_{60} layer up to 80 nm. Beyond this range, further increases in the C_{60} layer thickness produced a drastic reduction in the MR ratio. Fortunately, a C_{60} layer thickness of 110 nm was successfully observed with an MR ratio of approximately 0.5% at room temperature in the devices. To the best of our knowledge, this thickness is the largest SDT length reported at room temperature in OSEs with vertical SV structures to date. We suggest that when compared with other reported OSEs, such as rubrene and Alq_3, the weaker spin-orbit coupling and hyperfine interaction of the C_{60} molecule [55, 56] significantly reduces spin flip in the organic film. Thus, a large SDT length was obtained for the C_{60} layer. Our results indicate that the SDT length relies strongly on the molecular configuration of the OSE.

To analyse the dependence of the MR ratio on the C_{60} layer thickness, the structures of the C_{60} layers were studied. X-ray diffraction measurements showed that the C_{60} layers of different thickness were structurally amorphous. The C_{60} layer thickness in our devices was far greater than the tunnelling barrier limit that has been reported thus far. Thus, the SDT within the C_{60} layer via hopping between localised neighbouring states was mainly considered. Our results can therefore be described by the Gaussian disorder model. Bobbert et al. [55] presented a theory for spin transport in disordered OSEs. The authors concluded that both the spin diffusion length (l_s) and mobility (μ) decreased with a decreasing electric field (E) ($E < 5[\sigma/ea]$). E was determined by the equation $E = U/d$, where U and d are the applied voltage and distance between two electrodes, respectively. In our SV devices, U was fixed at 30 mV and d was determined by the C_{60} layer thickness. Thus, as the C_{60} layer thickness increased, E decreased, as shown in Fig. 17.17b. As a result, l_s and μ decreased. Bobber et al. also suggested that l_s and μ increase with decreasing disorder strength. The activation energy (E_a) is a direct reflection of the amount of disorder strength [57]. The E_a of C_{60} was estimated from the temperature dependence of the resistance and is plotted in Fig. 17.17b. E_a clearly decreased with increasing C_{60} thickness, indicating that the disorder strength of the C_{60} layer decreased. This result indicates that l_s and μ increased with increasing C_{60} thickness. The competition between these two factors (E and disorder) likely leads to a maximum for both l_s and μ. As l_s and μ increase, the dependence of MR on the magnetisation alignments of the two electrodes becomes stronger, which is the origin of the MR effect [55]. Consequently, a maximum of $\Delta R/R_p$ can appear with increasing C_{60} layer thickness. The change ratios for l_s and μ were distinctly different for different values of E [55]. This observation could explain the different increasing slopes of $\Delta R/R_p$ for C_{60} layer thicknesses of 10–40 nm and 40–80 nm. The decrease in E_a implies that the molecular interactions between C_{60} molecules may vary. In our organic SV devices, E_a was smaller than the values reported in the literature [57], likely due to the dependence of the mobility on the charge carrier density induced by different devices [57, 58].

The evaporation temperature of the C_{60} layers was 723 K for the organic SV devices reported above. We found that the evaporation temperature of the C_{60} layer played an essential role in modifying the SDT properties. Organic SV devices with an 80-nm-thick C_{60} layer were also fabricated by evaporating C_{60} at 673 K and

773 K. Both series of devices showed MR ratios of approximately 1.7% at room temperature. These MR ratios were three times smaller than that of the device with C_{60} evaporated at 723 K. Note that the R_p of the devices with C_{60} evaporated at 673 K was approximately 12.5 kΩ, which was approximately two times smaller than that of the devices with C_{60} evaporated at 723 K. In contrast, the R_p of the device with C_{60} evaporated at 773 K was nearly 1 MΩ. The resistance was inversely proportional to the carrier mobility. The decrease of R_p at 673 K did not lead to an increase in the MR ratio, indicating that the MR ratio was not solely affected by the carrier mobility, in agreement with the analysis for Fig. 17.17a, b as well as the literature [59].

The roughness and growth rates of the C_{60} films for different evaporation temperatures are plotted in Fig. 17.17c. The roughness for all films was approximately 0.5 nm, which is less than the diameter of a C_{60} molecule (approximately 1 nm), indicating that the films were extremely smooth. However, the growth rates were significantly different, corresponding to 0.015 nm/s, 0.045 nm/s, and 0.22 nm/s for the films evaporated at 673 K, 723 K, and 773 K, respectively. These differences can be attributed to the different kinetic energies for the C_{60} molecules obtained at each evaporation temperature. The kinetic energy of the molecules clearly affected the thermal equilibrium of the individual C_{60} molecules when the final film formed. Thus, the molecular distance and intermolecular interactions in organic films will vary for different temperatures. The intermolecular distance and interactions may affect the orbital overlap of the π electrons in OSEs [60] that determine the transport efficiency of the charge carriers. Previous studies have also reported that temperature affects the lattice fluctuations [61, 62] and intramolecular interactions [63] of C_{60}, which may cause variations in the σ-orbital. The σ-orbital affects the SDT length in organic OSEs [59]. Moreover, the C_{60} molecule exhibits a series of energy levels [61]. The evaporation temperature may affect the distribution of the energy levels in a C_{60} layer. This energy redistribution can reduce the matching of energy levels between the C_{60} molecules for hopping transport, thus decreasing the hopping efficiency of the C_{60} layer. Meanwhile, the energy redistribution may also change the interactions between the C_{60} energy levels and electrodes at the interfaces, leading to lower spin injection and/or detection efficiencies. This study has shown that the microstructure in organic SV devices is closely correlated with magnetotransport, in agreement with the literature [48]. Based on an optimised evaporation temperature for the C_{60} layer at 723 K, we deduced that an efficient SDT pathway exists in a SV device with a C_{60} layer of 80 nm for achieving a high MR ratio in which spin electrons are efficiently injected [64] into the C_{60} layer and hopping transport occurs with very a small spin flip. The most recent theoretical investigation predicts that an SDT length of more than 430 nm for C_{60} can be achieved at room temperature [65]. In the organic SV devices considered here, the C_{60} layer had an amorphous structure. Thus, a larger SDT length for C_{60} layers at room temperature can be expected when C_{60} forms a single-crystalline structure.

17.4 TMR Effect in Organic Spin Valve

If the organic layer is thin enough, e.g., bellow 20 nm, quantum transport mechanism will appear. That is, spin transport follows the tunneling mechanism, rather than diffusion. Moreover, single-step tunneling is dominated when the organic layer thickness is below 5 nm and multiple-step tunneling becomes popular with a thicker layer.

Rubrene has a high carrier mobility that is larger by one order of magnitude than that of 6 T and by five orders of magnitude than that of Alq_3. Moreover, the admixture parameter for spin–orbit coupling in rubrene is approximately one order of magnitude smaller than those of 6 T and Alq_3 and a relatively large spin diffusion length has been predicted in rubrene [65]. MR effect was evaluated by applying magnetic field along the direction of Fe_3O_4 strip. The MR curves for the rubrene-based spin valves were measured at different temperatures (300 K, 250 K, 200 K, and 150 K). Figure 17.18a, b show the MR curves for the devices with 2-nm-and 6-nm-thick rubrene layers, respectively. The MR ratio was approximately 6% for the devices with a rubrene-layer thickness of 2 nm. This is one of the highest MR ratios at room temperature in rubrene-based spin valves reported to date [19–27]. Upon reducing the measurement temperature, we observed that the width of the MR curves increased mainly due to increase in the coercivity of the Fe_3O_4 electrode. Meanwhile, the MR ratios were also enhanced with decreasing measurement temperatures. The MR ratios reached values of up to 7.2%, 8%, and 11% for temperatures of 250 K, 200 K, and 150 K, respectively. The increase in the MR ratio basically originates from the decrease of spin scattering at lower temperatures. The temperature dependences of the MR ratios and MR curves for device with the 6-nm-thick rubrene layer were similar to those for the device with the 2-nm-thick rubrene layer. However, the MR ratio of the device with the 6-nm rubrene layer was smaller than that of 2-nm rubrene-layer device. The former exhibits MR ratios of 3.3%, 4.1%, 7%, and 9% corresponding to the temperature at 300 K, 250 K, 200 K and 150 K, respectively. Without rubrene layer, MR effect does not appear in the spin valve of Fe_3O_4/Al-O/Co/Al with the same size as rubrene-based spin valve device. This can be understood by considering the low conductivity of Fe_3O_4 compared to metal.

The current–voltage curves for the devices with rubrene layer (6 nm and 20 nm) were measured at 300 K and 150 K. Nonlinear current–voltage dependences were observed, as shown in Fig. 17.19a, b. Moreover, with increasing rubrene-layer thickness, the nonlinear response of current–voltage dependences became stronger because of increase in the device resistance. Figure 17.19c depicts the clear bias-voltage dependence of MR ratios measured at 300 K (black circles) and 150 K (red squares) for a device with a 6-nm-thick rubrene layer. Figure 17.19d shows the conductance versus bias curves for a device with a 20-nm-thick rubrene layer measured at 300 K (black curve) and 150 K (red curve). A primary difference was the downshift of the curve at 150 K because of the increase of the resistance.

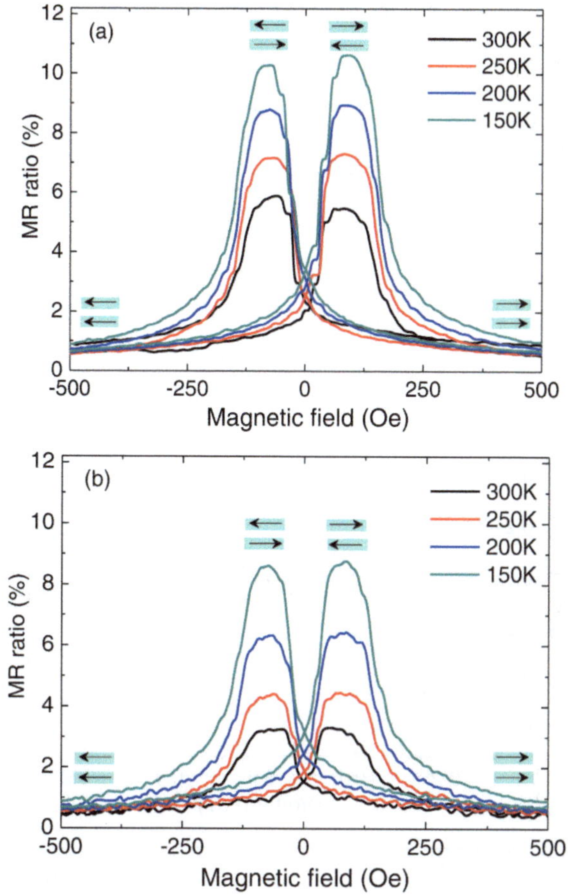

Fig. 17.18 Magnetoresistance curves for organic spin valve devices composed of (**a**) 2-nm-thickand (**b**) 6-nm-thickrubrene layers. The devices were measured at 300 K, 250 K, 200 K and 150 K, respectively. The arrows illustrate the magnetization alignments in the anti-parallel and parallel directions. The bias voltage for the measurement was fixed at 5 mV (from Zhang, et al. ACS Appl. Mater. Interfaces. 2015,7,4685–4692)

The observation was similar with the results reported previously using a 15-nm-thick rubrene layer [66].

To investigate the spin-dependent electron transfer behaviors, a series of rubrene-based devices with different rubrene-layer thicknesses (2, 5, 10, and 20 nm) were fabricated, and the MR effects of these devices were measured at 300 K, 250 K, 200 K, and 150 K, as depicted in Fig. 17.20. The MR ratio decreased monotonously with increasing rubrene-layer thickness, as shown in Fig. 17.20a. The largest MR ratio was approximately 6% for a rubrene-layer-thickness of 2 nm at room temperature, and the MR ratio decreased to 0.6% for devices with a 20-nm-thick rubrene layer. The MR ratio dependence on the

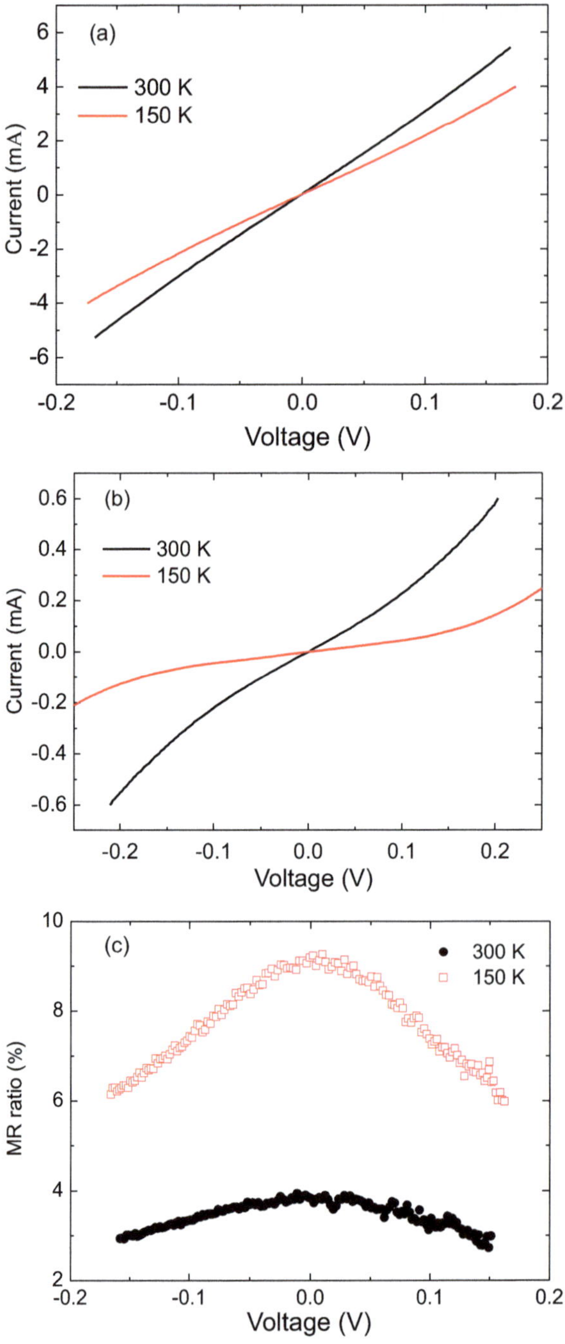

Fig. 17.19 Nonlinear current–voltage curves for spin valves with (**a**) 6-nm-thick and (**b**) 20-nm-thick rubrene layers, which were measured at 300 K and 150 K, respectively. (**c**) Bias-voltage dependence of MR ratios at 300 K (*black circles*) and 150 K (*red squares*) for a device with of a 6-nm-thick rubrene layer. (**d**) Conductance versus bias curves for a device with a 20-nm-thick rubrene layer measured at 300 K (*black curve*) and 150 K (*red curve*). (from Zhang, et al. ACS Appl. Mater. Interfaces. 2015,7,4685–4692)

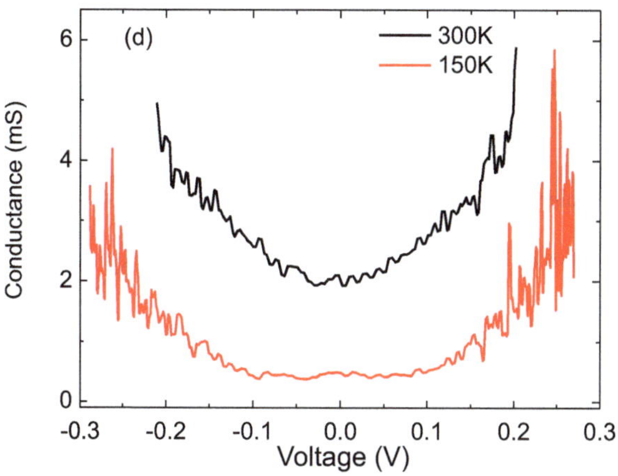

Fig. 17.19 (continued)

measurement temperature is plotted in Fig. 17.20b. The MR ratio for devices with 2-nm- and 6-nm-thick rubrene layers exhibited a sharper increase with decreasing measurement temperature in comparison with those of devices with thicker rubrene layers of 12 nm and 20 nm. The devices with rubrene layer thickness below 2 nm usually failed. This may be due to a discontinuous film of thinner rubrene layer or a thin "ill-defined layer" [13] for rubrene/Co interface. In the following two paragraphs, we will try to discuss the possible spin transport mechanism in these OSVs.

Several theoretical methods have been proposed to analyze the related factors affecting on the spin-dependent transport mechanisms, tunneling or diffusion, in disordered OSs [67–69]. It has been demonstrated that the site-energy disorder and/or positional disorder, spin mixing, and the exchange coupling between localized polarons contribute to the spin-dependent transport. These theories have promoted an understanding of the spin transport behavior in organic spintronics. In contrast, there is still no generally accepted experimental method to clearly clarify the spin-dependent transport mechanism in OSVs. Therefore, it is very difficult to illustrate the exact spin electron transport mechanism in the present rubrene-based devices. Here, we speculate on the possible spin transport mechanism. In general, the thickness of the spacer layer in spin valves, around several nanometers, is in the tunneling transport range. With increasing spacer-layer thickness, the transport behaviors become complicated, and the tunneling mechanism changes from single-step tunneling to multiple-step tunneling. For spacer-layer thickness over several tens of nanometers, diffusion mechanisms play a dominant role. In this regard, Schoonus and co-authors have reported a decrease in the MR ratio with increasing Alq_3 layer thickness ranging from 1 nm to 4 nm. They hypothesized that this decrease was caused by a transition in the spin-dependent transport from single-step tunneling to multiple-step tunneling with increasing Alq_3 layer thickness. This explanation agrees well with the reports in Alq_3 based spin valves from Moodera's group [23] and Coey's group [70]. In C_{60}-based spin valve, multiple-step tunneling mechanism

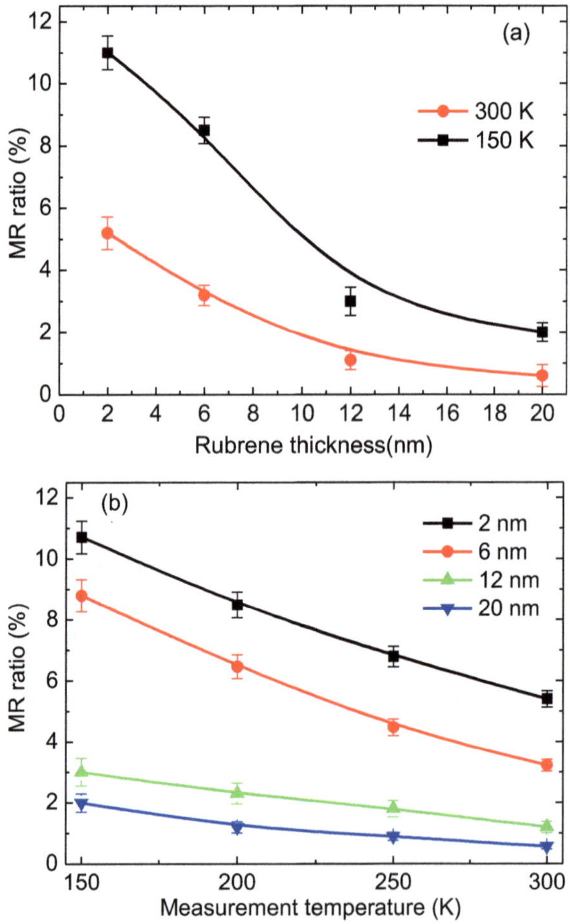

Fig. 17.20 (**a**) Dependence of the magnetoresistance ratios on the rubrene-layer thickness measured at 300 K (*red circles*) and 150 K (*black squares*). (**b**) Measurement temperature dependence of the magnetoresistance ratios for devices with different rubrene-layer thicknesses. The lines serve as a visual guide (from Zhang, et al. ACS Appl. Mater. Interfaces. 2015,7,4685–4692)

occurs when the organic layer thickness is over ~4 nm [69, 71]. In the present rubrene-based devices, single-step tunneling is most probably dominant in the devices with a 2-nm-thick rubrene layer. Upon increasing the rubrene thickness, the multiple-step tunneling processes become stronger, thereby resulting in a decrease in the MR ratios.

The defects appear in the amorphous rubrene-layer, which will also affect the spin-dependent transport behaviors [66, 68]. This effect could occurs in our OSVs and may be stronger with increasing rubrene layer thickness. The observed temperature dependence of the MR ratios for the spin valves with 2-nm- and 6-nm-thick rubrene layers differs from those for spin valves with the 12-nm and 20-nm-thick rubrene layers, as shown in Fig. 17.20b. This is probably due to change in the spin-dependent transport mechanism based on the competition of above factors.

For inorganic spin valves, the MR ratio can usually be increased by post-annealing. The resulting increase is mainly due to decrease in spin scattering by increasing the interface quality after annealing. Consequently, we annealed the rubrene-based spin valves in vacuum for 6 minutes at 363 K to enhance the MR ratios. After the sample device was cooled to room temperature, its MR curve was measured again. Figure 17.21a shows the results measured at 300 K for the device with a 6-nm-thick rubrene layer before and after annealing. We note that the shape of the MR curve after annealing is same as that before annealing. However, it is also noteworthy that the MR ratio reduces from 3.3% to 1.9% after annealing. We also measured the MR curves at low temperatures, and their MR ratios are plotted in Fig. 17.21b. We find that the MR ratios decrease after annealing. Further, it is noted that the device resistances for both parallel and anti-parallel states increase after annealing compared to the corresponding values before annealing, as summarized in Fig. 17.5c. This indicates that the spin scattering becomes stronger after annealing compared with that before annealing. The rubrene layers annealed with the same condition were still of amorphous structure checked by the XRD measurement, which is owing to the low annealing temperature. It should also be noted that this temperature is hard to change the interface quality for Fe_3O_4/Al-O layers because the bilayer film was directly annealed at 623 K after deposition.

In order to further explore the possible reasons for the increase in spin scattering, atomic force microscopy was utilized to inspect the rubrene films. Figure 17.22a, b show the morphology of the rubrene films before and after annealing. We note that the surface of the film is very flat and uniformly smooth before annealing. But it becomes very rough after annealing. We observed that the roughness of the film increased from 0.25 nm to 0.41 nm, and the maximum height between peak and valley increased from 3 nm to 5 nm after annealing. Consequently, we deduced that the increased roughness after annealing increases the electron scattering at the interface, which probably leads to the observed decrease in the MR ratios. The increase of spin scattering reduces the electron number through the devices, agreeing with the increase of device resistances, as shown in Fig. 17.21c. It should be noted that the increased roughness is mainly attributed to the inhomogeneous growth of rubrene in the film during annealing. A theoretical calculation made by Yu predicted that the spin diffusion length could be over 430 nm in rubrene film [65]. For actual rubrene devices, small spin transport (diffusion and tunneling) lengths were generally observed [72–75]. In this report, MR response disappears for the device with rubrene layer thickness over 20 nm, agreeing with the these experimental results. The device performance is related to the micro-structure of film, such as molecular arrangement [76] and the impurity band in the organic layer [66, 68, 73]. In this regard, the fabrication technique should be further explored to obtain high-quality films for OSVs. If a rubrene film with single-crystal structure was fabricated and used for spin devices, the spin transport length and MR response could be significantly enhanced.

The Alq_3-based spin valves were prepared using Fe_3O_4/Al-O films as the bottom layer after annealing at 623 K. An MR curve for a device with an Alq_3 layer thickness of 5 nm is shown in Fig. 17.23a. It was shown that the MR ratio was

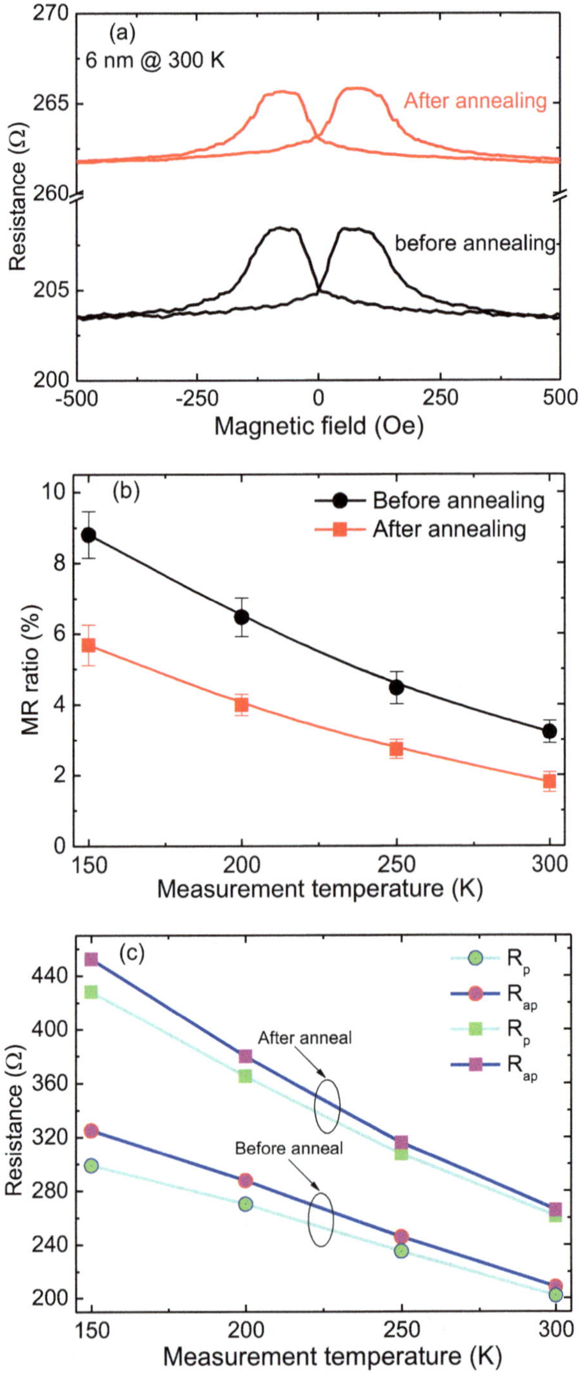

Fig. 17.21 (**a**) Magnetoresistance curves measured at 300 K for spin valves with a 6-nm-thick rubrene layer before annealing (*black lines*) and after annealing (*red lines*). Measurement temperature dependence of (**b**) magnetoresistance ratios and (**c**) resistances for the device before and after annealing. Resistances for parallel and anti-parallel states are denoted as R_p and R_{ap}, respectively. The lines serve as a visual guide (from Zhang, et al. ACS Appl. Mater. Interfaces. 2015,7,4685–4692)

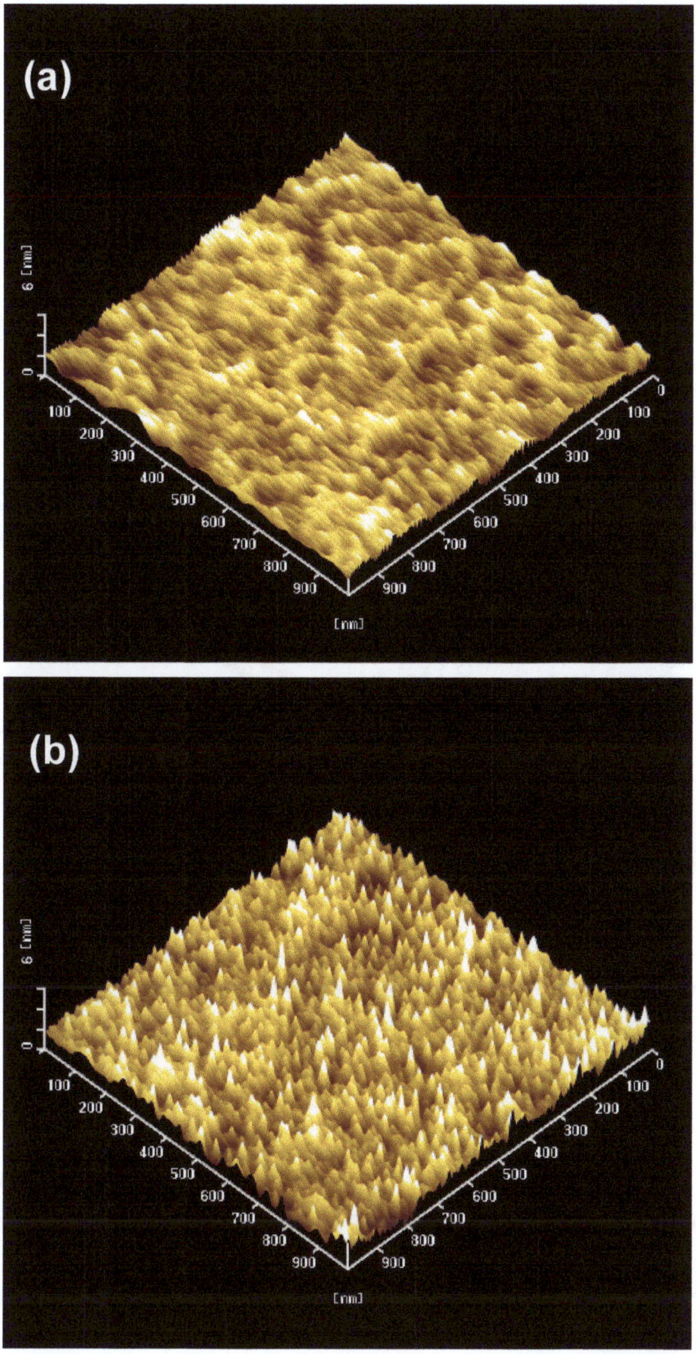

Fig. 17.22 Atomic force microscopy images of rubrene layer (100 nm) deposited on MgO/Fe_3O_4/Al-O (**a**) before and (**b**) after annealing (from Zhang, et al. ACS Appl. Mater. Interfaces. 2015,7,4685–4692)

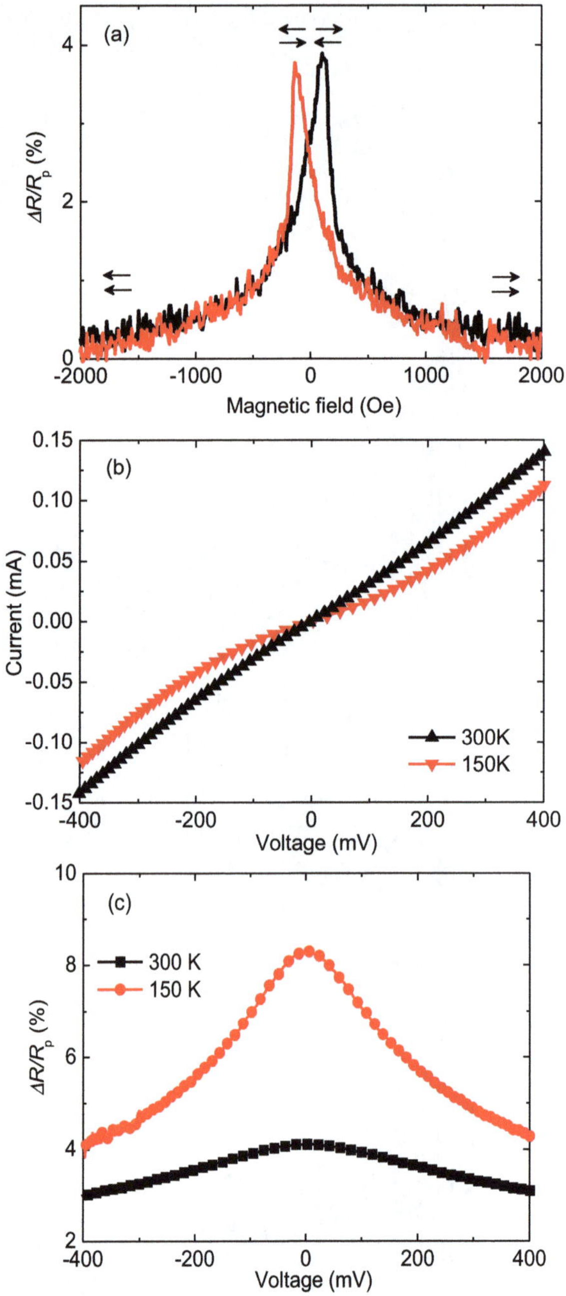

Fig. 17.23 (**a**) Magnetoresistance curves for the organic spin valve device composed of a 5-nm-thick Alq_3 layer measured at 300 K. The external magnetic field was swept from positive to negative indicated by the *red curves* and from negative to positive indicated by the *black curves*. The arrows illustrate the magnetization alignments in the anti-parallel and parallel directions. The bias voltage was fixed at 30 mV. (**b**) Nonlinear current-voltage curves and (**c**) the bias-voltage dependence of the magnetoresistance ratios were measured at 300 K and 150 K for the same device. (from Zhang, et al. J. Appl. Phys. 115, 172,608 (2014))

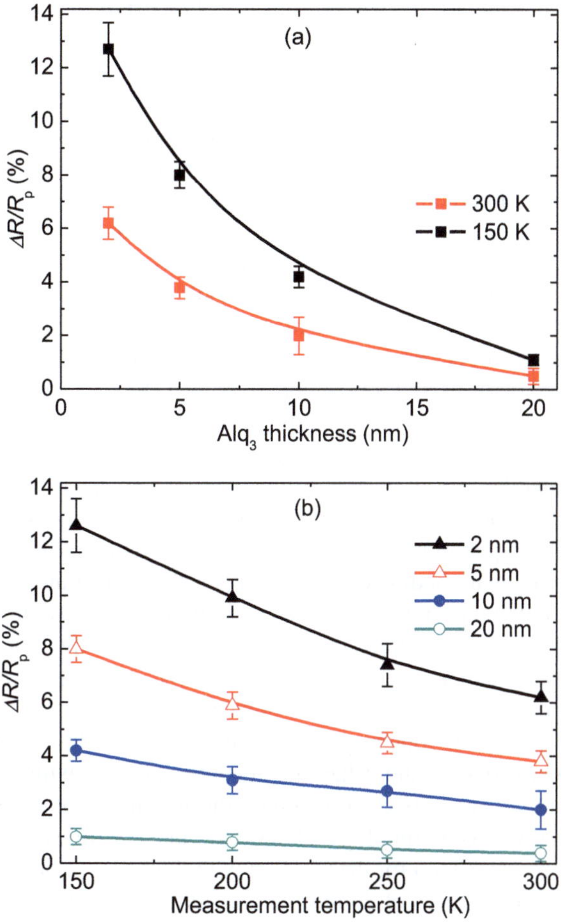

Fig. 17.24 (**a**) Dependence of the magnetoresistance ratios on the Alq_3 layer thickness measured at 300 K (*red squares*) and 150 K (*black squares*). (**b**) Measurement temperature dependence of the magnetoresistance ratios for the devices with different Alq_3 layer thicknesses. The bias voltage was fixed at 30 mV. The lines were inserted as a visual guide (from Zhang, et al. J. Appl. Phys. 115, 172,608 (2014))

around 4% at room temperature. The current-voltage curves were measured at 300 K and 150 K. Figure 17.23b shows that a nonlinear current-voltage curve was clearly observed at 300 K, which became stronger at 150 K. The bias-voltage dependence of the MR ratio is presented in Fig. 17.23c, showing a weak bias-voltage dependence at 300 K. In contrast, a strong bias-voltage dependence on the MR ratio appeared at 150 K, which showed that the MR ratio could be greater than 8% at zero bias-voltage.

A series of Alq_3-based devices with different Alq_3 layer thicknesses (2, 5, 10 and 20 nm) were fabricated and their MR effects were measured at 300 K, 250 K, 200 K and 150 K, as plotted in Fig. 17.24. The MR ratio decreased monotonously with increasing Alq_3 layer thickness, as shown in Fig. 17.24a. The largest MR ratio was over 6% for an Alq_3 layer thickness of 2 nm at room temperature, and the MR ratio decreased to 0.4% for devices with a 20-nm-thick Alq_3 layer. The MR ratio dependence on the measurement temperature is depicted in Fig. 17.24b. The MR ratio for devices with a 2 nm Alq_3 layer showed a quicker increase with decreasing measurement temperature in comparison with the devices with a thicker Alq_3 layer. In the present Alq_3-based devices, single-step tunneling likely dominated in the devices with a 2-nm-thick Alq_3 layer. Upon increasing the Alq_3 thickness, the multiple-step tunneling processes become stronger, resulting in a decrease in the MR ratios. This is similar with the results in that of rubrene case.

All-carbon materials, such as carbon nanotube [77], graphene [78] and fullerene [79], have been extensively investigated for their abundant properties in electricity, catalysts, magnetism and optical transition [80]. Owing to the lack of hydrogen element, all-carbon molecules should show much weaker hyperfine interaction than that of other organic semiconductors with hydrogen. This implies that their spin transport is very likely more efficient. Nanotube and graphene were investigated for spin transport at low temperatures [81–84]. In contrast, C_{60} molecule caught more attentions in spintronics due to the successful operation at room temperature and morphology effect of C_{60} film on MR ratio was investigated [85]. As an important member of fullerene family, C_{70} is also an electron acceptor and could be used for spintronics. C_{70} molecule can be envisioned by adding a ring of five hexagons along the equatorial plane of C_{60}. In our opinion, fullerenes based spin devices are excellent candidates to investigate the spin transport behaviors in organic semiconductors and could get an insight of spin transport mechanism, compared to that in carbon nanotube and graphene. This investigation contributes greatly to understand the criteria for organic semiconductors selection and design to improve the performance in spin based devices.

A small roughness and high uniform of film is a precondition to ensure the performance and reliability for OSV devices with stacking structure. AFM technique was utilized to evaluate the roughness of fullerene films. Figure 17.25a, b show the morphology of the C_{60} and C_{70} films, respectively. The roughness is around 0.9 nm for the C_{60} film with thickness of 50 nm, and 2.1 nm for the C_{70} film with thickness of 80 nm. As a result, the roughness of both fullerenes films with a

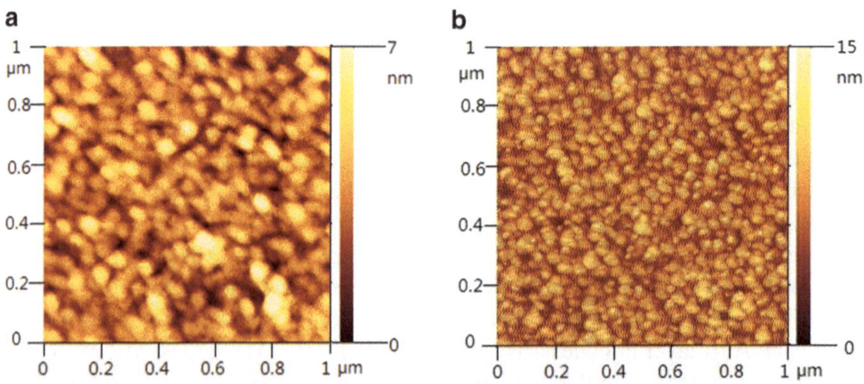

Fig. 17.25 Atomic force microscopy images for C_{60} (**a**) and C_{70} (**b**) films with the thickness of 50 nm and 80 nm, respectively (from Zhang, et al. Carbon. 2016, 106, 202–207)

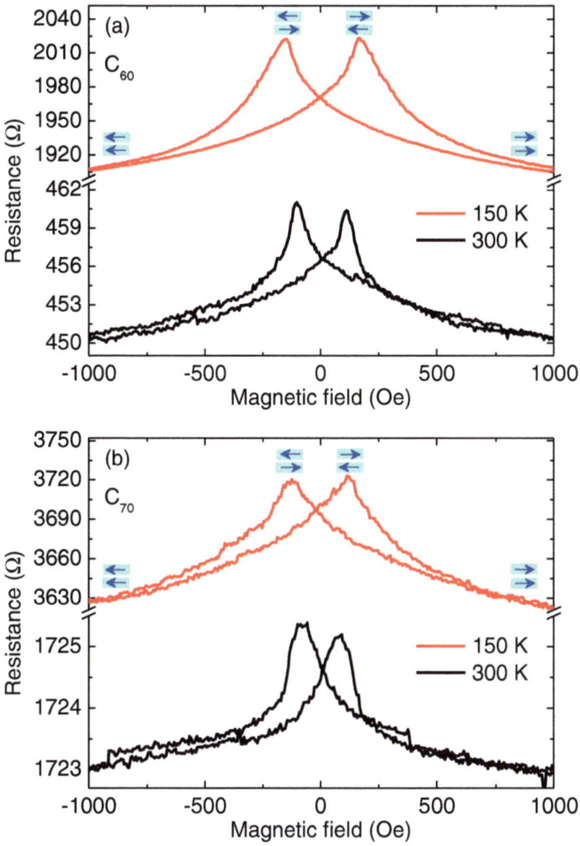

Fig. 17.26 Magnetoresistance curves for C_{60} (**a**) and C_{70} (**b**) based spin valves measured at 300 K (*black line*) and 150 K (*red line*). The arrows illustrate the magnetization alignments in the anti-parallel and parallel directions. The bias voltage for the measurement was fixed at 5 mV (from Zhang, et al. Carbon. 2016, 106, 202–207)

thickness of 5 nm could be estimated only around 0.1 nm. The C_{60} molecular is of a symmetric structure with the diameter of 0.71 nm [80]. The lengths of the short and long axes of ellipsoidal C_{70} molecule are 0.712 nm and 0.796 nm, respectively [80]. So the roughness of both fullerenes films (5 nm in thick) is smaller than that of C_{60} or C_{70} molecule size, which is very excellent to fabricate OSVs with a vertical structure.

MR effects are investigated at both room and low temperatures for both fullerenes devices. As shown in Fig. 17.26, the MR steps clearly appeared for both C_{60} and C_{70} based devices at around ±105 Oe at 300 K owing to the anti-parallel magnetization alignments of the Fe_3O_4 and Co electrodes. With reducing the measurement temperature to 150 K, the width of the MR curves become larger

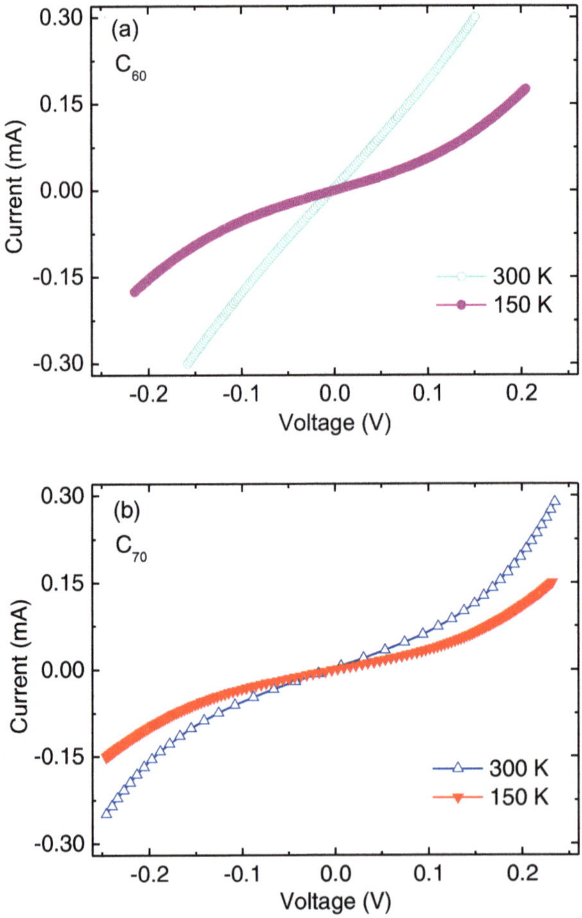

Fig. 17.27 Current-voltage curves for C_{60} (**a**) and C_{70} (**b**) based spin valves measured at 300 K and 150 K, respectively (from Zhang, et al. Carbon. 2016, 106, 202–207)

and the MR steps were shifted to about ±130 Oe. This is mainly due to the increase in the coercivity of the Fe_3O_4 electrode at low temperature.

Analysis of the current–voltage characteristics as a function of measurement temperature confirmed that the electrical transport across both fullerenes layers occur by tunneling. As shown in Fig. 17.27, nonlinear current–voltage curves were observed at 300 K and 150 K, as expected for tunnel junctions. C_{70} based OSV devices show a stronger nonlinear than that in C_{60} based OSVs at each temperature because of large device resistance (see Fig. 17.26). With decreasing measurement temperature, the nonlinear response of current–voltage dependences became stronger owing to the increase for device resistance in both fullerenes based devices at low temperature.

We calculated the MR ratio according to the following equation: $\Delta R/R_p = (R_{ap}/R_p - 1) \times 100\%$, the value for R_p was obtained with a magnetic field of 1000 Oe. The calculated results from 150 K to 300 K were ploted in Fig. 17.28. As reducing

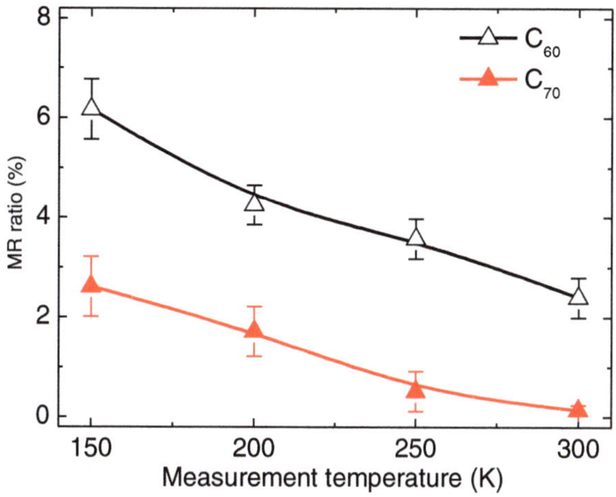

Fig. 17.28 Magnetoresistance ratios dependence on measurement temperature for C_{60} (*empty triangles*) and C_{70} (*solid triangles*) based spin valves (from Zhang, et al. Carbon. 2016, 106, 202–207)

measurement temperature, the MR ratios increased in both fullerenes based devices due to the reduction of spin scattering. The MR ratios for C_{60} based OSVs are much larger than that of C_{70} based devices. It is found that the MR ratio at 300 K was approximately 3% for the C_{60} based tunneling device, which is one of the highest MR ratios reported to date in organic spintronics. However, MR ratio is only 0.3% for the C_{70} based device at 300 K. This means the difference of MR ratios for both fullerenes devices is one order of magnitude at 300 K. It is also noted that C_{70} based devices show a larger device resistance than that in C_{60} based devices (see Fig. 17.26), as summarized in Fig. 17.29.

We will discuss the main reason for the smaller MR ratios in C_{70} devices compared to that in C_{60} OSVs. First, both fullerenes films with a 5 nm thick show a very small roughness, which is in the range of molecular scale, as discussed from Fig. 17.25. This means the electron scattering at the C_{70}/Co interface should be similar to that of C_{60}/Co interface. Second, the highest occupied molecular orbital and lowest unoccupied molecular orbital levels for C_{60} molecules are almost identical with those in C_{70} molecules. So the interface barriers for electron transport in both fullerenes based devices are also not the dominant reason for their difference of MR response. It has been pointed out in theory that molecular configuration significantly affects the spin conserved transport efficiency in organic solids [59]. It should be noted that the molecular configurations of both fullerenes are quite different. C_{60} molecule is spherical with high symmetry, but C_{70} molecule is ellipsoidal with low symmetry. This could induce a different electron states and their distributions in both fullerenes. Thus, C_{60} is isotropic and C_{70} is anisotropic. Even though C_{70} molecules tend to stand along the long axis, their lying alignment was also confirmed [86, 87]. As shown in Fig. 17.25, a large variety of particle size

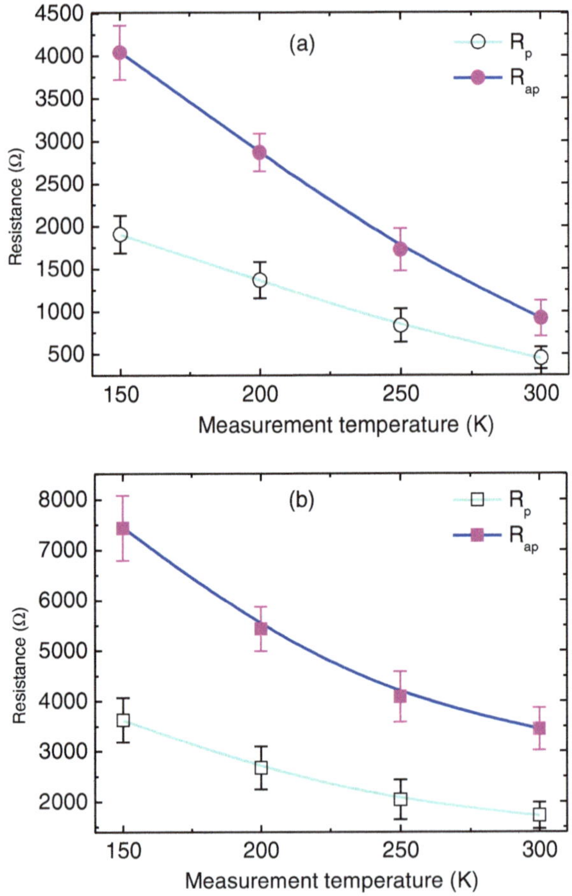

Fig. 17.29 Resistance dependence on measurement temperature for C_{60} (**a**) and C_{70} (**b**) based spin valves. Resistances for parallel and antiparallel states are denoted as R_p and R_{ap}, respectively. R_{ap} is enlarged by 2 times to show clearly (from Zhang, et al. Carbon. 2016, 106, 202–207)

for C_{70} film compared to that in C_{60} film likely reflects the co-exist of both alignments due to orientation disorder in C_{70} film. Moreover, C_{70} film shows far less long-range order than that in C_{60} film [87]. The stronger anisotropic and disorder of orientation in C_{70} film could lead to a larger energy loss for carriers between transport states compared to that in C_{60} film [56]. The dependence of MR response on spin flip in disordered organic solids has been investigated in theory. One step tunneling usually occurs for the spacer layer thickness within 2 nm in spin valves. A multistep tunneling transport process is mainly considered in the present fullerene layers with 5 nm thickness. Corresponding, in the OSV devices, polarized electron can easily move from one C_{60} energy state to the next C_{60} energy state with a small spin flip, compared with C_{70} film with strongly anisotropy, as illustrated in Fig. 17.30. In contrast to C_{60} based devices, the increase of spin flip reduces the transport efficiency of electron tunneling in C_{70} based OSVs, leading to a smaller MR ratios and larger device resistances, which agree with the observation in Fig. 17.28 and in Fig. 17.29, respectively.

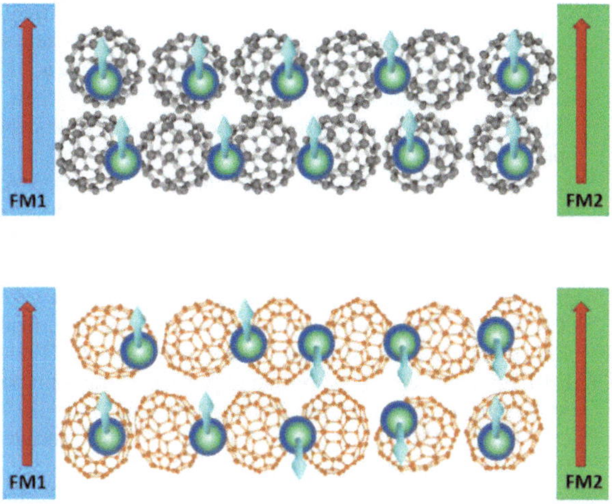

Fig. 17.30 Illustration of spin conserved electron transport in C_{60} (*up*) and C_{70} (*down*) layers. Solid ball (*dark blue*) depicts the electron and a blue arrow shows the spin direction. The *red arrow* shows magnetization alignment of ferromagnetic (FM) electrodes (from Zhang, et al. Carbon. 2016, 106, 202–207)

17.5 Summary

In summary, organic spintronics is an emerging and potential platform for future electronic devices. Investigations have been advanced to a range of materials combination of organic semiconductors and ferromagnetic electrodes to enhance device performance. Interface coupling between molecular layer and ferromagnetic metal plays important roles in determining spin injection/detection efficiency of traditionally organic spin devices and constructing molecular spin memory devices. Both giant magnetoresistance and tunneling magnetoresistance effects were also discussed to understand spin conserved electron transport behaviors in organic semiconductors. The configuration and composition of molecules contributes greatly to improve the performance in spin based devices. However, the research is still on going for actual applications. MR ratio is small for possible commercial development compared to their inorganic counterpart. Possible solutions include the design and synthesize novel molecules for efficiency spin transport and fabrication of a molecular film crystal with low defects. Another issue to be overcome in future is to develop organic magnetic materials as ferromagnetic electrode to make a fully organic spin devices.

References

1. V. Dediu, M. Murgia, F.C. Matacotta, C. Taliani, S. Barbanera, Room temperature spin polarized injection in organic semiconductors. Solid State Commun. **122**, 181–184 (2002)
2. Z.H. Xiong, D. Wu, Z.V. Vardeny, J. Shi, Giant magnetoresistance in organic spin-valves. Nature **427**, 821–824 (2004)

3. C. Barraud, P. Seneor, R. Mattana, S. Fusil, K. Bouzehouane, C. Deranlot, P. Graziosi, L. Hueso, I. Bergenti, V. Dediu, F. Petroff, A. Fert, Unravelling the role of the interface for spin injection into organic semiconductors. Nat. Phys. **6**, 615–620 (2010)
4. S. Steil, N. Großmann, M. Laux, A. Ruffing, D. Steil, M. Wiesenmayer, S. Mathias, O.L.A. Monti, M. Cinchetti, M. Aeschlimann, Spin-dependent trapping of electrons at spinterfaces. Nat. Phys. **9**, 242–247 (2013)
5. K.V. Raman, A.M. Kamerbeek, A. Mukherjee, N. Atodiresei, T.K. Sen, P. Lazic, V. Caciuc, R. Michel, D. Stalke, S.K. Mandal, S. Blugel, M. Munzenberg, J.S. Moodera, Interface-engineered templates for molecular spin memory devices. Nature **493**, 509–513 (2013)
6. S.W. Jiang, S. Liu, P. Wang, Z.Z. Luan, X.D. Tao, H.F. Ding, D. Wu, Exchange-dominated pure spin current transport in Alq_3 molecules. Phys. Rev. Lett. **115**, 086601 (2015)
7. A.K. Singh, J. Eom, Negative magnetoresistance in a vertical single-layer graphene spin valve at room temperature. ACS Appl. Mater. Interfaces **6**, 2493–2496 (2014)
8. F. Li, Effect of substrate temperature on the spin transport properties in C_{60}-based spin valves. ACS Appl. Mater. Interfaces **5**, 8099–8104 (2013)
9. D.L. Sun, E. Ehrenfreund, Z.V. Vardeny, The first decade of organic spintronics research. Chem. Commun. **50**, 1781–1793 (2014)
10. S. Shi, Z. Sun, A. Bedoya-Pinto, P. Graziosi, X. Li, X. Liu, M. Fahlman, Hybrid interface states and spin polarization at ferromagnetic metal–organic heterojunctions: interface engineering for efficient spin injection in organic spintronics. Adv. Funct. Mater. **24**, 4812–4821 (2014)
11. T.D. Nguyen, G. Hukic-Markosian, F.J. Wang, L. Wojcik, X.-G. Li, E. Ehrenfreund, Z.V. Vardeny, Isotope effect in spin response of π-conjugated polymer films and devices. Nat. Mater. **9**, 345–352 (2010)
12. X.M. Zhang, S. Mizukami, Q.L. Ma, T. Kubota, M. Oogane, H. Naganuma, Y. Ando, T. Miyazaki, Spin-dependent transport behavior in C_{60} and Alq_3 based spin valves with a magnetite electrode. J. Appl. Phys. **115**, 172608 (2014)
13. V. Dediu, L.E. Hueso, I. Bergenti, A. Riminucci, F. Borgatti, P. Graziosi, C. Newby, F. Casoli, M.P. De Jong, C. Taliani, Y. Zhan, Room-temperature spintronic effects in Alq_3-based hybrid devices. Phys. Rev. B **78**, 115203 (2008)
14. V. Dediu, L.E. Hueso, I. Bergenti, C. Taliani, Spin routes in organic semiconductors. Nat. Mater. **8**, 707–716 (2009)
15. Y.H. Zheng, F. Wudl, Organic spin transporting materials: present and future. J. Mater. Chem. A **2**, 48–57 (2014)
16. J. Devkota, R. Geng, R. Subedi, T. Nguyen, *Organic Spin Valves: A Review* (Adv. Funct, Mater, 2016). doi:10.1002/adfm.201504209
17. T.D. Nguyen, E. Ehrenfreund, Z.V. Vardeny, Spin-polarized light-emitting diode based on an organic bipolar spin valve. Science **337**, 204–209 (2012)
18. F. Al Ma'Mari, T. Moorsom, G. Teobaldi, W. Deacon, T. Prokscha, H. Luetkens, S. Lee, G.E. Sterbinsky, D.A. Arena, et al., Beating the Stoner criterion using molecular interfaces. Nature **524**, 69–73 (2015)
19. C. Barraud, P. Seneor, R. Mattana, S. Fusil, K. Bouzehouane, C. Deranlot, P. Graziosi, L. Hueso, I. Bergenti, V. Dediu, F. Petroff, A. Fert, Unravelling the role of the interface for spin injection into organic semiconductors. Nat. Phys. **6**, 615 (2010)
20. S. Majumdar, R. Laiho, P. Laukkanen, I.J. Vayrynen, H.S. Majumdar, R. Osterbacka, Application of regioregular polythiophene in spintronic devices: effect of interface. Appl. Phys. Lett. **89**, 122114 (2006)
21. D. Sun, L. Yin, C. Sun, H. Guo, Z. Gai, X.G. Zhang, T.Z. Ward, Z. Cheng, J. Shen, Giant magnetoresistance in organic spin valves. Phys. Rev. Lett. **104**, 236602 (2010)
22. J.H. Shim, K.V. Raman, Y.J. Park, T.S. Santos, G.X. Miao, B. Satpati, J.S. Moodera, Large spin diffusion length in an amorphous organic semiconductor. Phys. Rev. Lett. **100**, 22603 (2008)

23. T.S. Santos, J.S. Lee, P. Migdal, I.C. Lekshmi, B. Satpati, J.S. Moodera, Room-temperature tunnel magnetoresistance and spin-polarized tunneling through an organic semiconductor barrier. Phys. Rev. Lett. **98**, 016601 (2007)
24. W.H. Xu, J. Brauer, G. Szulczewski, M.S. Driver, A.N. Caruso, Electronic, magnetic, and physical structure of cobalt deposited on aluminum tris (8-hydroxy quinoline). Appl. Phys. Lett. **94**, 233302 (2009)
25. Y.Q. Zhan, E. Holmstroöm, R. Liza´rraga, O. Eriksson, X.J. Liu, F.H. Li, E. Carlegrim, S. Stafstroöm, M. Fahlman, Efficient spin injection through exchange coupling at organic semiconductor/ferromagnet heterojunctions. Adv. Mater. **22**, 1626 (2010)
26. T.L.A. Tran, P.K.J. Wong, M.P. de Jong, W.G. van der Wiel, Y.Q. Zhan, M. Fahlman, Hybridization-induced oscillatory magnetic polarization of C(60) orbitals at the C(60)/Fe (001) interface. Appl. Phys. Lett. **98**, 222505 (2011)
27. D. Çakır, D.M. Otalvaro, G. Brocks, Magnetoresistance in multilayer fullerene spin valves: a first-principles study. Phys. Rev. B **90**, 245404 (2014)
28. P.K. Johnny Wong, W. Zhang, G. Laan, M.P. de Jong, Hybridizationinduced charge rebalancing at the weakly interactive C60/Fe3O4(001) spinterface. Org. Electron. **29**, 39–43 (2016)
29. A.J. Drew et al., Direct measurement of the electronic spin diffusion length in a fully functional organic spin valve by low-energy muon spin rotation. Nat. Mater. **8**, 109–114 (2009)
30. M. Cinchetti et al., Determination of spin injection and transport in a ferromagnet/organic semiconductor heterojunction by two-photon photoemission. Nat. Mater. **8**, 115–119 (2009)
31. PRL. 81, 5229 (1998)
32. S.P. Park, S.S. Kim, J.H. Kim, C.N. Whang, S. Im, Optical and luminescence characteristics of thermally evaporated pentacene films on Si. Appl. Phys. Lett. **80**, 2872 (2002)
33. T. Kambayashi, H. Ohta, H. Hoshi, M. Hirano, H. Hosono, H. Takezoe, K. Ishikawa, Organic semiconductors for device applications: current trends and future prospects. Cryst. Growth Des. **5**, 143 (2005)
34. J.P. Velev, P.A. Dowben, E.Y. Tsymbal, S.J. Jenkins, A.N. Caruso, Interface effects in spin-polarized metal/insulator layered structures. Surf. Sci. Rep. **63**, 400 (2008)
35. Y.-L. Chan, Y.-J. Hung, C.-H. Wang, Y.-C. Lin, C.-Y. Chiu, Y.-L. Lai, H.-T. Chang, C.-H. Lee, Y.J. Hsu, D.H. Wei, Magnetic response of an ultrathin cobalt film in contact with an organic pentacene layer. Phys. Rev. Lett. **104**, 177204 (2010)
36. D.H. Wei, Y.-L. Chan, Y.-J. Hung, C.-H. Wang, Y.-C. Lin, Y.-L. Lai, H.-T. Chang, C.-H. Lee, Y.J. Hsu, Magnetic disparities at the interfaces of Co–pentacene–Co hybrid structures. Synth. Met. **161**, 581–585 (2011)
37. N. Atodiresei, J. Brede, P. Lazic, V. Caciuc, G. Hoffmann, R. Wiesendanger, S. Blugel, Design of the local spin polarization at the organic-ferromagnetic interface. Phys. Rev. Lett. **105**, 066601 (2010)
38. A.A. Sidorenko, C. Pernechele, P. Lupo, M. Ghidini, M. Solzi, R. De Renzi, I. Bergenti, P. Graziosi, V. Dediu, L. Hueso, A.T. Hindmarch, Interface effects on an ultrathin Co film in multilayers based on the organic semiconductor Alq_3. Appl. Phys. Lett. **97**, 162509 (2010)
39. A.N. Caruso, D.L. Schulz, P.A. Dowben, Metal hybridization and electronic structure of tris (8-hydroxyquinolato) aluminum (Alq_3). Chem. Phys. Lett. **413**, 321 (2005)
40. J.M. Baik, Y. Shon, S.J. Lee, Y.H. Jeong, T.W. Kang, J.-L. Lee, Electronic structure and magnetism in transition metals doped 8-hydroxy-quinoline aluminum. J. Am. Chem. Soc. **130**, 13522 (2008)
41. Y. Matsumoto, S. Sakai, Y. Takagi, T. Nakagawa, T. Yokoyama, T. Shimada, S. Mitani, H. Naramoto, Y. Maeda, X-ray absorption spectroscopy and magnetic circular dichroism in codeposited C_{60}–Co films with giant tunnel magnetoresistance. Chem. Phys. Lett. **470**, 244 (2009)

42. M. Cummings, S. Gliga, B. Lukanov, E.I. Altman, M. Bode, E.V. Barrera, Surface interactions of molecular C_{60} and impact on Ni(100) and Co(0001) film growth: a scanning tunneling microscopy study. Surf. Sci. **605**, 72 (2011)
43. K.V. Raman, J.S. Moodera, Materials chemistry: a magnetic facelift for non-magnetic metals. Nature **524**, 42–43 (2015)
44. J.J.H.M. Schoonus, P.G.E. Lumens, W. Wagemans, J.T. Kohlhepp, P.A. Bobbert, H.J.M. Swagten, B. Koopmans, Magnetoresistance in hybrid organic spin valves at the onset of multiple-step tunneling. Phys. Rev. Lett. **103**, 146601 (2009)
45. H.-J. Jang, K.P. Pernstich, D.J. Gundlach, O.D. Jurchescu, C.A. Richter, Observation of spin-polarized electron transport in Alq_3 by using a low work function metal. Appl. Phys. Lett. **101**, 102412 (2012)
46. L. Schulz, L. Nuccio, M. Willis, P. Desai, P. Shakya, T. Kreouzis, V.K. Malik, C. Bernhard, F.L. Pratt, N.A. Morley, A. Suter, G.J. Nieuwenhuys, T. Prokscha, E. Morenzoni, W.P. Gillin, A.J. Drew, Engineering spin propagation across a hybrid organic/inorganic interface using a polar layer. Nat. Mater. **10**, 39 (2011)
47. F.C. Wang, Z.H. Xiong, D. Wu, J. Shi, Z.V. Vardeny, Organic spintronics: the case of Fe/Alq_3/Co spin-valve devices. Synth. Met. **155**, 172 (2005)
48. Y.H. Liu, S.M. Watson, T. Lee, J.M. Gorham, H.E. Katz, J.A. Borchers, H.D. Fairbrother, D. H. Reich, Correlation between microstructure and magnetotransport in organic semiconductor spin-valve structures. Phys. Rev. B **79**, 075312 (2009a)
49. F.J. Wang, C.G. Yang, Z. Valy Vardeny, X.G. Li, Spin response in organic spin valves based on $La_{2/3}$ $Sr_{1/3}$ MnO_3 electrodes. Phys. Rev. B **75**, 245324 (2007)
50. Y.H. Liu, T. Lee, H.E. Katz, D.H. Reich, Effects of carrier mobility and morphology in organic semiconductor spin valves. J. Appl. Phys. **105**, 07C708 (2009)
51. A. Yanase, K. Siratori, Band structure in the high temperature phase of Fe_3O_4. J. Phys. Soc. Jpn. **53**, 312 (1984)
52. Z. Zhang, S. Satpathy, Electron states, magnetism, and the Verwey transition in magnetite. Phys. Rev. B **44**, 13319 (1991)
53. Y.S. Dedkov, U. Rüdiger, G. Güntherodt, Evidence for the half-metallic ferromagnetic state of Fe_3O_4 by spin-resolved photoelectron spectroscopy. Phys. Rev. B **65**, 064417 (2002)
54. P.A. Bobbert, W. Wagemans, F.W.A. van Oost, B. Koopmans, M. Wohlgenannt, Theory for spin diffusion in disordered organic semiconductors. Phys. Rev. Lett. **102**, 156604 (2009)
55. T.D.G. Nguyen et al., Isotope effect in spin response of π-conjugated polymer films and devices. Nat. Mater. **9**, 345–352 (2010)
56. J.J. Kwiatkowski, J.M. Frost, J. Nelson, The effect of morphology on electron field-effect mobility in disordered C60 thin films. Nano Lett. **9**, 1085–1090 (2009)
57. C. Tanase, E.J. Meijer, P.W.M. Blom, D.M. de Leeuw, Unification of the hole transport in polymeric field-effect transistors and light-emitting diodes. Phys. Rev. Lett. **91**, 216601 (2003)
58. A. Pivrikas, M. Ullah, H. Sitter, N.S. Sariciftci, Electric field dependent activation energy of electron transport in fullerene diodes and field effect transistors: Gill's law. Appl. Phys. Lett. **98**, 092114 (2011)
59. Z.G. Yu, Spin-orbit coupling, spin relaxation, and spin diffusion in organic solids. Phys. Rev. Lett. **106**, 106602 (2011)
60. C.-C. Chu et al., Self-assembly of supramolecular fullerene ribbons via hydrogen-bonding interactions and their impact on fullerene electronic interactions and charge carrier mobility. J. Am. Chem. Soc. **132**(36), 12717–12723 (2010)
61. Q.-M. Zhang, J.-Y. Yi, J. Bernholc, Structure and dynamics of solid C60. Phys. Rev. Lett. **66**, 2633–2636 (1991)
62. K. Harigaya, S. Abe, Optical-absorption spectra in fullerenes C60 and C70: effects of coulomb interactions, lattice fluctuations, and anisotropy. Phys. Rev. B **49**, 16746–16752 (1994)
63. L. Schulz et al., Importance of intramolecular electron spin relaxation in small molecule semiconductors. Phys. Rev. B **84**, 085209 (2011)

64. R. Fiederling et al., Injection and detection of a spin-polarized current in a light-emitting diode. Nature **402**, 787–790 (1999)
65. Z.G. Yu, Spin-orbit coupling and its effects in organic solids. Phys. Rev. B **85**, 115201 (2012)
66. R. Lin, F. Wang, J. Rybicki, M. Wohlgenannt, K.A. Hutchinson, Distinguishing between tunneling and injection regimes of ferromagnet/organic semiconductor/ferromagnet junctions. Phys. Rev. B **81**, 195214 (2010)
67. N.J. Harmon, M.E. Flatté, Distinguishing spin relaxation mechanisms in organic semiconductors. Phys. Rev. Lett. **110**, 176602 (2013)
68. Z.G. Yu, Impurity-band transport in organic spin valves. Nat. Commun. **5**, 4842 (2014)
69. T. Lan Anh Tran, T. Quyen Le, J.G.M. Sanderink, W.G. van der Wiel, M.P. De Jong, The multistep tunneling analogue of conductivity mismatching organic spin valves. Adv. Funct. Mater. **22**, 1180–1189 (2012)
70. G. Szulczewski, H. Tokuc, K. Oguz, J.M.D. Coey, Magnetoresistance in magnetic tunnel junctions with an organic barrier and a MgO spin filter. Appl. Phys. Lett. **95**, 202506 (2009)
71. M. Gobbi, F. Golmar, R. Llopis, F. Casanova, L.E. Hueso, Room-temperature spin transport in C60-based spin valves. Adv. Mater. **23**, 1609–1613 (2011)
72. J.-W. Yoo, C.-Y. Chen, H.W. Jang, C.W. Bark, V.N. Prigodin, C.B. Eom, A.J. Epstein, Spin injection/detection using an organic-based magnetic semiconductor. Nat. Mater. **9**, 638–642 (2010)
73. J.-W. Yoo, H.W. Jang, V.N. Prigodin, C. Kao, C.B. Eom, A.J. Epstein, Giant magnetoresistance in ferromagnet/organic semiconductor/ferromagnet heterojunctions. Phys. Rev. B **80**, 205207 (2009)
74. B. Li, C.Y. Kao, Y. Lu, J.-W. Yoo, V.N. Prigodin, A.J. Epstein, Room-temperature organic-based spin polarizer. Appl. Phys. Lett. **99**, 153503 (2011)
75. B. Li, C.Y. Kao, J.-W. Yoo, V.N. Prigodin, A.J. Epstein, Magnetoresistance in an all-organic-based spin valve. Adv. Mater. **23**, 3382–3386 (2011)
76. T. Vehoff, B. Baumeier, A. Troisi, D. Andrienko, Charge transport in organic crystals: role of disorder and topological connectivity. J. Am. Chem. Soc. **132**, 11702–11708 (2010)
77. S. Iijima, Helical microtubules of graphitic carbon. Nature **354**, 56–58 (1991)
78. K.S. Novoselov, A.K. Geim, S.V. Morozov, D. Jiang, Y. Zhang, S.V. Dubonos, I.V. Grigorieva, A.A. Firsov, Electric field effect in atomically thin carbon films. Science **306**, 666–669 (2004)
79. H.W. Kroto, J.R. Heath, S.C. Obrien, R.F. Curl, R.E. Smalley, C60: Buckminsterfullerene. Nature **318**, 162–163 (1985)
80. S. Okubo, T. Okazaki, K. Hirose-Takai, K. Suenaga, S. Okada, S. Bandow, S. Iijima, Electronic structures of single-walled carbon nanotubes encapsulating ellipsoidal C70. J. Am. Chem. Soc. **132**, 15252–15258 (2010)
81. K. Tsukagoshi, B.W. Alphenaar, H. Ago, Coherent transport of electron spin in a ferromagnetically contacted carbon nanotube. Nature **401**, 572–574 (1999)
82. L.E. Hueso et al., Transformation of spin information into large electrical signals using carbon nanotubes. Nature **445**, 410–413 (2007)
83. Y.-H. Zhou, J. Zeng, K.-Q. Chen, Spin filtering effect and magnetoresistance in zigzag 6, 6, 12-graphyne nanoribbon system. Carbon **76**, 175–182 (2014)
84. P.K. Johnny Wong, W. Zhang, G. Laan, M.P. de Jong, Hybridization-induced charge rebalancing at the weakly interactive C60/Fe3O4(001) spinterface. Org. Electron. **29**, 39–43 (2016)
85. T.D. Nguyen, F.J. Wang, X.-G. Li, E. Ehrenfreund, Z.V. Vardeny, Spin diffusion in fullerene-based devices: morphology effect. Phys. Rev. B **87**, 075205 (2013)
86. Y.J. Guo, N. Karasawa, W.A. Goddard III, Prediction of fullerene packing in C60 and C70 crystals. Nature **351**, 464–467 (1991)
87. D.R. McKenzie, C.A. Davis, D.J.H. Cockayne, D.A. Muller, A.M. Vassallo, The structure of C70 molecule. Nature **355**, 622–624 (1992)

MIX
Papier aus verantwortungsvollen Quellen
Paper from responsible sources
FSC® C105338

If you have any concerns about our products,
you can contact us on
ProductSafety@springernature.com

In case Publisher is established outside the EU,
the EU authorized representative is:
Springer Nature Customer Service Center GmbH
Europaplatz 3, 69115 Heidelberg, Germany

Printed by Libri Plureos GmbH
in Hamburg, Germany